GCSE Biology

third edition

GCSE

Biology

third edition

D. G. Mackean

Hodder Murray

A MEMBER OF THE HODDER HEADLINE GROUP

First published in 1986
by Hodder Murray, an imprint of Hodder Education
part of Hachette Livre UK
338 Euston Road
London NW1 3BH

Second edition 1995

Third edition 2002

Reprinted 2002, 2003, 2004 (twice), 2006 (twice), 2007, 2008

Layouts by Jenny Fleet
Original illustrations by Don Mackean, prepared and adapted by Wearset Ltd
with additional illustrations by Ethan Danielson, Richard Draper, Mike Humphries
Natural history artwork by Chris Etheridge
The full-colour illustrations on pages 270–73 are by Pamela Haddon
Cover design by John Townson/Creation

Typeset in 11½/13pt Bembo by Wearset Ltd, Boldon, Tyne and Wear
Printed and bound in Dubai by Oriental Press

A catalogue entry for this title is available from the British Library

ISBN-13: 978-0-719-58615-6

Acknowledgements

I should like to thank Mr D. Hayward and Dr C. J. Clegg, who read parts of the draft of the Third Edition, for their helpful and constructive suggestions. I am also grateful to Sophie Hartley for her assiduous photo research, and Chris Wyard and Michelle Sadler for their efficient and perceptive editing.

Examination questions

Exam questions have been reproduced with kind permission from the following examination boards:

AQA (Assessment and Qualifications Alliance)
CCEA (Northern Ireland Council for the Curriculum, Examinations and Assessment)
Edexcel
IGCSE (produced by University of Cambridge Local Examinations Syndicate/Cambridge International Examinations)
OCR (Oxford, Cambridge and RSA Examinations)
WJEC (Welsh Joint Education Committee)

Artwork figure acknowledgements

Figure 2.3 From *Principles of Plant Physiology* by Bonner and Galston. Copyright © 1952 by W. H. Freeman and Company. Used with permission.
Table p. 25 From *Textbook of Physiology* by Emslie-Smith, Paterson, Scratcherd and Read, reproduced by permission of the publisher Churchill Livingstone.
Figure 4.9 Data from Brierly, *Plant Physiology*, 1954, reproduced by kind permission of the publisher, the Association for Science Education.
Figure 5.15 From Dr S. B. Verma and Professor N. J. Rosenburg, *Agriculture and the atmospheric carbon dioxide build-up (Span*, 22 February 1979).
Figure 10.6 From *World Resources Report 1998–9*, by permission of the World Resources Institute.
Figure 12.27 From *Smoking or Health: a report of the Royal College of Physicians*, 1977, Pitman Medical Publishing Co. Ltd.
Figure 13.10 From *Smoking and Health Now: a report of the Royal College of Physicians*, Pitman Medical Publishing Co. Ltd.
Figure 16.20 From G. W. Corner, *The Hormones in Human Reproduction*. Copyright © 1942 by Princeton University Press. Reprinted/reproduced by permission of Princeton University Press.
Figure 20.2 Copyright © John Besford, 1984. Reprinted from *Good Mouthkeeping: or how to save your children's teeth and your own while you're about it* by John Besford (2nd edn, 1984) by permission of Oxford University Press.
Figure 23.15 After Grove and Newell, *Animal Biology*, 6th edn, University Tutorial Press, 1961.
Figure 24.13 Courtesy of the Department of Forensic Genetics, University of Copenhagen.
Figure 24.14 From Cellmark Diagnostics, Blacklands Way, Abingdon Business Park, Abington, Oxfordshire.
Figure 25.16 Reprinted with the permission of Prentice Hall/Pearson Higher Education from the Macmillan College text *Communities and Ecosystems 2/E* by Robert H. Whittaker. Copyright © 1975.
Figure 26.16 From W. E. Shewell-Cooper, *The ABC of Soils*, University Tutorial Press.
Figure 26.17 From Clive A. Edwards, *Soil Pollutants and Soil Animals*, copyright © 1969 by Scientific American Inc. All rights reserved.
Figure 26.25 From J. E. Hansen and S. Ledeboff in *New Scientist*, 22 October 1985.
Figures 29.1, 29.2 and 29.4 From Lewis and Taylor, *Introduction to Experimental Ecology*, 1966. Academic Press, USA.
Figure 29.6 Reproduced with permission from James Bonner, *The World's People and the World's Food Supply*, 1980, Carolina Biology Readers Series. Copyright © Carolina Biological Supply Company, Burlington, North Carolina, USA.
Figure 29.7 From F. M. Burnett, *Natural History of Infectious Disease*, 3rd edn, 1962, with permission of Cambridge University Press.
Figure 34.2 Copyright © N. Tinbergen, 1989. Reprinted from *The Study of Instinct* by N. Tinbergen (1989) by permission of Oxford University Press.
Figure 35.2 After Tanner, *Growth and Adolescence*, 2nd edn, 1962, reproduced by permission of Blackwell Publishing Ltd.
Figure 35.4 After C. M. Jackson.
Figure 36.2b From V. B. Wigglesworth, *Principles of Insect Physiology*, 1965, Methuen.
Figure 37.11 After Mast, 1911.
Figure 39.1 After Brian Jones, *Introduction to Human and Social Biology*, 2nd edn, 1985, John Murray.
Figure 39.2 From *World Resources Report 1998–9*, by permission of the World Resources Institute.
Figure 39.6 After Brian Jones, *Introduction to Human and Social Biology*, 2nd edn, 1985, John Murray.
Fig 41.11 From B. Holmes, Tough treaty to police global fisheries, *New Scientist*, August 1995, by permission of New Scientist.

Photo acknowledgements

Front cover Steve Knell/BBC wild; **back cover:** Peter Oxford/ BBC Wild; **p.1** David Parker/Science Photo Library; **p.2** Biophoto Associates; **p.3** Biophoto Associates; **p.5** Ed Reschke, Peter Arnold Inc./ Science Photo Library; **p.9** Biophoto Associates; **p.10** Biophoto Associates; **p.13** Biophoto Associates; **p.18** D.G. Mackean; **p.31** Martyn F. Chillmaid/Science Photo Library; **p.32** D.G. Mackean; **p.33** J.C. Revy/Science Photo Library; **p.39** Heather Angel; **p.42** © Dr Tim Wheeler, University of Reading; **p.45** Crighton Thomas Creative; **p.49** Hans Reinhard/Bruce Coleman; **p.51** Sidney Moulds/Science Photo Library; **p.52** Dr Geoff Holroyd/Lancaster University; **p.53** Gene Cox; **p.55** Biophoto Associates; **p.56** Biophoto Associates; **p.57** D.G. Mackean; **p.60** © Breck P. Kent/ Earth Scenes/Oxford Scientific Films; **p.62** Bruce Coleman; **p.68** D.G. Mackean; **p.69** © 1998 J.E. Swedberg/Ardea; **p.70** Bob Gibbons/Ardea; **p.72** *both* D.G. Mackean; **p.73** Kim Taylor/Bruce Coleman; **p.75** Bruce Coleman; **p.81** *l & c* D.G. Mackean, *r* Rosenfeld Images Ltd/Science Photo Library; **p.85** AVS Photography, Loughborough University; **p.89** *tl & bl* © MMB/The Anthony Blake Photo Library, *tr & br* © Milk Marque/The Anthony Blake Photo Library; **p.94** John Townson/Creation; **p.102** David Scharf/Science Photo Library; **p.109** Biophoto Associates; **p.113** Biophoto Associates; **p.117** Andrew Syred/Science Photo Library; **p.119** *l* Biophoto Associates, *r* Jerry Mason/Science Photo Library; **p.120** *t* Philip Harris Education, *b* Biophoto Associates/Science Photo Library; **p.124** Biophoto Associates; **p.126** Philip Harris Education; **p.128** *t & bl* Biophoto Associates, *br* BSIP, Alexandre/Science Photo Library; **p.132** Biophoto Associates; **p.135** Y. Beaulieu, Publiphoto Diffusion/Science Photo Library; **p.137** Biophoto Associates; **p.138** Mike Hewitt/Allsport; **p.142** John Walsh/Science Photo Library; **p.143** Biophoto Associates; **p.144** London Fertility Centre; **p.145** Professor W. J. Hamilton; **p.146** Maternity Center Assoc, New York; **p.147** Keith/Custom Medical Stock Photo/Science Photo Library; **p.148** *t* © Jenny Matthews/Format, *b* © Judy Harrison/Format; **p.163** © Colorsport; **p.167** © Mike Benyon/BBC Natural History Unit; **p.168** Biophoto Associates; **p.169** Biophoto Associates; **p.171** Biophoto Associates; **p.173** Brian J. Myers/ Allsport; **p.174** Sheila Terry/Science Photo Library; **p.176** Sporting Pictures; **p.181** Jane Burton/Bruce Coleman; **p.183** Manfred Kage/ Science Photo Library; **p.186** Biophoto Associates; **p.188** Dr A. Lesk, Laboratory of Molecular Biology, Cambridge/Science Photo Library; **p.195** Philip Harris Education; **p.198** Simon Fraser/Science Photo

Library; **p.200** Horticultural Research International, East Malling; **p.203** Biophoto Associates; **p.204** Kim Taylor/Bruce Coleman; **p.205** *tl & br* Biophoto Associates, *tr* Hans Reinhard/Bruce Coleman; **p.212** Sir Ralph Riley; **p.214** *t* © Javed Jafferji/Oxford Scientific Films, *bl* Philippe Plailly/Science Photo Library, *br* Julia Kamlish/ Science Photo Library; **p.215** *l* Cystic Fibrosis Trust, *r* Crighton Thomas Creative; **p.216** Zeneca Seeds; **p.217** courtesy of Syngenta; **p.218** Science Pictures Limited/Science Photo Library; **p.220** Biophoto Associates; **p.223** Tony Heald/BBC Wild; **p.225** *tl & tr* © Dr. D.P. Wilson/FLPA, *b* © Mike Lane/NHPA; **p.227** *tl & tr* Colin Green, *b* Bob Gibbons/Ardea; **p.230** D.G. Mackean; **p.231** © Image Quest 3-D/NHPA; **p.233** Marcelo Brodsky/Science Photo Library; **p.235** *tl* R.F. Porter/Ardea, *tr* © Environmental Investigation Agency, *c* Bob Gibbons/Ardea, *b* Last Resort; **p.236** *tl & b* Crighton Thomas Creative, *tr* Biophoto Associates; **p.238** *t* © Eyal Bartov/Oxford Scientific Films, *b* © Alan Williams/ NHPA; **p.239** Biophoto Associates; **p.240** *l* Simon Fraser/Science Photo Library, *r* ADAS Aerial Photograph Unit – Crown Copyright, Cambridge; **p.242** *t* J. Swedberg/Ardea, *b* Ecoscene/Michael Cuthbert; **p.243** *t* Thomas Nilson/JVZ/Science Photo Library, *b* Simon Fraser/Science Photo Library; **p.244** BSIP, M.I.G/Baeza/Science Photo Library; **p.247** © Joanna Van Gruisen/Ardea; **p.248** *tl* © Jack A. Bailey/Ardea, *tr* Paul Van Gaalen/Bruce Coleman, *bl* Luiz Claudio Marigo/Bruce Coleman, *br* Chris Gomersall/Bruce Coleman; **p.249** *t & b* © David Tipling/Oxford Scientific Films, *c* © Ecoscene/ Sally Morgan; **p.250** *t* Crighton Thomas Creative, *c* © David Woodfall/ NHPA, *b* E. Mickleburgh/Ardea; **p.251** *t* © Sean Sprague/Panos, *bl* Alex Bartel/Science Photo Library, *br* Prof. David Hall/Science Photo Library; **p.252** James Holmes/Zedcor/Science Photo Library; **p.254** *tl* Ecosphere Associates Inc., Tuscon, Arizona, *tr* © David Keith Jones/Images of Africa Photobank, *b* © Stefan Meyers/Ardea; **p.255** © Peggy Heard/FLPA; **p.256** Chris Martin Bahr/Ardea; **p.257** Bruce Coleman; **p.260** © Stephen Warman/Oxford Scientific Films; **p.263** Mark Edwards/Still Pictures; **p.267** © Sue Scott/ Oxford Scientific Films; **p.269** *t* Heather Angel, *b* National Institute of Agricultural Botany, Cambridge; **p.274** Sinclair Stammers/Science Photo Library; **p.278** Robert Maier/Bruce Coleman; **p.279** *t* Dr Alan Beaumont, *b* Heather Angel; **p.280** *tl & bl* © David M. Dennis/ Oxford Scientific Films, *tr* Biophoto Associates, *br* Heather Angel; **p.283** Biophoto Associates; **p.287** *tl* Jens Rydell/Bruce Coleman, *tr & br* Heather Angel, *bl* Nigel Cattln/Holt; **p.288** Biophoto Associates; **p.289** *t* Samuel Ashfield/Science Photo Library, *b* Dr David Hockley/ National Institute for Biological Standards and Control, Potters Bar, Hertfordshire; **p.290** *t & b* Biophoto Associates; **p.294** *tl* Heather Angel, *bl* D.G. Mackean, *r* © N. A. Callow/NHPA; **p.297** Dr Jeremy Burgess/ Science Photo Library; **p.299** *t* © David M. Dennis/Oxford Scientific Films, *b* Heather Angel; **p.302** *tl* Kim Taylor/Bruce Coleman, *tr* Heather Angel, *b* Jane Burton/Bruce Coleman; **p.303** *t* Heather Angel, *b* Science Pictures Limited/Science Photo Library; **p.304** *t* P. Morris/Ardea, *b* Heather Angel; **p.307** London Scientific Films/Oxford Scientific Films; **p.308** *l* © G.I. Bernard/NHPA, *r* Biophoto Associates; **p.315** Colorsport; **p.316** *tl & cl* D.G. Mackean, *bl* © Kathie Atkinson/Oxford Scientific Films, *r* J.C. Revy/ Science Photo Library; **p.317** D.G. Mackean; **p.318** D.G. Mackean; **p.319** D.G. Mackean; **p.320** D.G. Mackean; **p.322** *t* © Mark Hamblin/ Oxford Scientific Films, *c & bl* Heather Angel, *br* Kim Taylor/Bruce Coleman; **p.323** *t* Hans Reinhard/Bruce Coleman, *b* Heather Angel; **p.325** Andy Crump, TDR, WHO/Science Photo Library; **p.327** *l* © The Anthony Blake Photo Library, *r* © Gerrit Buntrock/ The Anthony Blake Photo Library; **p.328** © The Anthony Blake Photo Library; **p.329** © Dr. Ariel Louwrier; **p.331** Thames Water; **p.332** Dr David J. Patterson/Science Photo Library; **p.338** Matt Meadows, Peter Arnold Inc./Science Photo Library; **p.340** *l* Tim Beddow/Science Photo Library, *r* courtesy Tesco; **p.341** © Mark Turner/The Anthony Blake Photo Library; **p.342** Saturn Stills/ Science Photo Library; **p.343** *t* John Townson/Creation, *b* courtesy Trailfinders Travel Centre; **p.345** © English Heritage Photo Library/ Jonathan Bailey; **p.354** Science Source/Science Photo Library; **p.355** A. Barrington Brown/Science Photo Library; **p.356** *l* F.C. Taylor/Fortean Picture Library, *r* Wellcome Library, London; **p.357** *l* John Townson/Creation, *c* Allsport Australia/ Allsport, *r* Scott, Barbour/Allsport; **p.359** Ardea; **p.360** *tl* B & C Alexander/NHPA, *tr* Ecoscene/Close, *c* Charles D. Winters/Science Photo Library, *b* Merck Nitrate test strips available from VWR International Ltd; **p.361** courtesy Tim Pickering.

t = top, *b* = bottom, *l* = left, *r* = right, *c* = centre

Every effort has been made to contact copyright holders, and the publishers apologise for any omissions which they will be pleased to rectify at the earliest opportunity.

Contents

Diversity of organisms

Micro-organisms and humans

Ideas and evidence

To the student

This is a textbook to help you in studying biology for the GCSE or IGCSE. You will be following the GCSE specification of only one examination group but this book contains the material needed by all the groups. For this reason, amongst others, it is not expected that you will need to study or learn everything in this book.

Furthermore, the emphasis in GCSE is on the ability to understand and use biological information rather than on committing it all to memory. However, you will still need to use a book of this sort to find the facts and explanations before you can demonstrate your understanding or apply the biological principles.

The questions included in the chapter are intended to test your understanding of the text you have just read. If you cannot answer the question straightaway, read that section of text again with the question in mind.

There is a checklist at the end of each chapter, summarising the important points covered.

The questions at the end of each section are selected from the specimen papers published by the GCSE examining groups as sample material for their examinations in 2003 and from the November 2000 papers of the IGCSE.

In many cases they are designed to test your ability to apply your biological knowledge. The questions may provide certain facts and ask you to make interpretations or suggest explanations. In such cases the factual information may not be covered in this textbook.

Looking up information

To find the information you need, use the index, the contents pages and the summary at the beginning of each chapter. If the word you want does not appear in the index, try a related word. For example, information about 'sight' might be listed under 'vision', 'eyes' or 'senses'.

Practical work

Given standard laboratory equipment, it should be possible for you to attempt any of the practical work described in the book, though you will probably not have time to do it all.

For this reason, it has been necessary to give the expected results of the experiments so that you can appreciate the design and purpose of the experiment, and consider alternative interpretations of the results, even if you have not been able to do the experiment yourself.

Third Edition

'Now it remains I acquaint you with what I have performed in this edition, which is either by mending what was amiss, or by adding such as formerly were wanting; some places I have helped by putting out ... the kinds in divers places where they were not very necessary, by this means to get more room for things more necessary'.

John Gerard, 1633, in the preface to the second edition of his Herbal.

This edition has been adapted to meet the specifications of the GCSE Biology and Double Award Science for the 2003 examinations.

A new chapter on Applied genetics includes genetic engineering, cloning, GM crops, DNA fingerprinting, the human genome project and stem cells. A chapter on Conservation has also been included.

The requirements of 'Ideas and evidence' have been met in two ways. Throughout the book there are references to controlled experiments, homeostasis, negative feedback, interpretation of experimental results, cause and correlation, and hypothesis testing. In addition, there are two new chapters; one dealing with the historical development of ideas and the other addressing some of the scientific principles underlying the gathering of evidence in Biology.

Some principles of biology

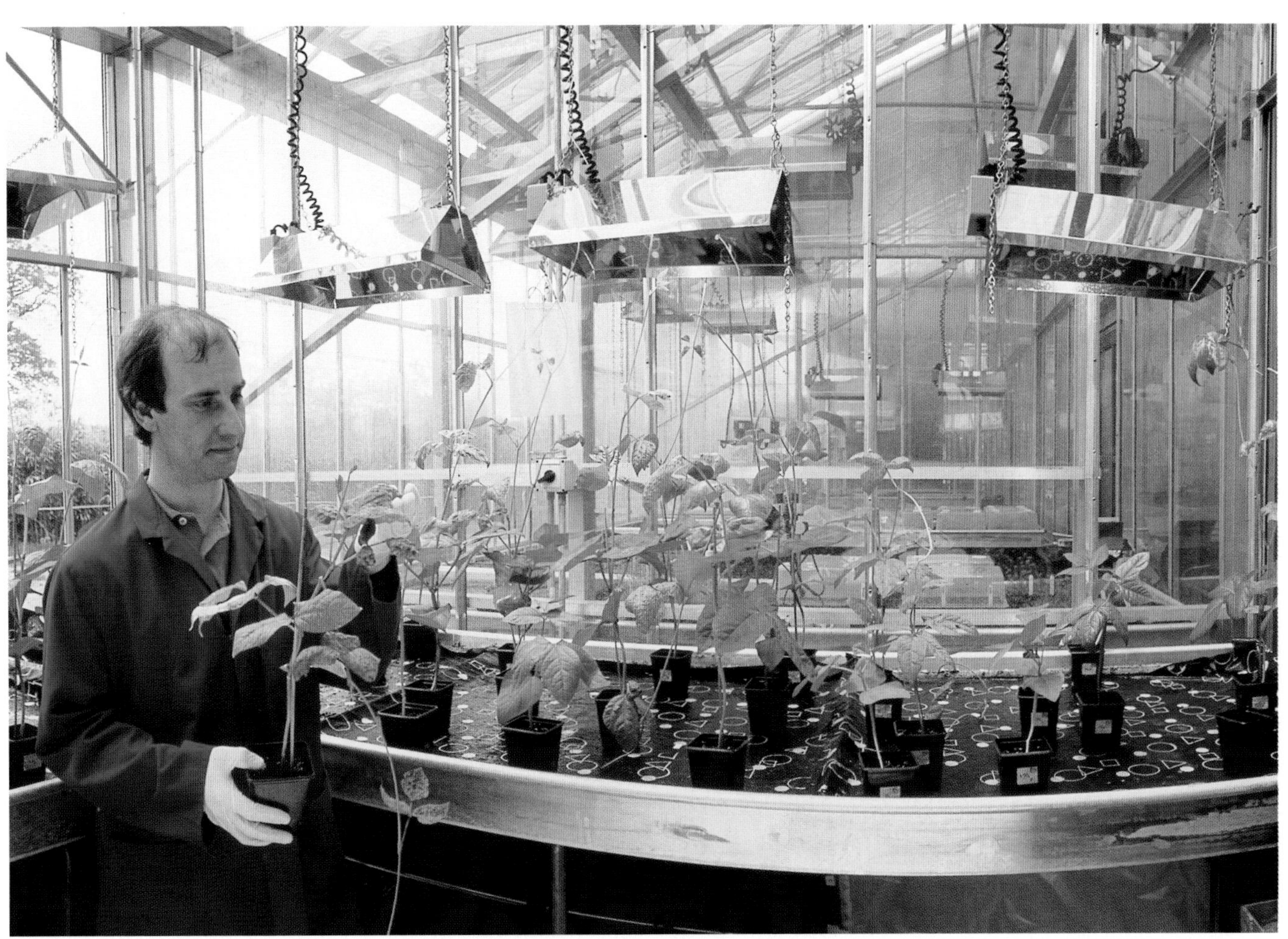

1 Cells and tissues

Cell structure
How tissues are studied to see cells: taking sections. Cell components. Plant cells.

Cell division and cell specialization
Cell division and growth. Specialization of cells for different functions.

Tissues and organs
Definitions and examples of tissues, organs and systems.

Practical work
Preparing, observing and drawing plant and animal cells.

Cell structure

If a very thin slice of a plant stem is cut and studied under a microscope, it can be seen that the stem consists of thousands of tiny, box-like structures. These structures are called **cells**. Figure 1.1 is a thin slice taken from the tip of a plant shoot and photographed through a microscope. Photographs like this are called **photomicrographs**. The one in Figure 1.1 is 60 times larger than life, so a cell which appears to be 2 mm long in the picture is only 0.03 mm long in life.

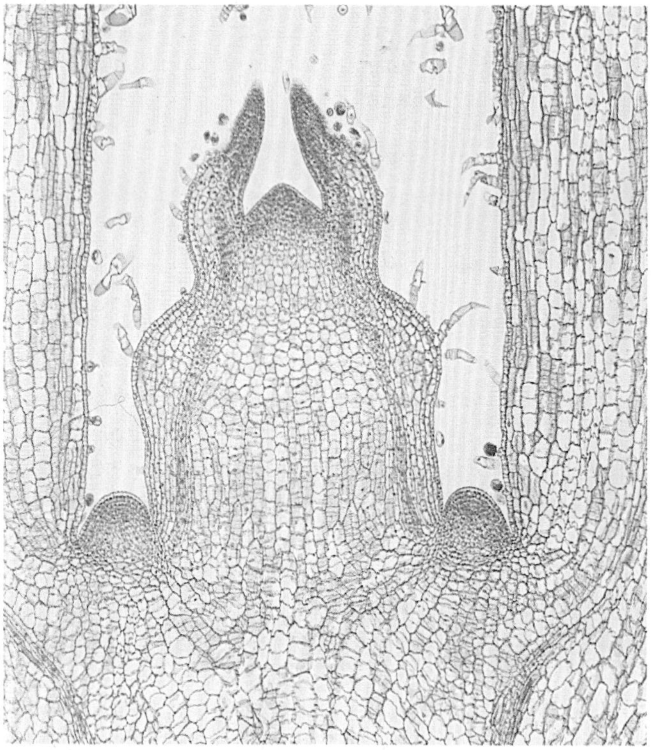

Figure 1.1 Longitudinal section through the tip of a plant shoot (× 60). The slice is only one cell thick, so light can pass through it and allow the cells to be seen clearly

Thin slices of this kind are called **sections**. If you cut *along the length* of the structure, you are taking a **longitudinal section** (Figure 1.2b). Figure 1.1 shows a longitudinal section, which passes through two small developing leaves near the tip of the shoot, and two larger leaves below them. The leaves, buds and stem are all made up of cells. If you cut *across* the structure, you make a **transverse section** (Figure 1.2a).

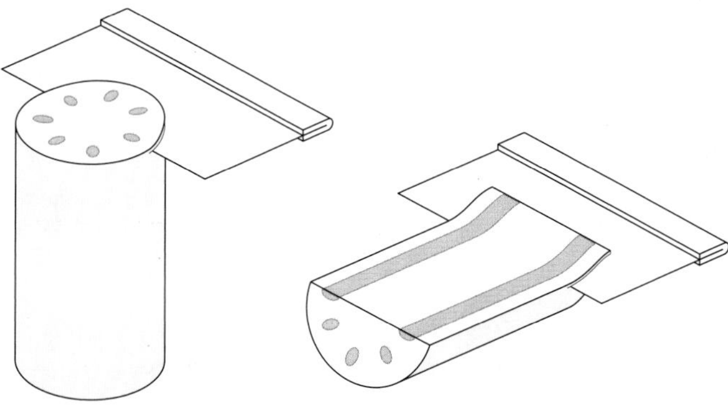

(a) transverse section **(b)** longitudinal section

Figure 1.2 Cutting sections of a plant stem

It is fairly easy to cut sections through plant structures just by using a razor blade. To cut sections of animal structures is more difficult because they are mostly soft and flexible. Pieces of skin, muscle or liver, for example, first have to be soaked in melted wax. When the wax goes solid it is then possible to cut thin sections. The wax is dissolved away after making the section.

When sections of animal structures are examined under the microscope, they, too, are seen to be made up of cells but they are much smaller than plant cells and need to be magnified more. The photomicrograph of kidney tissue in Figure 1.3 has been magnified 700 times to show the cells clearly. The sections are often treated with dyes, called 'stains', in order to show up the structures inside the cells more clearly.

Making sections is not the only way to study cells. Thin strips of plant tissue, only one cell thick, can be pulled off stems or leaves (Experiment 1, p. 9). Plant or animal tissue can be squashed or smeared on a microscope slide (Experiment 2, p. 10) or treated with chemicals to separate the cells before studying them.

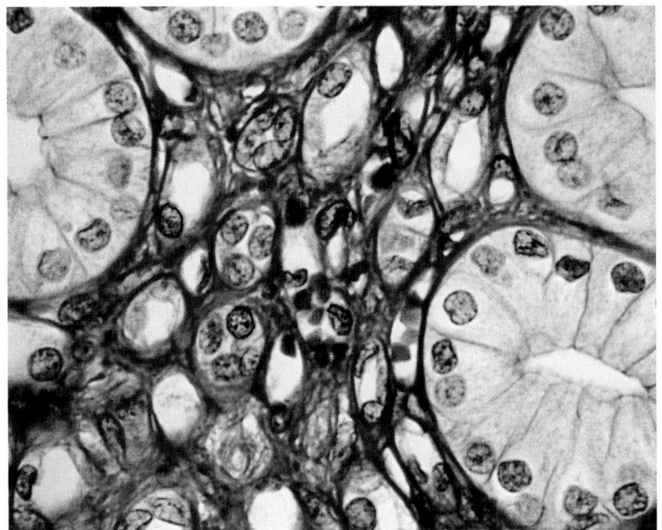

Figure 1.3 Transverse section through a kidney tubule (×700). A section through a tube will look like a ring (see Figure 1.12b on p. 7). In this case, each 'ring' consists of about 12 cells

There is no such thing as a typical plant or animal cell because cells vary a great deal in their size and shape depending on their function. Nevertheless, it is possible to make a drawing like Figure 1.4 to show features which are present in most cells. *All cells* have a **cell membrane**, which is a thin boundary enclosing the **cytoplasm**. Most cells have a **nucleus**.

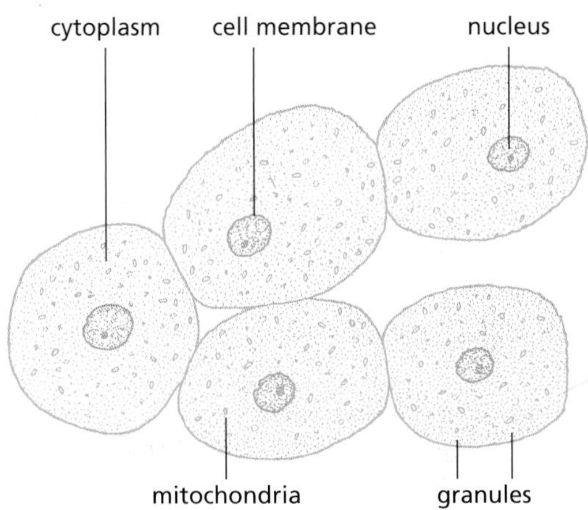

Figure 1.4 A group of liver cells. These cells have all the characteristics of animal cells

Cytoplasm

Under the ordinary microscope (light microscope), cytoplasm looks like a thick liquid with particles in it. In plant cells it may be seen to be flowing about. The particles may be food reserves such as oil droplets or granules of starch. Other particles are structures which have particular functions in the cytoplasm. These structures are the **organelles**. Examples are the **ribosomes**, which build up the cell's proteins (see p. 11) and the **mitochondria**, which generate energy for the cell's living processes (see p. 20).

When studied at much higher magnifications with the **electron microscope**, the cytoplasm no longer looks like a structureless jelly but appears to be organized into a complex system of membranes and vacuoles.

In the cytoplasm, a great many chemical reactions are taking place which keep the cell alive by providing energy and making substances that the cell needs (see pp. 11 and 20).

The liquid part of cytoplasm is about 90 per cent water with molecules of salts and sugars dissolved in it. Suspended in this solution there are larger molecules of fats (lipids) and proteins (see pp. 11–12). Lipids and proteins may be used to build up the cell structures, e.g. the membranes. Some of the proteins are **enzymes** (p. 14). Enzymes control the rate and type of chemical reactions which take place in the cells. Some enzymes are attached to the membrane systems of the cell, whereas others float freely in the liquid part of the cytoplasm.

Cell membrane

This is a thin layer of cytoplasm round the outside of the cell. It stops the cell contents from escaping and also controls the substances which are allowed to enter and leave the cell. In general, oxygen, food and water are allowed to enter; waste products are allowed to leave and harmful substances are kept out. In this way the cell membrane maintains the structure and chemical reactions of the cytoplasm.

Nucleus (plural = nuclei)

Most cells contain one nucleus, which is usually seen as a rounded structure enclosed in a membrane and embedded in the cytoplasm. In drawings of cells, the nucleus may be shown darker than the cytoplasm because, in prepared sections, it takes up certain stains more strongly than the cytoplasm. The function of the nucleus is to control the type and quantity of enzymes produced by the cytoplasm. In this way it regulates the chemical changes which take place in the cell. As a result, the nucleus determines what the cell will be, e.g. a blood cell, a liver cell, a muscle cell or a nerve cell.

The nucleus also controls cell division as shown in Figure 1.8 on p. 5. A cell without a nucleus cannot reproduce. Inside the nucleus are thread-like structures called **chromosomes**, which can be seen most easily at the time when the cell is dividing. (See p. 182 for a fuller account of chromosomes.)

Mitochondria

The mitochondria are tiny organelles present in plant and animal cells. They may be spherical, rod-like or elongated. They are most numerous in regions of rapid chemical activity and are responsible for producing energy from food substances (see 'Respiration', p. 19).

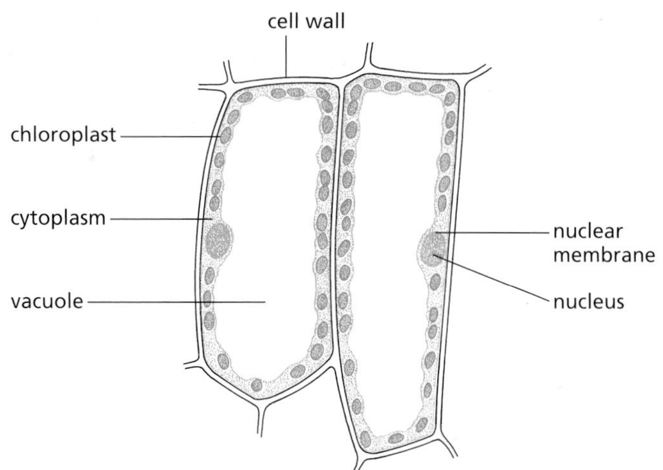

Figure 1.5 Palisade cells from a leaf

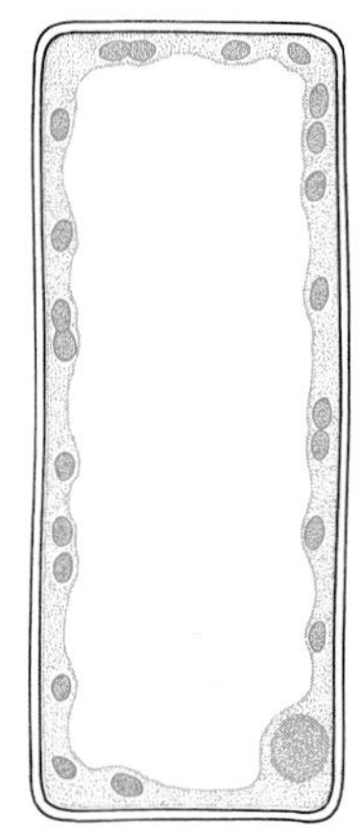

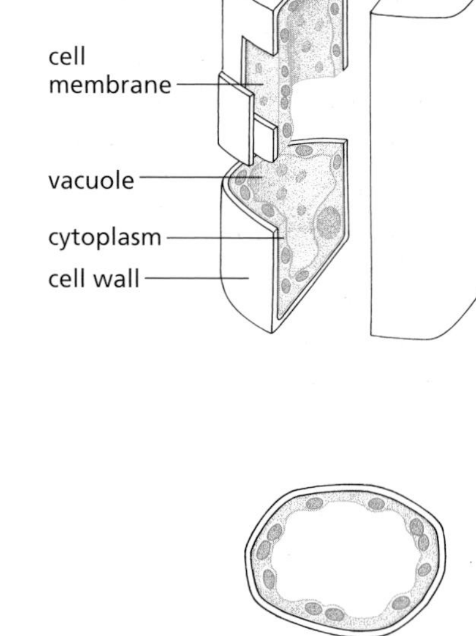

(a) longitudinal section **(b)** transverse section

Figure 1.6 Structure of a palisade cell. It is important to remember that, although cells look flat in sections or in thin strips of tissue, they are in fact three-dimensional and may seem to have different shapes according to the direction in which the section is cut. If the cell is cut across it will look like (b); if cut longitudinally it will look like (a)

Plant cells

A few generalized animal cells are represented by Figure 1.4, while Figure 1.5 is a drawing of two palisade cells from a plant leaf (see pp. 51 and 53).

Plant cells differ from animal cells in several ways:

1 Outside the cell membrane they all have a **cell wall** which contains cellulose and other compounds. It is non-living and allows water and dissolved substances to pass through. The cell wall is not selective like the cell membrane. (Note that plant cells *do* have a cell membrane but it is not easy to see or draw because it is pressed against the inside of the cell wall. See Figure 1.6.)

   Under the microscope, plant cells are quite distinct and easy to see because of their cell walls. In Figure 1.1 it is only the cell walls (and in some cases the nuclei) which can be seen. Each plant cell has its own cell wall but the boundary between two cells side by side does not usually show up clearly. Cells next to each other therefore appear to be sharing the same cell wall.

2 Most mature plant cells have a large, fluid-filled space called a **vacuole**. The vacuole contains **cell sap**, a watery solution of sugars, salts and sometimes pigments. This large, central vacuole pushes the cytoplasm aside so that it forms just a thin lining inside the cell wall. It is the outward pressure of the vacuole on the cytoplasm and cell wall which makes plant cells and their tissues firm (see p. 30). Animal cells may sometimes have small vacuoles in their cytoplasm but they are usually produced to do a particular job and are not permanent.

3 In the cytoplasm of plant cells are many organelles called **plastids** which are not present in animal cells. If they contain the green substance **chlorophyll**, the organelles are called **chloroplasts** (see p. 38). Colourless plastids usually contain starch, which is used as a food store.

The shape of a cell when seen in a transverse section may be quite different when the same cell is seen in a longitudinal section and Figure 1.6 shows why this is so. Figures 6.10b and 6.10c on p. 55 show the appearance of cells in a stem vein as seen in transverse and longitudinal section.

Questions

1 **a** What structures are usually present in all cells, whether they are from an animal or from a plant?
  b What structures are present in plant cells but not in animal cells?

2 What cell structure is largely responsible for controlling the entry and exit of substances into or out of the cell?

3 In what way does the red blood cell shown in Figure 12.1 on p. 108 differ from most other animal cells?

4 How does a cell membrane differ from a cell wall?

5 Why does the cell shown in Figure 1.6b appear to have no nucleus?

6 **a** In order to see cells clearly in a section of plant tissue, would you have to magnify the tissue
  (i) $\times 5$,
  (ii) $\times 10$,
  (iii) $\times 100$ or
  (iv) $\times 1000$?
  b What is the approximate width (in mm) of one of the largest cells in Figure 1.3?

7 In Figure 1.3, the cell membranes are not always clear. Why is it still possible to decide roughly how many cells there are in each tubule section?

Cell division and cell specialization

Cell division

When plants and animals grow, their cells increase in number by dividing. Typical growing regions are the ends of bones, layers of cells in the skin, root tips and buds (Figure 1.10). Each cell divides to produce two daughter cells. Both daughter cells may divide again, but usually one of the cells grows and changes its shape and structure and becomes adapted to do one particular job – in other words, it becomes **specialized** (Figure 1.7). At the same time it loses its ability to divide any more. The other cell is still able to divide and so continue the growth of the tissue. **Growth** is, therefore, the result of cell division, followed by cell enlargement and, in many cases, cell specialization.

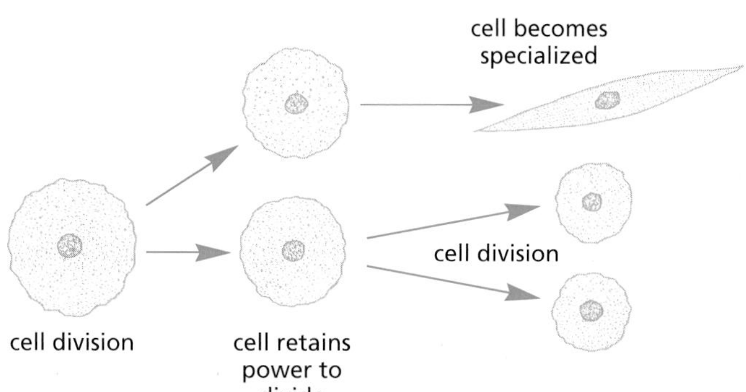

Figure 1.7 Cell division and specialization. Cells which retain the ability to divide are sometimes called **stem cells**

The process of cell division in an animal cell is shown in Figure 1.8. The events in a plant cell are shown in Figures 1.9 and 1.10. Because of the cell wall, the cytoplasm cannot simply pinch off in the middle, and a new wall has to be laid down between the two daughter cells. Also a new vacuole has to form.

Organelles such as mitochondria and chloroplasts are able to divide and are shared more or less equally between the daughter cells at cell division.

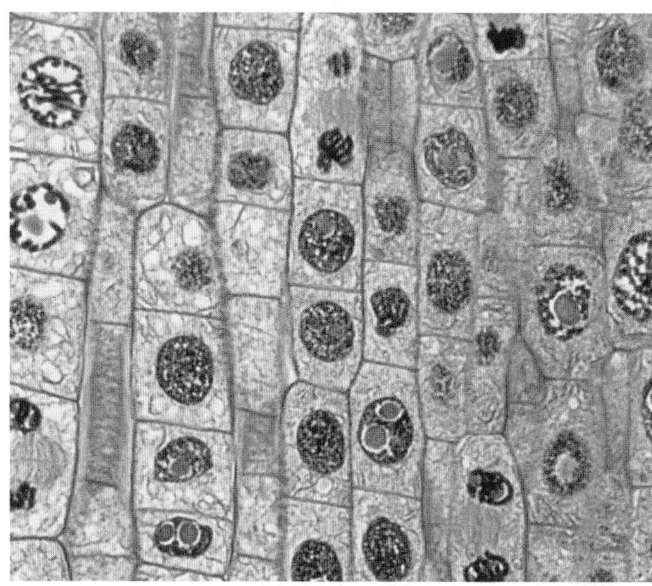

Figure 1.10 Cell division in an onion root tip (×300). The nuclei are stained blue. Most of the cells have just completed cell division

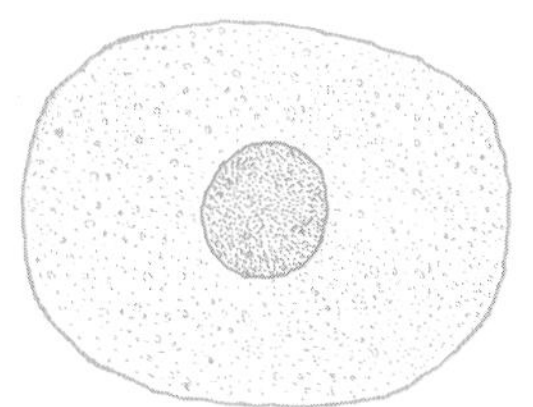

(a) Animal cell about to divide.

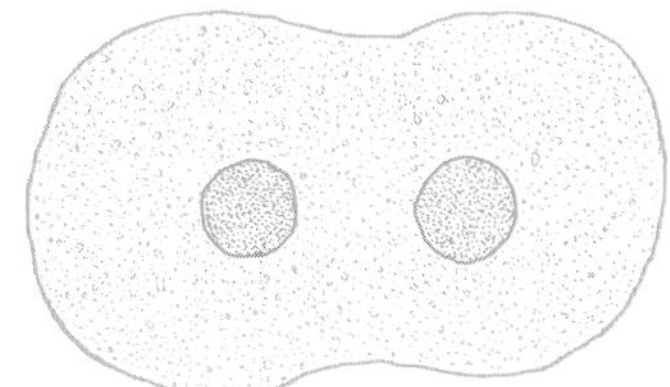

(b) The nucleus divides first.

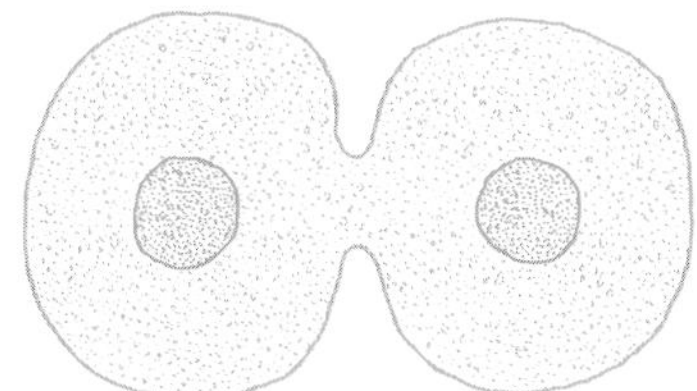

(c) The daughter nuclei separate and the cytoplasm pinches off between the nuclei.

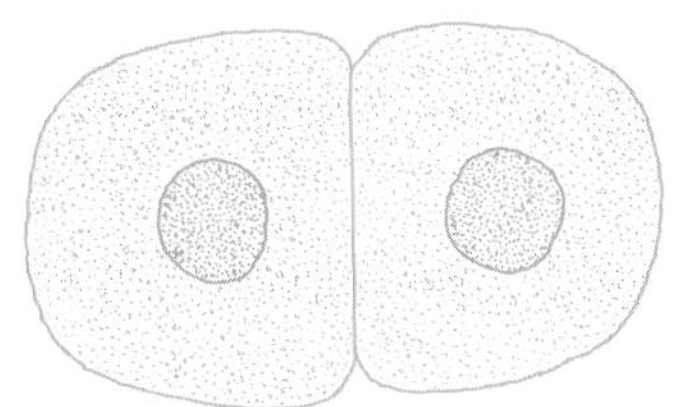

(d) Two cells are formed – one may keep the ability to divide, and the other may become specialized.

Figure 1.8 Cell division in an animal cell

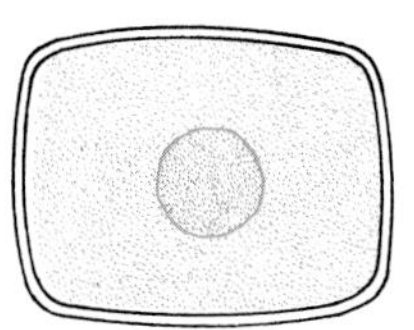

(a) A plant cell about to divide has a large nucleus and no vacuole.

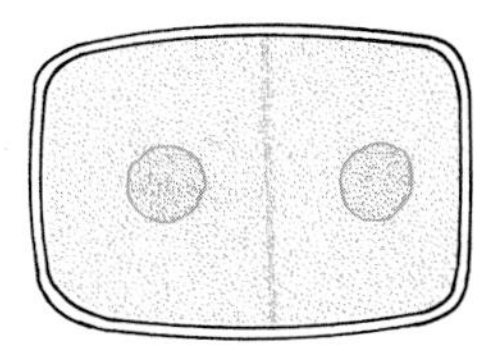

(b) The nucleus divides first. A new cell wall develops and separates the two cells.

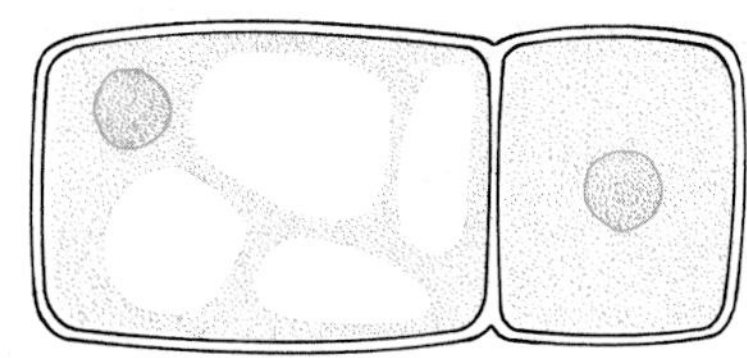

(c) The cytoplasm adds layers of cellulose on each side of the new cell wall. Vacuoles form in the cytoplasm of one cell.

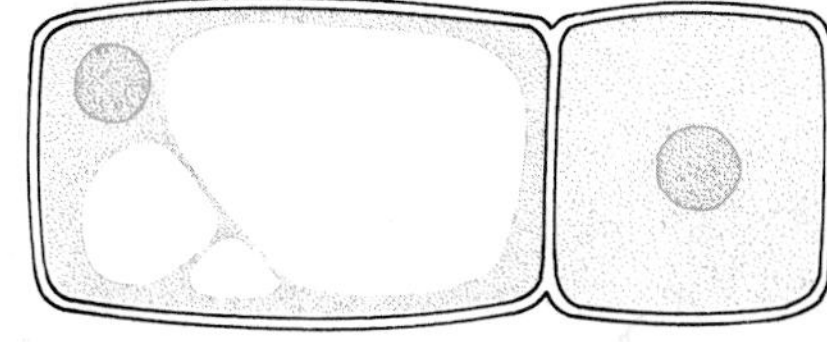

(d) The vacuoles join up to form one vacuole. This takes in water and makes the cell bigger. The other cell will divide again.

Figure 1.9 Cell division in a plant cell

Specialization of cells

Most cells, when they have finished dividing and growing, become specialized (Figure 1.11). This means that:

- They do one particular job.
- They develop a distinct shape.
- Special kinds of chemical change take place in their cytoplasm. The changes in shape and chemical reactions enable the cell to carry out its special function. Nerve cells and guard cells are examples of specialized cells.

Nerve cells (Figure 1.11e):

- Conduct electrical impulses to and from the brain.
- Some of them are very long and connect distant parts of the body to the spinal cord and brain.
- Their chemical reactions cause the impulses to travel along the fibre.

Root hair cells (Figure 1.11d):

- Absorb water and mineral salts from the soil.
- The hair-like projection penetrates between the soil particles and offers a large absorbing surface.
- The cell membrane is able to control which dissolved substances enter the cell.

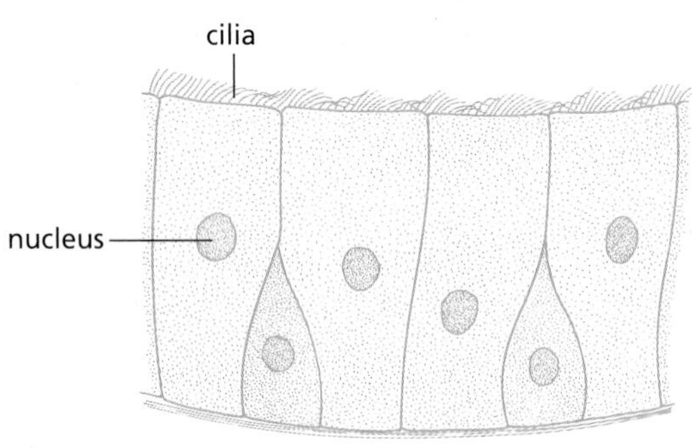

(a) ciliated cells
These form the lining of the nose and windpipe, and the tiny cytoplasmic 'hairs', called cilia, are in a continually flicking movement keeping up a stream of fluid (mucus) that carries dust and bacteria away from the lungs.

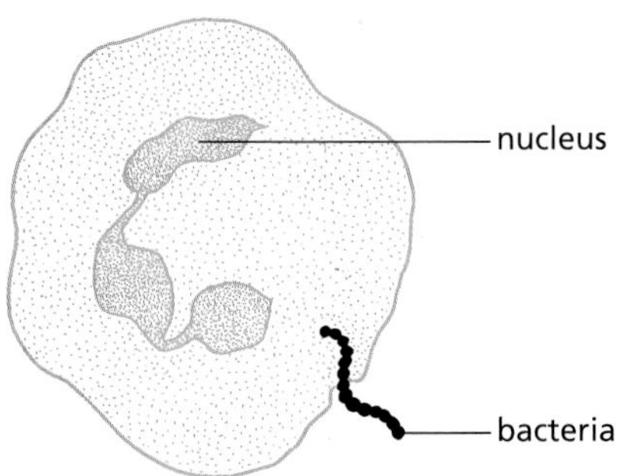

(b) white blood cell
Occurs in the bloodstream and is specialized for engulfing harmful bacteria. It is able to change its shape and move about, even through the walls of blood vessels into the surrounding tissues.

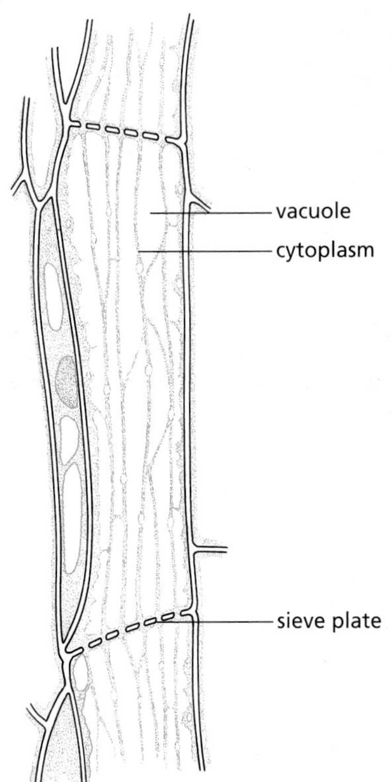

(c) food-conducting cell in a plant (phloem cell)
Long cells, joined end to end, and where they meet, perforations occur in the walls. Through these holes the cytoplasm of one cell communicates with the next. Dissolved food is thought to pass through the holes during its transport through the stem.

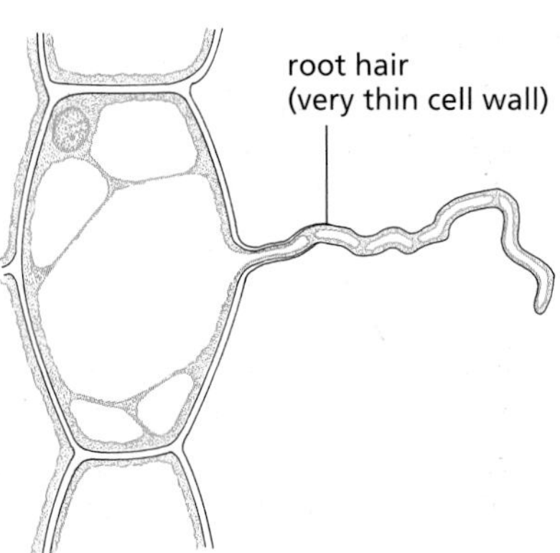

(d) root hair cell
These cells, in their thousands, form the outer layer of young roots and present a vast surface for absorbing water and mineral salts. (*See* Figures 6.14, 7.6 on pp. 57, 63.)

nerve fibre

cell body

nucleus

(e) nerve cell
Specialized for conducting impulses of an electrical nature along the fibre. The fibres may be very long, e.g. from the foot to the spinal column.

Figure 1.11 Specialized cells (not to scale)

The specialization of cells to carry out particular functions in an organism is sometimes referred to as **'division of labour'** within the organism. Similarly, the special functions of mitochondria, ribosomes and other cell organelles may be termed 'division of labour' within the cell.

Questions

1 Select from the following events and put them in the correct order for cell division in
(i) animal cells,
(ii) plant cells:
a cytoplasm divides,
b vacuole forms in one cell,
c new cell wall separates cells,
d nucleus divides.

2 Look at Figure 6.2 on p. 51.
a Whereabouts in a leaf are the food-carrying cells?
b What other specialized cells are there in the leaf?

Tissues and organs

There are some microscopic organisms that consist of one cell only and can carry out all the processes necessary for their survival (see p. 268). The cells of the larger plants and animals cannot survive on their own. A muscle cell could not obtain its own food and oxygen. Other specialized cells have to provide the food and oxygen needed for the muscle cell to live. Unless these cells are grouped together in large numbers and made to work together, they cannot exist for long.

Tissues

A tissue such as bone, nerve or muscle in animals, and epidermis, phloem or pith (p. 54) in plants, is made up of many hundreds of cells of a few types. The cells of each type have similar structures and functions so that the tissue itself can be said to have a particular function, e.g. nerves conduct impulses, phloem carries food in plants. Figure 1.12 shows how some cells are arranged to form simple tissues.

Organs

Organs consist of several tissues grouped together to make a structure with a special function. For example, the stomach is an organ which contains tissues made from epithelial cells, gland cells and muscle cells. These cells are supplied with food and oxygen brought by blood vessels. The stomach also has a nerve supply. The heart, lungs, intestines, brain and eyes are further examples of organs in animals. In flowering plants, the root, stem and leaves are the organs. The tissues of the leaf are epidermis, palisade tissue, spongy tissue, xylem and phloem (see pp. 39 and 50–3).

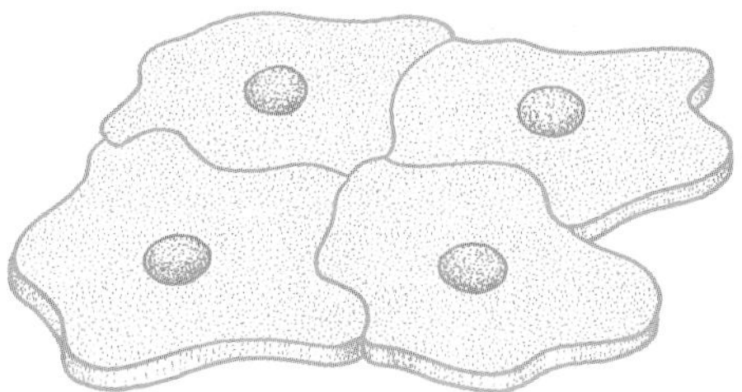

(a) cells forming an epithelium
A thin layer of tissue, e.g. the lining of the mouth cavity. Different types of epithelium form the internal lining of the windpipe, air passages, food canal, etc., and protect these organs from physical or chemical damage.

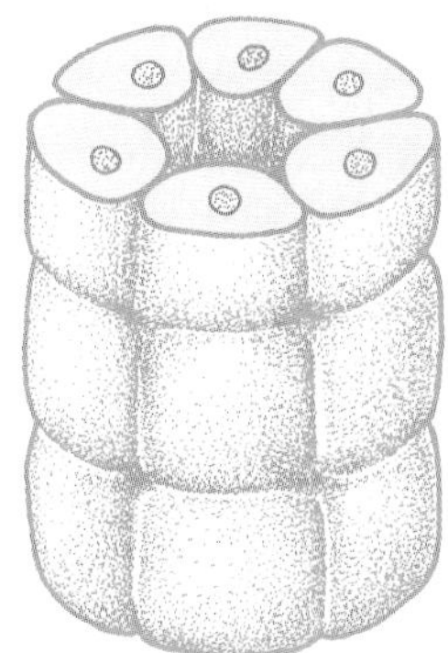

(b) cells forming a small tube
e.g. a kidney tubule (see p. 132). Tubules such as this carry liquids from one part of an organ to another.

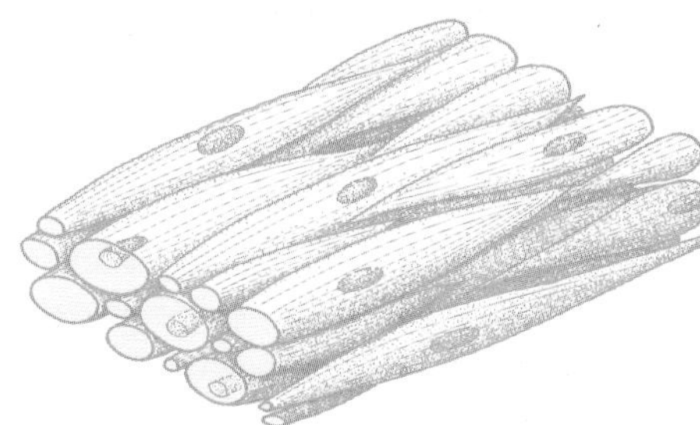

(c) one kind of muscle cell
Forms a sheet of muscle tissue. Blood vessels, nerve fibres and connective tissues will also be present. Contractions of this kind of muscle help to move food along the food canal or to close down small blood vessels.

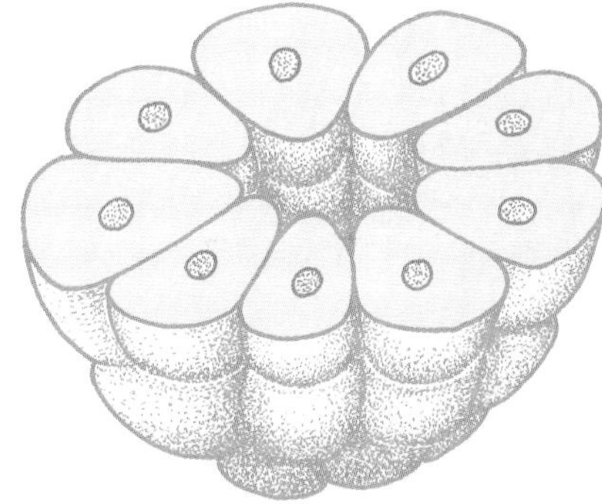

(d) cells forming part of a gland
The cells make chemicals which are released into the central space and carried away by a tubule such as shown in **(b)**. Hundreds of cell groups like this would form a gland like the salivary gland.

Figure 1.12 How cells form tissues

System

A system usually refers to a group of organs whose functions are closely related. For example, the heart and blood vessels make up the **circulatory system**; the brain, spinal cord and nerves make up the **nervous system** (Figure 1.13). In a flowering plant, the stem, leaves and buds make up a system called the shoot (p. 50).

Organism

An organism is formed by the organs and systems working together to produce an independent plant or animal.

An example in the human body of how cells, tissues and organs are related is shown in Figure 1.14.

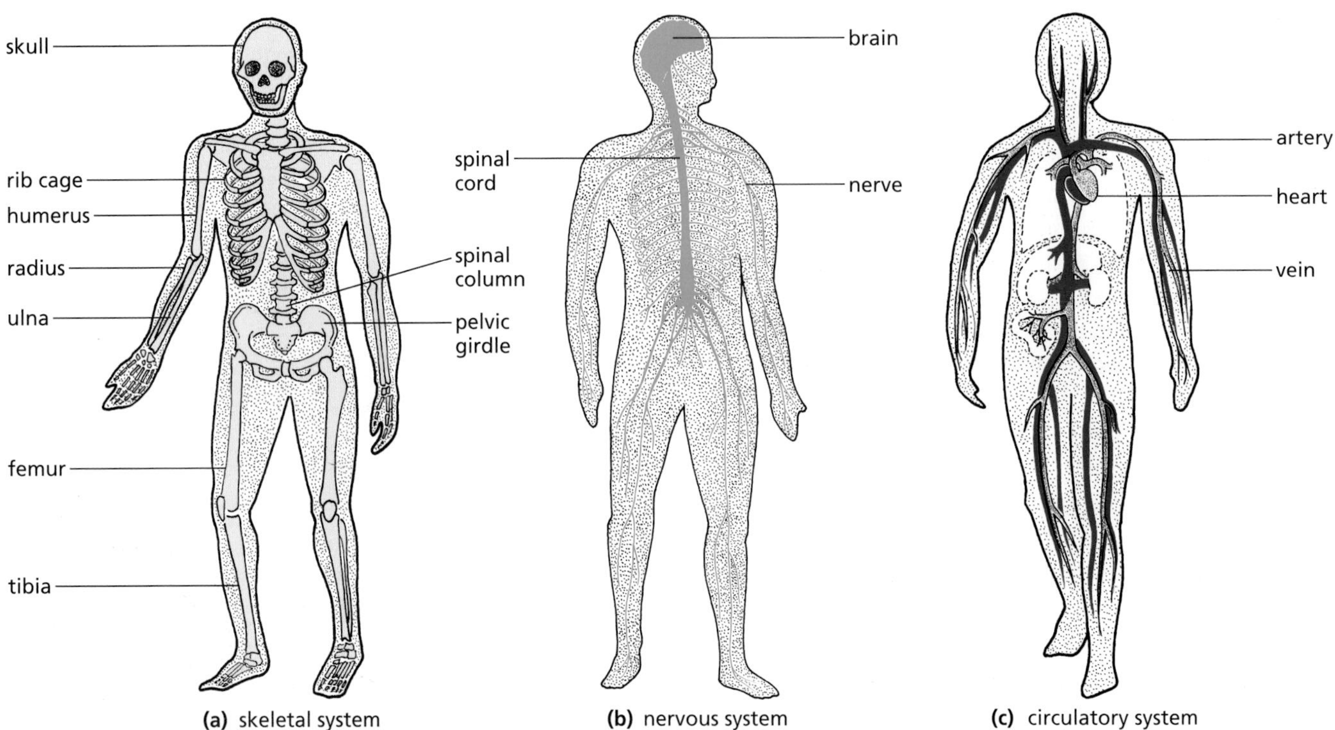

Figure 1.13 Three examples of systems in the human body

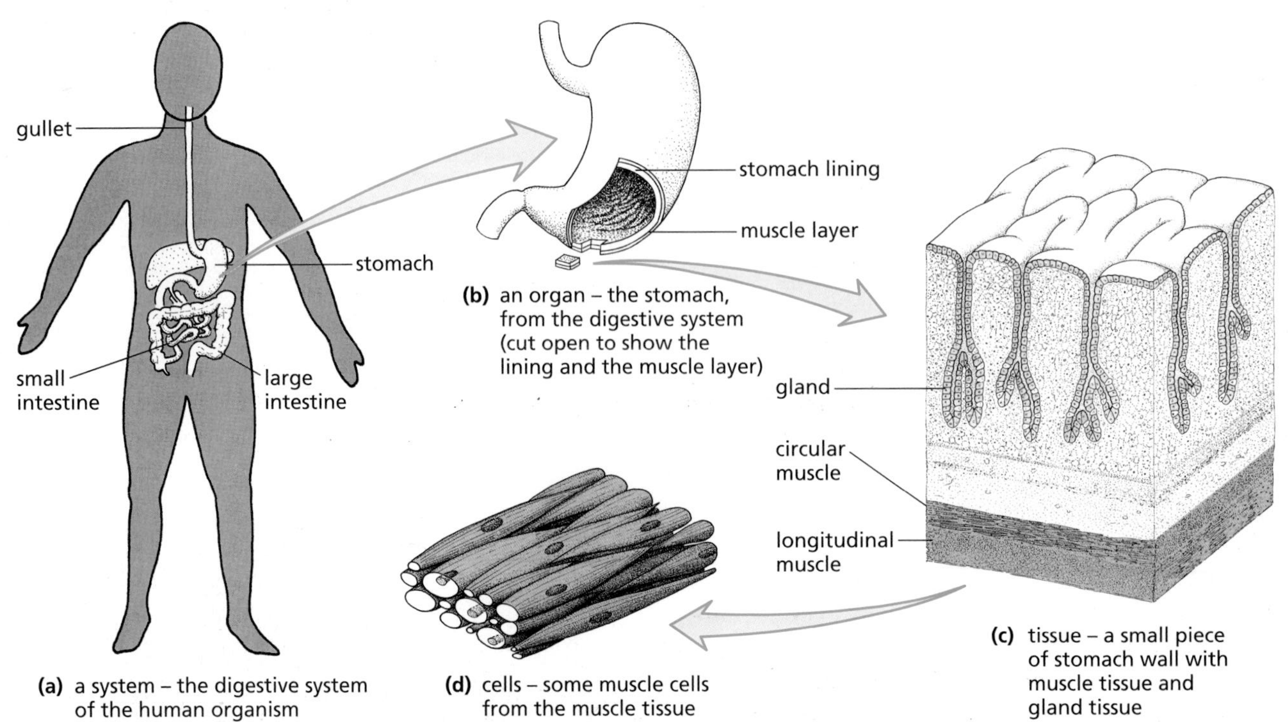

Figure 1.14 An example of how cells, tissues and organs are related

Tissue culture

It is possible to take samples of developing animal tissues, separate the cells by means of enzymes and make the cells grow and divide in shallow dishes containing a nutrient solution. This technique is called **tissue culture**. The cells do not become specialized but may move about, establish contact with each other and eventually cover the floor of the culture dish with a layer, one cell thick. At this point, they stop dividing unless they are separated and transferred to fresh culture vessels. Even so, most mammal cells cease to reproduce after about 20 divisions.

These tissue cultures are used to study cells and cell division, to test out new drugs and vaccines, to culture viruses or to test the effect of possible harmful chemicals. For some tests and experiments, tissue cultures are able to take the place of laboratory animals.

Large-scale tissue cultures are being developed in order to obtain cell chemicals which may be useful in medical treatments.

Plant tissue culture is described on p. 81.

Questions

1 Say whether you think the following are cells, tissues, organs or organisms: lungs (p. 123), root hair (p. 56), mesophyll (p. 53), multi-polar neurone (p. 164).

2 What tissues are shown in the following drawings: Figure 36.13 on p. 315; Figure 19.9 on p. 166?

3 Look at Figure 11.5 on p. 99.
 a Which of the structures do you think are organs?
 b What system is represented by the drawing?

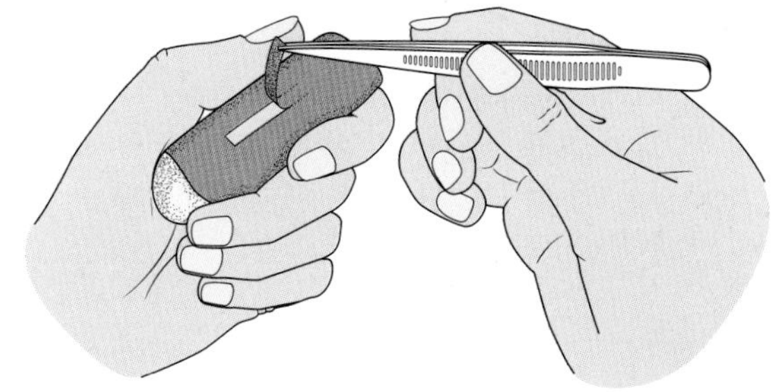

(a) peel a strip of red epidermis from a piece of rhubarb skin

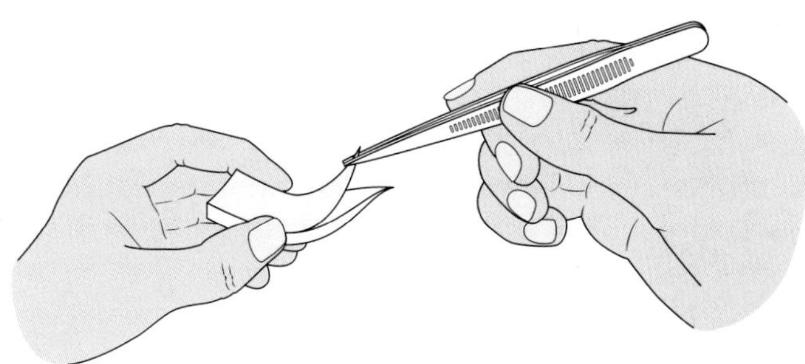

(b) alternatively, peel the epidermis from the inside of an onion scale

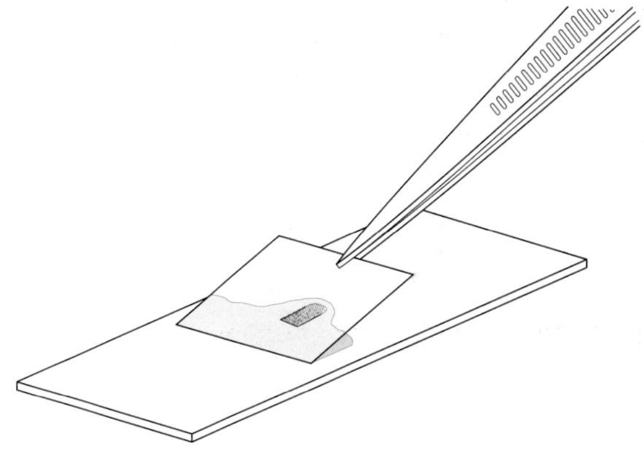

(c) place the epidermis in a drop of water or weak iodine solution on a slide and carefully lower a cover-slip over it

Figure 1.15 Looking at plant cells

Practical work

1 Plant cells

The outer layer of cells (epidermis) from a stem or an onion scale can be stripped off as shown in Figure 1.15a and b. A piece of onion or a rhubarb stalk is particularly suitable for this. A small piece of this epidermis is placed in a drop of weak iodine solution on a microscope slide and covered with a cover-slip (Figure 1.15c). The tissue is then studied under the microscope. The iodine will stain the cell nuclei pale yellow and the starch grains will stain blue. If red epidermis from rhubarb stalk is used, you will see the red cell sap in the vacuoles.

To see chloroplasts, use fine forceps to pull a leaf from a moss plant, mount it in water on a slide as before and examine it with the high power objective of the microscope (Figure 1.16). At the base of the leaf, the chloroplasts are less densely packed and, therefore, easier to see.

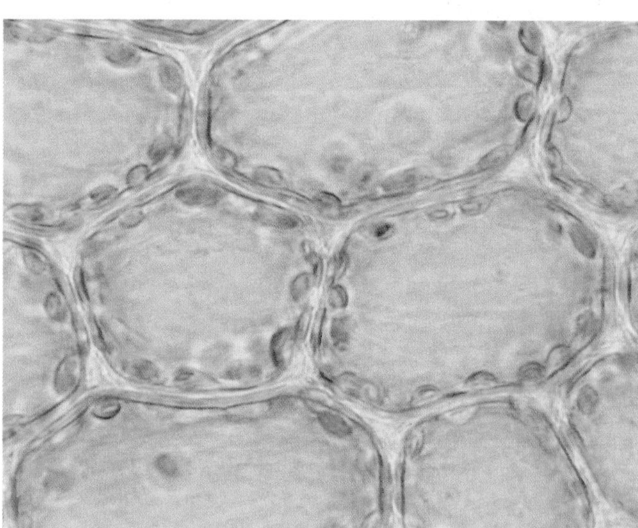

Figure 1.16 Cells in a moss leaf (× 500). The vacuole occupies most of the space in each cell. The chloroplasts are confined to the layer of cytoplasm lining the cell wall

2 Animal cells

Note The Department of Education and Science recommends that schools no longer use the technique which involves studying the epithelial cells which appear in a smear taken from the inside of the cheek, because of the very small risk of transmitting the AIDS virus. Some Local Education Authorities may therefore forbid the use of this technique in their schools, but the Institute of Biology suggests that if the following procedure is adopted the risk is negligible (*Biologist*, 35 (4) p. 211, September 1988).

Cotton buds from a freshly opened pack are rubbed lightly on the inside of the cheek and gums. The buds are rubbed on to clean slides and then dropped into a container of absolute alcohol. The smear on the slide is covered with a few drops of methylene blue solution before being examined under the microscope (Figure 1.17). The slides are placed in laboratory disinfectant before washing.

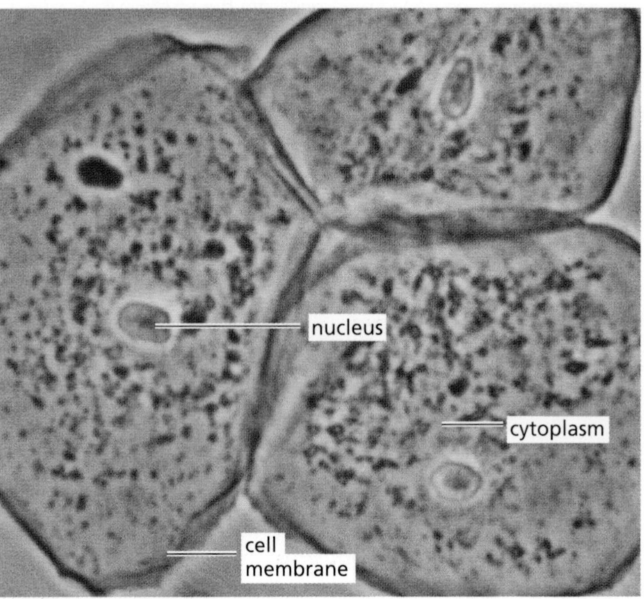

Figure 1.17 Cells from the lining epithelium of the cheek (× 1500)

An alternative method of obtaining cells is to press some 'Sellotape' on to a well-washed wrist. When the tape is removed and studied under the microscope, cells with nuclei can be seen. A few drops of methylene blue solution will stain the cells and make the nuclei more distinct.

Checklist

- Nearly all plants and animals are made up of thousands or millions of microscopic cells.
- All cells contain cytoplasm enclosed in a cell membrane.
- Most cells have a nucleus.
- Cytoplasm contains organelles such as mitochondria, chloroplasts and ribosomes.
- Many chemical reactions take place in the cytoplasm to keep the cell alive.
- The nucleus directs the chemical reactions in the cell and also controls cell division.
- Plant cells have a cellulose cell wall and a large central vacuole.
- Cells are often specialized in their shapes and activities to carry out particular jobs.
- Large numbers of similar cells packed together form a tissue.
- Different tissues arranged together form organs.
- A group of related organs makes up a system.

2 The chemicals of living cells

Cell physiology
Description.

Chemical components of cells
Water, proteins, lipids, carbohydrates, salts, ions, vitamins; their chemical structure and role in the cell. Interconversion of substances in cells.

Enzymes
Definition. Methods of action. Effects of temperature and pH. Enzyme specificity. Intra- and extracellular enzymes.

Practical work
Experiments with enzymes: catalase, starch phosphorylase, pH and temperature.

Cell physiology

The term 'physiology' refers to all the normal functions that take place in a living organism. Digestion of food, circulation of blood and contraction of muscles are some aspects of human physiology. Absorption of water by roots, production of food in the leaves, and growth of shoots towards light are examples of plant physiology. The next three chapters are concerned with physiological events in individual cells.

The physiology of a whole organism is, in some ways, the sum of the physiology of its component cells. If all cells need a supply of oxygen then the whole organism must take in oxygen. Cells need chemical substances to make new cytoplasm and to produce energy. Therefore the organism must take in food to supply the cells with these substances. Of course, it is not quite as simple as this; most cells have specialized functions (p. 6) and so have differing needs. However, all cells need water, oxygen, salts and food substances and all cells consist of water, proteins, lipids, carbohydrates, salts and vitamins or their derivatives.

Chemical components of cells

Water

Most cells contain about 75 per cent of water and will die if their water content falls much below this. Water is a good solvent and many substances move about the cells in a watery solution. Water molecules take part in a great many vital chemical reactions. For example, in green plants, water combines with carbon dioxide to form sugar (see 'Photosynthesis', p. 35). In animals, water helps to break down and dissolve food molecules (see 'Digestion', p. 97).

The physical and chemical properties of water differ from those of most other liquids but make it uniquely effective in supporting living activities. For example, water has a high capacity for heat (high thermal capacity). This means that it can absorb a lot of heat without its temperature rising to levels which damage the proteins in the cytoplasm (see below). However, because water freezes at 0 °C most cells are damaged if their temperature falls below this and ice crystals form in the cytoplasm. (Oddly enough, rapid freezing of cells in liquid nitrogen at below −196 °C does not harm them.)

Proteins

Some proteins contribute to the structures of the cell, e.g. to the cell membranes, the mitochondria, ribosomes and chromosomes. These proteins are called **structural proteins**.

There is another group of proteins called **enzymes**. Enzymes are present in the membrane systems, in the mitochondria, in special vacuoles and in the fluid part of the cytoplasm. Enzymes control the chemical reactions which keep the cell alive (see p. 14).

Although there are many different types of protein, all contain carbon, hydrogen, oxygen and nitrogen, and many contain sulphur. Their molecules are made up of long chains of simpler chemicals called **amino acids**. There are about 20 different amino acids in animal proteins, including alanine, leucine, valine, glutamine, cysteine, glycine and lysine. A small protein molecule might be made up from a chain consisting of a hundred or so amino acids, e.g. glycine–valine–valine–cysteine–leucine–glutamine–, etc. Each type of protein has its amino acids arranged in a particular order.

The chain of amino acids in a protein takes up a particular shape as a result of cross-linkages. Cross-linkages form between amino acids which are not neighbours, as shown in Figure 2.1. The shape of a protein molecule has a very important effect on its reactions with substances, as explained in 'Enzymes' on p. 14.

When a protein is heated to temperatures over 50 °C, the cross-linkages in its molecules break down; the protein molecules lose their shape and will not usually regain it even when cooled. The protein is said to have been **denatured**. Because the shape of the molecules has been altered, the protein will have lost its original properties.

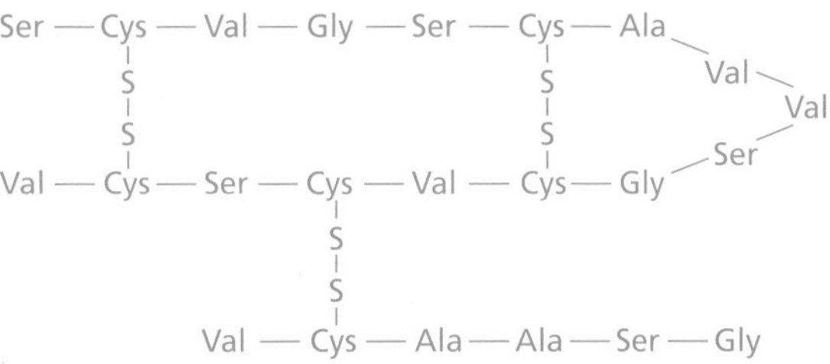

Figure 2.1 A small imaginary protein made from only five different kinds of amino acid. Note that cross-linkage occurs between cysteine molecules with the aid of sulphur atoms

Egg-white is a protein. When it is heated, its molecules change shape and the egg-white goes from a clear, runny liquid to a white solid and cannot be changed back again. The egg-white protein, albumen, has been denatured by heat.

Proteins form enzymes and many of the structures in the cell, so if they are denatured the enzymes and the cell structures will stop working and the cell will die. Whole organisms may survive for a time above 50 °C depending on the temperature, the period of exposure and the proportion of the cells which are damaged.

Lipids

Lipids are oils or fats and substances related to or derived from them. Fats are formed from carbon, hydrogen and oxygen only. A molecule of fat is made up of three molecules of an organic acid, called a **fatty acid**, combined with one molecule of **glycerol**.

$$\text{glycerol}\left\{\begin{array}{l} H_2{-}C{-}O{-} \\ \quad\ \ | \\ \ H{-}C{-}O{-} \\ \quad\ \ | \\ H_2{-}C{-}O{-} \end{array}\right. \begin{array}{l} \text{stearic acid} \\ \\ \text{oleic acid} \\ \\ \text{palmitic acid} \end{array}\left.\vphantom{\begin{array}{l} H \\ | \\ H \\ | \\ H \end{array}}\right\}\text{fatty acids}$$

Lipids form part of the cell membrane and the internal membranes of the cell such as the nuclear membrane. Droplets of fat or oil form a source of energy when stored in the cytoplasm.

Carbohydrates

These may be simple, soluble sugars or complex materials like starch and cellulose, but all carbohydrates contain carbon, hydrogen and oxygen only. A commonly occurring simple sugar is **glucose**, whose chemical formula is $C_6H_{12}O_6$.

The glucose molecule is often in the form of a ring, represented as

CH_2OH … O … H, HO, OH, H, OH, OH (ring structure)

or more simply as

Two molecules of glucose can be combined to form a molecule of maltose $C_{12}H_{22}O_{11}$ (p.14).

Sugars with a single carbon ring are called **monosaccharides**, e.g. glucose and fructose. Those sugars with two carbon rings in their molecules are called **disaccharides**, e.g. maltose and sucrose. Mono- and disaccharides are readily soluble in water.

When many glucose molecules are joined together, the carbohydrate is called a **polysaccharide**. **Glycogen** (Figure 2.2) is a polysaccharide which forms a food storage substance in many animal cells. The **starch** molecule is made up of hundreds of glucose molecules joined together to form long chains. Starch is an important storage substance in the plastids of plant cells. **Cellulose** consists of even longer chains of glucose molecules. The chain molecules are grouped together to form microscopic fibres, which are laid down in layers to form the cell wall in plant cells (Figures 2.3 and 2.4).

Polysaccharides are not readily soluble in water.

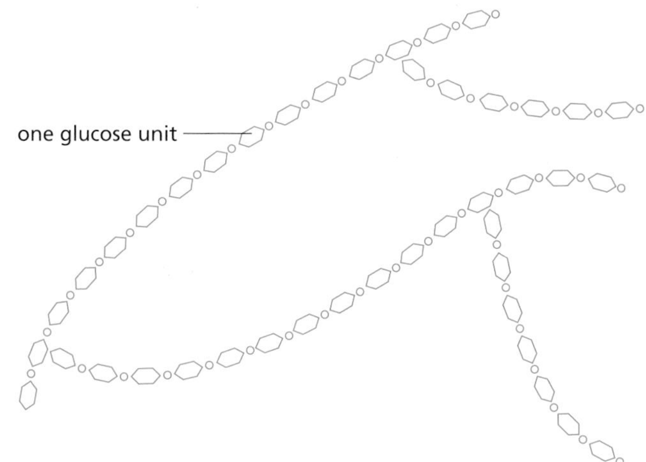

Figure 2.2 Part of a glycogen molecule

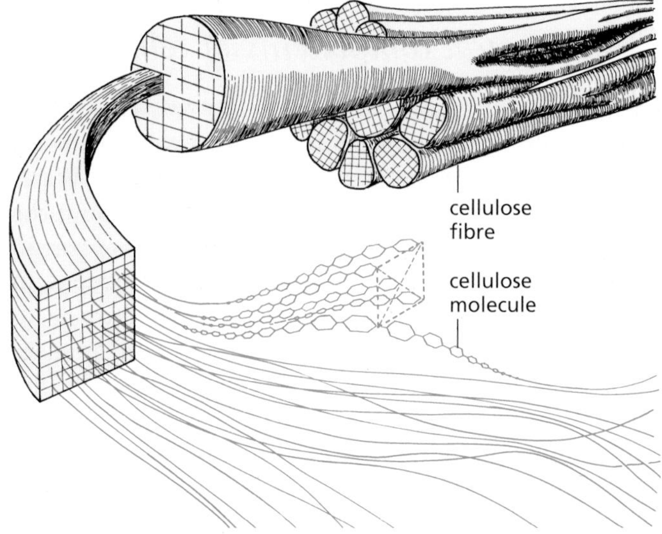

Figure 2.3 Cellulose. Plant cell walls are composed of long, interwoven and interconnected cellulose fibres which are large enough to be seen with the electron microscope. Each fibre is made up of many long-chain cellulose molecules

Figure 2.4 Electron micrograph of a plant cell wall ($\times$ 20 000) showing the cellulose fibres

Salts

Salts are present in cells in the form of their ions, e.g. sodium chloride in a cell will exist as sodium ions (Na^+) and chlorine ions (Cl^-). The ions may be free to move about in the water of the cell or attached to other molecules such as the proteins or lipids.

Ions take part in and influence many chemical reactions in the cell, e.g. phosphate ions (PO_4^{3-}) are essential for energy transfer reactions. Ions are also involved in determining how much water enters or leaves the cell (see 'Osmosis', p. 29). Calcium, potassium and sodium ions are particularly important in chemical changes related to electrical activities of a cell, e.g. responding to stimuli or conducting nerve impulses. A shortage or excess of ions in the cells upsets their physiology and affects the normal functioning of the whole organism.

Vitamins

This is a category of substances which, in their chemical structure at least, have little in common. Plants can make their own vitamins. Animals have to obtain many of their vitamins ready-made. Vitamins, or substances derived from them, play a part in chemical reactions in cells – for example those which involve a transfer of energy from one compound to another. If cells are not supplied with vitamins or the substances needed to make them, the cell physiology is thrown out of order and the whole organism suffers.

Synthesis and conversion in cells

Cells are able to build up (synthesize) or break down their proteins, lipids and carbohydrates, or change one to the other. For example, animal cells synthesize glycogen from glucose by joining glucose molecules together (Figure 2.2); plant cells synthesize starch and cellulose from glucose. All cells can make proteins from amino acids and they can build up fats from glycerol and fatty acids. Animal cells can change carbohydrates to lipids, and lipids to carbohydrates; they can also change proteins to carbohydrates but they cannot make proteins unless they are supplied with amino acids. Plant cells, on the other hand, can make their own amino acids starting from sugars and salts. The cells in the green parts of plants can even make glucose starting from only carbon dioxide and water (see pp. 35–46).

Questions

1 Why do you think it is essential for us to include proteins in our diet?

2 Which substances are particularly important in
a forming the structures of the cell and
b storing food in the cell?

3 Which of the following – sugar, butter, iron sulphate, starch, sodium chloride, olive oil, cellulose – are
a carbohydrates,
b salts,
c lipids?

4 What additional elements are needed to convert a carbohydrate into a protein? Where do you think plants get these elements from?

5 What chemical changes can occur in a plant cell that do not take place in animal cells?

Enzymes

Enzymes are proteins that act as **catalysts**. They are made in all living cells. A catalyst is a chemical substance which speeds up a reaction but does not get used up during the reaction. One enzyme can be used many times over (Figure 2.5).

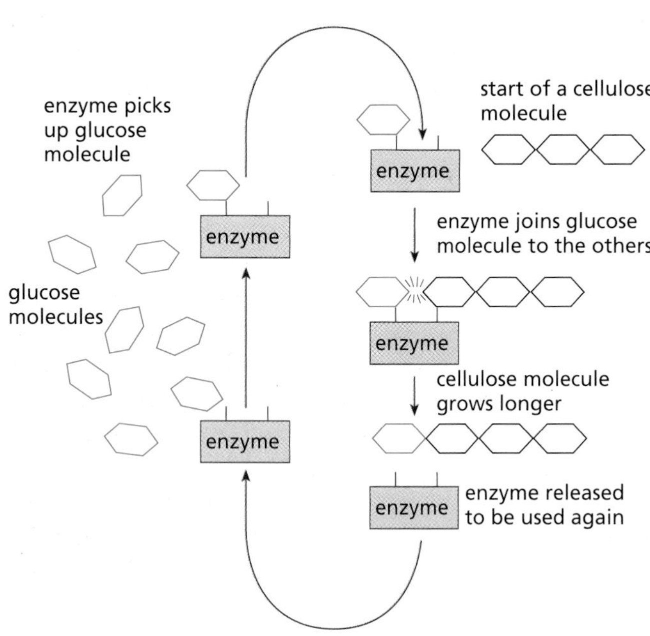

Figure 2.5 Building up a cellulose molecule

How an enzyme molecule might work to join two other molecules together and so form a more complicated substance is shown by the diagram in Figure 2.6.

An example of an enzyme-controlled reaction such as this is the joining up of two glucose molecules to form a molecule of maltose.

$$\underset{\text{glucose}}{C_6H_{12}O_6} + \underset{\text{glucose}}{C_6H_{12}O_6} \xrightarrow{\text{enzyme}} \underset{\text{maltose}}{C_{12}H_{22}O_{11}} + \underset{\text{water}}{H_2O}$$

In a similar way, hundreds of glucose molecules might be joined together, end to end, to form a long molecule of starch to be stored in the plastid of a plant cell. The glucose molecules might also be built up into a molecule of cellulose to be added to the cell wall. Protein molecules are built up by enzymes which join together tens or hundreds of amino acid molecules. These proteins are added to the cell membrane, to the cytoplasm or to the nucleus of the cell. They may also become the proteins which act as enzymes.

Reactions in which large molecules are built up from smaller molecules are called **anabolic** reactions (Figure 2.6a).

After the new substance has been formed, the enzyme is set free to start another reaction. Molecules of the two substances might have combined without the enzyme being present but they would have done so very slowly. By bringing the substances close together, the enzyme molecule makes the reaction take place much more rapidly. A chemical reaction which would take hours or days to happen on its own takes only a few seconds when the right enzyme is present.

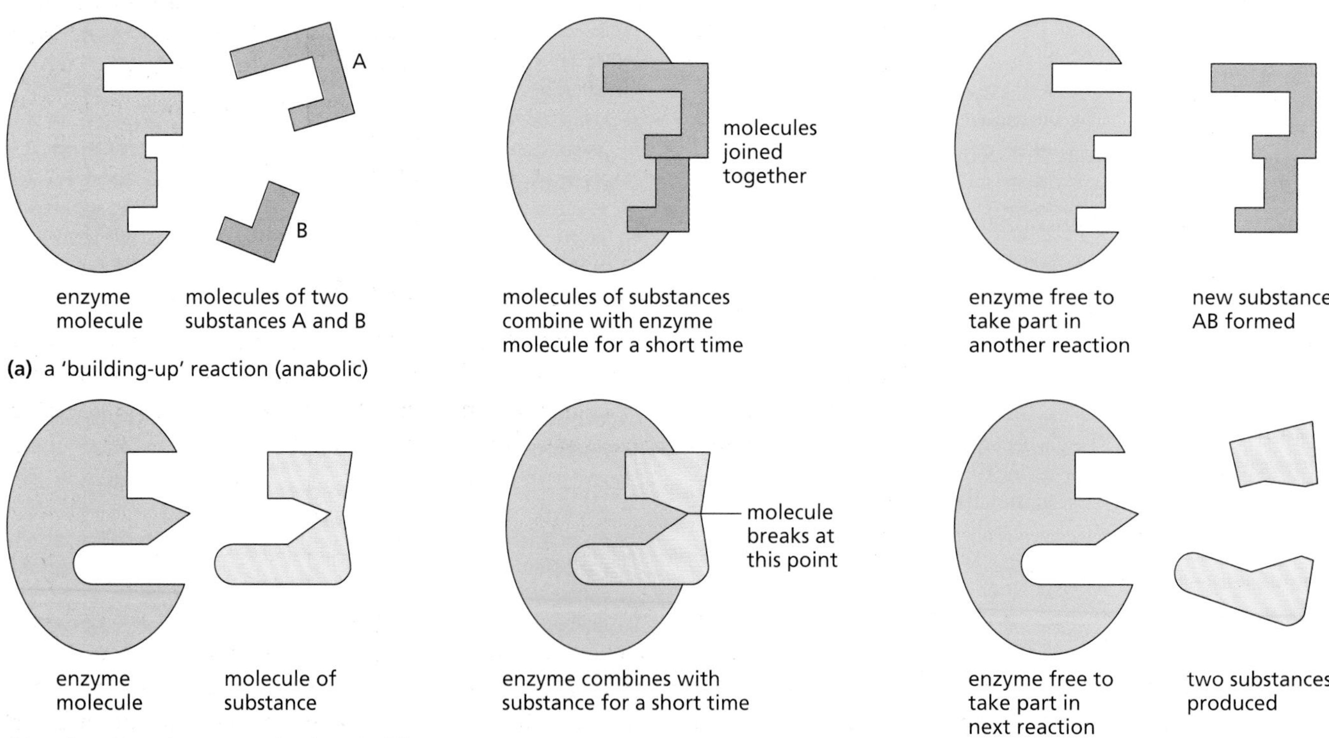

Figure 2.6 Possible explanation of enzyme action (the 'lock and key' model)

Figure 2.6b shows an enzyme speeding up a chemical change but this time it is a reaction in which the molecule of a substance is split into smaller molecules. If starch is mixed with water it will break down very slowly to sugar, taking several years. In your saliva there is an enzyme called **amylase** which can break down starch to sugar in minutes or seconds. In cells, many of the 'breaking-down' enzymes are helping to break down glucose to carbon dioxide and water in order to produce energy (see p. 19).

Reactions which split large molecules into smaller ones are called **catabolic** reactions.

Enzymes and temperature

A rise in temperature increases the rate of most chemical reactions; a fall in temperature slows them down. In many cases a rise of 10 °C will double the rate of reaction in a cell. This is equally true for enzyme-controlled reactions, but above 50 °C the enzymes, being proteins, are denatured and stop working.

Figure 2.6 shows how the shape of an enzyme molecule could be very important if it has to fit the substances on which it acts. Above 50 °C the shapes of enzymes are changed and the enzymes can no longer combine with the substances.

This is one of the reasons why organisms may be killed by prolonged exposure to high temperatures. The enzymes in their cells are denatured and the chemical reactions proceed too slowly to maintain life.

One way to test whether a substance is an enzyme is to heat it to boiling point. If it can still carry out its reactions after this, it cannot be an enzyme. This technique is used as a 'control' (see p. 23) in enzyme experiments.

Enzymes and pH

Acid or alkaline conditions alter the chemical properties of proteins, including enzymes. Most enzymes work best at a particular level of acidity or alkalinity (pH). The protein-digesting enzyme in your stomach, for example, works well at an acidity of pH 2. At this pH, the enzyme amylase, from your saliva, cannot work at all. Inside the cells, most enzymes will work best in neutral conditions (pH 7). The pH or temperature at which an enzyme works best is often called its **optimum** pH or temperature.

Although changes in pH affect the activity of enzymes, these effects are usually reversible, i.e. an enzyme which is inactivated by a low pH will resume its normal activity when its optimum pH is restored. Extremes of pH, however, may denature some enzymes irreversibly.

Enzymes are specific

This means simply that an enzyme which normally acts on one substance will not act on a different one. Figure 2.6a shows how the shape of an enzyme could decide what substances it combines with. The enzyme in Figure 2.6a has a shape which exactly fits the substances on which it acts, but would not fit the substance in Figure 2.6b. Thus, an enzyme which breaks down starch to maltose will not also break down proteins to amino acids. Also, if a reaction takes place in stages, e.g.

starch → maltose (stage 1)
maltose → glucose (stage 2)

a different enzyme is needed for each stage.

The names of enzymes usually end with **-ase** and they are named according to the substance on which they act, or the reaction which they promote. For example, an enzyme which acts on proteins may be called a **protease**; one which removes hydrogen from a substance is a **dehydrogenase**.

The substance on which an enzyme acts is called its **substrate**. Thus, the enzyme **sucrase** acts on the substrate **sucrose** to produce the monosaccharides glucose and fructose.

Rates of enzyme reactions

As explained above, the rate of an enzyme-controlled reaction depends on the temperature and pH. It also depends on the concentrations of the enzyme and its substrate. The more enzyme molecules produced by a cell, the faster the reaction will proceed, provided there are enough substrate molecules available. Similarly, an increase in the substrate concentration will speed up the reaction if there are enough enzyme molecules to cope with the additional substrate.

Intra- and extracellular enzymes

All enzymes are made inside cells. Most of them remain inside the cell to speed up reactions in the cytoplasm and nucleus. These are called **intracellular enzymes** ('intra' means 'inside'). In a few cases, the enzymes made in the cells are let out of the cell to do their work outside. These are **extracellular enzymes** ('extra' means 'outside'). Fungi (p. 286) and bacteria (p. 283) release extracellular enzymes in order to digest their food. A mould growing on a piece of bread releases starch-digesting enzymes into the bread and absorbs the soluble sugars which the enzyme produces from the bread. In the digestive systems of animals (p. 97), extracellular enzymes are released into the stomach and intestines in order to digest the food.

Practical work

Tests for proteins, fats and carbohydrates are described on pp. 95–6. Experiments on the digestive enzymes amylase and pepsin are described on pp. 105–6.

1 Extracting and testing an enzyme from living cells

In this experiment, the enzyme to be extracted and tested is **catalase**, and the substrate is hydrogen peroxide (H_2O_2). Certain reactions in the cell produce hydrogen peroxide, which is poisonous. Catalase renders the hydrogen peroxide harmless by breaking it down to water and oxygen.

$$2H_2O_2 \xrightarrow{\text{catalase}} 2H_2O + O_2$$

Grind a small piece of liver with about 20 cm^3 water and a little sand, in a mortar. This will break open the liver cells and release their contents. Filter the mixture and share it between two test-tubes, A and B. The filtrate will contain a great variety of substances dissolved out from the cytoplasm of the liver cells, including many enzymes. Because enzymes are specific, however, only one of these, catalase, will act on hydrogen peroxide. Add some drops of the filtrate from test-tube A to a few cm^3 hydrogen peroxide in a test-tube. You will see a vigorous reaction as the hydrogen peroxide breaks down to produce oxygen. (The oxygen can be tested with a glowing splint.)

Now boil the filtrate in tube B for about 30 seconds. Add a few drops of the boiled filtrate to a fresh lot of hydrogen peroxide. There will be no reaction because boiling has denatured the catalase.

Next, shake a little manganese(IV) oxide powder in a test-tube with some water and pour this into some hydrogen peroxide. There will be a vigorous reaction similar to the one with the liver extract. If you now boil some manganese(IV) oxide with water and add this to hydrogen peroxide, the reaction will still occur. Manganese(IV) oxide is a catalyst but it is not an enzyme because heating has not altered its catalytic properties.

The experiment can be repeated with a piece of potato to compare its catalase content with that of the liver. The piece of potato should be about the same size as the liver sample.

2 The effect of temperature on an enzyme reaction

Amylase is an enzyme which breaks down starch to a sugar (maltose).

Draw up 5 cm^3 of 5 per cent amylase solution in a plastic syringe (or graduated pipette) and place 1 cm^3 in each of 3 test-tubes labelled A, B and C. Rinse the syringe and use it to place 5 cm^3 of a 1 per cent starch solution in each of 3 test-tubes labelled 1, 2 and 3.

To each of tubes 1 to 3, add 6 drops only of dilute iodine solution using a dropping pipette.

Prepare three water baths by half filling beakers or jars with:

1. ice and water, adding ice during the experiment to keep the temperature at about 10 °C;
2. water from the cold tap at about 20 °C;
3. warm water at about 35 °C by mixing hot and cold water.

Place tubes 1 and A in the cold water bath, tubes 2 and B in the water at room temperature, and tubes 3 and C in the warm water. Leave them for 5 minutes to reach the temperature of the water (Figure 2.7).

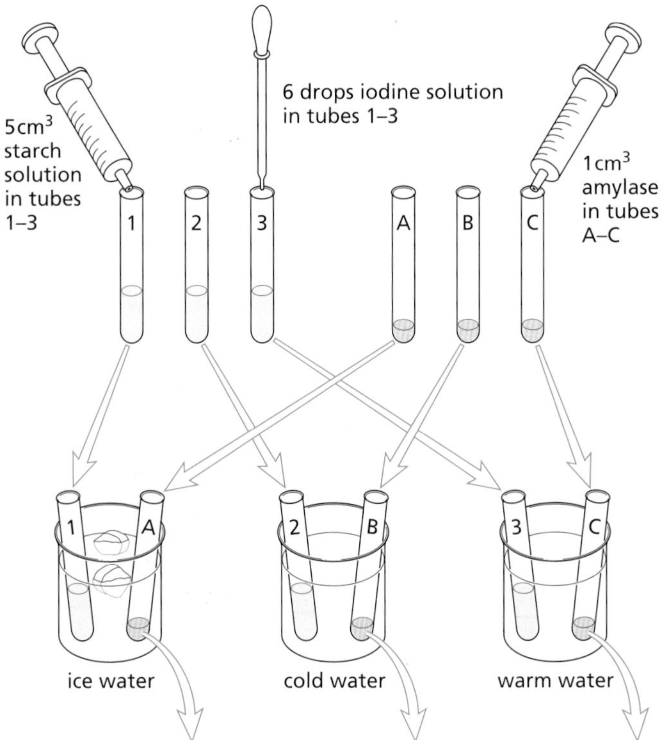

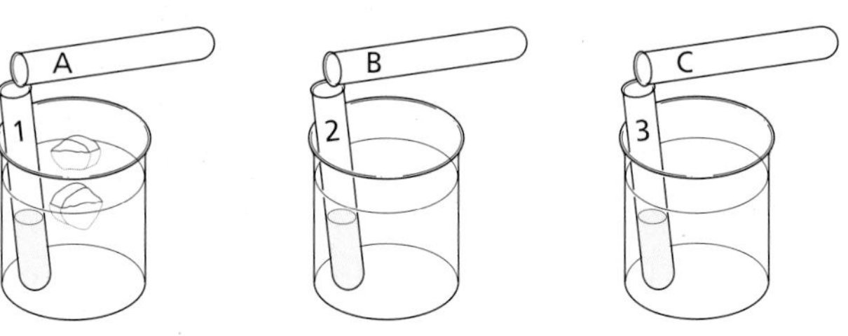

Figure 2.7 The effect of temperature on an enzyme reaction

After 5 minutes, take the temperature of each water bath, then pour the amylase from tube A into the starch solution in tube 1 and return tube 1 to the water bath. Repeat this with tubes 2 and B, and 3 and C.

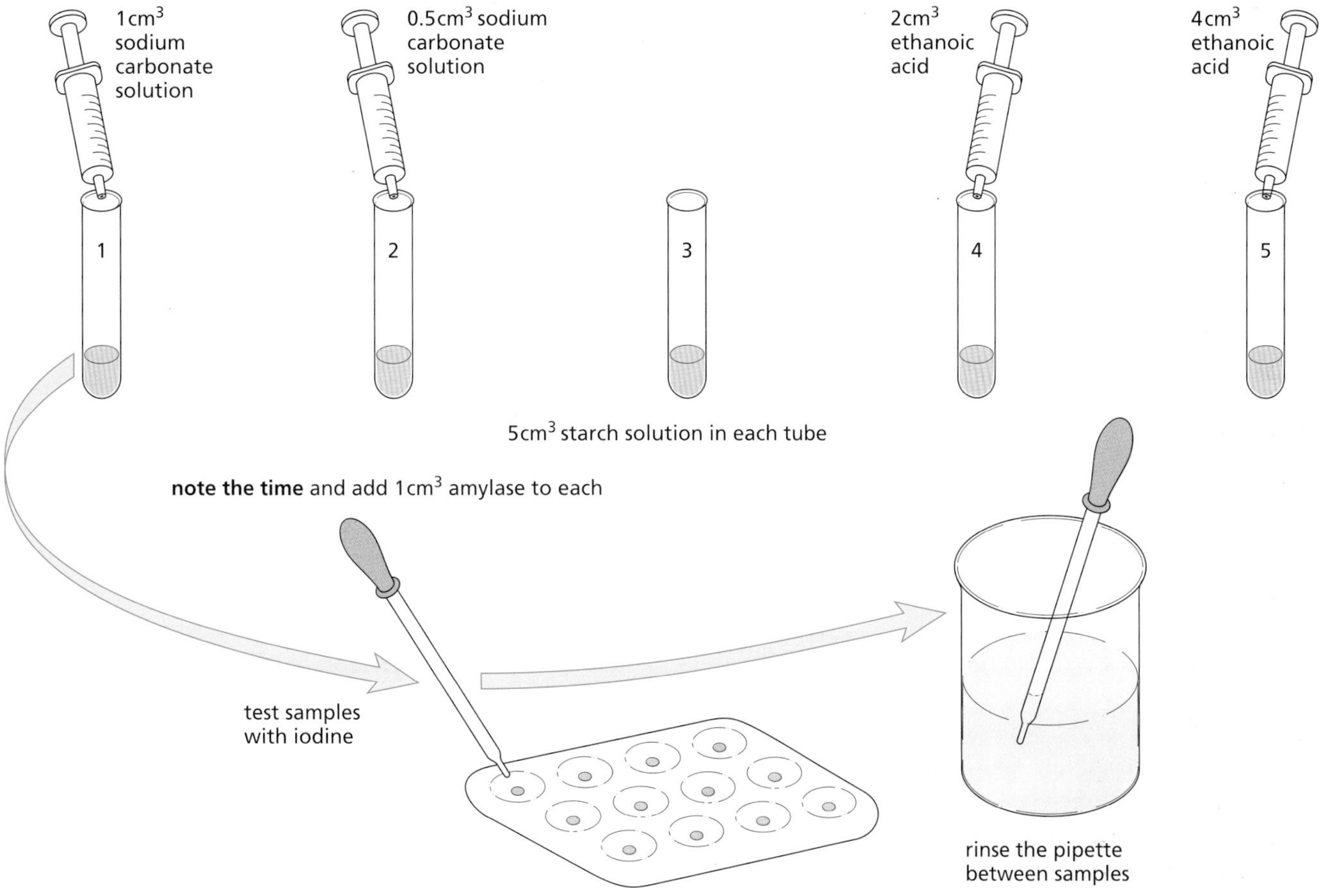

Figure 2.8 The effect of pH on an enzyme reaction

As the amylase breaks down the starch, it will cause the blue colour to disappear. Make a note of how long this takes in each case and answer the following questions:

a At what temperature did the amylase break down starch most rapidly?

b What do you think would have been the result if a fourth water bath at 90 °C had been used?

3 The effect of pH on an enzyme reaction

Label 5 test-tubes 1 to 5 and use a plastic syringe (or graduated pipette) to place 5 cm^3 of a 1 per cent starch solution in each tube. Add acid or alkali to each tube as indicated in the table below. Rinse the syringe when changing from sodium carbonate to acid.

Tube	Chemical	Approximate pH	
1	1 cm^3 sodium carbonate solution (M/20)	9	(alkaline)
2	0.5 cm^3 sodium carbonate solution (M/20)	7–8	(slightly alkaline)
3	nothing	6–7	(neutral)
4	2 cm^3 ethanoic (acetic) acid (M/10)	6	(slightly acid)
5	4 cm^3 ethanoic (acetic) acid (M/10)	3	(acid)

Place several rows of iodine solution drops in a cavity tile.

Draw up 5 cm^3 of 5 per cent amylase solution in a clean syringe and place 1 cm^3 in each tube. Shake the tubes and note the time (Figure 2.8).

Use a clean dropping pipette to remove a small sample from each tube in turn and let 1 drop fall on to one of the iodine drops in the cavity tile. Rinse the pipette in a beaker of water between each sample. Keep on sampling in this way.

When any of the samples fails to give a blue colour, this means that the starch in that tube has been completely broken down to sugar by the amylase. Note the time when this happens for each tube and stop taking samples from that tube. Do not continue sampling for more than about 15 minutes, but put a drop from each tube on to a piece of pH paper and compare the colour produced with a colour chart of pH values.

Answer the following questions:

a At what pH did the enzyme, amylase, work most rapidly?

b Is this its optimum pH?

c Explain why you might have expected the result which you got.

d Your stomach pH is about 2. Would you expect starch digestion to take place in the stomach?

Questions

1 Which of the following statements apply both to enzymes and to any other catalysts:
   a Their activity is stopped by high temperature.
   b They speed up chemical reactions.
   c They are proteins.
   d They are not used up during the reaction.

2 How would you expect the rate of an enzyme-controlled reaction to change if the temperature was raised
   a from 20 °C to 30 °C,
   b from 35 °C to 55 °C?
   Explain your answers.

3 There are cells in your salivary glands which can make an extracellular enzyme, amylase. Would you expect these cells to make intracellular enzymes as well? Explain your answer.

4 Apple cells contain an enzyme which turns the tissues brown when an apple is peeled and left for a time. Boiled apple does not go brown (Figure 2.9). Explain why the boiled apple behaves differently.

Figure 2.9 Enzyme activity in an apple. Slice A has been freshly cut. B and C were cut 2 days earlier but C was dipped immediately in boiling water for 1 minute

5 Suppose that the enzyme in Figure 2.6a is joining a glycine molecule (p. 11) to a valine molecule. Suggest why the same enzyme will not join glycine to serine.

Enzymes in industry

Brewing, baking and cheese-making are examples of industries which, for many years, have made use of enzymes on a large scale. In the case of brewing and baking it is the sugar-fermenting enzymes in living yeast cells which are exploited. These enzymes convert sugar to alcohol and carbon dioxide. It is the bubbles of carbon dioxide which make the dough 'rise' before baking and so give the bread a 'light' texture. In brewing, it is mainly the alcohol which is wanted but the carbon dioxide gives a 'sparkle' to beers and sparkling wine.

Brewing and baking exploit enzymes in living cells but for cheese-making an enzyme was extracted from calves' stomachs (it is now produced by genetically engineered bacteria, see p. 213). The enzyme is called **rennin** (the genetically engineered product is **chymosin**); it clots milk in the first stages of making cheese.

Of about 2000 known enzymes, about 50 are used commercially in large quantities. The advantages of using enzymes are:

- they work at low temperatures, so saving energy;
- they are not corrosive, as are the strong acids and alkalis that would otherwise be needed to carry out the chemical changes;
- they are specific, i.e. they act on only one substrate, so the process can be carefully controlled, and unwanted by-products avoided.

Two-thirds of commercial enzyme production is taken up by protein-digesting enzymes (proteases) and fat-digesting enzymes (lipases) for 'biological' washing powders, and by starch-digesting enzymes (amylases) for the food and textile industries. The proteases and lipases in washing powders help to remove organic stains such as blood, gravy or fruit juice. The industrial amylases are used to convert starch to glucose and fructose for sweeteners, or to destarch fabrics in the course of processing.

The commercial production and use of enzymes is described more fully on p. 330.

Checklist

- Living matter is made up of water, proteins, lipids, carbohydrates, salts and vitamins.
- Proteins are built up from amino acids joined together by chemical bonds.
- In different proteins the 20 or so amino acids are in different proportions and arranged in different sequences.
- Proteins are denatured by heat and some chemicals.
- Lipids include fats, fatty acids and oils.
- Fats are made from fatty acids and glycerol.
- Proteins and lipids form the membranes outside and inside the cell.
- Enzymes are proteins which catalyse chemical reactions in the cell.
- Enzymes are affected by pH and temperature and are denatured above 50 °C.
- Different enzymes may accelerate reactions which build up or break down molecules.
- Each enzyme acts on only one substance (breaking down), or a pair of substances (building up).
- The substance on which an enzyme acts is called the substrate.

3 Energy from respiration

Respiration
Definition. Aerobic and anaerobic respiration. Metabolism.

Practical work
Principles of the experimental design. Uptake of oxygen, output of carbon dioxide, temperature rise, anaerobic respiration.

Controlled experiments
Description. Clinical trials.
Hypotheses and hypothesis testing in scientific investigations. Use of carbon-14.

Respiration

Most of the processes taking place in cells need energy to make them happen. Building up proteins from amino acids or making starch from glucose needs energy. When muscle cells contract, or nerve cells conduct electrical impulses or plant cells form cell walls, they use energy. This energy comes from the food which cells take in. The food mainly used for energy in cells is glucose.

The process by which energy is produced from food is called **respiration**.

Respiration is a chemical process which takes place in cells. It must not be confused with the process of breathing, which is also sometimes called 'respiration'. To make the difference quite clear, the chemical process in cells is sometimes called **cellular respiration**, **internal respiration** or **tissue respiration**. The use of the word 'respiration' for breathing is best avoided altogether.

Aerobic respiration

The word **aerobic** means that oxygen is needed for this chemical reaction. The food molecules are combined with oxygen. The process is called **oxidation** and the food is said to be **oxidized**. All food molecules contain carbon, hydrogen and oxygen atoms. The process of oxidation converts the carbon to carbon dioxide (CO_2) and the hydrogen to water (H_2O) and, at the same time, sets free energy which the cell can use to drive other reactions.

Aerobic respiration can be summed up by the equation

$$\underset{\text{glucose}}{C_6H_{12}O_6} + \underset{\text{oxygen}}{6O_2} \xrightarrow{\text{enzymes}} \underset{\text{carbon dioxide}}{6CO_2} + \underset{\text{water}}{6H_2O} + \underset{\text{energy}}{2830\,kJ}$$

The 2830 kilojoules (kJ) is the amount of energy you would get by completely oxidizing 180 grams of glucose to carbon dioxide and water. In the cells, the energy is not released all at once. The oxidation takes place in a series of small steps and not in one jump as the equation suggests. Each small step needs its own enzyme and at each stage a little energy is released (Figure 3.1).

Although the energy is used for the processes mentioned above, some of it always appears as heat. In 'warm-blooded' animals (birds and mammals) some of this heat is retained to keep up their body temperature.

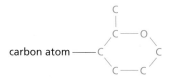

(a) molecule of glucose (H and O atoms not all shown)

(b) the enzyme attacks and breaks the glucose molecule into two 3-carbon molecules

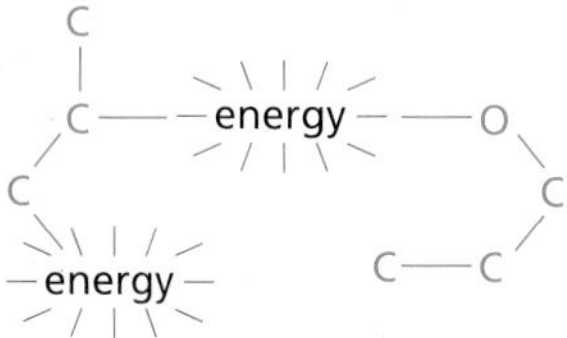

(c) this breakdown sets free energy

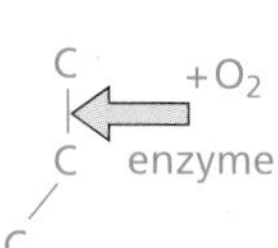

(d) each 3-carbon molecule is broken down to carbon dioxide

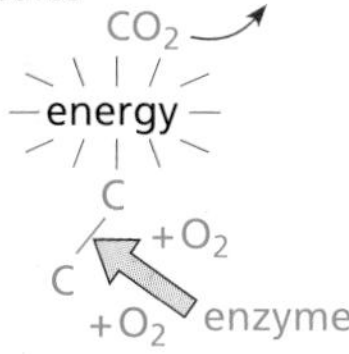

(e) more energy is released and CO_2 is produced

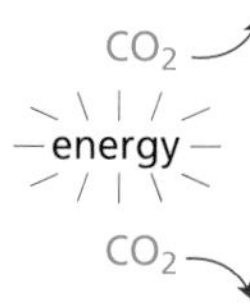

(f) the glucose has been completely oxidized to carbon dioxide (and water), and all the energy released

Figure 3.1 Aerobic respiration

In 'cold-blooded' animals (e.g. reptiles and fish) the heat may build up for a time and allow the animal to move about faster. In plants the heat is lost to the surroundings (by conduction, convection and evaporation) as fast as it is produced.

Mitochondria

It is in the mitochondria that the chemistry of aerobic respiration takes place. The mitochondria generate a compound called ATP, which is used by the cell as the source of energy for driving other chemical reactions in the cytoplasm and nucleus.

Anaerobic respiration

The word **anaerobic** means 'in the absence of oxygen'. In this process, energy is still released from food by breaking it down chemically but the reactions do not use oxygen though they do often produce carbon dioxide. A common example is the action of yeast on sugar solution to produce alcohol. The sugar is not completely oxidized to carbon dioxide and water but converted to carbon dioxide and alcohol. This process is called **fermentation** and is shown by the following equation:

$$\underset{\text{glucose}}{C_6H_{12}O_6} \xrightarrow{\text{enzymes}} \underset{\text{alcohol}}{2C_2H_5OH} + \underset{\text{carbon dioxide}}{2CO_2} + \underset{\text{energy}}{118\,\text{kJ}}$$

The processes of brewing and bread-making rely on anaerobic respiration by yeast (p. 326). As with aerobic respiration, the reaction takes place in small steps and needs several different enzymes. The yeast uses the energy for its growth and living activities, but you can see from the equation that less energy is produced by anaerobic respiration than in aerobic respiration. This is because the alcohol still contains a great deal of energy which the yeast is unable to use.

In animals, the first stages of respiration in muscle cells are anaerobic and produce **pyruvic acid** (the equivalent of the yeast's alcohol). Only later on is the pyruvic acid completely oxidized to carbon dioxide and water.

$$\text{glucose} \xrightarrow{\text{enzymes}} \text{pyruvic acid} \xrightarrow{\text{enzymes and oxygen}} CO_2 + H_2O$$

ANAEROBIC STAGE AEROBIC STAGE

During exercise pyruvic acid may build up in a muscle faster than it can be oxidized. In this case it is turned into **lactic acid** and removed in the bloodstream. On reaching the liver, some of the lactic acid is oxidized to carbon dioxide and water, using up oxygen in the process. After exercise has stopped, a high level of oxygen consumption may persist until the excess of lactic acid is oxidized. This build-up of lactic acid which is oxidized later is said to create an **oxygen debt**.

Accumulation of lactic acid in the muscles may be one of the causes of muscular fatigue. It is sometimes claimed that lactic acid also causes cramp but this is not established.

Metabolism

All the chemical changes taking place inside a cell or a living organism are called its **metabolism**. The minimum turnover of energy needed simply to keep an organism alive, without movement or growth, is called the **basal metabolism**. Our basal metabolism maintains vital processes such as breathing, heart beat, digestion and excretion.

The processes which break substances down are sometimes called **catabolism**. Respiration is an example of catabolism in which carbohydrates are broken down to carbon dioxide and water. Chemical reactions which build up substances are called **anabolism**. Building up a protein from amino acids is an example of anabolism. The energy released by the **catabolic** process of respiration is used to drive the **anabolic** reactions which build up proteins.

You may have heard of anabolic steroids, in connection with drug-taking by athletes. These chemicals do reduce the rate of protein breakdown and may enhance the build-up of certain proteins. However, their effects are complicated and not fully understood, they have undesirable side-effects and their use contravenes athletics codes (p. 173).

Questions

1 a If, in one word, you had to say what respiration was about, which word would you choose from this list: breathing, energy, oxygen, cells, food?
b In which parts of a living organism does respiration take place?

2 a What are the main differences between aerobic and anaerobic respiration?
b What is the difference between aerobic and anaerobic respiration in the amount of energy released from one molecule of glucose?

3 What chemical substances
a from outside the cell,
b from inside the cell, must be provided for aerobic respiration to take place?
c What are the products of aerobic respiration?

4 Victims of drowning who have stopped breathing are sometimes revived by a process called 'artificial respiration'. Why would a biologist object to the use of this expression? ('Resuscitation' is a better word to use.)

5 Why do you think your breathing rate and heart rate stay high for some time after completing a spell of vigorous exercise?

6 Table 10.2 on p. 90 shows energy consumption in different conditions. Which figure do you think represents 'basal metabolism'?

Practical work

Experiments on respiration

If you look below at the chemical equation which represents aerobic respiration you will see that a tissue or an organism which is respiring should be (a) using up food, (b) using up oxygen, (c) giving off carbon dioxide, (d) giving off water and (e) releasing energy which can be used for other processes.

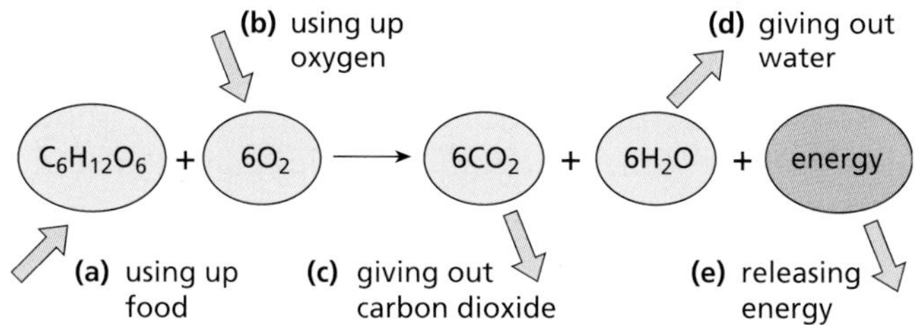

If we wish to test whether aerobic respiration is taking place:

- '(d) giving out water' is not a good test because non-living material will give off water vapour if it is wet to start with.
- '(a) using up food' can be tested by seeing if an organism loses weight. This is not so easy as it seems because most organisms lose weight as a result of evaporation of water and this may have nothing to do with respiration. It is the decrease in 'dry weight' which must be measured (p. 307).
- (b), (c) and (e) are fairly easy to demonstrate either with whole organisms or with pieces of living tissue as in Experiments 1 to 4.

Seeds are often used as the living organisms because when they start to grow (germinate) there is a high level of chemical activity in the cells. The seeds are easy to obtain and to handle and they fit into small-scale apparatus. In some cases blowfly maggots can be used as animal material.

1 Using up oxygen during respiration

The apparatus in Figure 3.2 is a **respirometer** (a 'respire meter'), which can measure the rate of respiration by seeing how quickly oxygen is taken up. Germinating seeds or blowfly larvae are placed in the test-tube and, as they use up the oxygen for respiration, the level of liquid in the delivery tubing will go up.

There are two drawbacks to this. One is that the organisms usually give out as much carbon dioxide as they take in oxygen. So there may be no change in the total amount of air in the test-tube and the liquid level will not move. This drawback is overcome by placing **soda-lime** in the test-tube. Soda-lime will absorb carbon dioxide as fast as the organisms give it out. So only the uptake of oxygen will affect the amount of air in the tube. The second drawback is that quite small changes in temperature will make the air in the test-tube expand or contract and so cause the liquid to rise or fall whether or not respiration is taking place. To overcome this, the test-tube is kept in a beaker of water (a water bath). The temperature of water changes far more slowly than that of air, so there will not be much change during a 30-minute experiment.

Control To show that it is a living process which uses up oxygen, a similar respirometer is prepared but containing an equal quantity of germinating seeds which have been killed by boiling. (If blowfly larvae are used, the control can consist of an equivalent volume of glass beads. This is not a very good control but is probably more acceptable than killing an equivalent number of larvae.)

The apparatus is finally set up as shown in Figure 3.2 and left for 30 minutes (10 minutes if blowfly larvae are used).

The capillary tube and reservoir of liquid are called a **manometer**.

Result The level of liquid in the experiment goes up more than in the control. The level in the control may not move at all.

Interpretation The rise of liquid in the delivery tubing shows that the living seedlings have taken up part of the air. It does not prove that it is oxygen which has been taken up. Oxygen seems the most likely gas, however, because (1) there is only 0.03 per cent carbon dioxide in the air to start with and (2) the other gas, nitrogen, is known to be less active than oxygen.

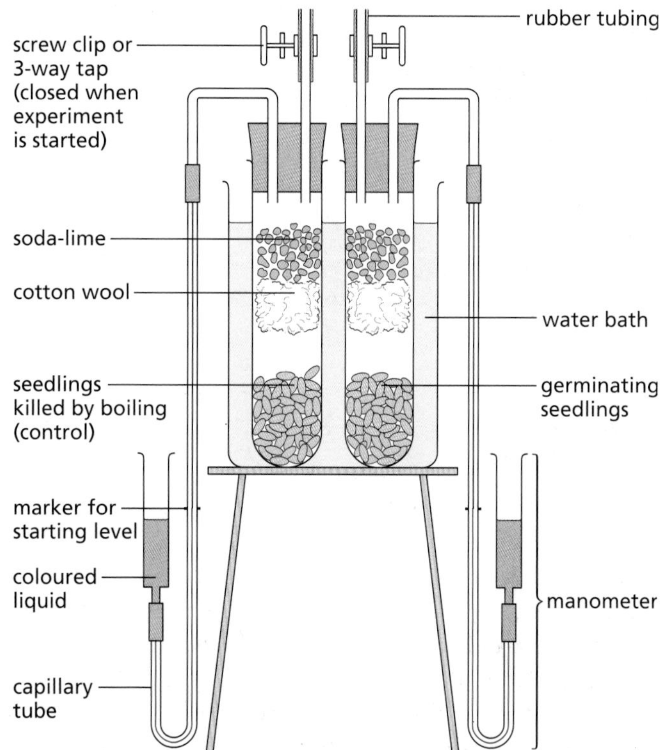

Figure 3.2 To see if oxygen is taken up in respiration

If the experiment is allowed to run for a long time, the uptake of oxygen could be checked at the end by placing a lighted splint in each test-tube in turn. If some of the oxygen has been removed by the living seedlings, the flame should go out more quickly than it does in the tube with dead seedlings.

2 Carbon dioxide from germinating seeds

Put some germinating wheat grains in a large test-tube. Cover the mouth of the tube with aluminium foil. After 15–20 minutes, take a sample of the air from the test-tube. Do this by pushing a glass tube attached to a 10 cm^3 plastic syringe through the foil and into the test-tube (Figure 3.3a). Withdraw the syringe plunger enough to fill the syringe with air from the test-tube. Now slowly bubble this air sample through a little clear lime water in a small test-tube (Figure 3.3b). Cover the mouth of the small test-tube and shake the lime water up.

Result The lime water will go milky.

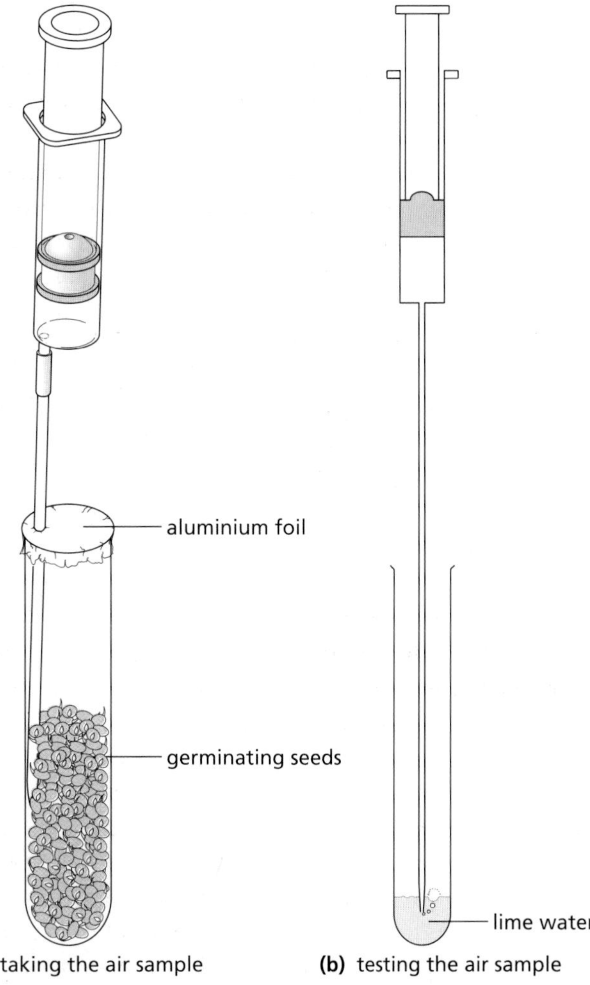

Figure 3.3 Production of carbon dioxide by germinating seeds

Interpretation Lime water turning milky is evidence of carbon dioxide but it could be argued that the carbon dioxide came from the air or that the seeds give off carbon dioxide whether or not they are respiring. The only way to disprove these arguments is to do a **control experiment** (see p. 23).

Control Boil some of the germinating wheat grains before starting the experiment. When you set up the experiment, put an equal amount of boiled wheat grains in a large test-tube and cover the mouth of the tube with aluminium foil exactly as you did for the living seeds. When you test the air from the living seeds, also test the air from the dead seeds. It should not turn the lime water milky. This means that the carbon dioxide did not come from the air, nor was it given off by the dead seeds. It must be a living process in the seeds which produces carbon dioxide and this process is likely to be respiration. However, since you stopped all living processes by boiling the seeds in the control experiment, you have not been able to prove that it was respiration rather than some other chemical change which produced the carbon dioxide.

An experiment to show carbon dioxide production by plants, using hydrogencarbonate indicator, is described on p. 40.

3 Releasing energy in respiration

Fill a small vacuum flask with wheat grains which have been soaked for 24 hours and rinsed in 1 per cent formalin (or domestic bleach diluted 1 + 4) for 5 minutes. These solutions will kill any bacteria or fungi on the surface of the grains. Kill an equal quantity of soaked grains by boiling them for 5 minutes. Cool the boiled seeds in cold tap water, rinse them in bleach or formalin for 5 minutes as before and then put them in a vacuum flask of the same size as the first one. This flask is the control.

Place a thermometer in each flask so that its bulb is in the middle of the seeds (Figure 3.4). Plug the mouth of each flask with cotton wool and leave both flasks for 2 days, noting the thermometer readings whenever possible.

Result The temperature in the flask with the living seeds will be 5–10 °C higher than that of the dead seeds.

Interpretation Provided that there are no signs of the living seeds going mouldy, the heat produced must have come from living processes in the seeds, because the dead seeds in the control did not give out any heat. There is no evidence that this process is respiration rather than any other chemical change but the result is what you would expect if respiration does produce energy.

For an experiment comparing the energy value of different food substances, see p. 96.

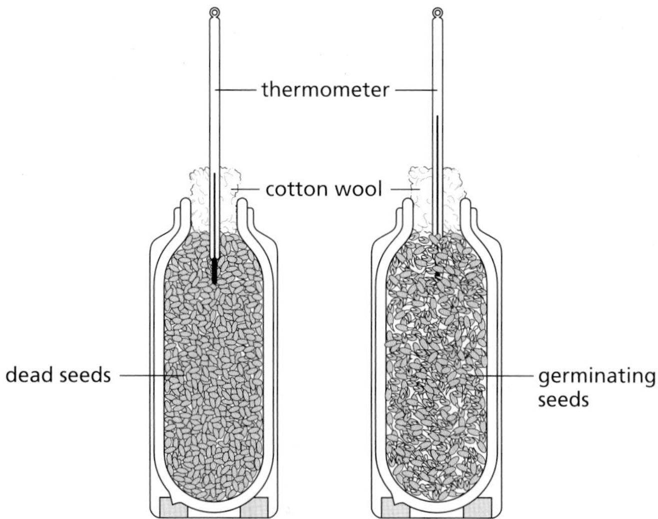

Figure 3.4 Energy release in germinating seeds

4 Anaerobic respiration in yeast

Boil some water to expel all the dissolved oxygen. When cool, use the boiled water to make up a 5 per cent solution of glucose and a 10 per cent suspension of dried yeast. Place $5\,cm^3$ of the glucose solution and $1\,cm^3$ of the yeast suspension in a test-tube and cover the mixture with a thin layer of liquid paraffin to exclude atmospheric oxygen. Fit a delivery tube as shown in Figure 3.5 and allow it to dip into clear lime water.

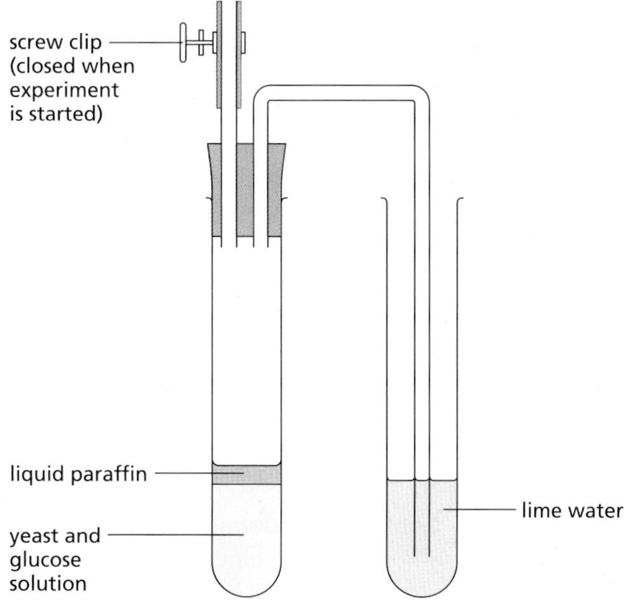

Figure 3.5 Anaerobic respiration in yeast

Result After 10–15 minutes, with gentle warming if necessary, there should be signs of fermentation in the yeast–glucose mixture and the bubbles of gas escaping through the lime water should turn it milky.

Interpretation The fact that the lime water goes milky shows that the yeast–glucose mixture is producing carbon dioxide. If we assume that the production of carbon dioxide is evidence of respiration, then it looks as if the yeast is respiring. In setting up the experiment, you took care to see that oxygen was removed from the glucose solution and the yeast suspension, and the liquid paraffin excluded air (including oxygen) from the mixture. Any respiration taking place must, therefore, be anaerobic (i.e. without oxygen).

Control It might be suggested that the carbon dioxide came from a chemical reaction between yeast and glucose (as between chalk and acid) which had nothing to do with respiration or any other living process. A control should, therefore, be set up using the same procedure as before but with yeast which has been killed by boiling. The failure, in this case, to produce carbon dioxide supports the claim that it was a living process in the yeast in the first experiment which produced the carbon dioxide.

Controlled experiments

In most biological experiments, a second experiment called a **control** is set up. This is to make sure that the results of the first experiment are due to the conditions being studied and not to some other cause which has been overlooked.

In the experiment in Figure 3.2, the liquid rising up the tube could have been the result of the test-tube cooling down, so making the air inside it contract. The identical experiment with dead seeds – the control – showed that the result was not due to a temperature change, because the level of liquid in the control did not move.

The term 'controlled experiment' refers to the fact that the experimenter (1) sets up a control and (2) controls the conditions in the experiment. In the experiment shown in Figure 3.2 the seeds are enclosed in a test-tube, and soda-lime is added. This makes sure that any uptake or output of oxygen will make the liquid go up or down, and that the output of carbon dioxide will not affect the results. The experimenter had controlled both the amount and the composition of the air available to the germinating seeds.

If you did an experiment to compare the growth of plants in the house and in a greenhouse, you could not be sure whether it was the extra light or the high temperature of the greenhouse which caused better growth. This would not, therefore, be a properly controlled experiment. You must alter only one condition (called a **variable**) at a time, either the light or the temperature, and then you can compare the results with the control experiment.

A properly controlled experiment, therefore, alters only one variable at a time and includes a control which shows that it is this condition and nothing else which gave the result.

Clinical trials

Before a new drug can be released for general use it has to undergo a series of clinical trials. These trials often take the form of controlled experiments. The regulatory body needs to be sure that the drug is effective and has no unwanted side-effects. If the drug were given to 100 patients and 60 of them showed signs of improvement, you could not be sure (a) that they would not have got better anyway and (b) that the psychological effect of receiving treatment was responsible for their improvement.

For this reason, a control has to be designed. It usually takes the form of giving half of the trial group the drug, while the other half, the control group, receives a 'dummy' pill that is similar in every respect to the real pill except that it lacks the active ingredient. The 'dummy' pill is called a **placebo**. The degree of recovery in the two groups can then be compared and analysed statistically. Of course, the two groups need to be as similar as possible with respect to age, sex and symptoms.

It may seem unethical to withhold treatment from the control group, but it has to be remembered that it is a new drug that is being tested, and its effectiveness is not yet known. In fact, in some cases, if the new drug produces a significant improvement in the test group, the trial is cut short and the control group is also given the drug.

Questions

1 Which of the following statements are true? If an organism is respiring you would expect it to be
a giving out carbon dioxide,
b losing heat,
c breaking down food,
d using up oxygen,
e gaining weight,
f moving about?

2 What was the purpose of the soda-lime in Experiment 1 on p. 21, and the lime water in Experiments 2 and 4 on pp. 22 and 23?

3 In an experiment like the one shown in Figure 3.2 on p. 21, the growing seeds took in 5 cm^3 oxygen and gave out 7 cm^3 carbon dioxide. How does the volume change
a if no soda-lime is present,
b if soda-lime is present?

4 The germinating seeds in Figure 3.4 on p. 23 will release the same amount of heat whether they are in a beaker or a vacuum flask. Why then is it necessary to use a vacuum flask for this experiment?

5 What is the purpose of the control in the experiment to show carbon dioxide production (Figure 3.3, p. 22)?

Hypothesis testing

You will have noticed that none of the experiments described above claim to have *proved* that respiration is taking place. The most we can claim is that they have not disproved the proposal that energy is produced from respiration. There are many reactions taking place in living organisms and, for all we know at this stage, some of them may be using oxygen or giving out carbon dioxide without releasing energy, i.e. they would not fit our definition of respiration.

This inability to 'prove' that a particular proposal is 'true' is not restricted to experiments on respiration. It is a feature of many scientific experiments. One way that science makes progress is by putting forward a **hypothesis**, making predictions from the hypothesis, and then testing these predictions by experiments. A hypothesis is an attempt to explain some event or observation using the information currently available. If an experiment's results do not confirm the predictions, the hypothesis must be abandoned or altered.

For example, biologists observing that living organisms take up oxygen might put forward the hypothesis that 'oxygen is used to convert food to carbon dioxide, so producing energy for movement, growth, reproduction, etc.'. This hypothesis can be tested by predicting that, '*if* the oxygen is used to oxidize food *then* an organism that takes up oxygen will also produce carbon dioxide'. Experiments 1 and 2 on pp. 21 and 22 test this and fulfil this prediction and, therefore, support the hypothesis. Looking at the equation for respiration, we might also predict that an organism which is respiring will produce carbon dioxide and take up oxygen. Experiment 4 with yeast, however, does not fulfil this prediction and so does not support the hypothesis as it stands, because here is an organism producing carbon dioxide without taking up oxygen. The hypothesis will have to be modified, e.g. 'energy is released from food by breaking it down to carbon dioxide; some organisms use oxygen for this process, others do not'.

There are still plenty of tests which we have not done. For example, we have not attempted to see whether it is food that is the source of energy and carbon dioxide. One way of doing this is to provide the organism with food, e.g. glucose, in which the carbon atoms are radioactive. Carbon-14 (^{14}C) is a radioactive form of carbon and can be detected by using a Geiger counter. If the organism produces radioactive carbon dioxide, it is reasonable to suppose that the carbon dioxide comes from the glucose.

$$C_6H_{12}O_6 + 6O_2 \longrightarrow 6CO_2 + 6H_2O + \text{energy}$$

This is **direct evidence** in support of the hypothesis. All the previous experiments have provided only **indirect evidence**.

Criteria for a good hypothesis

A good hypothesis must:

- explain *all* aspects of the observation;
- be the simplest possible explanation;
- be expressed in such a way that predictions can be made from it;
- be testable by experiment.

Questions

1 The experiment with yeast on p. 23 supported the claim that anaerobic respiration was taking place. The experiment was repeated using unboiled water and without the liquid paraffin. Fermentation still took place and carbon dioxide was produced. Does this mean that the design or the interpretation of the first experiment was wrong? Explain your answer.

2 Twenty seeds are placed on soaked cotton wool in a closed glass dish and after 5 days in the light 15 of the seeds had germinated. If the experiment is intended to see if light is needed for germination, which of the following would be a suitable control:

a exactly the same set-up but with dead seeds;
b the same set-up but with 50 seeds;
c an identical experiment but with 20 seeds of a different species;
d an identical experiment but left in darkness for 5 days?

3 Certain bacteria which live in sulphurous springs in areas of volcanic activity take up hydrogen sulphide (H_2S) and produce sulphates ($-SO_4$). Put forward a hypothesis to account for this chemical activity. Suggest one way of testing your hypothesis.

4 The table below shows the energy used up each day either as kilojoules per kilogram of body mass or as kilojoules per square metre of body surface.

		kJ per day	
	Mass/kg	per kg body mass	per m^2 body surface
pig	128.0	80	4510
man	64.3	134	4360
dog	15.2	216	4347
mouse	0.018	2736	4971

Reprinted from *Textbook of Physiology*, Emslie-Smith, Paterson, Scratcherd and Read, by permission of the publisher Churchill Livingstone, 1988

a According to the table, what is the total amount of energy used each day by
(i) a man,
(ii) a mouse?
b Which of these two shows a greater rate of respiration in its body cells?
c Why, do you think, is there so little difference in the energy expenditure per square metre of body surface?

Checklist

- Respiration is the process in cells which releases energy from food.
- Aerobic respiration needs oxygen; anaerobic respiration does not.
- The oxidation of food produces carbon dioxide as well as releasing energy.
- Experiments to investigate respiration try to detect uptake of oxygen, production of carbon dioxide, release of energy as heat or a reduction in dry weight.
- In a controlled experiment, the scientist tries to alter only one condition at a time, and sets up a control to check this.
- A control is a second experiment, identical to the first experiment except for the one condition being investigated.
- The control is designed to show that only the condition under investigation is responsible for the results.
- Experiments are designed to test predictions made from hypotheses; they cannot 'prove' a hypothesis.

4 How substances get in and out of cells

Diffusion
Explanation. Diffusion into and out of cells. Rates of diffusion. Controlled diffusion. Surface area. Endo- and exocytosis.
Active transport

Osmosis
Definition. Explanation. Artificial partially permeable membranes; the cell membrane. Water potential. Osmosis in animal and plant cells.
Practical work
Experiments on turgor, osmotic flow, plasmolysis, selective permeability (dialysis), turgor in potato tissue.

Cells need food materials which they can oxidize for energy or use to build up their cell structures. They also need salts and water which play a part in chemical reactions in the cell. Finally, they need to get rid of substances such as carbon dioxide, which, if they accumulated in the cell, would upset some of the chemical reactions and even poison the cell.

Substances may pass through the cell membrane either passively by diffusion, or actively by some form of active transport.

Diffusion

The molecules of a gas such as oxygen are moving about all the time. So are the molecules of a liquid, or a substance such as sugar dissolved in water. As a result of this movement, the molecules spread themselves out evenly to fill all the available space (Figure 4.1).

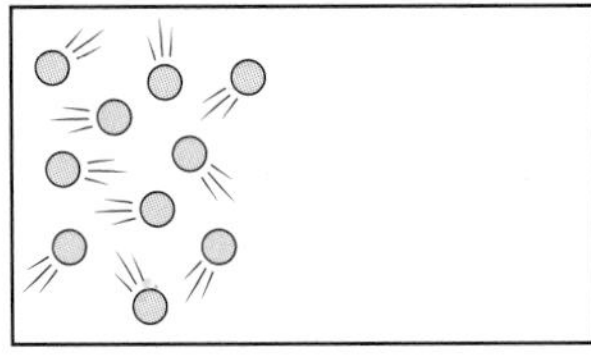

molecules moving about

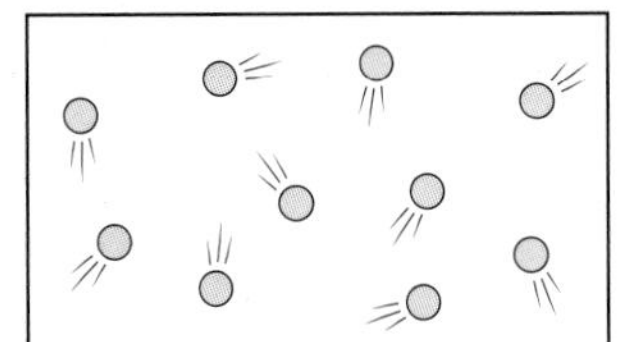

become evenly distributed

Figure 4.1 Diffusion

This process is called **diffusion**. One effect of diffusion is that the molecules of a gas, a liquid or a dissolved substance will move from a region where there are a lot of them (i.e. concentrated) to regions where there are few of them (i.e. less concentrated) until the concentration everywhere is the same. Figure 4.2a is a diagram of a cell with a high concentration of molecules (e.g. oxygen) outside and a low concentration inside. The effect of this difference in concentration is to make the molecules diffuse into the cell until the concentration inside and outside is the same (Figure 4.2b).

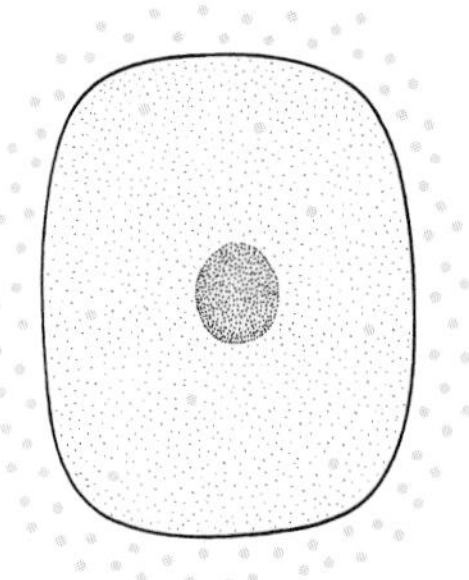

(a) greater concentration outside cell

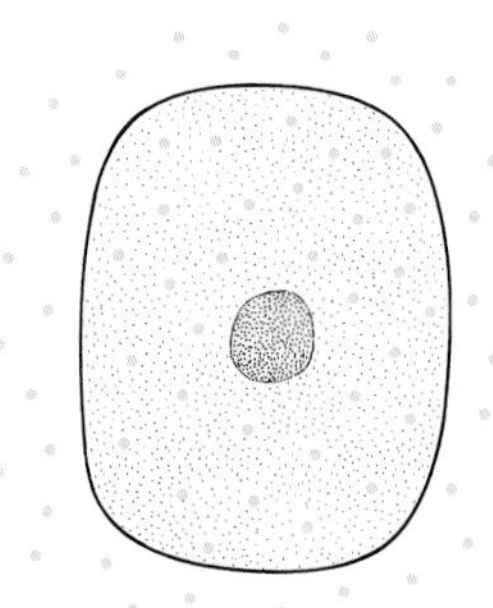

(b) concentrations equal on both sides of the cell membrane

Figure 4.2 Molecules entering a cell by diffusion

Whether this will happen or not depends on whether the cell membrane will let the molecules through. Small molecules such as water (H_2O), carbon dioxide (CO_2) and oxygen (O_2) can pass through the cell membrane fairly easily. So diffusion tends to equalize the concentration of these molecules inside and outside the cell all the time.

When a cell uses up oxygen for its aerobic respiration, the concentration of oxygen inside the cell falls and so oxygen molecules diffuse into the cell until the concentration is raised again. During tissue respiration, carbon dioxide is produced and so its concentration inside the cell goes up. Once again diffusion takes place, but this time the molecules move out of the cell. In this way, diffusion can explain how a cell takes in its oxygen and gets rid of its carbon dioxide.

Rates of diffusion

The speed with which a substance diffuses through a cell wall or cell membrane will depend on temperature, pressure and many other conditions including (1) the distance it has to diffuse, (2) its concentration inside and outside the cell and (3) the size of its molecules or ions.

- Cell membranes are all about the same thickness (about 0.007 µm) but plant cell walls vary in their thickness and permeability. Generally speaking, the thicker the wall, the slower is the rate of diffusion.

- The bigger the difference in concentration of a substance on either side of a membrane, the faster it will tend to diffuse. The difference is called a **concentration gradient** or **diffusion gradient** (Figure 4.3). If a substance on one side of a membrane is steadily removed, the diffusion gradient is maintained. When oxygen molecules enter a red cell they combine with a chemical (haemoglobin) which takes them out of solution. Thus the concentration of free oxygen molecules inside the cell is kept very low and the diffusion gradient for oxygen is maintained.

Figure 4.3 Diffusion gradient

- In general, the larger the molecules or ions, the slower they diffuse. However, many ions and molecules in solution attract water molecules around them (see p. 29) and so their effective size is greatly increased. It may not be possible to predict the rate of diffusion from the molecular size alone.

Controlled diffusion

Although for any one substance, the rate of diffusion through a cell membrane depends partly on the concentration gradient, the rate is often faster or slower than expected. Water diffuses more slowly and amino acids diffuse more rapidly through a membrane than might be expected. In some cases this is thought to happen because the ions or molecules can pass through the membrane only by means of special pores. These pores may be few in number or they may be open or closed in different conditions.

In other cases, the movement of a substance may be speeded up by an enzyme working in the cell membrane. So it seems that 'simple passive' diffusion, even of water molecules, may not be so simple or so passive after all, where cell membranes are concerned.

When a molecule gets inside a cell there are a great many structures and processes which may move it from where it enters to where it is needed. Simple diffusion is unlikely to play a very significant part in this movement.

Surface area

If 100 molecules diffuse through 1 mm^2 of a membrane in 1 minute, it is reasonable to suppose that an area of 2 mm^2 will allow twice as many through in the same time. Thus the rate of diffusion into a cell will depend on the cell's surface area. The greater the surface area, the faster is the total diffusion. Cells which are involved in rapid absorption, e.g. in the kidney or the intestine, often have their 'free' surface membrane formed into hundreds of tiny projections called **microvilli** (Figure 4.4) which increase the absorbing surface.

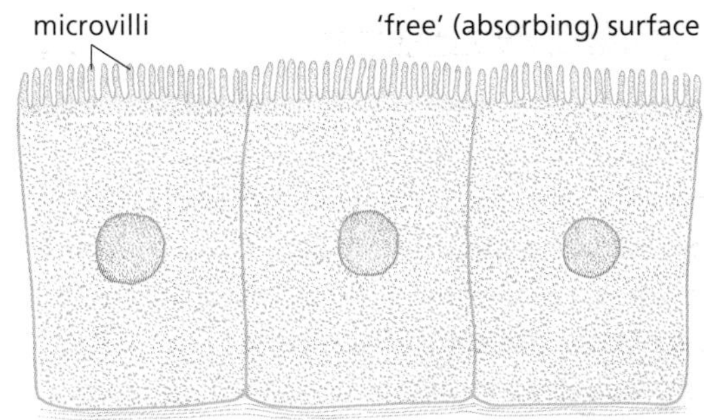

Figure 4.4 Microvilli

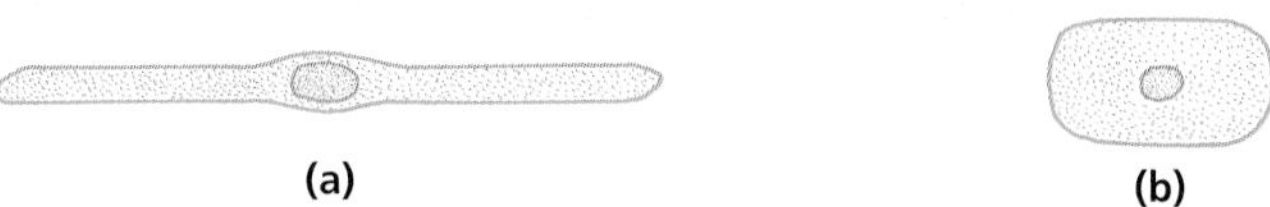

Figure 4.5 Surface area. The cells have the same volume but **(a)** has a much greater surface area

The shape of a cell will also affect the surface area. For example, the cell in Figure 4.5a has a greater surface area than that in Figure 4.5b, even though they each have the same volume.

Endo- and exocytosis

Some cells can take in (**endocytosis**) or expel (**exocytosis**) solid particles or drops of fluid through the cell membrane. Endocytosis occurs in single-celled 'animals' such as *Paramecium* (p. 268) when they feed, or in certain white blood cells (phagocytes, p. 108) when they engulf bacteria, a process called **phagocytosis** (Figure 4.6). Exocytosis takes place in the cells of some glands. A secretion, e.g. a digestive enzyme, forms vacuoles or granules in the cytoplasm and these are expelled through the cell membrane to do their work outside the cell (Figure 4.7).

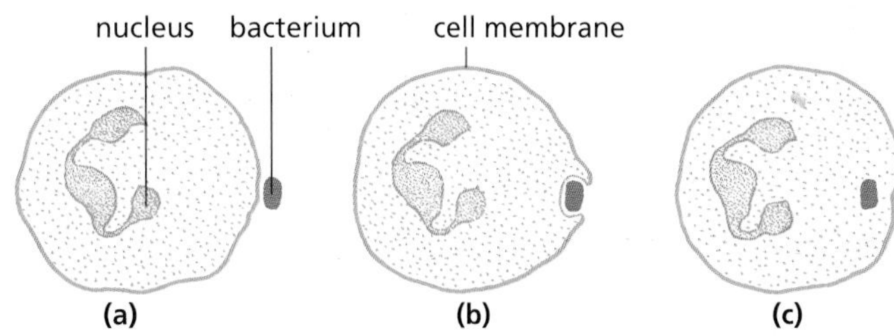

Figure 4.6 Endocytosis (phagocytosis) in a white blood cell

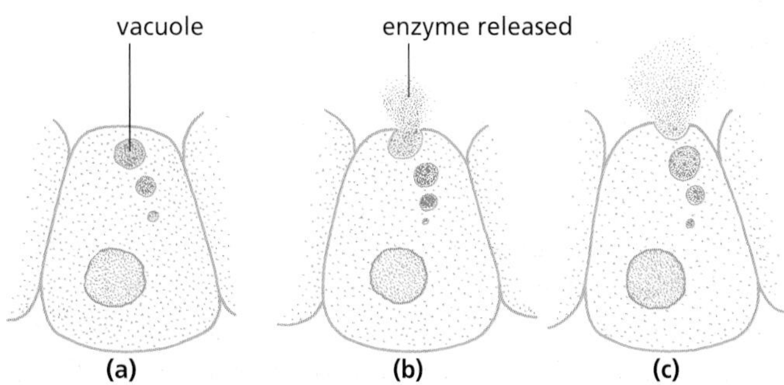

Figure 4.7 Exocytosis in a gland cell

Active transport

If diffusion were the only method by which a cell could take in substances, it would have no control over what went in or out. Anything that was more concentrated outside would diffuse into the cell whether it was harmful or not. Substances which the cell needed would diffuse out as soon as their concentration inside rose above that outside the cell. The cell membrane, however, has a great deal of control over the substances which enter and leave the cell.

In some cases, substances are taken into or expelled from the cell against the concentration gradient. For example, sodium ions may continue to pass out of a cell even though the concentration outside is greater than inside. The processes by which such reverse concentrations are produced are not fully understood and may be quite different for different substances but are all generally described as **active transport**. The cells lining the small intestine take up glucose by active transport (p. 102).

Anything which interferes with respiration, e.g. lack of oxygen or glucose, prevents active transport taking place (Figure 4.8). Thus it seems that active transport needs a supply of energy from respiration.

In some cases, a combination of active transport and controlled diffusion seems to occur. For example, sodium ions are thought to get into a cell by diffusion through special pores in the membrane and are expelled by a form of active transport. The reversed diffusion gradient for sodium ions created in this way is very important in the conduction of nerve impulses in nerve cells.

Plants need to absorb mineral salts from the soil, but these salts are in very dilute solution. Active transport enables the cells of the plant roots to take up salts from this dilute solution against the concentration gradient.

Questions

1 Look at Figure 12.3 on p. 109. If the symbol O_2 represents an oxygen molecule, explain why oxygen is entering the cells drawn on the left but leaving the cells on the right.

2 Look at Figure 13.9 on p. 126 representing one of the small air pockets (an alveolus) which form the lung.
   a Suggest a reason why the oxygen and carbon dioxide are diffusing in opposite directions.
   b What might happen to the rate of diffusion if the blood flow were to speed up?

3 List the ways in which a cell membrane might regulate the flow of substances into the cell.

4 What is your interpretation of the results shown by the graph in Figure 4.9?

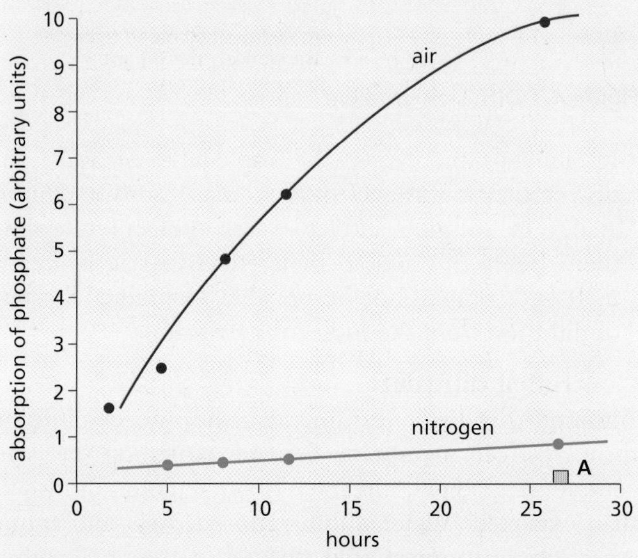

Figure 4.9 Absorption of phosphate ions in air and in nitrogen by roots of beech. **A** represents the concentration of phosphate in external solution

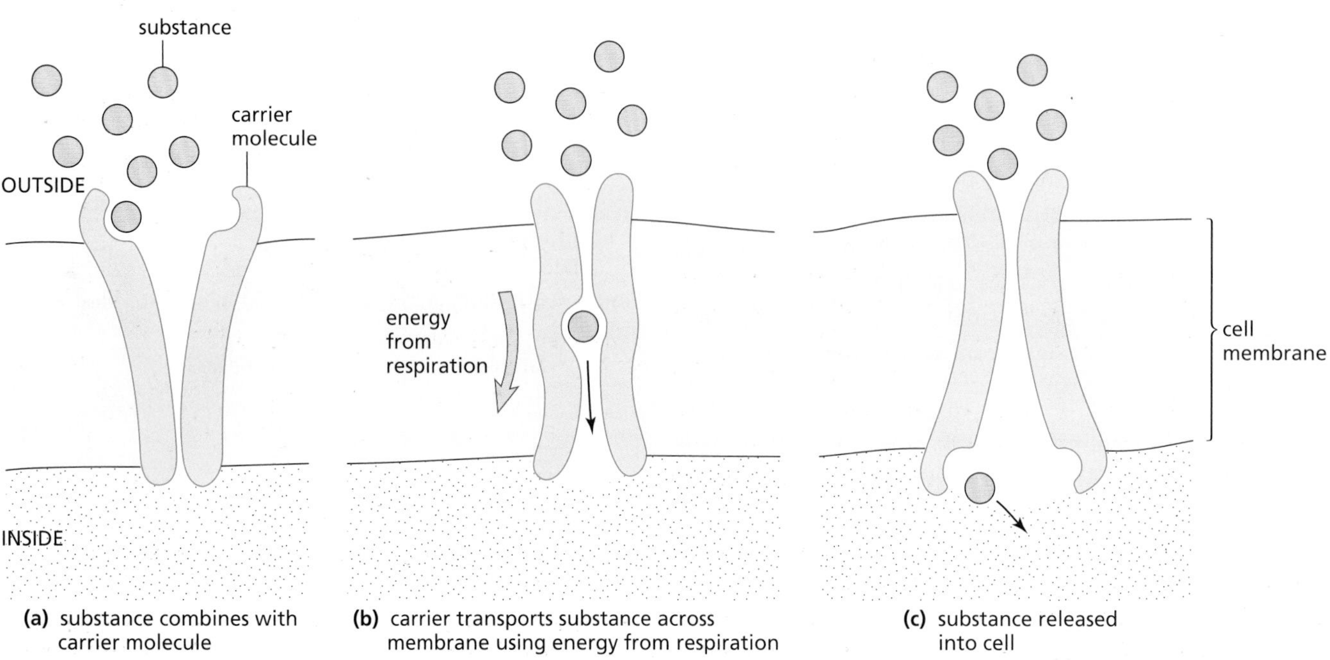

Figure 4.8 Theoretical model to explain active transport

Osmosis

If a dilute solution is separated from a concentrated solution by a **partially permeable** membrane, water diffuses across the membrane from the dilute to the concentrated solution. This is known as **osmosis** (Figure 4.10).

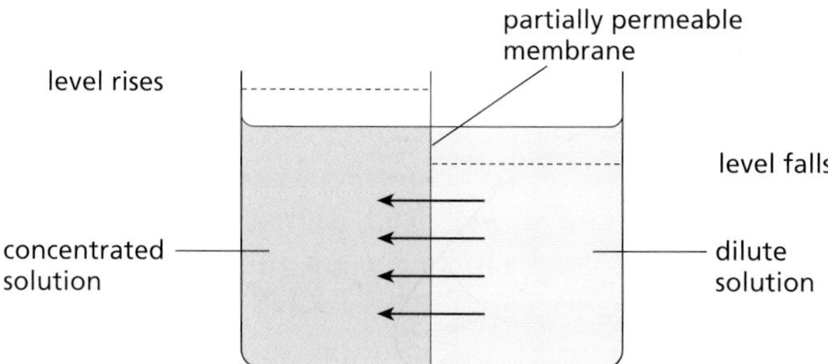

Figure 4.10 Osmosis. Water will diffuse from the dilute to the concentrated solution through the partially permeable membrane. As a result, the liquid level will rise on the left and fall on the right

A partially permeable membrane is porous but allows water to pass through it more rapidly than dissolved substances.

Since a dilute solution contains, in effect, more water molecules than a concentrated solution, there is a diffusion gradient which favours the passage of water from the dilute to the concentrated solution.

In living cells, the cell membrane is partially permeable and the cytoplasm and vacuole (in plant cells) contain dissolved substances. As a consequence, water tends to diffuse into cells by osmosis if they are surrounded by a weak solution, e.g. fresh water. If the cells are surrounded by a stronger solution, e.g. sea water, the cells may lose water by osmosis.

These effects are described more fully on p. 30.

Explanation of osmosis

When a substance such as sugar dissolves in water, the sugar molecules attract some of the water molecules and stop them moving freely. This, in effect, reduces the concentration of water molecules. In Figure 4.11 the sugar molecules on the right have 'captured' half the water molecules. There are more free water molecules on the left of the membrane than on the right, so water will diffuse more rapidly from left to right across the membrane than from right to left.

The partially permeable membrane does not act like a sieve, in this case. The sugar molecules can diffuse from right to left but, because they are bigger and surrounded by a cloud of water molecules, they diffuse more slowly than the water (Figure 4.12).

Artificial partially permeable membranes are made from cellulose acetate in sheets or tubes and used for **dialysis** (p. 34) rather than for osmosis. The pore size can be adjusted during manufacture so that large molecules cannot get through at all.

The **cell membrane** behaves like a partially permeable membrane. The partial permeability may depend on pores in the cell membrane but the processes involved are far more complicated than in an artificial membrane and depend on the structure of the membrane and on living processes in the cytoplasm (see p. 27). The cell membrane contains lipids and proteins. Anything which denatures proteins, e.g. heat, also destroys the structure and the partially permeable properties of a cell membrane. If this happens, the cell will die as essential substances diffuse out of the cell and harmful chemicals diffuse in.

Water potential

The **water potential** of a solution is a measure of whether it is likely to lose or gain water molecules from another solution. A dilute solution, with its high proportion of free water molecules, is said to have a higher water potential than a concentrated solution, because water will flow from the dilute to the concentrated solution (from a high potential to a low potential). Pure water has the highest possible water potential because water molecules will flow from it to any other aqueous solution, no matter how dilute.

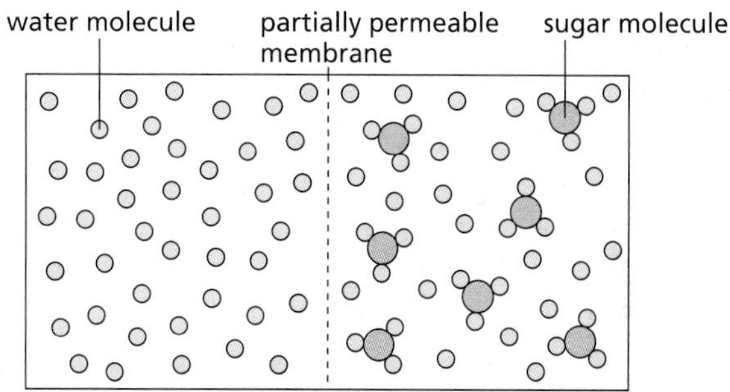

Figure 4.11 The diffusion gradient for water. There are more free water molecules on the left, so more will diffuse from left to right than in the other direction. Sugar molecules will diffuse more slowly from right to left

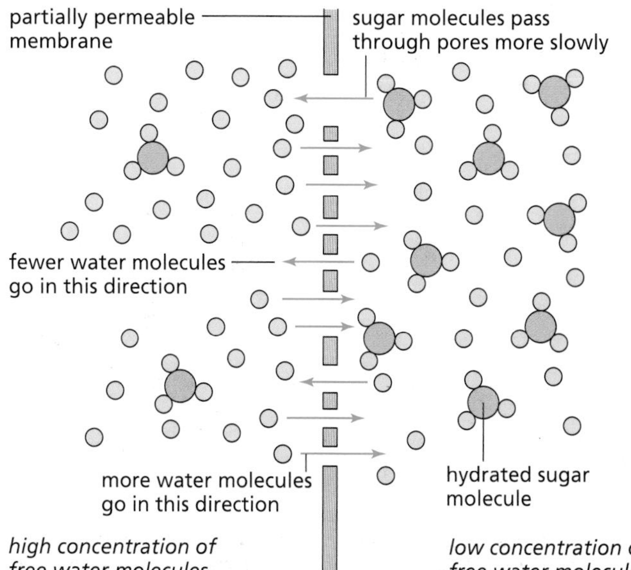

Figure 4.12 The diffusion theory of osmosis

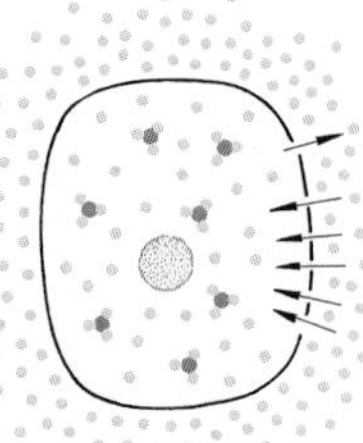

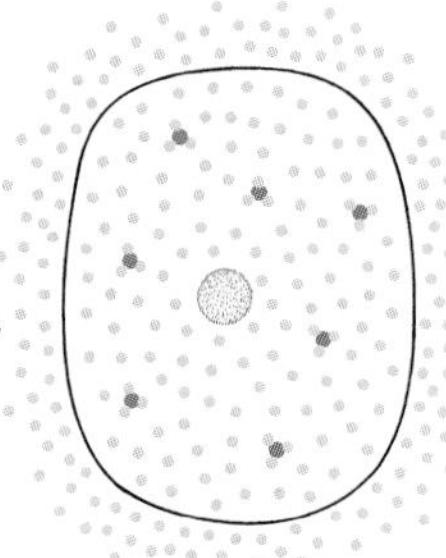

(a) There is a higher concentration of free water molecules outside the cell than inside, so water diffuses into the cell.

(b) The extra water makes the cell swell up.

(Note that molecules are really far too small to be seen at this magnification.)

Figure 4.13 Osmosis in an animal cell

Animal cells

In Figure 4.13 an animal cell is shown very simply. The coloured circles represent molecules in the cytoplasm. They may be sugar, salt or protein molecules. The grey circles represent water molecules.

The cell is shown surrounded by pure water. Nothing is dissolved in the water; it has 100 per cent concentration of water molecules. So the concentration of free water molecules outside the cell is greater than that inside and, therefore, water will diffuse into the cell by osmosis.

The membrane allows water to go through either way. So in our example, water can move into or out of the cell.

The cell membrane is partially permeable to most of the substances dissolved in the cytoplasm. So although the concentration of these substances inside may be high, they cannot diffuse freely out of the cell.

The water molecules move into and out of the cell, but because there are more of them on the outside, they will move in faster than they move out. The liquid outside the cell does not have to be 100 per cent pure water. As long as the concentration of water outside is higher than that inside, water will diffuse in by osmosis.

Water entering the cell will make it swell up, and unless the extra water is expelled in some way the cell will burst.

Conversely, if the cells are surrounded by a solution which is more concentrated than the cytoplasm, water will pass out of the cell by osmosis and the cell will shrink. Excessive uptake or loss of water by osmosis may damage cells.

For this reason, it is very important that the cells in an animal's body are surrounded by a liquid which has the same concentration as the liquid inside the cells. The outside liquid is called 'tissue fluid' (see p. 113) and its concentration depends on the concentration of the blood. In vertebrate animals the blood's concentration is monitored by the brain and adjusted by the kidneys, as described on p. 133.

By keeping the blood concentration within narrow limits, the concentration of tissue fluid remains more or less constant (see pp. 116 and 135) and the cells are not bloated by taking in too much water, or dehydrated by losing too much.

Plant cells

The cytoplasm of a plant cell and the cell sap in its vacuole contain salts, sugars and proteins which effectively reduce the concentration of free water molecules inside the cell. The cell wall is freely permeable to water and dissolved substances but the cell membrane of the cytoplasm is partially permeable. If a plant cell is surrounded by water or a solution more dilute than its contents, water will pass into the vacuole by osmosis. The vacuole will expand and press outwards on the cytoplasm and cell wall. The cell wall of a mature plant cell cannot be stretched, so there comes a time when the inflow of water is resisted by the unstretchable cell wall (Figure 4.14).

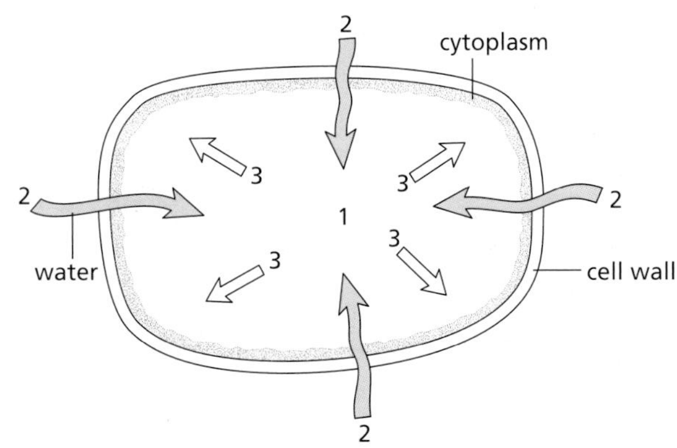

1 since there is effectively a lower concentration of water in the cell sap
2 water diffuses into the vacuole
3 and makes it push out against the cell wall

Figure 4.14 Osmosis in a plant cell

This has a similar effect to inflating a soft bicycle tyre. The tyre represents the firm cell wall, the floppy inner tube is like the cytoplasm and the air inside corresponds to the vacuole. If enough air is pumped in, it pushes the inner tube against the tyre and makes the tyre hard. A plant cell with the vacuole pushing out on the cell wall is said to be **turgid** and the vacuole is exerting **turgor pressure** on the cell wall.

If all the cells in a leaf and stem are turgid, the stem will be firm and upright and the leaves held out straight. If the vacuoles lose water for any reason, the cells will lose their turgor and become **flaccid** (Experiment 3, p. 32). A leaf with flaccid cells will be limp and the stem will droop. A plant which loses water to this extent is said to be 'wilting' (Figure 4.15).

(a) plant wilting
(b) plant recovered after watering

Figure 4.15 Wilting

Questions

1. A 10 per cent solution of copper sulphate is separated by a partially permeable membrane from a 5 per cent solution of copper sulphate. Will water diffuse from the 10 per cent to the 5 per cent solution, or from the 5 per cent to the 10 per cent solution?
2. If a fresh beetroot is cut up, the pieces washed in water and then left for an hour in a beaker of water, little or no red pigment escapes from the cells into the water. If the beetroot is boiled first, the pigment does escape into the water. Bearing in mind the properties of a living cell membrane, offer an explanation for this difference.
3. When doing experiments with animal tissues they are usually bathed in Ringer's solution, which has a concentration similar to that of blood or tissue fluid. Why do you think this is necessary?
4. Why does a dissolved substance reduce the number of 'free' water molecules in a solution?
5. When a plant leaf is in daylight, its cells make sugar from carbon dioxide and water (see p. 35). The sugar is at once turned into starch and deposited in plastids. What is the osmotic advantage of doing this? (Sugar is soluble in water; starch is not. See p. 12.)

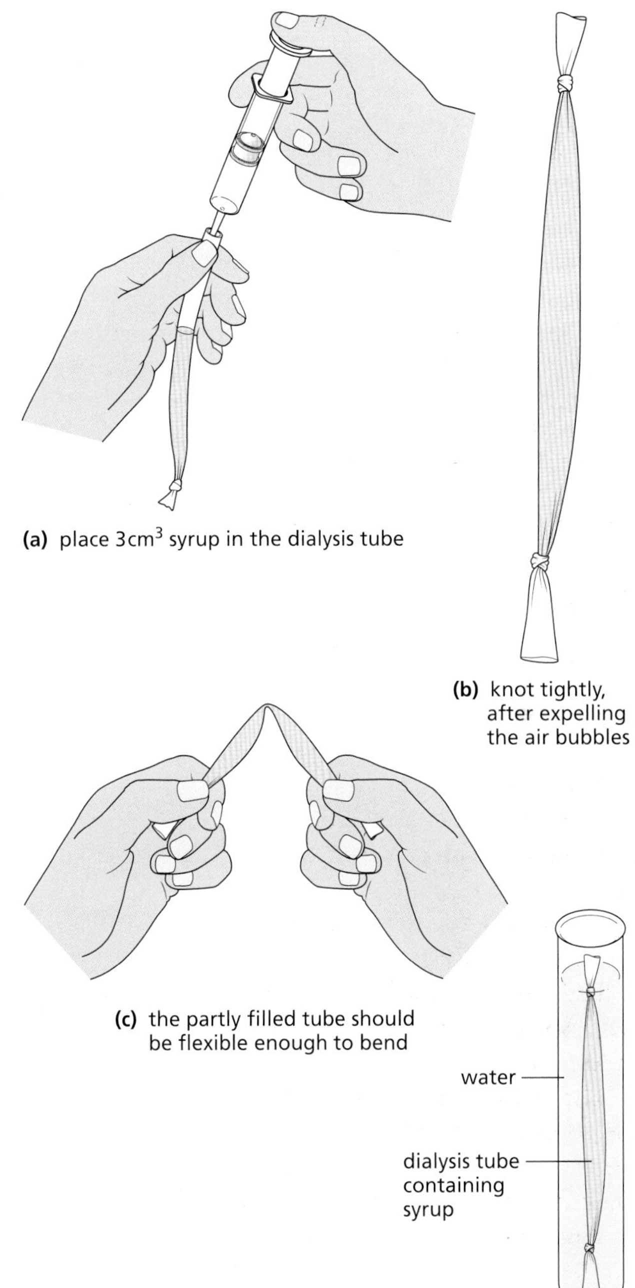
(a) place 3 cm³ syrup in the dialysis tube
(b) knot tightly, after expelling the air bubbles
(c) the partly filled tube should be flexible enough to bend
water
dialysis tube containing syrup

Figure 4.16 Experiment to illustrate turgor in a plant cell

Practical work

Experiments on osmosis and dialysis

Experiments 1, 2 and 5 use 'Visking' dialysis tubing. It is made from cellulose and is partially permeable, allowing water molecules to diffuse through freely, but restricting the passage of dissolved substances to varying extents. It is used in kidney dialysis machines because it lets the small molecules of harmful waste products (e.g. urea, p. 134) out of the blood but retains the blood cells and large protein molecules.

1 Osmosis and turgor

Take a 20 cm length of dialysis tubing which has been soaked in water and tie a knot tightly at one end. Place $3\,cm^3$ of a strong sugar solution in the tubing using a plastic syringe (Figure 4.16a) and then knot the open end of the tube (Figure 4.16b). The partly filled tube should be quite floppy (Figure 4.16c). Place the tubing in a test-tube of water for 30–45 minutes. After this time, remove the dialysis tubing from the water and look for any changes in how it looks or feels.

Result The tubing will now be firm, distended by the solution inside.

Interpretation The dialysis tubing is partially permeable and the solution inside has fewer free water molecules than outside. Water has, therefore, diffused in and increased the volume and the pressure of the solution inside.

This is a crude model of what is thought to happen to a plant cell when it becomes turgid. The sugar solution represents the cell sap and the dialysis tubing represents the cell membrane and cell wall combined.

2 Osmosis and water flow

Tie a knot in one end of a piece of soaked dialysis tubing and fill it with sugar solution as in the previous experiment but, this time, add a little coloured dye. Then fit it over the end of a length of capillary tubing and hold it in place with an elastic band. Push the capillary tubing into the dialysis tubing until the sugar solution enters the capillary. Now clamp the capillary tubing so that the dialysis tubing is totally immersed in a beaker of water as shown in Figure 4.17. Watch the level of liquid in the capillary tubing over the next 10 or 15 minutes.

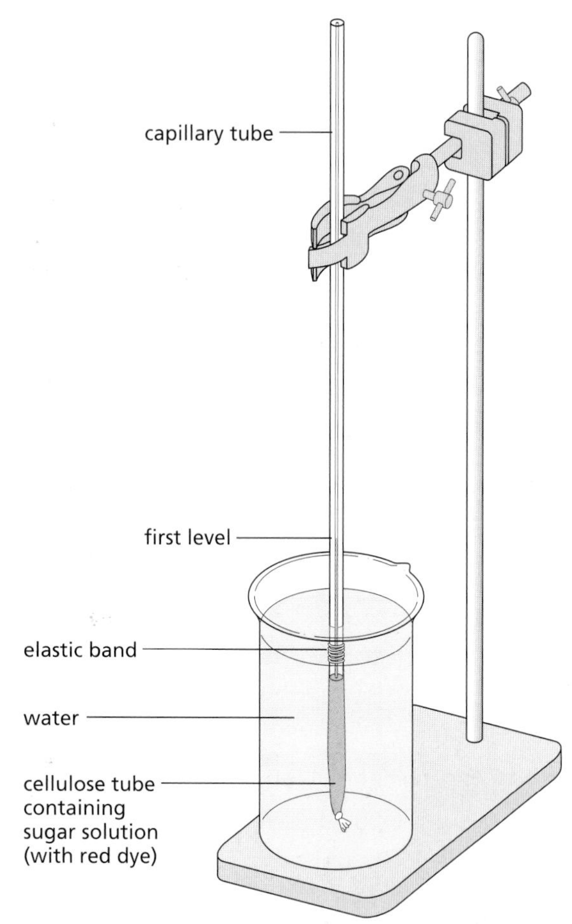

Figure 4.17 Demonstration of osmosis

Result The level of liquid in the capillary tube will be seen to rise.

Interpretation Water must be passing into the sugar solution from the beaker. This is what you would expect when a concentrated solution is separated from water by a partially permeable membrane. The results are similar to those in Experiment 1 but instead of the expanding solution distending the dialysis tube, it escapes up the capillary.

A process similar to this might be partially responsible for moving water from the roots to the stem of a plant.

3 Plasmolysis

Peel a small piece of epidermis (outer layer of cells) from a red area of a rhubarb stalk (see Figure 1.15a, p. 9). Place the epidermis on a slide with a drop of water and cover with a cover-slip (see Figure 1.15c, p. 9). Place a 30 per cent solution of sugar at one edge of the cover-slip with a pipette and then draw the solution under the cover-slip by placing a piece of blotting-paper on the opposite side (Figure 4.18). As you are doing this, study a small group of cells under the microscope and watch for any changes in their appearance (Figure 4.20).

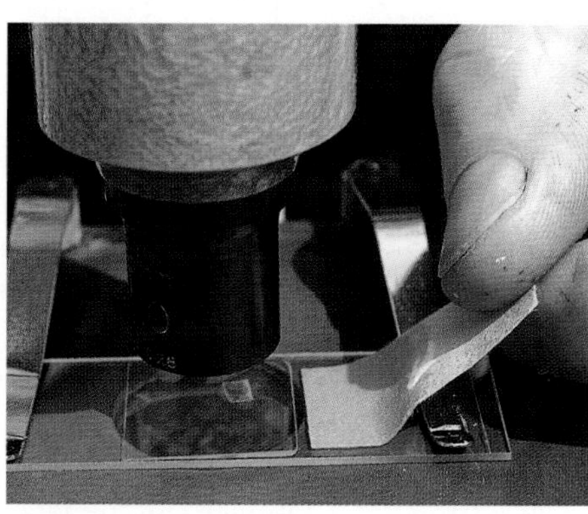

Figure 4.18 Changing the water for sugar solution

Result The red cell sap will appear to shrink and get darker and pull the cytoplasm away from the cell wall leaving clear spaces. (It will not be possible to see the cytoplasm but its presence can be inferred from the fact that the red cell sap seems to have a distinct outer boundary in those places where it has separated from the cell wall.)

Interpretation The interpretation in terms of osmosis is outlined in Figure 4.19. The cells are said to be **plasmolysed**.

The plasmolysis can be reversed by drawing water under the cover-slip in the same way that you drew the sugar solution under. It may need two or three lots of water to flush out all the sugar. If you watch a group of cells, you should see their vacuoles expanding to fill the cells once again.

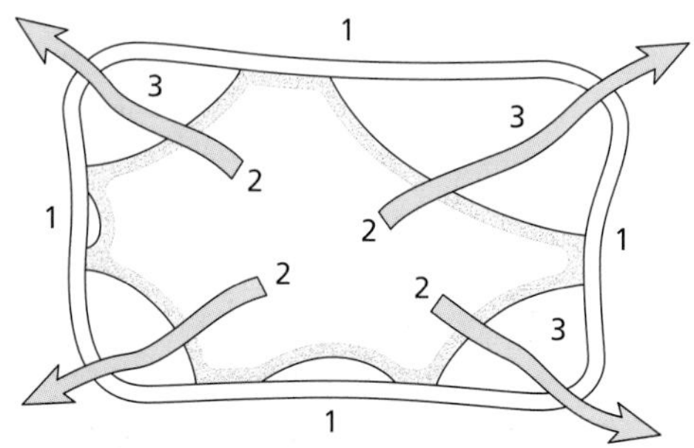

1 the solution outside the cell is more concentrated than the cell sap
2 water diffuses out of the vacuole
3 the vacuole shrinks, pulling the cytoplasm away from the cell wall, leaving the cell flaccid

Figure 4.19 Plasmolysis

Rhubarb is used for this experiment because the coloured cell sap shows up. If rhubarb is not available the epidermis from a red onion scale can be used with results similar to those in Figure 4.20.

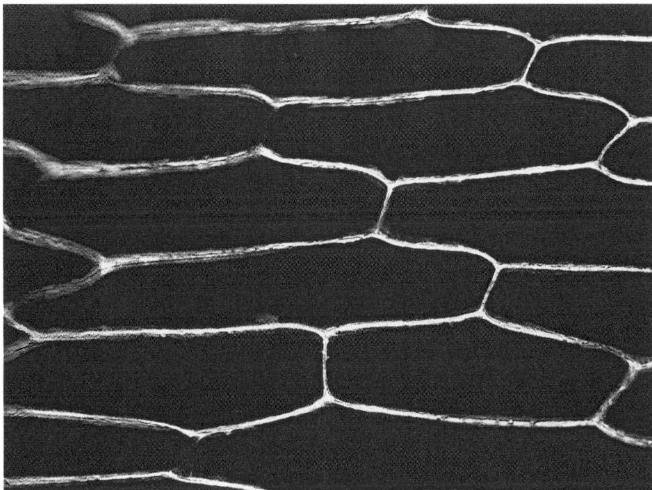

Figure 4.20 **(a)** Turgid cells (× 100). The cells are in a strip of epidermis from a red onion scale. The cytoplasm is pressed against the inside of the cell wall by the vacuole

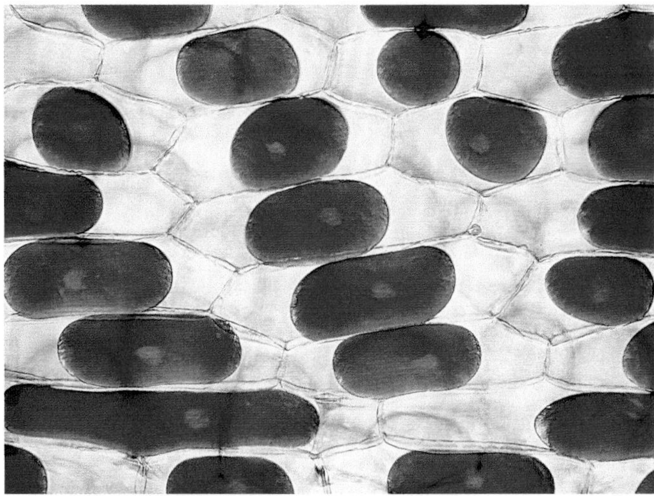

(b) Plasmolysed cells (× 100). The same cells as they appear after treatment with salt solution. The vacuole has lost water by osmosis, shrunk and pulled the cytoplasm away from the cell wall

4 Turgor in potato tissue

Push a No. 4 or No. 5 cork borer into a large potato. *Caution* Do not hold the potato in your hand but use a board as in Figure 4.21a.

Push the potato tissue out of the cork borer using a pencil as in Figure 4.21b. Prepare a number of potato cylinders in this way and choose the two longest (at least 50 mm). Cut these two accurately to the same length, e.g. 50, 60 or 70 mm. Measure carefully.

Label two test-tubes A and B and place a potato cylinder in each. Cover the potato tissue in tube A with water; cover the tissue in B with a 20 per cent sugar solution. Leave the tubes for a day.

After this time, remove the cylinder from tube A and measure its length. Notice also whether it is firm or flabby. Repeat this for the potato in tube B, but rinse it in water before measuring it.

Result The cylinder from tube A should have gained a millimetre or two and feel firm. The cylinder from tube B should be a millimetre or two shorter and feel flabby.

Interpretation If the potato cells were not fully turgid at the beginning of the experiment, they would take up water by osmosis (tube A), and cause an increase in length.

In tube B, the sugar solution is stronger than the cell sap of the potato cells, so these cells will lose water by osmosis. The cells will lose their turgor and the potato cylinder will become flabby and shorter.

An alternative to measuring the potato cores is to weigh them before and after the 24 hours' immersion in water or sugar solution. The core in tube A should gain weight and that in tube B should lose weight. It is important to blot the cores dry with a paper towel before each weighing.

Whichever method is used, it is a good idea to pool the results of the whole class since the changes may be quite small. A gain in length of 1 or 2 mm might be due to an error in measurement, but if most of the class record an increase in length, then experimental error is unlikely to be the cause.

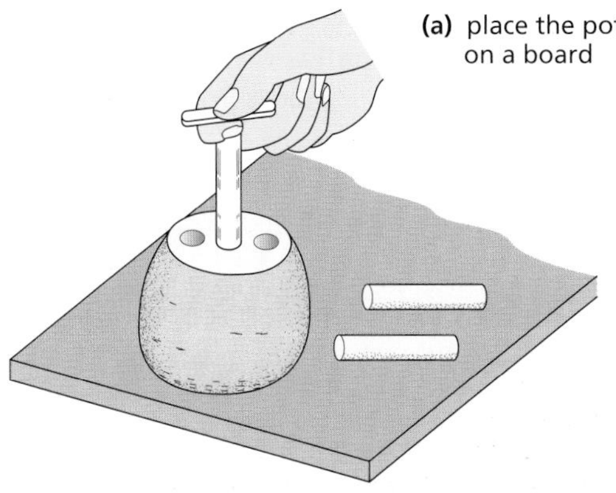

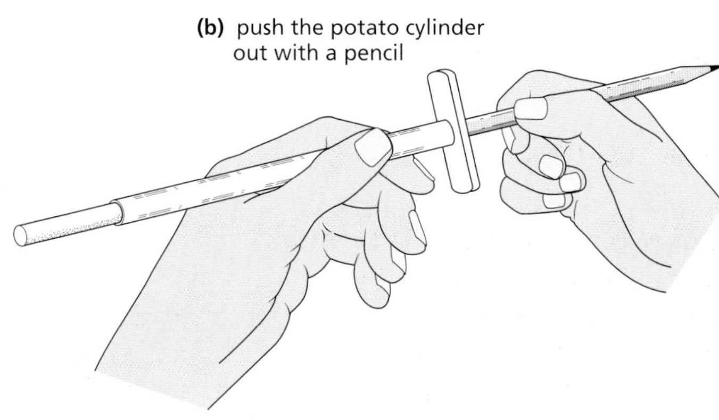

(a) place the potato on a board

(b) push the potato cylinder out with a pencil

Figure 4.21 Obtaining cylinders of potato tissue

5 Partial permeability (dialysis)

Take a 15 cm length of dialysis tubing which has been soaked in water and tie a knot tightly at one end. Use a dropping pipette to partly fill the tubing with 1 per cent starch solution. Put the tubing in a test-tube and hold it in place with an elastic band as shown in Figure 4.22. Rinse the tubing and test-tube under the tap to remove all traces of starch solution from the outside of the dialysis tube.

Fill the test-tube with water and add a few drops of iodine solution to colour the water yellow. Leave for 10–15 minutes.

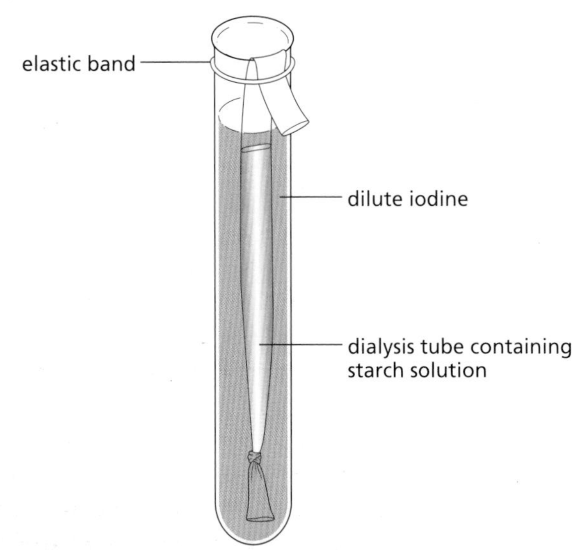

Figure 4.22 Dialysis

Result The starch inside the dialysis tubing goes blue but the iodine outside stays yellow.

Interpretation The blue colour is characteristic of the reaction which takes place between starch and iodine, and is used as a test for starch (see p. 95). The results show that iodine molecules have passed through the dialysis tubing into the starch but the starch molecules have not moved out into the iodine. This is what we would expect if the dialysis tubing were partially permeable on the basis of its pore size. Starch molecules are very large (see p. 12) and probably cannot get through the pores. Iodine molecules are much smaller and can, therefore, get through.

Note This experiment illustrates the process of dialysis rather than osmosis. The movement of water is not necessarily involved and the pore size of the membrane makes it genuinely partially permeable with respect to iodine and starch.

Questions

1. In Experiment 2 (Figure 4.17), what do you think would happen
   - **a** if a much stronger sugar solution was placed in the cellulose tube?
   - **b** if the beaker contained a weak sugar solution instead of water?
   - **c** if the sugar solution was in the beaker and the water was in the cellulose tube?
2. In Experiment 1 (Figure 4.16), what might happen if the cellulose tube filled with sugar solution was left in the water for several hours?
3. In Experiment 3, Figure 4.19 explains why the vacuole shrinks. Give a brief explanation of why it swells up again when the cell is surrounded by water.
4. An alternative interpretation of the results of Experiment 5 might be that the dialysis tubing allowed molecules (of any size) to pass in but not out. Describe an experiment to test this possibility and say what results you would expect
   - **a** if it were correct and
   - **b** if it were false.
5. In Experiment 2 on p. 32, the column of liquid accumulating in the capillary tube exerts an ever-increasing pressure on the solution in the dialysis tube. Bearing this in mind and assuming a very long capillary, at what stage would you expect the net flow of water from the beaker into the dialysis tubing to cease?

Checklist

- Diffusion is the result of molecules of liquid, gas or dissolved solid moving about.
- The molecules of a substance diffuse from a region where they are very concentrated to a region where they are less concentrated.
- Substances may enter cells by simple diffusion, controlled diffusion, active transport or endocytosis.
- Osmosis is the diffusion of water through a partially permeable membrane.
- Water diffuses from a dilute solution of salt or sugar to a concentrated solution because the concentrated solution contains fewer free water molecules.
- Cell membranes are partially permeable and cytoplasm and cell sap contain many substances in solution.
- Cells take up water from dilute solutions but lose water to concentrated solutions because of osmosis.
- Osmosis maintains turgor in plant cells.

5 Photosynthesis and nutrition in plants

Photosynthesis

All living organisms need food. They need it as a source of raw materials to build new cells and tissues as they grow. They also need food as a source of energy. Food is a kind of 'fuel' which drives essential living processes and brings about chemical changes (see pp. 19 and 86). Animals take in food, digest it, and use the digested products to build their tissues or to produce energy.

Plants also need energy and raw materials but, apart from a few insect-eating species, plants do not appear to take in food. The most likely source of their raw materials would appear to be the soil. However, experiments show that the weight gained by a growing plant is far greater than the weight lost by the soil it is growing in. So there must be additional sources of raw materials.

These additional sources can only be water and air. A hypothesis to explain the source of food in a plant is that it **makes it** from air, water and soil salts. Carbohydrates (p. 12) contain the elements carbon, hydrogen and oxygen, as in glucose ($C_6H_{12}O_6$). The carbon and oxygen could be supplied by carbon dioxide (CO_2) from the air, and the hydrogen could come from the water (H_2O) in the soil. The nitrogen and sulphur needed for making proteins (p. 11) could come from nitrates and sulphates in the soil.

This building-up of complex food molecules from simpler substances is called **synthesis** and it needs enzymes and energy to make it happen. The enzymes are present in the plant's cells and the energy for the first stages in the synthesis comes from sunlight. The process is, therefore, called **photosynthesis** ('photo' means 'light'). There is evidence to suggest that the green substance **chlorophyll**, in the chloroplasts of plant cells, plays a part in photosynthesis. Chlorophyll absorbs sunlight and makes the energy from sunlight available for chemical reactions. Thus, in effect, the function of chlorophyll is to convert light energy to chemical energy.

Our working hypothesis for photosynthesis is, therefore, the **building-up of food compounds from carbon dioxide and water by green plants using energy from sunlight which is absorbed by chlorophyll**.

A chemical equation for photosynthesis would be

$$\underset{\text{carbon dioxide}}{6CO_2} + \underset{\text{water}}{6H_2O} \xrightarrow{\text{light energy}} \underset{\text{glucose}}{C_6H_{12}O_6} + \underset{\text{oxygen}}{6O_2}$$

In order to keep the equation simple, glucose is shown as the food compound produced. This does not imply that it is the only substance synthesized by photosynthesis.

Practical work

Experiments to test photosynthesis

The design of biological experiments was discussed on p. 23 and this should be revised before studying the next section. It would also be helpful to read the section on 'Hypothesis testing' (p. 24) if you have not already done so.

A **hypothesis** is an attempt to explain certain observations. In this case the hypothesis is that plants make their food by photosynthesis. The equation given on p. 36 is one way of stating the hypothesis and is used here to show how it might be tested.

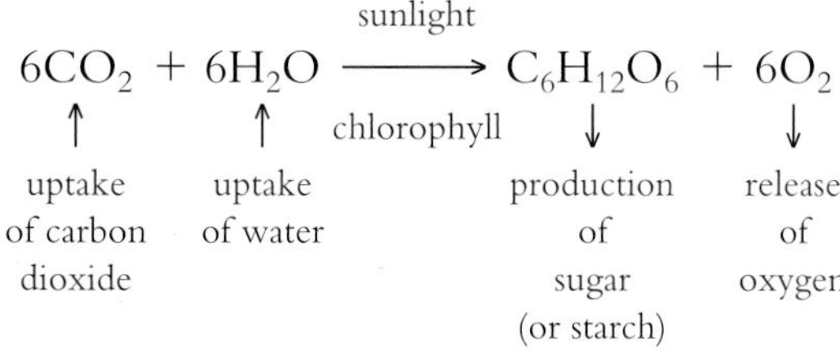

If photosynthesis is going on in a plant, then the leaves should be producing sugars. In many leaves, as fast as sugar is produced, it is turned into starch. Since it is easier to test for starch than for sugar, we regard the production of starch in a leaf as evidence that photosynthesis has taken place.

The first three experiments described below are designed to see if the leaf can make starch without chlorophyll, sunlight, or carbon dioxide, in turn. If the photosynthesis story is sound, then the lack of any one of these three conditions should stop photosynthesis, and so stop the production of starch. But, if starch production continues, then the hypothesis is no good and must be altered or rejected.

In designing the experiments, it is very important to make sure that only *one* variable is altered. If, for example, the method of keeping light from a leaf also cuts off its carbon dioxide supply, it would be impossible to decide whether it was the lack of light or lack of carbon dioxide which stopped the production of starch. To make sure that the experimental design has not altered more than one variable, a **control** is set up in each case. This is an identical situation, except that the condition missing from the experiment, e.g. light, carbon dioxide or chlorophyll, is present in the control (see p. 23).

Destarching a plant

If the production of starch is your evidence that photosynthesis is taking place, then you must make sure that the leaf does not contain any starch at the beginning of the experiment. This is done by **destarching** the leaves. It is not possible to remove the starch chemically, without damaging the leaves, so a plant is destarched simply by leaving it in darkness for 2 or 3 days. Potted plants are destarched by leaving them in a dark cupboard for a few days. In the darkness, any starch in the leaves will be changed to sugar and carried away from the leaves to other parts of the plant. For plants in the open, the experiment is set up on the day before the test. During the night, most of the starch will be removed from the leaves. Better still, wrap the leaves in aluminium foil for 2 days while they are still on the plant. Then test one of the leaves to see that no starch is present.

Testing a leaf for starch

Iodine solution (yellow) and starch (white) form a deep blue colour when they mix. The test for starch, therefore, is to add iodine solution to a leaf to see if it goes blue. First, however, the leaf has to be treated as follows:

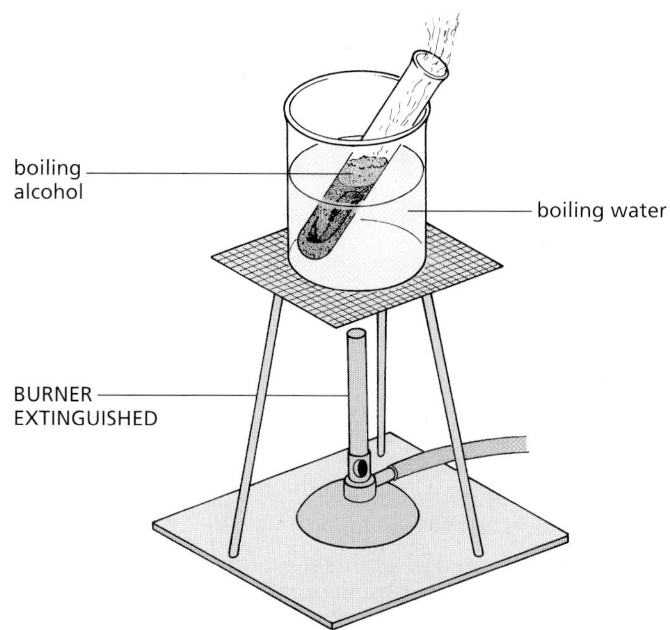

Figure 5.1 To remove chlorophyll from a leaf

1 Heat some water to boiling point in a beaker and then **turn out the Bunsen flame**.
2 Use forceps to dip a leaf in the hot water for about 30 seconds. This kills the cytoplasm, denatures the enzymes and makes the leaf more permeable to iodine solution.
3 Push the leaf to the bottom of a test-tube and cover it with alcohol (ethanol). Place the tube in the hot water (Figure 5.1). The alcohol will boil and dissolve out most of the chlorophyll. This makes colour changes with iodine easier to see.
4 Pour the green alcohol into a spare beaker, remove the leaf and dip it once more into the hot water to soften it.
5 Spread the decolorized leaf flat on a white tile and drop iodine solution on to it. The parts containing starch will turn blue; parts without starch will stain brown or yellow with iodine.

1 Is chlorophyll necessary for photosynthesis?

It is not possible to remove chlorophyll from a leaf without killing it, and so a **variegated** leaf, which has chlorophyll only in patches, is used. A leaf of this kind is shown in Figure 5.2a. The white part of the leaf serves as the experiment, because it lacks chlorophyll, while the green part with chlorophyll is the control. After being destarched, the leaf – still on the plant – is exposed to daylight for a few hours. Remove a leaf from the plant; draw it carefully to show where the chlorophyll is (i.e. the green parts), and test it for starch as described above.

Result Only the parts that were previously green turn blue with iodine. The parts that were white stain brown (Figure 5.2b).

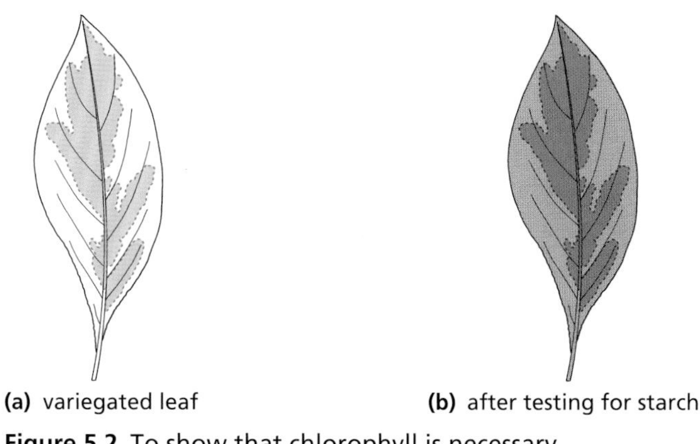

Figure 5.2 To show that chlorophyll is necessary

Interpretation Since starch is present only in the parts which originally contained chlorophyll, it seems reasonable to suppose that chlorophyll is needed for photosynthesis.

It must be remembered, however, that there are other possible interpretations which this experiment has not ruled out; for example, starch could be made in the green parts and sugar in the white parts. Such alternative explanations could be tested by further experiments.

2 Is light necessary for photosynthesis?

Cut a simple shape from a piece of aluminium foil to make a stencil and attach it to a destarched leaf (Figure 5.3a). After 4 to 6 hours of daylight, remove the leaf and test it for starch.

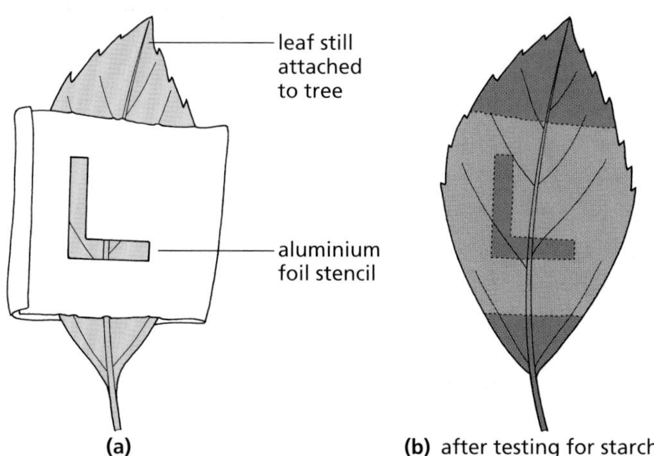

Figure 5.3 To show that light is necessary

Result Only the areas which had received light go blue with iodine (Figure 5.3b).

Interpretation As starch has not formed in the areas which received no light, it seems that light is needed for starch formation and thus for photosynthesis.

You could argue that the aluminium foil had stopped carbon dioxide from entering the leaf and that it was shortage of carbon dioxide rather than absence of light which prevented photosynthesis taking place. A further control could be designed, using transparent material instead of aluminium foil for the stencil.

3 Is carbon dioxide needed for photosynthesis?

Water two destarched potted plants and enclose their shoots in polythene bags. In one pot place a dish of soda-lime to absorb the carbon dioxide from the air (the experiment). In the other place a dish of sodium hydrogencarbonate solution to produce carbon dioxide (the control), as shown in Figure 5.4. Place both plants in the light for several hours and then test a leaf from each for starch.

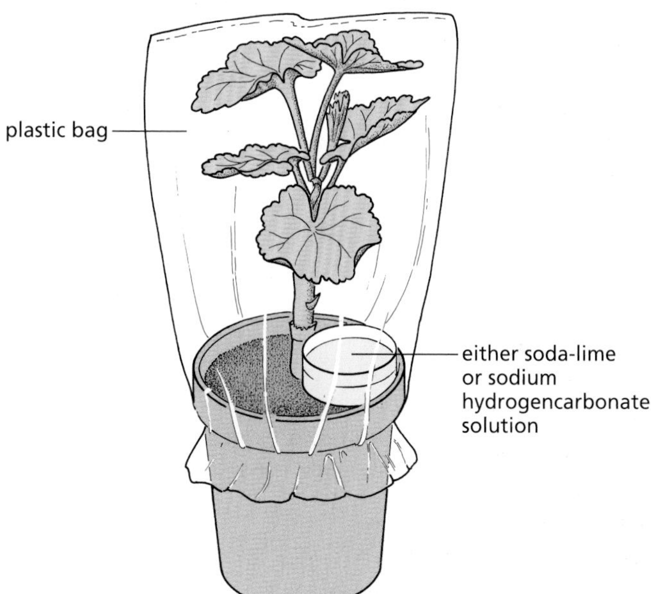

Figure 5.4 To show that carbon dioxide is necessary

Result The leaf which had no carbon dioxide does not turn blue. The one from the polythene bag containing carbon dioxide does turn blue.

Interpretation The fact that starch was made in the leaves which had carbon dioxide, but not in the leaves which had no carbon dioxide, suggests that this gas must be necessary for photosynthesis. The control rules out the possibility that high humidity or high temperature in the plastic bag prevents normal photosynthesis.

4 Is oxygen produced during photosynthesis?

Place a short-stemmed funnel over some Canadian pondweed in a beaker of water. Fill a test-tube with water and place it upside-down over the funnel stem (Figure 5.5, overleaf). (The funnel is raised above the bottom of the beaker to allow the water to circulate.) Place the apparatus in sunlight. Bubbles of gas should appear from the cut stems and collect in the test-tube. Set up a control in a similar way but place it in a dark cupboard. When sufficient gas has collected from the plant in the light, remove the test-tube and insert a glowing splint.

Result The glowing splint bursts into flames.

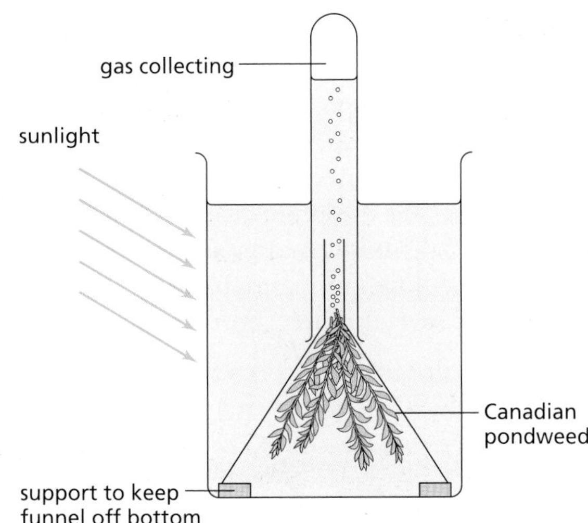

Figure 5.5 To show that oxygen is produced

Interpretation The relighting of a glowing splint does not prove that the gas collected in the test-tube is *pure* oxygen, but it does show that it contains extra oxygen and this must have come from the plant. The oxygen is given off only in the light.

Note that water contains dissolved oxygen, carbon dioxide and nitrogen. These gases may diffuse in or out of the bubbles as they pass through the water and collect in the test-tube. The composition of the gas in the test-tube may not be the same as that in the bubbles leaving the plant.

Controls When setting up an experiment and a control, which of the two procedures constitutes the 'control' depends on the way the prediction is worded. For example, if the prediction is that 'in the absence of light, the pondweed will not produce oxygen', then the 'control' is the plant in the light. If the prediction is that 'the pondweed in the light will produce oxygen', then the 'control' is the plant in darkness. As far as the results and interpretation are concerned, it does not matter which is the 'control' and which is the 'experiment'.

The results of these four experiments support the hypothesis of photosynthesis as stated on p. 35 and as represented by the equation. Starch formation (our evidence for photosynthesis) does not take place in the absence of light, chlorophyll or carbon dioxide, and oxygen production occurs only in the light.

If starch or oxygen production had occurred in the absence of any one of these conditions, we should have to change our hypothesis about the way plants obtain their food. Bear in mind, however, that although our results support the photosynthesis theory, they do not prove it. For example, it is now known that many stages in the production of sugar and starch from carbon dioxide do not need light (the 'dark' reaction).

Questions

1 Which of the following are needed for starch production in a leaf
- a carbon dioxide,
- b oxygen,
- c nitrates,
- d water,
- e chlorophyll,
- f soil,
- g light?

2 In Experiment 1 (concerning the need for chlorophyll), why was it not necessary to set up a separate control experiment?

3 What is meant by 'destarching' a leaf? Why is it necessary to destarch leaves before setting up some of the photosynthesis experiments?

4 In Experiment 3 (concerning the need for carbon dioxide), what were the functions of
- a the soda-lime,
- b the sodium hydrogencarbonate,
- c the polythene bag?

5 Why do you think a pondweed, rather than a land plant, is used for Experiment 4 (concerning production of oxygen)? In what way might this choice make the results less useful?

6 A green plant makes sugar from carbon dioxide and water. Why do we not try the experiment of depriving a plant of water to see if that stops photosynthesis?

7 Does the method of destarching a plant take for granted the results of Experiment 2? Explain your answer.

The process of photosynthesis

Although the details of photosynthesis vary in different plants, the hypothesis as stated in this chapter has stood up to many years of experimental testing and is universally accepted. The next section describes how photosynthesis takes place in a plant.

The process takes place mainly in the cells of the leaves (Figure 5.6) and is summarized in Figure 5.7. In land plants water is absorbed from the soil by the roots and carried in the water vessels of the veins, up the stem to the leaf. Carbon dioxide is absorbed from the air through the stomata (pores in the leaf, see p. 52). In the leaf cells, the carbon dioxide and water are combined to make sugar; the energy for this reaction comes from sunlight which has been absorbed by the green pigment **chlorophyll**. The chlorophyll is present in the **chloroplasts** of the leaf cells and it is inside the chloroplasts that the reaction takes place. Chloroplasts (Figure 5.7d) are small, green structures present in the cytoplasm of the leaf cells. Chlorophyll is the substance which gives leaves and stems their green colour. It is able to absorb energy from light and use it to split water molecules into hydrogen and oxygen (the 'light' reaction). The oxygen escapes from the leaf and the hydrogen molecules are added to carbon dioxide molecules to form sugar (the 'dark' reaction).

Figure 5.6 All the reactions involved in producing food take place in the leaves. Notice how little the leaves overlap

There are four types of chlorophyll which may be present in various proportions in different species. There are also a number of photosynthetic pigments, other than chlorophyll, which may mask the colour of chlorophyll even when it is present, e.g. the brown and red pigments which occur in certain seaweeds.

Questions

1 What substances must a plant take in, in order to carry on photosynthesis? Where does it get each of these substances from?

2 Look at Figure 6.7a on p. 53. Identify the palisade cells, the spongy mesophyll cells and the cells of the epidermis. In which of these would you expect photosynthesis to occur
a most rapidly,
b least rapidly,
c not at all?
Explain your answer.

3 a What provides a plant with energy for photosynthesis?
b What chemical process provides a plant with energy to carry on all other living activities?

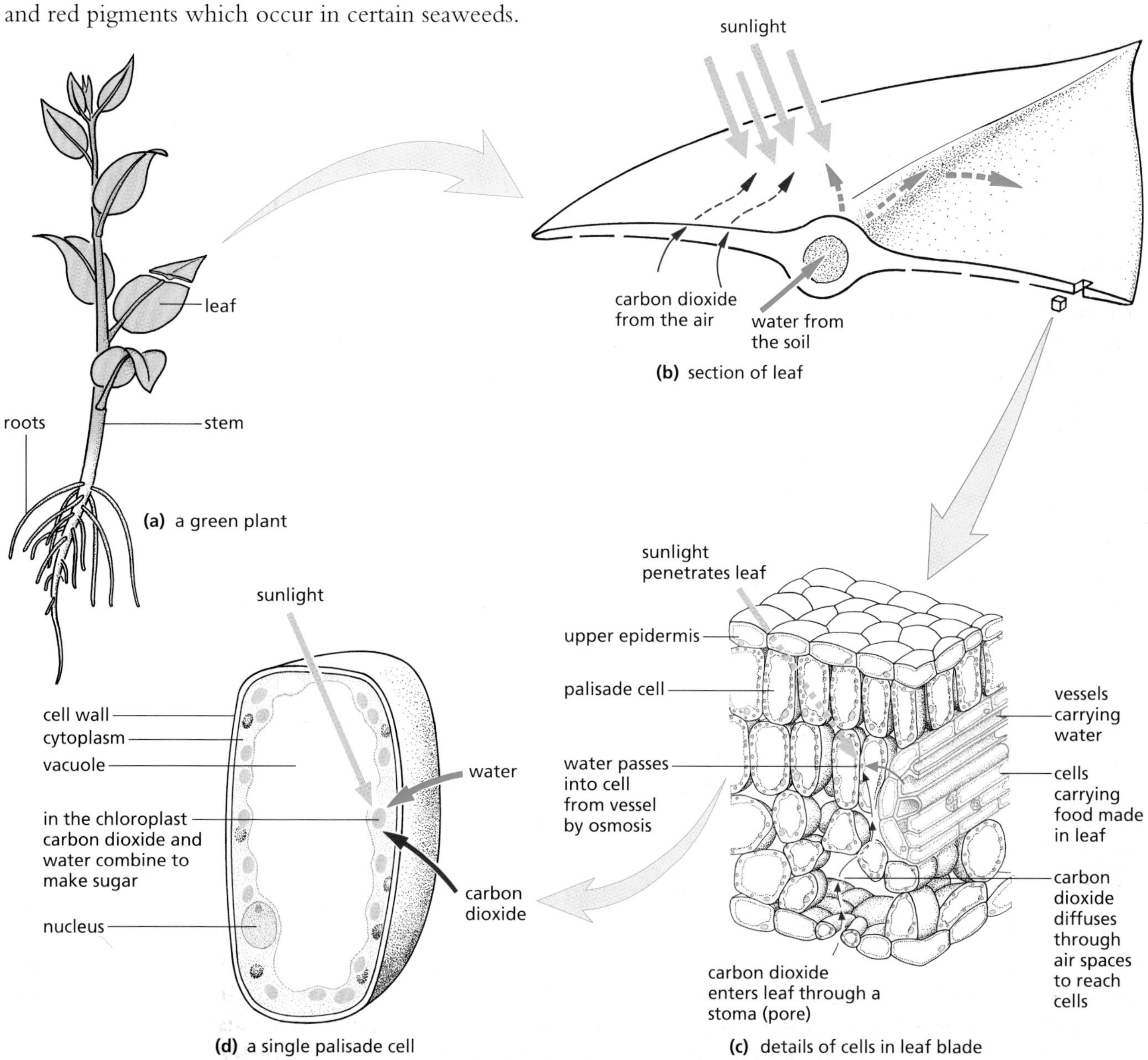

leaf
roots
stem
(a) a green plant

sunlight
carbon dioxide from the air
water from the soil
(b) section of leaf

sunlight penetrates leaf
upper epidermis
palisade cell
water passes into cell from vessel by osmosis
vessels carrying water
cells carrying food made in leaf
carbon dioxide diffuses through air spaces to reach cells
carbon dioxide enters leaf through a stoma (pore)
(c) details of cells in leaf blade

sunlight
cell wall
cytoplasm
vacuole
in the chloroplast carbon dioxide and water combine to make sugar
nucleus
water
carbon dioxide
(d) a single palisade cell

Figure 5.7 Photosynthesis in a leaf

Gaseous exchange in plants

Air contains the gases nitrogen, oxygen, carbon dioxide and water vapour. Plants and animals take in or give out these last three gases and this process is called **gaseous exchange**.

You can see from the equation for photosynthesis (p. 35) that one of its products is oxygen. Therefore, in daylight, when photosynthesis is going on in green plants, they will be taking in carbon dioxide and giving out oxygen. This exchange of gases is the opposite of that resulting from respiration (p. 19) but it must not be thought that green plants do not respire. The energy they need for all their living processes – apart from photosynthesis – comes from respiration and this is going on all the time, using up oxygen and producing carbon dioxide.

During the daylight hours, plants are photosynthesizing as well as respiring, so that all the carbon dioxide produced by respiration is used up by photosynthesis. At the same time, all the oxygen needed by respiration is provided by photosynthesis. Only when the rate of photosynthesis is faster than the rate of respiration will carbon dioxide be taken in and the excess oxygen be given out (Figure 5.8).

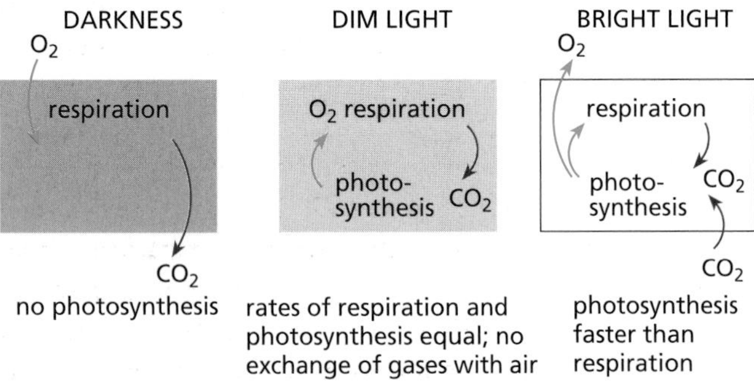

Figure 5.8 Respiration and photosynthesis

Compensation point

As the light intensity increases during the morning and fades during the evening, there will be a time when the rate of photosynthesis exactly matches the rate of respiration. At this point, there will be no net intake or output of carbon dioxide or oxygen. This is the **compensation point**. The sugar produced by photosynthesis exactly compensates for the sugar broken down by respiration.

Practical work

5 Gaseous exchange during photosynthesis

Wash three test-tubes first with tap water, then with distilled water and finally with hydrogencarbonate indicator. Then place 2 cm^3 hydrogencarbonate indicator in each tube.

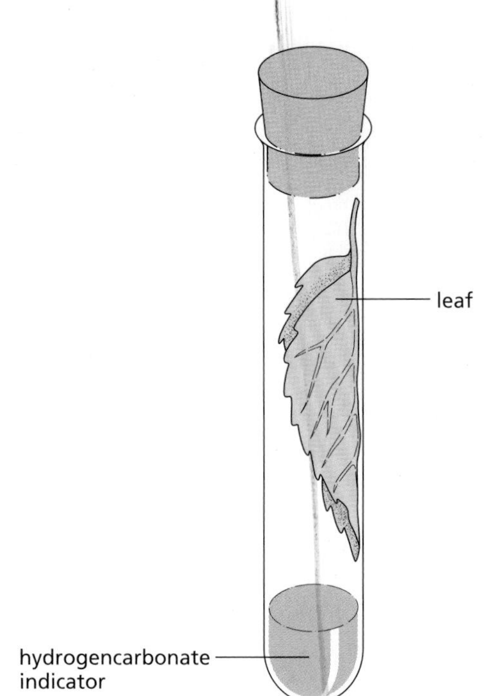

Figure 5.9 Gaseous exchange during photosynthesis and respiration

Place a green leaf in tubes 1 and 2 so that it is held against the walls of the tube and does not touch the indicator (Figure 5.9). Close the three tubes with bungs. Cover tube 1 with aluminium foil and place all three in a rack in direct sunlight, or a few centimetres from a bench lamp, for about 40 minutes.

Result The indicator (which was originally orange) should not change colour in tube 3, the control; that in tube 1, with the leaf in darkness, should turn yellow; and in tube 2 with the illuminated leaf, the indicator should be scarlet or purple.

Interpretation Hydrogencarbonate indicator is a mixture of dilute sodium hydrogencarbonate solution with the dyes cresol red and thymol blue. It is a pH indicator in equilibrium with the atmospheric carbon dioxide, i.e. its original colour represents the acidity produced by the carbon dioxide in the air. Increase in atmospheric carbon dioxide makes it more acid and it changes colour from orange to yellow. Decrease in atmospheric carbon dioxide makes it less acid and causes a colour change to red or purple.

The results, therefore, provide evidence that in darkness (tube 1) leaves produce carbon dioxide (from respiration), while in light (tube 2) they use up more carbon dioxide in photosynthesis than they produce in respiration. Tube 3 is the control, showing that it is the presence of the leaf which causes a change in the atmosphere in the test-tube.

The experiment can be criticized on the grounds that the hydrogencarbonate indicator is not a specific test for carbon dioxide but will respond to any change in acidity or alkalinity. In tube 2 there would be the same change in colour if the leaf produced an alkaline gas such as ammonia, and in tube 1 any acid gas produced by the leaf would turn the indicator yellow. However, a knowledge of the metabolism of the leaf suggests that these are less likely events than changes in the carbon dioxide concentration.

Questions

1 What gases would you expect a leaf to be
(i) taking in,
(ii) giving out,
a in bright sunlight,
b in darkness?

2 Measurements on a leaf show that it is giving out carbon dioxide and taking in oxygen. Does this prove that photosynthesis is *not* going on in the leaf? Explain your answer.

3 How could you adapt the experiment with hydrogencarbonate indicator to find the light intensity which corresponded to the compensation point?
How would you expect the compensation points to differ between plants growing in a wood and those growing in a field?

Adaptation of leaves for photosynthesis

When biologists say that something is **adapted**, they mean that its structure is well suited to its function. The detailed structure of the leaf is described on pp. 50 to 53, and although there are wide variations in leaf shape the following general statements apply to a great many leaves, and are illustrated in Figures 5.7b and c.

- Their broad, flat shape offers a large surface area for absorption of sunlight and carbon dioxide.
- Most leaves are thin and the carbon dioxide has to diffuse across short distances to reach the inner cells.
- The large spaces between cells inside the leaf provide an easy passage through which carbon dioxide can diffuse.
- There are many stomata (pores) in the lower surface of the leaf. These allow the exchange of carbon dioxide and oxygen with the air outside.
- There are more chloroplasts in the upper (palisade) cells than in the lower (spongy mesophyll) cells. The palisade cells, being on the upper surface, will receive most sunlight and this will reach the chloroplasts without being absorbed by too many cell walls.
- The branching network of veins provides a good water supply to the photosynthesizing cells. No cell is very far from a water-conducting vessel in one of these veins.

Although photosynthesis takes place mainly in the leaves, any part of the plant which contains chlorophyll will photosynthesize. Many plants have green stems in which photosynthesis takes place.

Effects of external factors on rate of photosynthesis

The rate of the light reaction will depend on the light intensity. The brighter the light, the faster will water molecules be split in the chloroplasts. The 'dark' reaction will be affected by temperature. A rise in temperature will increase the rate at which carbon dioxide is combined with hydrogen to make carbohydrate.

Limiting factors

If you look at Figure 5.10a (overleaf), you will see that an increase in light intensity does indeed speed up photosynthesis, but only up to a point. Beyond that point, any further increase in light intensity has only a small effect. This limit on the rate of increase could be because all available chloroplasts are fully occupied in light absorption. So, no matter how much the light intensity increases, no more light can be absorbed and used. Alternatively, the limit could be imposed by the fact that there is not enough carbon dioxide in the air to cope with the increased supply of hydrogen atoms produced by the light reaction. Or, it may be that low temperature is restricting the rate of the 'dark' reaction.

Figure 5.10b (overleaf) shows that, if the temperature of a plant is raised, then the effect of increased illumination is not limited so much. Thus, in Figure 5.10a, it seems likely that the increase in the rate of photosynthesis could have been limited by the temperature. Any one of the external factors – temperature, light intensity or carbon dioxide concentration – may limit the effects of the other two. A temperature rise may cause photosynthesis to speed up, but only to the point where the light intensity limits further increase. In such conditions, the external factor which restricts the effect of the others is called the **limiting factor**.

Since there is only 0.03 per cent of carbon dioxide in the air, it might seem that shortage of carbon dioxide could be an important limiting factor. Indeed, experiments do show that an increase in carbon dioxide concentration does allow a faster rate of photosynthesis. However, recent work in plant physiology has shown that the extra carbon dioxide affects reactions other than photosynthesis.

The main effect of extra carbon dioxide is to slow down the rate of oxidation of sugar by a process called **photorespiration** and this produces the same effect as an increase in photosynthesis.

Although carbon dioxide concentration limits photosynthesis only indirectly, artificially high levels of carbon dioxide in greenhouses do effectively increase yields of crops (Figure 5.11, overleaf).

Greenhouses also maintain a higher temperature and so reduce the effect of low temperature as a limiting factor.

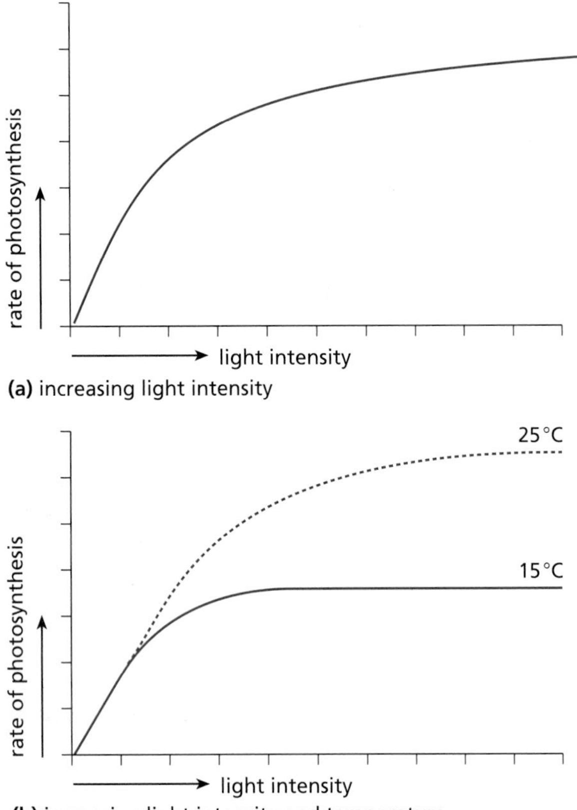

Figure 5.10 Limiting factors in photosynthesis

Figure 5.11 Carrot plants grown in increasing concentrations of carbon dioxide from left to right

The concept of limiting factors does not apply only to photosynthesis. Adding fertilizer to the soil, for example, may increase crop yields, but only up to the point where the roots can take up all the nutrients and the plant can build them into proteins, etc. The uptake of mineral ions is limited by the absorbing area of the roots, rates of respiration, aeration of the soil and availability of carbohydrates from photosynthesis.

Currently there is debate about whether athletic performance is limited by the ability of the heart and lungs to supply oxygenated blood to muscles, or by the ability of the muscles to take up and use the oxygen.

The role of the stomata

The stomata (p. 52) in a leaf may affect the rate of photosynthesis according to whether they are open or closed. When photosynthesis is taking place, carbon dioxide in the leaf is being used up and its concentration falls. At low concentrations of carbon dioxide, the stomata will open. Thus, when photosynthesis is most rapid, the stomata are likely to be open, allowing carbon dioxide to diffuse into the leaf. When the light intensity falls, photosynthesis will slow down and the build-up of carbon dioxide from respiration will make the stomata close. In this way, the stomata are normally regulated by the rate of photosynthesis rather than photosynthesis being limited by the stomata. However, if the stomata close during the daytime as a result of excessive water loss from the leaf, their closure will restrict photosynthesis by preventing the inward diffusion of atmospheric carbon dioxide.

Normally the stomata are open in the daytime and closed at night. Their closure at night, when intake of carbon dioxide is not necessary, reduces the loss of water vapour from the leaf (p. 61).

Practical work

6 The effect of light intensity on the rate of photosynthesis

Fill a beaker or jar with tap-water and add about 5 cm^3 saturated sodium hydrogencarbonate solution. (This is to maintain a good supply of carbon dioxide.)

Select a pondweed shoot about 5–10 cm long. Partly prise open a small paper-clip and slide it over the tip of the shoot and drop the shoot into the jar of water. The paper-clip should hold the shoot under water as shown in Figure 5.12.

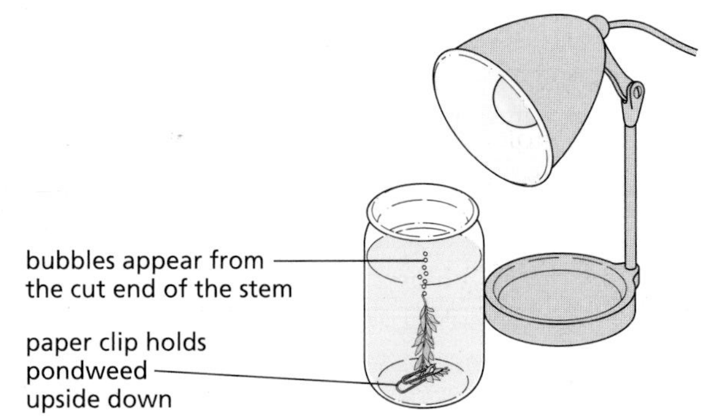

Figure 5.12 Light intensity and oxygen production

Switch on a bench lamp and bring it close to the jar. After a minute or two, bubbles should appear from the cut end of the stem. If they do not, try a different piece of pondweed.

When the bubbles are appearing steadily, switch off the bench lamp and observe any change in the production of bubbles.

Now place the lamp about 25 cm from the jar. Switch on the lamp and try to count the number of bubbles coming off in a minute. Move the lamp to about 10 cm from the jar and count the bubbles again.

If the bubbles appear too rapidly to count, try tapping a pen or pencil on a sheet of paper at the same rate as the bubbles appear and get your partner to slide the paper slowly along for 15 seconds. Then count the dots (Figure 5.13).

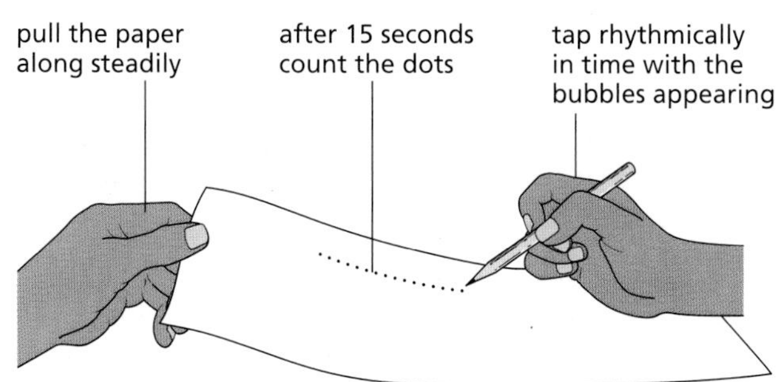

Figure 5.13 Estimating the rate of bubble production

Result When the light is switched off, the bubbling should stop. The rate of bubbling should be faster when the lamp is closer to the plant.

Interpretation Assuming that the bubbles contain oxygen produced by photosynthesis, it seems that an increase in light intensity produces an increase in the rate of photosynthesis. We are assuming also that the bubbles do not change in size during the experiment. A fast stream of small bubbles might represent the same volume of gas as a slow stream of large bubbles.

The plant's use of photosynthetic products

The glucose molecules produced by photosynthesis are quickly built up into starch molecules and added to the growing starch granules in the chloroplast. If the glucose concentration was allowed to increase in the mesophyll cells of the leaf, it could disturb the osmotic balance between the cells (p. 30). Starch is a relatively insoluble compound and so does not alter the osmotic potential of the cell contents.

The starch, however, is steadily broken down to sucrose (p. 12) and this soluble sugar is transported out of the cell into the food-carrying cells (see p. 55) of the leaf veins. These veins will distribute the sucrose to all parts of the plant which do not photosynthesize, e.g. the growing buds, the ripening fruits, the roots and the underground storage organs.

The cells in these regions will use the sucrose in a variety of ways (Figure 5.14).

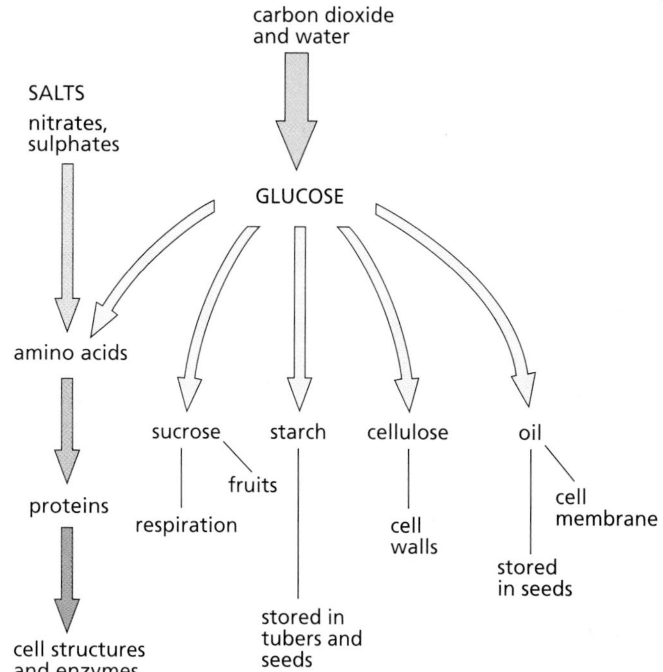

Figure 5.14 Green plants can make all the materials they need from carbon dioxide, water and salts

Respiration

The sugar can be used to provide energy. It is oxidized by respiration (p. 19) to carbon dioxide and water, and the energy released is used to drive other chemical reactions such as the building-up of proteins described below.

Storage

Sugar which is not needed for respiration is turned into starch and stored. Some plants store it as starch grains in the cells of their stems or roots. Other plants such as the potato or parsnip have special storage organs (tubers) for holding the reserves of starch (p. 80). Sugar may be stored in the fruits of some plants; grapes, for example, contain a large amount of glucose.

Synthesis of other substances

As well as sugars for energy and starch for storage, the plant needs cellulose for its cell walls, lipids for its cell membranes, proteins for its cytoplasm and pigments for its flower petals, etc. All these substances are built up (synthesized) from the sugar molecules and other molecules produced in photosynthesis.

By joining hundreds of glucose molecules together, the long-chain molecules of cellulose (Figure 2.3, p. 12) are built up and added to the cell walls.

Amino acids (see p. 11) are made by combining **nitrogen** with sugar molecules or smaller carbohydrate molecules. These amino acids are then joined together to make the proteins which form the enzymes and the cytoplasm of the cell. The nitrogen for this synthesis comes from **nitrates** which are absorbed from the soil by the roots.

Some proteins also need **sulphur** molecules and these are absorbed from the soil in the form of **sulphates** (SO_4). **Phosphorus** is needed for DNA (p. 186) and for reactions involving energy release. It is taken up as **phosphates** (PO_4).

The chlorophyll molecule needs **magnesium** (Mg). This metallic element is also obtained in salts from the soil (see the salts listed under 'Water cultures', below).

Many other elements, e.g. iron, manganese, boron, are also needed in very small quantities for healthy growth. These are often referred to as **trace elements**.

The metallic and non-metallic elements are all taken up in the form of their ions (p. 13).

All these chemical processes, such as the uptake of salts and the building-up of proteins, need energy from respiration to make them happen.

Questions

1 What substances does a green plant need to take in, to make
  a sugar,
  b proteins?
  What must be present in the cells to make reactions **a** and **b** work?

2 A molecule of carbon dioxide enters a leaf cell at 4 p.m. and leaves the same cell at 6 p.m. What is likely to have happened to the carbon dioxide molecule during the 2 hours it was in the leaf cell?

3 In a partially controlled environment such as a greenhouse
  a how could you alter the external factors to obtain maximum photosynthesis,
  b which of these alterations might not be cost effective?

4 Figure 5.15 is a graph showing the average daily change in the carbon dioxide concentration, 1 metre above an agricultural crop in July. From what you have learned about photosynthesis and respiration, try to explain the changes in the carbon dioxide concentration.

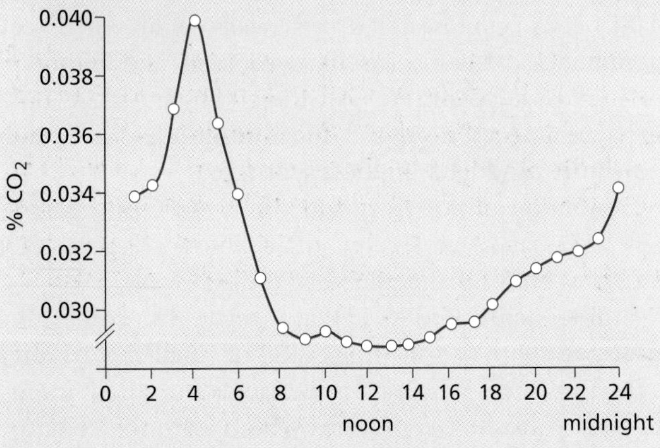

Figure 5.15 Daily changes in concentration of carbon dioxide 1 metre above a plant crop

Sources of mineral elements

The substances mentioned above (nitrates, phosphates, etc.) are often referred to as 'mineral salts', or 'mineral elements', (nitrogen, phosphorus, etc.). If any of these mineral elements is lacking, or deficient, in the soil then the plants may show visible deficiency symptoms.

Many slow-growing wild plants will show no deficiency symptoms even on poor soils. Fast-growing crop plants, on the other hand, will show distinct deficiency symptoms though these will vary according to the species of plant.

If nitrate ions are in short supply, the plant will show stunted growth and yellowish leaves. Lack of phosphate leads to poor root growth and dark green or purplish leaves with red coloration round the veins. Potassium deficiency causes discoloured and mottled leaves.

Farmers and gardeners can recognize these symptoms and take steps to replace the missing minerals.

The mineral elements needed by plants are absorbed from the soil in the form of salts. For example, a plant's needs for potassium (K) and nitrogen (N) might be met by absorbing the ions of the salt **potassium nitrate** (KNO_3). Salts like this come originally from rocks which have been broken down to form the soil. They are continually being taken up from the soil by plants or washed out of the soil by rain. They are replaced partly from the dead remains of plants and animals. When these organisms die and their bodies decay, the salts they contain are released back into the soil. This process is explained in some detail, for nitrates, on p. 229.

In arable farming, the ground is ploughed and whatever is grown is removed. There are no dead plants left to decay and replace the mineral salts. The farmer must replace them by spreading animal manure, sewage sludge or artificial fertilizers in measured quantities over the land.

Three manufactured fertilizers in common use are ammonium nitrate, superphosphate and compound NPK.

Ammonium nitrate (NH_4NO_3)

The formula shows that ammonium nitrate is a rich source of nitrogen but no other plant nutrients. It is sometimes mixed with calcium carbonate to form a compound fertilizer such as 'Nitro-chalk'.

Superphosphates

These fertilizers are mixtures of minerals. They all contain calcium and phosphate and some have sulphate as well.

Compound NPK fertilizer

'N' is the chemical symbol for nitrogen, 'P' for phosphorus and 'K' for potassium. NPK fertilizers are made by mixing ammonium sulphate, ammonium phosphate and potassium chloride in varying proportions. They provide the ions of nitrate, phosphate and potassium, which are the ones most likely to be below the optimum level in an agricultural soil.

Water cultures

It is possible to demonstrate the importance of the various mineral elements by growing plants in water cultures. A full water culture is a solution containing the salts which provide all the necessary elements for healthy growth, e.g.

- potassium nitrate for potassium and nitrogen;
- magnesium sulphate for magnesium and sulphur;
- potassium phosphate for potassium and phosphorus;
- calcium nitrate for calcium and nitrogen.

From these elements, plus the carbon dioxide, water and sunlight needed for photosynthesis, a green plant can make all the substances it needs for a healthy existence.

Some branches of horticulture, e.g. growing of glasshouse crops, make use of water cultures on a large scale. Tomatoes may be grown with their roots in flat polythene tubes. The appropriate water culture solution is pumped along these tubes (Figure 5.16). This method has the advantage that the yield is increased and the need to sterilize the soil each year, to destroy pests, is eliminated. This kind of technique is sometimes described as hydroponics or soil-less culture.

Figure 5.16 Soil-less culture. The tomato plants are growing in a nutrient solution circulated through troughs of polythene. The network of roots can be seen in the polythene

Practical work

7 The importance of different mineral elements

Place wheat seedlings in test-tubes containing water cultures as shown in Figure 5.17. Cover the tubes with aluminium foil to keep out light and so stop green algae from growing in the solution (p. 279). Some of the solutions have one of the elements missing. For example, in the list of chemicals on the left, magnesium chloride is used instead of magnesium sulphate and so the solution will lack sulphur. In a similar way, solutions lacking nitrogen, potassium and phosphorus can be prepared.

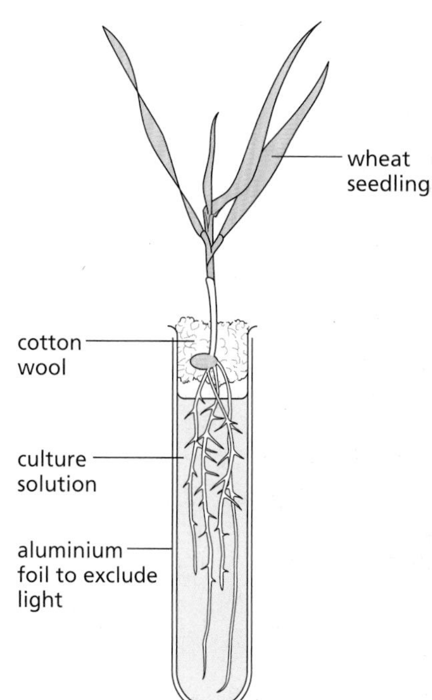

Figure 5.17 To set up a water culture

Leave the seedlings to grow in these solutions for a few weeks, keeping the tubes topped up with distilled water.

Result The kind of result which might be expected from wheat seedlings is shown in Figure 5.18 (overleaf). Generally, the plants in a complete culture will be tall and sturdy, with large, dark green leaves. The plants lacking nitrogen will usually be stunted and have small, pale leaves. In the absence of magnesium, chlorophyll cannot be made, and these plants will be small with yellow leaves.

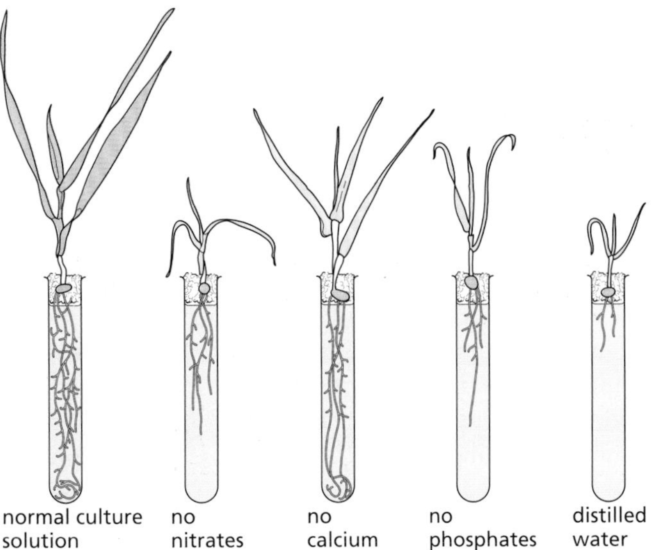

Figure 5.18 Result of water culture experiment

Interpretation The healthy plant in the full culture is the control and shows that this method of raising plants does not affect them. The other, less healthy plants show that a full range of mineral elements is necessary for normal growth.

Quantitative results Although the effects of mineral deficiency can usually be seen simply by looking at the wheat seedlings, it is better if actual measurements are made.

The height of the shoot, or the total length of all the leaves on one plant, can be measured. The total root length can also be measured, though this is difficult if root growth is profuse.

Alternatively, the dry weight (see p. 307) of the shoots and roots, can be measured. In this case, it is best to pool the results of several experiments. All the shoots from the complete culture are placed in a labelled container, all those from the 'no nitrate' culture solution are placed in another container, and so on for all the plants from the different solutions. The shoots are then dried at 110 °C for 24 hours and weighed. The same procedure can be carried out for the roots.

You would expect the roots and shoots from the complete culture to weigh more than those from the nutrient-deficient cultures.

Questions

1 What salts would you put in a water culture which is to contain *no* nitrogen?

2 What mineral elements do you think are provided by **a** bone meal (p. 152), **b** dried blood (p. 108)?

3 How can a floating pond plant, such as duckweed, survive without having its roots in soil?

4 In the water culture experiment, why should
a lack of nitrate,
b lack of phosphate cause reduced growth?

5 Figure 5.19 shows the increased yield of winter wheat in response to adding more nitrogenous fertilizer.
a If the applied nitrogen is doubled from 50 to 100 kg per hectare, how much extra wheat does the farmer get?
b If the applied nitrogen is doubled from 100 to 200 kg per hectare, how much extra wheat is obtained?
c What sort of calculations will a farmer need to make before deciding to increase the applied nitrogen from 150 to 200 kg per hectare?

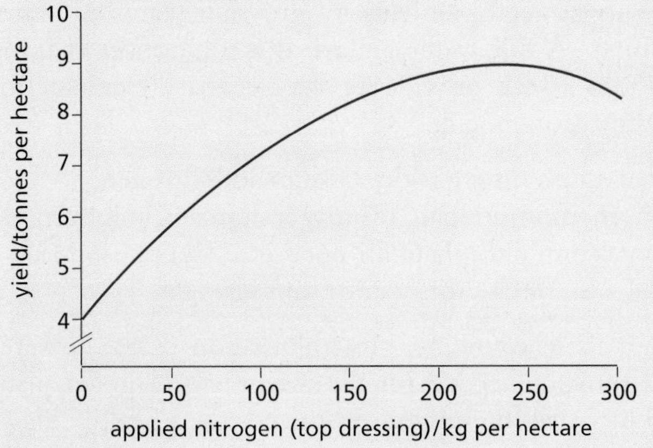

Figure 5.19

Checklist

- Photosynthesis is the way plants make their food.
- They combine carbon dioxide and water to make sugar.
- To do this, they need energy from sunlight, which is absorbed by chlorophyll.
- Chlorophyll converts light energy to chemical energy.
- The equation to represent photosynthesis is

$$6CO_2 + 6H_2O \xrightarrow[\text{absorbed by chlorophyll}]{\text{energy from sunlight}} C_6H_{12}O_6 + 6O_2$$

- Plant leaves are adapted for the process of photosynthesis by being broad and thin, with many chloroplasts in their cells.
- From the sugar made by photosynthesis, a plant can make all the other substances it needs, provided it has a supply of mineral salts like nitrate, phosphate and potassium.
- In daylight, respiration and photosynthesis will be taking place in a leaf; in darkness, only respiration will be taking place.
- In daylight a plant will be taking in carbon dioxide and giving out oxygen.
- In darkness a plant will be taking in oxygen and giving out carbon dioxide.
- Experiments to test photosynthesis are designed to exclude light, or carbon dioxide, or chlorophyll, to see if the plant can still produce starch.
- The rate of photosynthesis may be limited by light intensity and temperature.
- Plants need a supply of mineral salts to make protein and other vital substances.

Some principles of biology
Examination questions

Do not write on this page. Where necessary copy drawings, tables or sentences.

1 This apparatus was used to measure the effect of temperature on the respiration rate of mealworms.

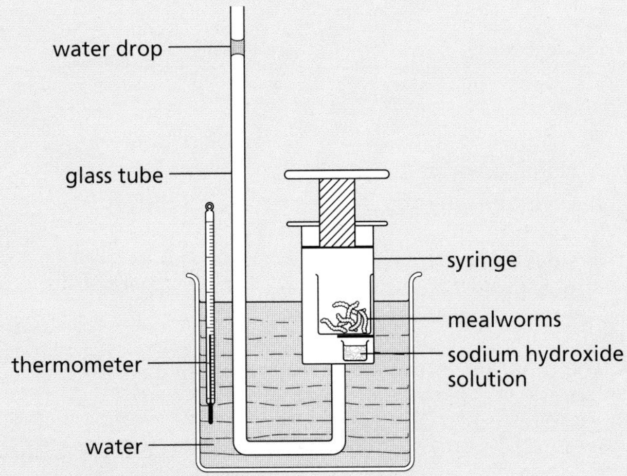

a (i) Name the gas absorbed by sodium hydroxide solution. (1)
(ii) Show on the diagram the direction of movement of the water drop. (1)
(iii) Give one difference between the apparatus shown and a suitable control apparatus. (1)

b The rate of respiration was measured at intervals from 20 °C to 50 °C. The graph shows the results of the investigation.

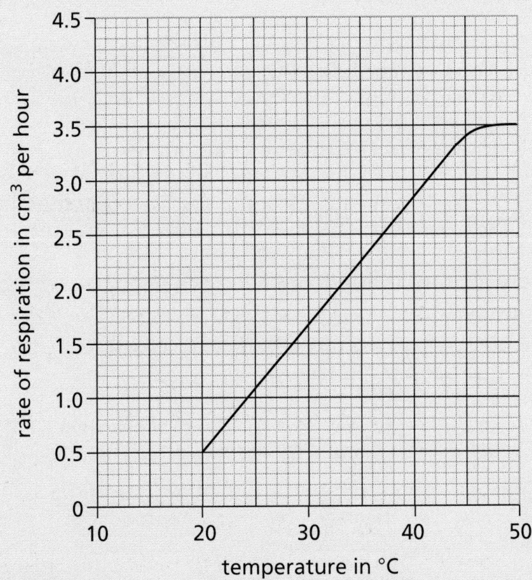

(i) Explain the results shown on the graph. (2)
(ii) What was the rate of respiration at 37 °C? (1)
(iii) Suggest what would happen if temperatures above 50 °C were used.
Give a reason for your answer. (2)

(Edexcel)

2 Cellulase and amylase are enzymes (carbohydrases) which break down different carbohydrates. Cellulase breaks down the cellulose in plant cell walls to sugar. Amylase breaks down starch to sugar. Some pupils carried out an investigation into the effect of these enzymes on corn (maize) which was mashed up in distilled water. The investigation was at room temperature (22 °C). The diagram below shows the apparatus used.

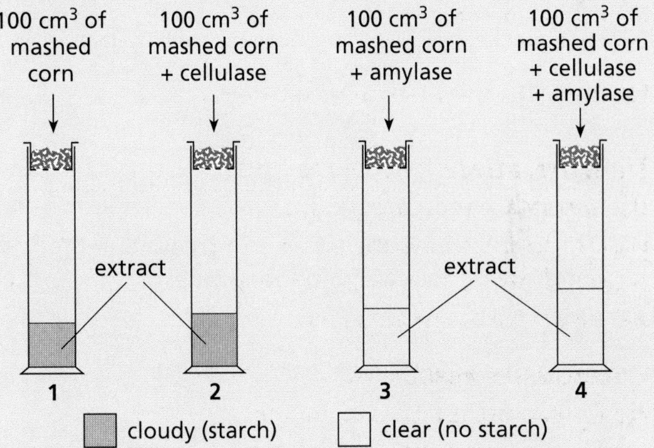

The volumes and concentrations of the enzymes were constant. The apparatus was left for 24 hours. Here are the results:

	Cylinder			
	1	2	3	4
Volume of extract (cm^3)	25	34	35	50

a Calculate the volume of mashed corn needed to give 1000 cm^3 of extract if **both** enzymes are used. (1)

b Refer to the appearance and volumes of the extracts and explain fully the results in the four cylinders. (9)

(WJEC)

3 Three potato cylinders were weighed and left in distilled water overnight. They were then reweighed.

before

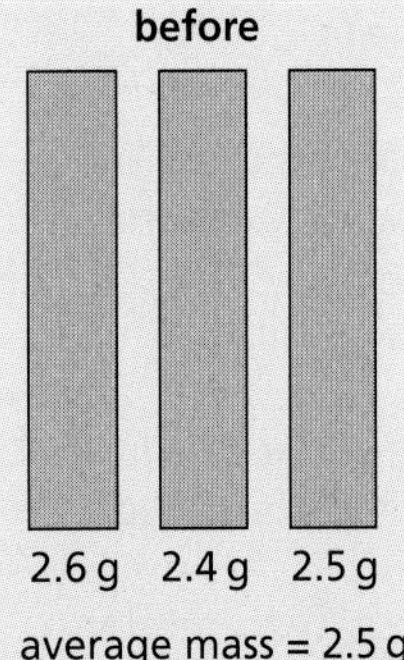

average mass = 2.5 g

after

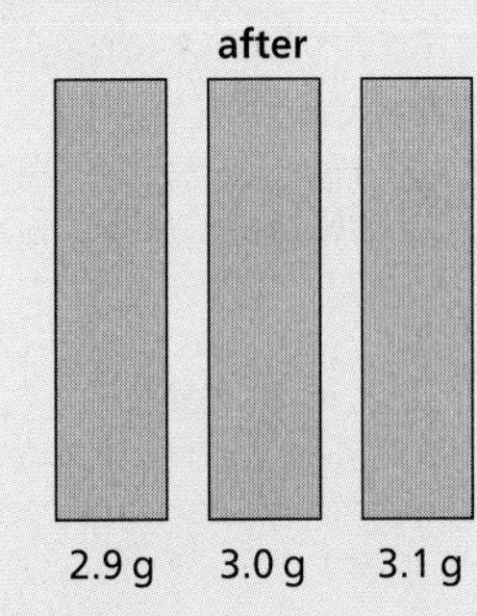

a Calculate the average mass of the potato cylinders after reweighing. Show your working. (2)

b Explain why the mass of the cylinders has increased. (2)

c Describe **one other** change, visible in the diagram, that has occurred in the potato cylinders. (1)

(CCEA)

4 a Finish the balanced symbol equation for photosynthesis.

$$6CO_2 + 6H_2O \xrightarrow[\text{chlorophyll}]{\text{light}} \_\_\_\_\_ + \_\_\_\_\_$$ (2)

b Market gardeners use automatic control mechanisms in their greenhouses.
The diagram shows a commercial greenhouse.
This greenhouse provides everything plants need for a high rate of photosynthesis.

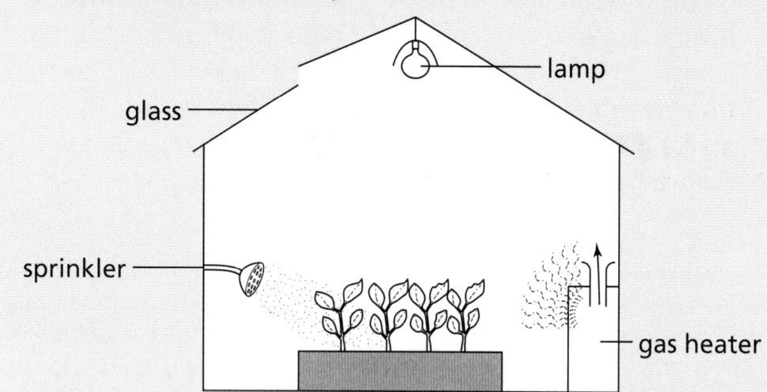

Explain two ways in which the gas heater could increase the rate of photosynthesis. (2)

c The graph shows the effect of increasing light intensity on the rate of photosynthesis.

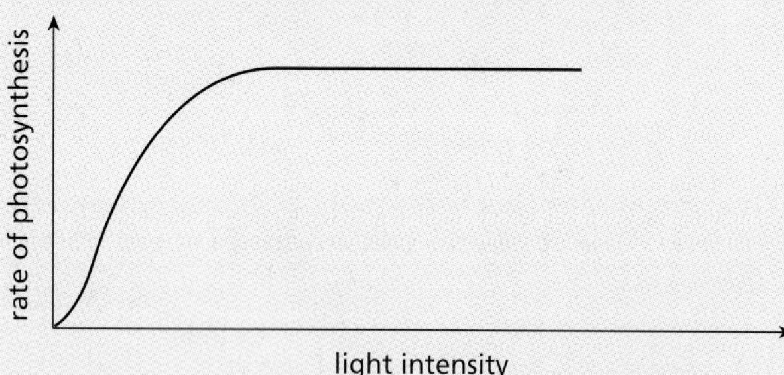

Explain why the rate of photosynthesis does not continue to increase as light intensity continues to increase. (2)

(OCR)

5 a The chemical equation for photosynthesis shown below is incomplete.

$$\underset{\text{water}}{6H_2O} + \_\_\_\_\_ \xrightarrow[\text{plant pigment}]{\text{energy}} \underset{\text{glucose}}{C_6H_{12}O_6} + \_\_\_\_\_$$

(i) Complete the equation in either symbols or words. (2)
(ii) State the source of energy for this reaction. (1)
(iii) Name the plant pigment necessary for this reaction. (1)
(iv) Which mineral is needed by a plant to form this pigment? (1)

b (i) Name the tissue in which the sugar produced by photosynthesis is carried to other parts of the plant. (1)
(ii) In many plants some of the sugar formed in photosynthesis is converted to starch for storage. Explain the advantage of storing starch rather than sugar. (2)
(iii) Name the carbohydrate, formed from sugar produced in photosynthesis, which is used to build cell walls. (1)

(IGCSE)

6 The diagram shows three cells taken from different parts of the body.

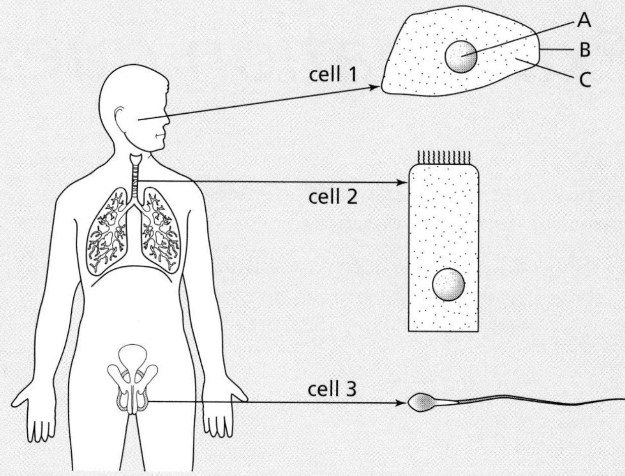

a Name parts A, B and C. (3)
b Give **one** way the structure of cell 2 differs from cell 1. (1)
c How is the structure of cell 3 adapted to its function? (2)

(CCEA)

7 The diagram shows a cell from a blade of grass.

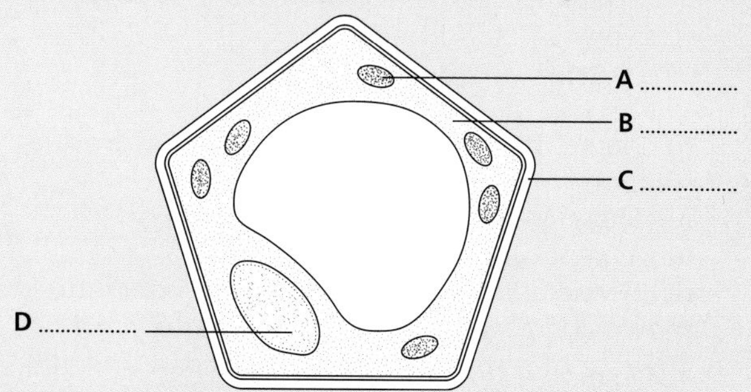

a On the diagram, use words from the list to name the parts labelled A–D.

cell membrane cell wall chloroplast cytoplasm nucleus vacuole (4)

b Name two parts of the grass cell which are not found in any of the cells in an animal. (2)

(AQA)

8 The human body contains organs which belong to organ systems.
Draw a straight line from each organ to the correct organ system.
Use each organ and organ system only once.
One has been done for you.

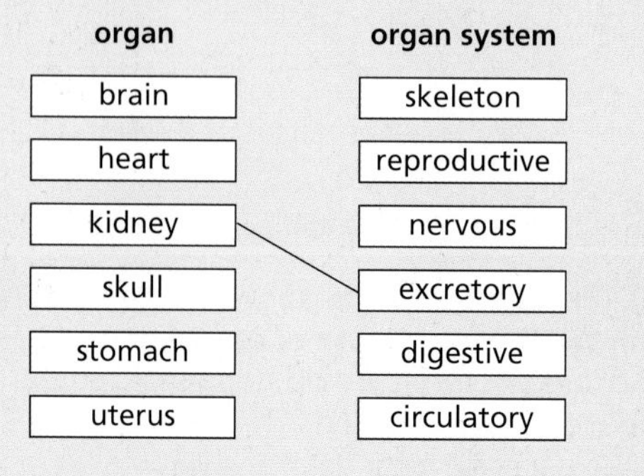

(4)

(OCR)

Flowering plants

6 Plant structure and function

Leaf
Epidermis, stomata, mesophyll, veins.
Stem
Epidermis, vascular bundles and their function.
Root
Outer layer and root hairs.

Structure and function
Practical work
Stomata. Tension in stems.

A young sycamore plant is shown in Figure 6.1. It is typical of many flowering plants in having a **root system** below the ground and a **shoot** above ground. The shoot consists of an upright stem, with leaves and buds. The buds on the side of the stem are called **lateral buds**. When they grow, they will produce branches. The bud at the tip of the shoot is the **terminal bud** and when it grows, it will continue the upward growth of the stem. The lateral buds and the terminal buds may also produce flowers.

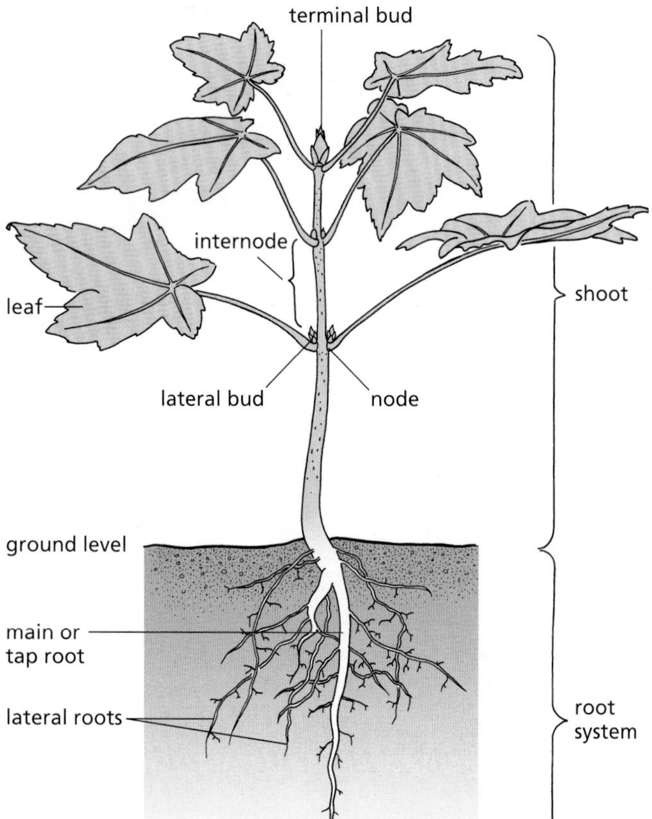

Figure 6.1 Structure of a typical flowering plant

The region of stem from which leaves and buds arise is called a **node**. The region of stem between two nodes is the **internode**.

The **leaves** make food by photosynthesis (p. 35) and pass it back to the stem.

The **stem** carries this food to all parts of the plant which need it and also carries water and dissolved salts from the roots to the leaves and flowers.

In addition, the stem supports and spaces out the leaves so that they can receive sunlight and absorb carbon dioxide, which they need for photosynthesis.

An upright stem also holds the flowers above the ground, helping the pollination by insects or the wind (p. 69). A tall stem may help in seed dispersal later on (p. 72).

The **roots** anchor the plant in the soil and prevent it falling over or being blown over by the wind. They also absorb the water and salts which the plant needs for making food in the leaves.

The structure and functions of the plant organs will be considered in more detail in this chapter.

Leaf

A typical leaf of a broad-leaved plant is shown in Figure 6.2a (Figure 6.2b shows a transverse section through the leaf). It is attached to the stem by a **leaf stalk**, which continues into the leaf as a **midrib**. Branching from the midrib is a network of veins which deliver water and salts to the leaf cells and carry away the food made by them.

As well as carrying food and water, the network of veins forms a kind of skeleton which supports the softer tissues of the leaf blade.

The **leaf blade** (or **lamina**) is broad and a vertical section through a small part of a leaf blade is shown in Figure 6.2c, and Figure 6.3 is a photograph of a leaf section under the microscope.

(a) whole leaf

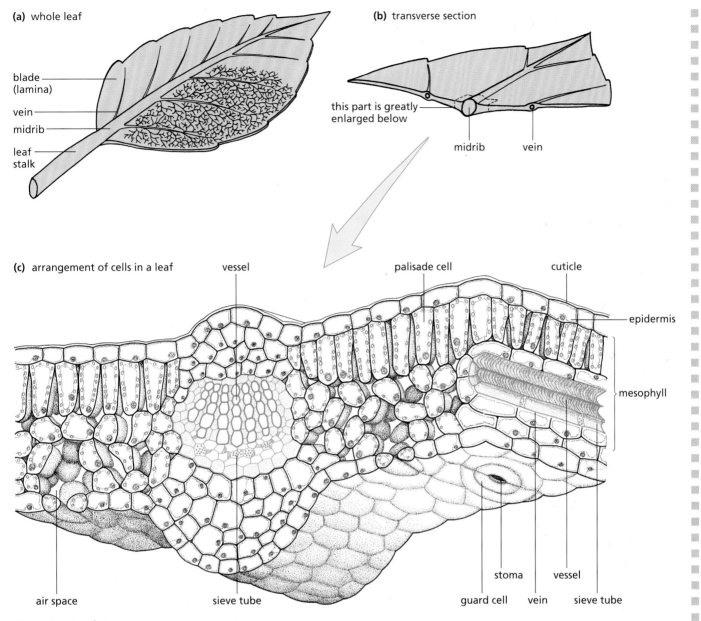

blade (lamina)
vein
midrib
leaf stalk

(b) transverse section

this part is greatly enlarged below
midrib
vein

(c) arrangement of cells in a leaf

vessel
palisade cell
cuticle
epidermis
mesophyll
air space
sieve tube
stoma
vessel
guard cell
vein
sieve tube

Figure 6.2 Leaf structure

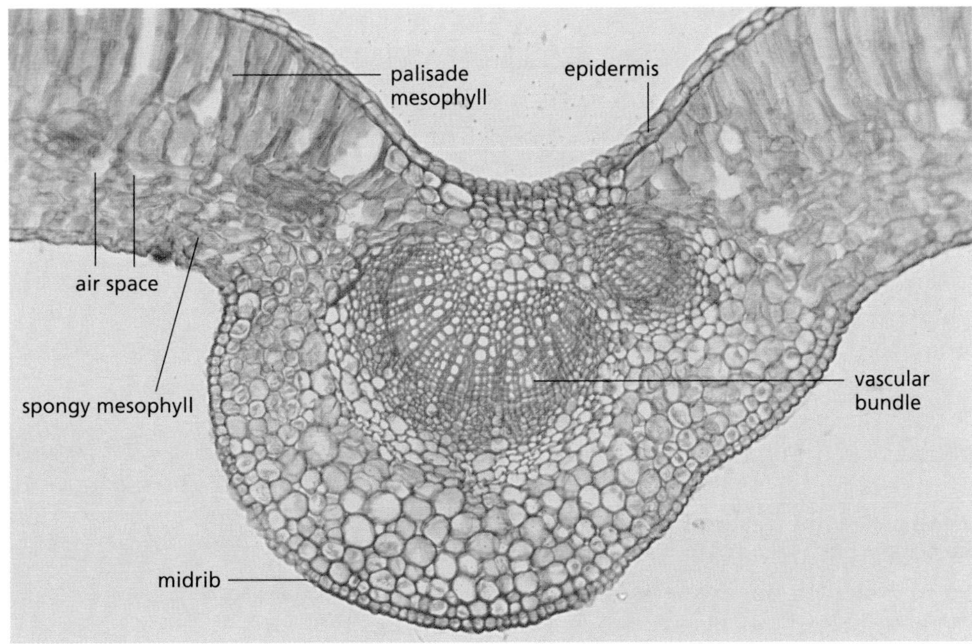

Figure 6.3 Transverse section through a leaf (×30)

Epidermis

The **epidermis** is a single layer of cells on the upper and lower surfaces of the leaf. The epidermis helps to keep the leaf's shape. The closely fitting cells (Figure 6.2c) reduce evaporation from the leaf and prevent bacteria and fungi from getting in. There is a thin waxy layer called the **cuticle** over the epidermis which helps to reduce water loss.

Stomata

In the leaf epidermis there are structures called **stomata** (singular = stoma). A stoma consists of a pair of **guard cells** (Figure 6.4) surrounding an opening, or stomatal pore. Changes in the turgor (p. 30) and shape of the guard cells can open or close the stomatal pore. In most dicotyledons (i.e. the broad-leaved plants; see p. 278), the stomata occur only in the lower epidermis. In monocotyledons (i.e. narrow-leaved plants such as grasses) the stomata are equally distributed on both sides of the leaf.

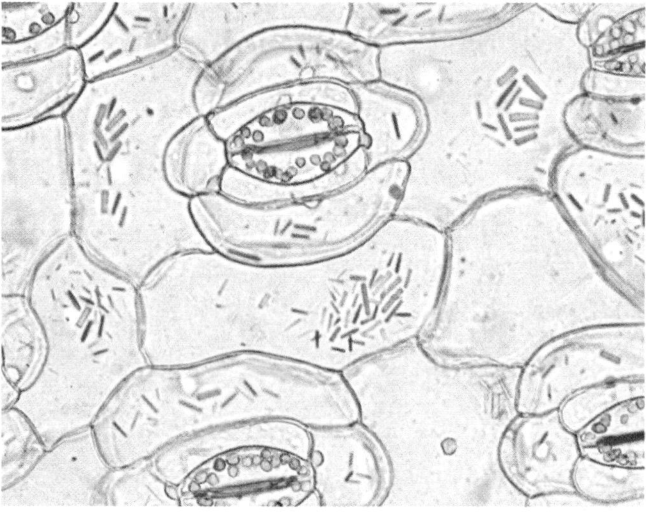

Figure 6.4 Stomata in the lower epidermis of a leaf (× 350)

In very general terms, stomata are open during the hours of daylight but closed during the evening and most of the night (Figure 6.5). This pattern, however, varies greatly with the plant species. A satisfactory explanation of stomatal rhythm has not been worked out, but when the stomata are open (i.e. mostly during daylight), they allow carbon dioxide to diffuse into the leaf where it is used for photosynthesis.

If the stomata close, the carbon dioxide supply to the leaf cells is virtually cut off and photosynthesis stops. However, in many species, the stomata are closed during the hours of darkness, when photosynthesis is not taking place anyway.

It seems, therefore, that stomata allow carbon dioxide into the leaf when photosynthesis is taking place and prevent excessive loss of water vapour (see pp. 60 and 61) when photosynthesis stops, but the story is likely to be more complicated than this.

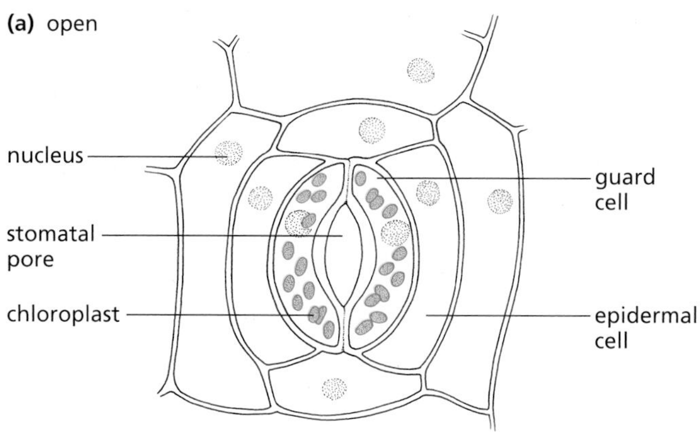

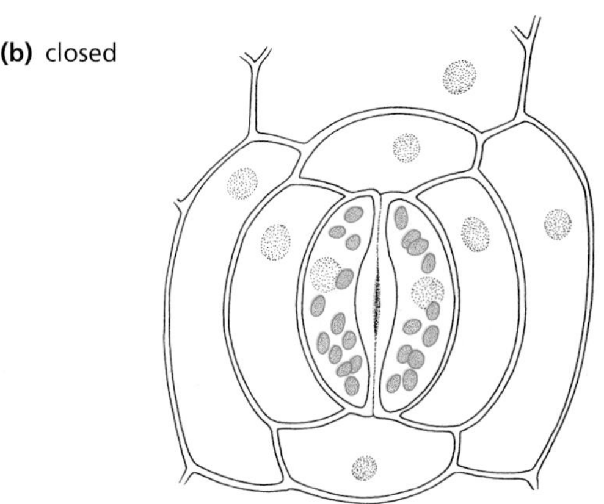

Figure 6.5 Stoma

The detailed mechanism by which stomata open and close is not fully understood, but it is known that, in the light, the potassium concentration in the guard cell vacuoles increases. This lowers the water potential (p. 29) of the cell sap and water enters the guard cells by osmosis (p. 29) from their neighbouring epidermal cells. This inflow of water raises the turgor pressure inside the guard cells.

The cell wall next to the stomatal pore is thicker than elsewhere in the cell and is less able to stretch (Figure 6.6). So, although the increased turgor tends to expand the whole guard cell, the thick inner wall cannot expand. This causes the guard cells to curve in such a way that the stomatal pore between them is opened.

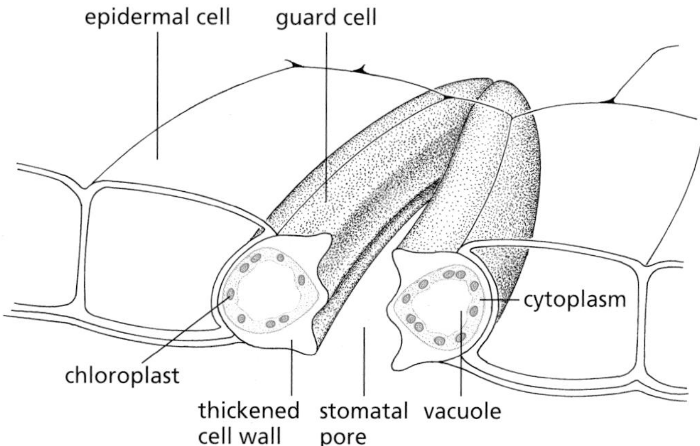

Figure 6.6 Structure of guard cells

When potassium ions leave the guard cell, the water potential rises, water passes out of the cells by osmosis, the turgor pressure falls and the guard cells straighten up and close the stoma.

Where the potassium ions come from and what triggers their movement into or out of the guard cells is still under active investigation.

You will notice from Figures 6.5 and 6.6 that the guard cells are the only epidermal cells containing chloroplasts. At one time it was thought that the chloroplasts built up sugar by photosynthesis during daylight, that the sugars made the cell sap more concentrated and so caused the increase in turgor. In fact, little or no photosynthesis takes place in these chloroplasts and their function has not been explained, though it is known that starch accumulates in them during the hours of darkness. In some species of plants, the guard cells have no chloroplasts.

Mesophyll

The tissue between the upper and lower epidermis is called **mesophyll** (Figure 6.2c). It consists of two zones: the upper, **palisade mesophyll** and the lower, **spongy mesophyll** (Figure 6.7). The palisade cells are usually long and contain many chloroplasts. The spongy mesophyll cells vary in shape and fit loosely together, leaving many air spaces between them (see Figure 5.7, p. 39).

The function of the palisade cells and – to a lesser extent – of the spongy mesophyll cells is to make food by photosynthesis. Their chloroplasts absorb sunlight and use its energy to join carbon dioxide and water molecules to make sugar molecules as described on p. 35.

In daylight, when photosynthesis is rapid, the mesophyll cells are using up carbon dioxide. As a result, the concentration of carbon dioxide in the air spaces falls to a low level and more carbon dioxide diffuses in (p. 26) from the outside air, through the stomata (Figure 6.7). This diffusion continues through the air spaces, up to the cells which are using carbon dioxide. These cells are also producing oxygen as a by-product of photosynthesis. When the concentration of oxygen in the air spaces rises, it diffuses out through the stomata.

Veins (vascular bundles)

The water needed for making sugar by photosynthesis is brought to the mesophyll cells by the **veins**. The mesophyll cells take in the water by osmosis (p. 29) because the concentration of free water molecules in a leaf cell, which contains sugars, will be less than the concentration of water in the water vessels of a vein. The branching network of leaf veins means that no cell is very far from a water supply.

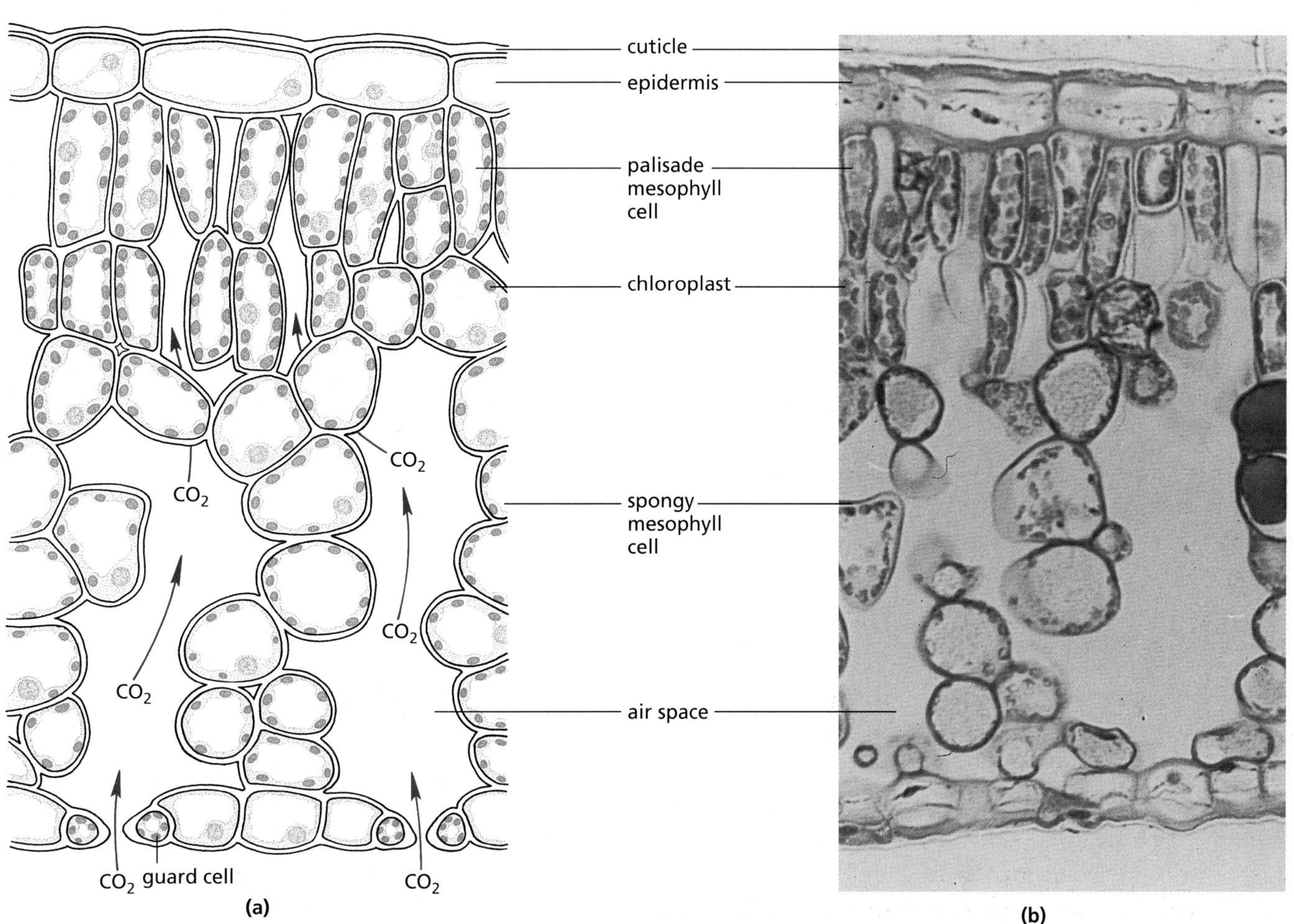

Figure 6.7 Vertical section through a leaf blade (×300)

The sugars made in the mesophyll cells are passed to the phloem cells (see below) of the veins, and these cells carry the sugars away from the leaf into the stem.

The ways in which a leaf is thought to be well adapted to its function of photosynthesis are listed on p. 41.

Questions

1 What are the functions of
  a the epidermis,
  b the mesophyll of a leaf?

2 Look at Figure 6.7. Why do you think that photosynthesis does not take place in the cells of the epidermis?

3 During bright sunlight, what gases are
  a passing out of the leaf through the stomata,
  b entering the leaf through the stomata?

4 What types of leaves do you know which do not have any midrib?

5 In some plants, the stomata close for a period at about midday. Suggest some possible advantages and disadvantages of this to the plant.

Stem

In Figure 6.8 a stem is shown cut across (transversely) and down its length (longitudinally) to show its internal structure.

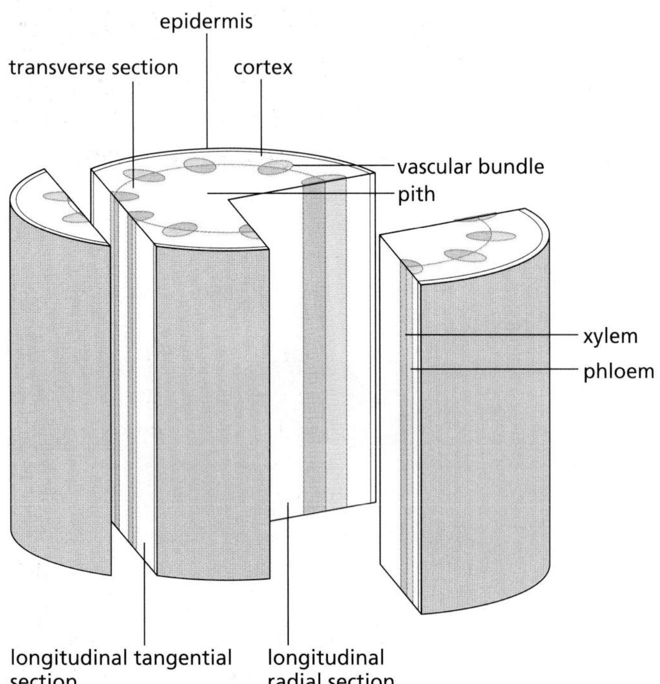

Figure 6.8 Structure of plant stem

Epidermis

Like the leaf epidermis, this is a single layer of cells which helps to keep the shape of the stem and cuts down the loss of water vapour. There are stomata in the epidermis which allow the tissues inside to take up oxygen and get rid of carbon dioxide. In woody stems, the epidermis is replaced by bark, which consists of many layers of dead cells.

Vascular bundles

These are made up of groups of specialized cells which conduct water, dissolved salts and food up or down the stem. The vascular bundles in the roots, stem, leaf stalks and leaf veins all connect up to form a transport system throughout the entire plant (Figure 6.9). The two main tissues in the vascular bundles are called **xylem** and **phloem** (Figure 6.10). Food substances travel in the phloem; water and salts travel mainly in the xylem. The cells in each tissue form elongated tubes called **vessels** (in the xylem) or **sieve tubes** (in the phloem) and they are surrounded and supported by other cells.

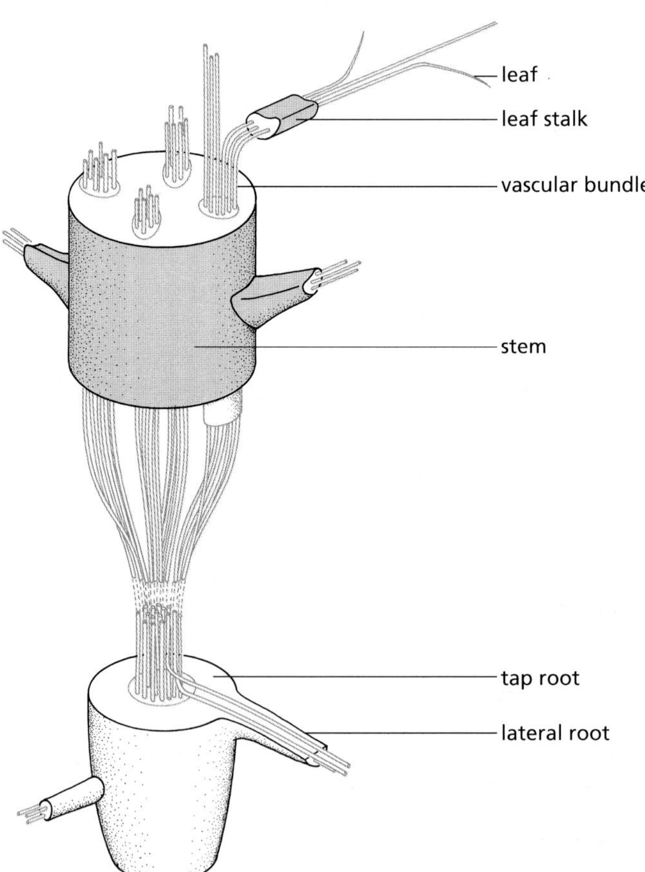

Figure 6.9 Distribution of veins from root to leaf

Vessels

The cells in the xylem which carry water become vessels. A vessel is made up of a series of long cells joined end to end (Figure 6.11a, p. 56). Once a region of the plant has ceased growing, the end walls of these cells are digested away to form a continuous, fine tube (Figure 6.10c). At the same time, the cell walls are thickened and impregnated with a substance called **lignin**, which makes the cell wall very strong and impermeable. Since these lignified cell walls prevent

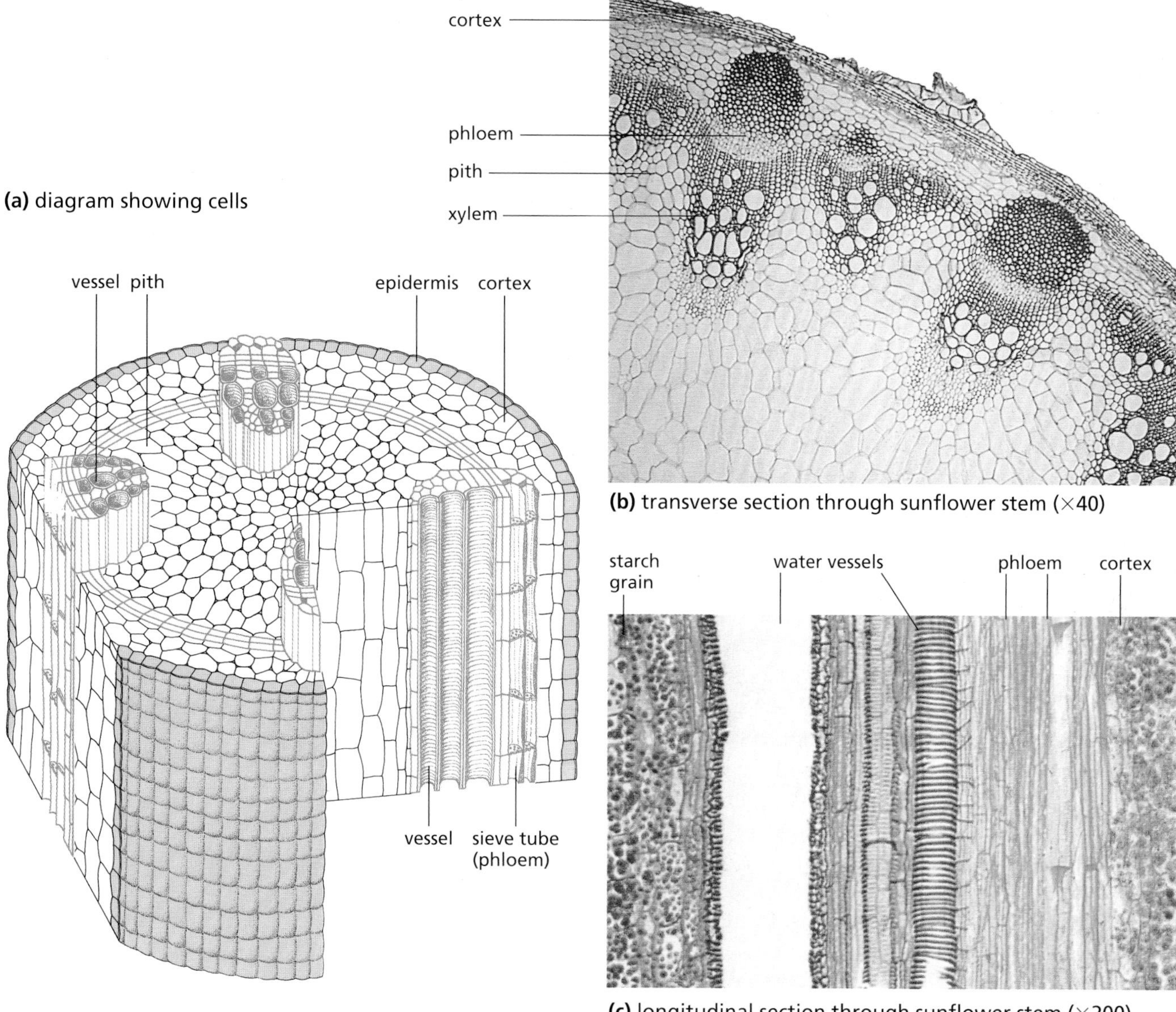

(a) diagram showing cells

(b) transverse section through sunflower stem (×40)

(c) longitudinal section through sunflower stem (×200)

Figure 6.10 Structure of plant stem

the free passage of water and nutrients, the cytoplasm dies. This does not affect the passage of water in the vessels. Xylem also contains many elongated, lignified supporting cells called **fibres**.

Sieve tubes

The conducting cells in the phloem remain alive and form sieve tubes. Like vessels, they are formed by vertical columns of cells (Figure 6.11b, overleaf). Perforations appear in the end walls, allowing substances to pass from cell to cell, but the cell walls are not lignified and the cell contents do not die, although they do lose their nuclei. The perforated end walls are called **sieve plates**.

Phloem contains supporting cells as well as sieve tubes.

Functions of vascular bundles

In general, water travels up the stem in the xylem from the roots to the leaves. Food may travel either up or down the stem in the phloem, from the leaves where it is made, to any part of the plant which is using or storing it.

Vascular bundles have a supporting function as well as a transport function, because they contain vessels, fibres and other thick-walled, lignified, elongated cells. In many stems, the vascular bundles are arranged in a cylinder, a little way in from the epidermis. This pattern of distribution helps the stem to resist the sideways bending forces caused by the wind. In a root, the vascular bundles are in the centre (Figure 6.12, overleaf) where they resist the pulling forces which the root is likely to experience when the shoot is being blown about by the wind.

The network of veins in many leaves supports the soft mesophyll tissues and resists stresses which could lead to tearing.

The methods by which water, salts and food are moved through the vessels and sieve tubes are discussed in Chapter 7 (p. 59).

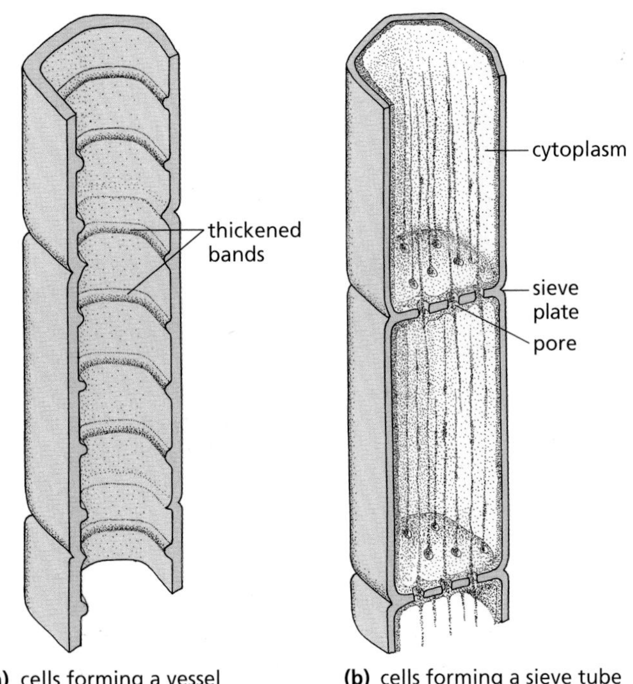

Figure 6.11 Conducting structures in a plant

Figure 6.12 Transverse section through a root (×40). Notice that the vascular tissue is in the centre. Some root hairs can be seen in the outer layer of cells

Cortex and pith

The tissue between the vascular bundles and the epidermis is called the **cortex**. Its cells often store starch. In green stems, the outer cortex cells contain chloroplasts and make food by photosynthesis. The central tissue of the stem is called **pith**. The cells of the pith and cortex act as packing tissues and help to support the stem in the same way that a lot of blown-up balloons packed tightly into a plastic bag would form quite a rigid structure.

Questions

1. Make a list of the types of cells or tissues you would expect to find in a vascular bundle.
2. What structures help to keep the stem's shape and upright position?
3. What are the differences between xylem and phloem
   a in structure,
   b in function?

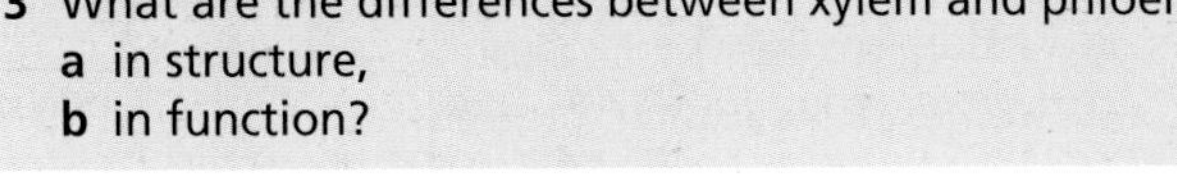

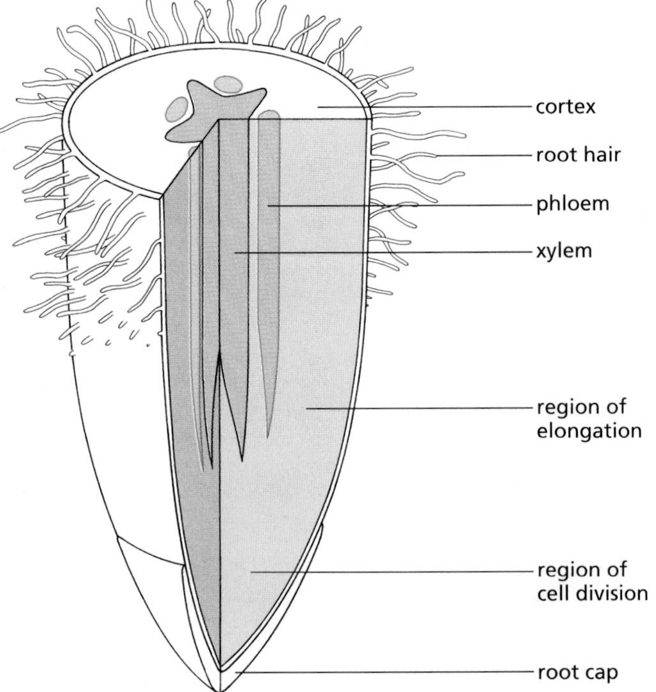

Figure 6.13 Root structure

Root

The internal structure of a typical root is shown in Figure 6.13. The vascular bundle is in the centre of the root (Figure 6.12), unlike the stem where the bundles form a cylinder in the cortex.

The xylem carries water and salts from the root to the stem. The phloem will bring food from the stem to the root, to provide the root cells with substances for their energy and growth.

Outer layer and root hairs

There is no distinct epidermis in a root. At the root tip are several layers of cells forming the **root cap**. These cells are continually replaced as fast as they are worn away when the root tip is pushed through the soil.

In a region above the root tip, where the root has just stopped growing, the cells of the outer layer produce tiny, tube-like outgrowths called **root hairs** (Figure 7.6, p. 63). These can just be seen as a downy layer on the roots of seedlings grown in moist air (Figure 6.14). In the soil, the root hairs grow between the soil particles and stick closely to them. The root hairs take up water from the soil by osmosis (see p. 63), and absorb mineral salts (as ions) by active transport (p. 28).

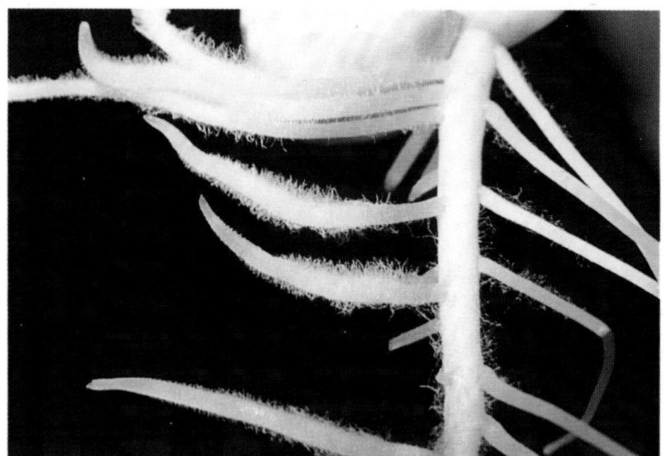

Figure 6.14 Root hairs (× 5) as they appear on a root grown in moist air

The large number of tiny root hairs greatly increases the absorbing surface of a root system. The surface area of the root system of a mature rye plant has been estimated at about 200 m². The additional surface provided by the root hairs was calculated to be 400 m².

Root hairs remain alive for only a short time. The region of root just below a root hair zone is producing new root hairs, while the root hairs at the top of the zone are shrivelling (Figure 6.15). Above the root hair zone, the cell walls of the outer layer become less permeable. This means that water cannot get in so easily.

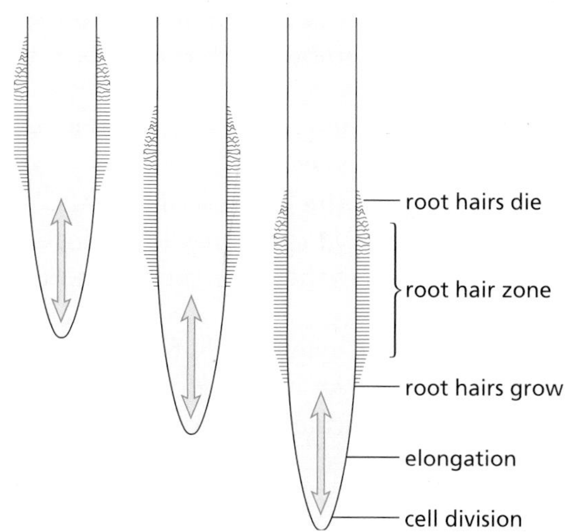

Figure 6.15 The root hair zone changes as the root grows

Tap root (Figure 6.16a)

When a seed germinates, a single root grows vertically down into the soil. Later, lateral roots grow from this at an acute angle outwards and downwards, and from these laterals other branches may arise. Where a main root is recognizable the arrangement is called a **tap-root system**.

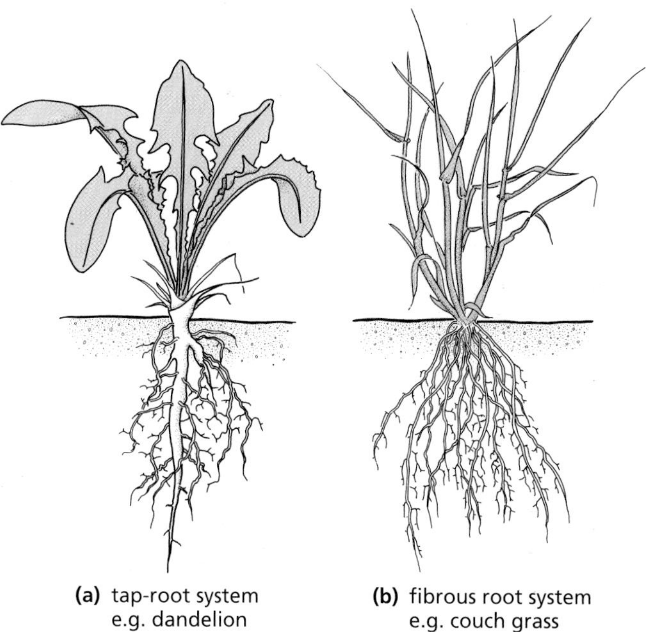

Figure 6.16 Types of root system

Fibrous root (Figure 6.16b)

When a seed of the grass and cereal group germinates, several roots grow out at the same time and laterals grow from them. There is no distinguishable main root, and it is called a **fibrous root system**.

Adventitious root

Where roots grow, not from a main root, but directly from the stem as they do in bulbs, corms, rhizomes or ivy, they are called **adventitious roots**, but such a system may also be described as a fibrous rooting system.

Questions

1 State briefly the functions of the following: xylem, palisade cell, root hair, root cap, stoma, epidermis.

2 If you were given a cylindrical structure cut from part of a plant, how could you tell whether it was a piece of stem or a piece of root
  a with the naked eye,
  b with the aid of a microscope or hand lens?

3 Describe the path taken by
  a a carbon dioxide molecule from the air and
  b a water molecule from the soil
  until they reach a mesophyll cell of a leaf to be made into sugar.

4 Why do you think that root hairs are produced only on the parts of the root system that have stopped growing?

5 Discuss whether you would expect to find a vascular bundle in a flower petal.

Structure and function

It is always tempting, when studying an organism, to ascribe some function to the structures which are being observed. In some cases this is easy. It is obvious, for example, that the function of a mammal's hind limb is locomotion because the limb can be seen in action.

The functions of internal organs are not so obvious and guesses about their function may be quite inaccurate. At one time it was thought that the arteries carried air because the arteries seen in the dissection of dead animals often contained no blood. The chloroplasts in the guard cells (p. 52) were assumed for a long time to be the site of photosynthesis and it was only after conducting experiments that this assumption was shown to be false.

The functions of tissues described in this chapter have been stated as if they were certainly known and without offering any evidence for these functions. In Chapter 7 experiments will be described which do provide some evidence for the statements.

In general, although function can be guessed at from studying anatomy of dead organisms, it cannot be confirmed without experiments to test the guesses (hypotheses). It is usually most unwise to assume a particular function from simply studying anatomy.

Practical work

1 Stomata

Strip off a piece of the lower epidermis of a rhubarb leaf and place it on a slide, in a little water, under the microscope. Draw a drop of 10 per cent salt solution under the cover-slip as shown in Figure 4.18 on p. 32. This solution withdraws water by osmosis from the guard cells, which lose their turgor. As a result the stomata close. Now flush out the salt solution by drawing water under the cover-slip; the guard cells take up water and become turgid.

2 To show the tension in stems

Partly remove a strip of epidermis from a rhubarb stalk as shown in Figure 6.17a. When you replace the epidermis in position it will have become too short, showing the shrinking stress that exists in the epidermis.

Push a cork borer into the pith and withdraw it without removing any tissue. The cylinder of pith so formed, freed from the constraint of the epidermis, expands and protrudes slightly. This shows the elongating tendency of the inner tissues (Figure 6.17b).

These two opposing stresses in the stem help to give it rigidity when the cells are turgid (p. 30).

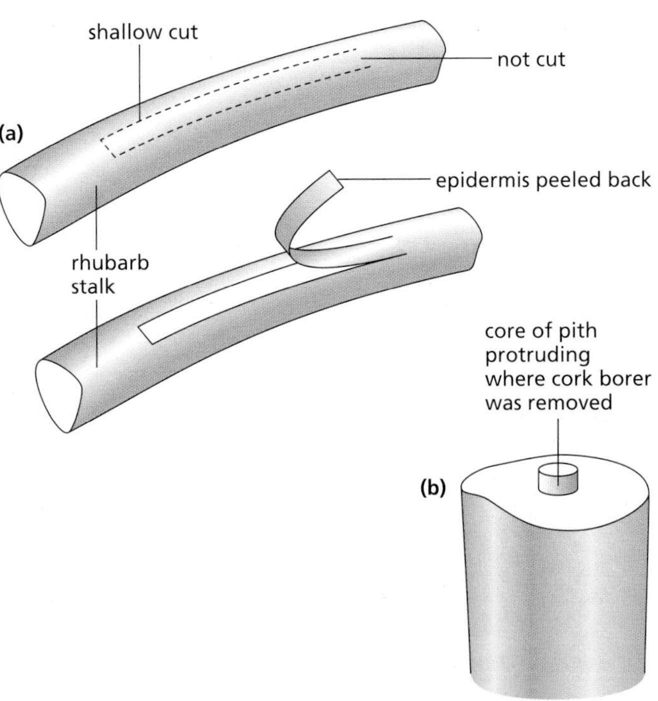

Figure 6.17 To show opposite tensions in pith and epidermis

Checklist

- The shoot of a plant consists of the stem, leaves, buds and flowers.
- The leaf makes food by photosynthesis in its mesophyll cells.
- The water for photosynthesis is carried in the leaf's veins.
- The carbon dioxide for photosynthesis enters the leaf through the stomata and diffuses through the air spaces in the leaf.
- Closure of the stomata stops the entry of carbon dioxide into a leaf but also reduces water loss.
- Sunlight is absorbed by the chloroplasts in the mesophyll cells.
- The food made in the leaf is carried away in the phloem cells.
- The stem supports the leaves and flowers.
- The stem contains vascular bundles (veins).
- The water vessels in the veins carry water up the stem to the leaves.
- The phloem in the veins carries food up or down the stem to wherever it is needed.
- The position of vascular bundles helps the stem to withstand sideways bending and the root to resist pulling forces.
- The roots hold the plant in the soil and absorb the water and mineral salts needed by the plant for making sugars and proteins.
- The root hairs make very close contact with soil particles and are the main route by which water and mineral salts enter the plant.

7 Transport in plants

Transport of water
Transpiration: its function and control. Rates of transpiration.

Transport of salts

Transport of food
Movement of solutes in the phloem.

Uptake of water and salts
Osmosis in the roots. Possible mechanisms for salt uptake.

Transport of gases
Diffusion through stomata and intercellular spaces.

Absorption by leaves

Practical work
Experiments on transport in plants: the potometer; rates of transpiration; role of vascular bundles; evaporation from a leaf.

Transport of water

Transpiration

The main force which draws water from the soil and through the plant is caused by a process called **transpiration**. Water evaporates from the leaves and causes a kind of 'suction' which pulls water up the stem (Figure 7.1). The water travels up the vessels in the vascular bundles (see Figure 6.9, p. 54) and this flow of water is called the **transpiration stream**. The cells in part of a leaf blade are shown in Figure 7.2. As explained on p. 30, the cell sap in each cell is exerting a turgor pressure outwards on the cell wall. This pressure forces some water out of the cell wall and into the air space between the cells. Here the water evaporates and the water vapour passes by diffusion through the air spaces in the mesophyll and out of the stomata. It is this loss of water vapour from the leaves which is called 'transpiration'.

The cell walls which are losing water in this way replace it by drawing water from the nearest vein. Most of this water travels along the cell walls without actually going inside the cells (Figure 7.3, overleaf). Thousands of leaf cells are evaporating water like this and drawing water to replace it from the xylem vessels in the veins. As a result, water is pulled through the xylem vessels and up the stem from the roots. This transpiration pull is strong enough to draw up water 50 metres or more in trees (Figure 7.4, overleaf).

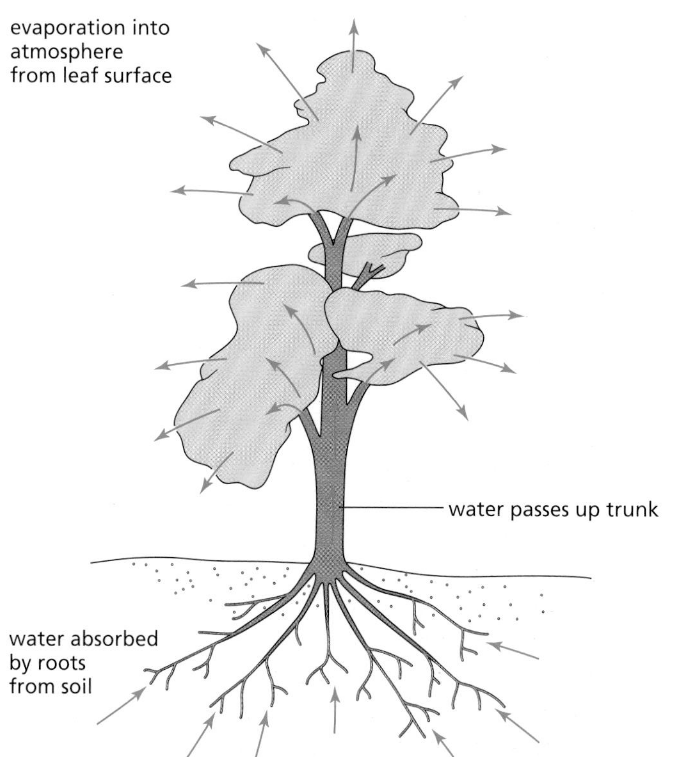

Figure 7.1 The transpiration stream

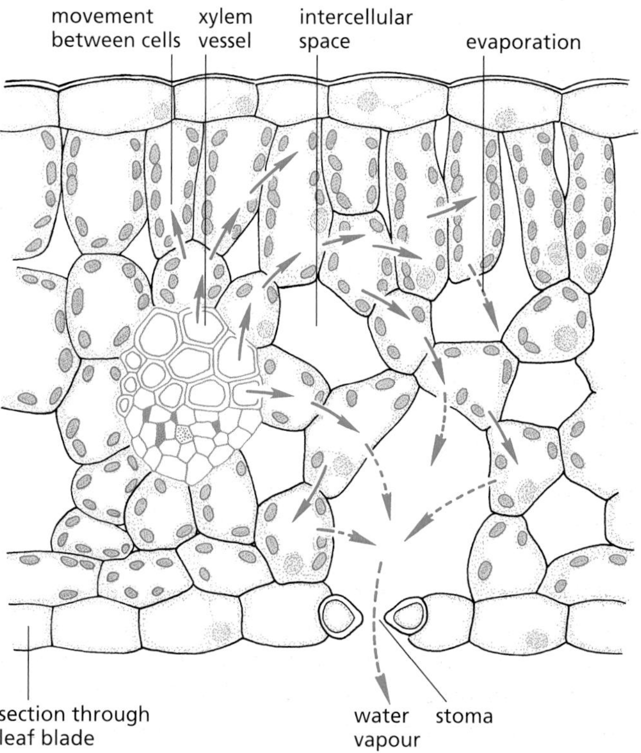

Figure 7.2 Movement of water through a leaf

In addition to the water passing along the cell walls, a small amount will pass right through the cells. When leaf cell A in Figure 7.3 loses water, its turgor pressure will fall. This fall in pressure allows the water in the cell wall to enter the vacuole and so restore the turgor pressure. In conditions of water shortage, cell A may be able to get water by osmosis from cell B more easily than B can get it from the xylem vessels. In this case, all the mesophyll cells will be losing water faster than they can absorb it from the vessels, and the leaf will wilt (see pp. 30–31).

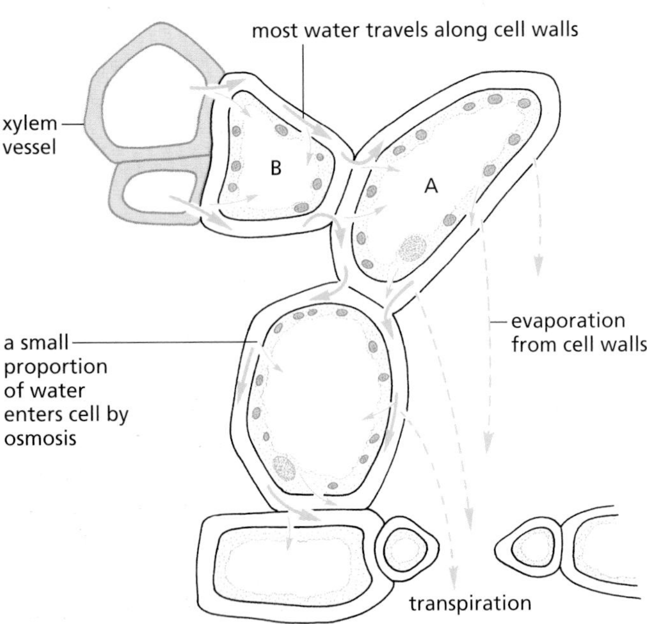
most water travels along cell walls
xylem vessel
B
A
evaporation from cell walls
a small proportion of water enters cell by osmosis
transpiration

Figure 7.3 Probable pathway of water through leaf cells

Importance of transpiration

A tree, on a hot day, may draw up hundreds of litres of water from the soil (Figure 7.4). Most of this water evaporates from the leaves; only a tiny fraction is retained for photosynthesis and to maintain the turgor of the cells. The advantage to the plant of this excessive evaporation is not clear. A rapid water flow may be needed to obtain sufficient mineral salts, which are in very dilute solution in the soil. Evaporation may also help to cool the leaf when it is exposed to intense sunlight.

Against the first possibility, it has to be pointed out that, in some cases, an increased transpiration rate does not increase the uptake of minerals.

The second possibility, the cooling effect, might be very important. A leaf exposed to direct sunlight will absorb heat and its temperature may rise to a level which could kill the cytoplasm. Water evaporating from a leaf absorbs its latent heat and cools the leaf down. This is probably one value of transpiration. However, there are plants whose stomata close at around midday, greatly reducing transpiration. How do these plants avoid overheating?

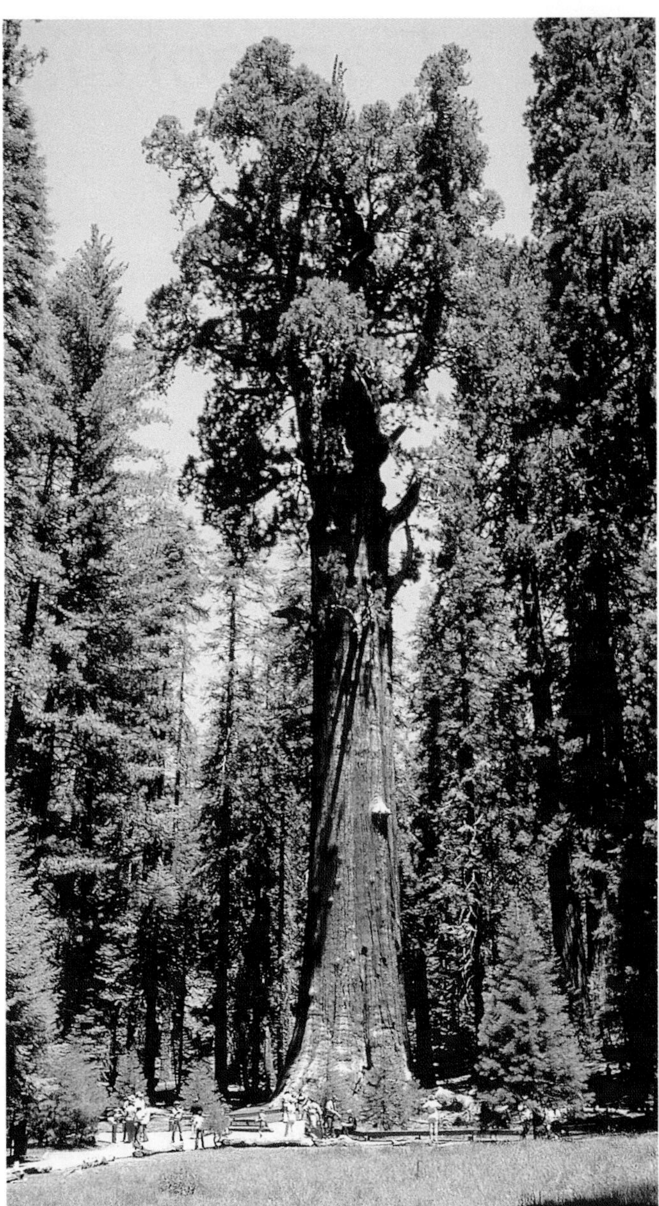

Figure 7.4 Californian redwoods. Some of these trees are over 100 metres tall. Transpiration from their leaves pulls hundreds of litres of water up the trunk

Many biologists regard transpiration as an inevitable consequence of photosynthesis. In order to photosynthesize, a leaf has to take in carbon dioxide from the air. The pathway which allows carbon dioxide in will also let water vapour out whether the plant needs to lose water or not. In all probability, plants have to maintain a careful balance between the optimum intake of carbon dioxide and a damaging loss of water. Plants achieve this balance in different ways, some of which are described on p. 61.

Question

1 Describe the pathway followed by a water molecule from the time it enters a plant root to the time it escapes into the atmosphere from a leaf.

The role of stomata

The opening and closing of stomata can be triggered by a variety of factors, principally light intensity, carbon dioxide concentration and humidity. These factors interact with each other. For example, a rise in light intensity will increase the rate of photosynthesis and so lower the carbon dioxide concentration in the leaf. These are the conditions you would expect to influence stomatal aperture if the stomata are to control the balance between loss of water vapour and uptake of carbon dioxide.

The stomata also react to water stress, i.e. if the leaf is losing water by transpiration faster than it is being taken up by the roots. Before wilting sets in, the stomata start to close. Although they do not prevent wilting, the stomata do seem to delay its onset.

Rate of transpiration

Transpiration is the evaporation of water from the leaves, so any change which increases or reduces evaporation will have the same effect on transpiration.

Light intensity

Light itself does not affect evaporation, but in daylight the stomata (p. 52) of the leaves are open. This allows the water vapour in the leaves to diffuse out into the atmosphere. At night, when the stomata close, transpiration is greatly reduced.

Generally speaking, then, transpiration speeds up when light intensity increases because the stomata respond to changes in light intensity.

Sunlight may also warm up the leaves and increase evaporation (see below).

Humidity

If the air is very humid, i.e. contains a great deal of water vapour, it can accept very little more from the plants and so transpiration slows down. In dry air, the diffusion of water vapour from the leaf to the atmosphere will be rapid.

Air movements

In still air, the region round a transpiring leaf will become saturated with water vapour so that no more can escape from the leaf. In these conditions, transpiration would slow down. In moving air, the water vapour will be swept away from the leaf as fast as it diffuses out. This will speed up transpiration.

Temperature

Warm air can hold more water vapour than cold air. Thus evaporation or transpiration will take place more rapidly into warm air.

Furthermore, when the sun shines on the leaves, they will absorb heat as well as light. This warms them up and increases the rate of evaporation of water.

Questions

1. What kind of climate and weather conditions do you think will cause a high rate of transpiration?
2. What would happen to the leaves of a plant which was losing water by transpiration faster than it was taking it up from the roots?
3. In what two ways does sunlight increase the rate of transpiration?
4. Apart from drawing water through the plant, what else may be drawn up by the transpiration stream?
5. Transpiration has been described in this chapter as if it took place only in leaves. What other parts of a plant might transpire?

Experiments which investigate the effect of some of these conditions on the rate of transpiration are described on p. 64 (Experiment 1).

Water movement in the xylem

You may have learned that you cannot draw water up by 'suction' to a height of more than about 10 metres. Many trees are taller than this yet they can draw up water effectively. The explanation offered is that, in long vertical columns of water in very thin tubes, the attractive forces between the water molecules are greater than the forces trying to separate them. So, in effect, the transpiration stream is pulling up thin threads of water which resist the tendency to break.

There are still problems, however. It is likely that the water columns in some of the vessels do have air breaks in them and yet the total water flow is not affected.

Evidence for the pathway of water

Experiment 3 on p. 66 uses a dye to show that, in a cut stem, the dye, and, therefore, presumably the water, travels in the vascular bundles. Closer examination with a microscope would show that it travels in the xylem vessels.

Removal of a ring of bark (which includes the phloem) does not affect the passage of water along a branch (Experiment 4). Killing parts of a branch by heat or poisons does not interrupt the flow of water, but anything which blocks the vessels does stop the flow.

The evidence all points to the non-living xylem vessels as the main route by which water passes from the soil to the leaves.

Adaptations to arid conditions

In both hot and cold climates, plants may suffer from water shortage. High temperatures accelerate evaporation from leaves. At very low temperatures the soil water becomes frozen and therefore unavailable to the roots of plants.

It is thought that the autumn leaf-fall of deciduous trees and shrubs is an essential adaptation to winter 'drought'. Loss of leaves removes virtually all evaporating surfaces at a time when water may become unavailable. Without leaves, however, the plants cannot make food by photosynthesis and so they enter a dormant condition in which metabolic activity is at a low level.

Evergreen trees, e.g. pine and spruce (p. 280), which survive in cold climates, have small, compact needle-like leaves. The small surface area of such leaves offers little resistance to high winds and can reduce the amount of water lost in transpiration. However, photosynthesis can continue whenever water is available.

Other adaptations to drought include leaves with thick waxy cuticles, leaves with few stomata often sunken below the level of the epidermis, hairy leaves or rolled-up leaves. These last four adaptations help to retain a static layer of air, saturated with water vapour, close to the leaf. This effectively reduces the diffusion gradient. Waxy cuticles reduce evaporation.

Cacti are adapted to hot, dry conditions in several ways. Often they have no leaves. Photosynthesis is carried on by a thick green stem which offers only a small surface area for evaporation. Cacti are succulent, i.e. they store water in their fleshy tissues and draw on this store for photosynthesis (Figure 7.5).

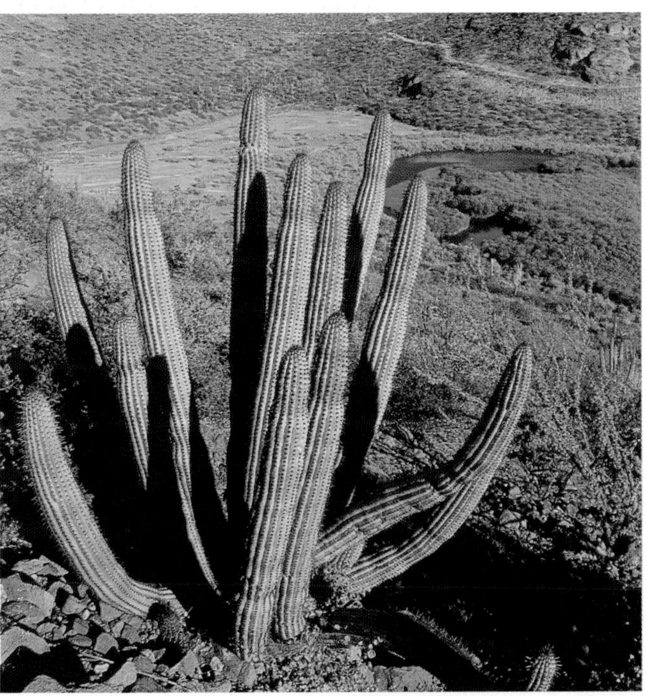

Figure 7.5 A succulent cactus growing in the Arizona desert

The stomata of many cacti are closed during the day, when temperatures are high, and open at night when evaporation is at a minimum. This strategy requires a slightly different form of photosynthesis. At night, carbon dioxide diffuses in through the open stomata and is 'fixed' (i.e. incorporated into) an organic acid. Little water vapour is lost at night. In the daytime the stomata are closed but the organic acid breaks down to yield carbon dioxide which is then built into sugars by photosynthesis. Closure of the stomata in daytime greatly reduces water loss.

Transport of salts

The liquid which travels in the xylem is not, in fact, pure water. It is a very dilute solution, containing from 0.1 to 1.0 per cent dissolved solids, mostly amino acids, other organic acids and mineral salts. The organic acids are made in the roots; the mineral salts come from the soil. The faster the flow in the transpiration stream, the more dilute is the xylem sap.

Experimental evidence suggests that salts are carried from the soil to the leaves mainly in the xylem vessels.

Transport of food

The xylem sap is always a very dilute solution, but the phloem sap may contain up to 25 per cent of dissolved solids, the bulk of which consists of sucrose and amino acids. There is a good deal of evidence to support the view that sucrose, amino acids and many other substances are transported in the phloem. This is called **translocation**.

The movement of water and salts in the xylem is always upwards, from soil to leaf, but in the phloem the solutes may be travelling up or down the stem. The carbohydrates made in the leaf during photosynthesis are converted to sucrose and carried out of the leaf to the stem. From here, the sucrose may pass upwards to growing buds and fruits or downwards to the roots and storage organs. All parts of a plant which cannot photosynthesize will need a supply of nutrients brought by the phloem. It is quite possible for substances to be travelling upwards and downwards at the same time in the phloem.

There is no doubt that substances travel in the sieve tubes (p. 55) of the phloem but the mechanism by which they are moved is not fully understood. We do know that translocation depends on living processes because anything which inhibits cell metabolism, e.g. poisons or high temperatures, also arrests translocation.

Questions

1 How do sieve tubes and vessels differ
  a in the substances they transport,
  b in the directions these substances are carried?

2 A complete ring of bark cut from round the circumference of a tree-trunk causes the tree to die. The xylem continues to carry water and salts to the leaves, which can make all the substances needed by the tree. So why does the tree die?

3 Make a list of all the non-photosynthetic parts of a plant which need a supply of sucrose and amino acids.

Uptake of water and salts

Uptake of water

The water tension developed in the vessels by a rapidly transpiring plant is thought to be sufficient to draw water through the root from the soil. The precise pathway taken by the water is the subject of some debate, but the path of least resistance seems to be in or between the cell walls rather than through the cells.

When transpiration is slow, e.g. at night-time or just before bud burst in a deciduous tree, then osmosis may play a more important part in the uptake of water. In Figure 7.6, showing a root hair in the soil, the cytoplasm of the root hair is partially permeable to water. The soil water is more dilute than the cell sap and so water passes by osmosis (p. 29) from the soil into the cell sap of the root hair cell. This flow of water into the root hair cell raises the cell's turgor pressure (p. 30). So water is forced out through the cell wall into the next cell and so on, right through the cortex of the root to the xylem vessels (Figure 7.7).

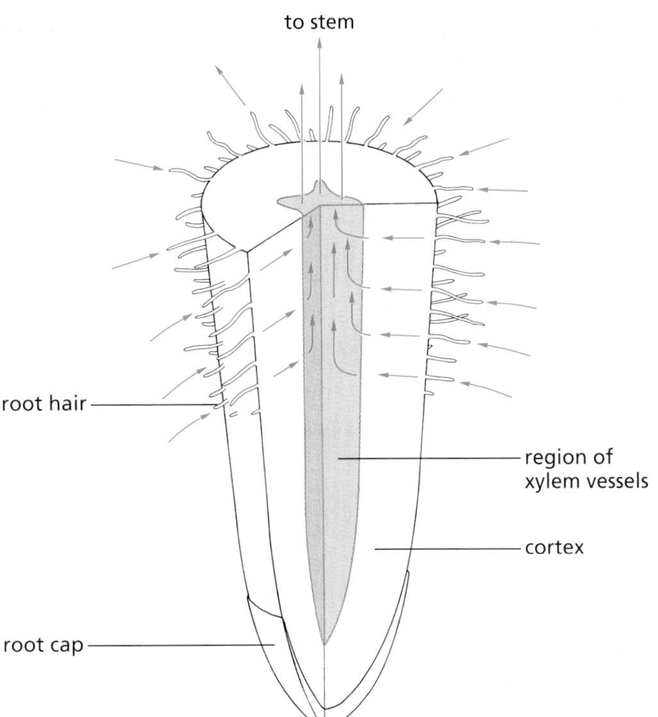

Figure 7.7 Diagrammatic section of root to show passage of water from the soil

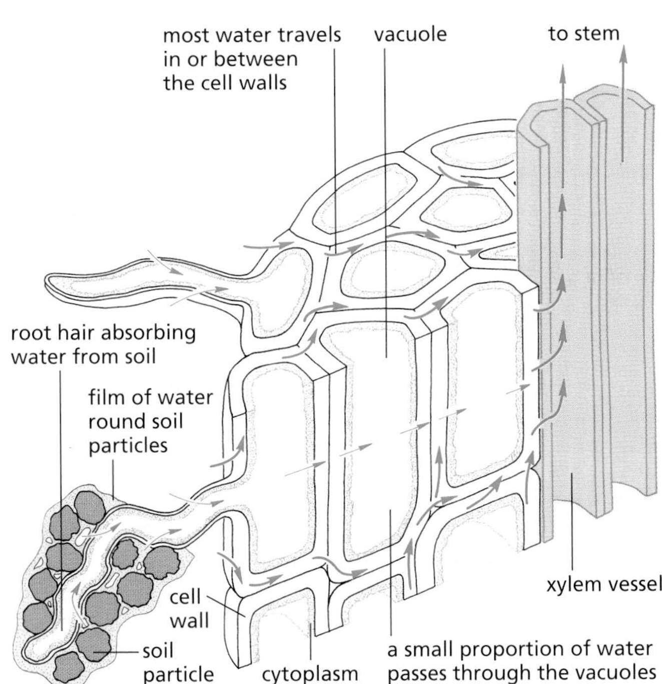

Figure 7.6 The probable pathways of water through a root

One problem for this explanation is that it has not been possible to demonstrate that there is an osmotic gradient across the root cortex which could produce this flow of water from cell to cell. Nevertheless, root pressure developed probably by osmosis does force water up the root system and into the stem.

Uptake of salts

The methods by which roots take up salts from the soil are not fully understood. Some salts may be carried in with the water drawn up by transpiration and pass mainly along the cell walls in the root cortex and into the xylem.

It may be that diffusion from a relatively high concentration in the soil to a lower concentration in the root cells accounts for uptake of some individual salts, but it has been shown (a) that salts can be taken from the soil even when their concentration is below that in the roots and (b) that anything which interferes with respiration impairs the uptake of salts. This suggests that active transport (p. 28) plays an important part in the uptake of salts.

The growing region of the root and the root hair zone (Figure 6.15, p. 57) seem to be most active in taking up salts. Most of the salts appear to be carried at first in the xylem vessels, though they soon appear in the phloem as well.

The salts are used by the plant's cells to build up essential molecules. Nitrates, for example, are combined with carbohydrates to make amino acids in the roots. These amino acids are used later to make proteins.

Questions

1. If root hairs take up water from the soil by osmosis, what would you expect to happen if so much nitrate fertilizer was put on the soil that the soil water became a stronger solution than the cell sap of the root hairs?
2. A plant's roots may take up water and salts less efficiently from a waterlogged soil than from a fairly dry soil. Revise 'Active transport' (p. 28) and suggest reasons for this.
3. Why do you think that, in a deciduous tree in spring, transpiration is negligible before bud burst?

Transport of gases

The process of diffusion described on p. 26 accounts for the movement of gases in and out of a plant. During respiration, oxygen is taken in and carbon dioxide given out. When photosynthesis is faster than respiration, carbon dioxide diffuses in and oxygen diffuses out (Figure 5.8, p. 40). In leaves and green stems, the gases enter and leave through the stomata (p. 52). Then they diffuse through the air spaces between the cells to reach all parts of the plant shoot. In woody plants, the stems have no stomata and the gases have to pass through small openings in the bark called **lenticels**.

Roots obtain their oxygen from the air spaces in the soil. Much of this oxygen will be dissolved in the soil water which enters the root through the growing region and the root hairs.

Absorption by leaves

Leaves are able to absorb certain substances if these are sprayed on to them. Mineral ions in solution can be absorbed through the cuticle or stomata and, for some crops, this is a method of applying 'fertilizer'. (The process is called **foliar feeding**.)

Some insecticide and fungicide sprays are absorbed through the leaves and translocated through the plant. Such pesticides are called **systemic** because they enter the plant's system. A caterpillar which ate part of a leaf treated with a systemic insecticide would be poisoned by the chemical in the cells of the leaf.

It is important, of course, that systemic pesticides are broken down to harmless compounds by the plant, long before its leaves or fruits are used for human consumption.

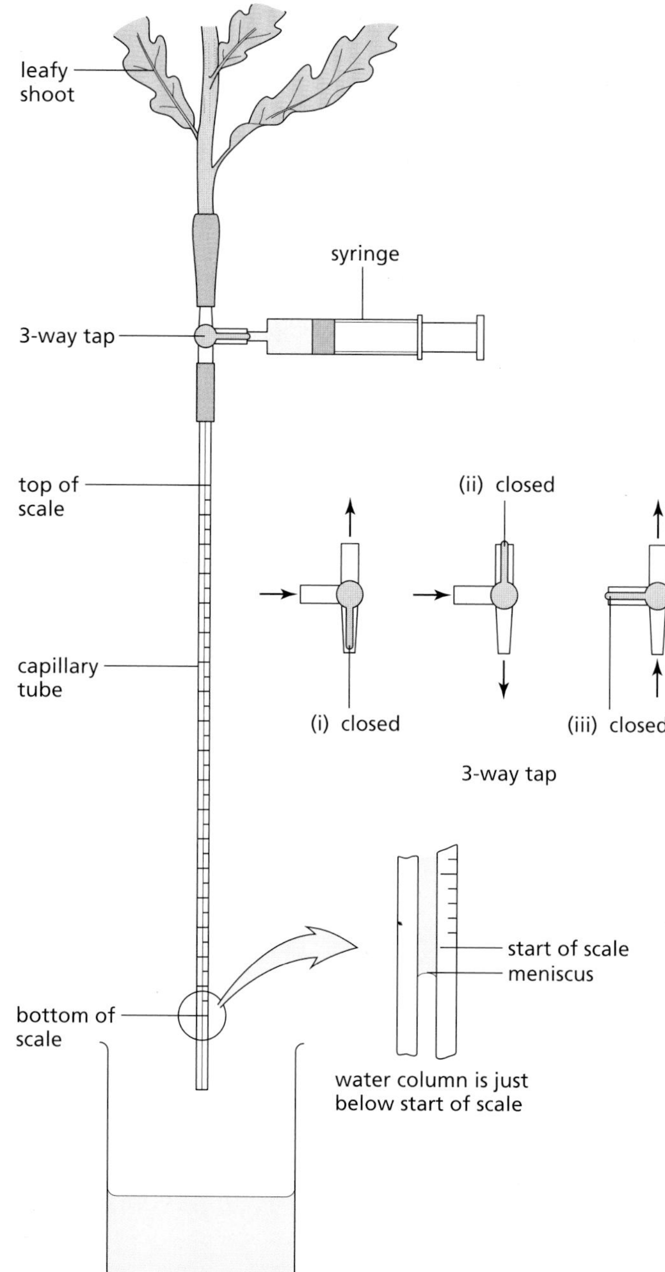

Figure 7.8 A potometer

Practical work

Experiments on transport in plants

1 Rates of water uptake in different conditions

The apparatus shown in Figure 7.8 is called a **potometer**. It is designed to measure the rate of uptake of water in a cut shoot.

Fill the syringe with water and attach it to the side arm of the 3-way tap. Turn the tap downwards (i) and press the syringe until water comes out of the rubber tubing at the top.

Collect a leafy shoot and push its stem into the rubber tubing as far as possible. Set up the apparatus in a part of the laboratory that is not receiving direct sunlight.

Turn the tap up (ii) and press the syringe till water comes out of the bottom of the capillary tube. Turn the tap horizontally (iii).

As the shoot transpires, it will draw water from the capillary tube and the level can be seen to rise. Record the distance moved by the water column in 30 seconds or a minute.

Turn the tap up and send the water column back to the bottom of the capillary. Turn the tap horizontally and make another measurement of the rate of uptake. In this way obtain the average of three readings.

The conditions can now be changed in one of the following ways:

1 Move the apparatus into sunlight or under a fluorescent lamp.
2 Blow air past the shoot with an electric fan or merely fan it with an exercise book.
3 Cover the shoot with a plastic bag.

After each change of conditions, take three more readings of the rate of uptake and notice whether they represent an increase or a decrease in the rate of transpiration.

You might expect results as follows:

1 An increase in light intensity should make the stomata open and allow more rapid transpiration.
2 Moving air should increase the rate of evaporation and, therefore, the rate of uptake.
3 The plastic bag will cause a rise in humidity round the leaves and suppress transpiration.

Ideally, you should change only one condition at a time. If you took the experiment outside, you would be changing the light intensity, the temperature and the air movement. When the rate of uptake increased, you would not know which of these three changes was mainly responsible.

To obtain reliable results, you should really keep taking readings until three of them are nearly the same. A change in conditions may take 10 or 15 minutes before it produces a new, steady rate of uptake. In practice, you may not have time to do this, but even your first three readings should indicate a trend towards increased or decreased uptake.

Limitations of the potometer

Although we use the potometer to compare rates of transpiration, it is really the rates of uptake that we are observing. Not all the water taken up will be transpired; some will be used in photosynthesis; some may be absorbed by cells to increase their turgor. However, these quantities are very small compared with the volume of water transpired and they can be disregarded.

The rate of uptake of a cut shoot may not reflect the rate in the intact plant. If the root system were present, it might offer resistance to the flow of water or it could be helping the flow by means of its root pressure.

2 To find which surface of a leaf loses more water vapour

Cut four leaves of about the same size from a plant (do not use an evergreen plant). Protect the bench with newspaper and then treat each leaf as follows:

a smear a thin layer of Vaseline (petroleum jelly) on the lower surface;
b smear Vaseline on the upper surface;
c smear Vaseline on both surfaces;
d no Vaseline on either surface.

Place a little Vaseline on the cut end of the leaf stalk and then suspend the four leaves from a retort stand with cotton threads for several days.

Result All the leaves will have shrivelled and curled up to some extent but the ones which lost most water will be the most shrivelled (Figure 7.9).

(a) lower surface

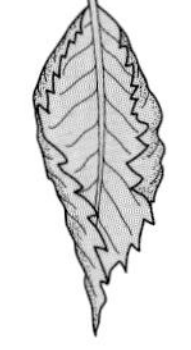

(b) upper surface

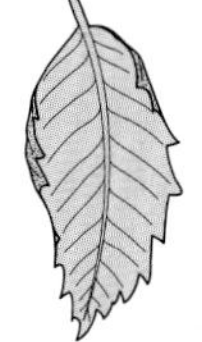

(c) both surfaces

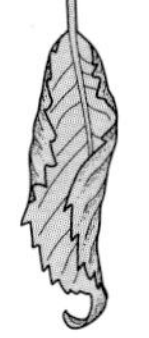

(d) neither surface

Figure 7.9 The results of evaporation from leaves subjected to different treatments

Interpretation The Vaseline prevents evaporation. The untreated leaf and the leaf with its upper surface sealed show the greatest degree of shrivelling, so it is from the lower surface that leaves evaporate most water.

More accurate results may be obtained by weighing the leaves at the start and the end of the experiment. It is best to group the leaves from the whole class into their respective batches and weigh each batch. Ideally, the weight loss should be expressed as a percentage of the initial weight.

More rapid results can be obtained by sticking small squares of blue cobalt chloride paper to the upper and lower surface of the same leaf using transparent adhesive tape (Figure 7.10). Cobalt chloride paper changes from blue to pink as it takes up moisture. By comparing the time taken for each square to go pink, the relative rates of evaporation from each surface can be compared.

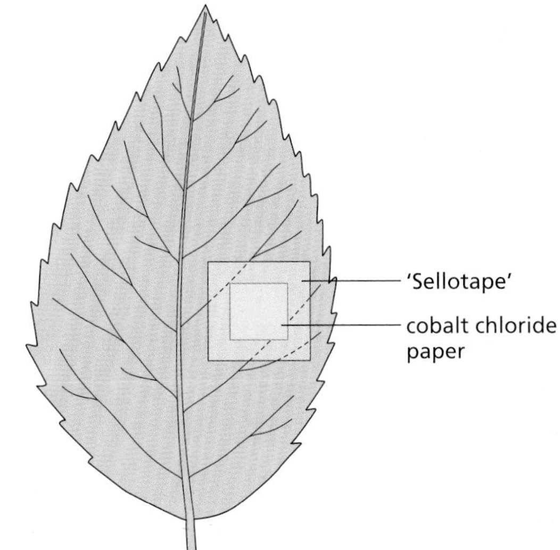

Figure 7.10 To find which surface of a leaf loses more water vapour

The results of either experiment can be correlated with the numbers of stomata on the upper and lower epidermis. This can be done by painting clear nail varnish over each surface and allowing it to dry. The varnish is then peeled off and examined under the microscope. The outlines of the guard cells can be seen and counted.

3 Transport in the vascular bundles

Place the shoots of several leafy plants in a solution of 1 per cent methylene blue. 'Busy Lizzie' (*Impatiens*) or celery stalks with leaves are usually effective. Leave the shoots in the light for 30 minutes or more.

Result In some cases, after this time, the blue dye will appear in the leaf veins. If some of the stems are cut across, the dye will be seen in the vascular bundles (see Figure 1.2, p. 2).

Interpretation These results show that the dye, and, therefore, probably also the water, travel up the stem in the vascular bundles. Closer study would show that they travel in the xylem vessels.

4 Transport of water in the xylem

Cut three leafy shoots from a deciduous tree or shrub. Each shoot should have about the same number of leaves.

On one twig remove a ring of bark about 5 mm wide, about 100 mm up from the cut base. With the second shoot, smear a layer of Vaseline over the cut base so that it blocks the vessels. The third twig is a control.

Place all three twigs in a jar with a little water. The water level must be below the region from which you removed the ring of bark. Leave the twigs where they can receive direct sunlight.

Result After an hour or two, you will probably find that the twig with blocked vessels shows signs of wilting. The other two twigs should still have turgid leaves.

Interpretation Removal of the bark (including the phloem) has not prevented water from reaching the leaves, but blocking the xylem vessels has. The vessels of the xylem, therefore, offer the most likely route for water passing up the stem.

Questions

1 A leafy shoot, plus the beaker of water in which it is placed, weighs 275 grams. Two hours later, it weighs 260 grams. An identical beaker of water, with no plant, loses 3 grams over the same period of time. What is the rate of transpiration of the shoot?

2 In Experiment 2, suggest
- **a** why forceps should be used to handle the cobalt chloride paper squares,
- **b** why it was water vapour from the leaf and not water vapour from the air which made the cobalt chloride paper change from blue to pink.

3 An interesting experiment is to use a syringe, attached to the cut end of a woody shoot, to force air through the xylem vessels. If the shoot is held under water while you press the syringe, would you expect to see air escaping from the stomata? If you cut away the top half of each leaf, what would you expect to see when the syringe is pressed? Explain your reasoning.

Checklist

- Transpiration is the evaporation of water from the leaves of a plant.
- Transpiration produces the force which draws water up the stem.
- The water travelling in the transpiration stream will contain dissolved salts.
- Closure of stomata and shedding of leaves may help to regulate the transpiration rate.
- The rate of transpiration is increased by sunlight, high temperature, low humidity and air movements.
- Root pressure forces water up the stem as a result of osmosis in the roots.
- Salts are taken up from the soil by roots, and are carried in the xylem vessels.
- Water and salts move up the stem from the roots to the leaves.
- Food made in the leaves moves up or down the stem in the phloem.
- Oxygen and carbon dioxide move in or out of the leaf by diffusion through the stomata.

8 Reproduction in flowering plants

Sexual reproduction

Flower structure
Wallflower and lupin.

Pollination
Cross- and self-pollination. Lupin. Adaptation to wind and insect pollination.

Fertilization and fruit formation
Fertilization: the pollen tube. Fruit and seed formation; the ovary forms the fruit.

Dispersal of fruit and seeds
Wind, animal and explosive methods.

Practical work
The growth of pollen tubes.

Seed structure and germination
Seed structure: French bean. Germination: stages in the French bean.

The importance of water, oxygen and temperature
Uses of water. Uses of oxygen. Importance of temperature. Germination and light

Practical work: conditions for germination
Experiments on the need for water, oxygen and suitable temperature. Controlling the variables; experimental design.

The importance of water, oxygen and temperature
Use of water in the seedling. Uses of oxygen. Importance of temperature. Germination and light. Dormancy.

Sexual reproduction

Flowers are reproductive structures; they contain the reproductive organs of the plant. The male organs are the **stamens**, which produce pollen. The female organs are the **carpels**. After fertilization, part of the carpel becomes the fruit of the plant and contains the seeds. In the flowers of most plants there are both stamens and carpels. These flowers are, therefore, male and female at the same time, a condition known as **bisexual** or **hermaphrodite**.

Some species of plants have unisexual flowers, i.e. any one flower will contain either stamens or carpels but not both. Sometimes both male and female flowers are present on the same plant, e.g. the hazel, which has male and female catkins on the same tree. In the willow tree, on the other hand, the male and female catkins are on different trees.

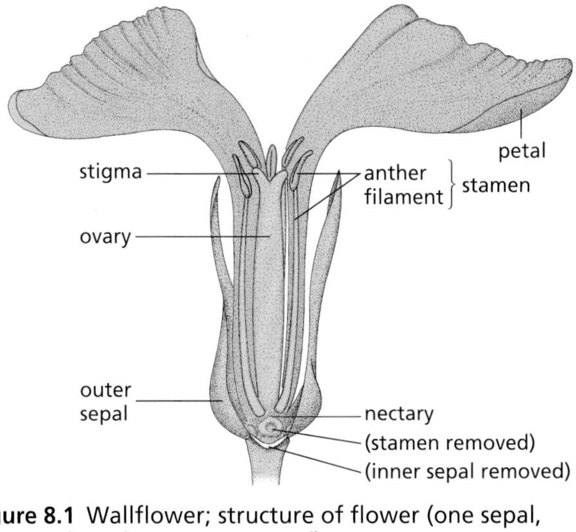

Figure 8.1 Wallflower; structure of flower (one sepal, two petals and stamen removed)

Flower structure

The basic structure of a flower is shown in Figures 8.1 and 8.2.

Petals

These are usually brightly coloured and sometimes scented. They are arranged in a circle (Figure 8.1) or a cylinder. Most flowers have from four to ten petals. Sometimes they are joined together to form a tube (Figures 8.3 and 8.4, overleaf) and the individual petals can no longer be distinguished. The colour and scent of the petals attract insects to the flower; the insects may bring about pollination (p. 69).

The flowers of grasses and many trees do not have petals but small, leaf-like structures which enclose the reproductive organs.

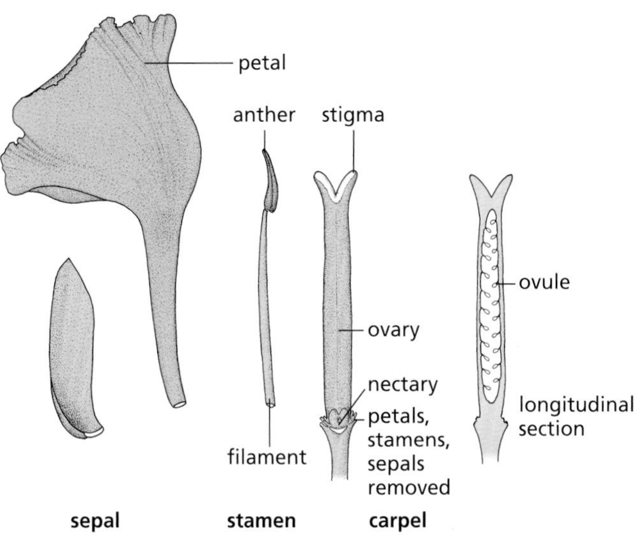

Figure 8.2 Floral parts of wallflower

Figure 8.3 Daffodil flower cut in half. The inner petals form a tube. Three stamens are visible round the long style and the ovary contains many ovules

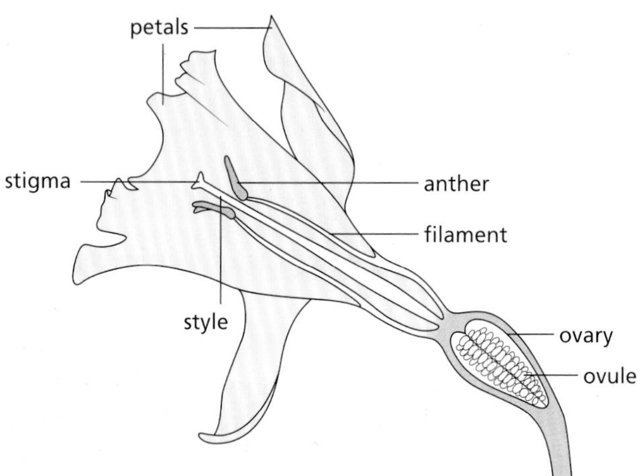

Figure 8.4 Daffodil flower. Outline drawing of Figure 8.3. In daffodils, lilies, tulips, etc. (monocots) there is no distinction between sepals and petals

Sepals

Outside the petals is a ring of sepals. They are often green and much smaller than the petals. They may protect the flower when it is in the bud.

Stamens

The stamens are the male reproductive organs of a flower. Each stamen has a stalk called the **filament**, with an **anther** on the end. Flowers such as the buttercup and blackberry have many stamens; others such as the tulip have a small number, often the same as, or double, the number of petals or sepals. Each anther consists of four **pollen sacs** in which the pollen grains are produced by cell division. When the anthers are ripe, the pollen sacs split open and release their pollen (see Figure 8.8, right).

Carpels

These are the female reproductive organs. Flowers such as the buttercup and blackberry have a large number of carpels while others, such as the lupin, have a single carpel. Each carpel consists of an **ovary**, bearing a **style** and a **stigma**.

Inside the ovary there are one or more **ovules**. Each blackberry ovary contains one ovule but the wallflower ovary contains several. The ovule will become a **seed**, and the whole ovary will become a **fruit**. (In biology, a fruit is the fertilized ovary of a flower, not necessarily something to eat.)

The style and stigma project from the top of the ovary. The stigma has a sticky surface and pollen grains will stick to it during pollination. The style may be quite short (e.g. wallflower, Figure 8.1) or very long (e.g. daffodil, Figures 8.3 and 8.4).

Receptacle

The flower structures just described are all attached to the expanded end of a flower stalk. This is called the **receptacle** and, in a few cases after fertilization, it becomes fleshy and edible, e.g. apple and pear.

Lupin

The lupin flower is shown in Figures 8.5–8.7. There are five sepals but these are joined together forming a short tube. The five petals are of different shapes and sizes. The uppermost, called the **standard**, is held vertically. Two petals at the sides are called **wings** and are partly joined together. Inside the wings are two more petals joined together to form a boat-shaped **keel**.

The single carpel is long, narrow and pod shaped, with about ten ovules in the ovary. The long style ends in a stigma just inside the pointed end of the keel. There are ten stamens: five long ones and five short ones. Their filaments are joined together at the base to form a sheath round the ovary.

The flowers of peas and beans are very similar to those of lupins.

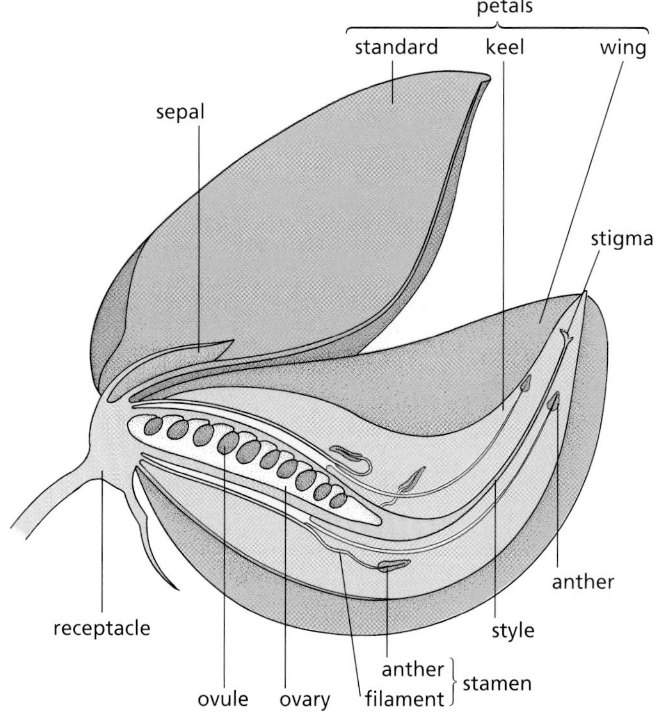

Figure 8.5 Half-flower of lupin

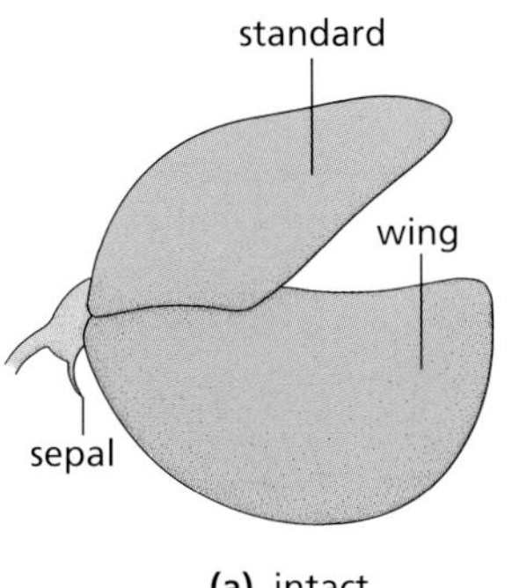

(a) intact

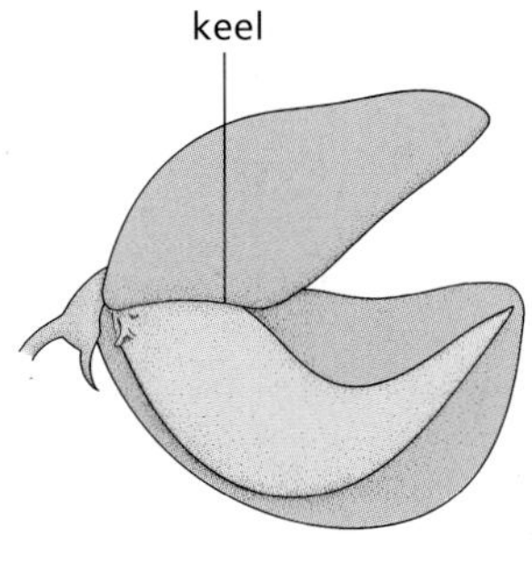

(b) one wing removed

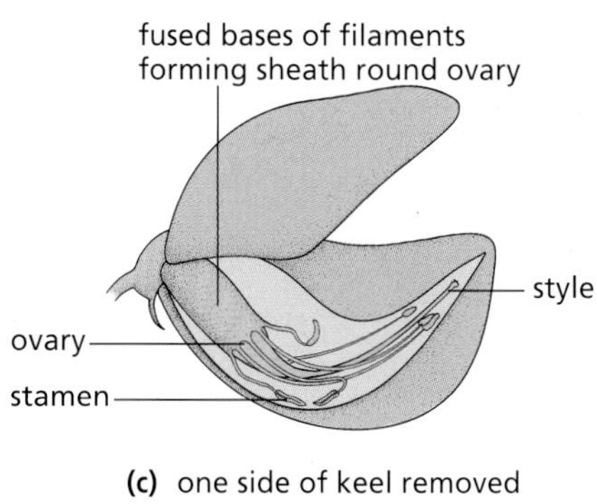

(c) one side of keel removed

Figure 8.6 Lupin flower dissected

The shoots or branches of a plant carrying groups of flowers are called **inflorescences**. The flowering shoots of the lupin in Figure 8.7 are inflorescences, each one carrying about a hundred individual flowers.

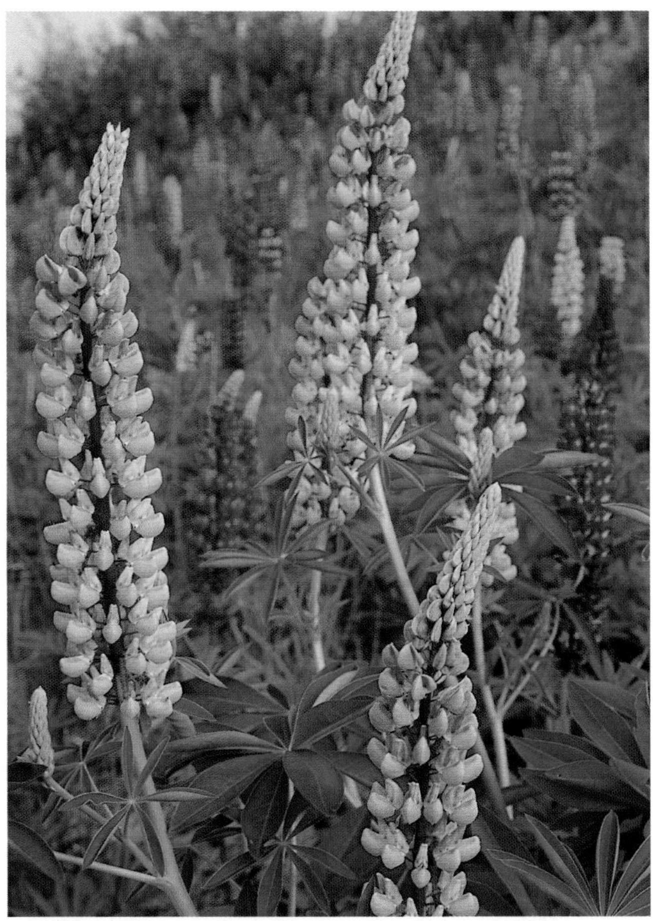

Figure 8.7 Lupin inflorescence. There are a hundred or more flowers in each inflorescence. The youngest flowers, at the top, have not yet opened. The oldest flowers are at the bottom and have already been pollinated

Questions

1 Working from outside to inside, list the parts of a bisexual flower.

2 What features of flowers might attract insects?

3 Which part of a flower becomes
a the seed,
b the fruit?

Pollination

The transfer of pollen from the anthers to the stigma is called 'pollination'. The anthers split open, exposing the microscopic pollen grains (Figure 8.8). The pollen grains are then carried away on the bodies of insects, or simply blown by the wind, and may land on the stigma of another flower. In **self-pollinating** plants, the pollen which reaches the stigma comes from the same flower or another flower on the same plant. In **cross-pollination**, the pollen is carried from the anthers of one flower to the stigma in a flower of another plant of the same species.

If a bee carried pollen from one of the younger flowers near the middle of a lupin plant to an older flower near the bottom, this would be self-pollination. If, however, the bee visited a separate lupin plant and pollinated its flowers, this would be cross-pollination.

The term 'cross-pollination', strictly speaking, should be applied only if there are genetic differences between the two plants involved. The flowers on a single plant all have the same genetic constitution. The flowers on plants growing from the same rhizome or rootstock (p. 79) will also have the same genetic constitution. Pollination between such flowers is little different from self-pollination in the same flower.

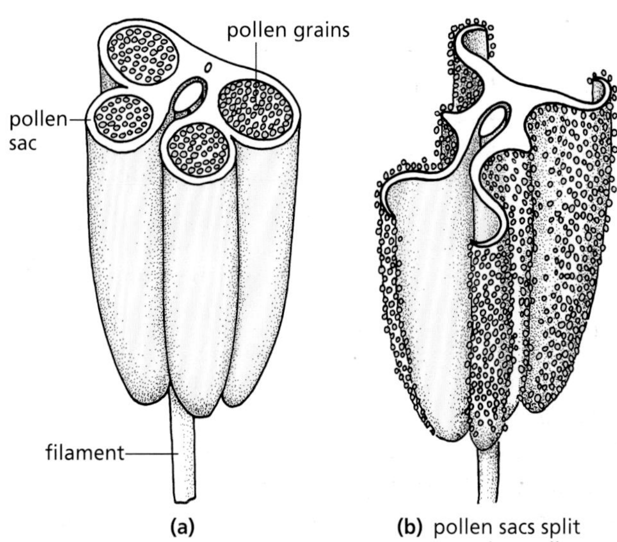

Figure 8.8 Structure of an anther (top cut off)

Pollination of the lupin

Lupin flowers have no nectar. The bees which visit them come to collect pollen, which they take back to the hive for food. Other members of the lupin family (Leguminosae), e.g. clover, do produce nectar.

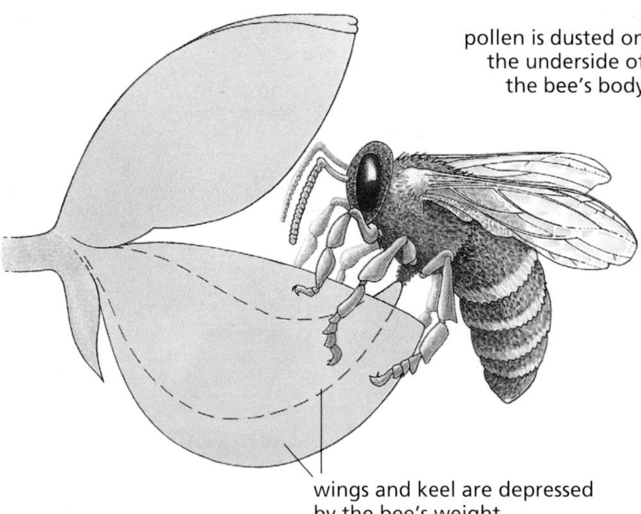
pollen is dusted on the underside of the bee's body
wings and keel are depressed by the bee's weight

Figure 8.9 Pollination of the lupin

The weight of the bee, when it lands on the flower's wings, pushes down these two petals and the petals of the keel. The pollen from the anthers has collected in the tip of the keel and, as the petals are pressed down, the stigma and long stamens push the pollen out from the keel on to the underside of the bee (Figure 8.9). The bee, with pollen grains sticking to its body, then flies to another flower. If this flower is older than the first one, it will already have lost its pollen. When the bee's weight pushes the keel down, only the stigma comes out and touches the insect's body, picking up pollen grains on its sticky surface.

Lupin and wallflower are examples of **insect-pollinated** flowers.

Wind pollination

Grasses, cereals and many trees are pollinated not by insects but by wind currents. The flowers are often quite small with inconspicuous, green, leaf-like bracts, rather than petals. They produce no nectar. The anthers and stigma are not enclosed by the bracts but are exposed to the air. The pollen grains, being light and smooth, may be carried long distances by the moving air and some of them will be trapped on the stigmas of other flowers.

In the grasses, at first, the feathery stigmas protrude from the flower, and pollen grains floating in the air are trapped by them. Later, the anthers hang outside the flower (Figure 8.10), the pollen sacs split, and the wind blows the pollen away. This sequence varies with species.

If the branches of a birch or hazel tree with ripe male catkins, or the flowers of the ornamental pampas grass, are shaken, a shower of pollen can easily be seen.

Figure 8.10 Grass flowers. Note that the anthers hang freely outside the bracts

Adaptation

Insect-pollinated flowers are considered to be adapted in various ways to their method of pollination. The term **'adaptation'** implies that, in the course of evolution, the structure and physiology of a flower have been modified in ways which improve the chances of successful pollination by insects.

Most **insect-pollinated flowers** have brightly coloured petals and scent, which attract a variety of insects. Some flowers produce nectar, which is also attractive to many insects. The dark lines ('honey guides') on petals are believed to help direct the insects to the nectar source and thus bring them into contact with the stamens and stigma.

These features are adaptations to insect pollination in general, but are not necessarily associated with any particular insect species. The various petal colours and the nectaries of the wallflower attract a variety of insects. Many flowers, however, have modifications which adapt them to pollination by only one type or species of insect. Flowers such as the honeysuckle, with narrow, deep petal tubes, are likely to be pollinated only by moths or butterflies whose long 'tongues' can reach down the tube to the nectar.

Tube-like flowers such as foxgloves need to be visited by fairly large insects to effect pollination. The petal tube is often lined with dense hairs which impede small insects that would take the nectar without pollinating the flower. A large bumble-bee, however, pushing into the petal tube, is forced to rub against the anthers and stigma.

Many tropical and sub-tropical flowers are adapted to pollination by birds or even by mammals, e.g. bats and mice.

Wind-pollinated flowers are adapted to their method of pollination by producing large quantities of light pollen, and having anthers and stigmas which project outside the flower. Many grasses have anthers which are not rigidly attached to the filaments and can be shaken by the wind. The stigmas of grasses are feathery and act as a net which traps passing pollen grains.

Questions

1 Put the following events in the correct order for pollination in a lupin plant.
   a Bee gets dusted with pollen.
   b Pollen is deposited on stigma.
   c Bee visits older flower.
   d Bee visits young flower.
   e Anthers split open.

2 Which of the following trees would you expect to be pollinated by insects: apple, hazel, oak, cherry, horse-chestnut, sycamore?

3 In what ways do you think
   a an antirrhinum flower,
   b a nasturtium flower
   are adapted to insect pollination?

4 In the course of evolution, some flowers may have become adapted to pollination by certain insect species. Discuss whether the insects are likely to have become adapted to the flowers. What sort of adaptation might you expect?

5 Draw up a table to contrast the features of a typical insect-pollinated flower with those of a typical wind-pollinated flower. Include features such as petals (or their equivalent), structure and position of anthers and stigmas, pollen and nectar.

Fertilization and fruit formation

Pollination is complete when pollen from an anther has landed on a stigma. If the flower is to produce seeds, pollination has to be followed by a process called **fertilization**. In all living organisms, fertilization happens when a male sex cell and a female sex cell meet and join together (they are said to **fuse** together). The cell which is formed by this fusion is called a **zygote** and develops into an embryo of an animal or a plant (Figure 8.11). The sex cells of all living organisms are called **gametes**.

In animals, the male gamete is the sperm and the female gamete is the egg or ovum (p. 140).

In flowering plants, the male gamete is in the pollen grain; the female gamete, called the **egg cell**, is in the ovule. For fertilization to occur, the nucleus of the male cell from the pollen grain has to reach the female nucleus of the egg cell in the ovule, and fuse with it. The following account explains how this happens.

Fertilization

The pollen grain absorbs liquid from the stigma and a microscopic **pollen tube** grows out of the grain. This tube grows down the style and into the ovary where it enters a small hole, the **micropyle**, in an ovule (Figure 8.12). The nucleus of the pollen grain travels down the pollen tube and enters the ovule. Here it combines with the nucleus of the egg cell. Each ovule in an ovary needs to be fertilized by a separate pollen grain.

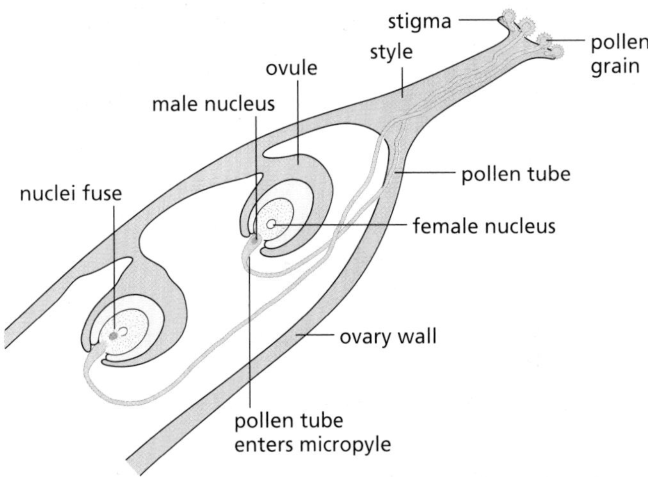

Figure 8.12 Diagram of fertilization showing pollen tube

Although pollination must occur before the ovule can be fertilized, pollination does not necessarily result in fertilization. A bee may visit many flowers on a Bramley apple tree, transferring pollen from one flower to the other. The Bramley, however, is 'self-sterile'; pollination with its own pollen will not result in fertilization. Pollination with pollen from a different variety of apple tree, for example a Worcester, can result in successful fertilization and fruit formation.

Fruit and seed formation

After the pollen and the egg nuclei have fused, the egg cell divides many times and produces a miniature plant called an **embryo**. This consists of a tiny root and shoot with two special leaves called **cotyledons**. In dicot plants (p. 278) food made in the leaves of the parent plant is carried in the phloem to the cotyledons.

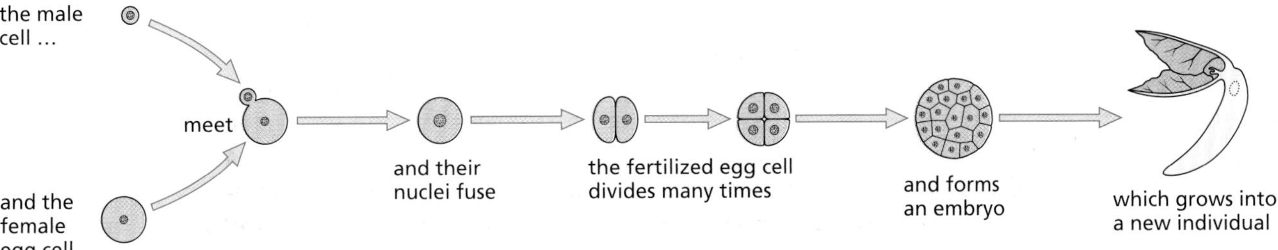

Figure 8.11 Fertilization. The male and female gametes fuse to form a zygote, which grows into a new individual

(a) tomato flowers
The petals of the older flowers are shrivelling

(b) after fertilization
The petals have dropped and the ovary is growing

(c) ripe fruit
The ovary has grown and ripened. The green sepals remain and the dried stigma is still attached

Figure 8.13 Tomato; fruit formation

The cotyledons eventually grow so large with this stored food that they completely enclose the embryo (see Figure 8.17, p. 74). In monocot plants (p. 278) the food store is laid down in a special tissue called endosperm which is outside the cotyledons. In both cases the outer wall of the ovule becomes thicker and harder, and forms the seed coat or **testa** (p. 74).

As the seeds grow, the ovary also becomes much larger and the petals and stamens shrivel and fall off (Figures 8.13b and 8.14). The ovary is now called a **fruit**. The biological definition of a fruit is a fertilized ovary; it is not necessarily edible. The lupin ovary forms a dry pod.

A plum is a good example of a fleshy, edible fruit. Tomatoes (Figure 8.13) and cucumbers are also fruits although they are classed as vegetables in the shops. Blackberries and raspberries are formed by many small fruits clustered together. In the apple and pear the edible part consists of the swollen receptacle surrounding and fused to the ovary wall.

Figure 8.14 Lupin flower after fertilization. The ovary (still with the style and stigma attached) has grown much larger than the flower and the petals have shrivelled

Questions

1 Which structures in a flower produce
  a the male gametes,
  b the female gametes?

2 In not more than two sentences, show that you understand the difference between pollination and fertilization.

3 In flowering plants
  a can pollination occur without fertilization,
  b can fertilization occur without pollination?

4 Which parts of a tomato flower
  a grow to form the fruit,
  b fall off after fertilization,
  c remain attached to the fruit?

5 Which of the following edible plant products – runner beans, peas, grapes, baked beans, marrow, rhubarb, tomatoes – do you think are, biologically,
  a fruits,
  b seeds,
  c neither?

Dispersal of fruits and seeds

When the seeds are mature, the whole fruit or the individual seeds fall from the parent plant to the ground and the seeds may then germinate (p. 75). In many plants, the fruits or seeds are adapted in such a way that they are carried a long distance from the parent plant (Figure 8.15a). This reduces competition for light and water between members of the same species. It may also result in plants growing in new places. The main adaptations are for dispersal by the wind and by animals but some plants have 'explosive' pods that scatter the seeds.

In the lupin, for example, the dry pod splits open suddenly, the two halves of the carpel curl back and the seeds are flicked out some distance from the parent plant (Figure 8.15b).

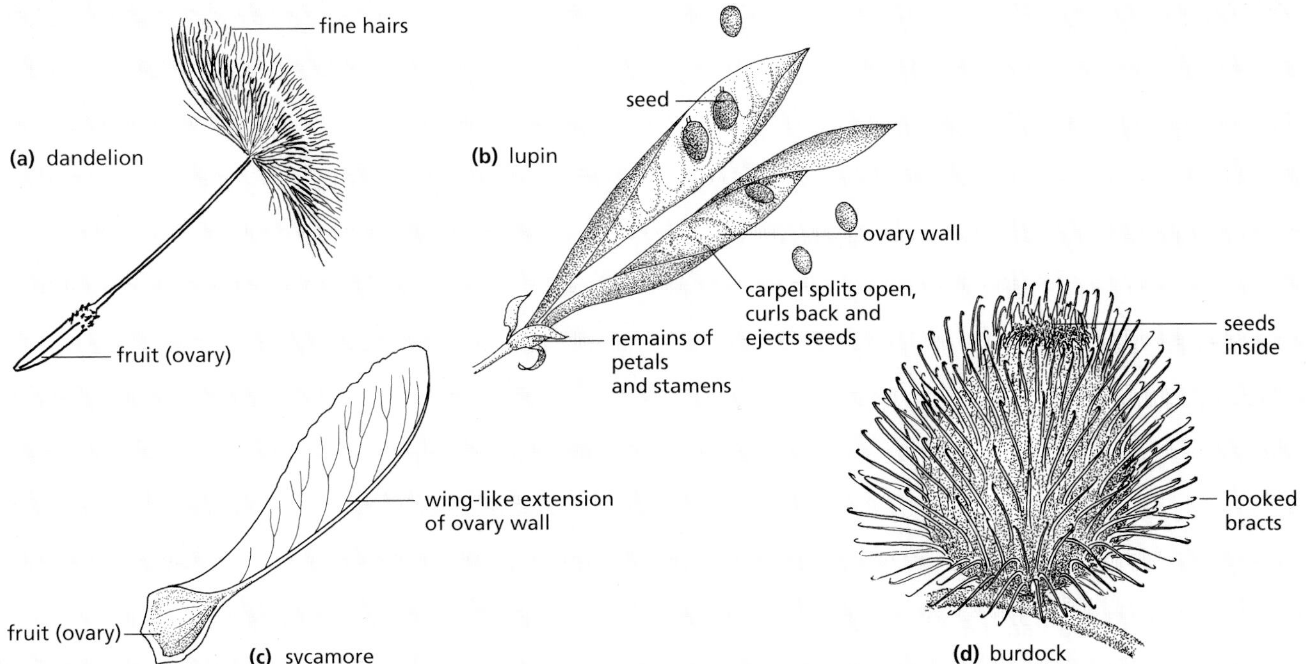

Figure 8.15 Fruit and seed dispersal

Wind dispersal

'Parachute' fruits and seeds

Clematis, thistles, willow herb and dandelion (Figure 8.15a) have seeds or fruits of this kind. Feathery hairs project from the fruit or seed and so increase its surface area. As a result, the seed 'floats' over long distances before sinking to the ground. It is, therefore, likely to be carried a long way from the parent plant by slight air currents.

'Winged' fruits

Fruits of the sycamore (Figure 8.15c) and ash trees have wing-like outgrowths from the ovary wall, or leaf-like structures on the flower stalk. These 'wings' cause the fruit to spin as it falls from the tree and so slow down its fall. This delay increases the chances of the fruit being carried away in air currents.

'Explosive' fruits

The pods of flowers in the pea family, e.g. gorse, broom, lupin and vetches, dry in the sun and shrivel. The tough fibres in the fruit wall shrink and set up a tension. When the fruit splits in half down two lines of weakness, the two halves curl back suddenly and flick out the seeds (Figure 8.15b).

Animal dispersal

Mammals and hooked fruits

The inflorescence of the burdock is surrounded by bracts which form hooks (Figure 8.15d). These hooks catch in the fur of passing mammals and the seeds fall out as the mammal moves about.

In other hooked fruits, e.g. agrimony, herb bennet and goose-grass, hooks develop from different parts of the flower.

Birds, mammals and succulent fruits

Fruits such as the blackberry and elderberry are eaten by birds and mammals (Figure 8.16). The hard pips containing the seed are not digested and so pass out with the droppings of the bird away from the parent plant. The soft texture and, in some cases, the bright colour of these fruits may be regarded as an adaptation to this method of dispersal.

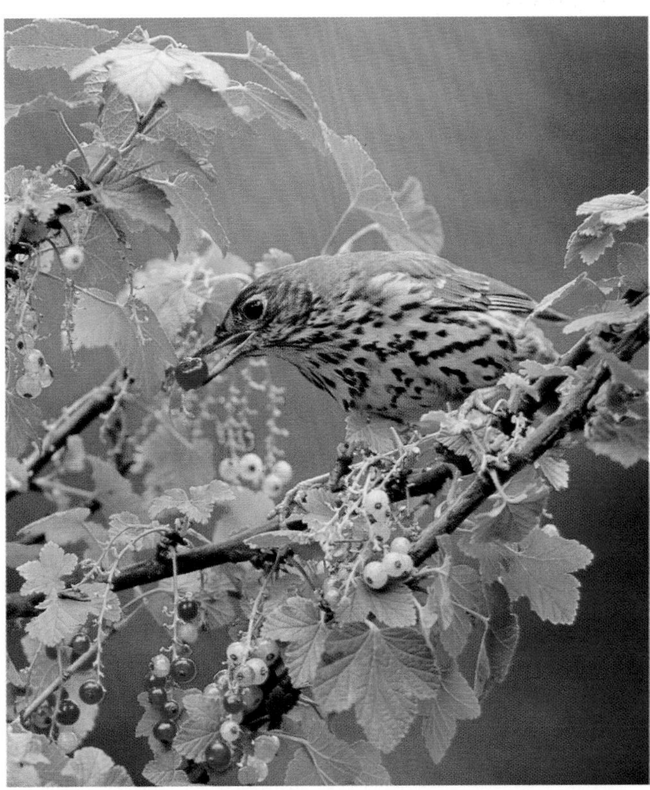

Figure 8.16 Animal dispersal. Thrush eating a red currant

Questions

1 What are the advantages of a plant dispersing its seeds over a wide area? What disadvantages might there be?

2 Which methods of dispersal are likely to result in seeds travelling the greatest distances? Explain your answer.

3 Students sometimes confuse 'wind pollination' and 'wind dispersal'. Write a sentence or two about each to make the difference clear.

Practical work

1 The growth of pollen tubes

1 Make a solution of 15 g sucrose and 0.1 g sodium borate in 100 cm^3 water. Put a drop of this solution on a cavity slide and scatter some pollen grains on the drop. This can be done by scraping an anther (which must already have opened to expose the pollen) with a mounted needle, or simply by touching the anther on the liquid drop.

Cover the drop with a cover-slip and examine the slide under the microscope at intervals of about 15 minutes. In some cases, pollen tubes may be seen growing from the grains.

Suitable plants include lily, narcissus, tulip, bluebell, lupin, wallflower, sweet pea or deadnettle, but a 15 per cent sucrose solution may not be equally suitable for all of them. It may be necessary to experiment with solutions ranging from 5 to 20 per cent.

2 Cut the stigma from a mature flower, e.g. honeysuckle, crocus, evening primrose or chickweed, and place it on a slide in a drop of 0.5 per cent methylene blue. Squash the stigma under a cover-slip (if the stigma is large, it may be safer to squash it between two slides), and leave it for 5 minutes.

Put a drop of water on one side of the slide, just touching the edge of the cover-slip, and draw it under the cover-slip by holding a piece of filter paper against the opposite edge (see Figure 4.18, p. 32). This will remove excess stain.

If the squash preparation is now examined under the microscope, pollen tubes may be seen growing between the spread-out cells of the stigma.

Seed structure and germination

The previous section described how a seed is formed from the ovule of a flower as a result of fertilization, and is then dispersed from the parent plant. If the seed lands in a suitable place it will germinate, i.e. grow into a mature plant.

Flowering plants can be divided into two major groups, the **monocotyledons** and the **dicotyledons**. A **cotyledon** is a modified leaf in a seed. The cotyledon plays a part in supplying food to the growing plant embryo. Monocotyledons (usually abbreviated to 'monocots') have only one of these seed leaves in the seed. Dicotyledons ('dicots') have two cotyledons in the seed and these two cotyledons store food.

Other characteristics and examples of monocots and dicots are discussed on p. 278.

To follow the process of germination, the structure of a dicot seed, in this case the French bean, will first be described.

Seed structure (French bean)

The seed (Figure 8.17) contains a miniature plant, the **embryo**, which consists of a root or **radicle**, and a shoot or **plumule**. The embryo is attached to two leaves called the **cotyledons**, which are swollen with stored food. This stored food, mainly starch, is used by the embryo when it starts to grow. The embryo and cotyledons are enclosed in a tough seed coat or **testa**. The **micropyle**, through which the pollen tube entered (p. 71), remains as a small hole in the testa and is an important route for the entry of water in some seeds. The **hilum** is the scar left where the seed was attached to the pod.

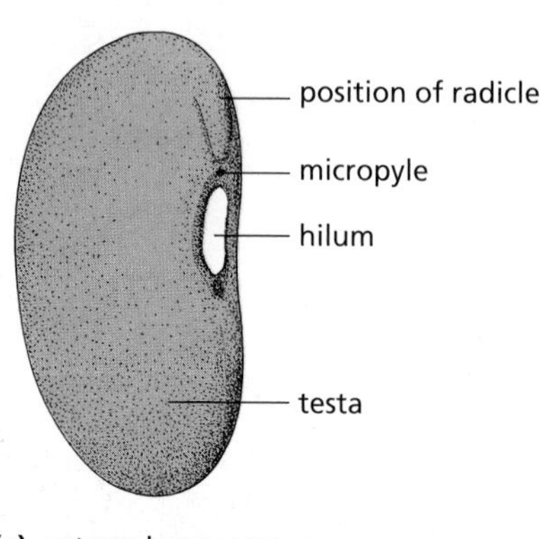

(a) external appearance

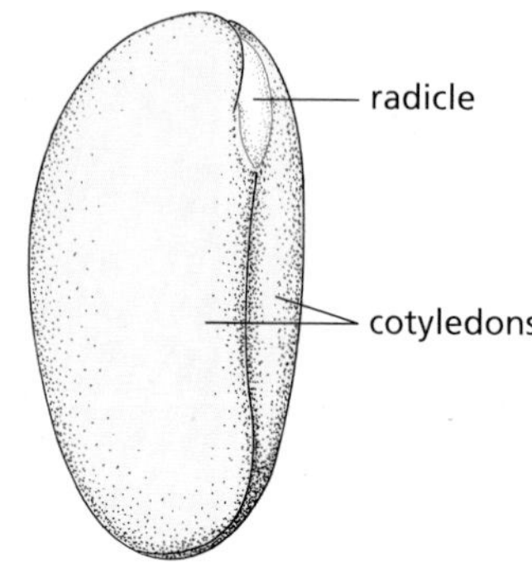

(b) testa removed

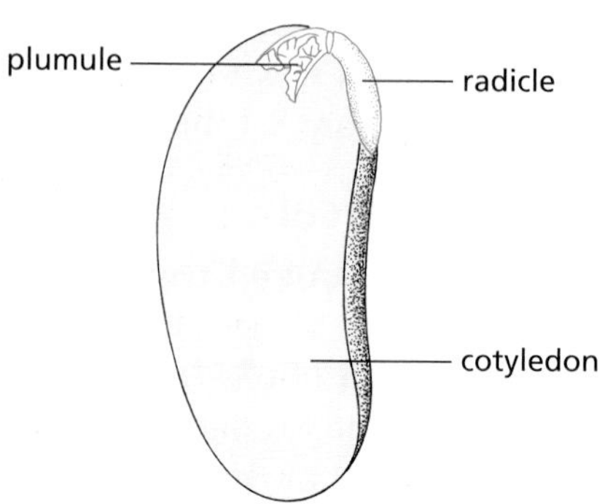

(c) one cotyledon removed

Figure 8.17 The French bean seed

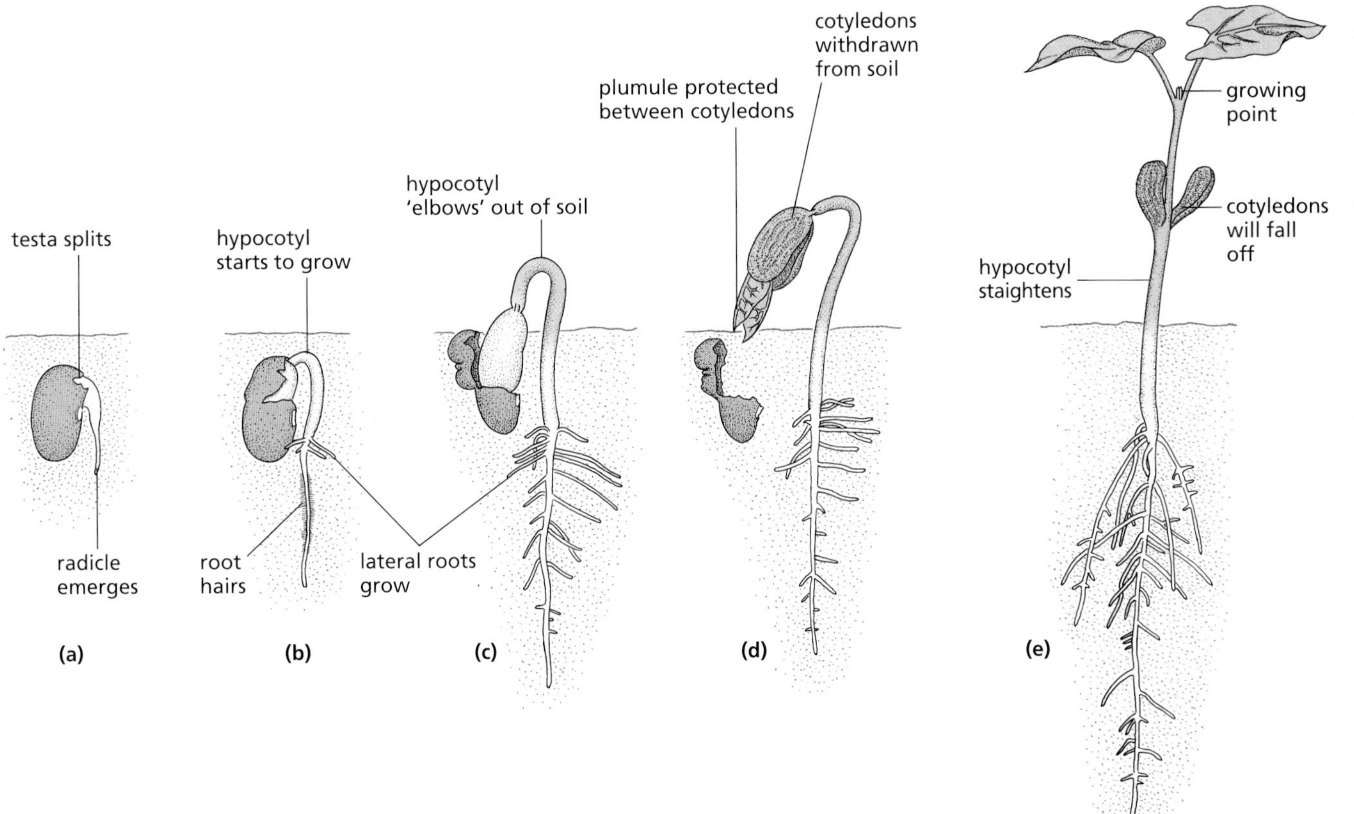

Figure 8.18 Germination of French bean

Germination

The stages of germination of a French bean are shown in Figure 8.18.

A seed just shed from its parent plant contains only 5–20 per cent water, compared with 80–90 per cent in mature plant tissues. Once in the soil, some seeds will absorb water and swell up, but will not necessarily start to germinate until other conditions are suitable (see pp. 76–8).

The radicle grows first and bursts through the testa (Figure 8.18a). The radicle continues to grow down into the soil, pushing its way between soil particles and small stones. Its tip is protected by the root cap (see p. 56). Branches, called **lateral roots**, grow out from the side of the main root and help to anchor it firmly in the soil. On the main root and the lateral roots, microscopic **root hairs** grow out. These are fine outgrowths from some of the outer cells. They make close contact with the soil particles and absorb water from the spaces between them (p. 63).

In the French bean a region of the embryo's stem, the **hypocotyl**, just above the radicle (Figure 8.18b), now starts to elongate. The radicle is by now firmly anchored in the soil, so the rapidly growing hypocotyl arches upwards through the soil, pulling the cotyledons with it (Figure 8.18c). Sometimes the cotyledons are pulled out of the testa, leaving it below the soil, and sometimes the cotyledons remain enclosed in the testa for a time. In either case, the plumule is well protected from damage while it is being pulled through the soil, because it is enclosed between the cotyledons (Figure 8.18d).

Once the cotyledons are above the soil, the hypocotyl straightens up and the leaves of the plumule open out (Figures 8.18e and 8. 19). Up to this point, all the food needed for making new cells and producing energy has come from the cotyledons.

The main type of food stored in the cotyledons is starch. Before this can be used by the growing shoot and root, the starch has to be turned into soluble sugar. In this form, it can be transported by the phloem cells. The change from starch to sugar in the cotyledons is brought about by enzymes which become active as soon as the seed starts to germinate. The cotyledons shrivel as their food reserve is used up, and they fall off altogether soon after they have been brought above the soil.

Figure 8.19 Germinating seeds. The brown testa is shed, the cotyledons turn green and the plumule expands

By now the plumule leaves have grown much larger, turned green and started to absorb sunlight and make their own food by photosynthesis (p. 35). Between the plumule leaves is a growing point which continues the upward growth of the stem and the production of new leaves. The embryo has now become an independent plant, absorbing water and mineral salts from the soil, carbon dioxide from the air and making food in its leaves.

Questions

1 What are the functions of
 a the radicle,
 b the plumule and
 c the cotyledons of a seed?

2 During germination of the broad bean, how are
 a the plumule,
 b the radicle protected from damage as they are forced through the soil?

3 List all the possible purposes for which a growing seedling might use the food stored in its cotyledons.

4 At what stage of development is a seedling able to stop depending on the cotyledons for its food?

5 What do you think are the advantages to a germinating seed of having its radicle growing some time before the shoot starts to grow?

Practical work

Experiments on the conditions for germination

The environmental conditions which might be expected to affect germination are temperature, light intensity and the availability of water and air. The relative importance of some of these conditions can be tested by the experiments which follow. The effects of these conditions are discussed on p. 77.

2 The need for water

Label three containers A, B and C and put dry cotton wool in the bottom of each. Place equal numbers of soaked seeds in all three. Leave A quite dry; add water to B to make the cotton wool moist; add water to C until all the seeds are completely covered (Figure 8.20). Put lids on the containers and leave them all at room temperature for a week.

Result The seeds in B will germinate normally. Those in A will not germinate. The seeds in C may have started to germinate but will probably not be as advanced as those in B and may have died and started to decay.

Interpretation Although water is necessary for germination, too much of it may prevent germination by cutting down the oxygen supply to the seed.

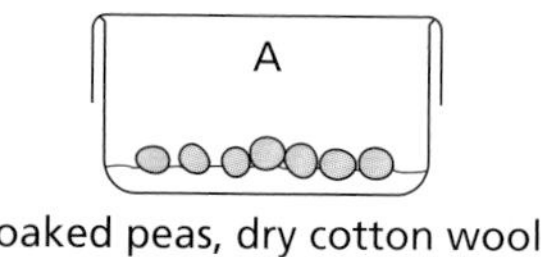

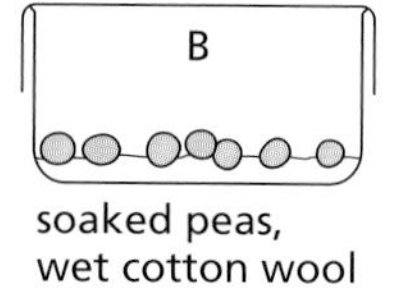

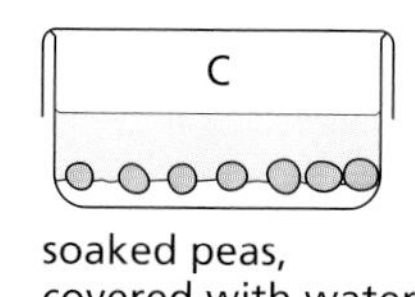

Figure 8.20 The need for water in germination

3 The need for oxygen

Set up the experiment as shown in Figure 8.21. (**CARE:** Pyrogallic acid and sodium hydroxide is a caustic mixture. Use eye shields, handle the liquids with care and report any spillage at once.)

If the moist cotton wool is rolled in some cress seeds, they will stick to it. The bungs must make an airtight seal in the flask and the cotton wool must not touch the solution. Pyrogallic acid and sodium hydroxide absorb oxygen from the air, so the cress seeds in flask A are deprived of oxygen. Flask B is the control (see p. 23). This is to show that germination can take place in these experimental conditions provided oxygen is present. Leave the flasks for a week at room temperature.

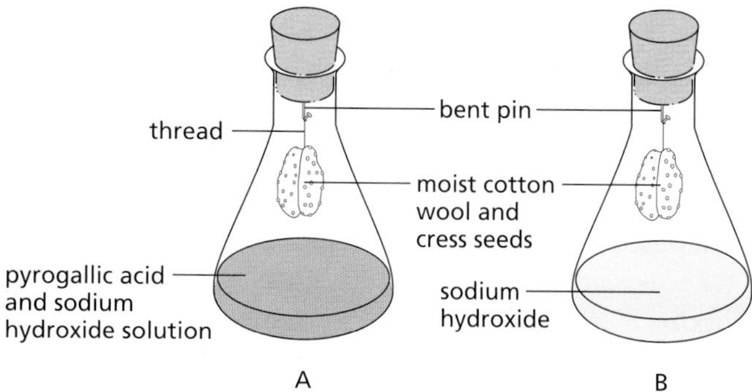

Figure 8.21 The need for oxygen

Result The seeds in flask B will germinate but there will be little or no germination in flask A.

Interpretation The main difference between flasks A and B is that A lacks oxygen. Since the seeds in this flask have not germinated, it looks as if oxygen is needed for germination.

To show that the chemicals in flask A had not killed the seeds, the cotton wool can be swapped from A to B. The seeds from A will now germinate.

Note Sodium hydroxide absorbs carbon dioxide from the air. The mixture (sodium hydroxide + pyrogallic acid) in flask A, therefore, absorbs both carbon dioxide and oxygen from the air in this flask. In the control flask B, the sodium hydroxide absorbs carbon dioxide but not oxygen. If the seeds in B germinate, it shows that lack of carbon dioxide did not affect them, whereas lack of oxygen did.

4 Temperature and germination

Soak some maize grains for a day and then roll them up in three strips of moist blotting paper as shown in Figure 8.22. Put the rolls into plastic bags. Place one in a refrigerator (about 4 °C), leave one upright in the room (about 20 °C) and put the third in a warm place such as over a radiator or, better, in an incubator set to 30 °C.

Because the seeds in the refrigerator will be in darkness, the other seeds must also be enclosed in a box or a cupboard, to exclude light. Otherwise it could be objected that it was lack of light rather than low temperature which affected germination.

After a week, examine the seedlings and measure the length of the roots and shoots.

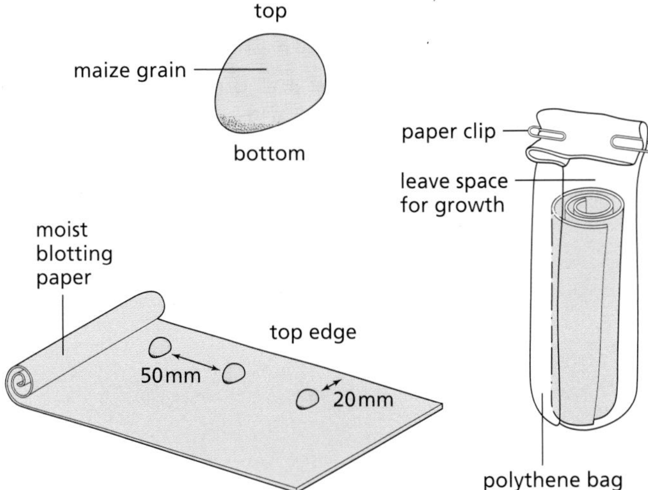

Figure 8.22 Temperature and germination. Roll the seeds in moist blotting-paper, and stand the rolls upright in plastic bags

Result The seedlings kept at 30 °C will be more advanced than those at room temperature. The grains in the refrigerator may not have started to germinate at all.

Interpretation Seeds will not germinate below a certain temperature. The higher the temperature, the faster the germination, at least up to 35–40 °C.

Questions

1 List the external conditions necessary for germination.

2 Do any of the results of Experiments 1–3 suggest whether or not light is necessary for germination? Explain.

3 a In Experiment 3 it could be argued that the low temperature of the refrigerator had killed the seeds and that this explains why they did not germinate. How could you check on this possibility?

b How could you modify Experiment 3, at least in theory, to find the minimum temperature at which maize seeds would germinate?

Controlling the variables

These experiments on germination illustrate one of the problems of designing biological experiments. You have to decide what conditions (the '**variables**') could influence the results and then try to change only one condition at a time. The dangers are that (1) some of the variables might not be controllable, (2) controlling some of the variables might also affect the condition you want to investigate, and (3) there might be a number of important variables you have not thought of.

1 In your germination experiments, you were unable to control the quality of the seeds, but had to assume that the differences between them would be small. If some of the seeds were dead or diseased, they would not germinate in any conditions and this could distort the results. This is one reason for using as large a sample as possible in the experiments.

2 You had to ensure that, when temperature was the variable, the exclusion of light from the seeds in the refrigerator was not an additional variable. This was done by putting all the seeds in darkness.

3 A variable you might not have considered could be the way the seeds were handled. Some seeds can be induced to germinate more successfully by scratching or chipping the testa.

The importance of water, oxygen and temperature

Use of water in the seedling

Most seeds, when first dispersed, contain very little water. In this dehydrated state, their metabolism is very slow and their food reserves are not used up. The dry seeds can also resist extremes of temperature and desiccation. Before the metabolic changes needed for germination can take place, seeds must absorb water.

Water is absorbed firstly through the micropyle, in some species, and then through the testa as a whole. Once the radicle has emerged, it will absorb water from the soil, particularly through the root hairs. The water which reaches the embryo and cotyledons is used to:

- activate the enzymes in the seed;
- help the conversion of stored starch to sugar, and proteins to amino acids (p. 11);
- transport the sugar in solution from the cotyledons to the growing regions;
- expand the vacuoles of new cells causing the root and shoot to grow and the leaves to expand;
- maintain the turgor (p. 30) of the cells and thus keep the shoot upright and the leaves expanded;
- provide the water needed for photosynthesis once the plumule and young leaves are above ground;
- transport salts from the soil to the shoot.

Uses of oxygen

In some seeds the testa is not very permeable to oxygen, and the early stages of germination are probably anaerobic (p. 20). The testa when soaked or split open allows oxygen to enter. The oxygen is used in aerobic respiration, which provides the energy for the many chemical changes involved in mobilizing the food reserves and making the new cytoplasm and cell walls of the growing seedling.

Importance of temperature

On p. 15 it was explained that a rise in temperature speeds up most chemical reactions, including those taking place in living organisms. Germination, therefore, occurs more rapidly at high temperatures, up to about 40 °C. Above 45 °C, the enzymes in the cells are denatured and the seedlings would be killed. Below certain temperatures (e.g. 0–4 °C) germination may not start at all in some seeds. However, there is considerable variation in the range of temperatures at which seeds of different species will germinate.

Germination and light

Since a great many cultivated plants are grown from seeds which are planted just below soil level, it seems obvious that light is not necessary for germination. There are some species, however, in which the seeds need some exposure to light before they will germinate, e.g. foxgloves and some varieties of lettuce. In all seedlings, once the shoot is above ground, light is necessary for photosynthesis.

Dormancy

When plants shed their seeds in summer and autumn, there is usually no shortage of water, oxygen and warmth. Yet, in a great many species, the seeds do not germinate until the following spring. These seeds are said to be **dormant**, i.e. there is some internal control mechanism which prevents immediate germination even though the external conditions are suitable.

If the seeds did germinate in the autumn, the seedlings might be killed by exposure to frost, snow and freezing conditions. Dormancy delays the period of germination so that adverse conditions are avoided.

The controlling mechanisms are very varied and are still the subject of investigation and discussion. The factors known to influence dormancy are plant growth substances (p. 319–20), the testa, low temperature and light, or a combination of these.

Questions

1 Describe the natural conditions in the soil that would be most favourable for germination. How could a gardener try to create these conditions?

Checklist

- Flowers contain the reproductive organs of plants.
- The stamens are the male organs. They produce pollen grains which contain the male gamete.
- The carpels are the female organs. They produce ovules which contain the female gamete and will form the seeds.
- The flowers of most plant species contain male and female organs. A few species have unisexual flowers.
- Brightly coloured petals attract insects, which pollinate the flower.
- Pollination is the transfer of pollen from the anthers of one flower to the stigma of the same or another plant.
- Pollination may be done by insects or by the wind.
- Flowers which are pollinated by insects are usually brightly coloured and have nectar.
- Flowers which are pollinated by the wind are usually small and green. Their stigmas and anthers hang outside the flower where they are exposed to air movements.
- Fertilization occurs when a pollen tube grows from a pollen grain into the ovary and up to an ovule. The pollen nucleus passes down the tube and fuses with the ovule nucleus.
- After fertilization, the ovary grows rapidly to become a fruit and the ovules become seeds.
- Seeds and fruits may be dispersed by the wind, by animals, or by an 'explosive' method.
- Dispersal scatters the seeds so that the plants growing from them are less likely to compete with each other and with their parent plant.
- A dicot seed consists of an embryo with two cotyledons enclosed in a seed coat (testa).
- The embryo consists of a small root (radicle) and shoot (plumule).
- The cotyledons contain the food store that the embryo will use when it starts to grow.
- The food stored in the cotyledons has to be made soluble by enzymes and transported to the growing regions.
- When germination takes place, (a) the radicle of the embryo grows out of the testa and down into the soil and (b) the plumule is pulled backwards out of the soil.
- In many seeds, the cotyledons are also brought above the soil. They photosynthesize for a while before falling off.
- Germination is influenced by temperature and the amount of water and oxygen available.
- Some seeds have a dormant period after being shed. During this period they will not germinate.

9 Asexual reproduction and cloning in plants

Vegetative propagation
Stolons; rhizomes; strawberry runners; bulbs and corms.
Food storage. Annuals, biennials and perennials.

Artificial propagation
Cuttings; tissue culture; clones.

Sexual reproduction and vegetative propagation compared

Vegetative propagation

In addition to reproducing by seeds, many plants are able to produce new individuals asexually, i.e. without gametes, pollination and fertilization. This form of **asexual reproduction** is also called **vegetative propagation**. When vegetative propagation takes place naturally, it usually results from the growth of a bud on a stem which is close to, or under, the soil. Instead of just making a branch, the bud produces a complete plant with roots, stem and leaves. When the old stem dies, the new plant is independent of the parent which produced it.

Stolons and rhizomes

The flowering shoots of plants such as the strawberry and the creeping buttercup are very short and, for the most part, below ground. The stems of shoots such as these are called **rootstocks**. The rootstocks bear leaves and flowers. After the main shoot has flowered, the lateral buds produce long shoots which grow horizontally over the ground (Figure 9.1). These shoots are called **stolons** (or 'runners'), and have only small scale-leaves at their nodes and very long internodes. At each node there is a bud which can produce not only a shoot, but roots as well. Thus a complete plant may develop and take root at the node, nourished for a time by food sent from the parent plant through the stolon. Eventually, the stolon dries up and withers, leaving an independent daughter plant growing a short distance away from the parent. In this way a strawberry plant can produce many daughter plants by vegetative propagation in addition to producing seeds.

In many plants, horizontal shoots arise from lateral buds near the stem base, and grow under the ground. Such underground horizontal stems are called **rhizomes**. At the nodes of the rhizome are buds which may develop to produce shoots above the ground. The shoots become independent plants when the connecting rhizome dies.

Many grasses propagate by rhizomes; the couch grass (Figure 9.2) is a good example. Even a small piece of rhizome, provided it has a bud, can produce a new plant.

In the bracken, the entire stem is horizontal and below ground. The bracken fronds you see in summer are produced from lateral buds on a rhizome many centimetres below the soil.

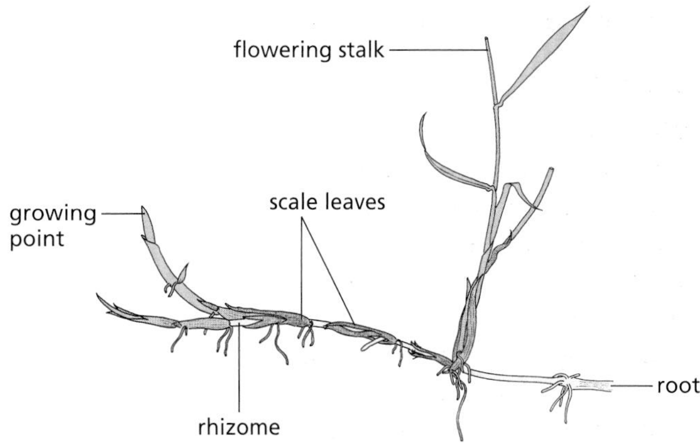

Figure 9.2 Couch grass rhizome

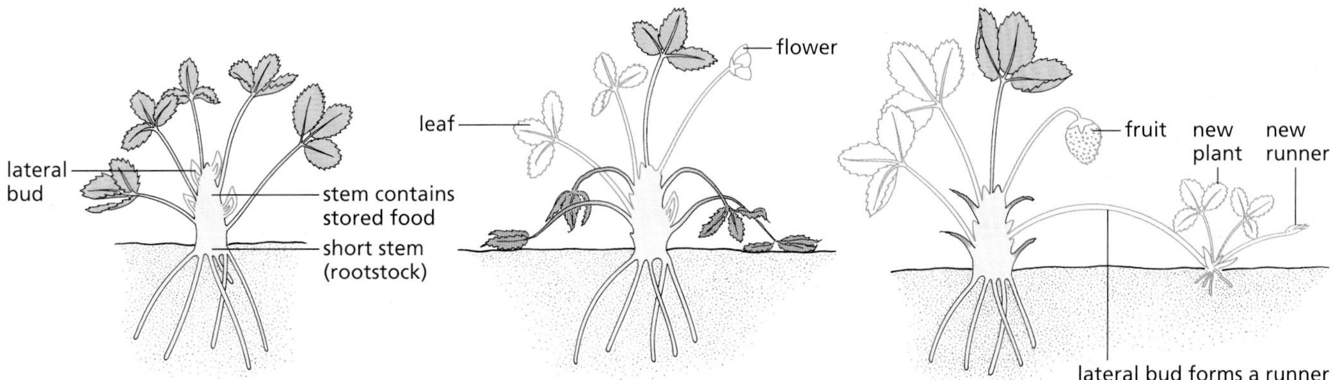

Figure 9.1 Strawberry runner developing from rootstock

Bulbs and corms

Bulbs such as those of the daffodil and snowdrop are very short shoots. The stem is only a few millimetres long and the leaves, which encircle the stem, are thick and fleshy with stored food.

In spring, the stored food is used by a rapidly growing terminal bud which produces a flowering stalk and a small number of leaves. During the growing season, food made in the leaves is sent to the leaf bases and stored. The leaf bases swell and form a new bulb ready for growth in the following year.

Vegetative reproduction occurs when some of the food is sent to a lateral bud as well as to the leaf bases. The lateral bud grows inside the parent bulb and, next year, will produce an independent plant (Figure 9.3).

The **corms** of crocuses and anemones have life cycles similar to those of bulbs but it is the stem, rather than the leaf bases, which swells with stored food. Vegetative reproduction takes place when a lateral bud on the short, fat stem grows into an independent plant.

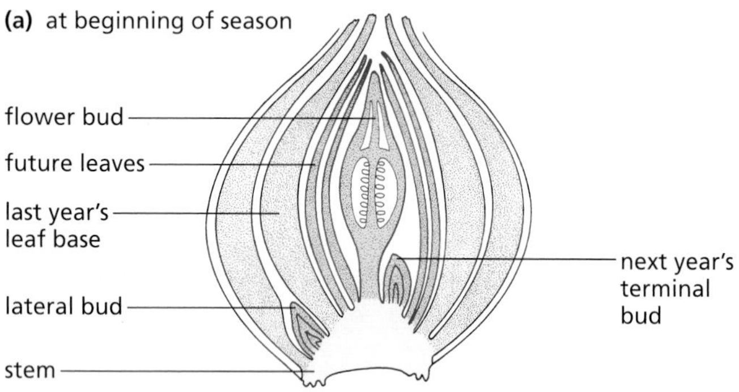

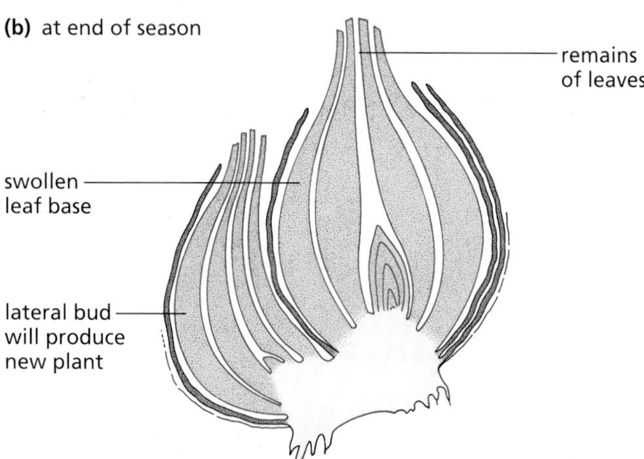

Figure 9.3 Daffodil bulb; vegetative reproduction

Questions

1. Plants can often be propagated from stems but rarely from roots. What features of shoots account for this difference?
2. The plants which survive a heath fire are often those which have a rhizome (e.g. bracken). Suggest a reason why this is so.

Food storage

In many cases the organs associated with asexual reproduction also serve as food stores.

Food in the storage organs enables very rapid growth in the spring. A great many of the spring and early summer plants have bulbs, corms, rhizomes or tubers: e.g. daffodil, snowdrop and bluebell, crocus and cuckoo pint, iris and lily-of-the-valley and lesser celandine.

Potatoes are **stem tubers**. Lateral buds at the base of the potato shoot produce underground shoots (rhizomes). These rhizomes swell up with stored starch and form tubers (Figure 9.4a). Because the tubers are stems, they carry buds. If the tubers are left in the ground or transplanted, the buds will produce shoots, using food stored in the tuber (Figure 9.4b). In this way, the potato plant can propagate vegetatively.

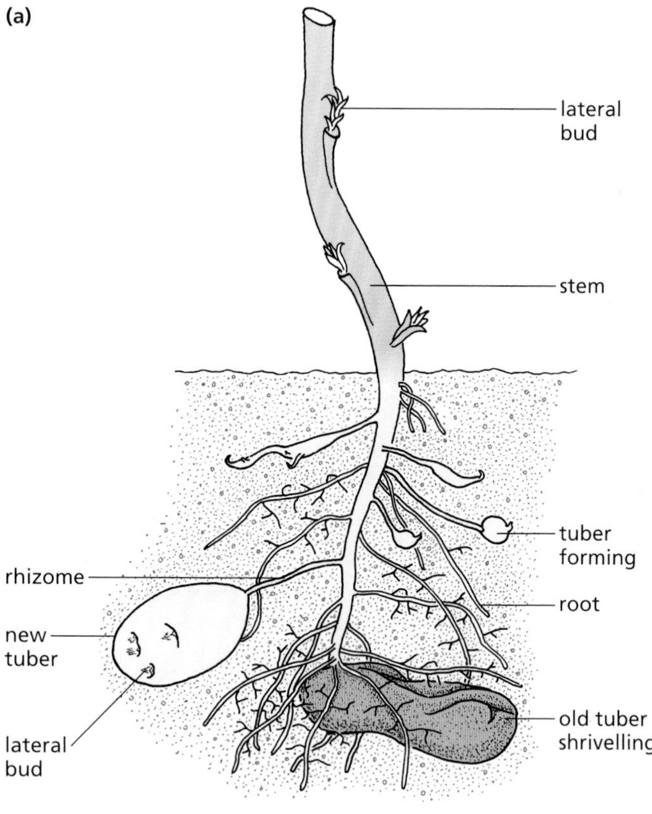

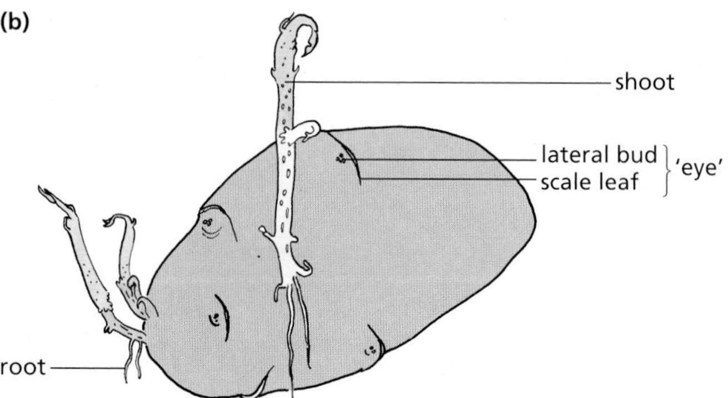

Figure 9.4 Stem tubers growing on potato plant and potato tuber sprouting

Carrots, parsnips and beetroot are examples of roots which are modified as storage organs. In the first year of growth, the leaves send food down to the tap root, which swells up to form a tuber. In the second year, the food store is used to promote rapid growth and the production of flowers and seeds. The parent plant then dies back.

This is an example of a **biennial** plant. Plants which continue to grow and reproduce asexually each year are called **perennials**. Plants which flower, produce seeds and then die are called **annuals**.

The food reserve in plant storage organs is often starch but in some cases it is glucose (as in the onion) or sucrose (as in sugar-beet).

In agriculture, humans have exploited many of these types of plant and bred them for bigger and more nutritious storage organs for their own consumption, e.g. onion, leek, potato, beetroot.

Artificial propagation

Agriculture and horticulture exploit vegetative reproduction in order to produce fresh stocks of plants. This can be done naturally, e.g. by planting potatoes, dividing up rootstocks or pegging down stolons at their nodes to make them take root. There are also methods which would not occur naturally in the plant's life cycle. Two methods of **artificial propagation** are by taking cuttings and by tissue culture.

Cuttings

It is possible to produce new individuals from certain plants by putting the cut end of a shoot into water or moist earth. Roots (Figure 9.5) grow from the base of the stem into the soil while the shoot continues to grow and produce leaves.

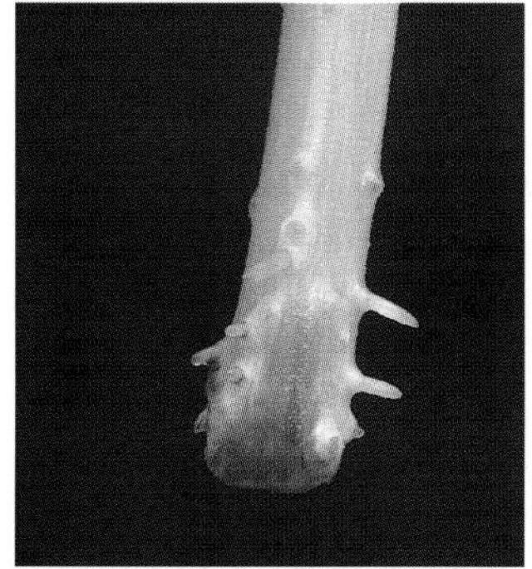

(a) roots developing from Busy Lizzie stem

(b) roots growing from coleus cutting

Figure 9.5 Rooted cuttings

In practice, the cut end of the stem may be treated with a rooting 'hormone' (see p. 310) to promote root growth, and evaporation from the shoot is reduced by covering it with polythene or a glass jar. Carnations, geraniums and chrysanthemums are commonly propagated from cuttings.

Tissue culture

Once a cell has become part of a tissue it usually loses the ability to reproduce. However, the nucleus of any cell in a plant still holds all the 'instructions' (genes, p. 183) for making a complete plant and in certain circumstances they can be brought back into action.

In laboratory conditions, single plant cells can be induced to divide and grow into complete plants. One technique is to take small pieces of plant tissue from a root or stem and treat it with enzymes to separate it into individual cells. The cells are then provided with particular plant 'hormones' which induce cell division and, eventually, the formation of roots, stems and leaves.

An alternative method is to start with a small piece of tissue and place it on a nutrient jelly (agar, p. 289). Cells in the tissue start to divide and produce many cells forming a shapeless mass called a **callus**. If the callus is then provided with the appropriate hormones it develops into a complete plant (Figure 9.6).

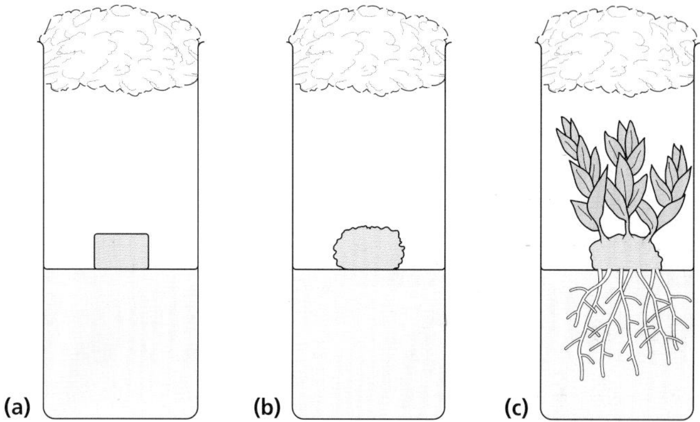

Figure 9.6 Propagation by tissue culture using nutrient jelly

Using the technique of tissue culture, large numbers of plants can be produced from small amounts of tissue (Figure 9.7) and they have the advantage of being free from fungal or bacterial infections. The plants produced in this way form **clones**, because they have been produced from a single parent plant.

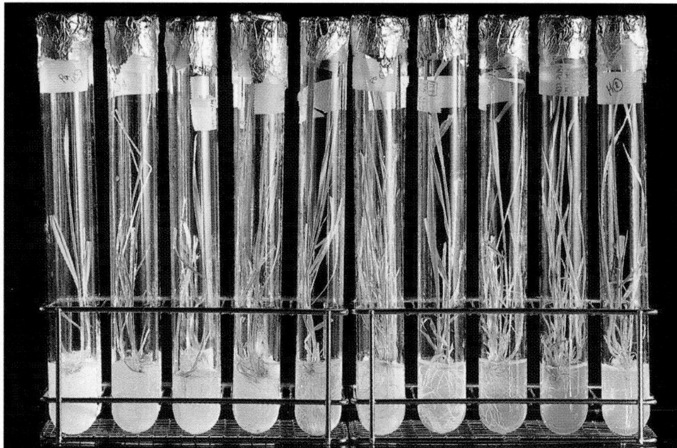

Figure 9.7 Tissue culture. Plants grown from small amounts of unspecialized tissue on an agar culture medium

Sexual reproduction and vegetative propagation compared

Variation

In sexual reproduction, the offspring are the result of the fusion of male and female gametes. These gametes may come from the same plant or from different plants of the same species. In either case, the production and subsequent fusion of gametes produce a good deal of variation among the offspring (p. 200). This may result from new combinations of characteristics, e.g. petal colour of one parent combined with fruit size of the other. It may also be the result of spontaneous changes in the gametes when they are produced (mutations, p. 189).

In vegetative propagation no gametes are involved and all the new plants are produced by cell division (mitosis, p. 182) from only one parent. Consequently they are genetically identical; there is no variation. A population of genetically identical individuals produced from a single parent is called a **clone** (p. 81).

A clone has the advantage that the good characteristics of the parent are passed on to all the offspring. This is a very useful feature in horticulture where the grower wants to preserve the quality of the produce. In natural conditions it might be a disadvantage, in the long term, if the climate or other conditions change. If a clone of vegetatively produced plants has no resistance to a particular disease, the whole population could be wiped out. In a population of plants which have been produced sexually, there is a chance that at least some of the offspring will have resistance to the disease. These plants will survive and produce further offspring with disease-resistance.

In fact, plants which reproduce asexually by vegetative propagation also reproduce sexually by flowers and seeds; so they enjoy the benefits of both methods.

Dispersal

The seeds produced as a result of sexual reproduction will be scattered over a relatively wide range. Some will land in inhospitable environments, perhaps lacking light or water. These seeds will fail to germinate. Nevertheless, most methods of seed dispersal result in some of the seeds establishing populations in new habitats.

A plant which reproduces vegetatively will already be growing in a favourable situation, so all the offspring will find themselves in a suitable environment. However, there is no vegetative dispersal mechanism and the plants will grow in dense colonies, competing with each other for water and minerals. The dense colonies, on the other hand, leave little room for competitors of other species.

As mentioned before, most plants which reproduce vegetatively also produce flowers and seeds. In this way they are able to colonize more distant habitats.

Food storage

The store of food in tubers, tap roots, bulbs, etc. enables the plants to grow rapidly as soon as conditions become favourable. Early growth enables the plant to flower and produce seeds before competition with other plants (for water, mineral salts and light) reaches its maximum. This must be particularly important in woods where, in summer, the leaf canopy prevents much light from reaching the ground and the tree roots tend to drain the soil of moisture over a wide area.

The seeds produced by sexual reproduction all contain some stored food but it is quickly used up during germination which produces only a miniature plant. It takes a long time for a seedling to become established and eventually produce seeds of its own.

Checklist

- Asexual reproduction occurs without gametes or fertilization.
- Plants reproduce asexually when some of their buds grow into new plants.
- Asexual reproduction in plants is called vegetative propagation.
- The stolon of the strawberry plant is a horizontal stem which grows above the ground, takes root at the nodes and produces new plants.
- The couch grass rhizome is a horizontal stem which grows below the ground and sends up shoots from its nodes.
- Bulbs are condensed shoots with circular fleshy leaves. Bulb-forming plants reproduce asexually from lateral buds.
- Rhizomes, corms, bulbs and tap roots may store food which is used to accelerate early growth.
- Vegetative propagation produces (genetically) identical individuals.
- A clone is a population of organisms produced asexually from a single parent.
- Artificial propagation from cuttings or grafts preserves the desirable characteristics of a crop plant.
- Whole plants can be produced from single cells or small pieces of tissue.

Flowering plants
Examination questions

Do not write on this page. Where necessary copy drawings, tables or sentences.

1 The diagram shows a plant.

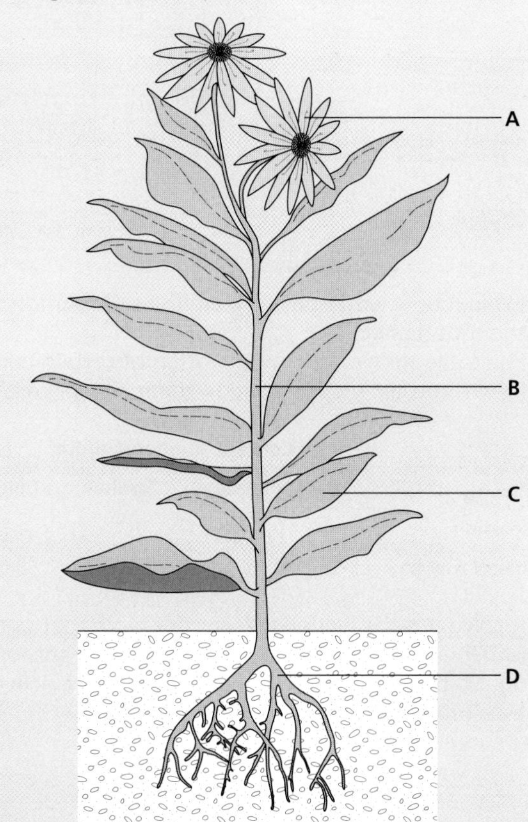

a Give **one** function of organs A, B and C. (3)
b Name the organs C and D. (2)
(CCEA)

2 The figure shows a section through a bean flower.

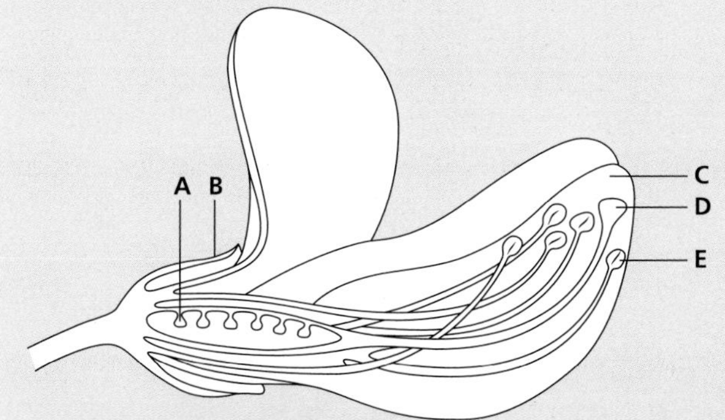

a Name the parts labelled A and B. (2)
b This flower is insect pollinated. Suggest how parts C, D and E help in the pollination of the flower. (3)
c After pollination the ovules develop into seeds. Describe the events which occur after pollination and which result in the formation of seeds. (4)
(IGCSE)

3 The diagram shows an experiment to investigate the uptake of water by roots.

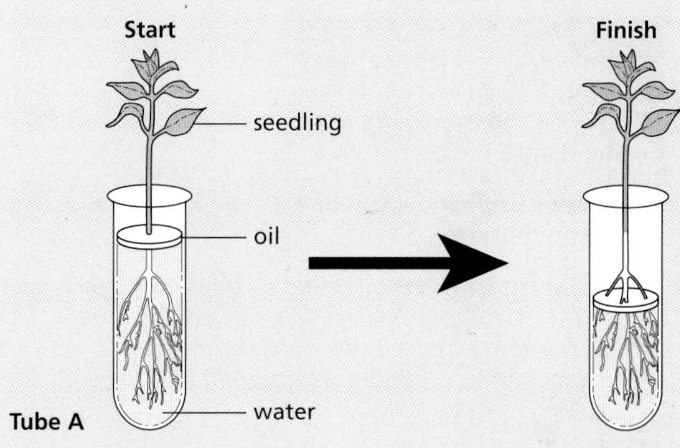

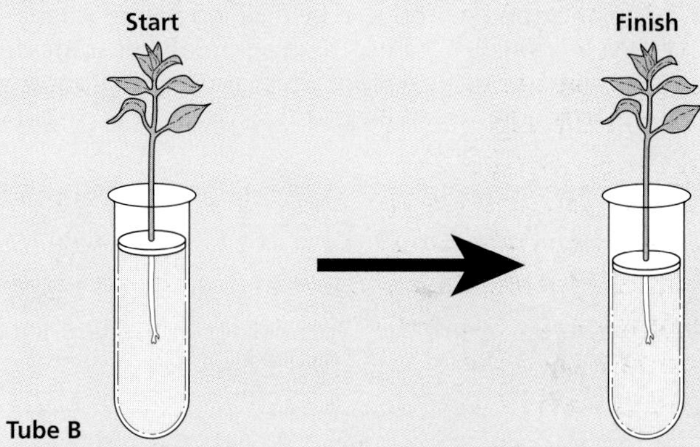

a In which tube does the water level drop most? Give **one** reason for this difference. (2)
b Describe **one** way in which roots are adapted to absorb water. (1)
c Name the tissue which transports water in plant seedlings. (1)
(CCEA)

4 A potted plant was left in a hot, brightly lit room for ten hours. The plant was not watered during this period. The drawings show how the mean width of stomata changed over the ten hour period.

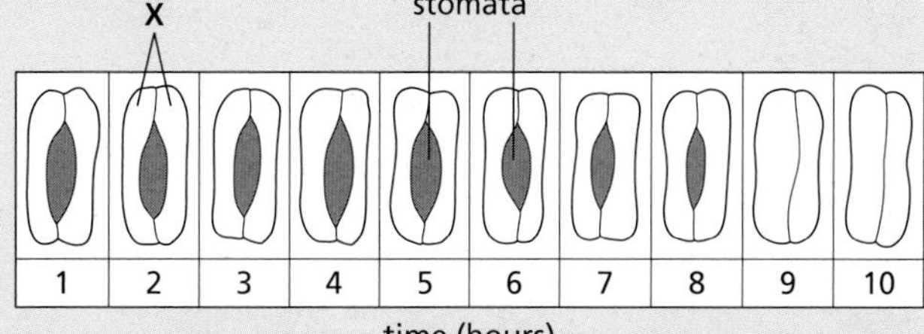

a Why do plants need stomata? (1)
b Name the cells labelled X on the drawing. (1)
c The width of the stomata changed over the ten hour period. Explain the advantage to the plant of this change. (2)
(AQA)

5 The diagram below shows a root hair cell.

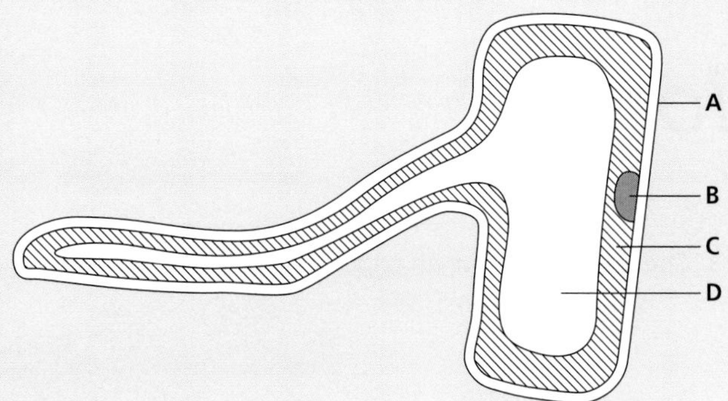
A
B
C
D

a (i) Use the following words to label A, B, C and D on the diagram.

vacuole cell wall cytoplasm nucleus cell membrane (4)

(ii) Which two parts shown on the diagram are found only in plant cells? (2)

(iii) Name the paired structures found in B. (1)

b Suggest why root hairs are so good at taking in water. (1)

(WJEC)

6 Lily plants can be produced by dividing parent plants into a few smaller plants. The diagram shows stages in the process of tissue culture which is used to produce very large numbers of lily plants.

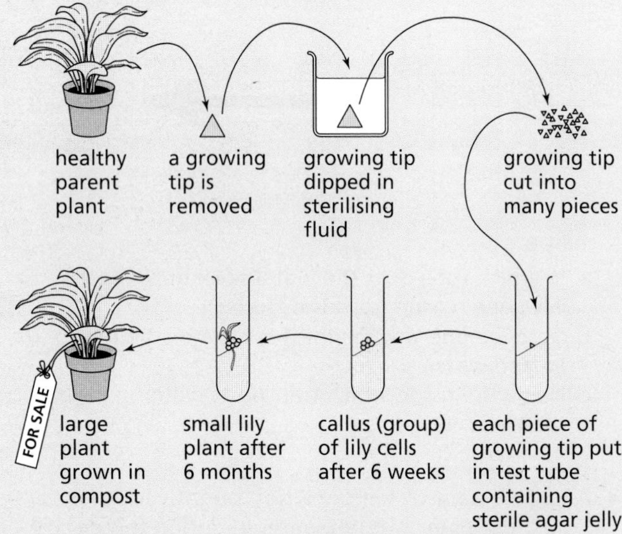
healthy parent plant
a growing tip is removed
growing tip dipped in sterilising fluid
growing tip cut into many pieces
large plant grown in compost
FOR SALE
small lily plant after 6 months
callus (group) of lily cells after 6 weeks
each piece of growing tip put in test tube containing sterile agar jelly

Use the diagram to help you explain the stages in the process of producing large numbers of lily plants. (7)

(Edexcel)

7 a Give a definition of osmosis. (4)

b What is the difference between osmosis and diffusion? (1)

c The diagram shows a section through a root hair cell.

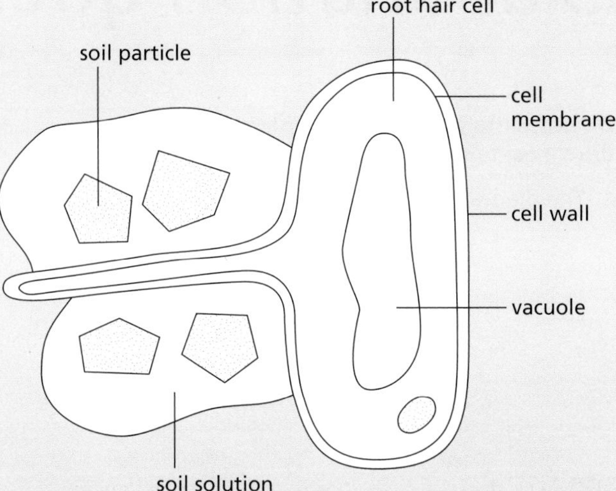
root hair cell
soil particle
cell membrane
cell wall
vacuole
soil solution

Explain how water passes from the soil solution to the root hair cell. (2)

d The table shows the concentration of certain mineral ions in the cell solution and in the root hair cells.

	Concentration mmol/dm^3		
	Potassium	**Sodium**	**Chloride**
Soil solution	0.1	1.1	1.3
Vacuole of root hair cell	93.0	51.0	58.0

(i) What is the evidence from this table that uptake of mineral ions from the soil is by active transport? (1)

(ii) Name a part of the human body where active transport takes place. (1)

(WJEC)

Human physiology

10 Food and diet

The need for food
Classes of food
Carbohydrates, proteins and fats.
Diet
Salts, vitamins, fibre and water.
Balanced diets
Energy and protein needs. Special needs.
Malnutrition
World food
Western diets
Preserving and processing
Additives.
Practical work
Food tests and energy from food.

The need for food

All living organisms need food. An important difference between plants and animals is that the green plants can make food in their leaves but animals have to take it in 'ready-made' by eating plants or the bodies of other animals. In all plants and animals, food is used as follows:

For growth

It provides the substances needed for making new cells and tissues.

As a source of energy

Energy is required for the chemical reactions which take place in living organisms to keep them alive. When food is broken down during respiration (see p. 19), the energy from the food is used for chemical reactions such as building complex molecules (pp. 12 and 13). In animals the energy is also used for activities such as movement, the heart beat and nerve impulses. Mammals and birds use energy to maintain their body temperature.

For replacement of worn and damaged tissues.

The substances provided by food are needed to replace – for example – the millions of our red blood cells that break down each day, and to replace the skin which is worn away, and to repair wounds.

Classes of food

There are three classes of food: carbohydrates, proteins and fats. The chemical structure of these substances is described on pp. 11 and 12.

Carbohydrates

Sugar and **starch** are important carbohydrates in our diet. Starch is abundant in potatoes, bread, maize, rice and other cereals. Sugar appears in our diet mainly as **sucrose** (table sugar) which is added to drinks and many prepared foods such as jam, biscuits and cakes. Glucose and fructose are sugars which occur naturally in many fruits and some vegetables.

Although all foods provide us with energy, carbohydrates are the cheapest and most readily available source of energy. They contain the elements carbon, hydrogen and oxygen (e.g. glucose is $C_6H_{12}O_6$). When carbohydrates are oxidized to provide energy by respiration they are broken down to carbon dioxide and water (see p. 19). One gram of carbohydrate can provide, on average, 16 kilojoules (kJ) of energy.

If we eat more carbohydrates than we need for our energy requirements, the excess is converted in the liver to either glycogen (see pp. 12, 103) or fat. The glycogen is stored in the liver and muscles; the fat is stored in fat depots in the abdomen, round the kidneys or under the skin (Figure 10.1).

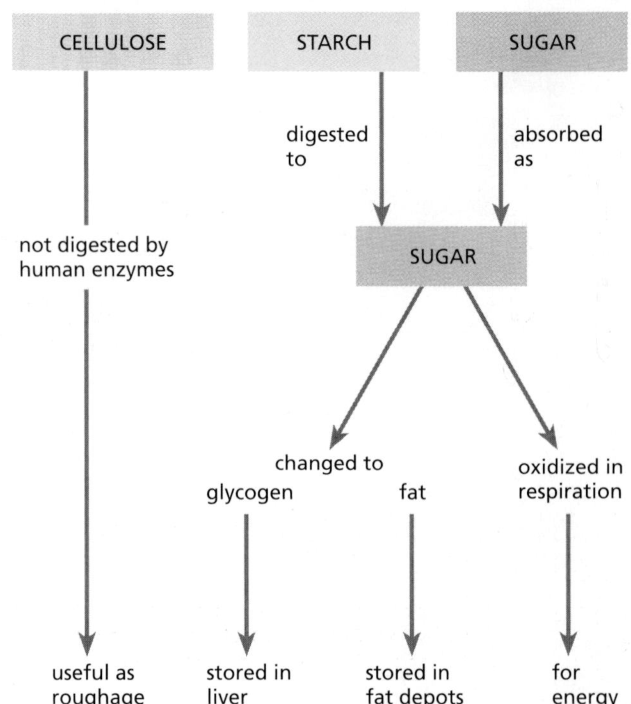

Figure 10.1 Digestion and use of carbohydrate

The **cellulose** in the cell walls of all plant tissues is a carbohydrate. We probably derive relatively little nourishment from cellulose but it is important in the diet as **fibre** (see p.88), which helps to maintain a healthy digestive system.

Proteins

Lean meat, fish, eggs, milk and cheese are important sources of animal protein. All plants contain some protein, but beans or cereals like wheat and maize are the best sources.

Proteins, when digested, provide the chemical substances needed to build cells and tissues, e.g. skin, muscle, blood and bones. Neither carbohydrates nor fats can do this and so it is essential to include some proteins in the diet.

Protein molecules consist of long chains of **amino acids** (see p. 11). When proteins are digested, the molecules are broken up into the constituent amino acids. The amino acids are absorbed into the bloodstream and used to build up different proteins. These proteins form part of the cytoplasm and enzymes of cells and tissues. Such a rearrangement of amino acids is shown in Figure 10.2.

The amino acids which are not used for making new tissues cannot be stored, but the liver removes their amino (—NH_2) groups and changes the residue to glycogen. The glycogen can be stored or oxidized to provide energy (p. 103). One gram of protein can provide 17 kJ of energy.

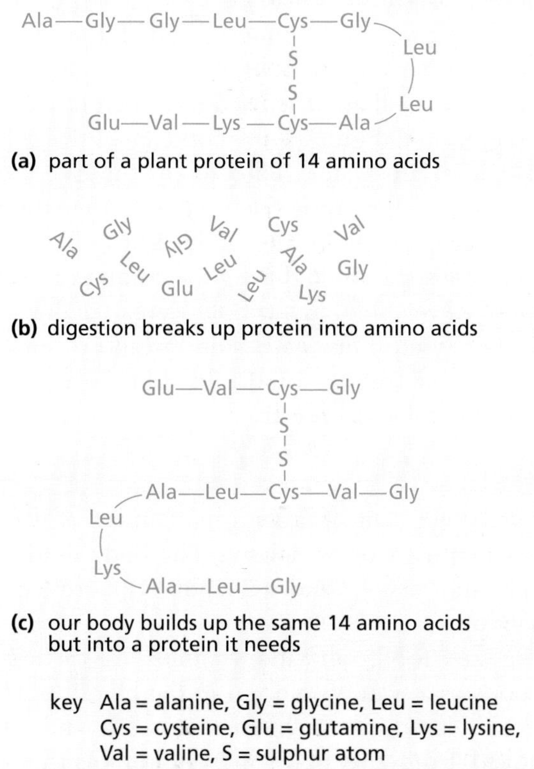

(a) part of a plant protein of 14 amino acids

(b) digestion breaks up protein into amino acids

(c) our body builds up the same 14 amino acids but into a protein it needs

key Ala = alanine, Gly = glycine, Leu = leucine
Cys = cysteine, Glu = glutamine, Lys = lysine,
Val = valine, S = sulphur atom

Figure 10.2 A model for digestion and use of a protein molecule

Chemically, proteins differ from both carbohydrates and fats because they contain nitrogen and sometimes sulphur as well as carbon, hydrogen and oxygen.

Fats

Animal fats are found in meat, milk, cheese, butter and egg-yolk. Plant fats occur as oils in fruits (e.g. palm oil) and seeds (e.g. sunflower seed oil), and are used for cooking and making margarine. Fats and oils are sometimes collectively called **lipids**.

Lipids are used in the cells of the body to form part of the cell membrane and other membrane systems (p. 3). Lipids can also be oxidized in respiration, to carbon dioxide and water. When used to provide energy in this way, 1 g fat gives 37 kJ of energy. This is more than twice as much energy as can be obtained from the same weight of carbohydrate or protein.

Fats can be stored in the body, so providing a means of long-term storage of energy in fat depots. The fatty tissue, **adipose tissue**, under the skin forms a layer which, if its blood supply is restricted, can reduce heat losses from the body.

Questions

1 What sources of protein-rich foods are available to a vegetarian who
 a will eat animal products but not meat itself,
 b will eat only plants and their products?

2 Why must all diets contain some protein?

3 Could you survive on a diet which contained no carbohydrate? Justify your answer.

4 In what sense can the fats in your diet be said to contribute to 'keeping you warm'?

5 How do proteins differ from fats (lipids) in
 a their chemical composition (pp. 11 and 12),
 b their energy value,
 c their role in the body?

6 Construct a flow chart for the digestion and use of proteins, similar to the one for carbohydrates in Figure 10.1.

Diet

In addition to proteins, carbohydrates and fats, the diet must include salts, vitamins, water and vegetable fibre (roughage). These substances are present in a balanced diet and do not normally have to be taken in separately.

Salts

These are sometimes called 'mineral salts' or just 'minerals'. Proteins, carbohydrates and fats provide the body with carbon, hydrogen, oxygen, nitrogen, sulphur and phosphorus but there are several more elements which the body needs and which occur as salts in the food we eat.

Iron

The red blood cells contain the pigment haemoglobin (p. 108). Part of the haemoglobin molecule contains iron and this plays an important part in carrying oxygen round the body. Millions of red cells break down each day and their iron is stored by the liver and used to make more haemoglobin. However, some iron is lost and adults need to take in about 15 mg each day. Iron is needed also in the muscles and for enzyme systems in all the body cells.

Red meat, especially liver and kidney, is the richest source of iron in the diet, but eggs, groundnuts, bread, spinach and other green vegetables are also important sources.

If the diet is deficient in iron, a person may suffer from some form of anaemia. Insufficient haemoglobin is made and the oxygen-carrying capacity of the blood is reduced.

Calcium

Calcium, in the form of calcium phosphate, is deposited in the bones and the teeth and makes them hard. It is present in blood plasma and plays an essential part in normal blood clotting (p. 117). Calcium is also needed for the chemical changes which make muscles contract and for the transmission of nerve impulses.

The richest sources of calcium are milk (liquid, skimmed or dried), and cheese, but calcium is present in most foods in small quantities and also in 'hard' water.

Many calcium salts are not soluble in water and may pass through the alimentary canal without being absorbed. Simply increasing the calcium in the diet may not have much effect unless the calcium is in the right form, the diet is balanced and the intestine is healthy. Vitamin D and bile salts (p. 104) are needed for efficient absorption.

Iodine

This is needed in only small quantities, but it forms an essential part of the molecule of **thyroxine**. Thyroxine is a hormone (p. 169) produced by the thyroid gland in the neck.

Specially rich sources of iodine are sea fish and shellfish but it is present in most vegetables, provided that the soil in which they grow is not deficient in the mineral. In some parts of the world, where soils have little iodine, potassium iodide may be added to table salt to bring the iodine in the diet to a satisfactory level.

Phosphorus

Phosphorus is needed for the calcium phosphate of bone, and also for DNA (p. 186). It is present in nearly all food but is particularly abundant in cheese, meat and fish.

Vitamins

All proteins are similar to each other in their chemical structure and so are all carbohydrates. Vitamins, on the other hand, are a group of organic substances quite unrelated to each other in their chemical structure.

The features shared by all vitamins are:

- They are not digested or broken down for energy.
- Mostly, they are not built into the body structures.
- They are essential in small quantities for health.
- They are needed for chemical reactions in the cells, working in association with enzymes.

Plants can make these vitamins in their leaves, but animals have to obtain many of them ready-made either from plants or from other animals.

If any one of the vitamins is missing or deficient in the diet, a vitamin-deficiency disease may develop. Such a disease can be cured, at least in the early stages, simply by adding the vitamin to the diet.

Fifteen or more vitamins have been identified and they are sometimes grouped into two classes: water-soluble or fat-soluble. The fat-soluble vitamins are found mostly in animal fats or vegetable oils, which is one reason why our diet should include some of these fats. The water-soluble vitamins are present in green leaves, fruits and cereal grains.

See Table 10.1 for details of some vitamins.

Dietary fibre (roughage)

When we eat vegetables and other fresh plant material, we take in a large quantity of plant cells. The cell walls of plants consist mainly of cellulose, but we do not have enzymes for digesting this substance. The result is that the plant cell walls reach the large intestine (colon) without being digested. This undigested part of the diet is called **fibre** or roughage. The colon contains many bacteria which can digest some of the substances in the plant cell walls to form fatty acids (see p. 12). Vegetable fibre, therefore, may supply some useful food material, but it has other important functions.

The fibre itself and the bacteria which multiply from feeding on it add bulk to the contents of the colon and help it to retain water. This softens the faeces (p. 102) and reduces the time needed for the undigested residues to pass out of the body. Both effects help to prevent constipation and keep the colon healthy.

Most vegetables and whole cereal grains contain fibre, but white flour and white bread do not contain much. Good sources of dietary fibre are vegetables, fruit and wholemeal bread.

Water

About 70 per cent of most tissue consists of water; it is an essential part of cytoplasm. The body fluids, blood, lymph and tissue fluid (Chapter 12) are composed mainly of water.

Digested food, salts and vitamins are carried round the body as a watery solution in the blood (p. 116) and excretory products such as excess salt and urea are removed from the body in solution by the kidneys (p. 133). Water thus acts as a solvent and as a transport medium for these substances.

Table 10.1 Vitamins

Name and source of vitamin	Diseases and symptoms caused by lack of vitamin	Notes
Retinol (vitamin A; fat-soluble): Liver, cheese, butter, margarine, milk, eggs. **Carotene** (vitamin A precursor; water-soluble). Fresh green leaves and carrots.	Reduced resistance to disease, particularly those which enter through the epithelium. Poor night vision. Cornea of eyes becomes dry and opaque leading to **keratomalacia** and blindness.	The yellow pigment, carotene, present in green leaves and carrots is turned into retinol by the body. Retinol forms part of the light-sensitive pigment in the retina (p. 159). Retinol is stored in the liver.
Ascorbic acid (vitamin C; water-soluble): Oranges, lemons, grapefruit, tomatoes, fresh green vegetables, potatoes.	Fibres in connective tissue of skin and blood vessels do not form properly, leading to bleeding under the skin, particularly at the joints, swollen, bleeding gums and poor healing of wounds. These are all symptoms of **scurvy**.	Possibly acts as a catalyst in cell respiration. Scurvy is only likely to occur when fresh food is not available. Cows' milk and milk powders contain little ascorbic acid so babies may need additional sources. Cannot be stored in the body; daily intake needed.
Calciferol (vitamin D; fat-soluble): Butter, milk, cheese, egg-yolk, liver, fish-liver oil.	Calcium is not deposited properly in the bones, causing **rickets** in young children because the bones remain soft and are deformed by the child's weight. Deficiency in adults causes **osteomalacia**; fractures are likely.	Calciferol helps the absorption of calcium from the intestine and the deposition of calcium salts in the bones. Natural fats in the skin are converted to a form of calciferol by sunlight.
The B vitamins There are ten or more water-soluble vitamins which occur together, particularly in whole cereals, peas and beans. A deficiency of any one of these vitamins is likely to occur only in communities living on restricted diets such as maize or milled rice. **Folic acid** is a B vitamin which, recently, has been shown to be effective in reducing the incidence of birth defects such as spina bifida. Women planning a pregnancy may be advised to take supplements of folic acid. It is present naturally in green vegetables, root vegetables and whole grain products.		
There are several other substances classed as vitamins, e.g. **riboflavin** (B_2), **tocopherol** (E), **phylloquinone**, but these are either (1) unlikely to be missing from the diet, or (2) their functions are not fully understood.		
Vitamin K plays a part in the blood-clotting process (p. 117). It is widely available in green vegetables and is also made by the bacteria living normally in the intestine. Consequently it is unlikely to be lacking except in people whose intestinal bacteria have been reduced by heavy doses of antibiotics.		

Figure 10.3 Examples of types of food in a balanced diet (see question 2)

Questions

1 Which tissues of the body need
 a iron, **b** glucose, **c** calcium, **d** protein, **e** iodine?

2 In Figure 10.3 some examples of the food that would give a balanced diet are shown. Consider each picture in turn and say what class of food or item of diet is mainly present. For example, the meat is mainly protein but will also contain some iron.

3 What is the value of leafy vegetables, such as cabbage and lettuce, in the diet?

4 Why is a diet consisting mainly of one type of food, e.g. rice or potatoes, likely to be unsatisfactory even if it is sufficient to meet our energy needs?

5 A zoologist is trying to find out whether rabbits need vitamin C in their diet. Assuming that a sufficiently large number of rabbits is used and adequate controls are applied, the best design of experiment would be to give the rabbits:
 a an artificial diet of pure protein, carbohydrate, fats, minerals and vitamins but lacking vitamin C;
 b an artificial diet as above but with extra vitamin C;
 c a natural diet of grass, carrots etc. but with added vitamin C;
 d natural food but of one kind only, e.g. exclusively grass or exclusively carrots.
 Justify your choice and say why you exclude the other alternatives.

Digestion is a process which uses water in a chemical reaction to break down insoluble substances to soluble ones (p. 98). These products then pass, in solution, into the bloodstream. In all cells there are many reactions in which water plays an essential part as a reactant and a solvent.

Since we lose water by evaporation, sweating, urinating and breathing, we have to make good this loss by taking in water with the diet.

Balanced diets

A balanced diet must contain enough carbohydrates and fats to meet our energy needs. It must also contain enough protein of the right kind to provide the essential amino acids to make new cells and tissues for growth or repair. The diet must also contain vitamins and mineral salts, plant fibre and water. The composition of four food samples is shown in Figure 10.4.

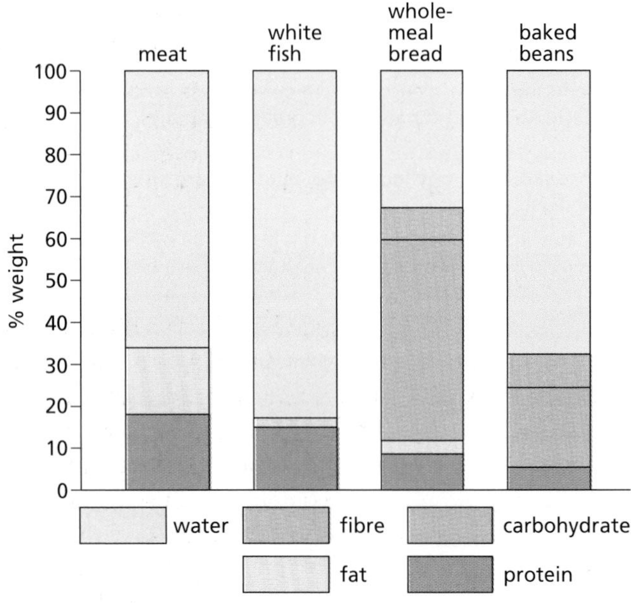

Figure 10.4 An analysis of four food samples
Note The percentage of water includes any salts and vitamins. There are wide variations in the composition of any given food sample according to its source and the method of preservation and cooking. 'White fish' (e.g. cod, haddock, plaice) contains only 0.5 per cent fat whereas herring and mackerel contain up to 14 per cent. White bread contains only 2–3 per cent fibre. Frying the food greatly adds to its fat content

Energy requirements

Energy can be obtained from carbohydrates, fats and proteins. The cheapest energy-giving food is usually carbohydrate; the greatest amount of energy is available in fats; proteins give about the same energy as carbohydrates but are expensive. Whatever mixture of carbohydrate, fat and protein makes up the diet, the total energy must be sufficient (1) to keep our internal body processes working (e.g. heart beating, breathing action), (2) to keep up our body temperature and (3) to meet the needs of work and other activities.

The amount of energy that can be obtained from food is measured in calories or joules. One gram of carbohydrate or protein can provide us with 16 or 17 kJ (kilojoules). A gram of fat can give 37 kJ. We need to obtain about 12 000 kJ of energy each day from our food. Table 10.2 shows how this figure is obtained. However, the figure will vary greatly according to our age, occupation and activity (Figure 10.5). It is fairly obvious that a person who does hard manual work, such as digging, will use more energy than someone who sits in an office.

Table 10.2 Energy requirements in kJ

8 hours asleep	2400
8 hours awake; relatively inactive physically	3000
8 hours physically active	6600
	Total 12 000

The 2400 kJ used during 8 hours' sleep represents the energy needed for **basal metabolism** which maintains the circulation, breathing, body temperature, brain function and essential chemical processes in the liver and other organs.

If the diet includes more food than is needed to supply the energy demands of the body, the surplus food is stored either as glycogen in the liver (see p. 103) or as fat below the skin and in the abdomen.

In 1984, a national study recommended that, for a balanced diet, 55 per cent of our energy intake should be made up of carbohydrate, 35 per cent of fat and 10 per cent of protein (though not necessarily used for energy). This means that a growing adolescent should take in about 63 g protein, 97 g fat and 344 g carbohydrate each day.

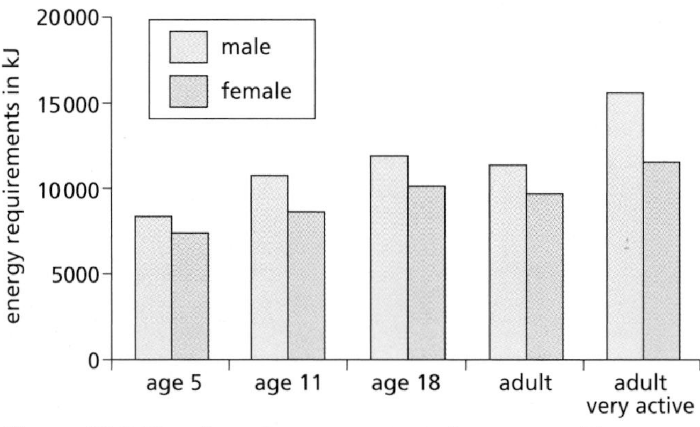

Figure 10.5 The changing energy requirements with age and activity

Protein requirements

As explained on p. 87, proteins are an essential part of the diet because they supply the amino acids needed to build up our own body structures. Estimates of how much protein we need have changed over the last few years. A recent WHO/FAO report recommended that an average person needs 0.57 g protein for every kilogram of body weight. That is, a 70 kg person would need $70 \times 0.57 = 39.9$, i.e. about 40 g protein per day.

This could be supplied by about 200 g (7 ounces) lean meat or 500 g bread but 2 kg potatoes would be needed to supply this much protein and even this will not contain all the essential amino acids.

Vegetarian and vegan diets

There is relatively less protein in food derived from plants than there is in animal products. **Vegetarians** and semivegetarians, who include dairy products, eggs and possibly fish in their diets, will obtain sufficient protein to meet their needs. **Vegans**, who eat no animal products, need to ensure that their diets include a good variety of cereals, peas, beans and nuts in order to obtain all the essential amino acids to build their body proteins.

Special needs

Pregnancy

A pregnant woman who is already receiving an adequate diet needs no extra food. Her body's metabolism will adapt to the demands of the growing baby. If, however, her diet is deficient in protein, calcium, iron, vitamin D or folic acid, she will need to increase her intake of these substances to meet the needs of the baby. The baby needs protein for making its tissues; calcium and vitamin D are needed for bone development, and iron is used to make the haemoglobin in its blood.

Lactation

'Lactation' means the production of breast milk for feeding the baby. The production of milk, rich in proteins and minerals, makes a large demand on the mother's resources. If her diet is already adequate, her metabolism will adjust to these demands. Otherwise, she may need to increase her intake of proteins, vitamins and calcium to produce milk of adequate quality and quantity.

Growing children

Most children up to the age of about 12 years need less food than adults, but they need more in proportion to their body weight. For example, an adult may need 0.57 g protein per kg body weight, but a 6–11-month baby needs 1.53 g per kg, and a 10-year-old child needs 0.8 g per kg for growth. In addition, children need extra calcium for growing bones, iron for their red blood cells, vitamin D to help calcify their bones and vitamin A for disease resistance.

Malnutrition

Malnutrition is often taken to mean simply not getting enough food, but it has a much wider meaning than this, including getting too much food or the wrong sort of food.

If the total intake of food is not sufficient to meet the body's need for energy, the body tissues themselves are broken down to provide the energy to stay alive. This leads to loss of weight, muscle wastage, weakness and ultimately starvation.

If food intake is drastically inadequate, it is likely that the diet will also be deficient in proteins, minerals and vitamins so that deficiency diseases such as anaemia and rickets (p. 89) also make an appearance. The victims of such malnutrition will also have reduced resistance to infectious diseases such as malaria or measles. Thus, the symptoms of malnutrition are usually the outcome of a variety of causes but all resulting from an inadequate diet.

Kwashiorkor (roughly = 'deposed child') is an example of protein-energy malnutrition (PEM) in the developing world. When a mother has her second baby, the first baby is weaned on to a starchy diet of yam, cassava or sweet potato, all of which have inadequate protein. The first baby then develops symptoms of kwashiorkor (dry skin, pot-belly, weakness and irritability). Protein deficiency is not the only cause of kwashiorkor. Infection, plant toxins, digestive failure or even psychological effects may be involved. The good news, however, is that it can often be cured or prevented by an intake of protein in the form of dried skimmed milk.

The **causes of malnutrition** can be famine due to drought or flood, soil erosion, wars, too little land for too many people, ignorance of proper dietary needs but, above all, poverty. Malnourished populations are often poor and cannot afford to buy enough nutritious food.

World food

The world population doubled in the last 30 years but food production, globally, rose even faster. The 'Green Revolution' of the 1960s greatly increased global food production by introducing high-yielding varieties of crops. However, these varieties needed a high input of fertilizer and the use of pesticides, so only the wealthy farmers could afford to use them. Moreover, since 1984, the yields are no longer rising fast enough to feed the growing population or keep pace with the loss of farmland due to erosion and urbanization.

It is estimated that, despite the global increase in food production, 15 per cent of the world population is undernourished and that 180 million children are underweight (Figure 10.6).

There are no obvious, easy or universal solutions to this situation. Genetically modified crops (p. 215) may hold out some hope but they are some way off. Redistribution of food from the wealthy to the poorer countries is not a practical proposition except in emergencies.

The strategies adopted need to be tailored to the needs and climate of individual countries. Crops suited to the region should be grown. Millet and sorghum grow far better in dry regions than do rice or wheat and need little or no irrigation. Cash crops such as coffee, tea or cotton can earn foreign currency but have no food value and do not feed the local population. It might be better to use the land, where suitable, to grow food crops.

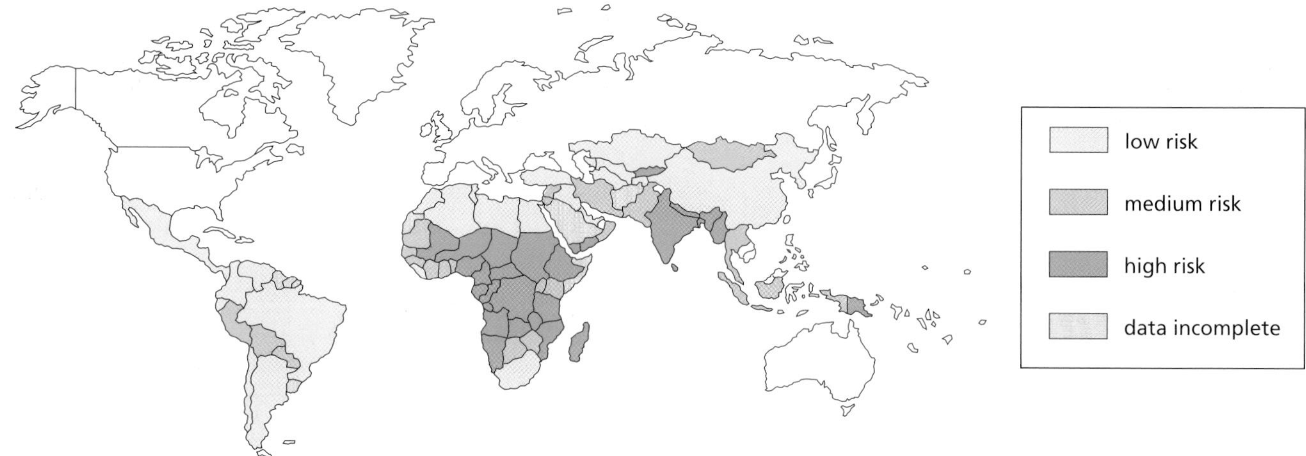

Figure 10.6 Countries with populations at risk of inadequate nutrition

The agricultural practices need to be sustainable and not result in erosion. Nearly one-third of the world's crop-growing land has had to be abandoned in the last 40 years because erosion has made it unproductive. Conservation of land, water and energy is essential for sustainable agriculture. A reduction in the growth of the world's population, if it could be achieved, would have a profound effect in reducing malnutrition.

Apart from the measures outlined above, lives could be saved by such simple and inexpensive steps as provision of regular vitamin and mineral supplements. It is estimated that about 30 million children are deficient in vitamin A. This deficiency leads to blindness and death if untreated.

Western diets

In the affluent societies, e.g. USA and Europe, there is no general shortage of food and most people can afford a diet with an adequate energy and protein content. So few people are undernourished. Eating too much food or food of the 'wrong' kind, however, leads to malnutrition of a different kind.

Refined sugar (sucrose)

This is a very concentrated source of energy. You can absorb a lot of sugar from biscuits, ice-cream, sweets, soft drinks, tinned fruits and sweet tea without ever feeling 'full up'. So you tend to take in more sugar than your body needs, which may lead to your becoming overweight or obese.

Sugar is also a major cause of tooth decay (p. 175).

Fats

Fatty deposits, called 'plaques', in the arteries can lead to coronary heart disease and strokes (p. 120). These plaques are formed from lipids and **cholesterol** combined with proteins (**low density lipoproteins** or **LDLs**). Although the liver makes LDLs, there is evidence to suggest that a high intake of fats, particularly animal fats, helps raise the level of LDLs in the blood and increase the risk of plaque formation.

Most animal fats are formed from **saturated fatty acids**, so called because of their molecular structure. Plant oils are formed from **unsaturated fatty acids** (polyunsaturates) and are thought less likely to cause fatty plaques in the arteries. For this reason, vegetable fats and certain margarines are considered, by some nutritionists, to be healthier than butter and cream. However, there is still much debate about the evidence for this.

Fibre

Many of the processed foods in Western diets contain too little fibre. White bread, for example, has had the fibre (bran) removed. Unprocessed foods, such as potatoes, vegetables and fruit, contain plenty of fibre. Food rich in fibre is usually bulky and makes you feel 'full up' so that you are unlikely to overeat.

As explained on p. 88, fibre helps prevent constipation and may also protect the intestine from cancer and other disorders.

Overweight and obesity

These are different degrees of the same disorder. If you take in more food than your body needs for energy, growth and replacement, the excess is converted to fat and stored in fat depots under the skin or in the abdomen.

Obese people are more likely to suffer from high blood pressure, coronary heart disease and diabetes. Having extra weight to carry also makes you reluctant to take exercise.

Why some people should be prone to obesity is unclear. There may be a genetic predisposition, in which the brain centre that responds to food intake may not signal when sufficient food has been taken; in some cases it may be the outcome of an infectious disease. Whatever the cause, the remedy is to reduce food intake to a level which matches but does not exceed the body's needs. Taking exercise helps, but it takes a great deal of exercise to 'burn off' even a small amount of surplus fat.

Questions

1 Make a list of the food and other substances that are needed to make up a balanced diet.

2 Why is it better to take in regular, small amounts of protein rather than to eat a large amount of protein at one meal? (Revise pp. 86–7.)

3 Select one food class and one mineral salt which are particularly important in the diet of *all three* of the following: pregnant woman, woman breast-feeding a baby, growing child.

4 a If you feel 'peckish' between meals, why is it better to eat an apple than a bar of chocolate?
  b If you are going to do a long-distance walk, why is it better to take chocolate bars than apples?

5 100 g of boiled potato will give you 240 kJ of energy, but 100 g of chips give you 900 kJ. Why do you think there is such a big difference?

6 Why should a 'high fibre' diet help to stop you putting on weight?

7 It is sometimes believed that a person who does hard, physical work needs to eat a lot of protein. Try to explain why this is not true.

8 How much protein would a 5 kg baby need each day?

9 From Table 10.2 on p. 90 work out the approximate minimum amount of energy needed each day to maintain your basal metabolism.

Preserving and processing

Food preservation

If food is kept for any length of time before it is eaten it may start to 'go off'. This may be because it is attacked by its own enzymes, oxidized by the air or, more important, decomposed by bacteria and fungi. All these processes make the food taste and smell unpleasant, but the greatest harm is likely to result from the fungi and bacteria.

Both these organisms may produce poisonous compounds (toxins) which make us ill if we eat them, e.g. *Salmonella* poisoning by bacteria. Cooking the food may kill the organisms but will not necessarily destroy the toxins they have already produced.

Methods of food preservation try to prevent the food's own enzymes from working and to stop the growth of fungi and bacteria.

Drying (dehydration)

Removal of water stops the enzymes from working and prevents the growth of bacteria and fungi. The food is usually dried under low air pressure. This makes the water evaporate without raising the food to a temperature at which it would start to cook. The dehydrated food is sealed in packets and kept at room temperature until it is reconstituted by adding water.

Refrigeration and freezing

Most refrigerators are kept at 4 °C. At this temperature bacteria reproduce only very slowly but they are not killed. The activity of any enzymes in the food is also slowed down. The temperature of −20 °C in the freezer stops the growth and reproduction of bacteria but it still does not kill them all.

Canning

Food is enclosed in cans which are sealed and heated to 120 °C to destroy enzymes and bacteria. Bacteria cannot enter the cans. So canned food can remain good to eat for years, provided the cans are not damaged or corroded.

Pasteurization

Milk is heated to 72 °C for 15 seconds then cooled rapidly to 7 °C (the 'flash' process). This does not sterilize the milk but does destroy bacteria which could cause diseases such as tuberculosis.

Ultra-high temperature (UHT)

Milk is heated under pressure to 132 °C for 1 second and sealed in cartons or plastic bottles. All the bacteria are killed and so the milk will keep for months. There is a slight change of flavour.

Irradiation

Food is sealed in plastic wrapping and exposed to low doses of X-rays or gamma-rays, which kill any bacteria or parasites in the food. The sealed wrapping prevents any more bacteria getting at the food. The food does not become radioactive because the radiation dose is too low.

The disadvantage is that the flavour, texture and colour of some foods may be altered and the vitamin content may be reduced (though this can be said of most methods of preservation). Of greater concern is the fact that the technique might be used to sterilize food which is already contaminated. The bacteria would be killed but any toxins they had produced would not be destroyed.

Irradiated food is used for patients whose immune systems have been suppressed but its general sale is not yet permitted in Britain.

Food additives

About 3500 different chemicals may be used by the processed food industry. These chemicals have no food value but are added to food (1) to stop it going bad, (2) to 'improve' its colour or (3) to alter or enhance its flavour.

Preservatives

There is always a time lag between harvesting a perishable food, processing it, packaging it and sending it to the food shops. The food also spends time on the shelf in the shop and in the kitchen. Thus, except for dried or frozen food, some chemical method of preservation is needed to stop bacterial growth.

Two chemical preservatives are sodium nitrite (E250) and sulphur dioxide (E220). Sodium nitrite is added to cured meat (ham or bacon, for example). It prevents the growth of bacteria, particularly *Clostridium botulinum*, which causes a deadly form of food poisoning called botulism. Sulphur dioxide is added to jams, fruit juices, beer and wine. It suppresses the growth of bacteria and fungi in these products.

Other additives

Of the 3500 additives, only about 1 per cent are preservatives. The rest are flavourings, colourings, stabilizers and bulking agents (Figure 10.7).

Flavouring

One widely used flavouring is monosodium glutamate, made from sugar-beet pulp and wheat protein. It is described as a 'flavour enhancer', and may achieve its effect by promoting the flow of saliva or stimulating the taste receptors on the tongue.

Sugar is a widely used flavouring agent. We expect to find it in cakes and biscuits. But it turns up in many unexpected places too – in instant baby cereal (34 per cent in certain brands), tomato sauce (14 per cent), frozen peas (7 per cent) and baked beans (5 per cent).

Colouring

Many substances (for example chlorophyll, saffron and turmeric) used for colouring food are from natural sources. This does not necessarily mean that they are harmless. Others are synthetic dyes. For example, tartrazine, sunset yellow and carmoisine are 'azo' dyes, and are known to cause allergic reactions in some people.

'Bulking agents'

Some substances, such as sodium polyphosphate (E450), are added to meat (ham and poultry particularly) as so-called 'tenderizers'. In fact they serve mainly to promote the absorption of water into the meat by osmosis, so that it weighs more.

Possible harmful effects

Only about 300 of the additives are regulated by law, but the food industry tests all the additives, usually by feeding them in large doses to animals. The results, however, are not necessarily applicable to humans. Some workers, for example, estimate that tests for cancer-causing properties may be only 37 per cent successful in animal trials.

There is little widely accepted evidence to show that food additives are harmful. Most allergic reactions to food are caused by naturally occurring substances in the food, though the yellow dye, tartrazine (E102), does cause an allergic reaction in a small number of people. Eczema in children, in some cases, has been relieved by eliminating artificial colour and flavour from their diets. The nitrates and nitrites used to 'cure' ham and bacon are sometimes suspected, without much evidence, as possible carcinogens (cancer-causers).

Most of us eat from 3 to 7 kg of additives each year with no obvious harm, but it is very difficult to know whether the additives have a long-term effect. The risks of cancer from eating cured meats are far lower than the risk of serious bacterial poisoning from eating unprocessed meat. Nevertheless, some people are uneasy that there are so many food additives, whose long-term effects are not known for sure, and which are not really essential for a safe and healthy diet.

THIN & CRIS
HAM & PINEAPPLE PIZZA
FRESH 9 INCH PIZZA
A thin and crispy pizza base topped with mozzarella che
honey roast ham, cheddar cheese and pineapple
INGREDIENTS
WHEAT FLOUR, WATER, MOZZARELLA CHEESE (13%), REFORMED HONEY ROAST HAM (10%) (WITH STABILISERS: DI-, TRI AND POLYPHOSPHATES; ANTIOXIDANT: SODIUM L-ASCORBATE; PRESERVATIVE: SODIUM NITRITE), CHEDDAR CHEESE (9%), PINEAPPLE (8%), TOMATO, VEGETABLE OIL, TOMATO PASTE, YEAST, SUGAR, SALT, MODIFIED MAIZE STARCH, DRIED PARSLEY, WHEY POWDER, PRESERVATIVE: CALCIUM PROPIONATE; VEGETABLE BOUILLON (WITH CITRIC ACID, FLAVOURINGS).
*CONTAINS MILK & WHEAT

Figure 10.7 Food labelling. The manufacturer must list all the ingredients including additives. Which of the ingredients would you regard as nutrients and which as additives?

Practical work

1 Food tests (Figure 10.8)

a Test for starch Shake a little starch powder in a test-tube with some cold water and then boil it to make a clear solution. When the solution is cold, add 3 or 4 drops of **iodine solution**. A dark blue colour should be produced.

b Test for glucose Heat a little glucose with some **Benedict's solution** in a test-tube. The heating is done by placing the test-tube in a beaker of boiling water (see Figure 10.8). The solution will change from clear blue to cloudy green, then yellow and finally to a red precipitate (deposit) of copper(I) oxide.

c Test for protein (Biuret test) To a 1 per cent solution of albumen (the protein of egg-white) add 5 cm^3 dilute sodium hydroxide (**CARE**: this solution is caustic), followed by 5 cm^3 1 per cent copper sulphate solution. A purple colour indicates protein.

d Test for fat Shake 2 drops of cooking oil with about 5 cm^3 ethanol in a dry test-tube until the fat dissolves. Pour this solution into a test-tube containing a few cm^3 water. A cloudy white emulsion will form. This shows that the solution contained some fat or oil.

e Test for vitamin C Draw up 2 cm^3 fresh lemon juice into a plastic syringe. Add this juice drop by drop to 2 cm^3 of a 0.1 per cent solution of DCPIP (a blue dye) in a test-tube. The DCPIP will become colourless quite suddenly as the juice is added. The amount of juice added from the syringe should be noted down. Repeat the experiment but with orange juice in the syringe. If it takes more orange juice than lemon juice to decolourize the DCPIP, the orange juice must contain less vitamin C.

2 Application of the food tests

The tests can be used on samples of food such as milk, potato, raisins, onion, beans, egg-yolk or peanuts to find what food materials are present. The solid samples are crushed in a mortar and shaken with warm water to extract the soluble products. Separate samples of the watery mixture of crushed food are tested for starch, glucose or protein as described above. To test for fats, the food must first be crushed in ethanol, not water, and then filtered. The clear filtrate is poured into water to see if it goes cloudy, indicating the presence of fats.

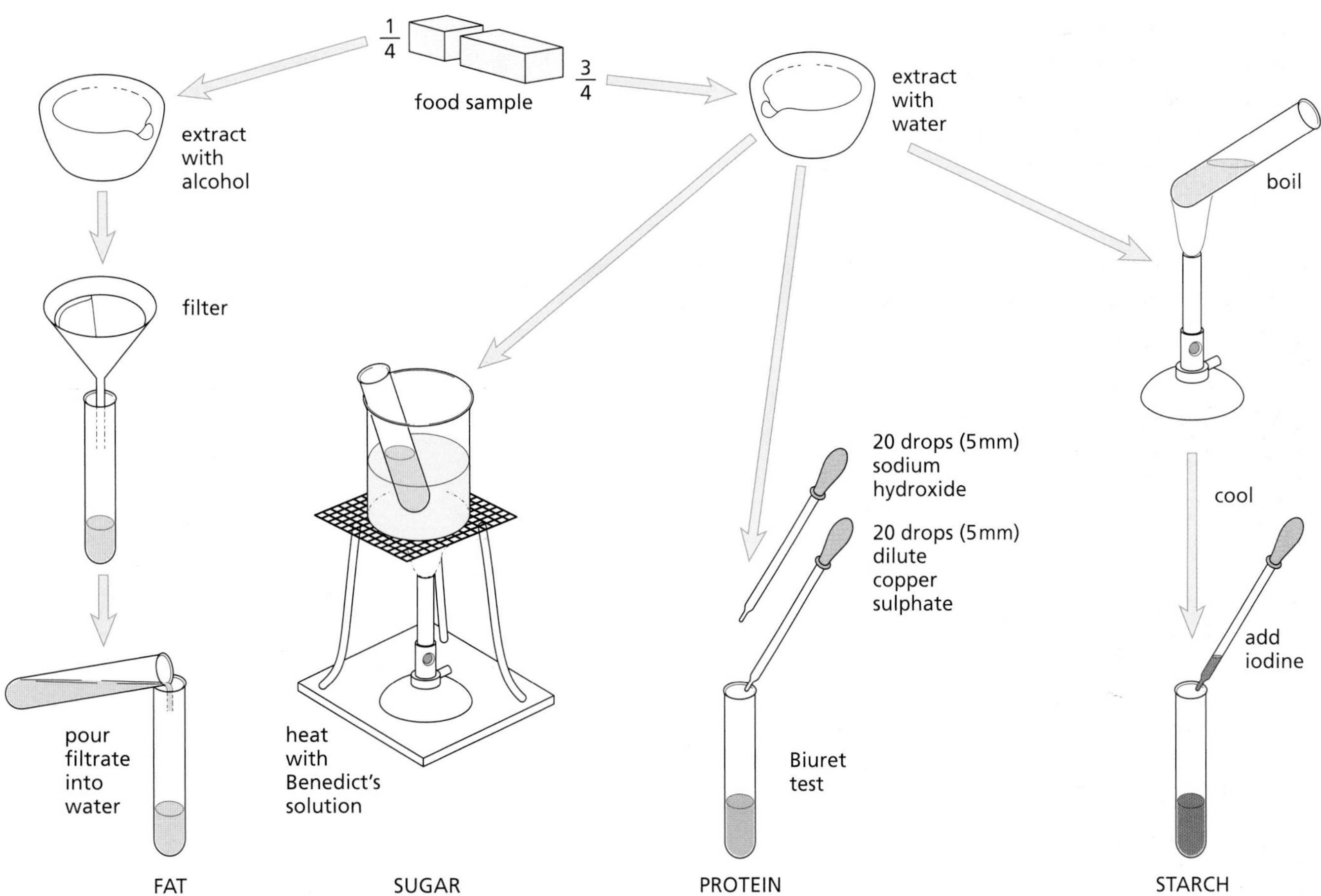

Figure 10.8 Food tests

3 Energy from food

Arrange the apparatus as shown in Figure 10.9. Use a measuring cylinder to place 100 cm^3 cold water in the metal can. With a thermometer, find the temperature of the water and make a note of it. In the nickel crucible or tin lid place 1 g sugar and heat it with the Bunsen flame until it begins to burn. As soon as it starts burning, slide the crucible under the can so that the flames heat the water. If the flame goes out, do not apply the Bunsen burner to the crucible while it is under the can, but return the crucible to the Bunsen flame to start the sugar burning again and replace the crucible beneath the can as soon as the sugar catches alight. When the sugar has finished burning and cannot be ignited again, gently stir the water in the can with the thermometer and record its new temperature.

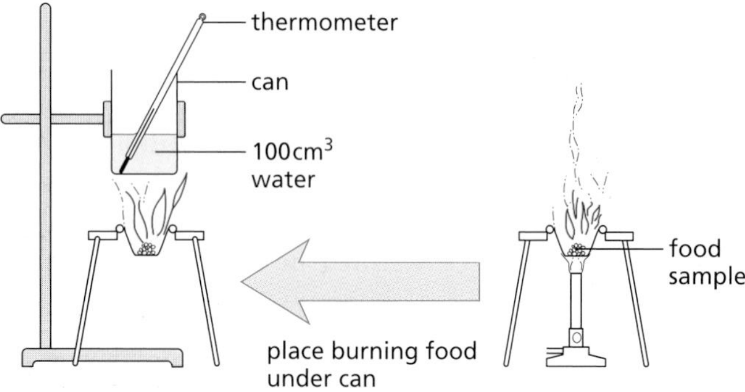

Figure 10.9 Energy from food

Calculate the rise in temperature by subtracting the first from the second temperature. Work out the quantity of energy transferred to the water from the burning sugar as follows:

4.2 joules raise 1 g water 1 °C
100 cm^3 cold water weighs 100 g
Let the rise in temperature be T °C

To raise 1 g water 1 °C needs 4.2 joules
∴ To raise 100 g water 1 °C needs 100 × 4.2 joules
∴ To raise 100 g water T °C needs T × 100 × 4.2 joules
∴ 1 g burning sugar produced 420 × T joules

The experiment may now be repeated using 1 g vegetable oil instead of sugar and replacing the warm water in the can with 100 cm^3 cold water.

Although the experiment is very inaccurate, the sources of error are more or less the same for both substances. Consequently, the results may be used for purposes of comparison.

Try to point out some of the faults in the design of the experiment.

Checklist

- Our diets must contain proteins, carbohydrates, fats, minerals, vitamins, fibre and water.
- Fats, carbohydrates and proteins provide energy.
- Proteins provide amino acids for the growth and replacement of the tissues.
- Mineral salts like calcium and iron are needed in tissues such as bone and blood.
- Vegetable fibre helps to maintain a healthy intestine.
- Adolescents and adults need about 10–12 000 kJ of energy each day from their food.
- Vitamins are essential in small quantities for chemical reactions in cells.
- The fat-soluble vitamins A and D occur mainly in animal products.
- Most cereals contain vitamins of the B group.
- Vitamin C occurs in certain fruits and in green leaves.
- Lack of vitamin A can lead to blindness; shortage of vitamin C causes scurvy; inadequate vitamin D causes rickets.
- Growing children have special dietary needs.
- Malnutrition is the result of taking in food which does not match the energy needs of the body, or is lacking in proteins, vitamins or minerals.
- About one in five people in the developing world are inadequately nourished.
- Increases in world food production may not keep pace with a growing population.
- Western diets often contain too much sugar and fat and too little fibre.
- Obesity results from taking in more food than the body needs for energy, growth or replacement.
- Methods of food preservation aim to stop enzymes working and to suppress growth of fungi and bacteria.

11 Digestion, absorption and use of food

The alimentary canal
Structure, function, peristalsis.

Digestion
Definition. Mouth, stomach, small intestine.

Absorption
Absorption in the small and large intestine.

Use of digested food
Glucose, fats, amino acids.

Storage of digested food
Glucose, fat, amino acids. Body weight.

The liver
Functions.

Practical work
Experiments on digestion. Salivary amylase, pepsin and lipase.

Feeding involves taking food into the mouth, chewing it and swallowing it down into the stomach. This satisfies our hunger, but for food to be of any use to the whole body it has first to be **digested**. This means that the solid food is dissolved. The soluble products then have to be **absorbed** into the bloodstream and carried by the blood all round the body. In this way, the blood delivers dissolved food to the living cells in all parts of the body such as the muscles, brain, heart and kidneys. This chapter describes how the food is digested and absorbed. Chapter 12 describes how the blood carries it round the body.

The alimentary canal

The **alimentary canal** is a tube, running through the body. Food is digested in the alimentary canal. The soluble products are absorbed and the indigestible residues expelled (egested). A simplified diagram of an alimentary canal is shown in Figure 11.1.

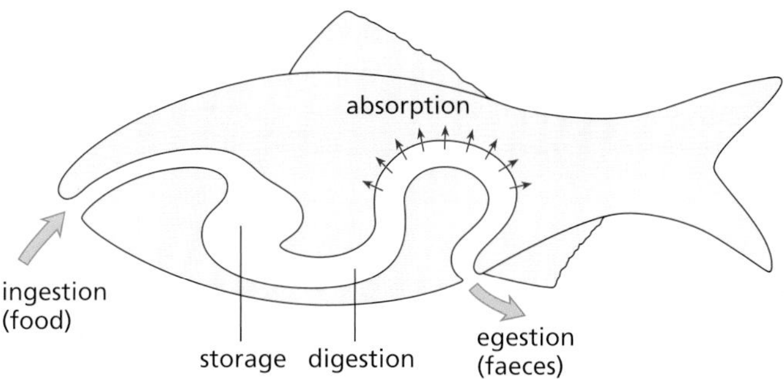

Figure 11.1 The alimentary canal (generalized)

The inside of the alimentary canal is lined with layers of cells forming what is called an **epithelium**. New cells in the epithelium are being produced all the time to replace the cells worn away by the movement of the food. There are also cells in the lining which produce **mucus**. Mucus is a slimy liquid that lubricates the lining of the canal and protects it from wear and tear. Mucus may also protect the lining from attack by the **digestive enzymes** which are released into the alimentary canal.

Some of the digestive enzymes are produced by cells in the lining of the alimentary canal, as in the stomach lining. Others are produced by **glands** which are outside the alimentary canal but pour their enzymes through tubes (called **ducts**) into the alimentary canal (Figure 11.2). The **salivary glands** (see Figure 11.6, p. 99) and the **pancreas** (see Figure 11.8, p. 100) are examples of such digestive glands.

The alimentary canal has a great many blood vessels in its walls, close to the lining. These bring oxygen needed by the cells and take away the carbon dioxide they produce. They also absorb the digested food from the alimentary canal.

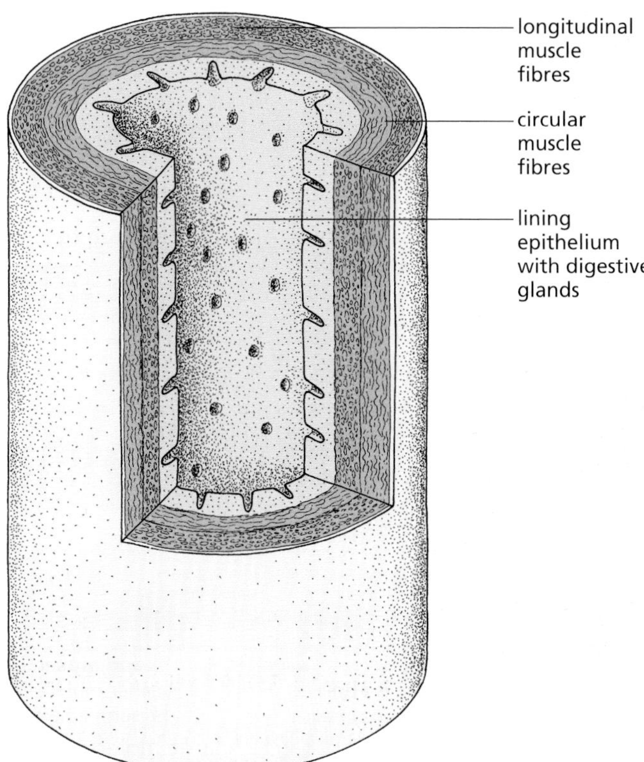

Figure 11.2 The general structure of the alimentary canal

Peristalsis

The alimentary canal has layers of muscle in its walls (Figure 11.2). The fibres of one layer of muscles run round the canal (**circular muscle**) and the others run along its length (**longitudinal muscle**). When the circular muscles in one region contract, they make the alimentary canal narrow in that region.

A contraction in one region of the alimentary canal is followed by another contraction just below it so that a wave of contraction passes along the canal pushing food in front of it. The wave of contraction, called **peristalsis**, is illustrated in Figure 11.3.

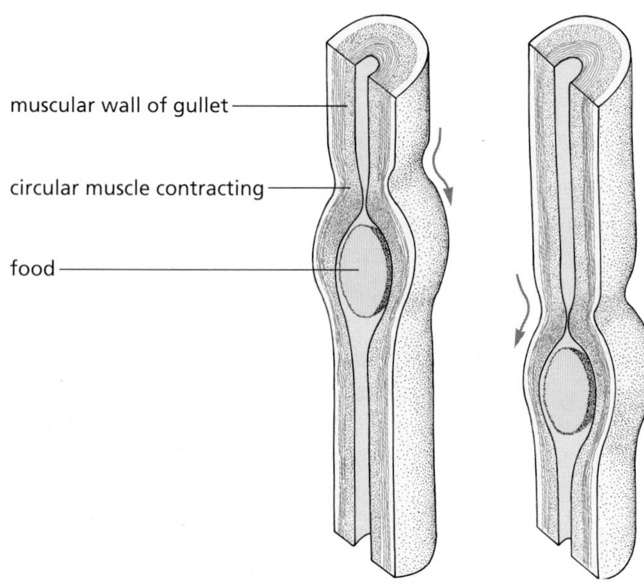
muscular wall of gullet
circular muscle contracting
food

Figure 11.3 Diagram to illustrate peristalsis

Questions

1 What three functions of the alimentary canal are shown in Figure 11.1?

2 Into what parts of the alimentary canal do
 a the pancreas,
 b the salivary glands, pour their digestive juices?

3 Starting from the inside, name the layers of tissue that make up the alimentary canal.

Digestion

Digestion is mainly a chemical process and consists of breaking down large molecules to small molecules. The large molecules are usually not soluble in water, while the smaller ones are. The small molecules can pass through the epithelium of the alimentary canal, through the walls of the blood vessels and into the blood.

Some food can be absorbed without digestion. The glucose in fruit juice, for example, could pass through the walls of the alimentary canal and enter the blood vessels. Most food, however, is solid and cannot get into blood vessels. Digestion is the process by which solid food is dissolved to make a solution.

The chemicals which dissolve the food are **enzymes**, described on p. 14. A protein might take 50 years to dissolve if just placed in water but is completely digested by enzymes in a few hours. All the solid starch in foods such as bread and potatoes is digested to **glucose**, which is soluble in water. The solid proteins in meat, egg and beans are digested to soluble substances called **amino acids**. Fats are digested to two soluble products called **glycerol** and **fatty acids** (see p. 12).

The chemical breakdown usually takes place in stages. For example, the starch molecule is made up of hundreds of carbon, hydrogen and oxygen atoms. The first stage of digestion breaks it down to a 12-carbon sugar molecule called **maltose**. The last stage of digestion breaks the maltose molecule into two 6-carbon sugar molecules called glucose (Figure 11.4). Protein molecules are digested first to smaller molecules called **peptides** and finally into completely soluble molecules called amino acids.

$$\text{starch} \rightarrow \text{maltose} \rightarrow \text{glucose}$$
$$\text{protein} \rightarrow \text{peptide} \rightarrow \text{amino acid}$$

These stages take place in different parts of the alimentary canal. The progress of food through the canal and the stages of digestion will now be described (Figures 11.5 and 11.6).

enzyme A (amylase)
enzyme A (amylase)
enzyme B (maltase)
enzyme B (maltase)
A large molecule (e.g. starch) . . .
. . . is attacked by enzymes . . .
. . . and broken into smaller molecules (e.g. the sugar, maltose) . . .
. . . which are attacked by different enzymes . . .
. . . and broken into even smaller molecules (e.g. the sugar, glucose)

Figure 11.4 Enzymes acting on starch. A large molecule (e.g. starch) is attacked by enzymes (A) and broken into smaller molecules (e.g. the sugar maltose) which are attacked by different enzymes (B) and broken into even smaller molecules (e.g. the sugar glucose)

The mouth

The act of taking food into the mouth is called **ingestion**. In the mouth, the food is chewed and mixed with **saliva**. The chewing breaks the food into pieces which can be swallowed and it also increases the surface area for the enzymes to work on later. Saliva is a digestive juice produced by three pairs of glands whose ducts lead into the mouth (Figure 11.6). It helps to lubricate the food and make the small pieces stick together. Saliva contains one enzyme, **salivary amylase** (sometimes called **ptyalin**), which acts on cooked starch and begins to break it down into maltose.

Strictly speaking, the 'mouth' is the aperture between the lips. The space inside, containing the tongue and teeth, is called the **buccal cavity**. Beyond the buccal cavity is the 'throat' or **pharynx**.

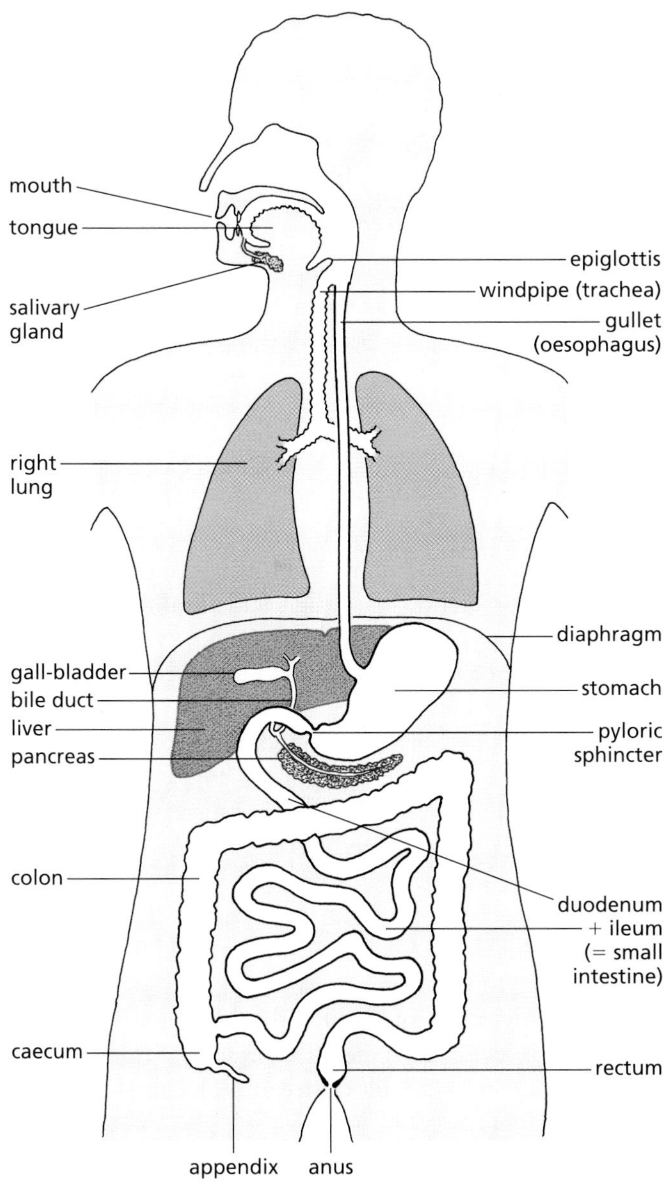

Figure 11.5 The alimentary canal

Swallowing

By studying Figure 11.6a, it can be seen that for food to enter the gullet (oesophagus), it has to pass over the windpipe. All the complicated actions which occur during swallowing ensure that food does not enter the windpipe and cause choking.

1 The tongue presses upwards and back against the roof of the mouth, forcing a pellet of food, called a **bolus**, to the back of the mouth.
2 The soft palate closes the nasal cavity at the back.
3 The larynx cartilage round the top of the windpipe is pulled upwards so that the opening of the windpipe (the **glottis**) lies under the back of the tongue.
4 The glottis is also partly closed by the contraction of a ring of muscle.
5 The **epiglottis**, a flap of cartilage (gristle) helps to prevent the food from going down the windpipe instead of the gullet.

The beginning of the swallowing action is voluntary, but once the food reaches the back of the mouth, swallowing becomes an automatic or reflex action. The food is forced into and down the oesophagus, or gullet, by peristalsis. This takes about 6 seconds with relatively solid food and then the food is admitted to the stomach. Liquid travels more rapidly down the gullet.

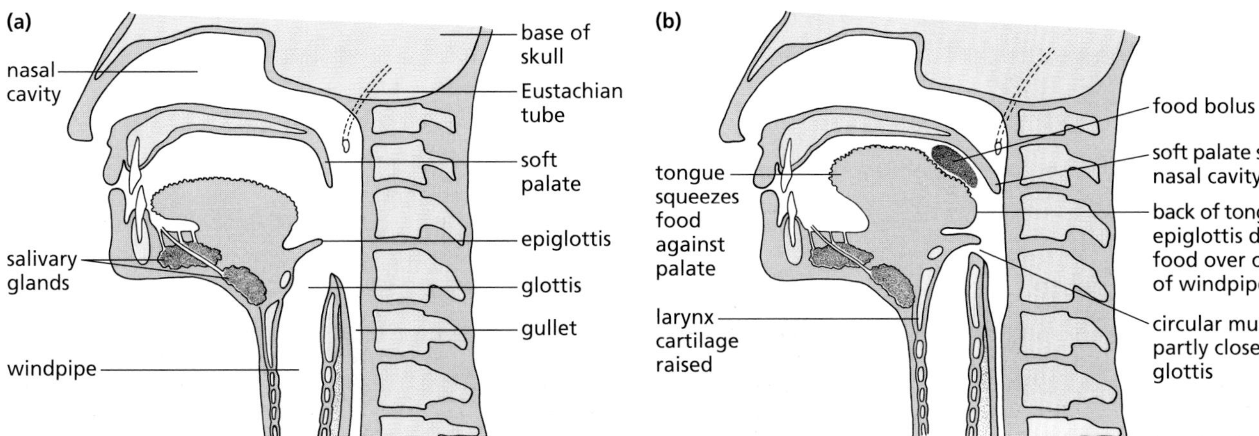

Figure 11.6 Section through head to show swallowing action

The stomach

The stomach has elastic walls which stretch as the food collects in it. The **pyloric sphincter** is a circular band of muscle at the lower end of the stomach which stops solid pieces of food from passing through. The main function of the stomach is to store the food from a meal, turn it into a liquid and release it in small quantities at a time to the rest of the alimentary canal.

Glands in the lining of the stomach (Figure 11.7) produce **gastric juice** containing the enzyme **pepsin**. Pepsin is a **protease** (or proteinase), i.e. it acts on proteins and breaks them down into soluble compounds called peptides. The stomach lining also produces hydrochloric acid which makes a weak solution in the gastric juice. This acid provides the best degree of acidity for pepsin to work in (see p. 15) and kills many of the bacteria taken in with the food.

The regular, peristaltic movements of the stomach, about once every 20 seconds, mix up the food and gastric juice into a creamy liquid. How long food remains in the stomach depends on its nature. Water may pass through in a few minutes; a meal of carbohydrate such as porridge may be held in the stomach for less than an hour, but a mixed meal containing protein and fat may be in the stomach for 1 or 2 hours.

The pyloric sphincter lets the liquid products of digestion pass, a little at a time, into the first part of the small intestine called the **duodenum**.

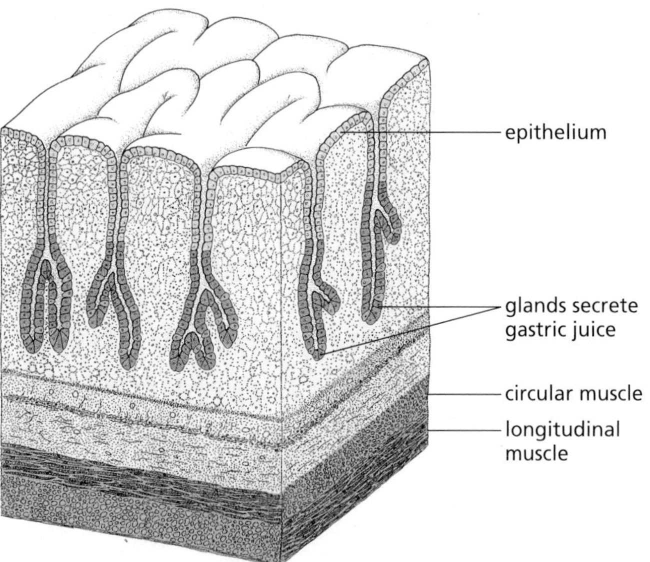

Figure 11.7 Diagram of section through stomach wall

The small intestine

A digestive juice from the pancreas (**pancreatic juice**) and bile from the liver are poured into the duodenum to act on food there. The pancreas is a digestive gland lying below the stomach (Figure 11.8). It makes a number of enzymes, which act on all classes of food.

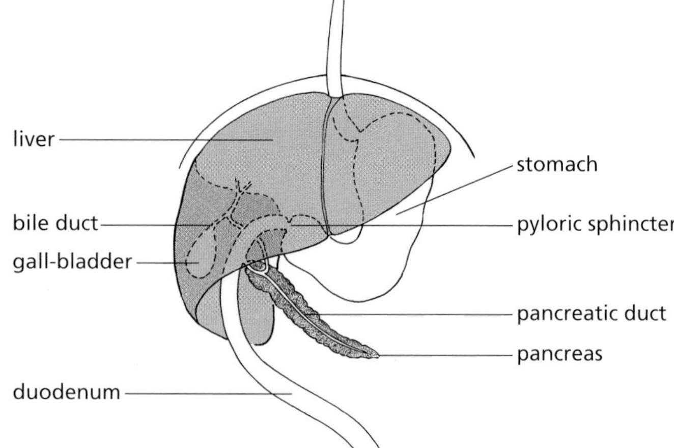

Figure 11.8 Relationship between stomach, liver and pancreas

There are several proteases which break down proteins to peptides and amino acids. **Pancreatic amylase** attacks starch and converts it to maltose. **Lipase** digests fats (lipids) to fatty acids and glycerol.

Pancreatic juice contains sodium hydrogencarbonate which partly neutralizes the acid liquid from the stomach. This is necessary because the enzymes of the pancreas do not work well in acid conditions.

Bile is a green, watery fluid made in the liver, stored in the gall-bladder and delivered to the duodenum by the bile duct (Figure 11.8). It contains no enzymes, but its green colour is caused by bile pigments which are formed from the breakdown of haemoglobin in the liver. Bile also contains bile salts which act on fats rather like a detergent. The bile salts **emulsify** the fats. That is, they break them up into small drops which are more easily digested by lipase.

All the digestible material is thus changed to soluble compounds which can pass through the lining of the intestine and into the bloodstream. The final products of digestion are:

food		**final products**
starch	→	glucose
proteins	→	amino acids
fats (lipids)	→	fatty acids and glycerol

The small intestine itself does not appear to liberate digestive enzymes. The structure labelled 'crypt' in Figure 11.10 on p. 102 is not a digestive gland, though some of its cells do produce mucus and other secretions. The main function of the crypts is to produce new epithelial cells (see 'Absorption', right) to replace those lost from the tips of the villi.

The epithelial cells of the villi contain, in their cell membranes, enzymes which complete the breakdown of sugars and peptides, before they pass through the cells on their way to the bloodstream. For example, the enzyme **maltase** converts the disaccharide maltose into the monosaccharide, glucose.

Prevention of self-digestion

The gland cells of the stomach and pancreas make protein-digesting enzymes (proteases) and yet the proteins of the cells which make these enzymes are not digested. One reason for this is that the proteases are secreted in an inactive form. Pepsin is produced as **pepsinogen** and does not become the active enzyme until it encounters the hydrochloric acid in the stomach. The lining of the stomach is protected from the action of pepsin probably by the layer of mucus.

Similarly, trypsin, one of the proteases from the pancreas, is secreted as the inactive **trypsinogen** and is activated by **enterokinase**, an enzyme secreted by the lining of the duodenum.

Control of secretion

The sight, smell and taste of food set off nerve impulses from the sense organs to the brain. The brain relays these impulses to the stomach and initiates gastric secretion. When the food reaches the stomach, it stimulates the stomach lining to produce a hormone (p. 169) called **gastrin**. This hormone circulates in the blood and, when it returns to the stomach in the bloodstream, it stimulates the gastric glands to continue secretion. Thus, gastric secretion is maintained all the time food is present.

In a similar way, the pancreas is affected first by nervous impulses and then by the hormone **secretin**. Secretin is released into the blood from cells in the duodenum when they are stimulated by the acid contents of the stomach. When secretin reaches the pancreas, it stimulates it to produce pancreatic juice.

Caecum and appendix

In some grass-eating animals (herbivores) such as horses and rabbits, the caecum and appendix are quite large. It is in these organs that digestion of plant cell walls takes place, largely as a result of bacterial activity.

In humans, the caecum and appendix are small structures, possibly without digestive function. The appendix, however, contains lymphoid tissue (p. 115) and may have an immunological function (p. 117).

Questions

1 Why can you not breathe while you are swallowing?

2 Why is it necessary for our food to be digested? Why do plants not need a digestive system? (See p. 35.)

3 In which parts of the alimentary canal is
 a starch
 b protein digested?

4 Study the characteristics of enzymes on pp. 14 and 15. In what ways does pepsin show the characteristics of an enzyme?

Absorption

The small intestine consists of the duodenum and the **ileum**. Nearly all the absorption of digested food takes place in the ileum, which is efficient at this for the following reasons:

- It is fairly long and presents a large absorbing surface to the digested food.
- Its internal surface is greatly increased by circular folds (Figure 11.9) bearing thousands of tiny projections called **villi** (singular = villus) (Figures 11.10 and 11.11, overleaf). These villi are about 0.5 mm long and may be finger-like or flattened in shape.
- The lining epithelium is very thin and the fluids can pass rapidly through it. The outer membrane of each epithelial cell has microvilli (p. 27) which increase by 20 times the exposed surface of the cell.
- There is a dense network of blood capillaries (tiny blood vessels, see p. 113) in each villus (Figure 11.10).

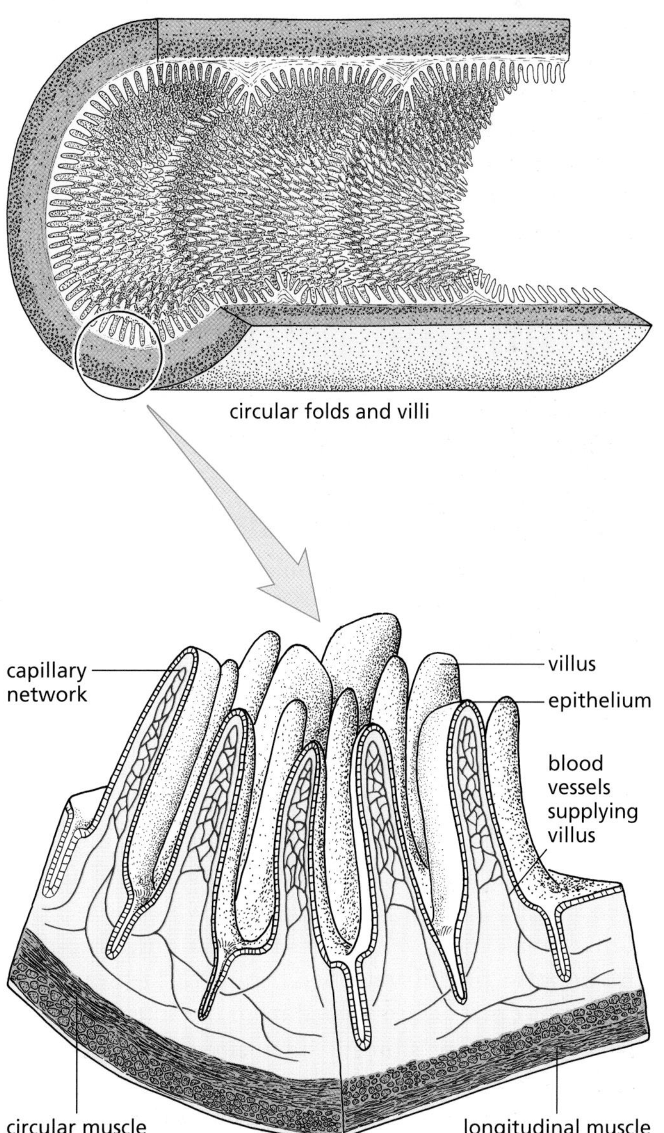

Figure 11.9 The absorbing surface of the ileum

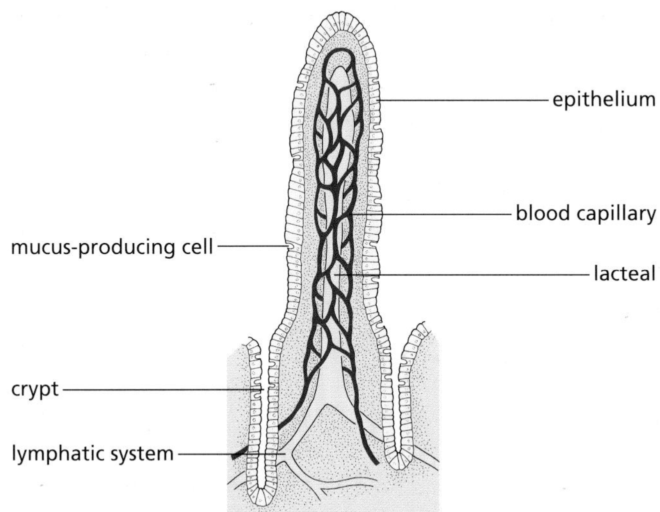

Figure 11.10 Structure of a single villus

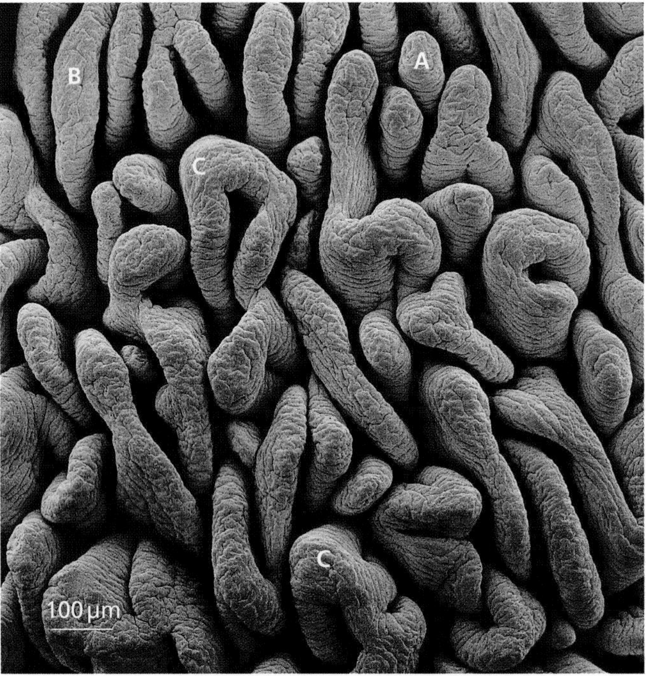

Figure 11.11 Scanning electron micrograph of the human intestinal lining (× 60). The villi are about 0.5 mm long. In the duodenum they are mostly leaf-like (C), but further towards the ileum they become narrower (B), and in the ileum they are mostly finger-like (A). This micrograph is of a region in the duodenum

The small molecules of the digested food, for example, glucose and amino acids, pass into the epithelial cells and then through the wall of the capillaries in the villus and into the bloodstream. They are then carried away in the capillaries, which join up to form veins. These veins unite to form one large vein, the **hepatic portal vein** (see Figure 12.11 on p. 112). This vein carries all the blood from the intestine to the liver, which may store or alter any of the digestion products. When these products are released from the liver, they enter the general blood circulation.

Some of the fatty acids and glycerol from the digestion of fats enter the blood capillaries of the villi. However, a large proportion of the fatty acids and glycerol may be combined to form fats again in the intestinal epithelium. These fats then pass into the **lacteals** (Figure 11.10). The fluid in the lacteals flows into the **lymphatic system,** which forms a network all over the body and eventually empties its contents into the bloodstream (see p. 115).

Absorption of the products of digestion and other dietary items is not just a matter of simple diffusion, except perhaps for alcohol and, sometimes, water. Although the mechanisms for transport across the intestinal epithelium have not been fully worked out, it seems likely that various forms of active transport (p. 28) are involved. Even water can cross the epithelium against an osmotic gradient (p. 29). Amino acids, sugars and salts are, almost certainly, taken up by active transport. Glucose, for example, crosses the epithelium faster than fructose although their rates of diffusion would be about the same.

Water-soluble vitamins may diffuse into the epithelium but fat-soluble vitamins are carried in the microscopic fat droplets that enter the cells. The ions of mineral salts are probably absorbed by active transport. Calcium ions need vitamin D for their effective absorption.

The epithelial cells of the villi are constantly being shed into the intestine. Rapid cell division in the epithelium of the crypts (Figure 11.10) replaces these lost cells. In effect there is a steady procession of epithelial cells moving up from the crypts to the villi.

The large intestine (colon and rectum)

The material passing into the large intestine consists of water with undigested matter, largely cellulose and vegetable fibres (roughage), mucus and dead cells from the lining of the alimentary canal. The large intestine secretes no enzymes but the bacteria in the colon digest part of the fibre to form fatty acids which the colon can absorb. Bile salts are absorbed and returned to the liver by the blood circulation. The colon also absorbs much of the water from the undigested residues. About 7 litres of digestive juices are poured into the alimentary canal each day. If the water from these was not absorbed by the ileum and colon, the body would soon be dehydrated.

The semi-solid waste, the **faeces** or 'stool', is passed into the rectum by peristalsis and is expelled at intervals through the anus. The residues may spend from 12 to 24 hours in the intestine. The act of expelling the faeces is called **egestion** or **defecation**.

Use of digested food

The products of digestion are carried round the body in the blood. From the blood, cells absorb and use glucose, fats and amino acids. This uptake and use of food is called **assimilation**.

Glucose

During respiration in the cells, glucose is oxidized to carbon dioxide and water (see p. 19). This reaction provides energy to drive the many chemical processes in the cells which result in, for example, the building-up of proteins, contraction of muscles or electrical changes in nerves.

Fats

These are built into cell membranes and other cell structures. Fats also form an important source of energy for cell metabolism. Fatty acids produced from stored fats, or taken in with the food, are oxidized in the cells to carbon dioxide and water. This releases energy for processes such as muscle contraction. Fats can provide twice as much energy as sugars.

Amino acids

These are absorbed by the cells and built up, with the aid of enzymes, into proteins. Some of the proteins will become plasma proteins in the blood (p. 109). Others may form structures such as the cell membrane or they may become enzymes which control the chemical activity within the cell. Amino acids not needed for making cell proteins are converted by the liver into glycogen which can then be used for energy.

Questions

1 What are the products of digestion of
  a starch,
  b protein,
  c fats, which are absorbed by the ileum?

2 What characteristics of the small intestine enable it to absorb digested food efficiently?

3 State briefly what happens to a protein molecule in food, from the time it is swallowed, to the time its products are built up into the cytoplasm of a muscle cell.

4 List the chemical changes which a starch molecule undergoes from the time it reaches the duodenum to the time its carbon atoms become part of carbon dioxide molecules. Say where in the body these changes occur.

Storage of digested food

If more food is taken in than the body needs for energy or for building tissues, such as bone or muscle, it is stored in one of the following ways.

Glucose

The sugar not required immediately for energy in the cells is changed in the liver to **glycogen**. The glycogen molecule is built up by combining many glucose molecules into a long-chain molecule similar to that of starch (see p. 12). Some of this insoluble glycogen is stored in the liver and the rest in the muscles. When the blood sugar falls below a certain level, the liver changes its glycogen back to glucose and releases it into the circulation. The muscle glycogen is not returned to the circulation but is used by muscle cells as a source of energy during muscular activity.

The glycogen in the liver is a 'short-term' store, sufficient for only about 6 hours. Excess glucose not stored as glycogen is converted to fat and stored in the fat depots.

Fat

Unlike glycogen, there is no limit to the amount of fat stored and because of its high energy value (p. 87) it is an effective 'long-term' store. The fat is stored in adipose tissue in the abdomen, round the kidneys and under the skin. These are the **fat depots**. The adipose tissue of the skin is shown in Figure 15.1 on p. 137.

Certain cells accumulate drops of fat in their cytoplasm. As these drops increase in size and number, they join together to form one large globule of fat in the middle of the cell, pushing the cytoplasm into a thin layer and the nucleus to one side (Figure 11.12). Groups of fat cells form adipose tissue.

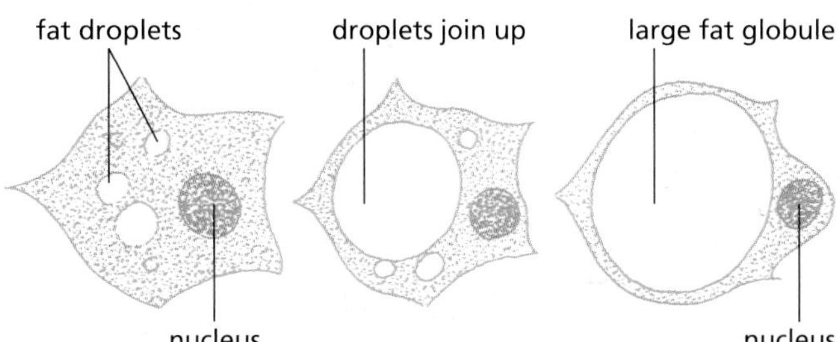

Figure 11.12 Accumulation of fat in a fat cell

Amino acids

Amino acids are not stored in the body. Those not used in protein formation are **deaminated** (see below). The protein of the liver and other tissues can act as a kind of protein store to maintain the protein level in the blood, but absence of protein in the diet soon leads to serious disorders.

All these conversions of one substance to another need specific enzymes to make them happen.

Body weight

The rate at which glucose is oxidized or changed into glycogen and fat is controlled by hormones (p. 169). When intake of carbohydrate and fat is more than enough to meet the energy requirements of the body, the surplus will be stored mainly as fat. Some people never seem to get fat no matter how much they eat, while others start to lay down fat when their intake only just exceeds their needs. Putting on weight is certainly the result of eating more food than the body needs, but it is not clear why people should differ so much in this respect. The explanation probably lies in the balance of hormones which, to some extent, is determined by heredity.

The liver

The liver has been mentioned several times in connection with the digestion, use and storage of food. This is only one aspect of its many important functions, some of which are listed below. It is a large, reddish-brown organ which lies just beneath the diaphragm and partly overlaps the stomach. All the blood from the blood vessels of the alimentary canal passes through the liver, which adjusts the composition of the blood before releasing it into the general circulation (Figure 11.13).

Regulation of blood sugar

After a meal, the liver removes excess glucose from the blood and stores it as glycogen. In the periods between meals, when the glucose concentration in the blood starts to fall, the liver converts some of its stored glycogen into glucose and releases it into the bloodstream. In this way, the concentration of sugar in the blood is kept at a fairly steady level.

The concentration of glucose in the blood of a person who has not eaten for 8 hours is usually between 90 and 100 mg/100 cm^3 blood. After a meal containing carbohydrate, the blood sugar level may rise to 140 mg/100 cm^3 but 2 hours later, the level returns to about 95 mg as the liver has converted the excess glucose to glycogen.

About 100 g glycogen is stored in the liver of a healthy man. If the concentration of glucose in the blood falls below about 80 mg/100 cm^3 blood, some of the glycogen stored in the liver is converted by enzyme action into glucose, which enters the circulation. If the blood sugar level rises above 160 mg/100 cm^3, glucose is excreted by the kidneys. A blood glucose level below 40 mg/100 cm^3 affects the brain cells adversely, leading to convulsions and coma. By helping to keep the glucose concentration between 80 and 150 mg, the liver prevents these undesirable effects and so contributes to the homeostasis (see below) of the body. (See Figure 12.11 on p. 112 for the circulatory supply to liver.)

Production of bile

Cells in the liver make bile continuously and this is stored in the gall-bladder until it is discharged through the bile duct into the duodenum. The green colour of the bile results from a pigment, **bilirubin**, which comes from the breakdown of haemoglobin from worn-out red blood cells.

The bile also contains bile salts which assist the digestion of fats as described on p. 100.

A large proportion of the bile salts is reabsorbed in the ileum along with the fats they have helped to emulsify. Bile salts are also absorbed in the colon.

Deamination

The amino acids not needed for making proteins are converted to glycogen in the liver. During this process, the nitrogen-containing, amino part (NH_2) of the amino acid is removed and changed to **urea**, which is later excreted by the kidneys (see p. 131).

When the -NH_2 group is removed from certain amino acids it forms ammonia, NH_3 (or, more strictly, the ammonium ion -NH_4^+). Ammonia is very poisonous to the body cells, and the liver converts it at once to urea, $(NH_2)_2CO$, which is a comparatively harmless substance.

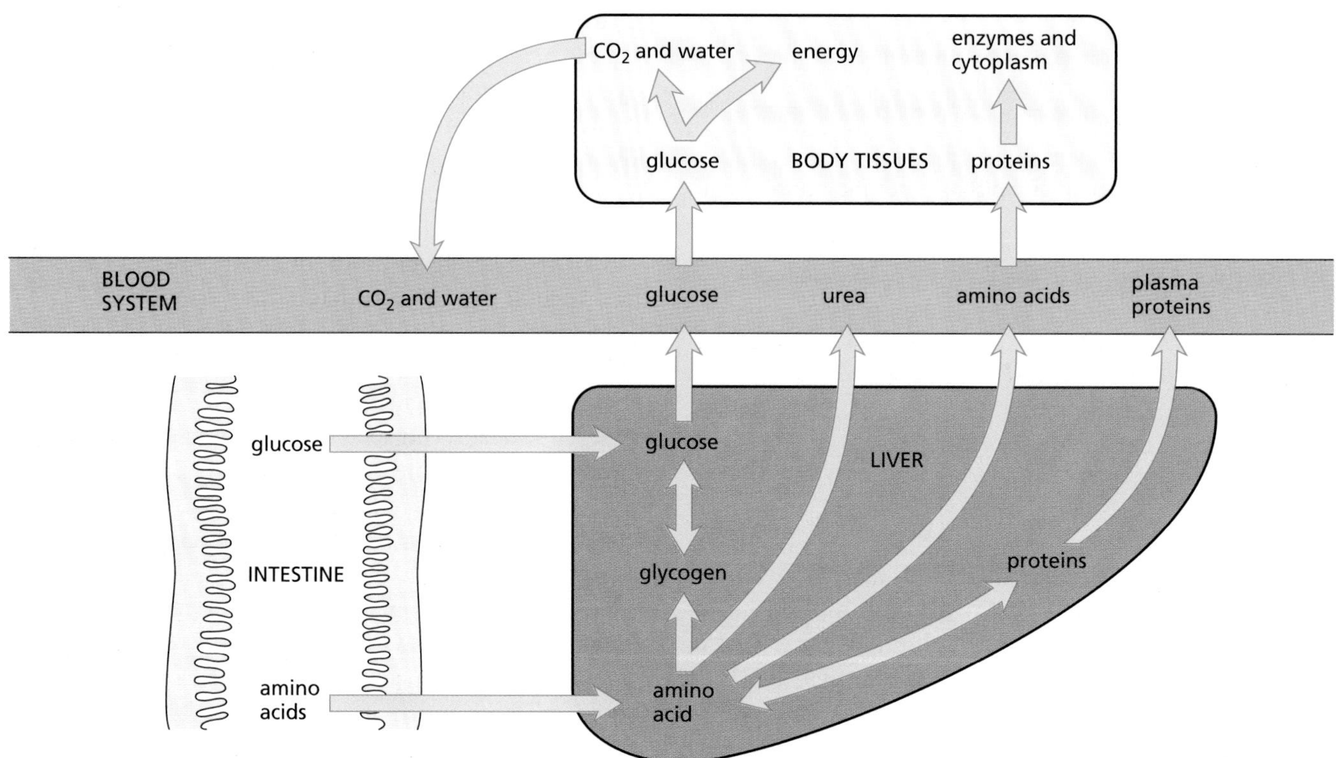

Figure 11.13 Some functions of the liver

Storage of iron
Millions of red blood cells break down every day. The iron from their haemoglobin (p. 108) is stored in the liver.

Manufacture of plasma proteins
The liver makes most of the proteins found in blood plasma, including fibrinogen, which plays an important part in the clotting action of the blood (p. 117).

Detoxication
Poisonous compounds, produced in the large intestine by the action of bacteria on amino acids, enter the blood, but on reaching the liver are converted to harmless substances, later excreted in the urine. Many other chemical substances normally present in the body or introduced as drugs are modified by the liver before being excreted by the kidneys. Hormones, for example, are converted to inactive compounds in the liver, so limiting their period of activity in the body.

Storage of vitamins
The fat-soluble vitamins A and D are stored in the liver. This is the reason why animal liver is a valuable source of these vitamins in the diet.

Questions

1. Explain how the liver exercises control over the substances coming from the intestine and entering the general blood circulation.
2. What contribution does the liver make to the process of digestion?
3. Suggest some ways in which extreme variations in blood composition might be harmful.

Homeostasis
A complete account of the functions of the liver would involve a very long list. It is most important, however, to realize that the one vital function of the liver, embodying all the details outlined above, is that it helps to maintain the concentration and composition of the body fluids, particularly the blood.

Within reason, a variation in the kind of food eaten will not produce changes in the composition of the blood.

If this **internal environment**, as it is called, were not so constant, the chemical changes that maintain life would become erratic and unpredictable so that with quite slight changes of diet or activity the whole organization might break down. The maintenance of the internal environment within narrow limits is called **homeostasis** and is discussed again on pp. 135, 139 and 172.

Practical work

1 The action of salivary amylase on starch
Rinse the mouth with water to remove traces of food. Collect saliva* in two test-tubes, labelled A and B, to a depth of about 15 mm (see Figure 11.14). Heat the saliva in tube B over a small flame until it boils for about 30 seconds and then cool the tube under the tap. Add about 2 cm^3 of a 2 per cent starch solution to each tube; shake each tube and leave them for 5 minutes.

Share the contents of tube A between two clean test-tubes. To one of these add some iodine solution. To the other add some Benedict's solution and heat in a water bath as described on p. 36. Test the contents of tube B in exactly the same way.

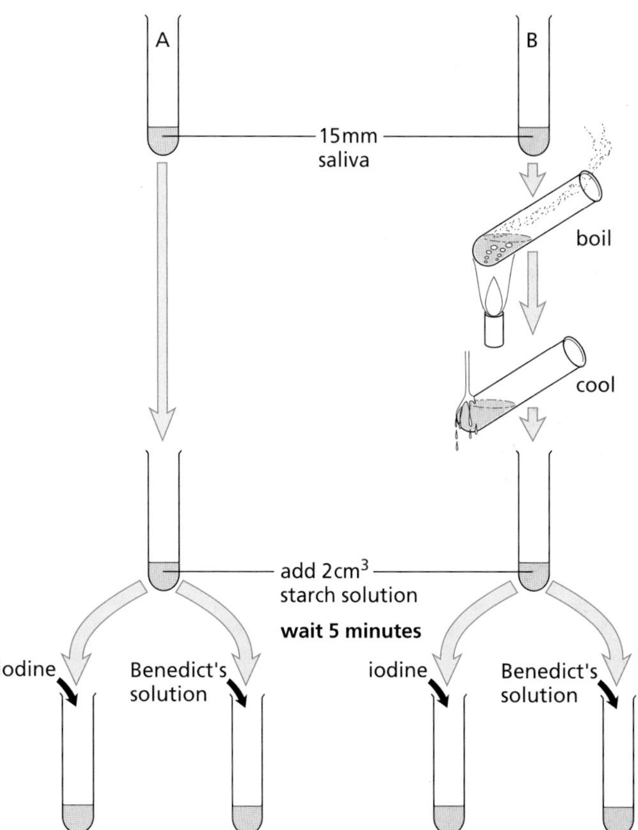

Figure 11.14 Salivary amylase acting on starch

Results The contents of tube A fail to give a blue colour with iodine, showing that the starch has gone. The other half of the contents, however, gives a red or orange precipitate with Benedict's solution, showing that sugar is present.

The contents of tube B still give a blue colour with iodine but do not form a red precipitate on heating with Benedict's solution.

*If there is some objection to using your own saliva, use a 5 per cent solution of commercially prepared amylase instead.

Interpretation The results with tube A suggest that something in saliva has converted starch into sugar. The fact that the boiled saliva in tube B fails to do this, suggests that it was an enzyme in saliva which brought about the change (see p. 15), because enzymes are proteins and are destroyed by boiling. If the boiled saliva had changed starch to sugar, it would have ruled out the possibility of an enzyme being responsible.

This interpretation assumes that it is something in saliva which changes starch into sugar. However, the results could equally well support the claim that starch can turn unboiled saliva into sugar. Our knowledge of (1) the chemical composition of starch and saliva and (2) the effect of heat on enzymes, makes the first interpretation more plausible.

2 The action of pepsin on egg-white protein

A cloudy suspension of egg-white is prepared by stirring the white of one egg into 500 cm^3 tap water, heating it to boiling point and filtering it through glass wool to remove the larger particles.

Label four test-tubes A, B, C and D and place 2 cm^3 egg-white suspension in each of them. Then add pepsin solution and/or dilute hydrochloric acid to the tubes as follows (Figure 11.15):

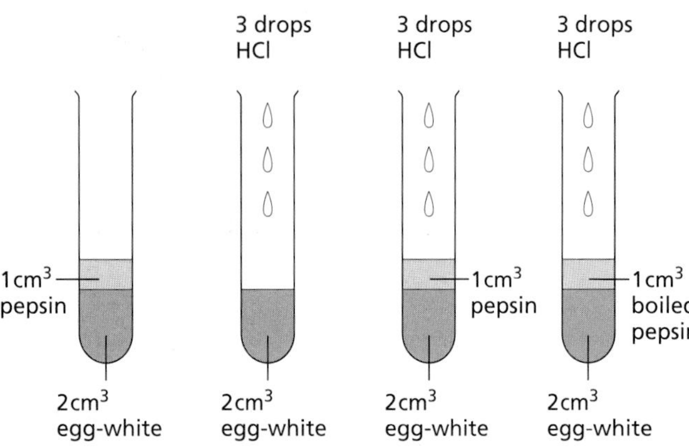

Figure 11.15 Pepsin acting on egg-white

A Egg-white suspension + 1 cm^3 pepsin solution (1 per cent)
B Egg-white suspension + 3 drops dilute hydrochloric acid (HCl)
C Egg-white suspension + 1 cm^3 pepsin + 3 drops HCl
D Egg-white suspension + 1 cm^3 boiled pepsin + 3 drops HCl

Place all four tubes in a beaker of warm water at 35 °C for 10–15 minutes.

Result The contents of tube C go clear. The rest remain cloudy.

Interpretation The change from a cloudy suspension to a clear solution shows that the solid particles of egg protein have been digested to soluble products. The failure of the other three tubes to give clear solutions shows that:

- Pepsin will only work in acid solutions.
- It is the pepsin and not the hydrochloric acid which does the digestion.
- Pepsin is an enzyme, because its activity is destroyed by boiling.

3 The action of lipase

Place 5 cm^3 milk and 7 cm^3 dilute (M/20) sodium carbonate solution in each of three test-tubes labelled 1 to 3 (Figure 11.16). Add 6 drops of phenolphthalein to each to turn the contents pink. Add 1 cm^3 of 3 per cent bile salts solution to tubes 2 and 3. Add 1 cm^3 of 5 per cent lipase solution to tubes 1 and 3, and an equal volume of boiled lipase to tube 2.

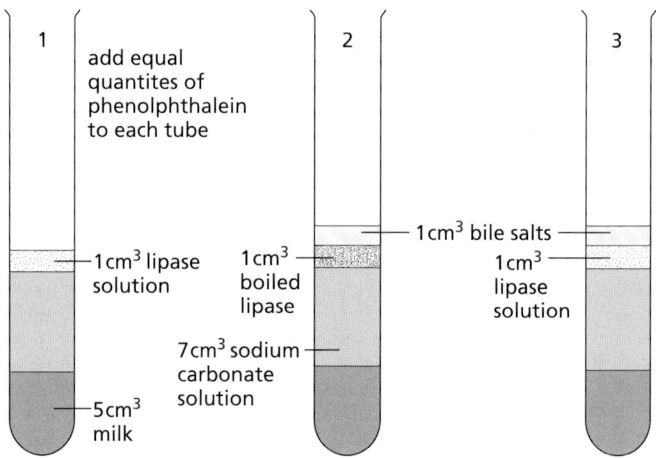

Figure 11.16 The action of lipase

Result In 10 minutes or less, the colour of the liquids in tubes 1 and 3 will change to white, tube 3 changing first. The liquid in tube 2 will remain pink.

Interpretation Lipase is an enzyme that digests fats to fatty acids and glycerol. When lipase acts on milk fats, the fatty acids so produced react with the alkaline sodium carbonate and make the solution more acid. In acid conditions the pH indicator, phenolphthalein, changes from pink to colourless. The presence of bile salts in tube 3 seems to accelerate the reaction, although bile salts with the denatured enzyme in tube 2 cannot bring about the change on their own.

For experiments investigating the effect of temperature and pH on enzyme action see pp. 16 and 17.

Questions

1 In experiments with enzymes, the control often involves the boiled enzyme. Suggest why this type of control is used.

2 In Experiment 2, why does the change from cloudy to clear suggest that digestion has occurred?

3 How would you modify Experiment 2 if you wanted to find the optimum temperature for the action of pepsin on egg-white?

4 Experiment 2 is really two experiments combined because there are two variables.
 a Identify the variables.
 b Which of the tubes could be the control?

5 It was suggested that an alternative interpretation of the result in Experiment 1 might be that starch has turned saliva into sugar. From what you know about starch, saliva and the design of the experiment, explain why this is a less acceptable interpretation.

6 Write down the menu for your breakfast and lunch (or supper). State the main food substances present in each item of the meal. State the final digestion product of each.

Checklist

- Digestion is the process which changes insoluble food into soluble substances (Table 11.1).
- Digestion takes place in the alimentary canal.
- The changes are brought about by chemicals called digestive enzymes.
- Maltose and sucrose are changed to glucose by enzymes in the epithelium of the villi.
- The ileum absorbs amino acids, glucose and fats.
- These are carried in the bloodstream first to the liver and then to all parts of the body.
- Internal folds, villi and microvilli greatly increase the absorbing surface of the small intestine.
- The digested food is used or stored in the following ways:
  Glucose is (1) oxidized for energy or (2) changed to glycogen or fat and stored.
  Amino acids are (1) built up into proteins or (2) deaminated to urea and glycogen and used for energy.
  Fats are (1) oxidized for energy or (2) stored.
- Glycogen in the liver and muscles acts as a short-term energy store; fat in the fat depots acts as a long-term energy store.
- The liver stores glycogen and changes it to glucose and releases it into the bloodstream to keep a steady level of blood sugar.
- The liver exercises control over many other aspects of blood composition and so helps maintain chemical stability in the body.

Table 11.1 Principal substances produced by digestion

Region of alimentary canal	Digestive gland	Digestive juice produced	Enzymes in the juice	Class of food acted upon	Substances produced
mouth	salivary glands	saliva	salivary amylase	starch	maltose
stomach	glands in stomach lining	gastric juice	pepsin	proteins	peptides
duodenum	pancreas	pancreatic juice	proteases	proteins and peptides	peptides and amino acids
			amylase	starch	maltose
			lipase	fats	fatty acids and glycerol

12 The blood circulatory system

Composition of blood
Blood cells and plasma.

The heart
Structure and function.

The circulation
Arteries, veins and capillaries. Blood pressure.

The lymphatic system
Spleen and thymus.

Functions of the blood
Homeostasis, transport and defence.

Antibodies and immunity
Vaccines. B and T lymphocytes.

Transplants and transfusions
ABO blood groups.

Coronary heart disease
Possible causes. Correlation and cause

The previous chapter explained how food is digested to amino acids, glucose, etc., which are absorbed in the small intestine. These substances are needed in all living cells in the body such as the brain cells, leg muscle cells and kidney cells. The substances are carried from the intestine to other parts of the body by the blood system. In a similar way, the oxygen taken in by the lungs is needed by all the cells and is carried round the body in the blood.

Composition of blood

Blood consists of **red cells**, **white cells** and **platelets** floating in a liquid called **plasma**. There are between 5 and 6 litres of blood in the body of an adult, and each cubic centimetre contains about 5 billion red cells.

Red cells

These are tiny, disc-like cells (Figures 12.1a and 12.2) which do not have nuclei. They are made of spongy cytoplasm enclosed in an elastic cell membrane. In their cytoplasm is the red pigment, **haemoglobin**, a protein combined with iron. Haemoglobin combines with oxygen in places where there is a high concentration of oxygen, to form **oxyhaemoglobin**. Oxyhaemoglobin is an unstable compound. It breaks down and releases its oxygen in places where the oxygen concentration is low (Figure 12.3). This makes haemoglobin very useful in carrying oxygen from the lungs to the tissues.

Blood which contains mainly oxyhaemoglobin is said to be **oxygenated**. Blood with little oxyhaemoglobin is called **deoxygenated**.

Each red cell lives for about 4 months, after which it breaks down. The red haemoglobin changes to a yellow pigment, bilirubin, which is excreted in the bile. The iron from the haemoglobin is stored in the liver. About 200 000 million red cells wear out and are replaced each day. This is about 1 per cent of the total.

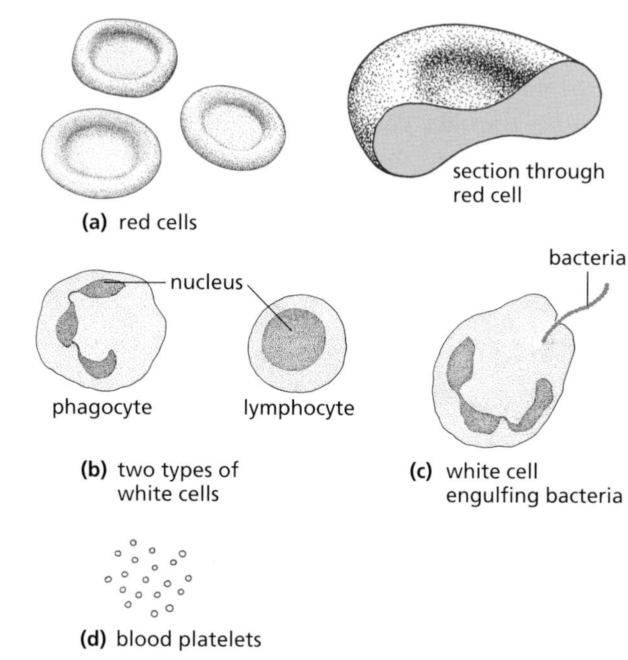

Figure 12.1 Blood cells

Red cells are made by the red bone marrow of certain bones in the skeleton – in the ribs, vertebrae and breastbone for example.

White cells

There are several different kinds of white cell (Figures 12.1b and 12.2). Most are larger than the red cells, and they all have a nucleus. There is 1 white cell to every 600 red cells and they are made in the same bone marrow that makes red cells. Many of them undergo a process of maturation and development in the thymus gland, lymph nodes or spleen (p. 115). The two most numerous types of white cells are **phagocytes** and **lymphocytes**.

The phagocytes can move about by a flowing action of their cytoplasm and can escape from the blood capillaries into the tissues by squeezing between the cells of the capillary walls. They collect at the site of an infection, engulfing (**ingesting**) and digesting harmful bacteria and cell debris (Figure 12.1c). In this way they prevent the spread of infection through the body.

One of the functions of lymphocytes is to produce antibodies (p. 117).

Platelets

These are pieces of special blood cells budded off in the red bone marrow. They help to clot the blood at wounds and so stop the bleeding (p. 117).

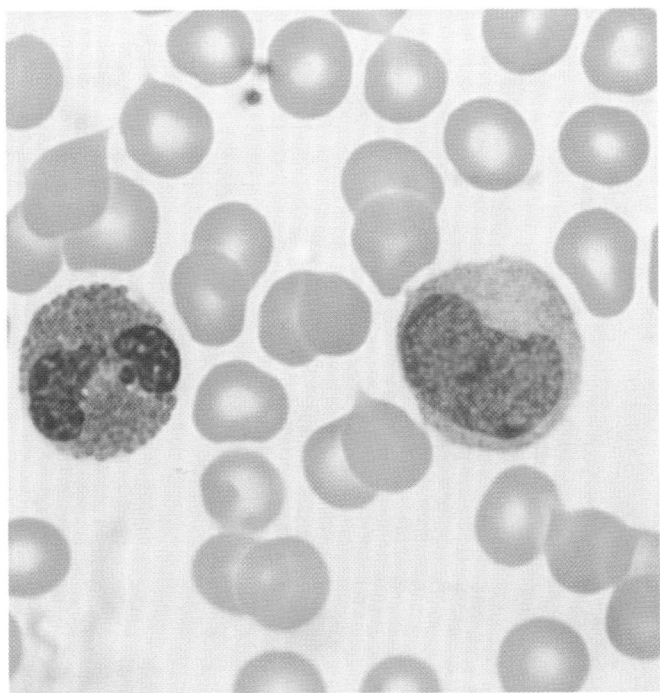

Figure 12.2 Red and white cells from human blood (×2500). The large nucleus can be clearly seen in the white cells

Plasma

The liquid part of the blood is called plasma. It is water with a large number of substances dissolved in it. The ions of sodium, potassium, calcium, chloride and hydrogen carbonate, for example, are present. Proteins such as fibrinogen, albumin and globulins constitute an important part of the plasma. Fibrinogen is needed for clotting (p. 117), and the globulin proteins include the antibodies which combat bacteria and other foreign matter (p. 117). The plasma will also contain varying amounts of food substances such as amino acids, glucose and lipids (fats). There may also be hormones (p. 169) present, depending on the activities taking place in the body. The excretory product, urea, is dissolved in the plasma.

The liver and kidneys keep the composition of the plasma more or less constant, but the amount of digested food, salts and water will vary within narrow limits according to food intake and body activities.

Questions

1. In what ways are white cells different from red cells in
   a their structure,
   b their function?
2. Where, in the body, would you expect haemoglobin to be combining with oxygen to form oxyhaemoglobin?
3. In what parts of the body would you expect oxyhaemoglobin to be breaking down to oxygen and haemoglobin?
4. Why is it important for oxyhaemoglobin to be an unstable compound, i.e. easily changed to oxygen and haemoglobin?

   What might be the effect on a person whose diet contained too little iron?

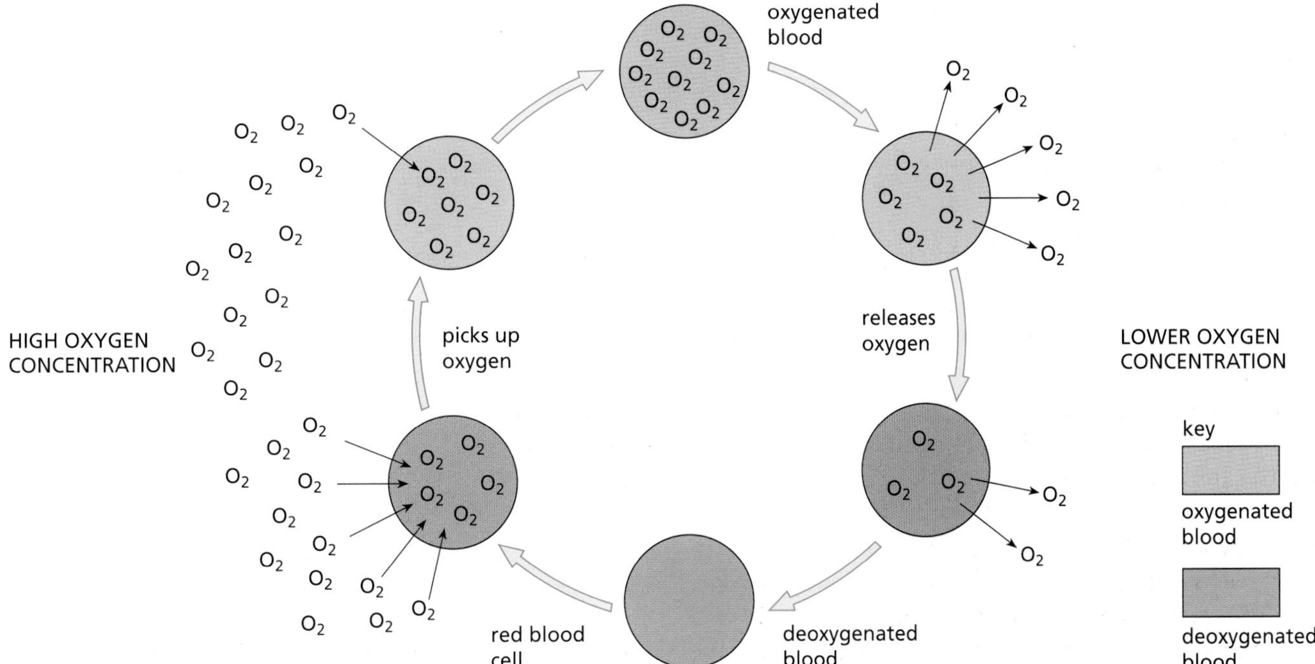

Figure 12.3 The function of the red cells

The heart

The heart pumps blood through the circulatory system all round the body. The appearance of the heart from the outside is shown in Figure 12.4, Figure 12.5 shows the left side cut open, while Figure 12.6 is a diagram of a vertical section to show its internal structure. Since the heart is seen as if in a dissection of a person facing you, the left side is drawn on the right.

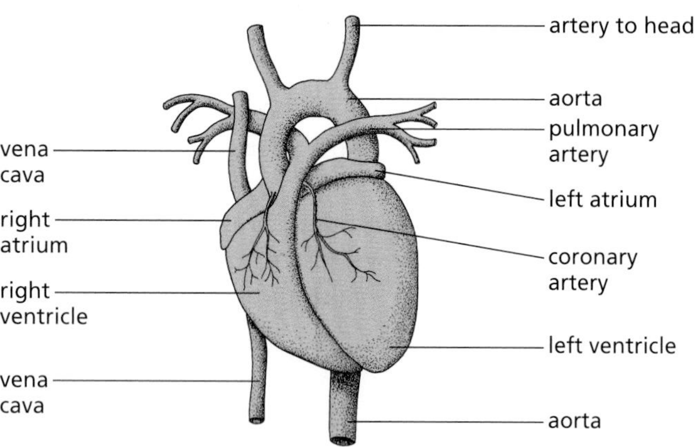

Figure 12.4 External view of the heart

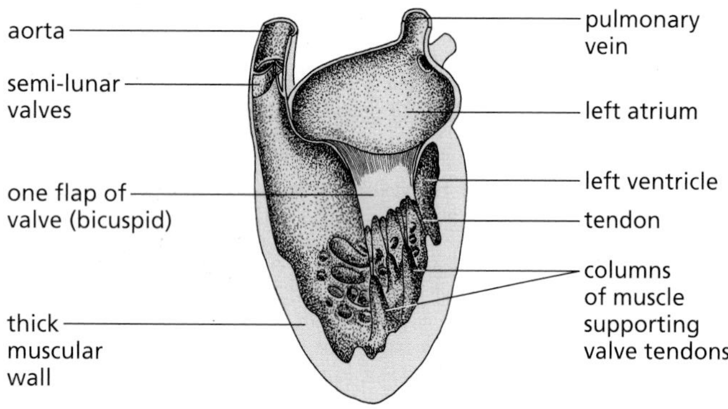

Figure 12.5 Diagram of the heart cut open (left side)

If you study Figure 12.6 you will see that there are four chambers. The upper, thin-walled chambers are the **atria** (singular = atrium) and each of these opens into a thick-walled chamber, the **ventricle**, below.

Blood enters the atria from large veins. The **pulmonary vein** brings oxygenated blood from the lungs into the left atrium. The **vena cava** brings deoxygenated blood from the body tissues into the right atrium. The blood passes from each atrium to its corresponding ventricle, and the ventricle pumps it out into the arteries.

The artery carrying oxygenated blood to the body from the left ventricle is the **aorta**. The **pulmonary artery** carries deoxygenated blood from the right ventricle to the lungs.

In pumping the blood, the muscle in the walls of the atria and ventricles contracts and relaxes (Figure 12.7). The walls of the atria contract first and force blood into the two ventricles. Then the ventricles contract and send blood into the arteries. The blood is stopped from flowing backwards by four sets of valves. Between the right atrium and the right ventricle is the **tricuspid** (= three flaps) valve. Between the left atrium and left ventricle is the **bicuspid** (= two flaps) valve. The flaps of these valves are shaped rather like parachutes, with 'strings' called **tendons** or **cords** to prevent their being turned inside out.

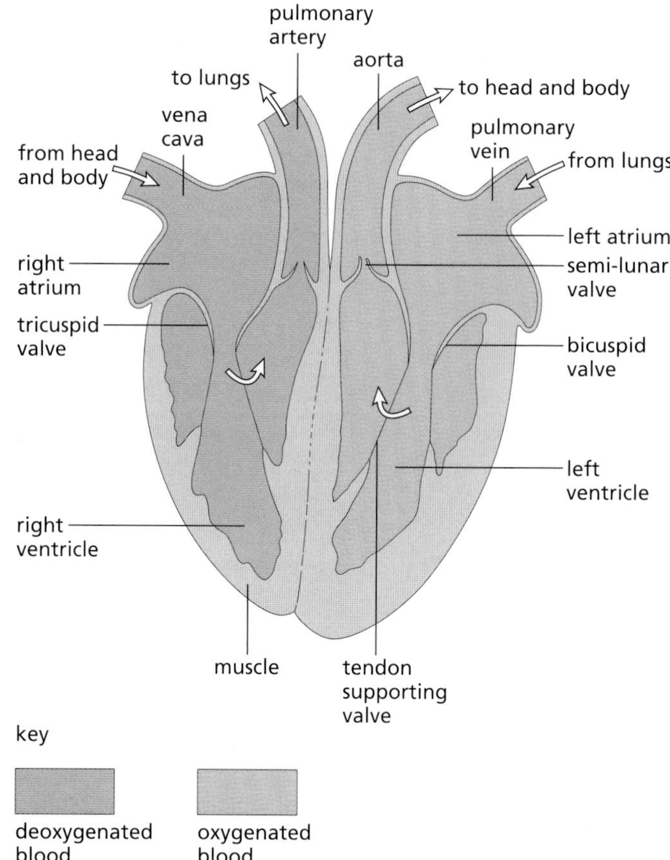

Figure 12.6 Diagram of the heart, vertical section

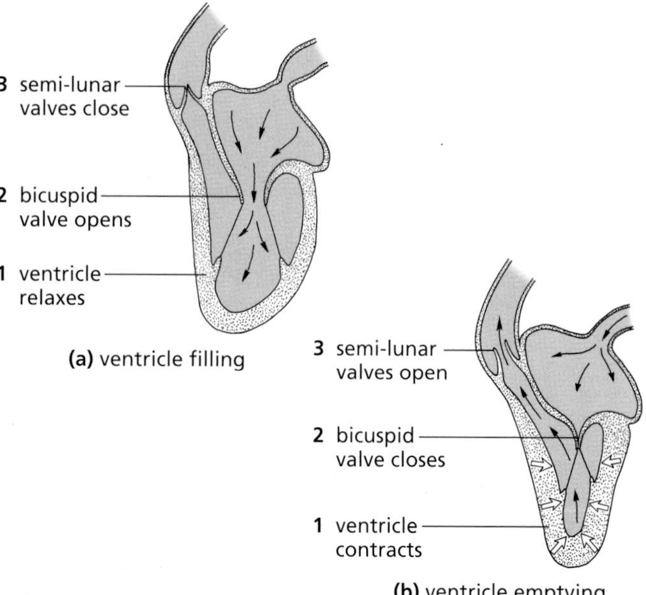

Figure 12.7 Diagram of heart beat (only the left side is shown)

In the pulmonary artery and aorta are the **semi-lunar** (= half-moon) valves. These each consist of three pockets which are pushed flat against the artery walls when blood flows one way. If blood tries to flow the other way, the 'pockets' fill up and meet in the middle to stop the flow of blood (Figure 12.8).

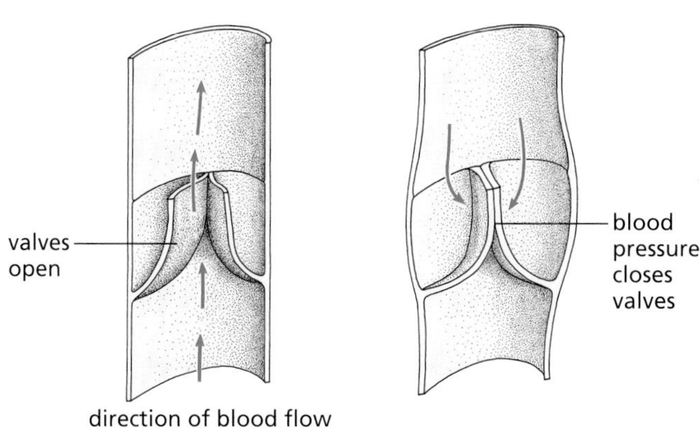

Figure 12.8 Action of the semi-lunar valves

When the ventricles contract, blood pressure closes the bicuspid and tricuspid valves and these prevent blood returning to the atria. When the ventricles relax, the blood pressure in the arteries closes the semi-lunar valves so preventing the return of blood to the ventricles.

The heart muscle is supplied with food and oxygen by the **coronary arteries** (Figure 12.4).

From the description above, it may seem that the ventricles are filled with blood as a result of the contraction of the atria. In fact, when the ventricles relax, their internal volume increases and they draw in blood from the pulmonary vein or vena cava through the relaxed atria. Atrial contraction then forces the final amount of blood into the ventricles just before ventricular contraction.

Control of the heart beat

Heart muscle has a natural rhythmic contraction of its own, about 40 contractions per minute. However, it is supplied by nerves which maintain a faster rate which can be adjusted to meet the body's needs for oxygen. At rest, the normal heart rate may lie between 50 and 100 beats per minute, according to age, sex and other factors. During exercise, the rate may increase to 200 per minute.

The heart beat is initiated by the '**pacemaker**', a small group of specialized muscle cells at the top of the right atrium. The pacemaker receives two sets of nerves from the brain. One group of nerves speeds up the heart rate and the other group slows it down. These nerves originate from a centre in the brain that receives an input from receptors (p. 158) in the circulatory system that are sensitive to blood pressure and levels of oxygen and carbon dioxide in the blood. If blood pressure rises, nervous impulses reduce the heart rate. A fall in blood pressure causes a rise in the rate. Reduced oxygen concentration or increased carbon dioxide in the blood also contribute to a faster rate. By this means, the heart rate is adjusted to meet the needs of the body at times of rest, exertion and excitement.

The hormone **adrenaline** (p. 170) also affects the heart rate. In conditions of excitement, activity or stress, adrenaline is released into the blood circulation from the adrenal glands. On reaching the heart it causes an increase in the rate and strength of the heart beat.

Questions

1 Which parts of the heart
   a pump blood into the arteries,
   b stop blood flowing the wrong way?

2 Put the following in the correct order
   a blood enters arteries,
   b ventricles contract,
   c atria contract,
   d ventricles relax,
   e blood enters ventricles,
   f semi-lunar valves close,
   g tri- and bicuspid valves close.

3 Why do you think that
   a the walls of the ventricles are more muscular than the walls of the atria and
   b the muscle of the left ventricle is thicker than that of the right ventricle?
   (Consult Figure 12.11.)

4 Which important veins are not shown in Figure 12.4?

5 Why is a person whose heart valves are damaged by disease unable to take part in active sport?

The circulation

The blood, pumped by the heart, travels all round the body in blood vessels. It leaves the heart in arteries and returns in veins. The route of the circulation is shown as a diagram in Figure 12.9 (overleaf). The blood passes twice through the heart during one complete circuit: once on its way to the body and again on its way to the lungs. The circulation through the lungs is called the **pulmonary** circulation; the circulation round the rest of the body is called the **systemic** circulation. On average, a red cell would go round the whole circulation in 45 seconds. A more detailed diagram of the circulation is shown in Figure 12.11 (overleaf).

Arteries

These are fairly wide vessels (Figures 12.10a, overleaf, and 12.15, p. 113) which carry blood from the heart to the limbs and organs of the body (Figure 12.17, p. 114). The blood in the arteries, except for the pulmonary arteries, is oxygenated.

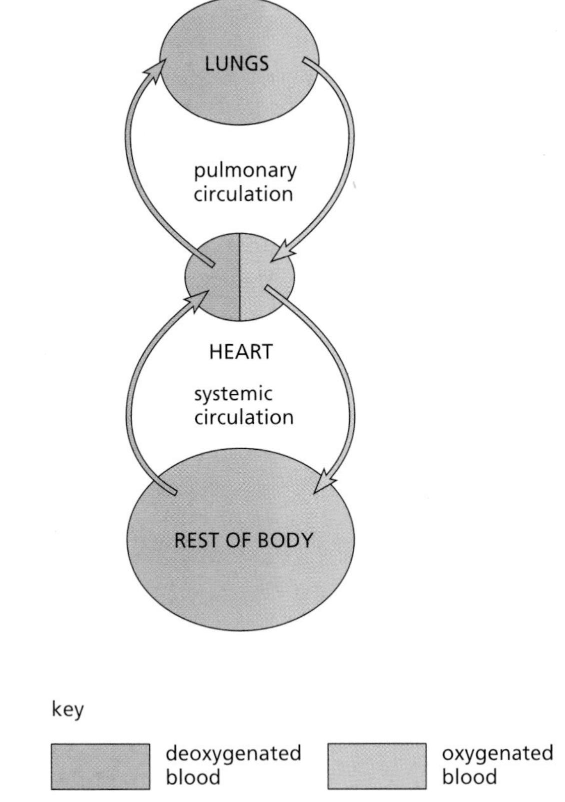

Figure 12.9 Blood circulation

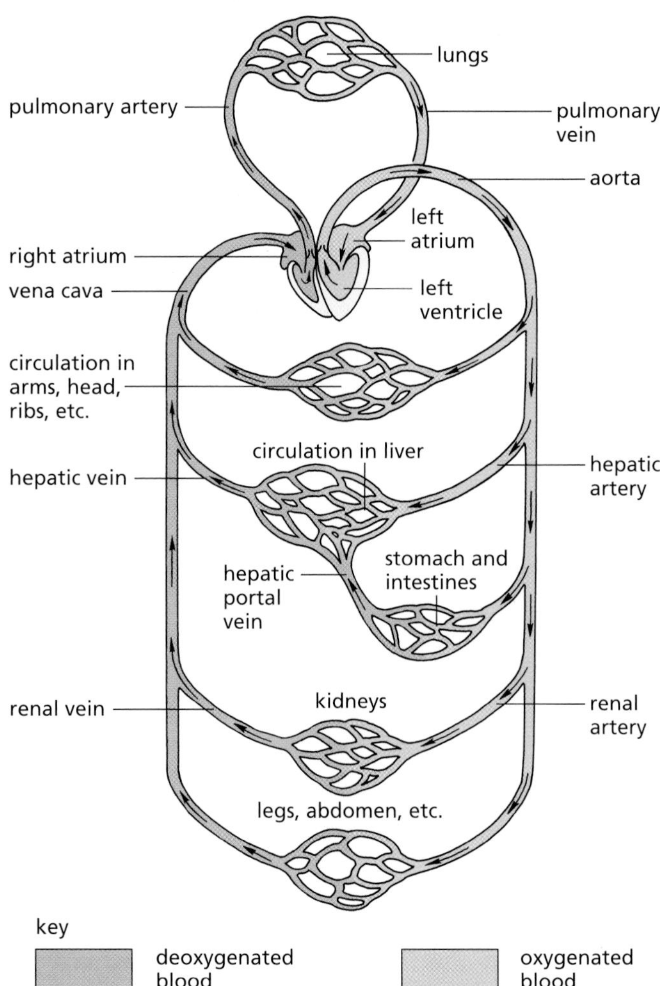

Figure 12.11 Diagram of human circulation

Arteries have elastic tissue and muscle fibres in their thick walls. The large arteries, near the heart, have a greater proportion of elastic tissue which allows these vessels to stand up to the surges of high pressure caused by the heart beat. The ripple of pressure which passes down an artery as a result of the heart beat can be felt as a 'pulse' when the artery is near the surface of the body. You can feel the pulse in your radial artery by pressing the finger-tips of one hand on the wrist of the other (Figure 12.12).

The arteries divide into smaller vessels called **arterioles**. The small arteries and the arterioles have proportionately less elastic tissue and more muscle fibres than the great arteries. When the muscle fibres of the arterioles contract, they make the vessels narrower and restrict the blood flow. In this way, the distribution of blood to different parts of the body can be regulated. (See p. 138 for an example of this.)

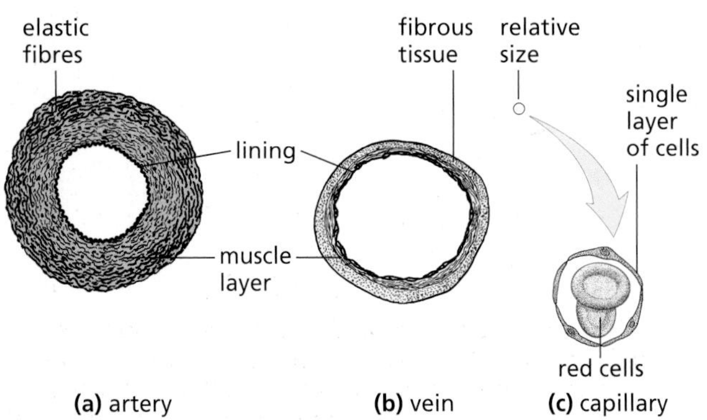

Figure 12.10 Blood vessels, transverse section

The arterioles divide repeatedly to form a branching network of microscopic vessels passing between the cells of every living tissue. These final branches are called **capillaries**.

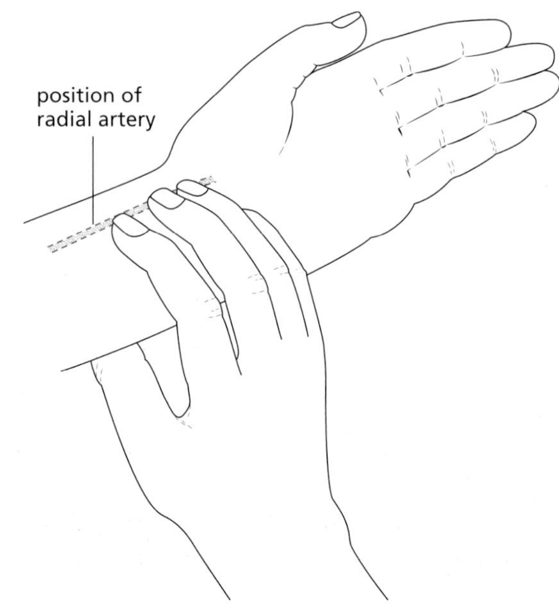

Figure 12.12 Taking the pulse

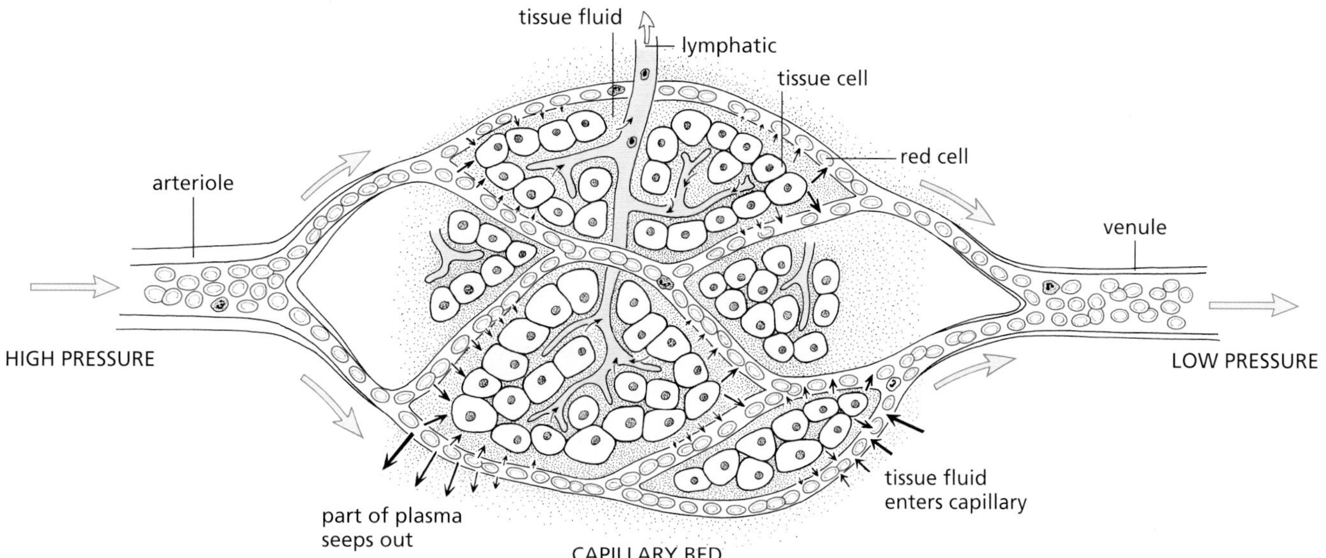

Figure 12.13 Relationship between capillaries, cells and lymphatics. The slow flow rate in the capillaries allows plenty of time for the exchange of oxygen, food, carbon dioxide and waste products

Capillaries

These are tiny vessels, often as little as 0.001 mm in diameter and with walls only one cell thick (Figures 12.10c and 12.14). Although the blood as a whole cannot escape from the capillary, the thin capillary walls allow some liquid to pass through, i.e. they are permeable. Blood pressure in the capillaries forces part of the plasma out through the walls. The fluid which escapes is not blood, nor plasma, but **tissue fluid**. Tissue fluid is similar to plasma but contains less protein. This fluid bathes all the living cells of the body and, since it contains dissolved food and oxygen from the blood, it supplies the cells with their needs (Figures 12.13 and 12.16, overleaf). The tissue fluid eventually seeps back into the capillaries, having given up its oxygen and dissolved food to the cells, but it has now received the waste products of the cells, such as carbon dioxide, which are carried away by the bloodstream.

The capillary network is so dense that no living cell is far from a supply of oxygen and food. The capillaries join up into larger vessels, called **venules**, which then combine to form **veins**.

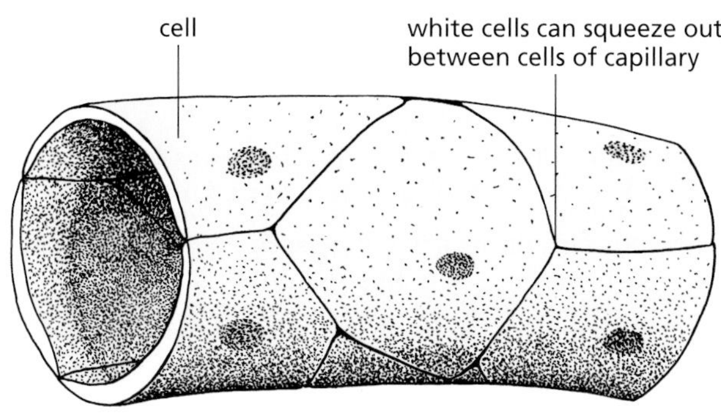

Figure 12.14 Diagram of blood capillary

Veins

Veins return blood from the tissues to the heart (Figure 12.17). The blood pressure in them is steady and is less than that in the arteries. They are wider and their walls are thinner, less elastic and less muscular than those of the arteries (Figures 12.10b and 12.15). They also have valves in them similar to the semi-lunar valves (Figure 12.8, p. 111).

Contraction of body muscles, particularly in the limbs, compresses the thin-walled veins. The valves in the veins prevent the blood flowing backwards when the vessels are compressed in this way. This assists the return of venous blood to the heart.

The blood in most veins is deoxygenated and contains less food but more carbon dioxide than the blood in most arteries. This is because respiring cells have used the oxygen and food and produced carbon dioxide (Figure 12.16, overleaf). The pulmonary veins, which return blood from the lungs to the heart, are an exception. They contain oxygenated blood and a reduced level of carbon dioxide.

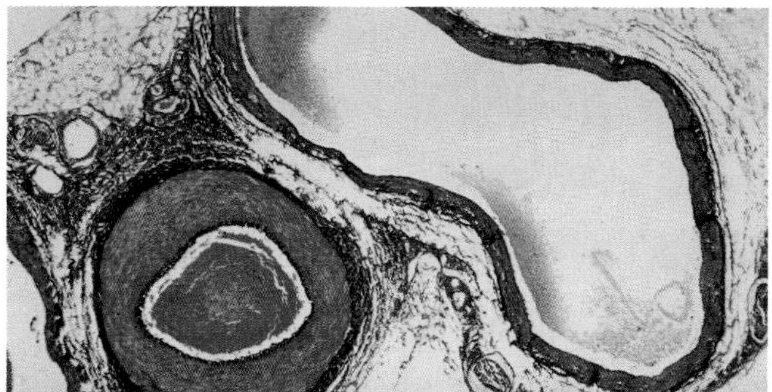

Figure 12.15 Transverse section through a vein and artery. The vein is on the right, the artery on the left. Notice that the wall of the artery is much thicker than that of the vein. The material filling the artery is formed from coagulated red blood cells. These are also visible in two regions of the vein

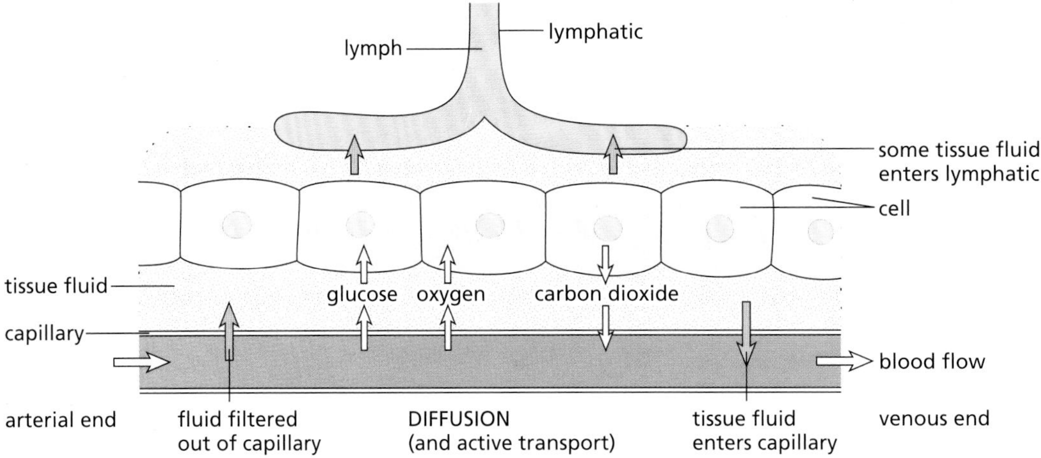

Figure 12.16 Blood, tissue fluid and lymph

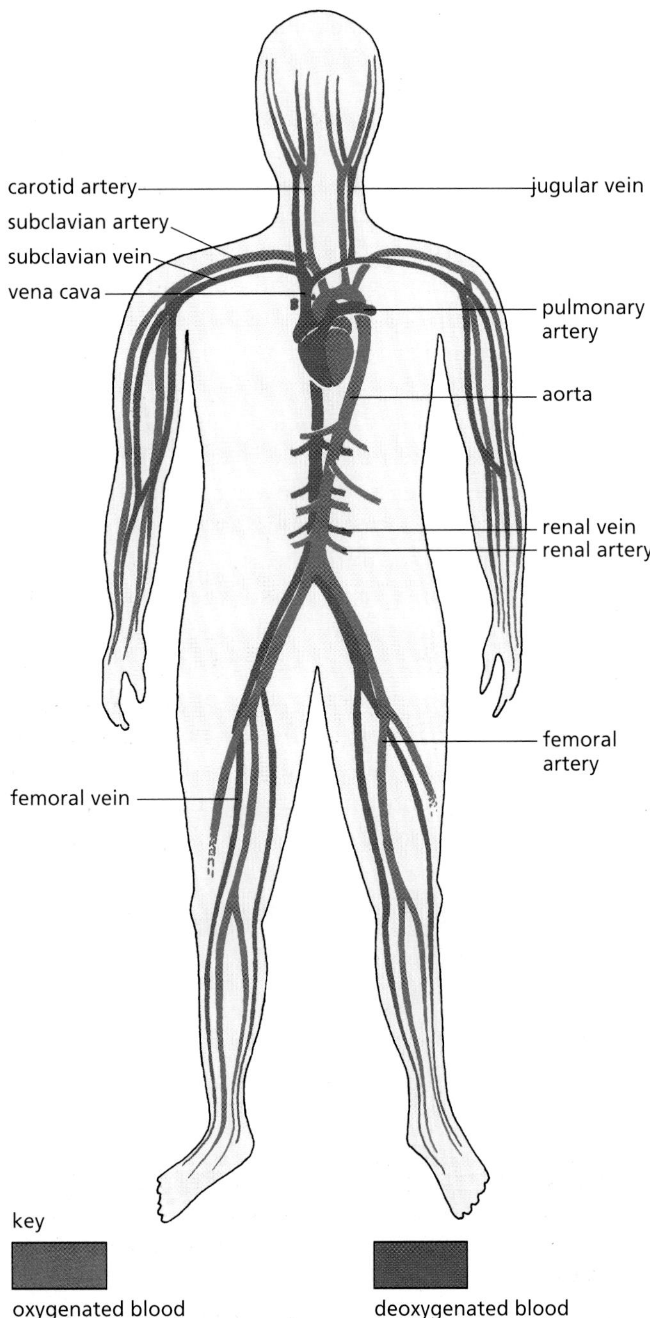

Figure 12.17 The main arteries and veins of the body

Blood pressure

The pumping action of the heart produces a pressure which drives blood round the circulatory system (Figure 12.17). In the arteries, the pressure fluctuates with the heart beat, and the pressure wave can be felt as a pulse. The millions of tiny capillaries offer resistance to the blood flow and, by the time the blood enters the veins, the surges due to the heart beat are lost and the blood pressure is greatly reduced.

Although blood pressure varies with age and activity, it is normally kept within specific limits by negative feedback (p. 172). The filtration process in the kidneys (p. 133) needs a fairly consistent blood pressure. If blood pressure falls significantly because, for example, of loss of blood or shock, then the kidneys may fail. Blood pressure consistently higher than normal increases the risk of heart disease or stroke (p. 120).

Questions

1 Starting from the left atrium, put the following in the correct order for circulation of the blood
- **a** left atrium,
- **b** vena cava,
- **c** aorta,
- **d** lungs,
- **e** pulmonary artery,
- **f** right atrium,
- **g** pulmonary vein,
- **h** right ventricle,
- **i** left ventricle.

2 Why is it incorrect to say 'all arteries carry oxygenated blood and all veins carry deoxygenated blood'?

3 How do veins differ from arteries in **a** their function, **b** their structure?

4 How do capillaries differ from other blood vessels in **a** their structure, **b** their function?

5 Why is it misleading to say that a person 'suffers from blood pressure'?

The lymphatic system

Not all the tissue fluid returns to the capillaries. Some of it enters blind-ended, thin-walled vessels called **lymphatics** (Figure 12.13, p. 113). The lymphatics from all parts of the body join up to make two large vessels which empty their contents into the blood system as shown in Figure 12.18.

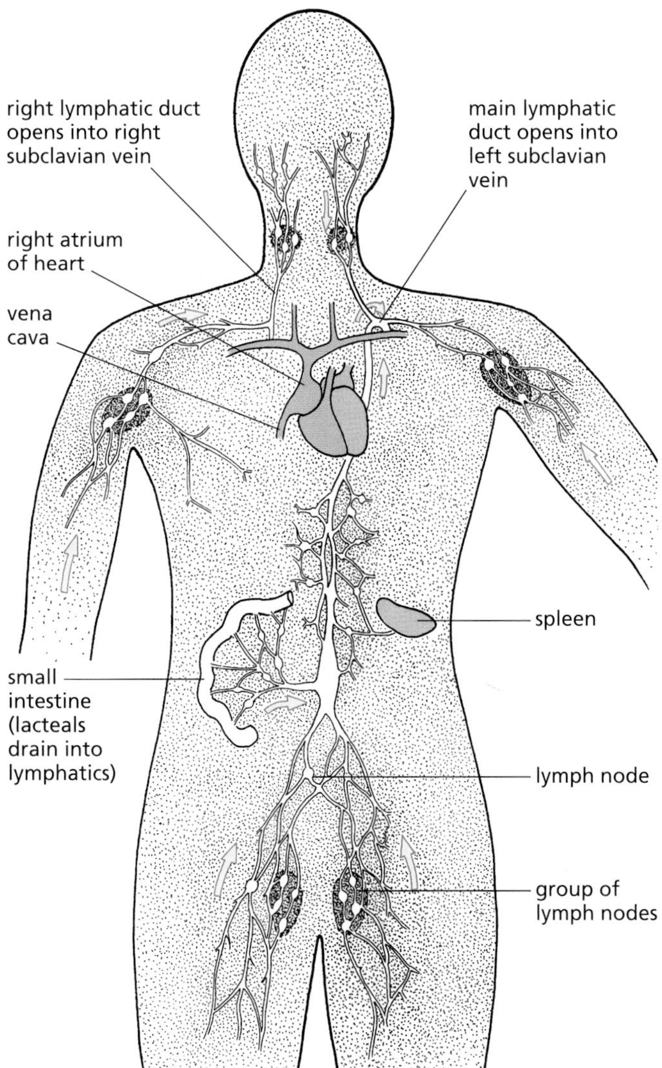

Figure 12.18 The main drainage routes of the lymphatic system

The lacteals from the villi in the small intestine (p. 102) join up with the lymphatic system, so most of the fats absorbed in the intestine reach the circulation by this route. The fluid in the lymphatic vessels is called **lymph** and is similar in composition to tissue fluid.

Some of the larger lymphatics can contract, but most of the lymph flow results from the vessels being compressed from time to time when the body muscles contract in movements such as walking or breathing. There are valves in the lymphatics (Figure 12.19) like those in the veins, so that when the lymphatics are squashed, the fluid in them is forced in one direction only: towards the heart.

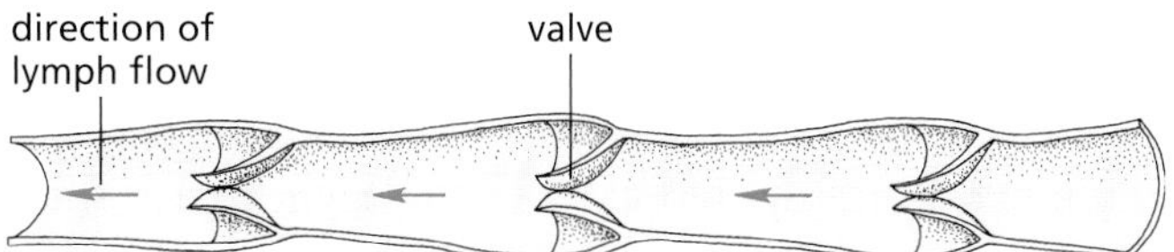

Figure 12.19 Lymphatic vessel cut open to show valves

At certain points in the lymphatic vessels there are swellings called **lymph nodes** (Figure 12.18). Lymphocytes are stored in the lymph nodes and released into the lymph to reach, eventually, the blood system. There are also phagocytes in the lymph nodes. If bacteria enter a wound and are not ingested by the white cells of the blood or lymph, they will be carried in the lymph to a lymph node and white cells there will ingest them. The lymph nodes thus form part of the body's defence system against infection.

Spleen

The spleen is the largest organ in the adult lymphatic system. It is a solid, deep-red body about 12 cm long and lies in the left side of the upper abdomen, between the lower ribs and the stomach. It contains lymphatics and blood vessels. Its main functions are to (1) remove worn-out red cells, bacteria and cell fragments from the blood, (2) produce lymphocytes and antibodies (p. 117).

If bacteria or their antigens (p. 117) reach the spleen, the lymphocytes there start to make antibodies against them.

Thymus

The thymus gland lies at the top of the thorax, partly over the heart and lungs. It is an important lymphoid organ particularly in the newborn where it controls the development of the spleen and lymph nodes. The thymus produces lymphocytes and is the main centre for providing immunity against harmful micro-organisms.

After puberty, the thymus becomes smaller but is still an important immunological organ. White cells from the bone marrow are stored in the thymus. Here they undergo cell division to produce a large population of lymphocytes (T lymphocytes) which can be 'programmed' to make antibodies against specific micro-organisms.

Questions

1. List the things you would expect to find if you analysed a sample of lymph.
2. Describe the course taken by a molecule of fat from the time it is absorbed in the small intestine to the time it reaches the liver to be oxidized for energy. (Use Figure 11.10 on p. 102, Figure 12.11 on p. 112 and Figure 12.18 on this page.)
3. Read pp. 117–18 and then explain why the spleen and thymus are described as 'immunological organs'.

Functions of the blood

It is convenient, at this point, to distinguish between the functions of the blood (1) as the agent replenishing the tissue fluid surrounding the cells, i.e. its role in homeostasis (p. 135), (2) as a circulatory transport system, and (3) as a defence mechanism against harmful bacteria, viruses and foreign proteins.

Homeostatic functions

All the cells of the body are bathed by tissue fluid which is derived from plasma. Tissue fluid supplies the cells with the food and oxygen necessary for their living chemistry. It also removes unwanted substances produced by the cell's metabolism.

The composition of the blood plasma is regulated by the liver and kidneys so that, within narrow limits, the living cells are soaked in a liquid of almost unvarying composition. This provides them with the environment they need and enables them to live and grow in the most favourable conditions. By delivering oxygen and nutrients to the tissue fluid and removing the excretory products, the blood fulfils a homeostatic function, maintaining the constancy of the internal environment. (See p. 135 for further details.)

Transport

Transport of oxygen from the lungs to the tissues

In the lungs, the concentration of oxygen is high and so the oxygen combines with the haemoglobin in the red cells, forming oxy-haemoglobin. The blood is now said to be **oxygenated**. When this oxygenated blood reaches tissues where oxygen is being used up, the oxyhaemoglobin breaks down and releases its oxygen to the tissues. Oxygenated blood is a bright red colour; **deoxygenated** blood is dark red.

Transport of carbon dioxide from the tissues to the lungs

The blood picks up carbon dioxide from actively respiring cells and carries it to the lungs. In the lungs, the carbon dioxide escapes from the blood and is breathed out (see p. 126).

The carbon dioxide is carried in the form of hydrogencarbonate ions (HCO_3^-). Some of the hydrogencarbonate is carried in the red cells, but most of it is dissolved in the plasma.

Transport of digested food from the intestine to the tissues

The soluble products of digestion pass into the capillaries of the villi lining the small intestine (p. 101). They are carried in solution by the plasma and, after passing through the liver, enter the main blood system. Glucose, salts, vitamins and some proteins pass out of the capillaries and into the tissue fluid. The cells bathed by this fluid take up the substances they need for their living processes.

Transport of nitrogenous waste from the liver to the kidneys

When the liver changes amino acids into glycogen (p. 104), the amino part of the molecules ($—NH_2$) is changed into the nitrogenous waste product, urea. This substance is carried away in the blood circulation. When the blood passes through the kidneys, much of the urea is removed and excreted (p. 133).

Transport of hormones

Hormones are chemicals made by certain glands in the body (see p. 169). The blood carries these chemicals from the glands which make them to the organs (target organs) where they affect the rate of activity (Table 12.1). For example, a hormone called **insulin**, made in the pancreas, is carried by the blood to the liver and controls how much glucose is stored as glycogen (p. 104).

Table 12.1 Transport by the blood system

Substance	From	To
oxygen	lungs	whole body
carbon dioxide	whole body	lungs
urea	liver	kidneys
hormones	glands	target organs
digested food	intestine	whole body
heat	abdomen and muscles	whole body

Note that the blood is not directed to a particular organ. A molecule of urea may go round the circulation many times before it enters the renal artery, by chance, and is removed by the kidneys.

Transport of heat

The limbs and head lose heat to the surrounding air. Chemical changes elsewhere in the body produce heat. The blood carries the heat from the warm places to the cold places and so helps to keep an even temperature in all regions. The regions which produce most heat are the abdominal organs, the brain and active muscles.

In addition, the blood system helps to control the body temperature by opening or closing blood vessels in the skin (see p. 138).

Question

1 What substance would the blood
  a gain,
  b lose, on passing through
    (i) the kidneys,
    (ii) the lungs,
    (iii) an active muscle? Remember that respiration (p.19) is taking place in all these organs.

Defence against infection

Clotting

When tissues are damaged and blood vessels cut, platelets clump together and block the smaller capillaries. The platelets and damaged cells at the wound also produce a substance which acts, through a series of enzymes, on the plasma protein called **fibrinogen**. As a result of this action, the fibrinogen is changed into **fibrin**, which forms a network of fibres across the wound. Red cells become trapped in this network and so form a blood clot. The clot not only stops further loss of blood, but also prevents the entry of harmful bacteria into the wound (Figures 12.20 and 12.21).

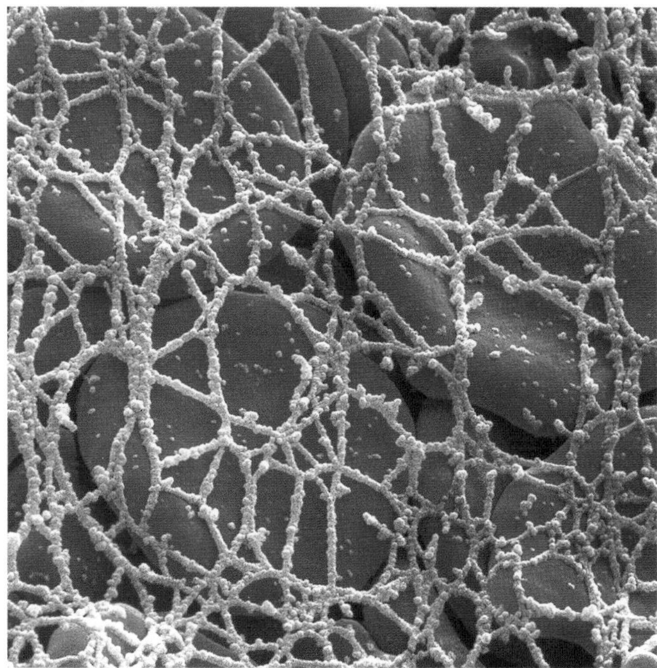

Figure 12.20 Red cells trapped in a fibrin network (×6500)

White cells

The phagocytes at the site of the wound, in the blood capillaries or in lymph nodes (p. 115) may ingest harmful bacteria and so stop them entering the general circulation. White cells can squeeze through the walls of capillary vessels and so attack bacteria which get into the tissues, even though the capillaries themselves are not damaged.

The lymphocytes produce antibodies (see below) against foreign proteins or cells.

Antibodies and immunity

On the surface of all cells there are chemical substances called **antigens**. Lymphocytes produce proteins called **antibodies**, which attack the antigens of bacteria or any alien cells or proteins which invade the body. The antibodies may attach to the surface of the bacteria and make it easier for the phagocytes to ingest them, they may clump the bacteria together or they may neutralize the poisonous proteins (**toxins**) that the bacteria produce.

Each antibody is very **specific**. This means that an antibody which attacks a typhoid bacterium will not affect a pneumonia bacterium. This is illustrated in the form of a diagram in Figure 12.22 (overleaf).

Some of the lymphocytes which produced the specific antibodies remain in the lymph nodes for some time and divide rapidly and make antibodies if the same antigen gets into the body again. This means that the body has become **immune** to the disease caused by the antigen and explains why, once you have recovered from measles or chickenpox, for example, you are very unlikely to catch the same disease again. This is called **natural acquired immunity**. You may inherit some forms of immunity or acquire antibodies from your mother's milk (p. 148). This is **innate immunity**.

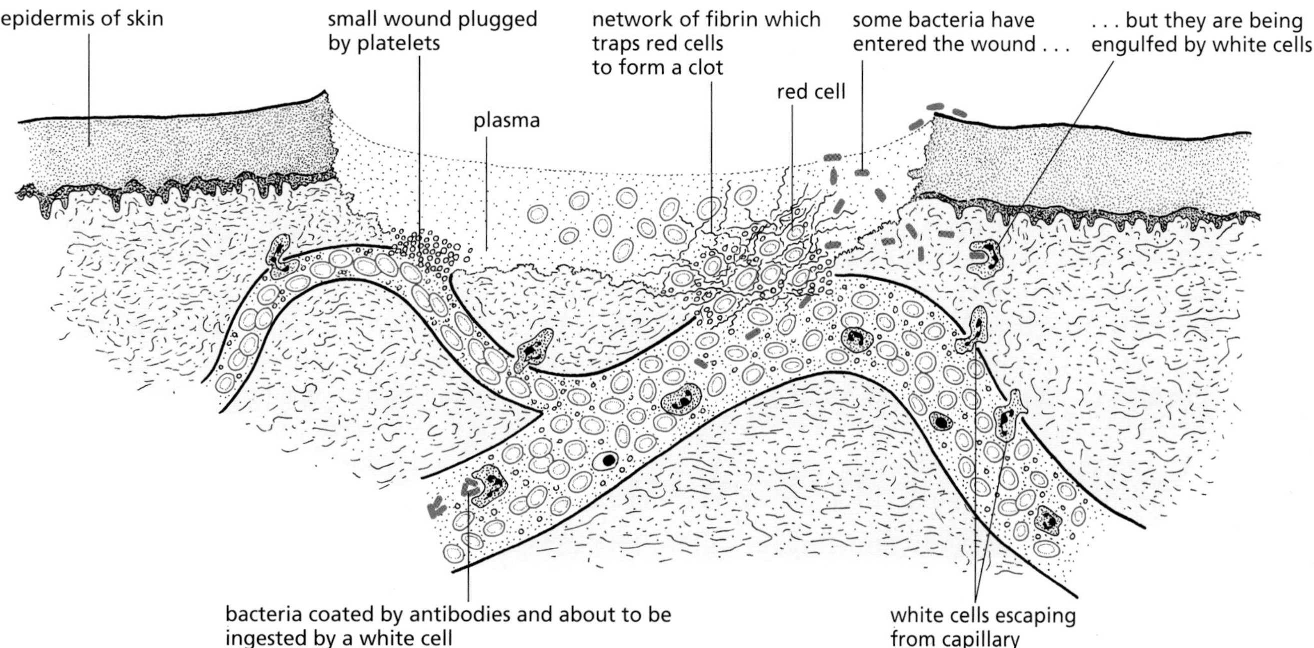

Figure 12.21 The defence against infection. An area of skin has been damaged and two capillaries broken open

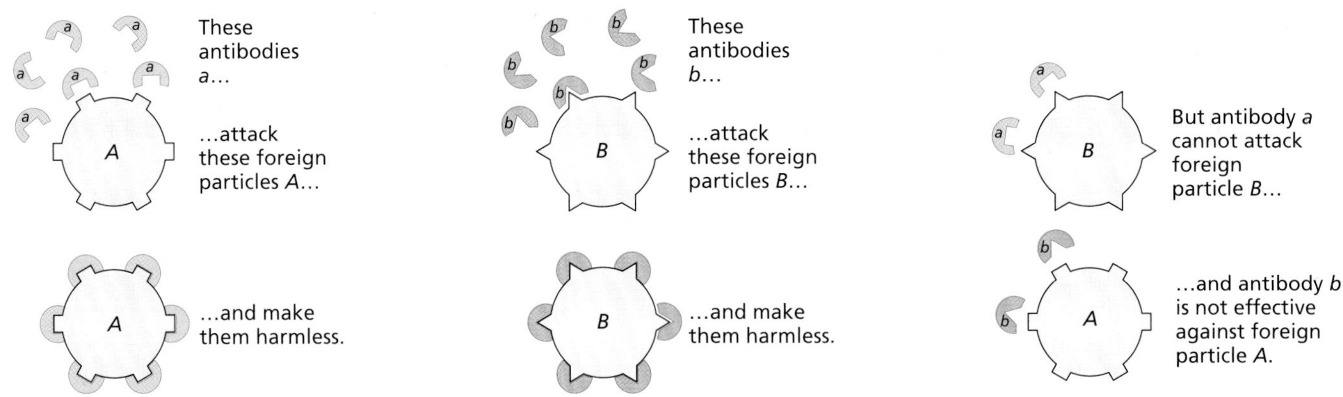

Figure 12.22 Antibodies are specific

When you are **inoculated** (vaccinated) against a disease, a harmless form of the bacteria or viruses is introduced into your body. The white cells make the correct antibodies, so that if the real micro-organisms get into the blood, the antibody is already present or very quickly made by the blood.

The material which is injected or swallowed is called a **vaccine** and is one of the following:

- a harmless form of the micro-organism, e.g. the BCG inoculation against tuberculosis and the Sabin oral vaccine against polio (oral, in this context, means 'taken by mouth');
- the killed micro-organisms, e.g. the Salk anti-polio vaccine and the whooping cough vaccine;
- a **toxoid**, i.e. the inactivated toxin from the bacteria, e.g. the diphtheria and tetanus vaccines. (A toxin is the poisonous substance produced by certain bacteria which causes the disease symptoms.)

The immunity produced by a vaccine is called **artificial acquired immunity**. Immunity acquired as a result of catching a disease or receiving a vaccine is called **active immunity** because the person produces his or her own antibodies.

Some diseases can be prevented or cured by injecting the patient with a serum from a person who has recovered from the disease. Serum is plasma with the fibrinogen removed. Sera are prepared from the plasma given by blood donors. People who have recently received an anti-tetanus inoculation will have made anti-tetanus antibodies in their blood. Some of these people volunteer to donate their blood, but their plasma is separated at once and the red cells returned to their circulation. The anti-tetanus antibodies are then extracted from the plasma and used to treat patients who are at risk of contracting tetanus, as a result of an accident, for example. In a similar way, antibodies against chickenpox and rabies can be produced.

The temporary immunity conferred by these methods is called **passive immunity**, because the antibodies have not been produced by the patient.

B and T lymphocytes

There are two main types of lymphocyte. Both types undergo rapid cell division in response to the presence of specific antigens but their functions are different (though interdependent). The B cells (from **B**one marrow) become short-lived **plasma** cells and produce antibodies which are released into the blood. These antibodies may attack antigens directly or stick to the surface membrane of infected or alien cells, e.g. cells carrying a virus, bacteria, cancer cells or transplanted cells.

'Killer' T cells (from the **T**hymus) have receptor molecules on their surface which attach them to these surface antibodies. The T cells then kill the cell by damaging its cell membrane.

'Helper' T cells stimulate the B cells to divide and produce antibodies. They also stimulate the phagocytes to ingest any cells carrying antibodies on their surface.

Some of the B cells remain in the lymph nodes as **memory cells**. These can reproduce swiftly and produce antibodies in response to any subsequent invasion of the body by the same foreign organism. This is called a 'secondary response'. The level of antibodies is much higher than in the 'primary response' and is the basis of acquired immunity to a disease.

Transplants and transfusions

Transplants

White cells, lymph nodes, the spleen and the thymus make up the **immune system**, because they have the ability to produce antibodies. When the immune system comes into play the body is said to be making an **immune response**. This response protects the body against invaders such as bacteria, viruses and foreign proteins.

All the body cells carry antigens on their surface. Apart from identical twins, no two people have the same antigens. They have different 'tissue types'. So, in transplant surgery, donated tissue is treated as 'foreign' by the immune system and attacked and destroyed by the white cells (i.e. 'rejected'). Skin grafts present fewer problems because the donated skin is from another part of the patient's own body so is recognized as 'self'.

Close relatives have similar, though not identical, tissue types and are therefore preferred as donors. Even so, to prevent donated organs such as hearts or kidneys being rejected, the patient is given **immunosuppressive drugs**. As the name implies, these drugs suppress the patient's immune response but, in doing so, leave him or her vulnerable to attack by disease-causing bacteria or viruses. In the recovery period, the patient may be kept in sterile condidtions.

Transfusions

If somebody loses a lot of blood as a result of an injury or surgical operation, he or she can be given a blood transfusion. Blood taken from a healthy person, the **donor**, is fed into one of the patient's veins. For a transfusion to be successful the blood type of the donor has to match the blood of the patient. If the two blood types do not match, the donor's red cells are clumped in the patient's blood vessels and cause serious harm. The red cells are clumped because they carry antigens on their cell membranes and if the blood types do not match, the antibodies in the patient's blood will act on the donor's red cells and clump them together (Figure 12.23).

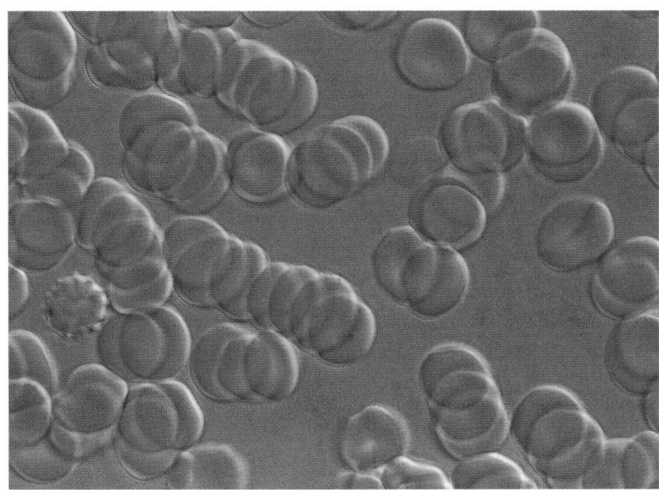

Figure 12.23 The red cells are being clumped (×1400)

For the purposes of transfusion, people can be put into one of four groups called group A, group B, group AB and group O. The red cells of group A people have antigen A on their cell membranes and the plasma contains anti-B antibodies. Since antibodies are specific, as explained on p. 118, the anti-B antibodies do not attack the A antigens on the red cells.

The red cells of group B people carry the B antigen and their plasma has the anti-A antibody. So if group A cells are introduced into a group B person, they will be clumped by the anti-A antibody. Group AB people have both A and B antigens on their cells but no A or B antibodies in their plasma. The red cells of group O people have neither A nor B antigens but their plasma contains both anti-A and anti-B antibodies (Table 12.2).

Table 12.2 Antigens and antibodies

Group	Antigen on cells	Antibody in plasma
A	A	anti-B
B	B	anti-A
AB	A and B	neither
O	neither	anti-A and anti-B

The red cells from group O people can be given to any other group because they have neither the A nor B antigens and so cannot be clumped. Group O people, on the other hand, can receive blood only from their own group because their plasma contains both anti-A and anti-B antibodies.

Group AB people, having neither anti-A nor anti-B antibodies in their plasma, can receive blood from any group. Table 12.3 shows the acceptable pattern of giving and receiving for the four groups.

Table 12.3 Blood transfusion

Group	Can donate blood to	Can receive blood from
A	A and AB	A and O
B	B and AB	B and O
AB	AB	all groups
O	all groups	O

A donor gives 420 cm^3 of blood from a vein in the arm (Figure 12.24). The blood is led into a sterilized bag containing sodium citrate which prevents clotting.

The blood is then stored at 5 °C for 10 days, or longer if glucose is added. Before blood is transfused, even though both groups are known, it is carefully tested against the patient's blood to make sure of a good match. Then it is fed into one of the patient's arm veins at the correct rate and temperature.

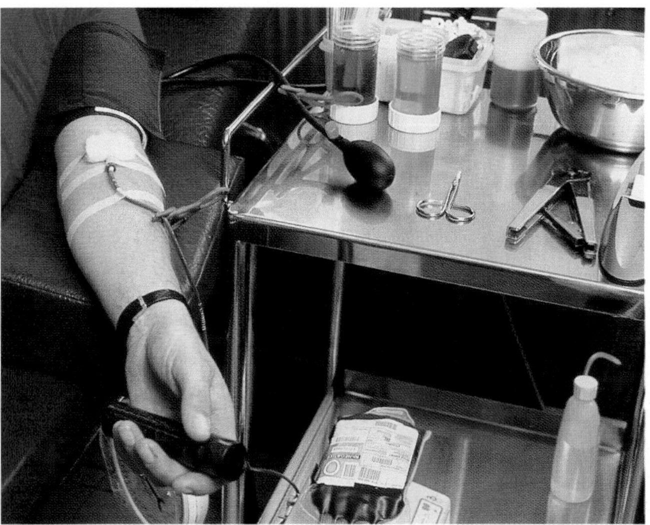

Figure 12.24 Blood donor. The veins in the upper arm are compressed by using an inflatable 'cuff'. The donor's blood is then tapped from a vein near the inside of the elbow. It takes 5–10 minutes to fill the bag

In a few hours, the donor will have made up his or her blood to the normal volume and in a week or two the red cells will have been replaced.

It is possible to find out a person's blood group by mixing a drop of his or her blood with anti-A serum and anti-B serum (Figure 12.25). Group AB cells will clump in both anti-sera; group O cells will clump in neither; A cells will clump only in anti-A; and B cells only in anti-B.

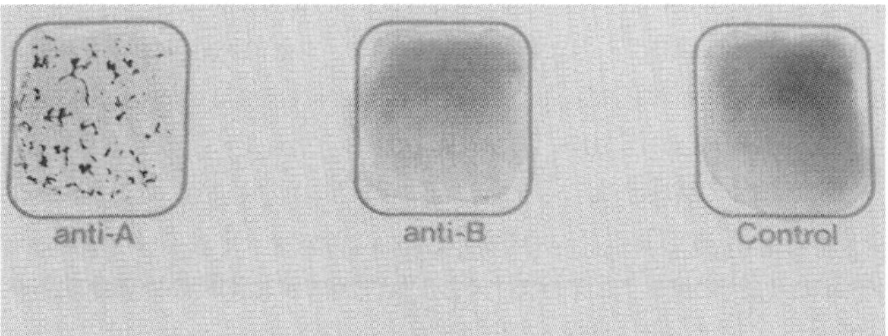

Figure 12.25 Blood group test. The subject's blood group is A because it has clumped in B serum

Questions

1 What part do white cells play in the defence of the body against infection?

2 Why is it necessary to inoculate people against a disease before they catch it rather than wait until they catch it?

3 Give an example of
a active immunity,
b passive immunity.

4 One of the ABO blood groups is sometimes called the 'universal donor'. Which group do you think this is and why?

5 A drop of a person's blood shows clumping in anti-B serum but not in anti-A. What is this person's blood group?

Coronary heart disease

In the lining of the large and medium arteries, deposits of a fatty substance, called **atheroma**, are laid down in patches. This happens to everyone and the patches get more numerous and extensive with age but until one of them actually blocks an important artery the effects are not noticed. It is not known how or why the deposits form. Some doctors think that fatty substances in the blood pass into the lining. Others believe that small blood clots form on damaged areas of the lining and are covered over by the atheroma patches. The patches may join up to form a continuous layer which reduces the internal diameter of the vessel (Figure 12.26).

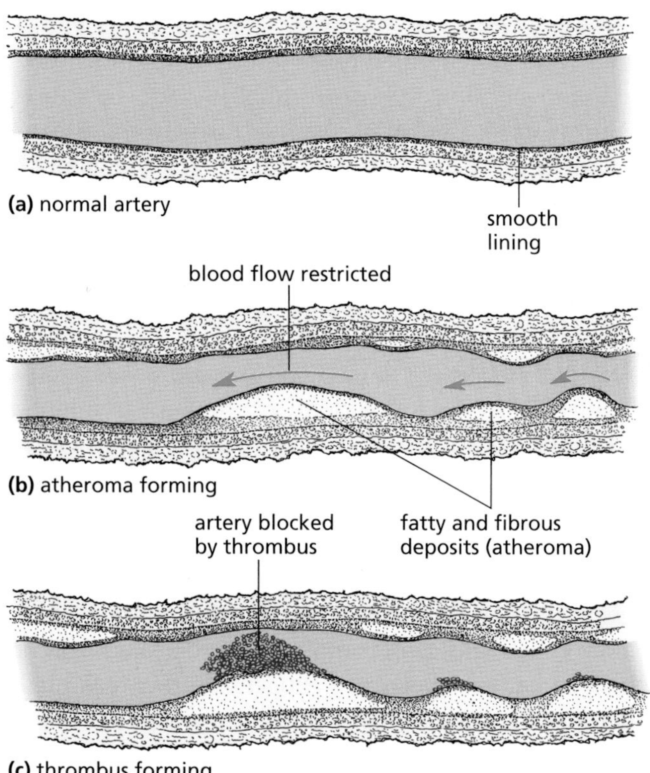

Figure 12.26 Atheroma and thrombus formation

The surface of a patch of atheroma sometimes becomes rough and causes fibrinogen in the plasma to deposit fibrin on it, so causing a blood clot (a **thrombus**) to form. If the blood clot blocks the **coronary artery** (Figure 12.4) which supplies the muscles of the ventricles with blood, it starves the muscles of oxygenated blood and the heart may stop beating. This is a severe heart attack from **coronary thrombosis**. A thrombus might form anywhere in the arterial system, but its effects in the coronary artery and in parts of the brain (strokes) are the most drastic.

In the early stages of coronary heart disease, the atheroma may partially block the coronary artery and reduce the blood supply to the heart (Figure 12.27). This can lead to **angina**, i.e. a pain in the chest which occurs during exercise or exertion. This is a warning to the person that he or she is at risk and should take precautions to avoid a coronary heart attack.

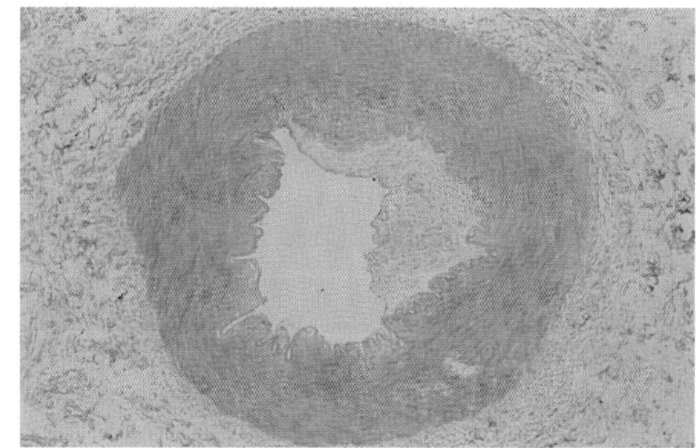

Figure 12.27 Atheroma partially blocking the coronary artery.

Possible causes

Atheroma and thrombus formation are the immediate causes of a heart attack but the long-term causes which give rise to these conditions are not well understood.

There is an inherited tendency towards the disease but the disease has increased very significantly in affluent countries in recent years. This makes us think that some features of 'Western' diets or lifestyles might be causing it. Although there is very little direct evidence, the main factors are thought to be smoking, a fatty diet, stress and lack of exercise.

Smoking

Statistical studies suggest that smokers are two to three times more likely to die from a heart attack than are non-smokers of a similar age (Figure 12.28). The carbon monoxide and other chemicals in cigarette smoke may damage the lining of the arteries, allowing atheroma to form, but there is not much direct evidence for this.

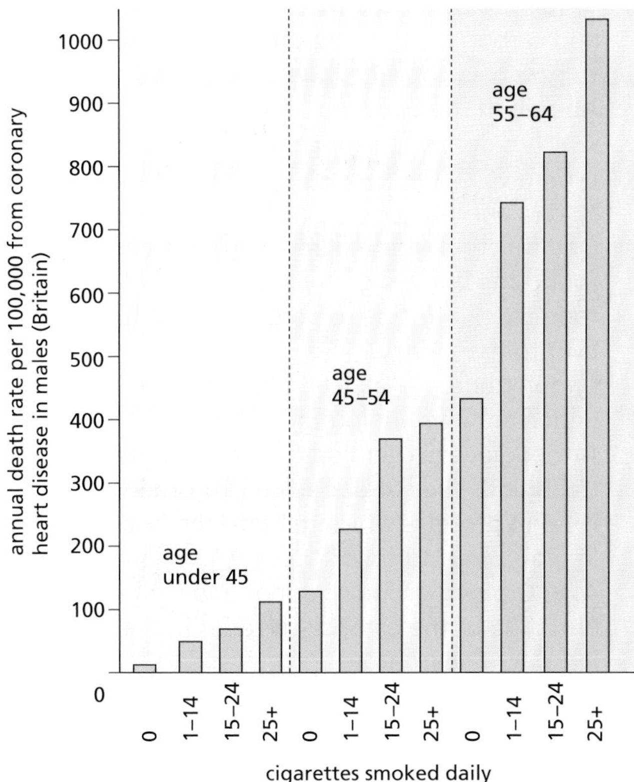

Figure 12.28 Smoking and heart disease. Obviously, as you get older you are more likely to die from a heart attack, but notice that, in any age group, the more you smoke the higher your chances of dying from heart disease

Fatty diet

The atheroma deposits contain cholesterol, which is present, combined with lipids and proteins, in the blood (p. 92). Cholesterol plays an essential part in our physiology, but it is known that people with high levels of blood cholesterol are more likely to suffer from heart attacks than people with low cholesterol levels.

Blood cholesterol can be influenced, to some extent, by the amount and type of fat in the diet. Many doctors and dieticians believe that animal fats (milk, cream, butter, cheese, egg yolk, fatty meat) are more likely to raise the blood cholesterol than are the vegetable oils which contain a high proportion of unsaturated fatty acids (p. 92). This is still a matter of some controversy.

Stress

Emotional stress often leads to a raised blood pressure. High blood pressure may increase the rate at which atheroma is formed in the arteries.

Lack of exercise

There is some evidence that regular, vigorous exercise reduces the chances of a heart attack. This could be the result of an improved coronary blood flow. A sluggish blood flow, resulting from lack of exercise, may allow atheroma to form in the arterial lining but, once again, the direct evidence for this is slim.

Correlation and cause

It is not possible or desirable to conduct experiments on humans to find out, more precisely, the causes of heart attack. The evidence has to be collected from long-term studies on populations of individuals, e.g. smokers and non-smokers. Statistical analysis of these studies will often show a **correlation**, e.g. more smokers, within a given age band, suffer heart attacks than do non-smokers of the same age. This correlation does *not* prove that smoking **causes** heart attacks. It could be argued that people who are already prone to heart attacks for other reasons (e.g. high blood pressure) are more likely to take up smoking. This may strike you as implausible, but until it can be shown that substances in tobacco smoke do cause an increase in atheroma, the correlation cannot be used on its own to claim a cause and effect.

Nevertheless, there are so many other correlations between smoking and ill-health (e.g. bronchitis, emphysema, lung cancer) that the circumstantial evidence against smoking is very strong.

Another example of a positive correlation is between the possession of a television set and heart disease. Nobody would seriously claim that television sets cause heart attacks. The correlation probably reflects an affluent way of life, associated with overeating, fatty diets, lack of exercise and other factors which may contribute to coronary heart disease.

Questions

1 a What positive steps could you take and
  b what things should you avoid, to reduce your risk of coronary heart disease in later life?

2 About 95 per cent of patients with disease of the leg arteries are cigarette smokers. Arterial disease of the leg is the most frequent cause of leg amputation.
  a Is there a correlation between smoking and leg amputation?
  b Does smoking cause leg amputation?
  c In what way could smoking be a possible cause of leg amputation?

3 Figure 12.29 shows the relative increase in the rates of four body processes in response to vigorous exercise.
  a How are the changes related physiologically to one another?
  b What other physiological changes are likely to occur during exercise?
  c Why do you think that the increase in blood flow in muscle is less than the total increase in the blood flow?

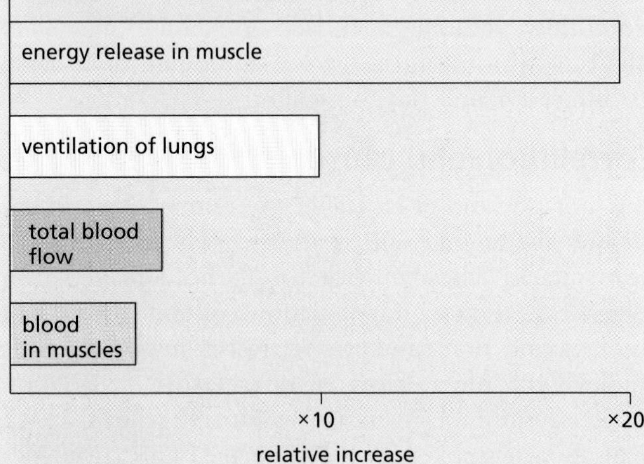

Figure 12.29

4 Figure 12.30 shows the changes in the levels of antibody in response to an inoculation of a vaccine, followed by a booster injection 3 weeks later. Use your knowledge of the immune reaction (pp. 117–18) to explain these changes.

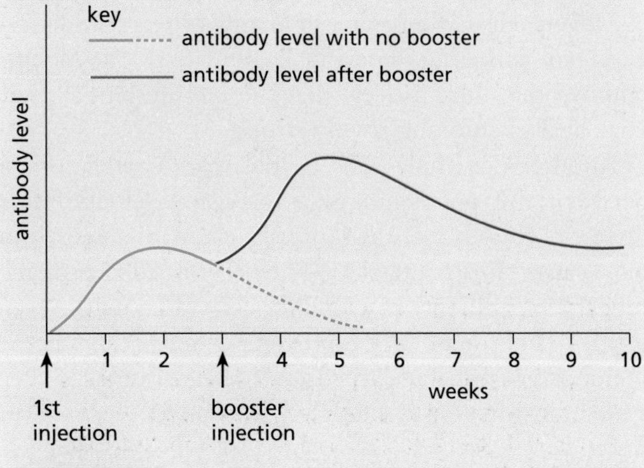

Figure 12.30

Checklist

- Blood consists of red cells, white cells and platelets suspended in plasma.
- Plasma contains water, proteins, salts, glucose and lipids.
- The red cells carry oxygen. The white cells attack bacteria.
- The heart is a muscular pump with valves, which sends blood round the circulatory system.
- The left side of the heart pumps oxygenated blood round the body.
- The right side of the heart pumps deoxygenated blood to the lungs.
- Blood pressure is essential in order to pump blood round the body.
- Arteries carry blood from the heart to the tissues.
- Veins return blood to the heart from the tissues.
- Capillaries form a network of tiny vessels in all tissues. Their thin walls allow dissolved food and oxygen to pass from the blood into the tissues, and carbon dioxide and other waste substances to pass back into the blood.
- All cells in the body are bathed in tissue fluid, which is derived from plasma.
- Lymph vessels return tissue fluid to the lymphatic system and finally into the blood system.
- One function of the blood is to carry substances round the body, e.g. oxygen from lungs to body, food from intestine to body and urea from the liver to the kidneys.
- Lymph nodes, the spleen and the thymus are important immunological organs.
- Antibodies are chemicals made by white cells in the blood. They attack any micro-organisms or foreign proteins which get into the body.
- In blood transfusions it is essential to match the A, B, O blood group of donor and recipient.
- Blockage of the coronary arteries in the heart leads to a heart attack.
- Smoking, fatty diets, stress and lack of exercise may contribute to heart disease.

13 Breathing

Lung structure
Air passages and alveoli.

Ventilation of the lungs
Inhaling, exhaling, lung capacity and breathing rate.

Gaseous exchange
Uptake of oxygen; removal of carbon dioxide.

Characteristics of respiratory surfaces
Surface area. Thickness. Ventilation. Capillaries.

Smoking
Effect of smoking on the lungs and circulatory system.

Practical work
The composition of exhaled air. Lung volume.

All the processes carried out by the body, such as movement, growth and reproduction, require energy. In animals, this energy can be obtained only from the food they eat. Before the energy can be used by the cells of the body, it must be set free from the chemicals of the food by a process called 'respiration' (see p. 19). Aerobic respiration needs a supply of oxygen and produces carbon dioxide as a waste product. All cells, therefore, must be supplied with oxygen and must be able to get rid of carbon dioxide.

In humans and other mammals, the oxygen is obtained from the air by means of the lungs. In the lungs, the oxygen dissolves in the blood and is carried to the tissues by the circulatory system (p. 112).

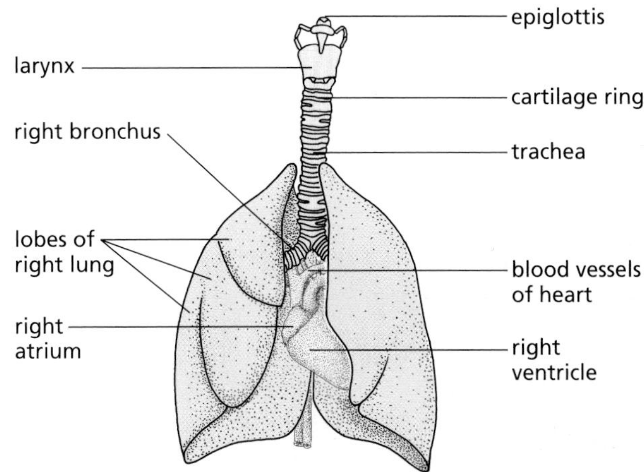

Figure 13.1 Diagram of lungs, showing position of heart

Lung structure

The lungs are enclosed in the thorax (chest region) (see Figure 11.5, p. 99). They have a spongy texture and can be expanded and compressed by movements of the thorax in such a way that air is sucked in and blown out. The lungs are joined to the back of the mouth by the windpipe or **trachea** (Figure 13.1). The trachea divides into two smaller tubes, called **bronchi** (singular = bronchus), which enter the lungs and divide into even smaller branches. When these branches are only about 0.2 mm in diameter, they are called **bronchioles** (Figure 13.2a). These fine branches end in a mass of little, thin-walled, pouch-like air sacs called **alveoli** (Figures 13.2b, c and 13.3, overleaf).

Rings of gristle (cartilage) stop the trachea and bronchi collapsing when we breathe in. The **epiglottis** and other structures at the top of the trachea stop food and drink from entering the air passages when we swallow (see p. 99).

The epithelium which lines the inside of the trachea, bronchi and bronchioles consists of ciliated cells (p. 6). There are also cells which secrete mucus. The mucus forms a thin film over the internal lining. Dust particles and bacteria become trapped in the sticky mucus film and the mucus is carried upwards, away from the lungs, by the flicking movements of the cilia. In this way, harmful particles are prevented from reaching the alveoli. When the mucus reaches the top of the trachea, it passes down the gullet during normal swallowing.

The alveoli have thin elastic walls, formed from a single-cell layer or **epithelium**. Beneath the epithelium is a dense network of capillaries (Figure 13.2c) supplied with deoxygenated blood (p. 108). This blood, from which the body has taken oxygen, is pumped from the right ventricle, through the pulmonary artery (see Figure 12.11, p. 112). In humans, there are about 350 million alveoli, with a total absorbing surface of about 90 m^2. This large absorbing surface makes it possible to take in oxygen and give out carbon dioxide at a rate to meet the body's needs.

Questions

1 Place the following structures in the order in which air will reach them when breathing in: bronchus, trachea, nasal cavity, alveolus.

2 One function of the small intestine is to absorb food (p. 101). One function of the lungs is to absorb oxygen. Point out the basic similarities in these two structures which help to speed up the process of absorption.

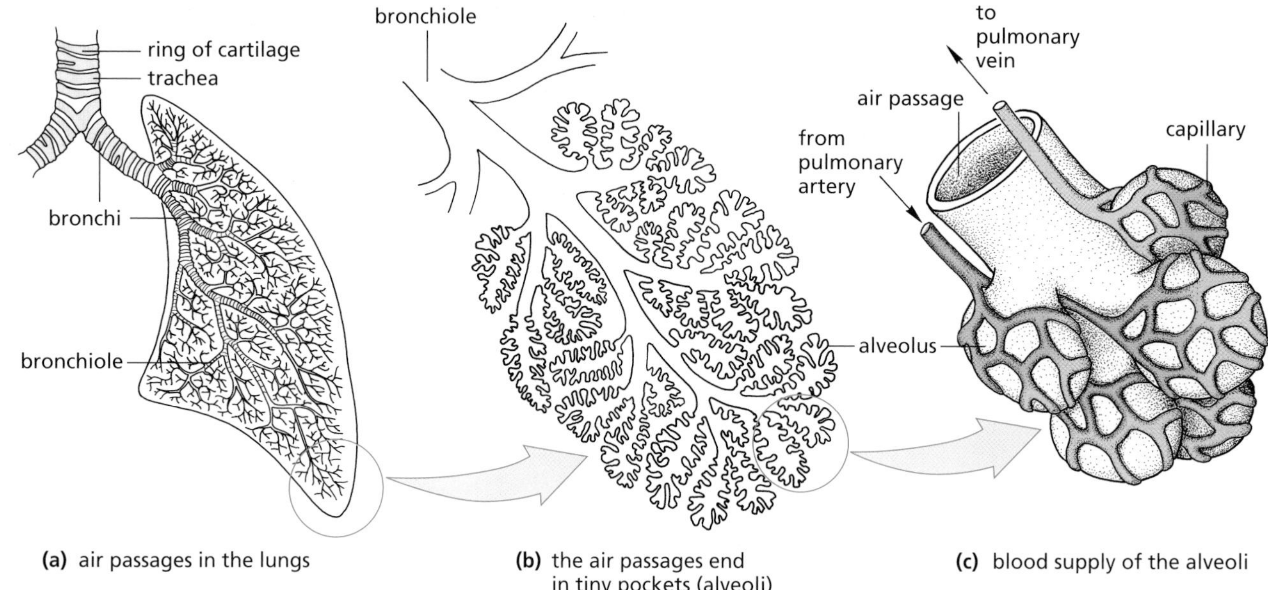

Figure 13.2 Lung structure

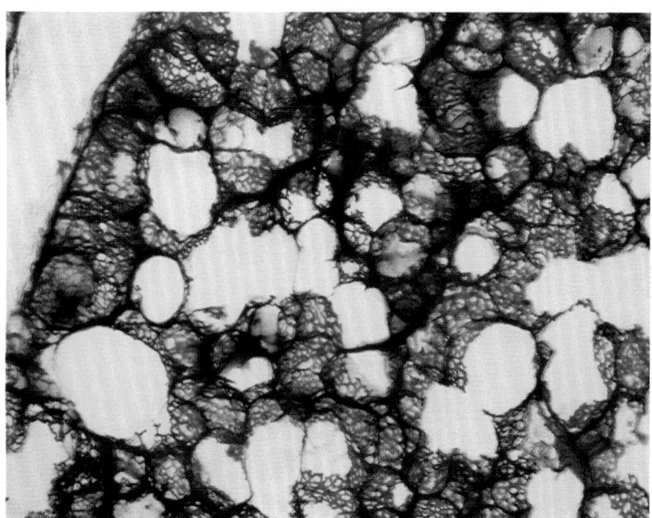

Figure 13.3 Small piece of lung tissue (×40). The capillaries have been injected with red and blue dye. The networks surrounding the alveoli can be seen

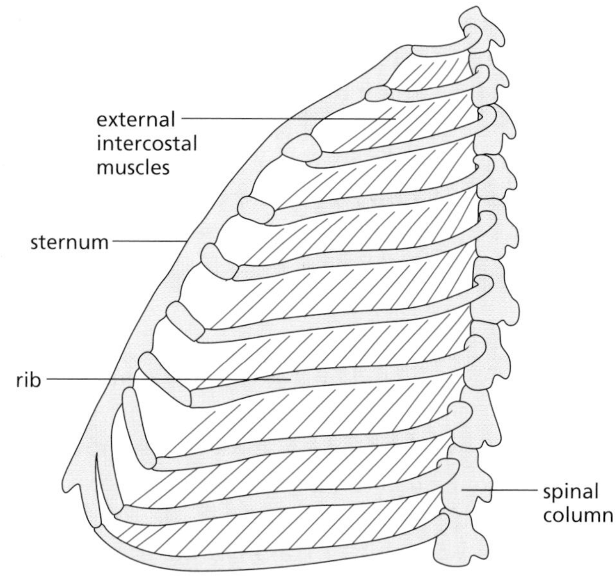

Figure 13.4 Rib cage seen from left side, showing external intercostal muscles

Ventilation of the lungs

The movement of air into and out of the lungs, called **ventilation**, renews the oxygen supply in the lungs and removes the surplus carbon dioxide. The lungs contain no muscle fibres and are made to expand and contract by movements of the ribs and diaphragm.

The **diaphragm** is a sheet of tissue which separates the thorax from the abdomen (see Figure 11.5, p. 99). When relaxed, it is domed slightly upwards. The ribs are moved by the **intercostal muscles**. The external intercostals (Figure 13.4) pull the ribs upwards and outwards. The internal intercostals pull them downwards and inwards. Figure 13.5 shows the contraction of the external intercostals making the ribs move upwards

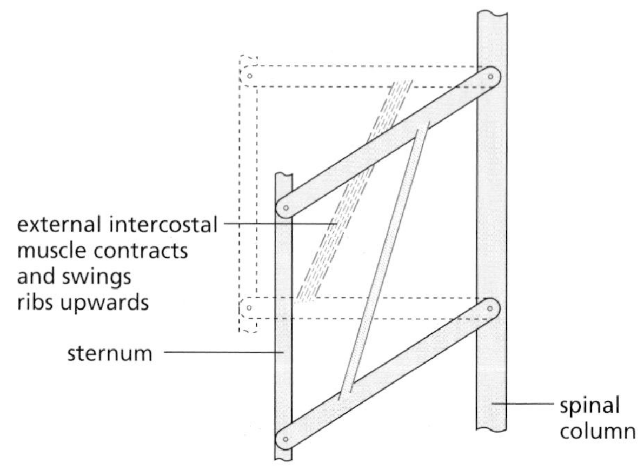

Figure 13.5 Model to show action of intercostal muscles

Inhaling

1 The diaphragm muscles contract and pull it down (Figure 13.6a).
2 The external intercostal muscles contract and pull the rib cage upwards and outwards (Figure 13.7a).

These two movements make the space in the thorax bigger, so forcing the lungs to expand and draw air in through the nose and trachea.

Exhaling

1 The diaphragm muscles relax, allowing the diaphragm to return to its domed shape (Figure 13.6b).
2 The internal intercostal muscles contract, pulling the ribs downwards (Figure 13.7b).

The lungs are elastic and shrink back to their relaxed size, forcing air out again.

The outside of the lungs and the inside of the thorax are lined with a smooth membrane called the **pleural membrane**. This produces a thin layer of liquid called **pleural fluid** which reduces the friction between the lungs and the inside of the thorax.

Lung capacity and breathing rate

The total volume of the lungs when fully inflated is about 5 litres in an adult. However, in quiet breathing, when asleep or at rest, you normally exchange only about 500 cm^3 (Figure 13.8, overleaf). During exercise you can take in and expel an extra 3 litres. There is a **residual volume** of 1.5 litres which cannot be expelled no matter how hard you breathe out.

At rest, you normally inhale and exhale about 16 times per minute. During exercise, the breathing rate may rise to 20 or 30 breaths per minute. The increased rate and depth of breathing during exercise allows more oxygen to dissolve in the blood and supply the active muscles. The extra carbon dioxide which the muscles put into the blood will be removed by the faster, deeper breathing. It is mainly the extra CO_2 in the blood reaching the brain which stimulates the increased rate of breathing.

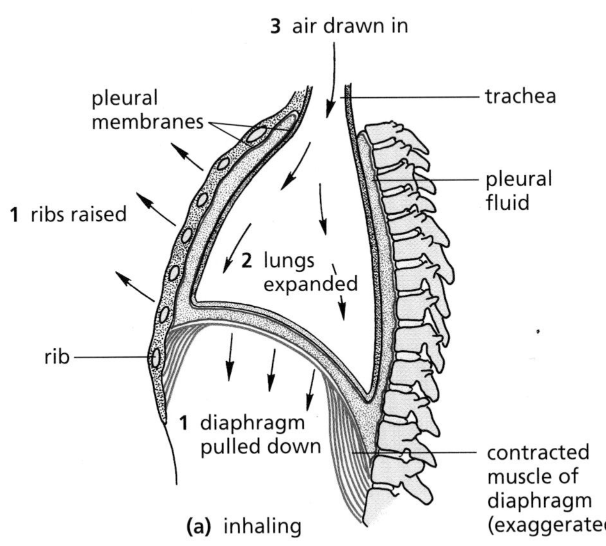

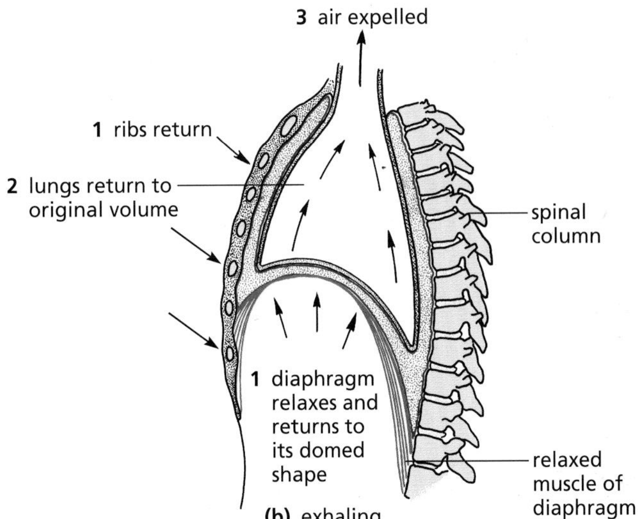

Figure 13.6 Diagrams of thorax to show mechanism of breathing

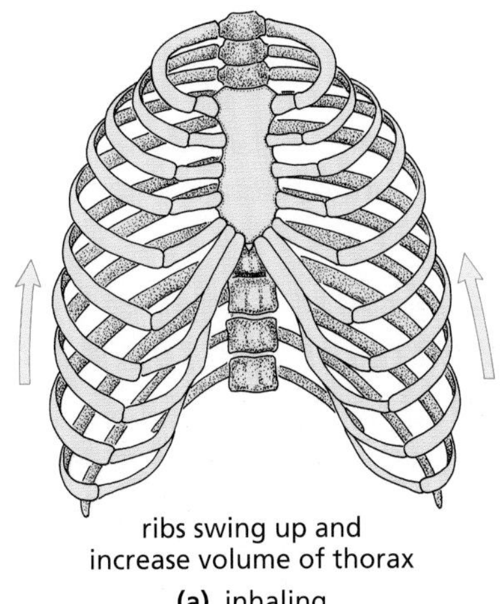

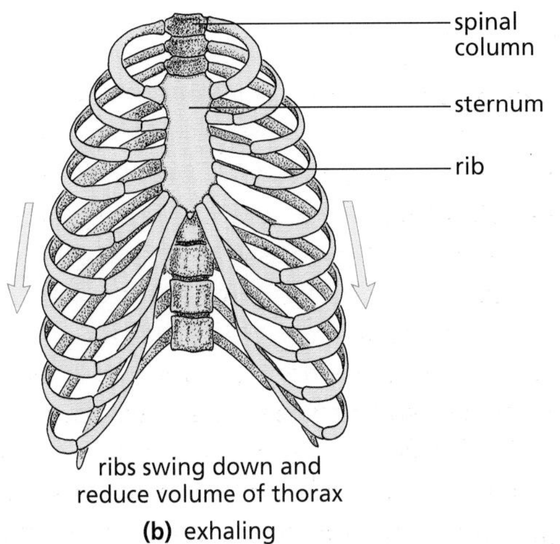

Figure 13.7 Movement of rib cage during breathing

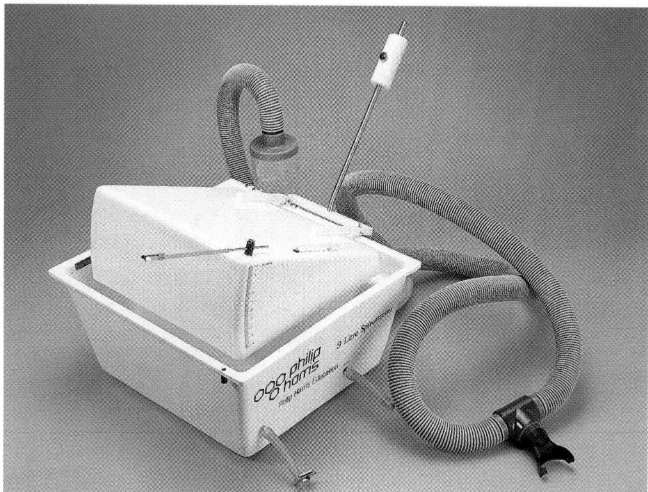

Figure 13.8 A spirometer. This instrument measures the volume of air breathed in and out of the lungs and can be used to measure oxygen consumption

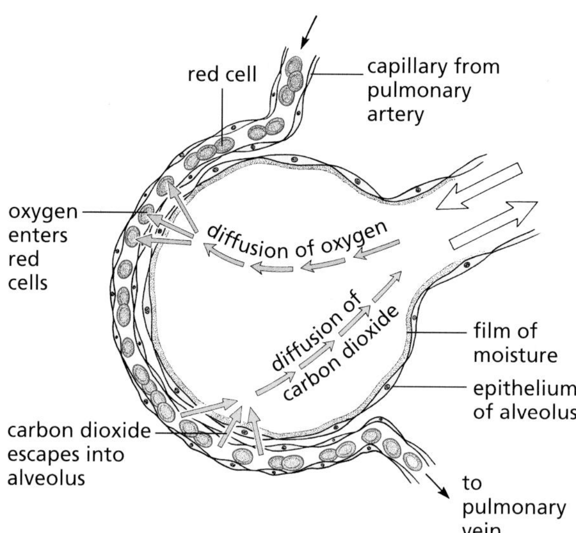

Figure 13.9 Gaseous exchange in the alveolus

Questions

1 What are the two principal muscular contractions which cause air to be inhaled?

2 Place the following in the correct order: lungs expand, ribs rise, air enters lungs, external intercostal muscles contract, thorax expands.

3 During inhalation, which parts of the lung structure would you expect to expand most?

Gaseous exchange

Ventilation refers to the movement of air into and out of the lungs. Gaseous exchange refers to the exchange of oxygen and carbon dioxide which takes place between the air and the blood vessels in the lungs (Figure 13.9).

The 1.5 litres of residual air in the alveoli is not exchanged during ventilation and oxygen has to reach the capillaries by the slower process of diffusion. Figure 13.9 shows how oxygen reaches the red blood cells and how carbon dioxide escapes from the blood.

The oxygen combines with the haemoglobin in the red blood cells, forming oxyhaemoglobin (p.108). The carbon dioxide in the plasma is released when the hydrogencarbonate ions ($-HCO_3$) break down to CO_2 and H_2O.

The capillaries carrying oxygenated blood from the alveoli join up to form the pulmonary vein (see Figure 12.11, p.112), which returns blood to the left atrium of the heart. From here it enters the left ventricle and is pumped all round the body, so supplying the tissues with oxygen.

The process of gaseous exchange in the alveoli does not remove all the oxygen from the air. The air breathed in contains about 21 per cent of oxygen; the air breathed out still contains 16 per cent of oxygen (see Table 13.1). The remaining 79 per cent of the air consists mainly of nitrogen, whose percentage composition does not change significantly during breathing.

Table 13.1 Changes in the composition of breathed air

	Inhaled %	Exhaled %
Oxygen	21	16
Carbon dioxide	0.04	4
Water vapour	variable	saturated

The lining of the alveoli is coated with a film of moisture in which the oxygen dissolves. Some of this moisture evaporates into the alveoli and saturates the air with water vapour. The air you breathe out, therefore, always contains a great deal more water vapour than the air you breathe in. The exhaled air is warmer as well, so in cold and temperate climates you lose heat to the atmosphere by breathing.

Sometimes the word **respiration** or **respiratory** is used in connection with breathing. The lungs, trachea and bronchi are called the **respiratory system**; a person's rate of breathing may be called his or her **respiration rate**. This use of the word should not be confused with the biological meaning of respiration, namely the release of energy in cells (p.19). This chemical process is sometimes called **tissue respiration** or **internal respiration** to distinguish it from breathing.

Characteristics of respiratory surfaces

The exchange of oxygen and carbon dioxide across a respiratory surface, as in the lungs or over the gills of a fish, depends on the diffusion of these two gases. Diffusion occurs more rapidly if (1) there is a large surface exposed to the gas, (2) the distance across which diffusion has to take place is small, (3) there is a big difference in the concentrations of the gas at two points brought about by ventilation and (4) there is a rich supply of blood capillaries.

Large surface

The presence of millions of alveoli in the lungs provides a very large surface for gaseous exchange. The many branching filaments in a fish's gills have the same effect.

Thin epithelium

There is only a two-cell layer, at the most, separating the air in the alveoli from the blood in the capillaries (Figure 13.9). Thus, the distance for diffusion is very short.

Ventilation

Ventilation of the lungs helps to maintain a steep diffusion gradient (p. 27), between the air at the end of the air passages and the alveolar air. The concentration of the oxygen in the air at the end of the air passages is high, because the air is constantly replaced by the breathing actions.

Capillary network

The continual removal of oxygen by the blood in the capillaries lining the alveoli keeps its concentration low. In this way, a steep diffusion gradient is maintained which favours the rapid diffusion of oxygen from the air passages to the alveolar lining.

The continual delivery of carbon dioxide from the blood into the alveoli, and its removal from the air passages by ventilation, similarly maintains a diffusion gradient which promotes the diffusion of carbon dioxide from the alveolar lining into the bronchioles.

The respiratory surfaces of land-dwelling mammals are invariably moist. Oxygen has to dissolve in the thin film of moisture before passing across the epithelium.

Questions

1 Try to make a clear distinction between 'respiration' (p. 19), 'gaseous exchange' and 'ventilation'. Say how one depends on the other.

2 Describe the path taken by a molecule of oxygen from the time it is breathed in through the nose, to the time it enters the heart in some oxygenated blood.

3 Figure 13.9 shows oxygen and carbon dioxide diffusing across an alveolus. What causes them to diffuse in opposite directions? (See p. 27.)

4 In 'mouth to mouth' resuscitation, air is breathed from the rescuer's lungs into the lungs of the person who has stopped breathing. How can this 'used' air help to revive the person?

5 What volume of air might you exchange in a minute
  a while resting,
  b during exercise?

Smoking

The short-term effects of smoking cause the bronchioles to constrict and the cilia lining the air passages to stop beating. The smoke also makes the lining produce more mucus. Nicotine, the addictive component of tobacco smoke, produces an increase in the rate of the heart beat and a rise in blood pressure. It may, in some cases cause an erratic and irregular heart beat.

The long-term effects of smoking may take many years to develop but they are severe, disabling and often lethal.

Lung cancer

Although all forms of air pollution are likely to increase the chances of lung cancer, many scientific studies show, beyond all reasonable doubt, that the vast increase in lung cancer (4000 per cent in the last century) is almost entirely due to cigarette smoking (Figure 13.10).

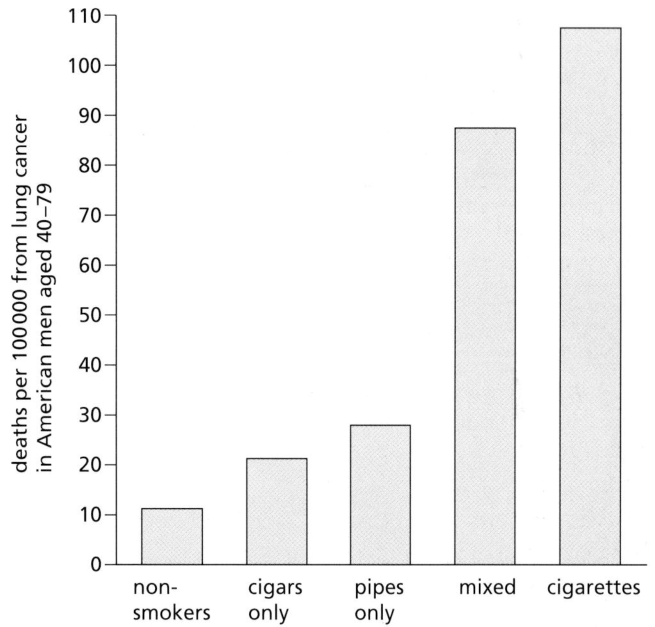

Figure 13.10 Smoking and lung cancer. Cigar and pipe smokers are probably at less risk because they often do not inhale. But notice that their death rate from lung cancer is still twice that of non-smokers

There are at least 17 substances in tobacco smoke known to cause cancer in experimental animals, and it is now thought that 90 per cent of lung cancer is caused by smoking. Table 13.2 shows the relationship between smoking cigarettes and the risk of developing lung cancer.

Table 13.2 Cigarette smoking and lung cancer

Number of cigarettes per day	Increased risk of lung cancer
1–14	×8
15–24	×13
25+	×25

Emphysema

Emphysema is a breakdown of the alveoli. The action of one or more of the substances in tobacco smoke weakens the walls of the alveoli. The irritant substances in the smoke cause a 'smokers' cough' and the coughing bursts some of the weakened alveoli. In time, the absorbing surface of the lungs is greatly reduced (Figure 13.11). Then the smoker cannot oxygenate his or her blood properly and the least exertion makes the person breathless and exhausted.

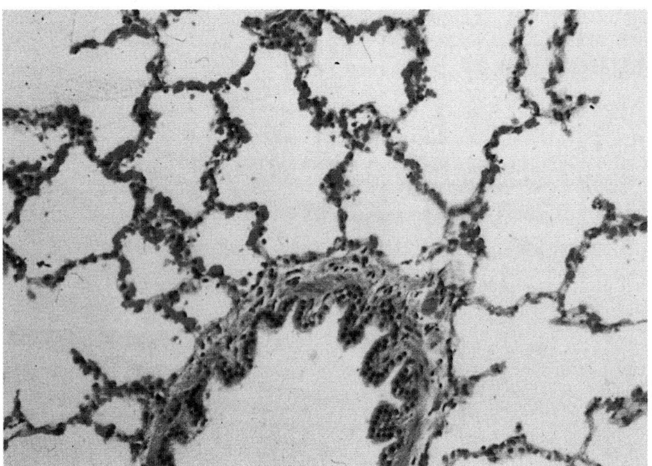

(a) Normal lung tissue showing a bronchiole and about 20 alveoli (×200)

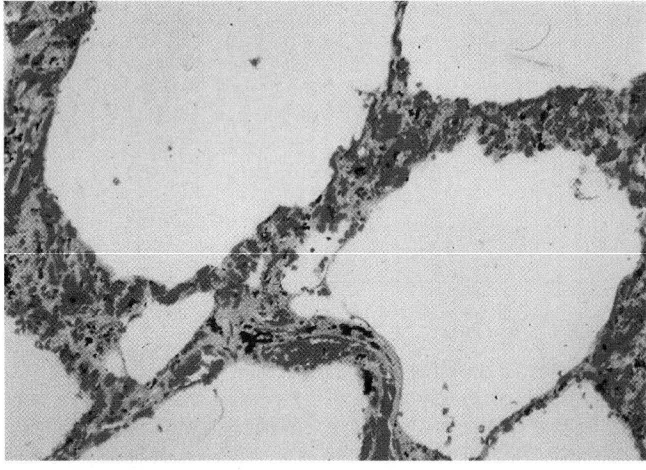

(b) Lung tissue from a person with emphysema. This is the same magnification as **(a)**. The alveoli have broken down leaving only about five air sacs which provide a much reduced absorbing surface

Figure 13.11 Emphysema

Chronic bronchitis

The smoke stops the cilia in the air passages from beating and so the irritant substances in the smoke and the excess mucus collect in the bronchi. This leads to the inflammation known as **bronchitis**. Over 95 per cent of people suffering from bronchitis are smokers and they have a 20 times greater chance of dying from bronchitis than non-smokers.

Heart disease

Coronary heart disease is the leading cause of death in most developed countries. It results from a blockage of coronary arteries by fatty deposits. This reduces the supply of oxygenated blood to the heart muscle and sooner or later leads to heart failure (see p. 120). High blood pressure, diets with too much animal fat, and lack of exercise are also thought to be causes of heart attack, but about a quarter of all deaths due to coronary heart disease are thought to be caused by smoking (Figure 12.27, p. 121).

The nicotine and carbon monoxide from cigarette smoke increase the tendency for the blood to clot and so block the coronary arteries, already partly blocked by fatty deposits. The carbon monoxide increases the rate at which the fatty material is deposited in the arteries.

Other risks

About 95 per cent of patients with disease of the leg arteries are cigarette smokers and this condition is the most frequent cause of leg amputations.

Strokes due to arterial disease in the brain are more frequent in smokers.

Cancer of the bladder, ulcers in the stomach and duodenum, tooth decay, gum disease and tuberculosis all occur more frequently in smokers.

Babies born to women who smoke during pregnancy are smaller than average, probably as a result of reduced oxygen supply caused by the carbon monoxide in the blood. In smokers, there is twice the frequency of miscarriages, a 50 per cent higher still-birth rate and a 26 per cent higher death rate of babies (Figure 13.12).

Figure 13.12 Cigarette smoke can harm the unborn baby. Pregnant women who smoke may have smaller babies and a higher chance of miscarriage or still-birth

A recent estimate is that one in every three smokers will die as a result of their smoking habits. Those who do not die at an early age will probably be seriously disabled by one of the conditions described above.

Passive smoking

It is not only the smokers themselves who are harmed by tobacco smoke. Non-smokers in the same room are also affected. One study has shown that children whose parents both smoke, breathe in as much nicotine as if they were themselves smoking 80 cigarettes a year.

Statistical studies also suggest that the non-smoking wives of smokers have an increased chance of lung cancer.

Reducing the risks

By giving up smoking, a person who smokes up to 20 cigarettes a day will, after 10 years, be at no greater risk than a non-smoker of the same age. The pipe or cigar smoker, provided he or she does not inhale, is at less risk than a cigarette smoker but still at greater risk than a non-smoker.

Correlations and causes

On p. 121 it was explained that a correlation between two variables does not prove that one of the variables causes the other. The fact that a higher risk of dying from lung cancer is correlated with heavy smoking does not actually prove that smoking is the cause of lung cancer. The alternative explanation is that people who become heavy smokers are, in some way, exposed to other potential causes of lung cancer, e.g. they live in areas of high air pollution or they have an inherited tendency to cancer of the lung. These alternatives are not very convincing, particularly when there is such an extensive list of ailments associated with smoking.

This is not to say that smoking is the only cause of lung cancer or that everyone who smokes will eventually develop lung cancer. There are likely to be complex interactions between life-styles, environments and genetic backgrounds which could lead, in some cases, to lung cancer. Smoking may be only a part, but a very important part, of these interactions.

Questions

1 What are
  a the immediate effects and
  b the long-term effects,
  of tobacco smoke on the trachea, bronchi and lungs?

2 Why does a regular smoker get out of breath sooner than a non-smoker of similar age and build?

3 If you smoke 20 cigarettes a day, by how much are your chances of getting lung cancer increased?

4 Apart from lung cancer, what other diseases are probably caused by smoking?

Practical work

1 Oxygen in exhaled air

Place a large screw-top jar on its side in a bowl of water (Figure 13.13a). Put a rubber tube in the mouth of the jar and then turn the jar upside-down, still full of water and with the rubber tube still in it. Start breathing out and when you feel your lungs must be about half empty, breathe the last part of the air down the rubber tubing so that the air collects in the upturned jar and fills it (Figure 13.13b). Put the screw top back on the jar under water, remove the jar from the bowl and place it upright on the bench.

Light the candle on the special wire holder (Figure 13.13c), remove the lid of the jar, lower the burning candle into the jar and count the number of seconds the candle stays alight. Now take a fresh jar, with ordinary air, and see how long the candle stays alight in this.

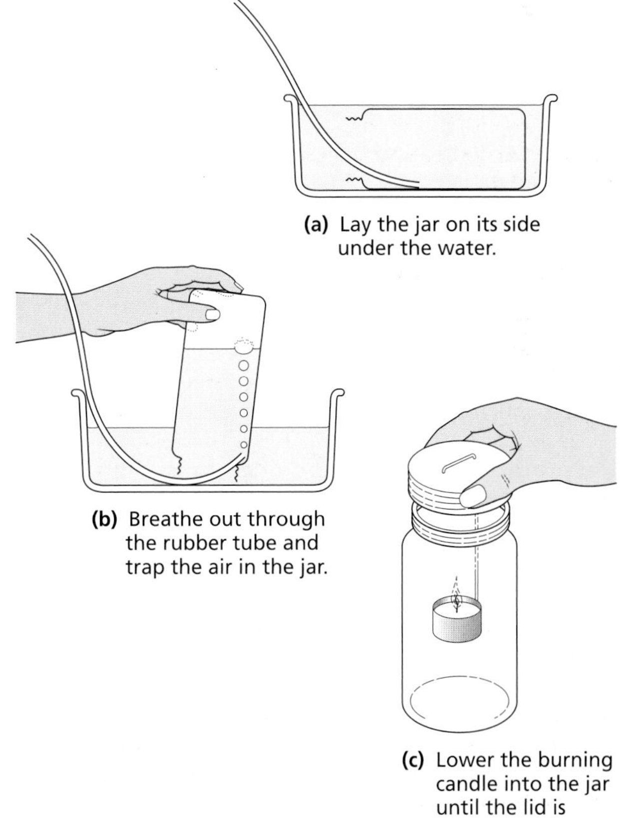

(a) Lay the jar on its side under the water.
(b) Breathe out through the rubber tube and trap the air in the jar.
(c) Lower the burning candle into the jar until the lid is resting on the rim.

Figure 13.13 Testing exhaled air for oxygen

Results The candle will burn for about 15–20 seconds in a large jar of ordinary air. In exhaled air it will go out in about 5 seconds.

Interpretation Burning needs oxygen. When the oxygen is used up, the flame goes out. It looks as if exhaled air contains much less oxygen than atmospheric air.

2 Carbon dioxide in exhaled air

Prepare two large test-tubes, A and B, as shown in Figure 13.14, each containing a little clear lime water. Put the ends of both rubber tubes at the same time in your mouth and breathe in and out gently through the tubes for about 15 seconds. Notice which tube is bubbling when you breathe out and which one bubbles when you breathe in.

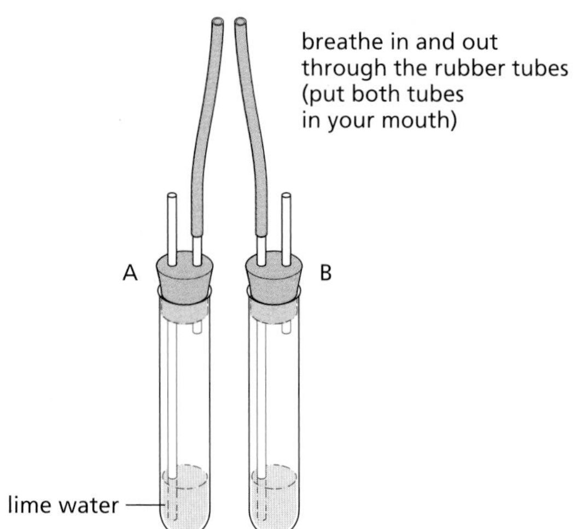

Figure 13.14 Comparing the carbon dioxide content of inhaled and exhaled air

If after 15 seconds there is no difference in the appearance of the lime water in the two tubes, continue breathing through them for another 15 seconds.

Results The lime water in tube B goes milky. The lime water in tube A stays clear.

Interpretation Carbon dioxide turns lime water milky. Exhaled air passes through tube B. Inhaled air passes through tube A. Exhaled air must, therefore, contain more carbon dioxide than inhaled air.

3 Volume of air in the lungs

Calibrate a large (5 litre) plastic bottle by filling it with water, half a litre at a time, and marking the water levels on the outside. Fill the bottle with water and put on the stopper. Put about 50 mm depth of water in a large plastic bowl. Hold the bottle upside-down with its neck under water and remove the screw top. Some of the water will run out but this does not matter. Push a rubber tube into the mouth of the bottle to position A, shown on the diagram (Figure 13.15). Take a deep breath and then exhale as much air as possible down the tubing into the bottle. The final water level inside the bottle will tell you how much air you can exchange in one deep breath.

Now push the rubber tubing further into the bottle, to position B (Figure 13.15), and blow out any water left in the tube. Support the bottle with your hand and breathe quietly in and out through the tube, keeping the water level inside and outside the bottle the same. This will give you an idea of how much air you exchange when breathing normally.

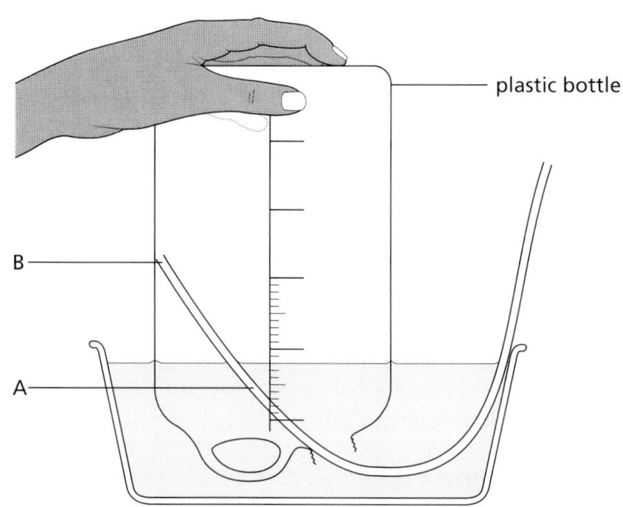

Figure 13.15 Measuring the volume of air exhaled from the lungs. (A) shows the position of the tube when measuring the maximum usable lung volume. (B) is the position for measuring the volume exchanged in quiet breathing

Checklist

- Ventilation is inhaling and exhaling air.
- The ribs, rib muscles and diaphragm make the lungs expand and contract. This causes inhaling and exhaling.
- Air is drawn into the lungs through the trachea, bronchi and bronchioles.
- The vast number of air pockets (alveoli) give the lungs an enormous internal surface area. This surface is moist and lined with capillaries.
- The blood in the capillaries picks up oxygen from the air in the alveoli and gives out carbon dioxide. This is called gaseous exchange.
- Ventilation exchanges the air in the air passages but not in the alveoli.
- Exchange of oxygen and carbon dioxide in the alveoli takes place by diffusion.
- The oxygen is carried round the body by the blood and used by the cells for their respiration.
- During exercise, the rate and depth of breathing increase. This supplies extra oxygen to the muscles and removes their excess carbon dioxide.
- Tobacco smoke causes the bronchioles to constrict, the cilia in their lining to stop beating and excessive mucus to be produced.
- Smoking is correlated with heart disease, bronchitis, emphysema and lung cancer.

14 Excretion and the kidneys

Excretion

Excretory organs
Lungs, kidneys, liver.

The kidneys
Structure. Function. Selective reabsorption.

Water balance and osmoregulation
Controlling the blood concentration.

The dialysis machine

Homeostasis
The stability of the internal environment.

Excretion

Many chemical reactions take place inside the cells of an organism in order to keep it alive. Some products of these reactions are poisonous and must be removed from the body. For example, the breakdown of glucose during respiration (p. 19) produces carbon dioxide. This is carried away by the blood and removed in the lungs. Excess amino acids are deaminated in the liver to form glycogen and **urea**, as explained on p. 104. The urea is removed from the tissues by the blood, and expelled by the kidneys.

Urea and similar waste products, like **uric acid**, from the breakdown of proteins, contain the element nitrogen. For this reason they are often called **nitrogenous waste products**.

During feeding, more water and salts are taken in with the food than are needed by the body. So these excess substances need to be removed as fast as they build up.

The hormones produced by the endocrine glands (p. 169) affect the rate at which various body systems work. Adrenaline, for example, speeds up the heart beat. When hormones have done their job, they are modified in the liver and excreted by the kidneys.

The nitrogenous waste products, excess salts and spent hormones are excreted by the kidneys as a watery solution called **urine**.

Excretion is the name given to the removal from the body of (1) the waste products of its chemical reactions, (2) the excess water and salts taken in with the diet and (3) spent hormones.

Excretion also includes the removal of drugs or other foreign substances taken into the alimentary canal and absorbed by the blood.

Questions

1 Write a list of the substances that are likely to be excreted from the body during the day.

2 Why do you think that urine analysis is an important part of medical diagnosis?

Excretory organs

Lungs

The lungs supply the body with oxygen, but they are also excretory organs because they get rid of carbon dioxide. They also lose a great deal of water vapour, but this loss is unavoidable and is not a method of controlling the water content of the body (Table 14.1).

Kidneys

The kidneys remove urea and other nitrogenous waste from the blood. They also expel excess water, salts, hormones (p. 169) and drugs.

Liver

The yellow/green bile pigment, bilirubin, is a breakdown product of haemoglobin (see p. 108). Bilirubin is excreted with the bile into the small intestine and expelled with the faeces. The pigment undergoes changes in the intestine and is largely responsible for the brown colour of the faeces.

Skin

Sweat consists of water, with sodium chloride and traces of urea dissolved in it. When you sweat, you will expel these substances from your body and so, in one sense, they are being excreted. However, sweating is a response to a rise in temperature and not to a change in the blood composition. In this sense, therefore, skin is not an excretory organ like the lungs and kidneys.

Table 14.1 Excretory products and incidental losses

	Excretory products	Incidental losses
Lungs	carbon dioxide	water
Kidneys	nitrogenous waste, water, salts, toxins, hormones, drugs	
Liver	bile pigments	
Skin		water, salt, urea

The kidneys

Structure

The two kidneys are fairly solid, oval structures. They are red-brown, enclosed in a transparent membrane and attached to the back of the abdominal cavity (Figure 14.1). The **renal artery** branches off from the aorta and brings oxygenated blood to them. The **renal vein** takes deoxygenated blood away from the kidneys to the vena cava (see Figure 12.11, p. 112). A tube, called the **ureter**, runs from each kidney to the bladder in the lower part of the abdomen.

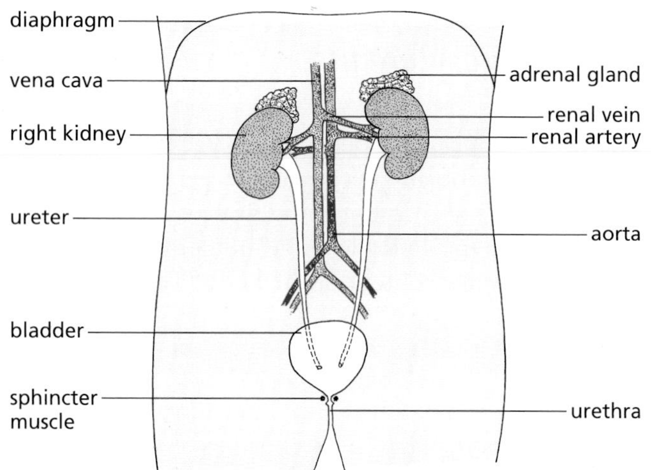

Figure 14.1 Position of the kidneys in the body

The kidney tissue consists of many capillaries and tiny tubes, called **renal tubules**, held together with connective tissue. If the kidney is cut down its length (sectioned), it is seen to have a dark, outer region called the **cortex** and a lighter, inner zone, the **medulla**. Where the ureter joins the kidney there is a space called the **pelvis** (Figure 14.2).

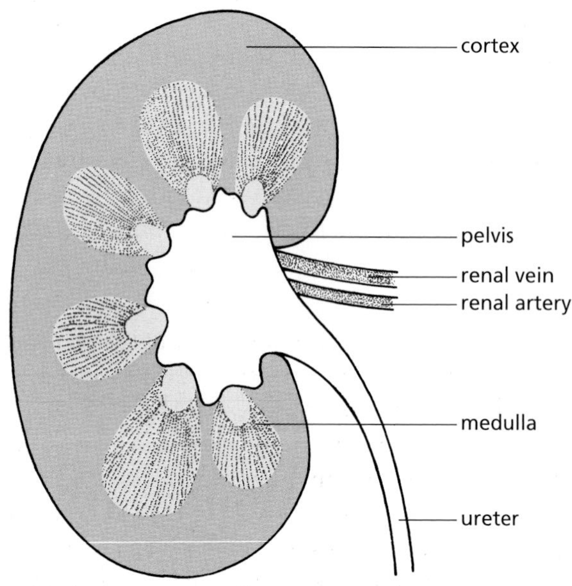

Figure 14.2 Section through the kidney to show regions

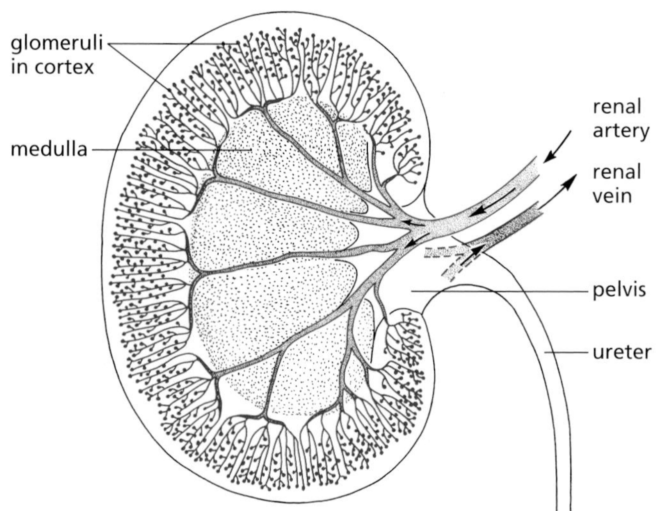

Figure 14.3 Section through kidney to show distribution of glomeruli

The renal artery divides up into a great many arterioles and capillaries, mostly in the cortex (Figure 14.3). Each arteriole leads to a **glomerulus**. This is a capillary repeatedly divided and coiled, making a knot of vessels (Figure 14.4). Each glomerulus is almost entirely surrounded by a cup-shaped organ called a **renal capsule**, which leads to a coiled **renal tubule**. This tubule, after a series of coils and loops, joins a **collecting duct** which passes through the medulla to open into the pelvis (Figure 14.5). There are thousands of glomeruli in the kidney cortex and the total surface area of their capillaries is very great.

A **nephron** is a single glomerulus with its renal capsule, renal tubule and blood capillaries (see Figure 14.6).

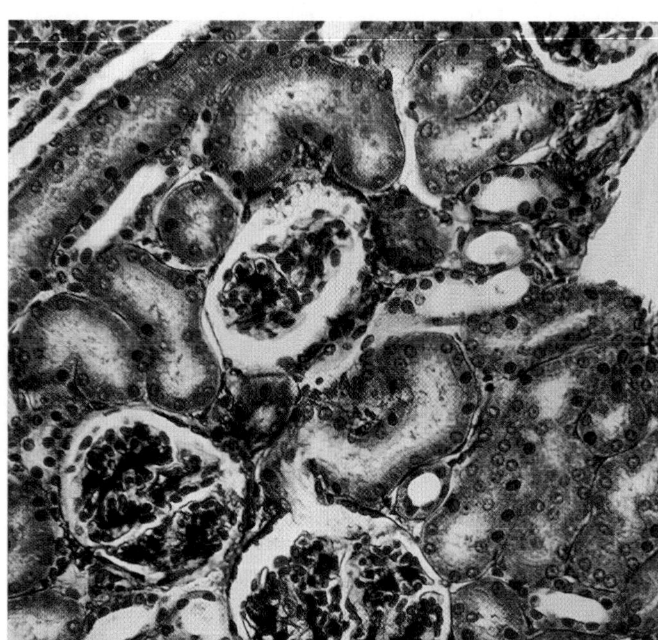

Figure 14.4 Glomeruli in the kidney cortex (× 300). The three glomeruli are surrounded by kidney tubules sectioned at different angles. The light space round each glomerulus represents the renal capsule

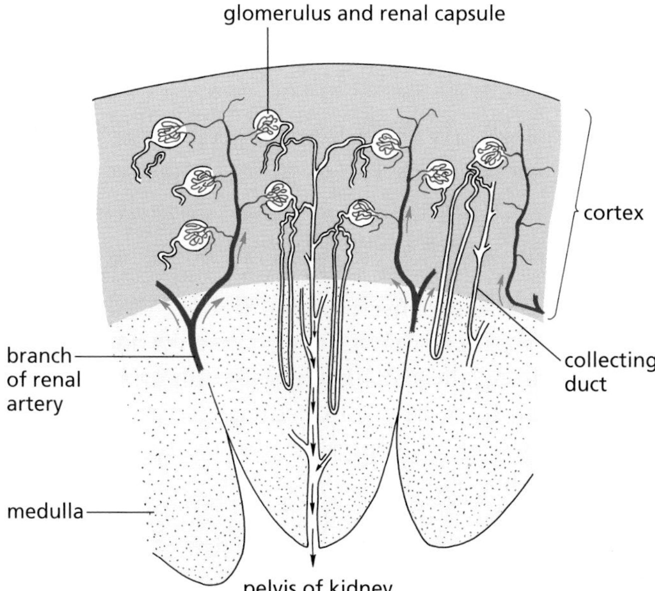

Figure 14.5 There are up to 4 million nephrons in a kidney. Only a few can be represented here, and not to scale

Function of the kidneys

The blood pressure in a glomerulus causes part of the blood plasma to leak through the capillary walls. The red blood cells and the plasma proteins are too big to pass out of the capillary, so the fluid that does filter through is plasma without the protein, i.e. similar to tissue fluid (see p. 113). The fluid thus consists mainly of water with dissolved salts, glucose, urea and uric acid. The process by which the fluid is filtered out of the blood by the glomerulus is called **ultrafiltration**.

The filtrate from the glomerulus collects in the renal capsule and trickles down the renal tubule (Figure 14.6). As it does so, the capillaries which surround the tubule absorb back into the blood those substances which the body needs. First, all the glucose is reabsorbed, with much of the water. Then some of the salts are taken back to keep the correct concentration in the blood. The process of absorbing back the substances needed by the body is called **selective reabsorption**.

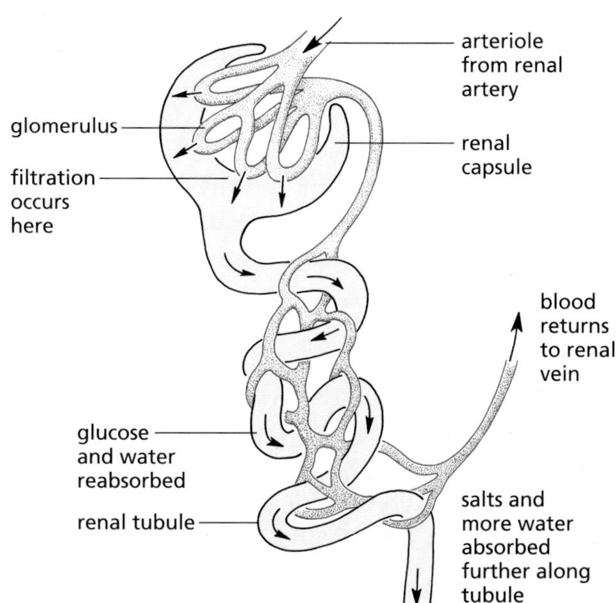

Figure 14.6 Part of a nephron (glomerulus, renal capsule and renal tubule)

Salts not needed by the body are left to pass on down the kidney tubule together with the urea and uric acid. So, these nitrogenous waste products, excess salts and water continue down the renal tube into the pelvis of the kidney. From here the fluid, now called **urine**, passes down the ureter to the bladder.

Table 14.2 shows some of the differences in composition between the blood plasma and the urine. The figures represent average values because the composition of the urine varies a great deal according to the diet, activity, temperature and intake of liquid.

Table 14.2 Composition of blood plasma and urine

	Plasma %	Urine %
Water	90–93	95.0
Urea	0.03	2.0
Uric acid	0.003	0.05
Ammonia	0.0001	0.05
Sodium	0.3	0.6
Potassium	0.02	0.15
Chloride	0.37	0.6
Phosphate	0.003	0.12

The **bladder** can expand to hold about 400 cm^3 urine. The urine cannot escape from the bladder because a band of circular muscle, called a sphincter, is contracted, so shutting off the exit. When this sphincter muscle relaxes, the muscular walls of the bladder expel the urine through the **urethra**. Adults can control this sphincter muscle and relax it only when they want to urinate. In babies, the sphincter relaxes by a reflex action (p. 167), set off by pressure in the bladder. By 3 years old, most children can control the sphincter voluntarily.

Water balance and osmoregulation

Your body gains water from food and drink. It loses water by evaporation, urination and defecation (p. 102). Evaporation from the skin takes place all the time but is particularly rapid when we sweat. Air from the lungs is saturated with water vapour which is lost to the atmosphere every time we exhale. Despite these gains and losses of water, the concentration of body fluids is kept within very narrow limits by the kidneys, which adjust the concentration of the blood flowing through them. If it is too dilute (i.e. has too much water), less water is reabsorbed from the renal tubules, leaving more to enter the bladder. After drinking a lot, a large volume of dilute urine is produced.

If the blood is too concentrated, more water is absorbed back into the blood from the kidney tubules. So, if the body is short of water, e.g. after sweating profusely, only a small quantity of concentrated urine is produced.

A rise in the blood concentration is thought to stimulate a 'thirst' centre in the brain. The drinking which follows this stimulation restores the blood to its correct concentration.

These regulatory processes keep the blood at a steady concentration and together are called **osmoregulation** because they regulate the osmotic strength (see p. 29) of the blood. Osmoregulation is one example of the process of **homeostasis**, which is described on p. 135.

Changes in the concentration of the blood are detected by an area in the brain called the **hypothalamus**. If the blood passing through the brain is too concentrated, the hypothalamus stimulates the pituitary gland beneath it to secrete into the blood a hormone (p. 169) called antidiuretic hormone (ADH). When this hormone reaches the kidneys, it causes the kidney tubules to absorb more water from the glomerular filtrate back into the blood. Thus the urine becomes more concentrated and the further loss of water from the blood is reduced. If blood passing through the hypothalamus is too dilute, production of ADH from the pituitary is suppressed and less water is absorbed from the glomerular filtrate.

Questions

1 Why should a fall in blood pressure sometimes lead to kidney failure?

2 In what ways would you expect the composition of blood in the renal vein to differ from that in the renal artery? (Remember that the cells in the kidney will be respiring.)

3 Where, in the urinary system, do the following take place (answer as precisely as possible): filtration, reabsorption, storage of urine, transport of urine, osmoregulation?

4 In hot weather, when you sweat a great deal, you urinate less often and the urine is a dark colour. In cold weather, when you sweat little, urination occurs more often and the urine is pale in colour. Use your knowledge of kidney function to explain these observations.

5 Trace the path taken by a molecule of urea from the time it is produced in the liver, to the time it leaves the body in the urine (see also p. 112).

6 The heart pumps about 5 litres of blood per minute. At rest, about one-quarter of the heart's output passes through the kidneys. About 180 litres of blood-derived liquid (glomerular filtrate) passes out of the glomeruli into the renal capsules each day. Use these figures to calculate the percentage volume of blood that is filtered by the kidneys.

The dialysis machine ('artificial kidney')

Kidney failure may result from an accident involving a drop in blood pressure, or from a disease of the kidneys. In the former case, recovery is usually spontaneous, but if it takes longer than 2 weeks, the patient may die as a result of a potassium imbalance in the blood, which causes heart failure. In the case of kidney disease, the patient can survive with only one kidney, but if both fail the patient's blood composition has to be regulated by a dialysis machine. Similarly, the accident victim can be kept alive on a dialysis machine until his or her blood pressure is restored.

In principle, a dialysis machine consists of a long cellulose tube coiled up in a water bath. The patient's blood is led from a vein in the arm and pumped through the cellulose (dialysis) tubing (Figures 14.7 and 14.8). The submicroscopic pores in the dialysis tubing allow small molecules, such as those of salts, glucose and urea, to leak out into the water bath. Blood cells and protein molecules are too large to get through the pores (see Experiment 5, p. 34). This stage is similar to the filtration process in the glomerulus.

To prevent a loss of glucose and essential salts from the blood, the liquid in the water bath consists of a solution of salts and sugar of the correct composition, so that only the substances above this concentration can diffuse out of the blood into the bathing solution. Thus, urea, uric acid and excess salts are removed.

The bathing solution is also kept at body temperature and is constantly changed as the unwanted blood solutes accumulate in it. The blood is then returned to the patient's arm vein.

A patient with total kidney failure has to spend 2 or 3 nights each week connected to the machine (Figure 14.8). With this treatment and a carefully controlled diet, the patient can lead a fairly normal life. A kidney transplant, however, is a better solution because the patient is not obliged to return to the dialysis machine.

The problem with kidney transplants is to find enough suitable donors of healthy kidneys and to prevent the transplanted kidney from being rejected.

The donor may be a close relative who is prepared to donate one of his or her kidneys (you can survive adequately with one kidney). Alternatively, the donated kidney may be taken from a healthy person who dies, for example, as a result of a road accident. People willing for their kidneys to be used after their death can carry a kidney donor card but the relatives must give their permission for the kidneys to be used.

The problem with rejection is that the body reacts to any transplanted cells or tissues as it does to all foreign proteins and produces lymphocytes which attack and destroy them (see p. 118). This rejection can

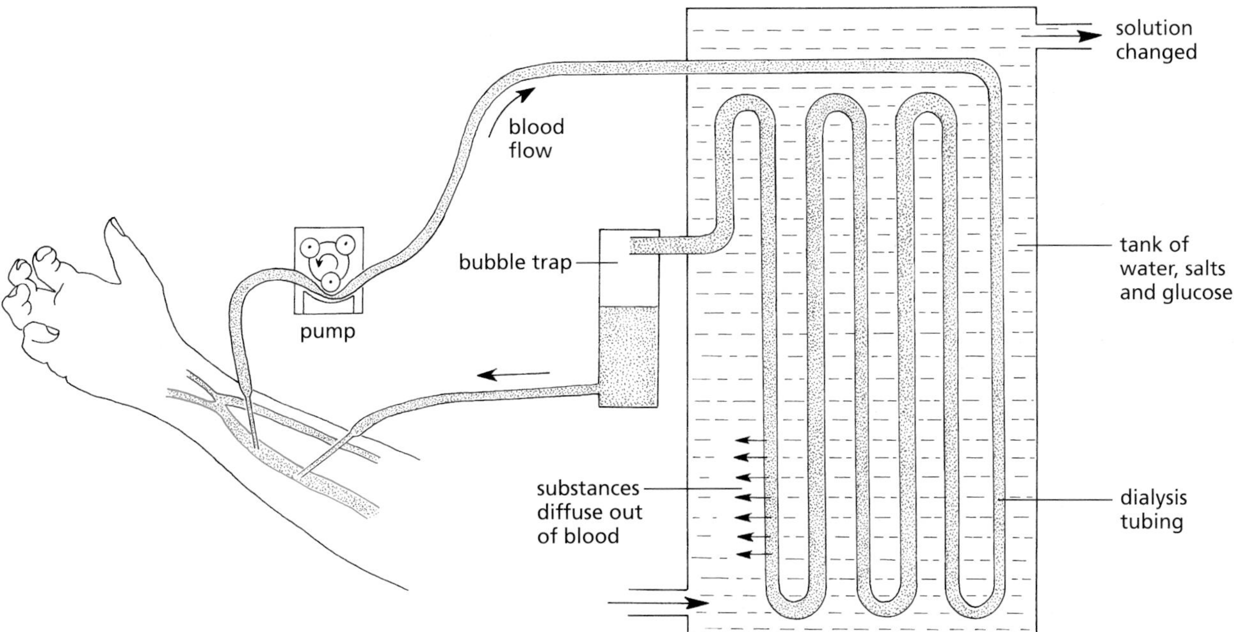

Figure 14.7 The principle of the kidney dialysis machine

be overcome by (1) choosing a donor whose tissues are as similar as possible to those of the patient, e.g. a close relative (p. 119), and (2) using immunosuppressive drugs which suppress the production of lymphocytes and their antibodies against the transplanted organ.

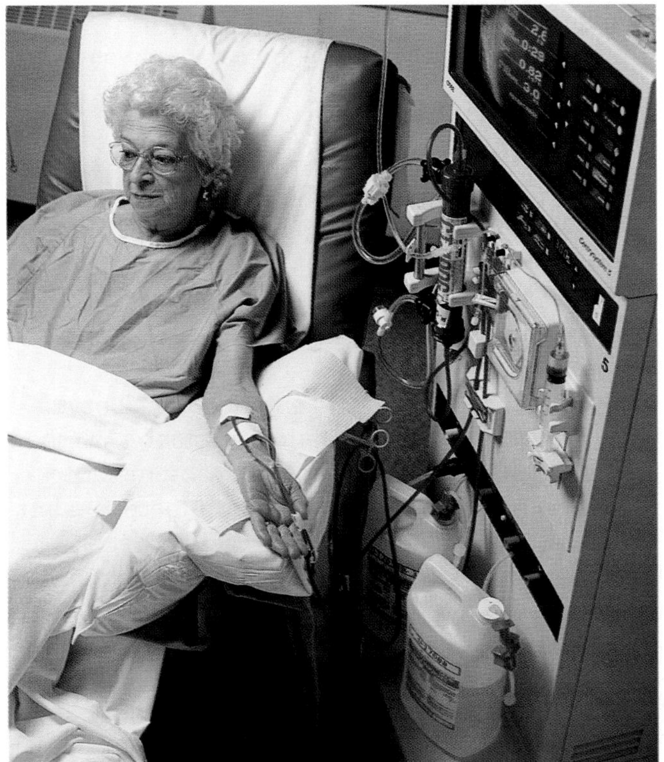

Figure 14.8 Kidney dialysis machine. The patient's blood is pumped to the dialyser, which removes urea and excess salts

Homeostasis

Homeostasis means 'staying similar'. It refers to the fact that the composition of the tissue fluid (p. 113) in the body is kept within narrow limits. The concentration, acidity and temperature of this fluid are being adjusted all the time to prevent any big changes.

On p. 14 it was explained that, in living cells, all the chemical reactions are controlled by enzymes. The enzymes are very sensitive to the conditions in which they work. A slight fall in temperature or a rise in acidity (p. 15) may slow down or stop an enzyme from working and thus prevent an important reaction from taking place in the cell.

The cell membrane controls the substances which enter and leave the cell, but it is the tissue fluid which supplies or removes these substances, and it is therefore important to keep the composition of the tissue fluid as steady as possible. If the tissue fluid were too concentrated, it would withdraw water from the cells by osmosis (p. 29) and the body would be dehydrated. If the tissue fluid were too dilute, the cells would take up too much water from it by osmosis and the tissues would become waterlogged and swollen.

Many systems in the body contribute to homeostasis (Figure 14.9). The obvious example is the kidneys, which remove substances that might poison the enzymes. The kidneys also control the level of salts, water and acids in the blood. The composition of the blood affects the tissue fluid which, in turn, affects the cells.

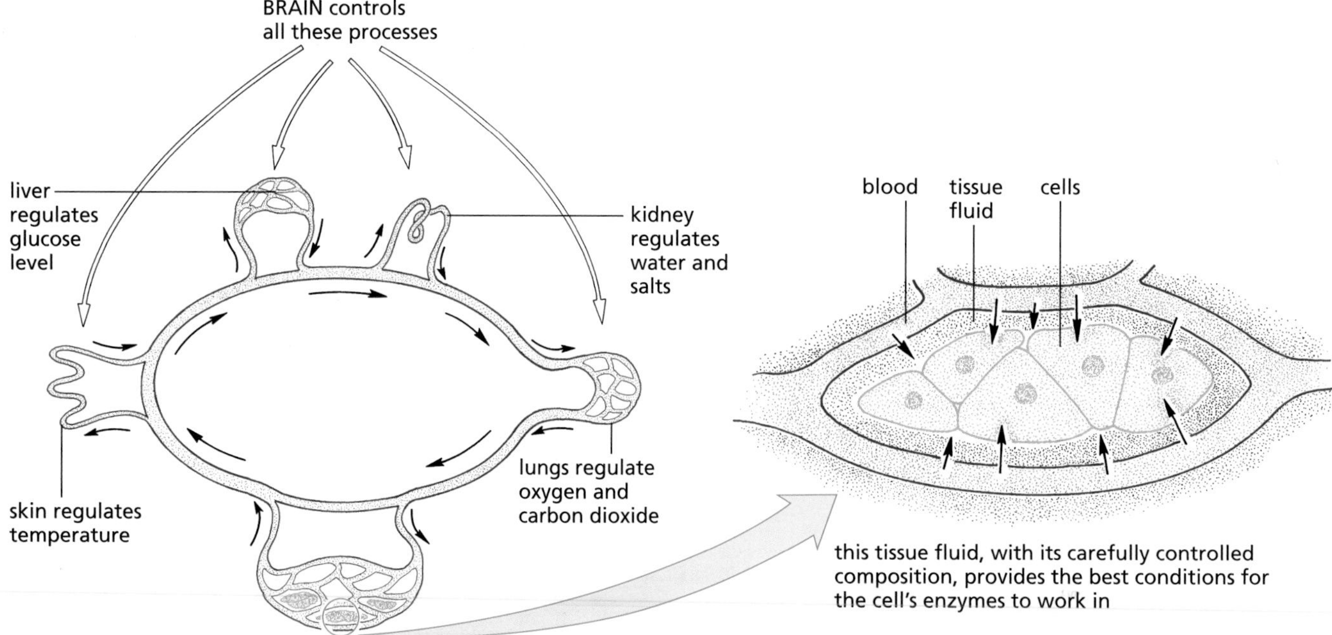

Figure 14.9 The homeostatic mechanisms of the body

Another example of a homeostatic organ is the liver, which regulates the level of glucose in the blood (p. 104). The liver stores any excess glucose as glycogen, or turns glycogen back into glucose if the concentration in the blood gets too low. The brain cells are very sensitive to the glucose concentration in the blood and if the level drops too far, they stop working properly, and the person becomes unconscious and will die unless glucose is injected into the blood system. This shows how important homeostasis is to the body.

The lungs (p. 123) play a part in homeostasis by keeping the concentrations of oxygen and carbon dioxide in the blood at the best level for the cells' chemical reactions, especially respiration.

The next chapter describes the way in which the skin regulates the temperature of the blood. If the cells were to get too cold, the chemical reactions would become too slow to maintain life. If they became too hot, the enzymes would be destroyed.

The brain has overall control of the homeostatic processes in the body. It checks the composition of the blood flowing through it and if it is too warm, too cold, too concentrated or has too little glucose, nerve impulses or hormones are sent to the organs concerned, causing them to make the necessary adjustments.

Questions

1 Where will the brain send nerve impulses or hormones if the blood flowing through it
 - **a** has too much water,
 - **b** contains too little glucose,
 - **c** is too warm,
 - **d** has too much carbon dioxide?

2 How does the dialysis machine
 - **a** resemble and
 - **b** differ from

 the nephron of a kidney in the way it functions?

Checklist

- Excretion is getting rid of unwanted substances produced by chemical reactions in the body or taken in with the diet.
- The lungs excrete carbon dioxide.
- The kidneys excrete urea, unwanted salts and excess water.
- Part of the blood plasma entering the kidneys is filtered out by the capillaries. Substances which the body needs, like glucose, are absorbed back into the blood. The unwanted substances are left to pass down the ureters into the bladder.
- The bladder stores urine, which is discharged at intervals.
- The kidneys help to keep the blood at a steady concentration by excreting excess salts and by adjusting the amounts of water (osmoregulation).
- The kidneys, lungs, liver and skin all help to keep the blood composition the same (homeostasis).

15 The skin, and temperature control

Skin structure
Epidermis, dermis, hair, sweat glands.

Skin function
Protection, sensitivity, temperature control.

Temperature control
Vasodilation, vasoconstriction, sweating.

Homeostasis and negative feedback

Skin structure (Figure 15.1)

In the **basal layer** some of the cells are continually dividing and pushing the older cells nearer the surface. Here they die and are shed at the same rate as they are replaced. The basal layer and the cells above it constitute the **epidermis**. The basal layer also contributes to the hair follicles. The dividing cells give rise to the hair.

There are specialized pigment cells in the basal layer and epidermis. These produce a black pigment, **melanin**, which gives the skin its colour. The more melanin, the darker is the skin.

The thickness of the epidermis and the abundance of hairs vary in different parts of the body (Figure 15.2).

The **dermis** contains connective tissue with hair follicles, sebaceous glands, sweat glands, blood vessels and nerve endings. There is usually a layer of adipose tissue (a fat depot) beneath the dermis.

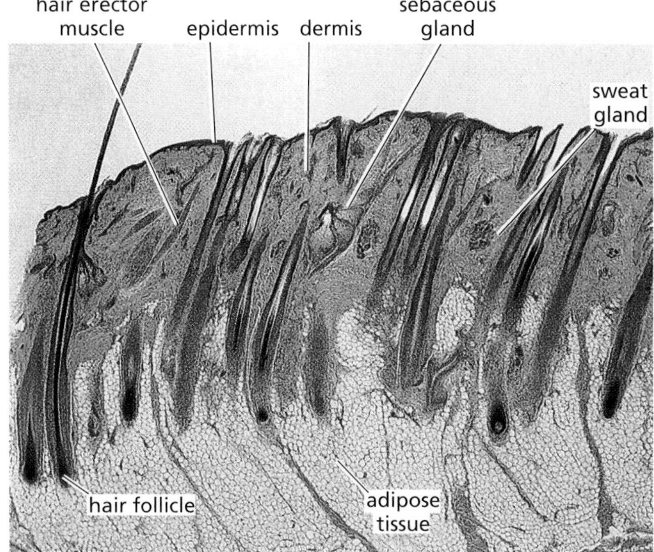

Figure 15.2 Section through hairy skin (×20)

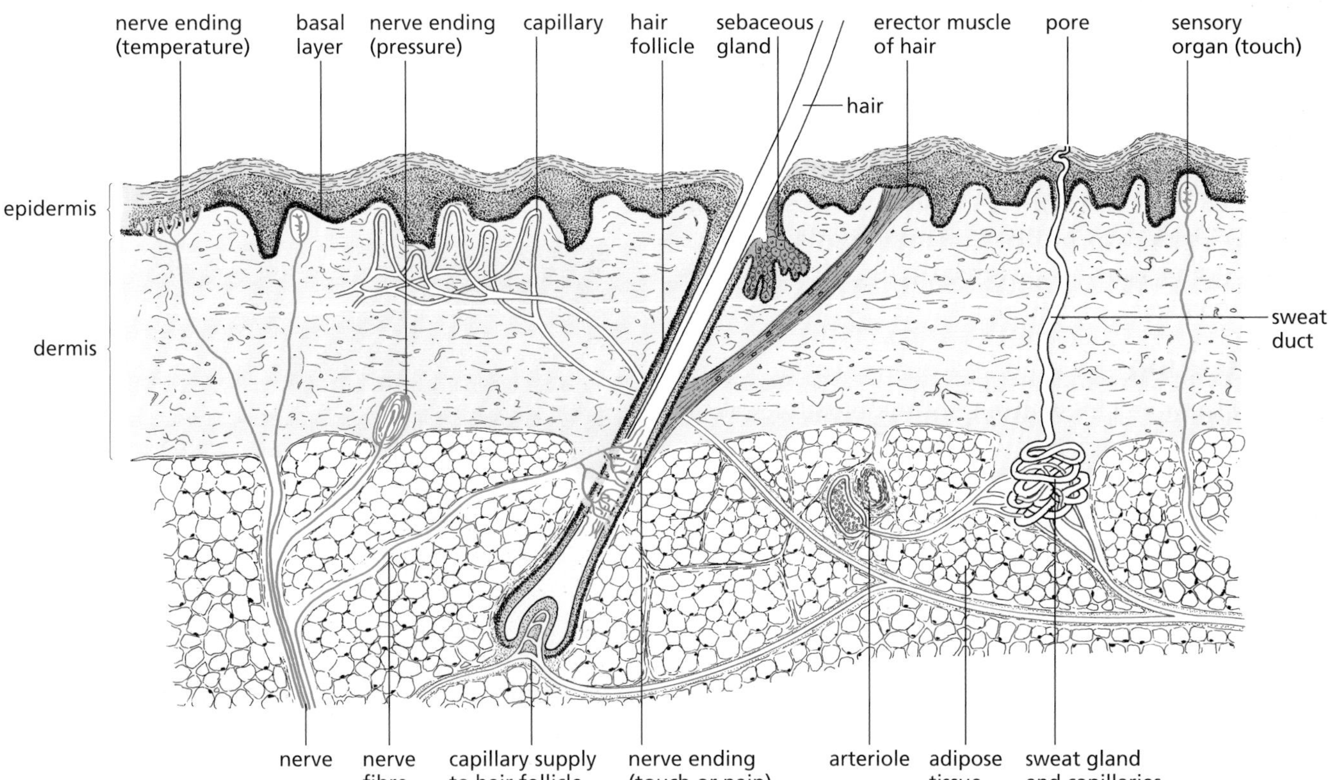

Figure 15.1 Generalized section through the skin

Skin function

Protection

The outermost layer of dead cells of the epidermis helps to reduce water loss and provides a barrier against bacteria. The pigment cells protect the skin from damage by the ultraviolet rays in sunlight. In white-skinned people, more melanin is produced in response to exposure to sunlight, giving rise to a tan.

Sensitivity

Scattered throughout the skin are a large numbers of tiny sense organs, which give rise to sensations of touch, pressure, heat, cold and pain. These make us aware of changes in our surroundings and enable us to take action to avoid damage, to recognize objects by touch and to manipulate objects with our hands.

Temperature regulation

The skin helps to keep the body temperature more or less constant. This is done by adjusting the flow of blood near the skin surface and by sweating. These processes are described more fully below.

Temperature control

Normal human body temperature varies between 35.8 °C and 37.7 °C. Temperatures below 34 °C or above 40 °C, if maintained for long, are considered dangerous. Different body regions, e.g. the hands, feet, head or internal organs, will be at different temperatures, but the **core** temperature, as measured with a thermometer under the tongue, will vary by only 1 or 2 degrees.

Heat is lost from the body surface by conduction, convection, radiation and evaporation. Heat is gained, internally, from the process of respiration (p. 19) in the tissues and, externally, from the surroundings or from the sun. These two processes are normally in balance but any imbalance is corrected by the methods described below.

Overheating

- *Vasodilation* – The widening of the blood vessels in the dermis allows more warm blood to flow near the skin surface and so lose more heat (Figure 15.3a).
- *Sweating* – The sweat glands secrete sweat on to the skin surface. When this layer of liquid evaporates, it takes heat (latent heat) from the body and cools it down (Figure 15.4).

Overcooling

- *Vasoconstriction* – Narrowing (constriction) of the blood vessels in the skin reduces the amount of warm blood flowing near the surface.
- *Sweat production stops* – Thus the heat lost by evaporation is reduced (Figure 15.3b).
- *Shivering* – Uncontrollable bursts of rapid muscular contraction in the limbs release heat as a result of respiration in the muscles.

In these ways, the body temperature remains at about 37 °C. We also control our temperature by adding or removing clothing or deliberately taking exercise.

Whether we feel hot or cold depends on the sensory nerve endings in the skin, which respond to heat loss or gain. You cannot consciously detect changes in your core temperature.

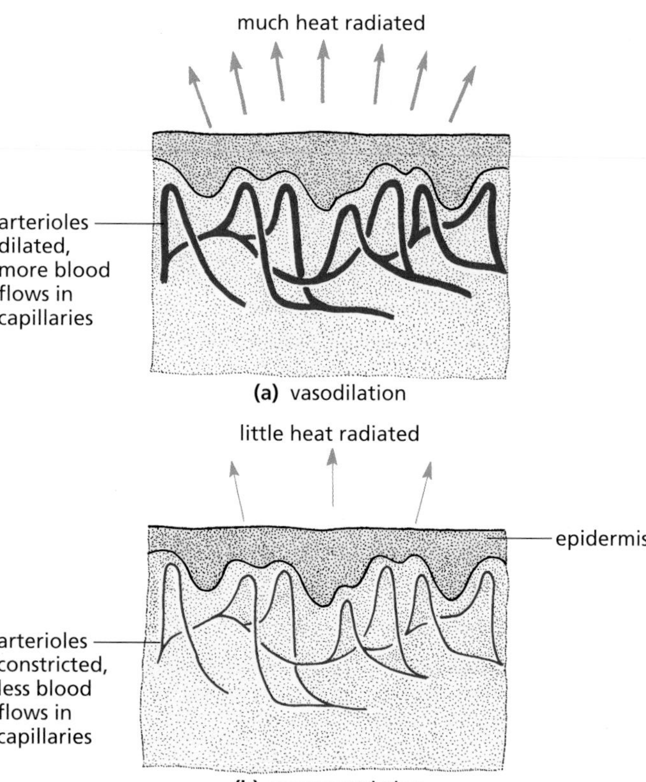

Figure 15.3 Vasodilation and vasoconstriction

Figure 15.4 Sweating. During vigorous activity the sweat evaporates from the skin and helps to cool the body. When the activity stops, continued evaporation of sweat may over-cool the body unless it is towelled off

Homeostasis and negative feedback

Temperature regulation is an example of homeostasis (p. 135). Maintenance of a constant body temperature ensures that vital chemical reactions continue at a predictable rate and do not speed up or slow down when the surrounding temperature changes. The constant-temperature or **homoiothermic** ('warm-blooded') animals, the birds and mammals, therefore have an advantage over the variable-temperature or **poikilothermic** ('cold-blooded') animals. Poikilotherms such as reptiles and insects can regulate their body temperature to some extent by, for example, basking in the sun or seeking shade. Nevertheless, if their body temperature falls, their vital chemistry slows down and their reactions become more sluggish. They are then more vulnerable to predators.

The 'price' that homoiotherms have to pay is the intake of enough food to maintain their body temperature, usually above that of their surroundings.

In the hypothalamus (p. 168) of a homoiotherm's brain, there is a thermoregulatory centre. This centre monitors the temperature of the blood passing through it and also receives sensory nerve impulses from temperature receptors in the skin. A rise in body temperature is detected by the thermoregulatory centre and it sends nerve impulses to the skin which result in vasodilation and sweating. Similarly, a fall in body temperature will be detected and will promote impulses which produce vasoconstriction and shivering.

This system of control is called **negative feedback**. The outgoing impulses counteract the effects which produced the incoming impulses. For example, a rise in temperature triggers responses which counteract the rise.

Questions

1 What conscious actions do we take to reduce the heat lost from the body?

2 What sort of chemical reaction in active muscle will produce heat? How does this heat get to other parts of the body? (See pp. 19 and 112.)

3 Draw up a balance sheet to show all the possible ways the human body can gain or lose heat. Make two columns, with 'Gains' on the left and 'Losses' on the right.

4 **a** Which structures in the skin of a furry mammal help to reduce heat loss?
b What changes take place in the skin of humans to reduce heat loss?

5 If your body temperature hardly changes at all, why do you sometimes feel hot and sometimes cold?

6 Sweating cools you down only if the sweat can evaporate.
a In what conditions might the sweat be unable to evaporate from your skin? (See p. 61.)
b What conditions might speed up the evaporation of sweat and so make you feel very cold?

7 Why, do you think, are most control systems examples of negative feedback rather than positive feedback?

Checklist

- Skin consists of an outer layer of epidermis and an inner dermis.
- The epidermis is growing all the time and has an outer layer of dead cells.
- The dermis contains the sweat glands, hair follicles, sense organs and capillaries.
- Skin (1) protects the body from bacteria and drying out, (2) contains sense organs which give us the sense of touch, warmth, cold and pain, and (3) controls the body temperature.
- Chemical activity in the body and muscular contractions produce heat.
- Heat is lost to the surroundings by conduction, convection, radiation and evaporation.
- If the body temperature rises too much, the skin cools it down by sweating and vasodilation.
- If the body loses too much heat, vasoconstriction and shivering help to keep it warm.

16 Human reproduction

Reproduction is the process of producing new individuals. Some single-celled creatures can reproduce by simply dividing into two. Some many-celled animals can produce offspring by a process of 'budding', in which part of their body breaks away and grows into a new individual. These are methods of asexual reproduction (see p.303).

Most animals reproduce sexually. The two sexes, male and female, each produce special types of reproductive cells, called **gametes**. The male gametes are the **sperms** (or **spermatozoa**) and the female gametes are the **ova** (singular = ovum) or eggs (Figure 16.1).

To produce a new individual, a sperm has to reach an ovum and join with it (**fuse** with it). The sperm nucleus then passes into the ovum and the two nuclei also fuse. This is called **fertilization**.

The cell formed after the fertilization of an ovum by a sperm is called a **zygote**. A zygote will grow by cell division to produce first an **embryo** and then a fully formed animal (Figure 16.2).

The male animal always produces a large number (millions) of sperms, while the female produces a smaller number of eggs. In some animals, such as fish and frogs, many eggs are fertilized and there are a large number of offspring. In mammals, a small number of eggs are fertilized, from one to twenty. In humans, usually only one egg is fertilized at a time; two eggs being fertilized produces (non-identical) twins (p. 146).

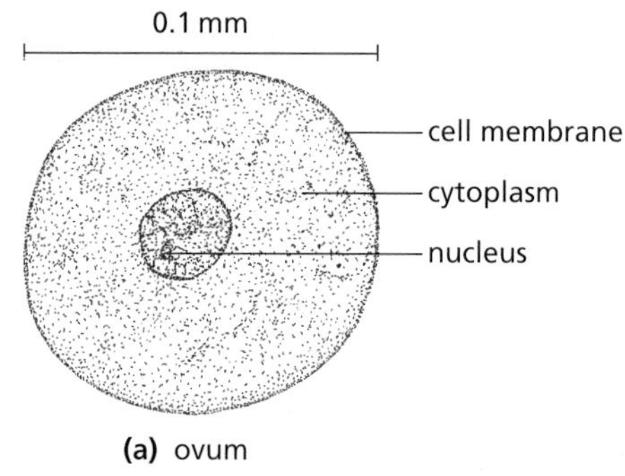

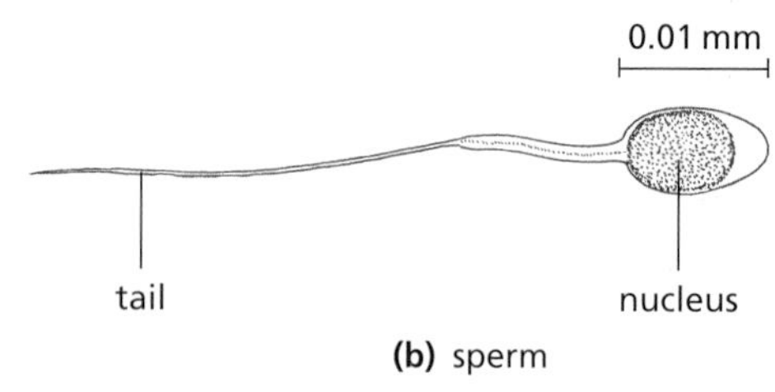

Figure 16.1 Human gametes

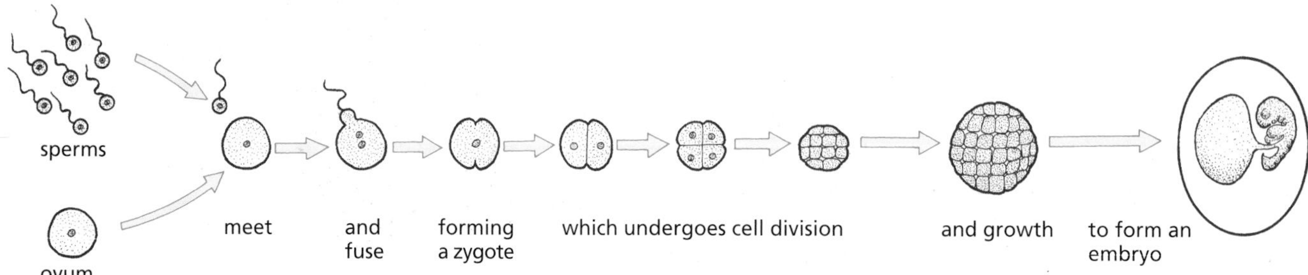

Figure 16.2 Fertilization and development

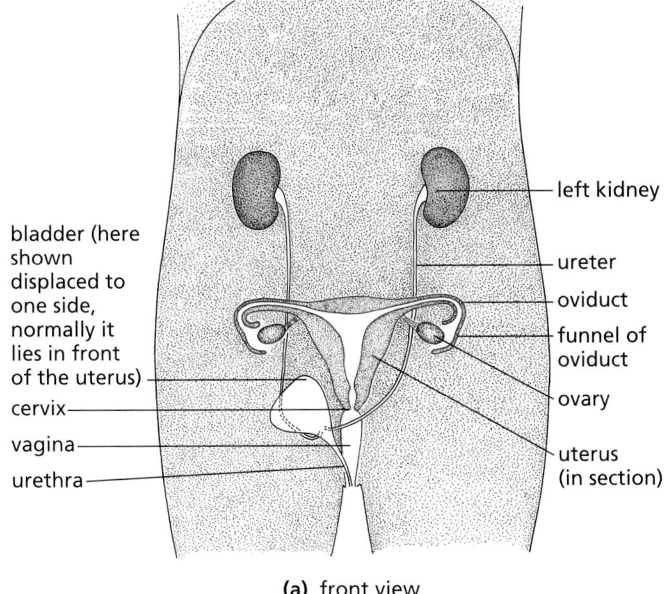

(a) front view

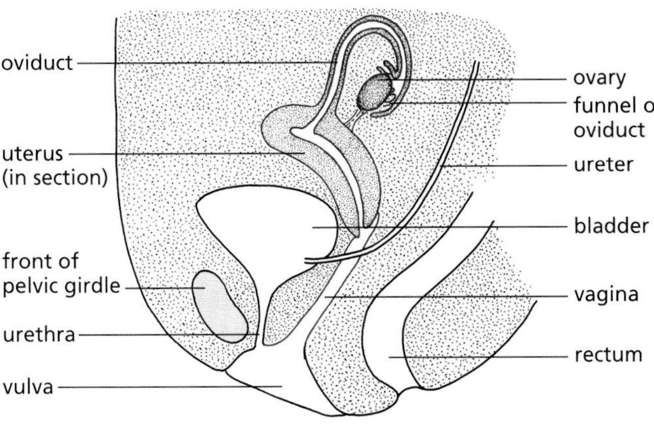

(b) vertical section

Figure 16.3 The female reproductive organs

To bring the sperms close enough to the ova for fertilization to take place, there is an act of mating or **copulation**. In mammals this act results in sperms from the male animal being injected into the female. The sperms swim inside the female's reproductive system and fertilize any eggs which are present. The zygote then grows into an embryo inside the body of the female.

The human reproductive system

Female

The eggs are produced from the female reproductive organs called **ovaries**. These are two whitish oval bodies, 3–4 cm long. They lie in the lower half of the abdomen, one on each side of the **uterus** (Figure 16.3a and b). Close to each ovary is the expanded, funnel-shaped opening of the **oviduct**, the tube down which the ova pass when released from the ovary. The oviduct is sometimes called the **Fallopian tube**.

The oviducts are narrow tubes that open into a wider tube, the uterus or womb, lower down in the abdomen. When there is no embryo developing in it, the uterus is only about 80 mm long. It leads to the outside through a muscular tube, the **vagina**. The **cervix** is a ring of muscle closing the lower end of the uterus where it joins the vagina. The urethra, from the bladder, opens into the **vulva** just in front of the vagina.

Male

Sperms are produced in the male reproductive organs (Figures 16.4 and 16.5), called the **testes** (singular = testis). These lie outside the abdominal cavity in a special sac called the **scrotum**. In this position they are kept at a temperature slightly below the rest of the body. This is the best temperature for sperm production.

The testes consist of a mass of sperm-producing tubes (Figure 16.6). These tubes join to form ducts leading to the **epididymis**, a coiled tube about 6 metres long on the outside of each testis. The epididymis, in turn, leads into a muscular **sperm duct**.

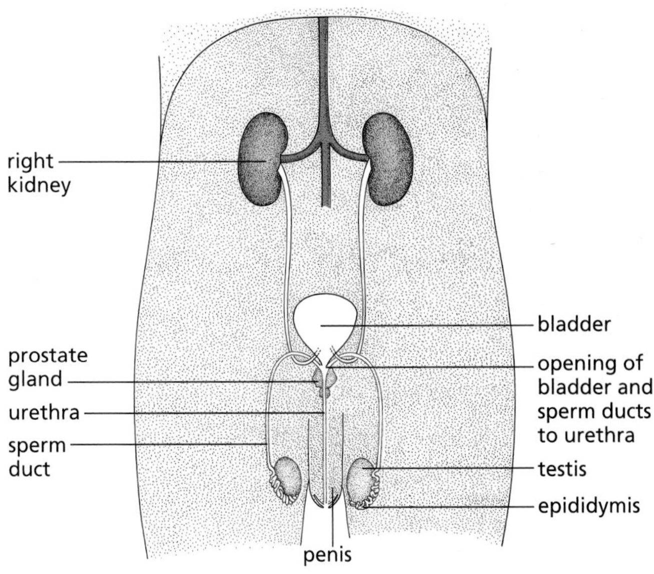

Figure 16.4 The male reproductive organs; front view

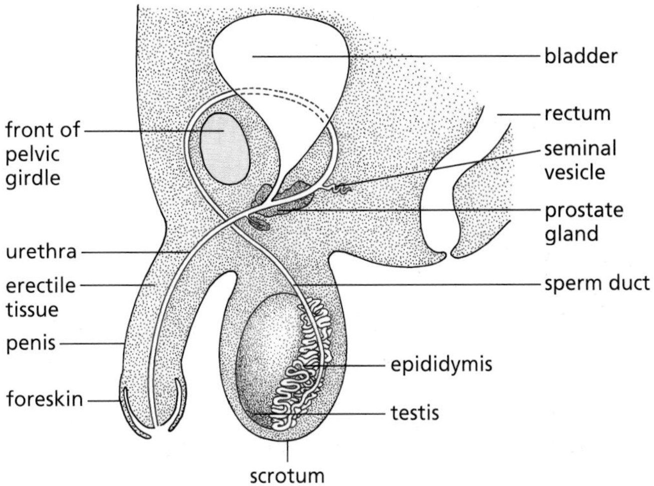

Figure 16.5 The male reproductive organs; side view

The two sperm ducts, one from each testis, open into the top of the urethra just after it leaves the bladder. A short, coiled tube called the **seminal vesicle** branches from each sperm duct just before it enters the **prostate gland**, which surrounds the urethra at this point.

The urethra passes through the **penis** and may conduct either urine or sperms at different times. The penis consists of connective tissue with many blood spaces in it. This is called **erectile tissue**.

Questions

1 How do sperms differ from ova in their structure (see Figure 16.1)?

2 List the structures, in the correct order, through which the sperms must pass from the time they are produced in the testis, to the time they leave the urethra.

3 What structures are shown in Figure 16.5 which are not shown in Figure 16.4?

4 In what ways does a zygote differ from any other cell in the body?

Production of gametes

Sperm production

The lining of the sperm-producing tubules in the testis consists of rapidly dividing cells (Figure 16.6). After a series of cell divisions, the cells grow long tails and become sperms (Figure 16.7) which pass into the epididymis.

During copulation, the epididymis and sperm ducts contract and force sperms out through the urethra. The prostate gland and seminal vesicle add fluid to the sperms. This fluid plus the sperms it contains is called **semen**, and the ejection of sperms through the penis is called **ejaculation**.

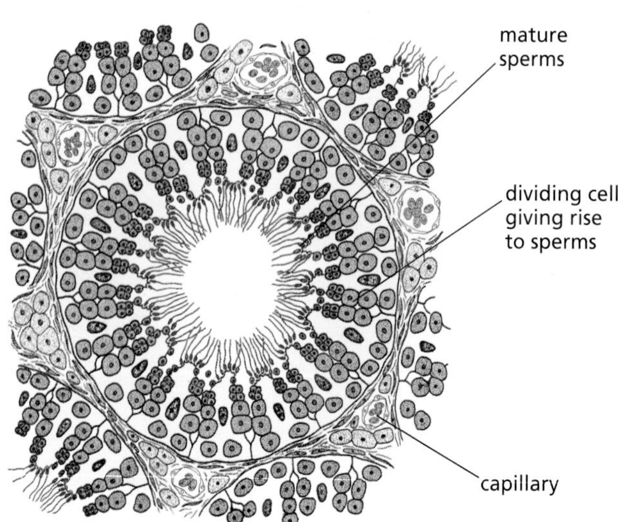

Figure 16.6 Section through sperm-producing tubules

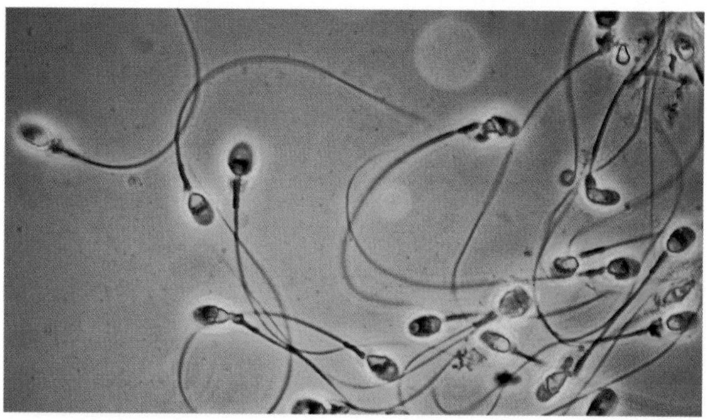

Figure 16.7 Human sperms (×800). The head of the sperm has a slightly different appearance when seen in 'side' view or in 'top' view

Ovulation

The egg cells (ova) are present in the ovary from the time of birth. No more are formed during the lifetime, but between the ages of 10 and 14 some of the egg cells start to mature and are released, one at a time about every 4 weeks from alternate ovaries. As each ovum matures, the cells round it divide rapidly and produce a fluid-filled sac. This sac is called a **follicle** (Figure 16.8) and, when mature, it projects from the surface of the ovary like a small blister (Figure 16.9). Finally, the follicle bursts and releases the ovum with its coating of cells into the funnel of the oviduct. This is called **ovulation**. From here, the ovum is wafted down the oviduct by the action of cilia (p. 6) in the lining of the tube. If the ovum meets sperm cells in the oviduct, it may be fertilized by one of them.

The released ovum is enclosed in a jelly-like coat called the **zona pellucida** and is still surrounded by a layer of follicle cells. Before fertilization can occur, sperms have to get through this layer of cells and the successful sperm has to penetrate the zona pellucida with the aid of enzymes secreted by the head of the sperm.

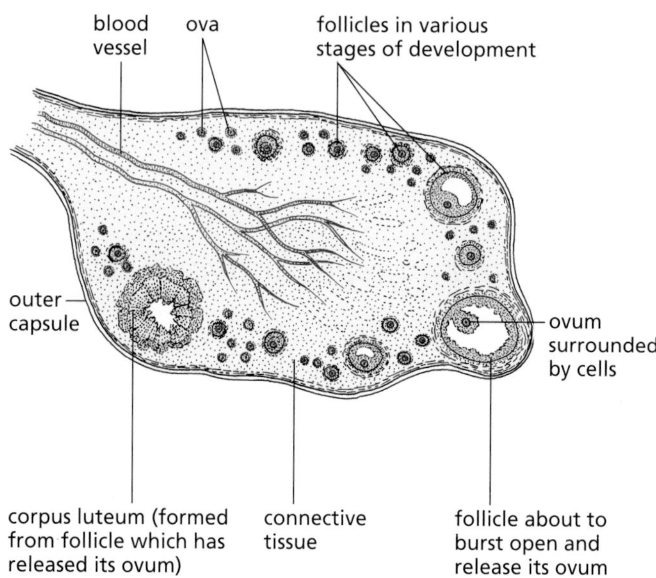

Figure 16.8 Section through an ovary

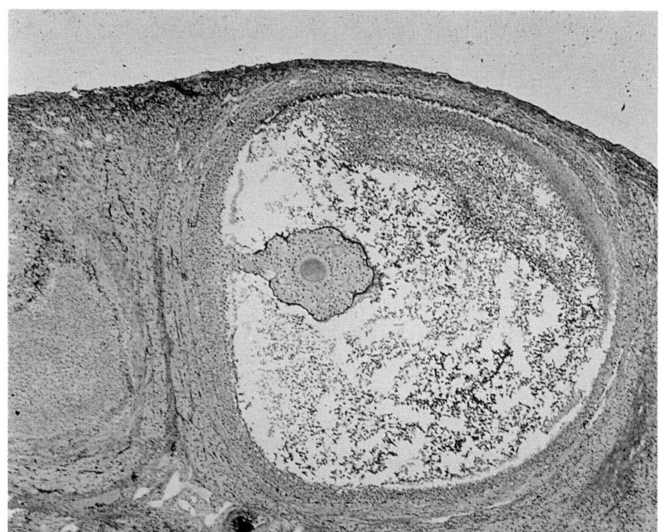

Figure 16.9 Mature follicle as seen in a section through part of an ovary (×30). The ovum is surrounded by follicle cells. These produce the fluid which occupies much of the space in the follicle

Mating and fertilization

Mating

Sexual arousal in the male results in an erection. That is, the penis becomes firm and erect as a result of blood flowing into the erectile tissue. Arousal in the female stimulates the lining of the vagina to produce mucus. This lubricates the vagina and makes it easy for the erect penis to enter.

In the act of copulation, the male inserts the penis into the female's vagina. The sensory stimulus (sensation) that this produces causes a reflex (p. 165) in the male which results in the ejaculation of semen into the top of the vagina.

The last paragraph is a very simple description of a biological event. In humans, however, the sex act has intense psychological and emotional importance. Most people feel a strong sexual drive, which has little to do with the need to reproduce. Sometimes the sex act is simply the meeting of an urgent physical need. Sometimes it is an experience that both man and woman enjoy together. At its 'highest' level it is both of these, and is also an expression of deeply felt affection within a lasting relationship.

Fertilization

The sperms swim through the cervix and into the uterus by wriggling movements of their tails. They pass through the uterus and enter the oviduct, but the method by which they do this is not known for certain. If there is an ovum in the oviduct, one of the sperms may bump into it and stick to its surface. The sperm then enters the cytoplasm of the ovum and the male nucleus of the sperm fuses with the female nucleus. This is the moment of fertilization and is shown in more detail in Figure 16.10. Although a single ejaculation may contain about five hundred million sperms, only a few hundred will reach the oviduct and only one will fertilize the ovum. The function of the others is not fully understood.

The released ovum is thought to survive for about 24 hours; the sperms might be able to fertilize an ovum for about 2 or 3 days. So there is only a short period of about 4 days each month when fertilization might occur. If this fertile period can be estimated accurately, it can be used either to achieve or to avoid fertilization (conception) (p.150).

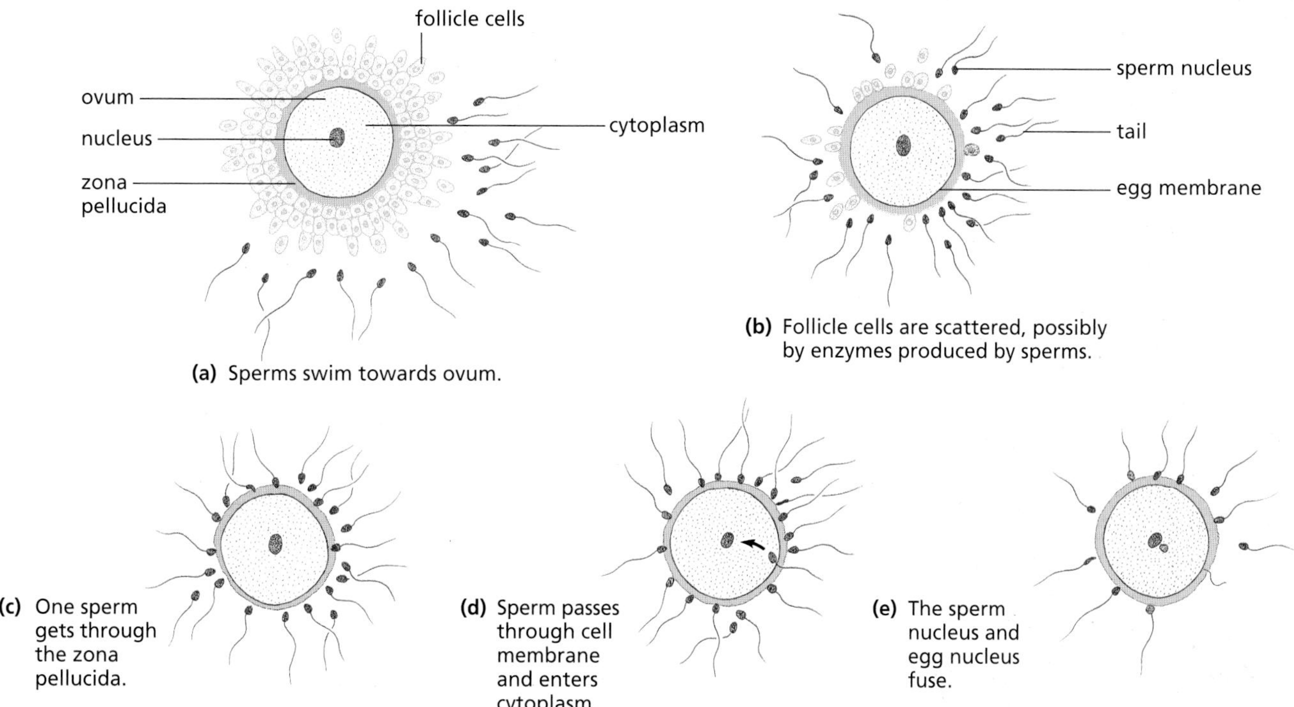
follicle cells
ovum
nucleus
zona pellucida
cytoplasm
(a) Sperms swim towards ovum.
sperm nucleus
tail
egg membrane
(b) Follicle cells are scattered, possibly by enzymes produced by sperms.
(c) One sperm gets through the zona pellucida.
(d) Sperm passes through cell membrane and enters cytoplasm.
(e) The sperm nucleus and egg nucleus fuse.

Figure 16.10 Fertilization of an ovum

Questions

1 If a woman starts ovulating at 13 and stops at 50,
  a how many ova are likely to be released from her ovaries,
  b about how many of these are likely to be fertilized?

2 List, in the correct order, the parts of the female reproductive system through which sperms must pass before reaching and fertilizing an ovum.

3 State exactly what happens at the moment of fertilization.

4 If mating takes place
  a 2 days before ovulation,
  b 2 days after ovulation, is fertilization likely to occur? Explain your answers.

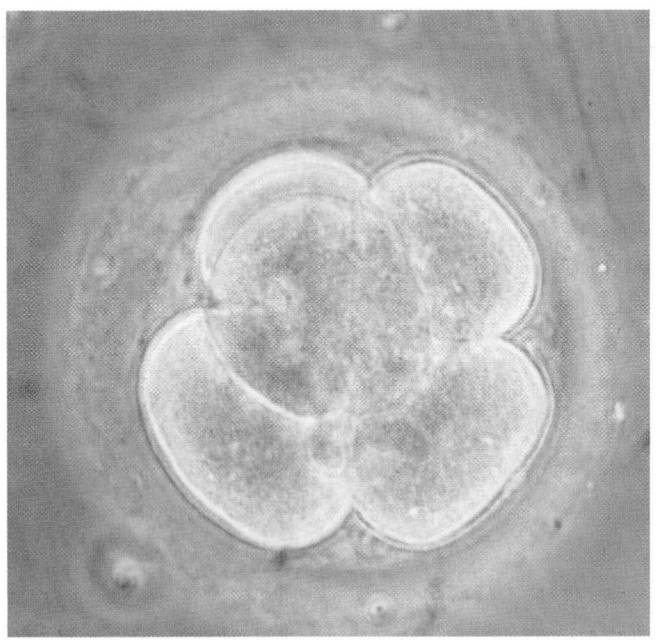

Figure 16.11 Human embryo at the 5-cell stage (× 230). The embryo is surrounded by the zona pellucida

Pregnancy and development

The fertilized ovum first divides into two cells. Each of these divides again, so producing four cells. The cells continue to divide in this way to produce a solid ball of cells (Figure 16.11), an early stage in the development of the **embryo**. This early embryo travels down the oviduct to the uterus. Here it sinks into the lining of the uterus, a process called **implantation** (Figure 16.12a). The embryo continues to grow and produces new cells which form tissues and organs (Figure 16.13). After 8 weeks, when all the organs are formed, the embryo is called a **fetus**. One of the first organs to form is the heart, which pumps blood round the body of the embryo.

As the embryo grows, the uterus enlarges to contain it. Inside the uterus the embryo becomes enclosed in a fluid-filled sac called the **amnion** or water sac, which protects it from damage and prevents unequal pressures from acting on it (Figure 16.12b and c). The oxygen and food needed to keep the embryo alive and growing are obtained from the mother's blood by means of a structure called the placenta.

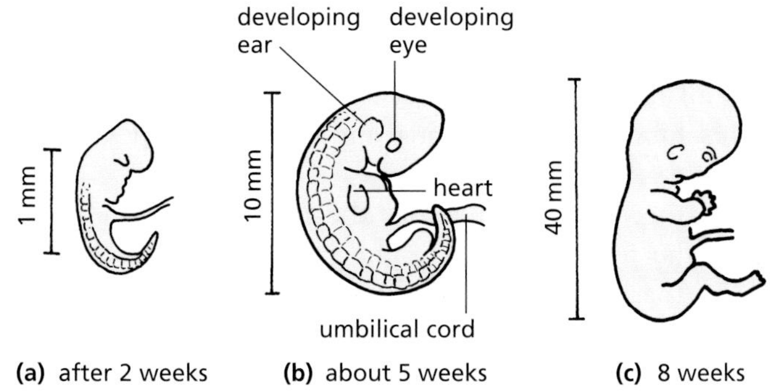

Figure 16.13 Human embryo: the first 8 weeks

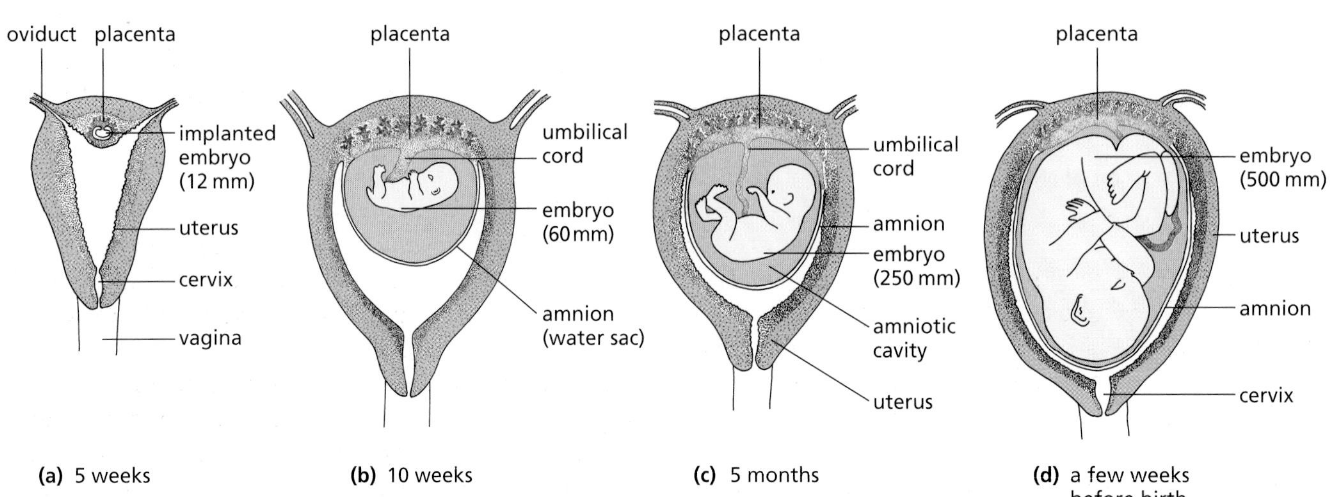

Figure 16.12 Growth and development in the uterus (not to scale)

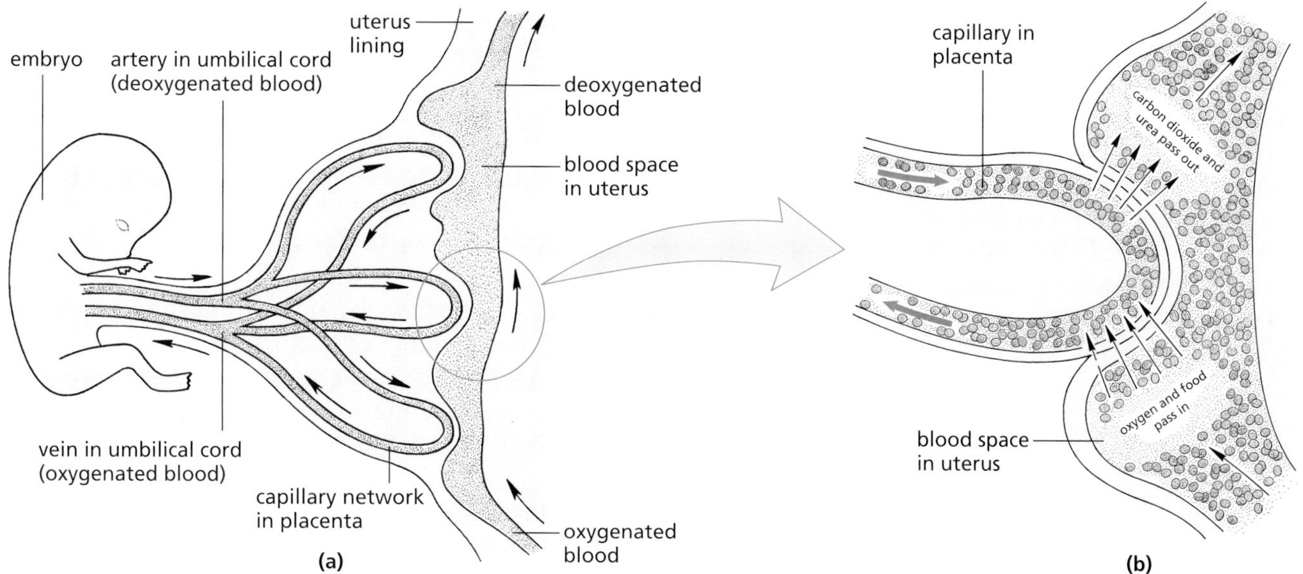

Figure 16.14 The exchange of substances between the blood of the embryo and the mother

Placenta

Soon after the ball of cells reaches the uterus, some of the cells, instead of forming the organs of the embryo, grow into a disc-like structure, the **placenta** (Figure 16.12c). The placenta becomes closely attached to the lining of the uterus and is attached to the embryo by a tube called the **umbilical cord** (Figure 16.12c). After a few weeks, the embryo's heart has developed and is circulating blood through the umbilical cord and placenta as well as through its own tissues. The blood vessels in the placenta are very close to the blood vessels in the uterus so that oxygen, glucose, amino acids and salts can pass from the mother's blood to the embryo's blood (Figure 16.14a). So the blood flowing in the umbilical vein from the placenta carries food and oxygen to be used by the living, growing tissues of the embryo. In a similar way, the carbon dioxide and urea in the embryo's blood escape from the vessels in the placenta and are carried away by the mother's blood in the uterus (Figure 16.14b). In this way the embryo gets rid of its excretory products.

There is no direct communication between the mother's blood system and the embryo's. The exchange of substances takes place across the thin walls of the blood vessels. In this way, the mother's blood pressure cannot damage the delicate vessels of the embryo and it is possible for the placenta to select the substances allowed to pass into the embryo's blood. The placenta can prevent some harmful substances in the mother's blood from reaching the embryo. It cannot prevent all of them, however, as is shown by the effects of cigarette smoke and alcohol described on p. 146.

Figure 16.15 shows the human embryo at 7 weeks surrounded by the amnion and placenta.

The placenta produces hormones, including oestrogens and progesterone. It is assumed that these hormones play an important part in maintaining the pregnancy and preparing for birth, but their precise function is not known. They may influence the development and activity of the muscle layers in the wall of the uterus and prepare the mammary glands in the breasts for milk production.

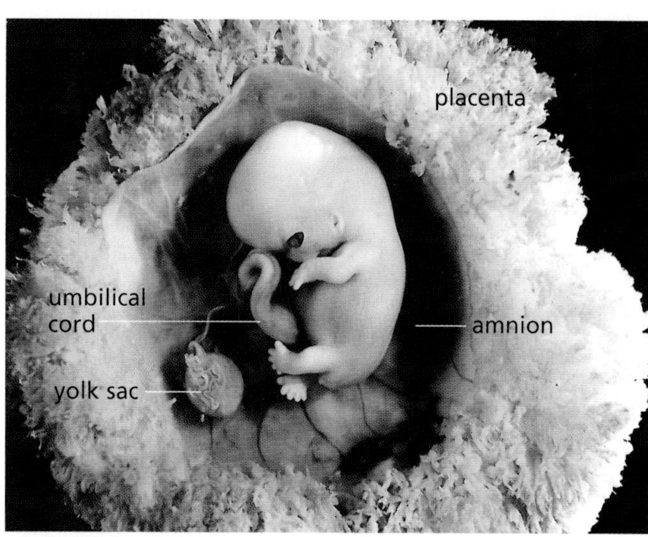

Figure 16.15 Human embryo, 7 weeks (× 1.5). The embryo is enclosed in the amnion. Its limbs, eye and ear-hole are clearly visible. The amnion is surrounded by the placenta; the fluffy-looking structures are the placental villi which penetrate into the lining of the uterus. The umbilical cord connects the embryo to the placenta

Antenatal care

'Antenatal' or 'prenatal' refers to the period before birth. Antenatal care is the way a woman should look after herself during pregnancy, so that the birth will be safe and her baby healthy.

The mother-to-be should make sure that she eats properly, and perhaps takes more iron and folic acid (p. 89) than she usually does to prevent anaemia. If her job is a light one, she may go on working for the first 6 months of pregnancy. She should not do heavy work, however, or repeated lifting or stooping.

Pregnant women who drink or smoke are more likely to have babies with low birth weights. These babies are more likely to be ill than babies of normal weight. Smoking may also make a miscarriage more likely. So a woman who smokes should give up smoking during her pregnancy. Heavy drinking is strongly suspected of damaging the developing brain of the fetus, so it is wise to avoid alcoholic drinks too.

During pregnancy, a woman should not take any drugs unless they are strictly necessary and prescribed by a doctor. In the 1950s, a drug called thalidomide was used to treat the bouts of early morning sickness which often occur in the first 3 months of pregnancy. Although tests had appeared to show the drug to be safe, it had not been tested on pregnant animals. About 20 per cent of pregnant women who took thalidomide had babies with deformed or missing limbs.

If a woman catches **rubella** (German measles) during the first 4 months of pregnancy, there is a danger that the virus may affect the fetus and cause abortion or stillbirth. Even if the baby is born alive, the virus may have caused defects of the eyes (cataracts), ears (deafness) or nervous system. All girls should be vaccinated against rubella to make sure that their bodies contain antibodies to the disease (see p. 117).

Twins

Sometimes a woman releases two ova when she ovulates. If both ova are fertilized, they may form twin embryos, each with its own placenta and amnion. Because the twins come from two separate ova, each fertilized by a different sperm, it is possible to have a boy and a girl. Twins formed in this way are called **fraternal twins**. Although they are both born within a few minutes of each other, they are no more alike than other brothers or sisters.

Another cause of twinning is when a single fertilized egg, during an early stage of cell division, forms two separate embryos. Sometimes these may share a placenta and amnion. Twins formed from a single ovum and sperm must be the same sex, because only one sperm (X or Y, see p. 196) fertilized the ovum. These 'one-egg' twins are sometimes called **identical twins** because, unlike fraternal twins, they will closely resemble each other in every respect.

Questions

1 In what ways will the composition of the blood in the umbilical vein differ from that in the umbilical artery?

2 An embryo is surrounded with fluid, its lungs are filled with fluid and it cannot breathe. Why doesn't it suffocate?

3 If a mother gives birth to twin boys, does this mean that they are identical twins? Explain.

Birth

From fertilization to birth takes about 38 weeks in humans. This is called the **gestation** period. A few weeks before the birth, the fetus has come to lie head downwards in the uterus, with its head just above the cervix (Figures 16.12d and 16.16). When birth starts, the uterus begins to contract rhythmically. This is the beginning of what is called 'labour'. These regular rhythmic contractions become stronger and more frequent. The opening of the cervix gradually widens enough to let the baby's head pass through and the contractions of the uterus are assisted by muscular contractions of the abdomen. The water sac breaks at some stage in labour and the fluid escapes through the vagina. Finally, the muscular contractions of the uterus and abdomen push the baby head-first through the widened cervix and vagina (Figure 16.17). The umbilical cord, which still connects the child to the placenta, is tied and cut. Later, the placenta breaks away from the uterus and is pushed out separately as the 'afterbirth'.

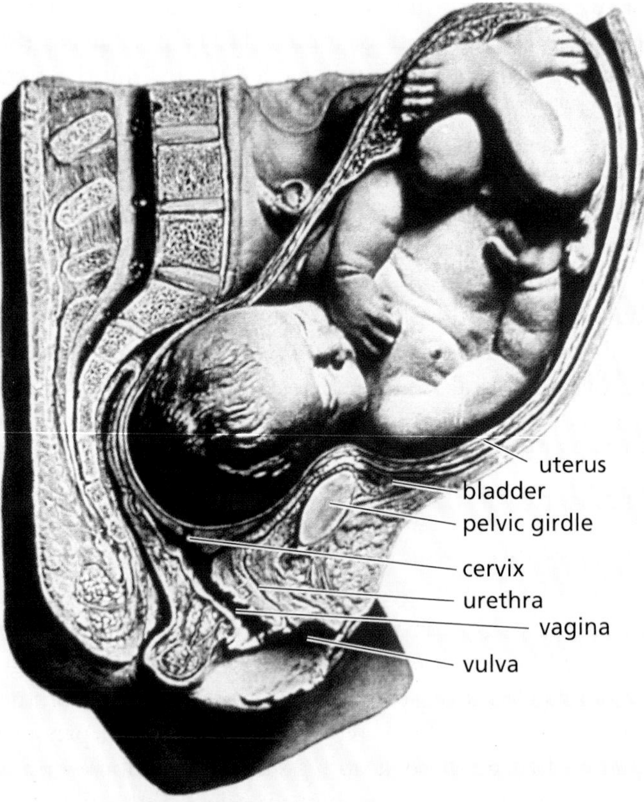

Figure 16.16 Model of human fetus just before birth. The cervix and vagina seem to provide narrow channels for the baby to pass through but they widen quite naturally during labour and delivery

The sudden fall in temperature felt by the newly born baby stimulates it to take its first breath and it usually cries. In a few days, the remains of the umbilical cord attached to the baby's abdomen shrivel and fall away, leaving a scar in the abdominal wall, called the navel.

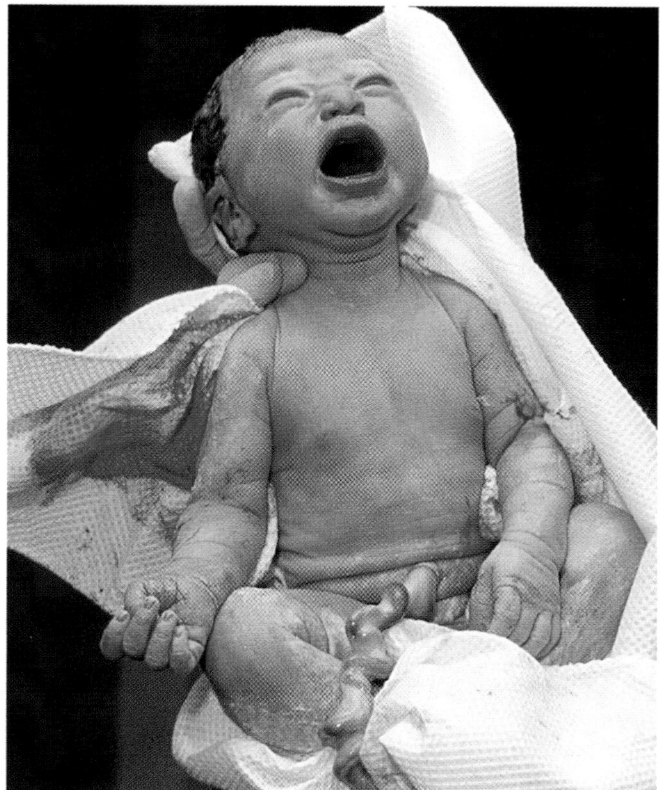

Figure 16.17 Delivery of a baby. The umbilical cord is still intact

Induced birth

Sometimes, when a pregnancy has lasted for more than 38 weeks or when examination shows that the placenta is not coping with the demands of the fetus, birth may be induced. This means that it is started artificially.

This is often done by carefully breaking the membrane of the amniotic sac. Another method is to inject a hormone, **oxytocin**, into the mother's veins. Either of these methods brings on the start of labour. Sometimes both are used together.

Amniocentesis

As a woman gets older, she is more likely to have a baby with Down's syndrome (p. 189). This can be detected fairly early in the pregnancy (14–15 weeks), by taking a sample of amniotic fluid. Then the chromosomes of the fetal cells floating in it are counted under the microscope. The cells of a Down's syndrome fetus have 47 chromosomes in their nuclei, instead of the normal 46. The mother of a Down's syndrome fetus can then decide whether to end her pregnancy rather than give birth to a mentally handicapped child.

In amniocentesis, a tube is inserted through the wall of the abdomen, through the uterus and into the amnion (Figure 22.7, p. 198). A local anaesthetic is used. In 1–2 per cent of cases, amniocentesis leads to a spontaneous abortion. The risk of this has to be balanced against the risk of having an affected child. For women over 40, amniocentesis is the lesser risk. Between 35 and 40, the risks are more evenly balanced.

The chromosomes of fetal cells can also be used to find out the sex of the fetus. This is useful where there is a family history of sex-linked genetic disease such as haemophilia (p. 197).

Analysis of the sample of amniotic fluid removed by amniocentesis can also detect certain chemicals which, if present, mean that the fetus has other defects.

A newer technique (chorionic villus sampling), takes a sample of the placental villi to look for abnormal chromosomes. One advantage is that the sampling tube can be passed through the cervix and does not enter the amnion. Placental sampling can also be carried out earlier in pregnancy than amniocentesis. So if an abnormality is found and the mother decides the pregnancy should be ended, the experience may be less distressing for her. The risks of this technique versus the possible benefit are still being studied and have recently been called into question.

Feeding and parental care

Within the first 24 hours after birth, the baby starts to suck at the breast. During pregnancy the mammary glands (breasts) enlarge as a result of an increase in the number of milk-secreting cells. No milk is secreted during pregnancy, but the hormones which start the birth process also act on the milk-secreting cells of the breasts. The breasts are stimulated to release milk by the first sucklings. The continued production of milk is under the control of hormones, but the amount of milk produced is related to the quantity taken by the child during suckling.

Milk contains the proteins, fats, sugar, vitamins and salts that babies need for their energy requirements and tissue-building, but there is too little iron present for the manufacture of haemoglobin. All the iron needed for the first weeks or months is stored in the liver of the fetus during gestation.

The liquid produced in the first few days is called **colostrum**. It is sticky and yellow, and contains more protein than the milk produced later. It also contains some of the mother's antibodies.

The mother's milk supply increases with the demands of the baby, up to 1 litre per day. It is gradually supplemented and eventually replaced entirely by solid food, a process known as **weaning**.

Cows' milk is not wholly suitable for human babies. It has more protein, sodium and phosphorus, and less sugar, vitamin A and vitamin C, than human milk. It is less easily digested than human milk. Manufacturers modify the components of dried cows' milk to resemble human milk more closely and this makes it more acceptable if the mother cannot breast-feed her baby.

Figure 16.18 Breast-feeding helps to establish an emotional bond between mother and baby

Cows' milk and proprietary dried milk both lack human antibodies, whereas the mother's milk contains antibodies to any diseases from which she has recovered. It also carries white cells which produce antibodies or ingest bacteria. These antibodies are important in defending the baby against infection at a time when its own immune responses are not fully developed. Breast-feeding provides milk free from bacteria, whereas bottle-feeding carries the risk of introducing bacteria which cause intestinal diseases. Breast-feeding also offers emotional and psychological benefits to both mother and baby (Figure 16.18).

Most young mammals are independent of their parents after a few weeks or months even though they may stay together as a family group. In humans, however, the young are dependent on their parents for food, clothing and shelter for many years. During this long period of dependence, the young learn to talk, read and write and learn a great variety of other skills that help them to survive and be self-sufficient (Figure 16.19).

Figure 16.19 Parental care. Human parental care includes a long period of education

Puberty and the menstrual cycle

Puberty

Although the ovaries of a young girl contain all the ova she will ever produce, they do not start to be released until she reaches an age of about 10–14 years. This stage in her life is known as **puberty**.

At about the same time as the first ovulation, the ovary also releases female sex hormones into the bloodstream. These hormones are called **oestrogens** and when they circulate round the body, they bring about the development of **secondary sexual characteristics**. In the girl these are the increased growth of the breasts, a widening of the hips and the growth of hair in the pubic region and in the armpits. There is also an increase in the size of the uterus and vagina. Once all these changes are complete, the girl is capable of having a baby.

Puberty in boys occurs at about the same age as in girls. The testes start to produce sperms for the first time and also release a hormone, called **testosterone**, into the bloodstream. The male secondary sexual characteristics, which begin to appear at puberty, are enlargement of the testes and penis, deepening of the voice, growth of hair in the pubic region, armpits, chest and, later on, the face. In both sexes there is a rapid increase in the rate of growth during puberty.

In addition to the physical changes at puberty, there are emotional and psychological changes associated with the transition from being a child to becoming an adult, i.e. the period of adolescence. Most people adjust to these changes smoothly and without problems. Sometimes, however, a conflict arises between having the status of a child and the sexuality and feelings of an adult.

The menstrual cycle

The ovaries release an ovum about every 4 weeks. As each follicle develops, the amount of oestrogens produced by the ovary increases. The oestrogens act on the uterus and cause its lining to become thicker and develop more blood vessels. These are changes which help an early embryo to implant as described on p. 144.

Two hormones, produced by the **pituitary gland** at the base of the brain (p. 171), promote ovulation. The hormones are **follicle-stimulating hormone** (FSH) and **luteinizing hormone** (LH). They act on a ripe follicle and stimulate maturation and release of the ovum.

Once the ovum has been released, the follicle which produced it develops into a solid body called the **corpus luteum**. This produces a hormone called **progesterone**, which affects the uterus lining in the same way as the oestrogens, making it grow thicker and produce more blood vessels.

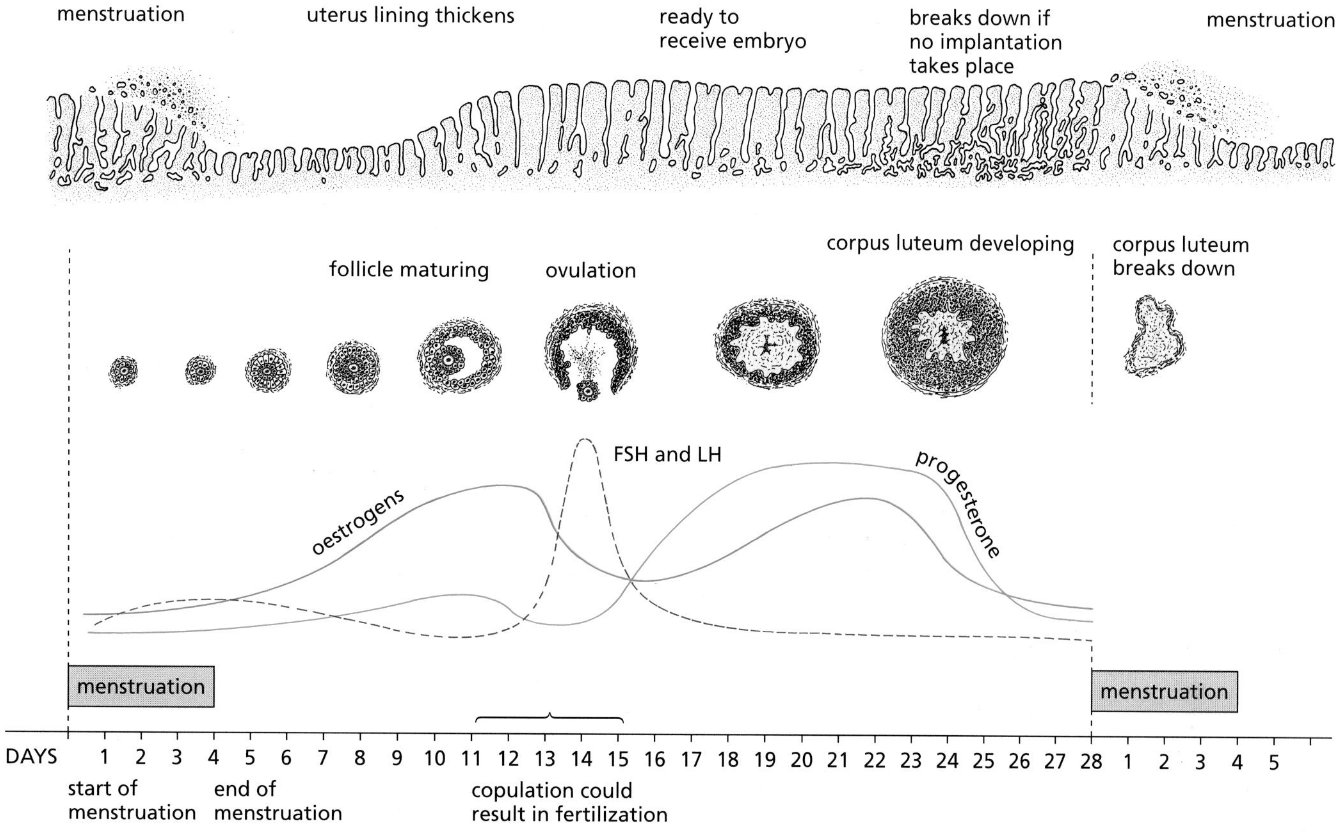

Figure 16.20 The menstrual cycle

If the ovum is fertilized, the corpus luteum continues to release progesterone and so keeps the uterus in a state suitable for implantation. If the ovum is not fertilized, the corpus luteum stops producing progesterone. As a result, the thickened lining of the uterus breaks down and loses blood which escapes through the cervix and vagina. This is known as a **menstrual period**. The appearance of the first menstrual period is one of the signs of puberty in girls. The events in the menstrual cycle are shown in Figure 16.20.

Menopause

Between the ages of 40 and 55, the ovaries cease to release ova or produce hormones. As a consequence, menstrual periods cease, the woman can no longer have children, and sexual desire is gradually reduced.

Questions

1. Apart from learning to talk, what other skills might young humans develop that would help them when they are no longer dependent on their parents in
   a an agricultural society,
   b an industrial society?
2. From the list of changes at puberty in girls, select those which are related to child-bearing and say what part you think they play.
3. One of the first signs of pregnancy is that the menstrual periods stop. Explain why you would expect this.

Infertility

About 85-90 per cent of couples trying for a baby achieve pregancy within a year. Those that do not may be subfertile or infertile.

Female infertility is usually caused by a failure to ovulate or a blockage or distortion of the oviducts. The latter can often be corrected by surgery; if this is not possible the couple may use '*in vitro*' fertilization (see p. 150). Failure to produce ova can be treated with **fertility drugs**. These drugs are similar to hormones and act by increasing the levels of FSH and LH. Administration of the drug is timed to promote ovulation to coincide with copulation.

Male infertility is caused by an inadequate quantity of sperms in the semen or by sperms which are insufficiently mobile to reach the oviducts. There are few effective treatments for this condition but pregnancy may be achieved by **artificial insemination** (AI). This involves injecting semen through a tube into the top of the uterus. In some cases, the husband's semen can be used but, more often, the semen is supplied by an anonymous donor.

With AI, the woman has the satisfaction of bearing her child rather than adopting, and 50 per cent of the child's genes are from the mother.

Apart from religious or moral objections, the disadvantages are that the child can never know his or her father and there may be legal problems about the legitimacy of the child in some countries.

In vitro fertilization ('test-tube babies')

'*In vitro*' means literally 'in glass' or, in other words, the fertilization is allowed to take place in laboratory glassware.

This technique may be employed where surgery cannot be used to repair blocked oviducts.

In vitro fertilization has received considerable publicity since the first 'test-tube' baby was born in 1978. The woman may be given fertility drugs which cause her ovaries to release several mature ova simultaneously. These ova are then collected by laparoscopy, i.e. they are sucked up in a fine tube inserted through the abdominal wall. The ova are then mixed with the husband's seminal fluid and watched under the microscope to see if cell division takes place. (Figure 16.11 on p. 144 is a photograph of such an '*in vitro*' fertilized ovum.)

One or more of the dividing zygotes are then introduced to the woman's uterus by means of a tube inserted through the cervix. Usually, only one (or none) of the zygotes develops though, occasionally, there are multiple births. The success rate for *in vitro* fertilization is between 12 and 40 per cent depending on how many embryos are transplanted. There is some controversy about the fate of the 'spare' embryos which are not returned to the uterus. Some people believe that since these embryos are potential human beings, they should not be destroyed or used for research. In some cases the 'spare' embryos have been frozen and used later if the first transplants did not work.

Family planning

As little as 4 weeks after giving birth, it is possible, though unlikely, that a woman may conceive again. Frequent breast-feeding may reduce the chances of conception. Nevertheless, it would be possible to have children at about 1-year intervals. Most people do not want, or cannot afford, to have as many children as this. All human communities, therefore, practise some form of birth control to space out births and limit the size of the family.

Natural methods of family planning

If it were possible to know exactly when ovulation occurred, intercourse could be avoided for 3–4 days before and 1 day after ovulation (see p. 149). At the moment, however, there is no simple, reliable way to recognize ovulation, though it is usually 12–16 days before the onset of the next menstrual period. By keeping careful records of the intervals between menstrual periods, it is possible to calculate a potentially fertile period of about 10 days in mid-cycle, when sexual intercourse should be avoided if children are not wanted.

On its own, this method is not very reliable but there are some physiological clues which help to make it more accurate. During or soon after ovulation, a woman's temperature rises about 0.5 °C. It is reasonable to assume that one day after the temperature returns to normal, a woman will be infertile. Another clue comes from the type of mucus secreted by the cervix and lining of the vagina. As the time for ovulation approaches, the mucus becomes more fluid. Women can learn to detect these changes and so calculate their fertile period.

By combining the 'calendar', 'temperature' and 'mucus' methods, it is possible to achieve about 80 per cent 'success', i.e. only 20 per cent unplanned pregnancies. Highly motivated couples may achieve better rates of success and, of course, it is a very helpful way of finding the fertile period for couples who do want to conceive.

Artificial methods of family planning

Contraception

The sheath or condom

A thin rubber sheath is placed on the erect penis before sexual intercourse. The sheath traps the sperms and prevents them from reaching the uterus.

The diaphragm

A thin rubber disc, placed in the vagina before intercourse, covers the cervix and stops sperms entering the uterus. Condoms and diaphragms, used in conjunction with chemicals that immobilize sperms, are about 95 per cent effective.

Spermicides

Spermicides are chemicals which, though harmless to the tissues, do kill or immobilize sperms. The spermicide, in the form of a cream, gel or foam, is placed in the vagina. On their own, spermicides are not very reliable but, in conjunction with condoms or diaphragms, they are effective.

Intra-uterine device (IUD)

A small metal or plastic strip bent into a loop or coil is inserted and retained in the uterus, where it probably prevents implantation of a fertilized ovum. It is about 98 per cent effective but there is a small risk of developing uterine infections, particularly if sexual relationships are promiscuous.

The contraceptive pill

The pill contains chemicals which have the same effect on the body as the hormones oestrogen and progesterone. When mixed in suitable proportions these hormones suppress ovulation (see 'Feedback', p. 172) and so prevent conception. The pills need to be taken each day for the 21 days between menstrual periods.

There are many varieties of contraceptive pill in which the relative proportions of oestrogen- and progesterone-like chemicals vary. They are 99 per cent effective but long-term use of some types may increase the risk of cancer of the breast and cervix.

Sterilization

Vasectomy

This is a simple and safe surgical operation in which the man's sperm ducts are cut and the ends sealed. This means that his semen contains the secretions of the prostate gland and seminal vesicle but no sperms and so cannot fertilize an ovum. Sexual desire, erection, copulation and ejaculation are quite unaffected.

The testis continues to produce sperms and testosterone. The sperms are removed by white cells as fast as they form. The testosterone ensures that there is no loss of masculinity.

The sperm ducts can be rejoined by surgery but this is not always successful.

Laparotomy

A woman may be sterilized by an operation in which her oviducts are tied, blocked or cut. The ovaries are unaffected. Sexual desire and menstruation continue as before, but sperms can no longer reach the ova. Ova are released, but break down in the upper part of the oviduct.

The operation cannot usually be reversed.

Checklist

- The male reproductive cells (gametes) are sperms. They are produced in the testes and expelled through the urethra and penis during mating.
- The female reproductive cells (gametes) are ova (eggs). They are produced in the ovaries. One is released each month. If sperms are present, the ovum may be fertilized as it passes down the oviduct to the uterus.
- Fertilization happens when a sperm enters an ovum and the sperm and egg nuclei join up (fuse).
- The fertilized ovum (zygote) divides into many cells and becomes embedded in the lining of the uterus. Here it grows into an embryo.
- The embryo gets its food and oxygen from its mother.
- The embryo's blood is pumped through blood vessels in the umbilical cord to the placenta, which is attached to the uterus lining. The embryo's blood comes very close to the mother's blood so that food and oxygen can be picked up and carbon dioxide and nitrogenous waste can be got rid of.
- When the embryo is fully grown, it is pushed out of the uterus through the vagina by contractions of the uterus and abdomen.
- Each month, the uterus lining thickens up in readiness to receive the fertilized ovum. If an ovum is not fertilized, the lining and some blood is lost through the vagina. This is menstruation.
- The release of ova and the development of an embryo are under the control of hormones like oestrogen, progesterone, follicle-stimulating hormone and luteinizing hormone.
- Twins may result from two ova being fertilized at the same time or from a zygote forming two embryos.
- At puberty, (1) the testes and ovaries start to produce mature gametes, (2) the secondary sexual characteristics develop.
- Human milk and breast-feeding are best for babies.
- Young humans are dependent on their parents for a long time. During this period much essential learning takes place.
- Female infertility may be relieved by surgery, fertility drugs or *in vitro* fertilization.
- Male infertility can be by-passed by artificial insemination.
- There are effective natural and artificial methods for spacing births and limiting the size of a family.

17 The skeleton, muscles and movement

Structure of the skeleton
Vertebral column, skull and limbs.

Functions of the skeleton
Support, protection and movement. Joints and muscle.

Muscles and movement
Muscles produce movement. Benefits of exercise. Injuries.

Practical work
Structure of bone.

Structure of the skeleton

A human skeleton is shown in Figure 17.1. It consists of a vertebral column (sometimes called the 'backbone', 'spine' or 'spinal column') which supports the skull. Twelve pairs of ribs are attached to the upper part of the vertebral column and the limbs are attached to it by means of **girdles**.

The hip girdle (pelvic girdle) is joined rigidly to the lower end of the vertebral column. The shoulder girdle (pectoral girdle) consists of a pair of collarbones and shoulder blades which are not rigidly fixed to the vertebral column but held in place by muscles.

The upper arm bone (humerus) fits into a socket in the shoulder blade; the thigh bone (femur) fits into a socket in the hip girdle.

Bone is very hard (i.e. it resists compression) and very strong (i.e. it resists bending). Despite these properties, it is a living tissue consisting of living cells plus strong fibres made from a protein called **collagen**, and crystals of calcium phosphate (and other minerals), which give bone its hardness.

Bone is penetrated by blood vessels which keep the cells alive and allow growth and repair to take place.

There are two main types of bone. The outside layer of bones is formed by a dense, hard **compact bone**. Inside this layer is **spongy bone**. Despite its name, spongy bone is not soft but consists of fine struts of bone with spaces in between. In the spaces are the cells which produce the red and white blood cells. The spongy bone forms the **red bone marrow**, found mainly in the heads of the limb bones and in the ribs and vertebrae. The shafts of the limb bones are filled with a soft, fatty marrow (Figure 17.2).

Vertebral column

The vertebral column forms the central supporting structure of the skeleton. It consists of 33 individual bones called **vertebrae** (singular = vertebra), separated by discs of fibrous cartilage. These discs allow the vertebrae to move slightly and so enable the vertebral column to bend backwards and forwards or from side to side.

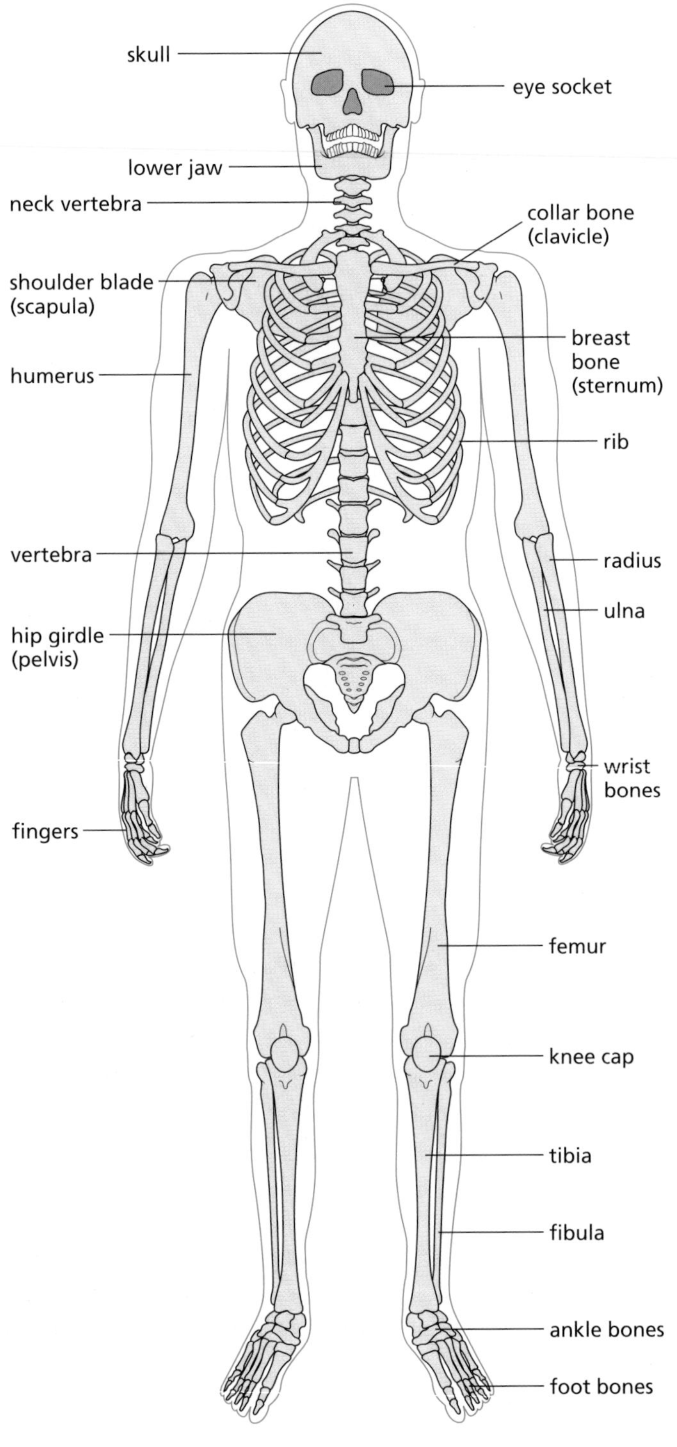

Figure 17.1 The skeleton

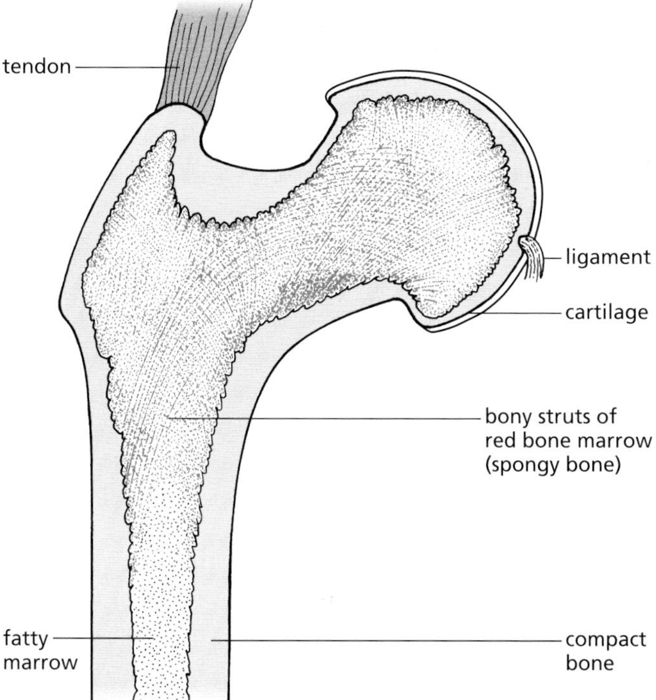

Figure 17.2 Section through head of femur

The spinal cord (p. 166) runs through an arch of bone (**neural arch**) formed by the vertebrae. In this position the cord is protected from damage.

The skull

This is made up of many bony plates joined together. It encloses and protects the brain and also carries and protects the main sense organs, the eyes, ears and nose. The upper jaw is fixed to the skull but the lower jaw is hinged to it in a way which allows chewing.

The base of the skull makes a joint with the top vertebra of the vertebral column. This joint allows the head to make nodding and rotational movements.

The limbs

Arm

The upper arm bone is the **humerus**. It is attached by a hinge joint to the lower arm bones, the **radius** and **ulna**. These two bones make a joint with a group of small wrist bones which in turn join to a series of five hand and finger bones. The ulna and radius can partly rotate round each other so that the hand can be held palm up or palm down.

Leg

The thigh bone or **femur** is attached at the hip to the pelvic girdle by a ball joint and at the knee it makes a hinge joint with the **tibia**. The **fibula** runs parallel to the tibia but does not form part of the knee joint. The ankle, foot and toe bones are similar to those of the wrist, hand and fingers.

Functions of the skeleton

Support

The skeleton holds the body off the ground and keeps its shape even when muscles are contracting to produce movement.

Protection

The brain is protected from injury by being enclosed in the skull. The heart, lungs and liver are protected by the rib cage, and the spinal cord is enclosed inside the neural arches of the vertebrae.

Movement

Many bones of the skeleton act as levers. When muscles pull on these bones, they produce movements such as the raising of the ribs during breathing (see p. 124), or the chewing action of the jaws. For a skeletal muscle to produce movement, both its ends need to have a firm attachment. The skeleton provides suitable points of attachment for the ends of muscles.

Production of blood cells

The red marrow of some bones, e.g. the vertebrae, ribs, breast bone and the heads of the limb bones, produce both red and white blood cells (p. 108).

Question

1 Study Figure 17.1 and then write the biological names of the following bones: upper arm bone, upper leg bone, hip bone, breastbone, 'backbone', lower arm bones.

Joints

Where two bones meet they form a joint. It may be a **fixed joint** as in the junction of the hip girdle and the vertebral column or a **movable joint** as in the knee. Two important types of movable joint already mentioned are the **ball and socket joints** of the hip (Figure 17.3a, overleaf) and the shoulder and the **hinge joints** of the elbow (Figure 17.3b, overleaf) and knee. The ball and socket joint allows movement forwards, backwards and sideways, whereas the hinge joint allows movement in only one direction.

Where the surfaces of the bones in a joint rub over each other, they are covered with smooth cartilage which reduces the friction between them. Friction is also reduced by a thin layer of lubricating fluid called **synovial fluid** (Figure 17.3b). (Movable joints are sometimes called **synovial joints**.) The bones forming the joint are held in place by tough bands of fibrous tissues called **ligaments** (Figure 17.3a). Ligaments keep the bones together but do not stop their various movements.

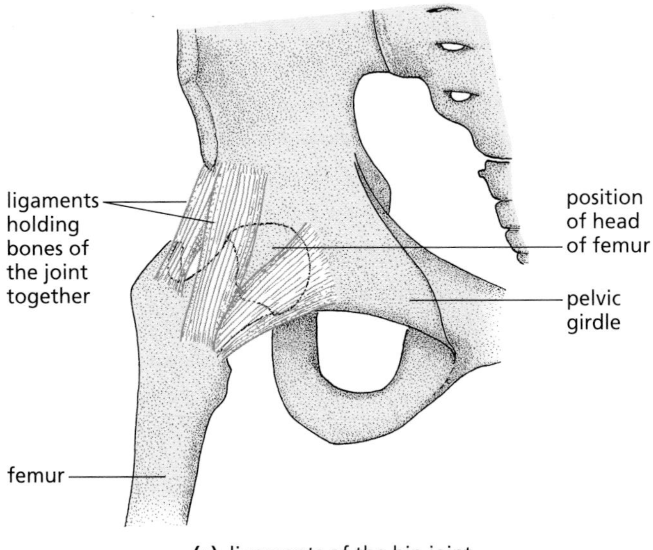

(a) ligaments of the hip joint

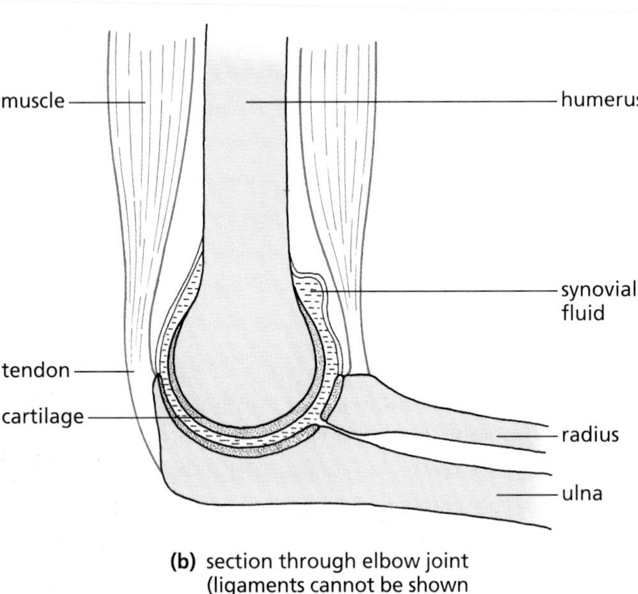

(b) section through elbow joint (ligaments cannot be shown in this sectional drawing)

Figure 17.3 The joints of the hip and elbow

The type of cartilage covering the surfaces of bones in a joint is firm but softer than bone. It contains fibres but no mineral salts. It forms a smooth, slippery surface.

Tendons and ligaments consist of tough collagen fibres, which makes them flexible but very strong and resistant to stretching.

Questions

1 Apart from the elbow and knee, what other joints in the body are like hinge joints?

2 Apart from its structural and mechanical functions, what other important function does the skeleton have?

3 Which parts of the skeleton are concerned with both protection and movement?

Muscle

There are three main types of muscle. One kind is called **skeletal muscle** (or striated, or voluntary muscle). Another kind is called **smooth muscle** (or unstriated, or involuntary muscle). A third kind occurs only in the heart.

Smooth muscle cells make layers of muscle tissue rather than distinct muscles, e.g. in the walls of the alimentary canal (p. 97), the uterus (p. 141) and the arterioles (p. 112).

Skeletal muscle

Skeletal muscle (Figure 17.4a) is made up of long fibres. Each fibre is formed from many cells, but the cells have fused together. The cell boundaries cannot be seen but the individual nuclei are still present (Figure 17.4c).

The muscle fibres are arranged in bundles which form distinct muscles (Figure 17.4b). Most of these are attached to bones, and produce movement as described below. Each muscle has a nerve supply. When a nerve impulse is sent to a muscle, it makes the muscle contract (i.e. get shorter and fatter). We can usually control most of our skeletal muscles. For this reason they are called **voluntary muscles**.

Muscle contraction

The fibres of skeletal muscle and the cells of smooth muscle have the special property of being able to contract, i.e. shorten, when stimulated by nerve impulses. However, the fibres and cells cannot elongate; they can only contract and relax. So, they have to be pulled back into their elongated shape by other muscles which work in the opposite direction (see 'Muscles and movement' below).

Muscles and movement

The ends of the limb muscles are drawn out into **tendons** which attach each end of the muscle to the skeleton (Figure 17.3b).

The way a muscle is attached to a limb to make it bend at the joint is shown in Figure 17.5. The tendon at one end is attached to a non-moving part of the skeleton while the tendon at the other end is attached to the movable bone close to the joint.

When the muscle contracts it pulls on the bones and makes one of them move. The position of the attachment means that a small contraction of the muscle will produce a large movement at the end of the limb. A model showing how the shortening of muscle can move a limb is given in Figure 17.6, and Figure 17.5 shows how a contraction of the **biceps muscle** bends (or **flexes**) the arm at the elbow, while the **triceps** straightens (or **extends**) the arm.

The non-moving end of the biceps is attached to the shoulder blade while the moving end is attached to the ulna, near the elbow joint.

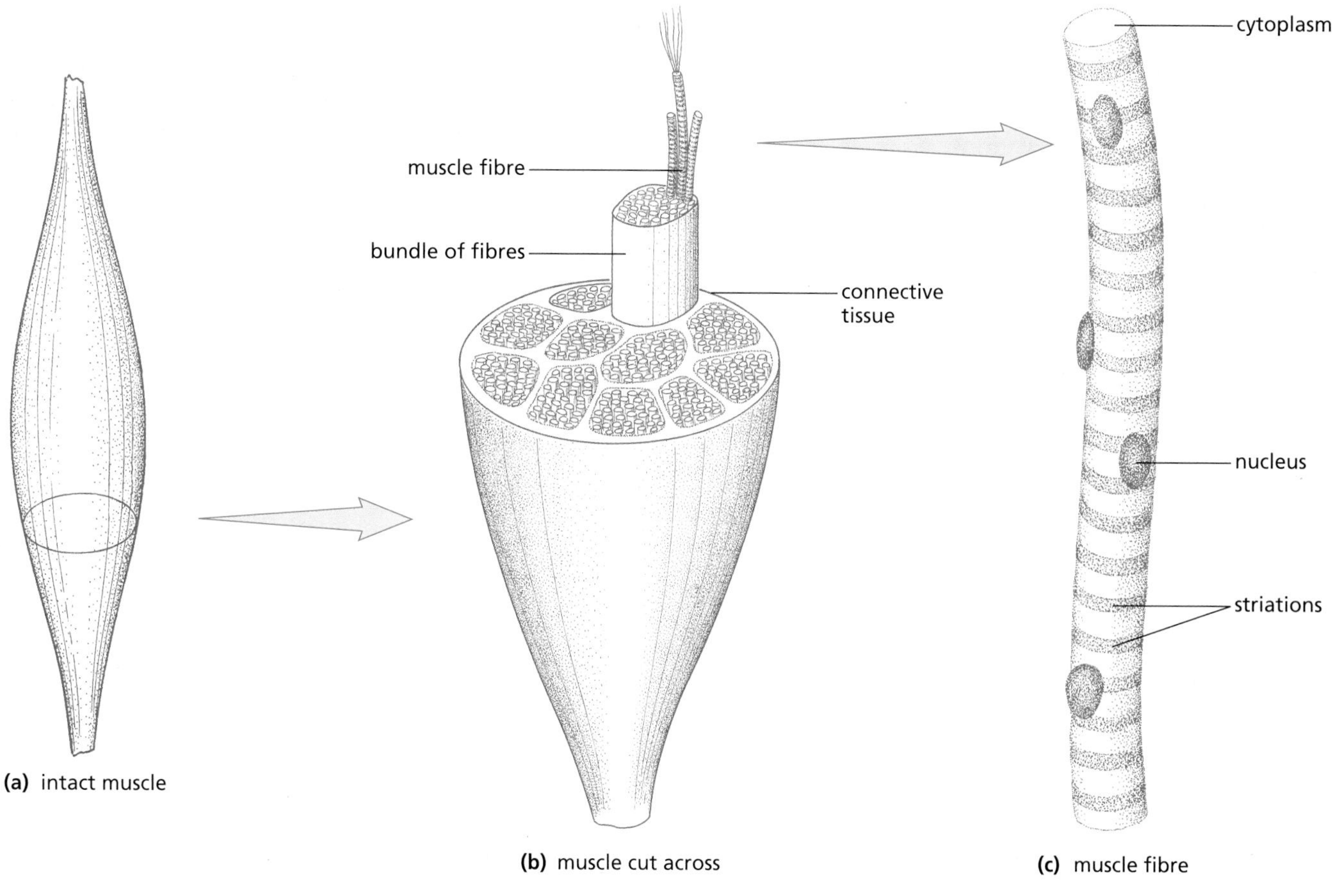

muscle fibre

bundle of fibres

connective tissue

cytoplasm

nucleus

striations

(a) intact muscle

(b) muscle cut across

(c) muscle fibre

Figure 17.4 Skeletal (striated) muscle

Limb muscles are usually arranged in pairs having opposite effects. This is because muscles can only shorten or relax, they cannot elongate, so the triceps is needed to pull the relaxed biceps back to its elongated shape after it has contracted. Pairs of muscles like this are called **antagonistic** muscles. Antagonistic muscles are also important in holding the limbs steady, both muscles keeping the same state of tension or 'tone'.

The contraction of muscles is controlled by nerve impulses. The brain sends out impulses in the nerves so that the muscles are made to contract or relax in the right order to make a movement. For example, when a muscle contracts to bend the limb, its antagonistic muscle must be kept in a relaxed state.

There are many muscular activities which bring about movements but do not result in locomotion. Chewing, breathing, throwing, swallowing and blinking are examples of such movements.

Energy for muscle contraction

The energy for muscle contraction comes from the respiration of lipids and carbohydrates, principally glycogen (p. 19).

Resting muscle has a store of glycogen but during exercise this has to be replenished from the glucose delivered to the muscle by its blood supply.

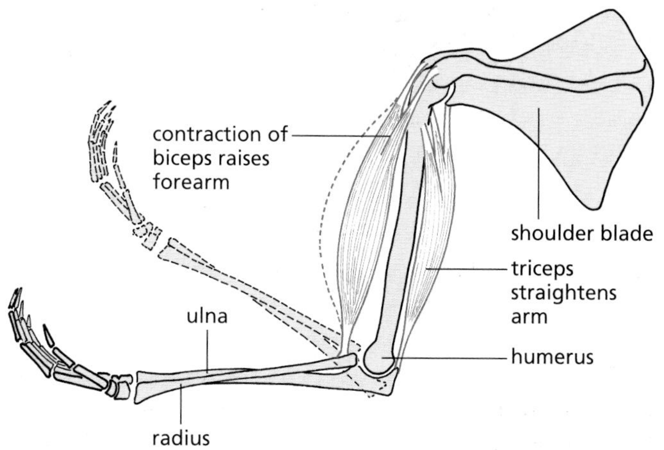

Figure 17.5 Antagonistic muscles of the forearm

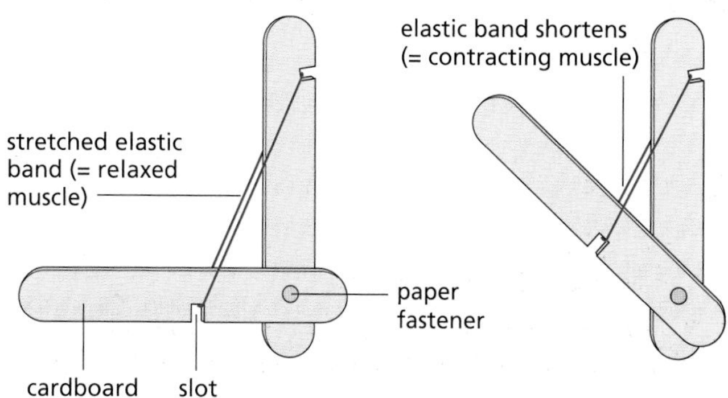

Figure 17.6 Model to show how muscles pull on bones to produce movement

Aerobic respiration needs a supply of oxygen, also carried by the blood and delivered to the muscle. In addition, the blood carries away the waste products of respiration, carbon dioxide, water and, in the case of anaerobic respiration, lactic acid (p. 20).

During exercise, there is an increase in the volume of blood (**stroke volume**) expelled at each beat of the heart, and in its rate of beating. The breathing rate and volume of air exchanged at each breath also increase. Both these changes increase the supply of glucose and oxygen to the active muscles and speed up the removal of the waste products.

During vigorous exercise, there is insufficient oxygen to meet the demands of the contracting muscle, which increasingly comes to depend on anaerobic respiration.

If the lactic acid produced by this form of respiration is not removed quickly enough, it may lead to muscle fatigue (though this has not been established for certain).

The heat generated by the chemical reactions in muscle contraction is carried away by the blood and may trigger the thermoregulatory centre to initiate vasodilation and sweating (p. 138).

Long-term benefits of exercise

Some of the evidence for the benefits of exercise comes from experiments on laboratory animals and some from observations and measurements on the general population.

Whatever benefits exercise may bring, they do not persist unless you exercise regularly. Taking part in sport at school will help to make you fit, but you will not stay fit unless you keep up a regular pattern of exercise.

Exercise does not need to be vigorous but should raise the heart and breathing rates and make you warm. The exercise needs to last 20–30 minutes and be repeated five times a week. Examples are brisk walking, jogging, team games, tennis, etc.

Regular exercise can produce the following physiological changes, which are beneficial:

- *The resting heart rate goes down.* The ventricles enlarge and the heart muscle grows stronger, so that the stroke volume is increased. This means that when you take exercise, your heart can deliver more blood to the muscles without its rate of beating rising too far. Coronary blood flow may possibly increase, either because the vessels widen or because more branches develop. But the evidence for this is not very strong. (There is no evidence that exercise has any beneficial effect on the lungs or breathing mechanism.)
- *The muscles used in exercise grow larger.* At first the muscle fibres grow thicker and then their number increases. The capillaries in the muscle develop more branches. Thus the muscles become stronger. But this only happens if the exercise makes them work to about 80 per cent of their maximum capacity.
- *More enzymes are made in the muscle tissue.* These are the enzymes needed for breaking down glucose, glycogen or fatty acids. Thus the muscle is able to take up oxygen and food more rapidly from the blood and increase the rate of energy production. The muscles can also store more glycogen.

All these changes increase your strength, and also your stamina and endurance. That means that you can take more vigorous exercise, for longer periods, without getting tired or out of breath. Weight training improves strength but not stamina. Walking improves stamina but not strength.

Only the muscles taking part in the exercise will grow stronger. So training for one sport probably won't make you better at another. Improved stamina is useful in all active sports, however.

- *Your ligaments and tendons become stronger.* This reduces the chance of injury during sudden vigorous activity. Back injury is less likely if the muscles and ligaments of the vertebral column are strong.
- *Your joints become more flexible*, giving a greater range of movement with a lower risk of 'pulling' a muscle or spraining a joint.

All these five changes help to postpone the effects of old age, but only if the pattern of exercise is maintained. Once regular exercise ceases, the muscles get thinner. All the improved physiological functions go back to their original level.

- *Protection from heart attacks.* Studies have been carried out on large groups of adults, comparing those who take regular exercise with those who do not. Some of these suggest that people who take regular exercise are less likely to suffer prematurely from heart disease. Other studies show no benefit, or even the reverse. On balance, however, it does seem likely that regular exercise throughout life reduces the chance of an early heart attack.

  The mechanism for this might be increased coronary circulation, a change in the ratio of lipids in the blood, a slowing down in the formation of atheroma (p. 120) in the arteries, a reduced tendency for the blood to form clots in the blood vessels, and long-term reduction in blood pressure. Increased efficiency of the muscles might reduce the strain on the heart.

  The evidence for these changes comes mainly from animal studies rather than from humans, and is not all clear-cut. However, even if exercise does not make you live longer, you will certainly stay fitter.
- *Well-being.* Most people agree that exercise, undertaken willingly, makes you 'feel good'. The 'scientific' evidence for this is lacking but it could be explained in physiological terms.

Questions

1 Say which of the following you would consider to be beneficial forms of exercise: running for a bus, walking 2 miles to school each day, a weekly game of squash, regular training for a football team. Give reasons for your answers.

2 Why do you have to exercise regularly in order to benefit?

3 Explain what part an oxygen debt (p. 20) might play in the time taken to recover from exercise.

Injuries to joints and muscles

Sprains

If a joint is forced beyond its normal degree of movement, some of the ligaments may be partially torn. This results in swelling, pain and a leakage of blood into the tissues (bruising).

A short period of immobilization followed by a return to normal movement as soon as possible is the recommended treatment. Turning your ankle is a common example of a sprain.

If a ligament is completely torn, a longer period of immobilization and surgical repair may be necessary.

Strains

Violent contraction of a muscle may cause some of the fibres to tear ('pulling' a muscle). This is the case in the 'ham string' injury of the posterior thigh muscle. Often, a long period of immobilization is needed for recovery.

Dislocations

This happens when the bones at a joint become displaced as a result of a violent movement. Usually the ligaments are damaged as well. The shallow ball and socket joint of the shoulder is vulnerable to dislocation. The bones have to be manipulated (by a qualified practitioner) back into place. Then the treatment is as for a severe sprain.

Fractures

When a limb bone is broken, the limb has to be restored to its correct shape and immobilized in a plaster cast (pod) until new bone grows and joins up the fractured ends.

Practical work

The structure of bone

Obtain two small bones, e.g. limb bones from a chicken. Place one of them in a test-tube or beaker, cover it with dilute hydrochloric acid and leave it for 24 hours, during which time the acid will dissolve most of the calcium salts in the bone. After 24 hours pour away the acid, wash the bone thoroughly with water and then try to bend it.

Take the second bone and hold one end of it in a hot Bunsen flame, heating it strongly for 2 minutes. At first the bone will char and then glow red as the fibrous organic material burns away, but it will retain its shape. Allow the bone to cool and then try crushing the heated end against the bench with the end of a pencil.

Result Both bones retain their shape after treatment, but the bone whose calcium salts have been dissolved in acid is rubbery and flexible because only the organic, fibrous connective tissue is left. The bone that had this fibrous tissue burned away is still hard but very brittle and easily shattered.

Interpretation The combination in bone of mineral salts and organic fibres produces a hard, strong and resilient structure.

Questions

1 What is the difference between the functions of a ligament and a tendon?

2 Where are the muscles which flex and extend your fingers?

Checklist

Skeleton

- The vertebral column ('backbone') is made up of 33 vertebrae.
- The vertebral column forms the main support for the body and also protects the spinal cord.
- The legs are attached to the vertebral column by the hip girdle.
- The shoulder blades and collar bones form the shoulder girdle.
- The arms are attached to the shoulder blades.
- The skull protects the brain, eyes and ears.
- The ribs protect the lungs, heart and liver, and also play a part in breathing.
- The limb joints are either ball and socket (e.g. hip and shoulder) or hinge (e.g. knee and elbow).
- The surfaces of the joints are covered with cartilage and lubricated with synovial fluid.

Muscles and movement

- Skeletal muscles are formed from fibres and can be consciously controlled.
- Smooth muscle is formed from layers of cells and cannot be consciously controlled.
- Limb muscles are attached to the bones by tendons.
- When the limb muscles contract, they pull on the bones and so bend and straighten the limb or move it forwards and backwards.
- Most limb muscles are arranged in antagonistic pairs, e.g. one bends and one straightens the limb.
- Exercise can have long-term benefits but only if maintained throughout life.

18 The senses

Types of receptor
Sight
Eye structure and function. Image formation. Accommodation. Control of light intensity.
Hearing
Ear structure and function.

Our senses make us aware of changes in our surroundings and in our own bodies. We have sense cells which respond to stimuli (singular = stimulus). A **stimulus** is a change in light, temperature, pressure, etc., which produces a reaction in a living organism. Structures which detect stimuli are called **receptors**. Some of these receptors are scattered through the skin while others are concentrated into special sense organs such as the eye and the ear.

The special property of sensory cells and sense organs is that they are able to convert one form of energy to another. Structures which can do this are called energy **transducers**. The eyes can convert light energy into the electrical energy of a nerve impulse. The ears convert the energy in sound vibrations into nerve impulses. The forms of energy which make up the stimuli may be very different, e.g. mechanical, chemical, light, but they are all transduced into pulses of electrical energy in the nerves.

When a receptor responds to a stimulus, it sends a nerve impulse to the brain which makes us aware of the sensation.

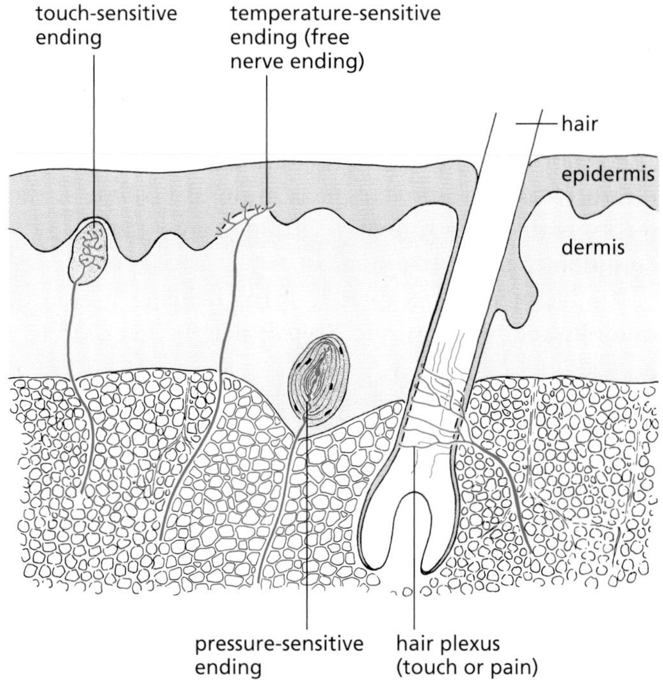

Figure 18.1 The sense organs of the skin (generalized diagram)

Types of receptor

Mechanoreceptors respond to mechanical stimuli such as touch, pressure or vibration. Figure 18.1 shows three types of mechanoreceptor in the skin. There are also internal receptors which respond to, for example, muscle movement (stretch receptors) or to changes in blood pressure. Specialized mechanoreceptors in the cochlea of the inner ear (see Figure 18.10, p. 162) respond to vibrations in the air transmitted by the ear drum and ear bones of the middle ear. These receptors give us our sense of sound. Mechanoreceptors in the semicircular canals and utriculus (p. 162) are stimulated by movement and changes in posture. They send nerve impulses to the brain which help us to maintain our balance.

Chemoreceptors respond to specific chemicals. The epithelium of the tongue carries four types of chemoreceptor which respond to the chemicals in food and drink, giving rise to the taste sensations of sweet, sour, salt and bitter. The nasal epithelium has a much greater variety of receptors that respond to chemicals in the air, and give us our sense of smell and flavour.

Temperature receptors are mostly in the skin (Figure 18.1) and respond to changes in temperature. They send nerve impulses to the brain which result in your feeling too hot or too cold and also set off adjustments to your heat gain or loss (p. 138).

Light receptors transduce the energy from light into the energy of nerve impulses and produce our sense of sight. The individual sensory cells are concentrated into the retina of the eye.

Generally, each type of nerve ending responds to only one kind of stimulus. For example, a heat receptor would send off a nerve impulse if its temperature were raised but not if it were touched.

Questions

1 What sensation would you expect to feel if a warm pin-head was pressed on to a touch receptor in your skin? Explain.

2 If a piece of ice is pressed on to the skin, which receptors are likely to send impulses to the brain?

3 Apart from the cells which detect chemicals, what other types of receptor must be present in the tongue?

Sight

The eye

The structure of the eye is shown in Figures 18.2 and 18.3. The **sclera** is the tough, white outer coating. The front part of the sclera is clear and allows light to enter the eye. This part is called the **cornea**. The **conjunctiva** is a thin epithelium which lines the inside of the eyelids and the front of the sclera, and is continuous with the epithelium of the cornea.

The eye contains a clear liquid whose outward pressure on the sclera keeps the spherical shape of the eyeball. The liquid behind the lens is jelly-like and called **vitreous humour**. The **aqueous humour** in front of the lens is watery.

The **lens** is a transparent structure, held in place by a ring of fibres called the **suspensory ligament**. Unlike the lens of a camera or a telescope, the eye lens is flexible and can change its shape. In front of the lens is a disc of tissue called the **iris**. It is the iris we refer to when we describe the colour of the eye as brown or blue. There is a hole in the centre of the iris called the **pupil**. This lets in light to the rest of the eye. The pupil looks black because all the light entering the eye is absorbed by the black pigment in the **choroid**. The choroid layer, which contains many blood vessels, lies between the retina and the sclera. In the front of the eyeball, it forms the iris and the **ciliary body** (see p. 161). The ciliary body produces aqueous humour.

The internal lining at the back of the eye is the **retina** and it consists of many thousands of cells which respond to light. When light falls on these cells, they send off nervous impulses which travel in nerve fibres, through the **optic nerve**, to the brain and so give rise to the sensation of sight.

Tear glands under the top eyelid produce tear fluid. This is a dilute solution of sodium chloride and sodium hydrogen carbonate. The fluid is spread over the eye surface by the blinking of the eyelids, keeping the surface moist and washing away any dust particles or foreign bodies. Tear fluid also contains an enzyme, **lysozyme**, which attacks bacteria.

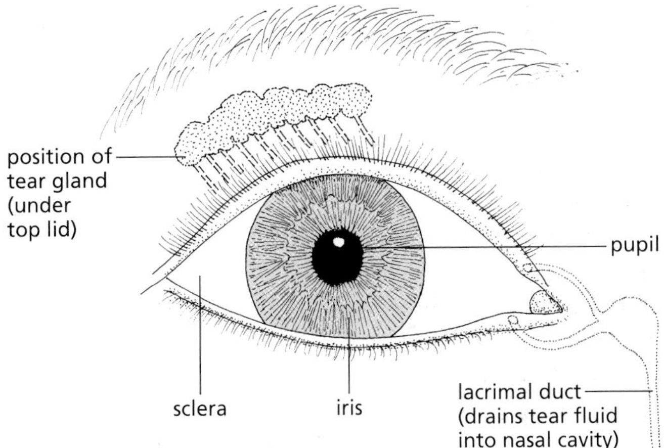

Figure 18.3 Appearance of right eye from the front

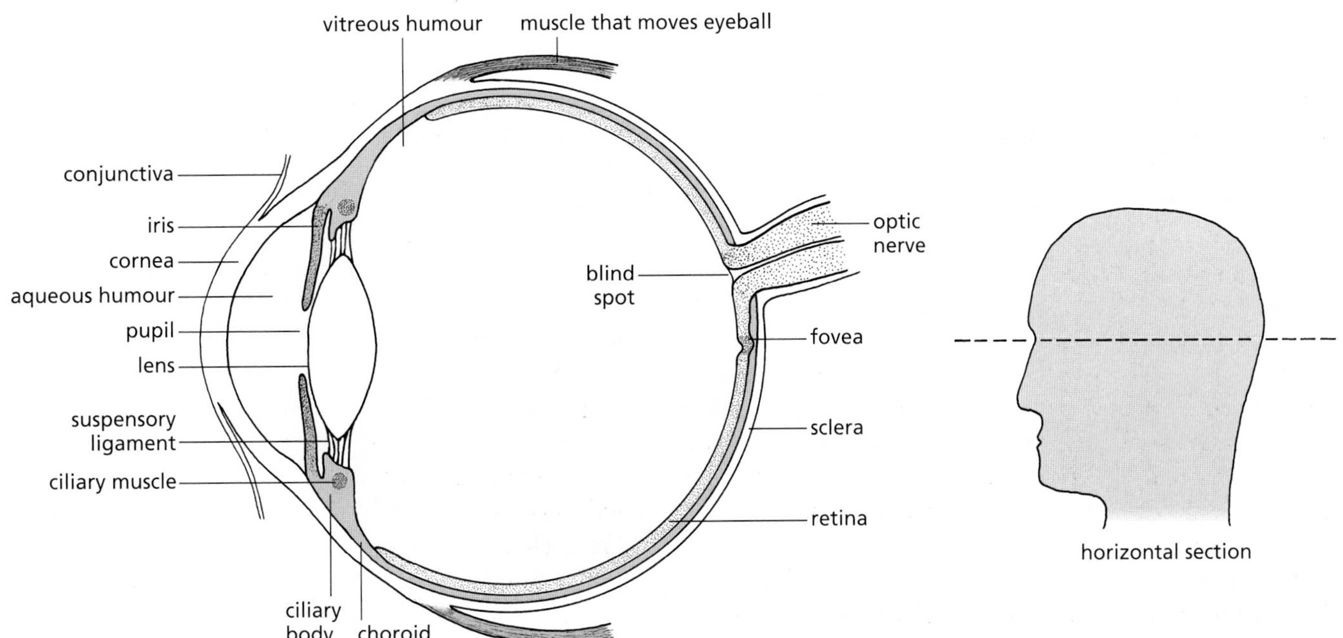

Figure 18.2 Horizontal section through left eye

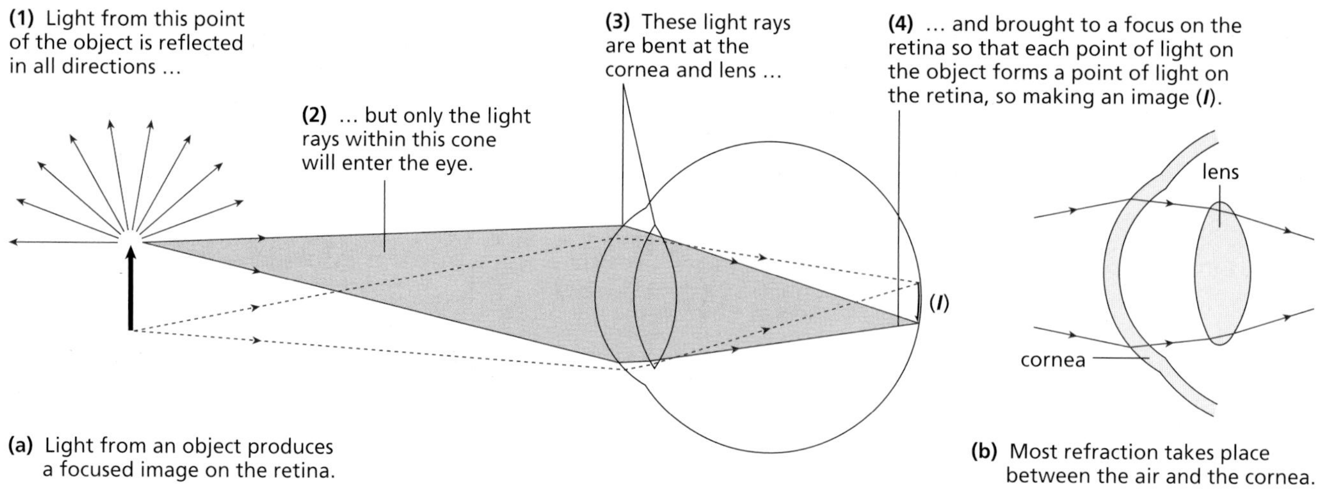

(1) Light from this point of the object is reflected in all directions ...
(2) ... but only the light rays within this cone will enter the eye.
(3) These light rays are bent at the cornea and lens ...
(4) ... and brought to a focus on the retina so that each point of light on the object forms a point of light on the retina, so making an image (I).
(I)
lens
cornea
(a) Light from an object produces a focused image on the retina.
(b) Most refraction takes place between the air and the cornea.

Figure 18.4 Image formation on the retina

Vision

Light from an object produces a focused **image** on the retina (like a 'picture' on a cinema screen) (Figures 18.4 and 18.5). The curved surfaces of the cornea and lens both 'bend' the light rays which enter the eye, in such a way that each 'point of light' from the object forms a 'point of light' on the retina. These points of light will form an image, upside-down and smaller than the object.

The cornea and the aqueous and vitreous humours are mainly responsible for the 'bending' (refraction) of light. The lens makes the final adjustments to the focus (Figure 18.4b).

The pattern of sensory cells stimulated by the image will produce a pattern of nerve impulses sent to the brain. The brain interprets this pattern, using past experience and learning, and forms an impression of the size, distance and upright nature of the object.

Retina

The millions of light-sensitive cells in the retina are of two kinds, the **rods** and the **cones** (according to shape). The cones enable us to distinguish colours, but the rods are more sensitive to low intensities of light. There are thought to be three types of cone cell. One type responds best to red light, one to green and one to blue. If all three types are equally stimulated we get the sensation of white. The cone cells are concentrated in a central part of the retina, called the **fovea** (Figure 18.3, p. 159), and when you study an object closely, you are making its image fall on the fovea.

Fovea

It is in the fovea that the image on the retina is analysed in detail. Only objects within a 2° cone from the eye form an image on the fovea. This means that only about two letters in any word on this page can be seen in detail. It is the constant scanning movements of the eye which enable you to build up an accurate 'picture' of a scene. The centre of the fovea contains only cones: it is here that colour discrimination occurs.

Blind spot

At the point where the optic nerve leaves the retina, there are no sensory cells and so no information reaches the brain about that part of the image which falls on this **blind spot** (Figure 18.6).

Figure 18.6 The blind spot. Hold the book about 50 cm away. Close the left eye and concentrate on the cross with the right eye. Slowly bring the book closer to the face. When the image of the dot falls on the blind spot it will seem to disappear

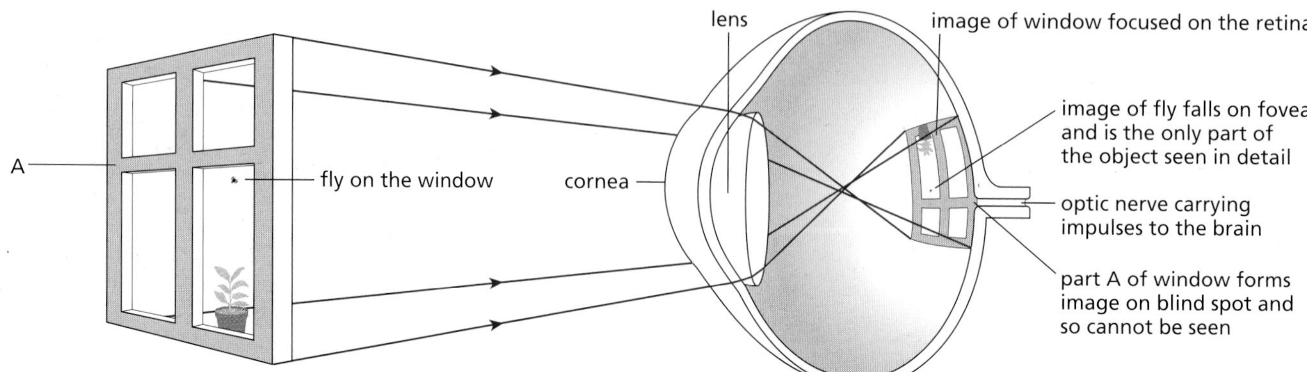

Figure 18.5 Image formation in the eye

Accommodation (focusing)

The eye can produce a focused image of either a near object or a distant object. To do this the lens changes its shape, becoming thinner for distant objects and fatter for near objects. This change in shape is caused by contracting or relaxing the **ciliary muscle** which forms a circular band of muscle in the ciliary body (Figures 18.7 and 18.9). When the ciliary muscle is relaxed, the outward pressure of the humours on the sclera pulls on the suspensory ligament and stretches the lens to its thin shape. The eye is now accommodated (i.e. focused) for distant objects (Figures 18.7a and 18.8a). To focus a near object, the ciliary muscle contracts to a smaller circle and this takes the tension out of the suspensory ligament (Figures 18.7b and 18.8b). The lens is elastic and flexible and so is able to change to its fatter shape. This shape is better at bending the light rays from a close object.

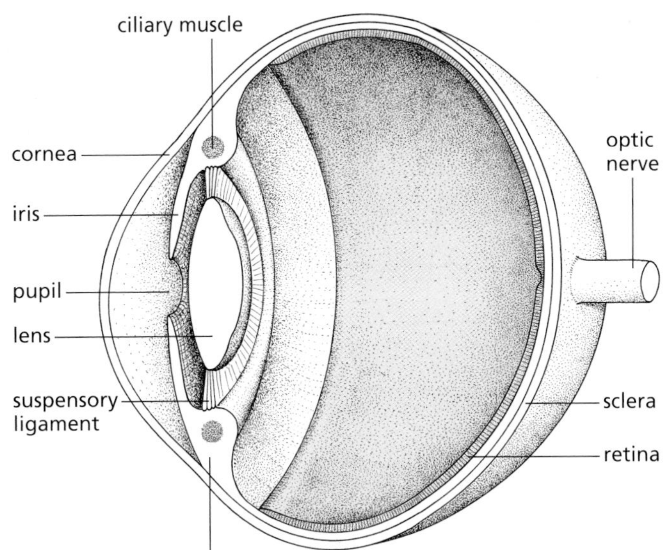

Figure 18.9 Vertical section through the left eye

Control of light intensity

The amount of light entering the eye is controlled by altering the size of the pupil. If the light intensity is high, it causes a contraction in a ring of muscle fibres (a sphincter) in the iris. This reduces the size of the pupil and cuts down the intensity of light entering the eye. High intensity light can damage the retina, so this reaction has a protective function.

In low light intensities, the sphincter muscle of the iris relaxes and muscle fibres running radially (i.e. like wheel spokes) contract. This makes the pupil enlarge and admits more light.

The change in size of the pupil is caused by an automatic reflex action (p. 165); you cannot control it consciously.

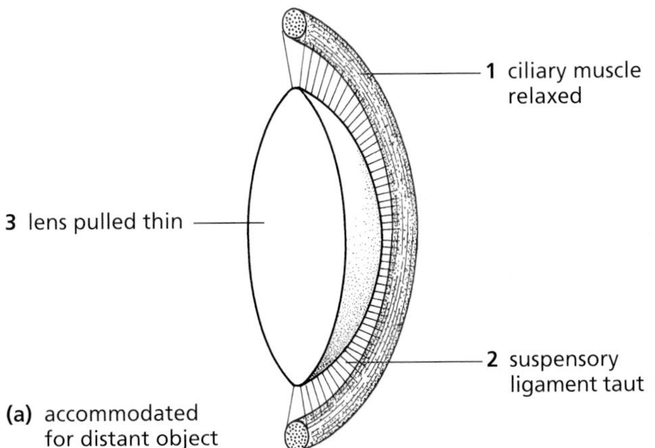

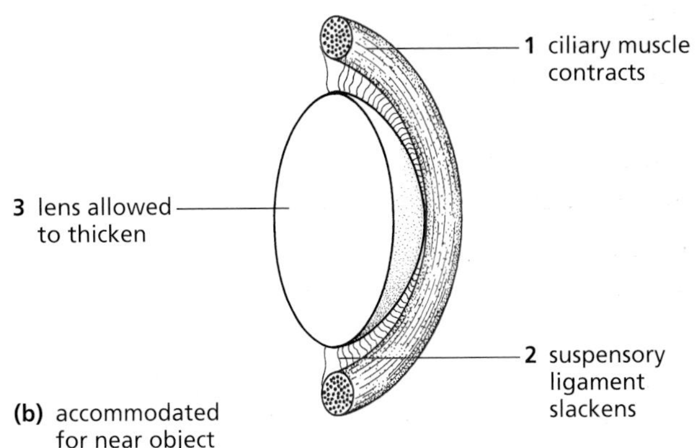

Figure 18.7 How accommodation is brought about

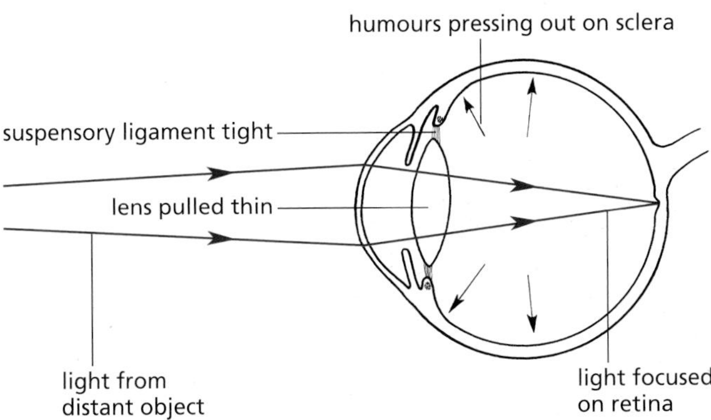

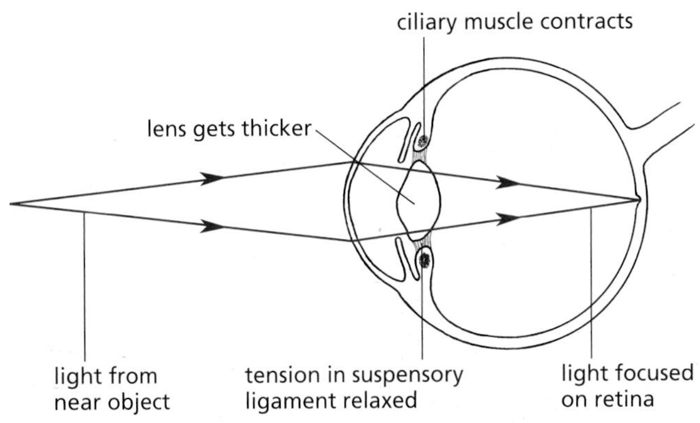

Figure 18.8 Accommodation

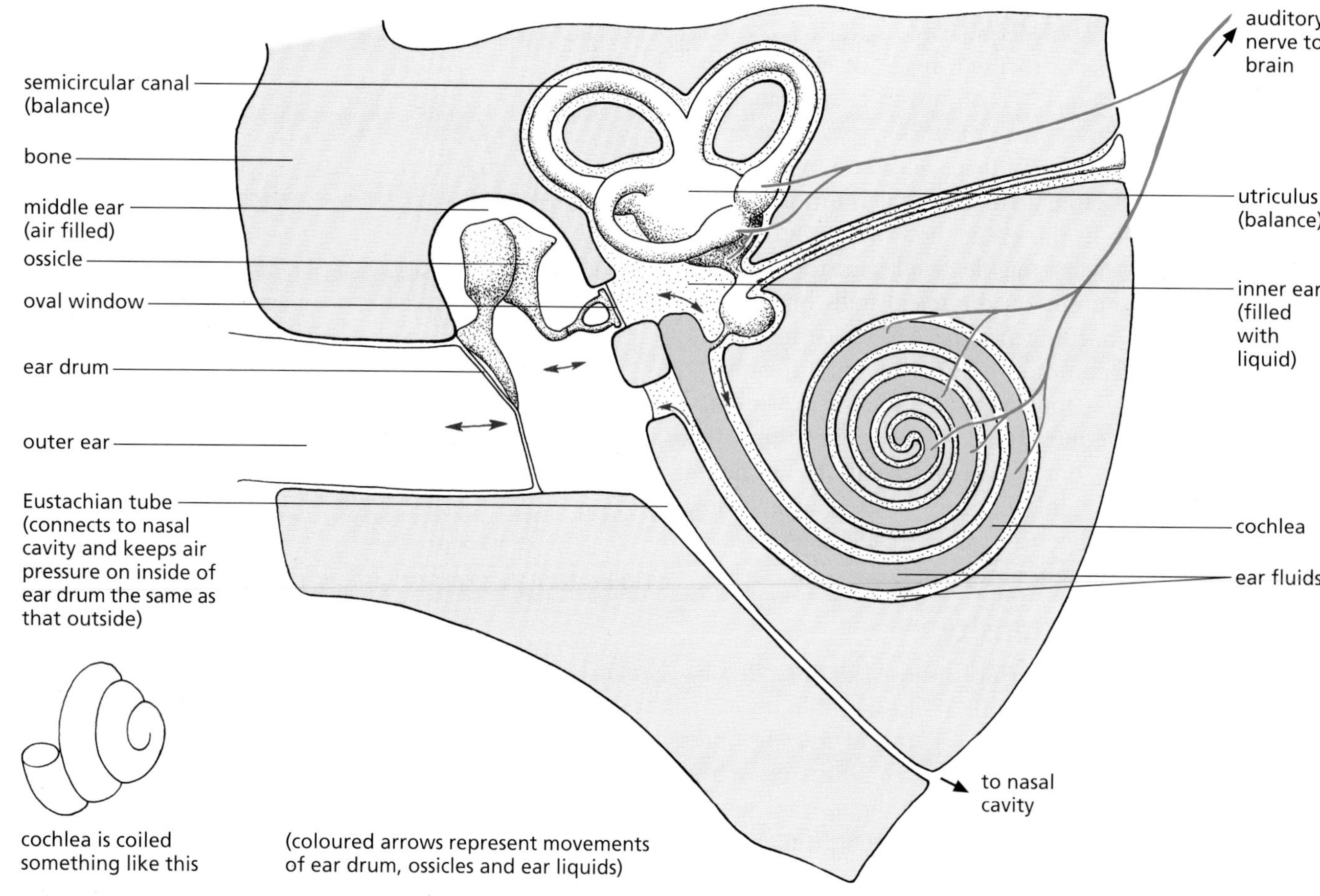

Figure 18.10 Diagram of the ear

Hearing

The hearing apparatus is enclosed in bone on each side of the skull, just behind the jaw hinge. Sound vibrations travel through a short, wide tube (the outer ear) and are converted to nerve impulses by the apparatus in the middle ear and inner ear. The structure of the ear is shown in Figure 18.10.

Outer ear

Sound is the name we give to the sensation we get as a result of vibrations in the air. These vibrations are pulses of compressed air. They enter the tube of the outer ear and hit the **ear drum**, a thin membrane like a drum-skin across the inner end of the tube. The air vibrations cause the ear drum to vibrate backwards and forwards. If there are 200 pulses of compressed air every second, the ear drum will move backwards and forwards at the same rate.

The vibrations are transmitted to the oval window by the chain of bones (ossicles) of the middle ear. The vibrations of the fluids in the inner ear stimulate sensory receptors in the **cochlea**, a fluid-filled coiled tube. Nerve impulses from these receptors pass to the brain in the **auditory nerve**. The first part of the cochlea reponds to high frequency sounds (e.g high notes). The later part responds to low frequency sounds (e.g. low notes).

Checklist

- Receptors respond to stimuli by sending nerve impulses to the brain.
- Stimuli may be, for example, light, chemicals, temperature, touch or pressure.
- Receptors respond to only one type of stimulus.
- The lens focuses light from the outside world to form a tiny image on the retina.
- The sensory cells of the retina are stimulated by the light and send nerve impulses to the brain.
- The brain interprets these nerve impulses and so gives us the sense of vision.
- The eye can focus on near or distant objects by changing the thickness of the lens.
- The ear drum is made to vibrate backwards and forwards by sound waves in the air.
- The vibrations are passed on to the inner ear by the ear bones.
- Sensory nerve endings in the inner ear respond to the vibrations and send impulses to the brain.
- The brain interprets these impulses as sound.

19 Co-ordination

Nervous system
Nerve cells, synapses, nervous impulses.

The reflex arc
The spinal cord. Reflexes.

Voluntary actions

The central nervous system
Spinal cord and brain.

The endocrine system
Compared with nervous system. Thyroid, adrenals, pancreas, diabetes, reproductive organs. Pituitary gland.

Homeostasis and feedback

Therapeutic and industrial uses of hormones

Performance-enhancing hormones

Co-ordination is the way all the organs and systems of the body are made to work efficiently together (Figure 19.1). If, for example, the leg muscles are being used for running, they will need extra supplies of glucose and oxygen. To meet this demand, the lungs breathe faster and deeper to obtain the extra oxygen and the heart pumps more rapidly to get the oxygen and glucose to the muscles more quickly.

The brain detects changes in the oxygen and carbon dioxide content of the blood and sends nervous impulses to the diaphragm, intercostal muscles and heart. In this example, the co-ordination of the systems is brought about by the **nervous system**.

Figure 19.1 Co-ordination. The badminton player's brain is receiving sensory impulses from her eyes, semicircular canals, utriculus and muscle stretch receptors. Using this information, the brain co-ordinates the muscles of her limbs so that even while running or leaping she can control her stroke

The extra supplies of glucose needed for running come from the liver. Glycogen in the liver is changed to glucose which is released into the bloodstream (p. 104). The conversion of glycogen to glucose is stimulated by, among other things, a chemical called adrenaline (p. 170). Co-ordination by chemicals is brought about by the **endocrine system**.

The nervous system works by sending electrical impulses along nerves. The endocrine system depends on the release of chemicals, called **hormones**, from **endocrine glands**. Hormones are carried by the bloodstream. For example, insulin (p. 171) is carried from the pancreas to the liver by the circulatory system.

The nervous system

The human nervous system is shown in Figure 19.2 (overleaf). The brain and spinal cord together form the **central nervous system**. Nerves carry electrical impulses from the central nervous system to all parts of the body, making muscles contract or glands produce enzymes or hormones.

Glands and muscles are called **effectors** because they go into action when they receive nerve impulses or hormones. The biceps muscle (p. 155) is an effector which flexes the arm; the salivary gland is an effector which produces saliva when it receives a nerve impulse from the brain.

The nerves also carry impulses back to the central nervous system from receptors in the sense organs of the body. These impulses from the eyes, ears, skin, etc. make us aware of changes in our surroundings or in ourselves. Nerve impulses from the sense organs to the central nervous system are called **sensory impulses**; those from the central nervous system to the effectors, resulting in action, are called **motor impulses**.

The nerves which connect the body to the central nervous system make up the **peripheral nervous system**.

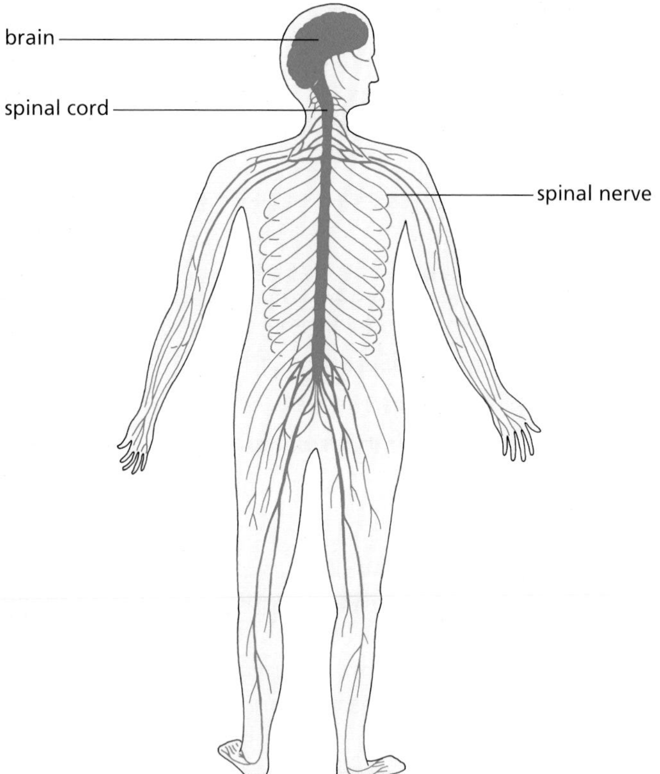

Figure 19.2 The human nervous system

Nerve cells (neurones)

The central nervous system and the peripheral nerves are made up of nerve cells, called **neurones**. Three types of neurone are shown in Figure 19.3. The **motor neurones** carry impulses from the central nervous system to muscles and glands. The **sensory neurones** carry impulses from the sense organs to the central nervous system. The **multi-polar neurones** are neither sensory nor motor but make connections to other neurones inside the central nervous system.

Each neurone has a **cell body** consisting of a nucleus surrounded by a little cytoplasm. Branching fibres, called **dendrites**, from the cell body make contact with other neurones. A long filament of cytoplasm, surrounded by an insulating sheath, runs from the cell body of the neurone. This filament is called a **nerve fibre** (Figure 19.3a and b). The cell bodies of the neurones are mostly located in the brain or in the spinal cord and it is the nerve fibres which run in the nerves. A **nerve** is easily visible, white, tough and stringy and consists of hundreds of microscopic nerve fibres bundled together (Figure 19.4). Most nerves will contain a mixture of sensory and motor fibres. So a nerve can carry many different impulses. These impulses will travel in one direction in sensory fibres and in the opposite direction in motor fibres.

Some of the nerve fibres are very long. The nerve fibres to the foot have their cell bodies in the spinal cord and the fibres run inside the nerves, without a break, to the skin of the toes or the muscles of the foot. A single nerve cell may have a fibre 1 m long.

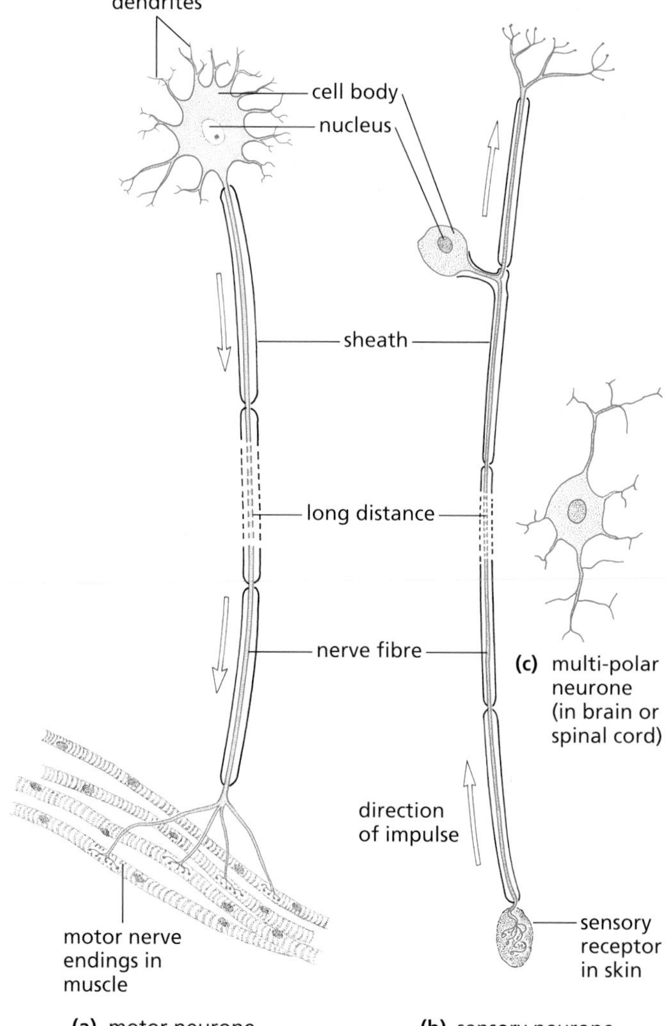

Figure 19.3 Nerve cells (neurones)

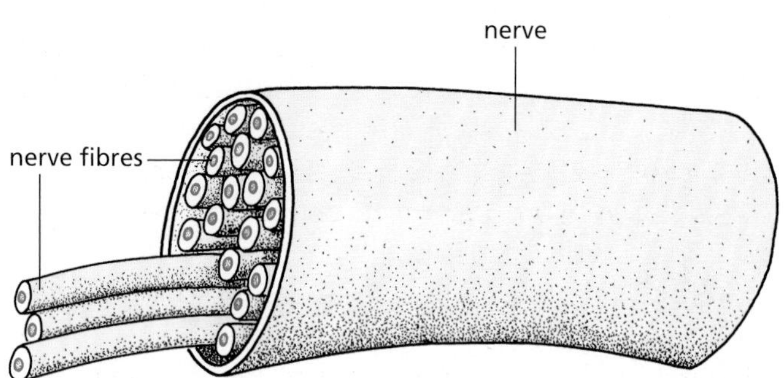

Figure 19.4 Nerve fibres grouped into a nerve

Questions

1 What is the difference between a nerve and a nerve fibre?

2 In what ways are sensory neurones and motor neurones similar
a in structure, **b** in function?
How do they differ?

3 Can
a a nerve fibre,
b a nerve, carry both sensory and motor impulses?
Explain your answers.

Synapse

Although nerve fibres are insulated, it is necessary for impulses to pass from one neurone to another. An impulse from the finger-tips has to pass through at least three neurones before reaching the brain and so produce a conscious sensation. The regions where impulses are able to cross from one neurone to the next are called **synapses**.

At a synapse, a branch at the end of one fibre is in close contact with the cell body or dendrite of another neurone (Figure 19.5). When an impulse arrives at the synapse, it releases a tiny amount of a chemical substance (a **neurotransmitter substance**) which sets off an impulse in the next neurone. Sometimes several impulses have to arrive at the synapse before enough transmitter substance is released to cause an impulse to be fired off in the next neurone.

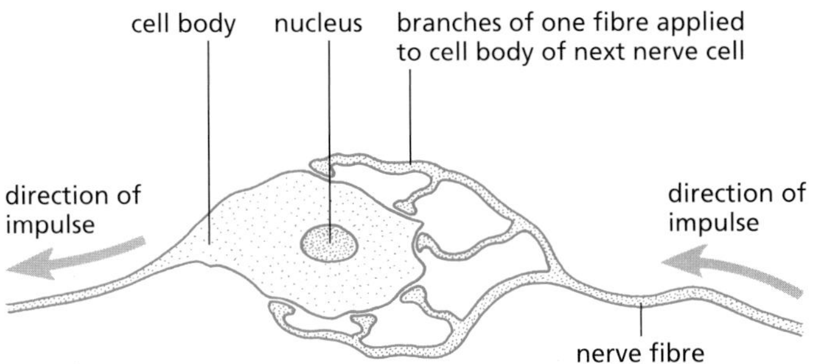

Figure 19.5 Diagram of synapses

The nerve impulse

The nerve fibres do not carry sensations like pain or cold. These sensations are felt only when a nerve impulse reaches the brain. The impulse itself is a series of electrical pulses which travel down the fibre. Each pulse lasts about 0.001 second and travels at speeds of up to 100 metres per second. All nerve impulses are similar; there is no difference between nerve impulses from the eyes, ears or hands.

We are able to tell where the sensory impulses have come from and what caused them only because the impulses are sent to different parts of the brain. The nerves from the eye go to the part of the brain concerned with sight. So when impulses are received in this area, the brain recognizes that they have come from the eyes and we 'see' something.

Questions

1. Look at Figure 19.7 (overleaf).
   a How many cell bodies are drawn?
   b How many synapses are shown?
   Look at Figure 19.9 and answer the same questions.
2. If you could intercept and 'listen to' the nerve impulses travelling in the spinal cord, could you tell which ones came from pain receptors and which from temperature receptors? Explain your answer.

The reflex arc

One of the simplest situations where impulses cross synapses to produce action is in the reflex arc. A **reflex action** is an automatic response to a stimulus. When a particle of dust touches the cornea of the eye, you will blink; you cannot prevent yourself from blinking. A particle of food touching the lining of the windpipe will set off a coughing reflex which cannot be suppressed. When a bright light shines in the eye, the pupil contracts (see p. 161). You cannot stop this reflex and you are not even aware that it is happening.

The nervous pathway for such reflexes is called a **reflex arc**. In Figure 19.6 the nervous pathway for a well-known reflex called the 'knee-jerk' reflex is shown.

One leg is crossed over the other and the muscles are totally relaxed. If the tendon just below the kneecap of the upper leg is tapped sharply, a reflex arc makes the thigh muscle contract and the lower part of the leg swings forward.

The pathway of this reflex arc is traced in Figure 19.7. Hitting the tendon stretches the muscle and stimulates a stretch receptor. The receptor sends off impulses in a sensory fibre. These sensory impulses travel in the nerve to the spinal cord.

In the central region of the spinal cord, the sensory fibre passes the impulse across a synapse to a motor neurone which conducts the impulse down the fibre, back to the thigh muscle. The arrival of the impulses at the muscle makes it contract, and jerk the lower part of the limb forward. You are aware that this is happening (which means that sensory impulses must be reaching the brain), but there is nothing you can do to stop it.

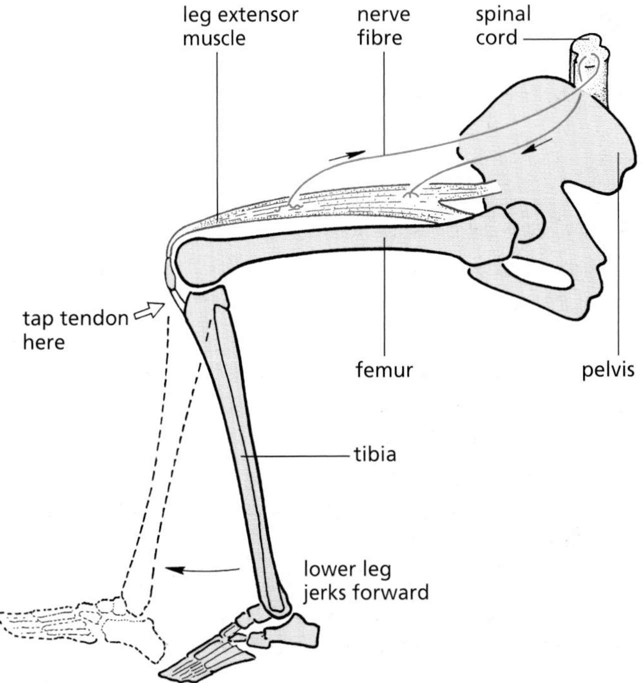

Figure 19.6 The reflex knee jerk

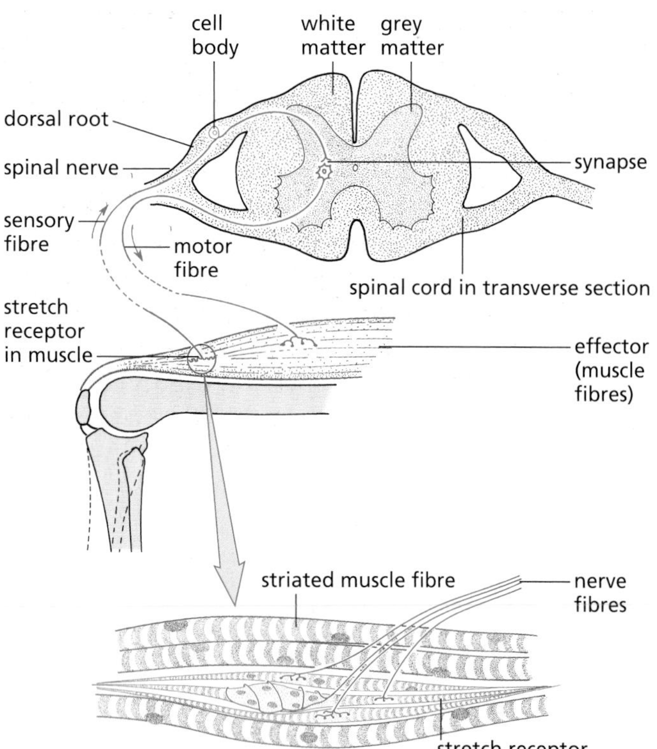

Figure 19.7 The reflex arc. This reflex arc needs only one synapse for making the response. Most reflex actions need many more synapses (i) to adjust other muscles in the body and (ii) to send impulses to the brain

The spinal cord

In Figure 19.7 the spinal cord is drawn in transverse section. The spinal nerve divides into two 'roots' at the point where it joins the spinal cord. All the sensory fibres enter through the **dorsal root** and the motor fibres all leave through the **ventral root**, but both kinds of fibre are contained in the same spinal nerve. This is like a group of insulated wires in the same electric cable. The cell bodies of all the sensory fibres are situated in the dorsal root and they make a bulge called a **ganglion** (Figure 19.8).

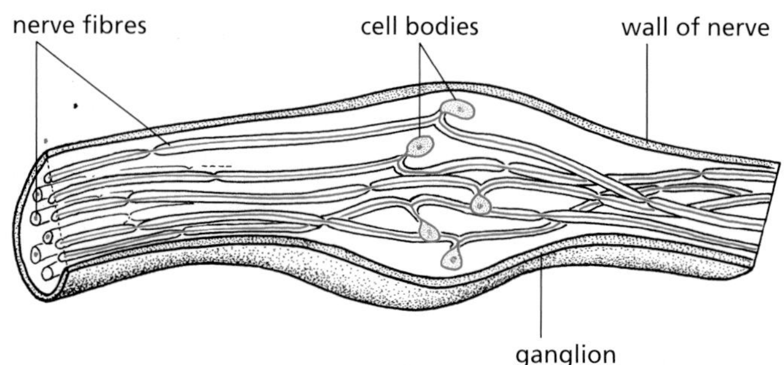

Figure 19.8 Cell bodies forming a ganglion

In even the simplest reflex action, many more nerve fibres, synapses and muscles are involved than are described here. In Figure 19.9 the reflex arc which would result in the hand being removed from a painful stimulus is illustrated. On the left side of the spinal cord, an incoming sensory fibre makes its first synapse with a **relay neurone**. This can pass the impulse on to many other motor neurones, although only one is shown in the diagram. On the right side of the spinal cord, some of the incoming sensory fibres are shown making synapses with neurones which send nerve fibres to the brain, thus keeping the brain informed about events in the body. Also, nerve fibres from the brain make synapses with motor neurones in the spinal cord so that 'commands' from the brain can be sent to muscles of the body.

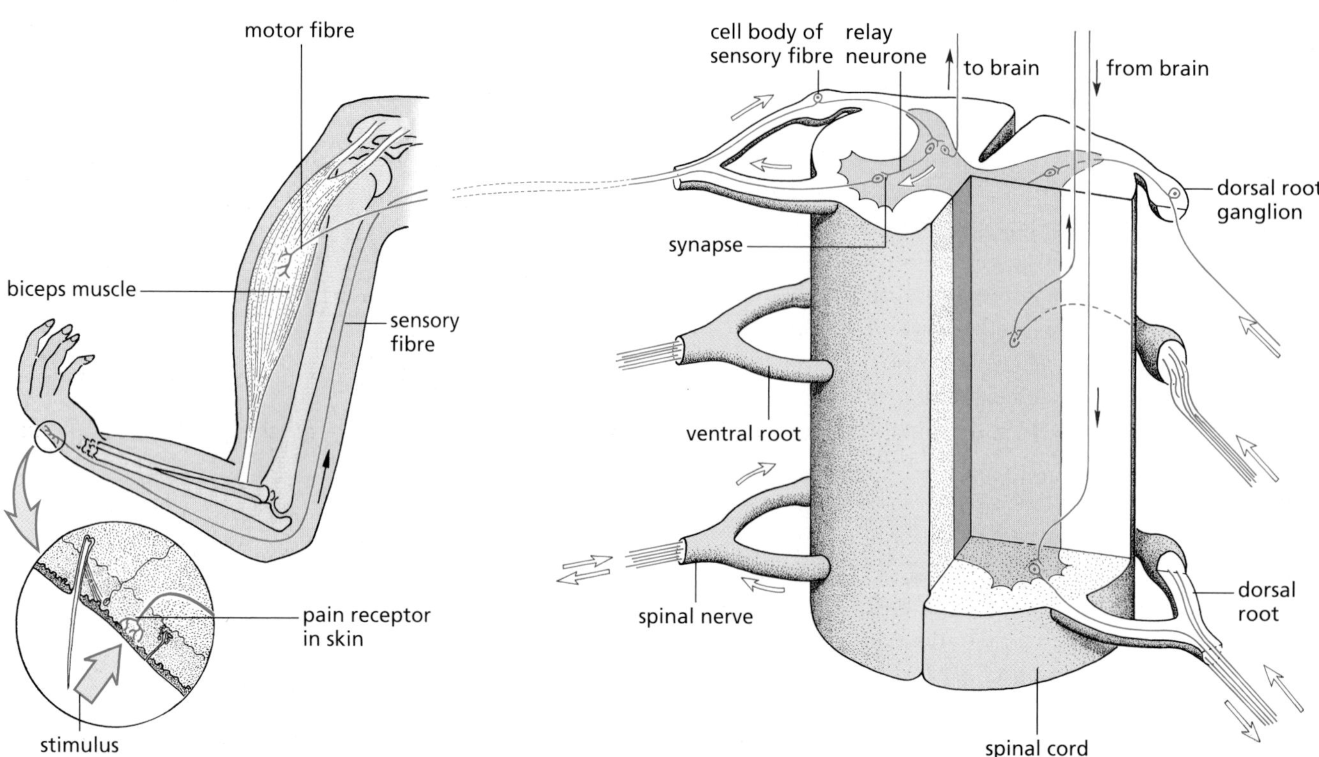

Figure 19.9 Reflex arc (withdrawal reflex)

Reflexes

The reflex just described is a **spinal reflex**. The brain, theoretically, is not needed for it to happen. Responses which take place in the head, such as blinking, coughing and iris contraction, have their reflex arcs in the brain, but may still not be consciously controlled.

Bright light stimulates the light-sensitive cells of the retina. The nerve impulses in the sensory fibres from these receptors travel through the optic nerve to the brain. In the mid-brain (p. 168) the fibres synapse with relay and motor fibres which carry impulses back through the optic nerve to the sphincter muscle of the iris and stimulate it to contract.

The reflex closure of the iris (p. 161) protects the retina from bright light; the withdrawal reflex removes the hand from a dangerously hot object; the coughing reflex dislodges a foreign particle from the windpipe. Thus, these reflexes have a protective function and all are **involuntary actions**.

There are many other reflexes going on inside our bodies. We are usually unaware of these, but they maintain our blood pressure, breathing rate, heart beat, etc. and so maintain the body processes.

Conditioned reflexes

At the beginning of the 20th century, a Russian scientist, **Ivan Pavlov**, studied the salivary reflex in dogs. If food is placed on a dog's tongue, the animal salivates (produces saliva). The stimulus of food on the chemoreceptors of the tongue sets off a reflex action, with the salivary glands as the effectors. Pavlov observed, however, that the sight of food alone produced the salivation response. This he described as a **conditioned reflex** since the nervous pathway originates in the eye and is not set off by the chemicals in the food.

He extended his experiments by ringing a bell every time food was presented. Eventually the dog salivated when the bell alone was rung, without food being presented. The dog had been **conditioned** to respond to the sound, a stimulus quite removed from the taste of food.

Conditioned reflexes may play a part in the early stages of learning. Day-old chicks will peck indiscriminately at all small objects on the ground. Some of these will be distasteful and the chick becomes conditioned to avoid them.

The early stages of learning to ride a bicycle may involve conditioned reflexes. The responses to stimuli which retain balance, normally involving the legs, become transferred to the arms controlling the handlebars.

Conditioning can produce some very complex behaviour (Figure 19.10) but learning usually involves a great deal more than a set of conditioned reflexes.

Figure 19.10 The blue tit has learned to pull out the matchstick to release the peanuts

Voluntary actions

A voluntary action starts in the brain. It may be the result of external events, such as seeing a book on the floor, but any resulting action, such as picking up the book, is entirely voluntary. Unlike a reflex action it does not happen automatically; you can decide whether or not you carry out the action.

The brain sends motor impulses down the spinal cord in the nerve fibres. These make synapses with motor fibres which enter spinal nerves and make connections to the sets of muscles needed to produce effective action. Many sets of muscles in the arms, legs and trunk would be brought into play in order to stoop and pick up the book, and impulses passing between the eyes, brain and arm would direct the hand to the right place and 'tell' the fingers when to close on the book.

One of the main functions of the brain is to co-ordinate these actions so that they happen in the right sequence and at the right time and place.

Questions

1 Put the following in the correct order for a simple reflex arc
- **a** impulse travels in motor fibre,
- **b** impulse travels in sensory fibre,
- **c** effector organ stimulated,
- **d** receptor organ stimulated,
- **e** impulse crosses synapse.

2 Which receptors and effectors are involved in the reflex actions of
- **a** sneezing,
- **b** blinking,
- **c** contraction of the iris?

3 Explain why the tongue may be considered to be both a receptor and an effector organ.

4 Discuss whether coughing is a voluntary or reflex action.

5 Give some examples of behaviour patterns in the domesticated dog which you would consider to be dependent on conditioned reflexes.

The central nervous system

Spinal cord

Like all other parts of the nervous system, the spinal cord consists of thousands of nerve cells. The structure of the spinal cord is shown in Figures 19.7, 19.9 and 19.11.

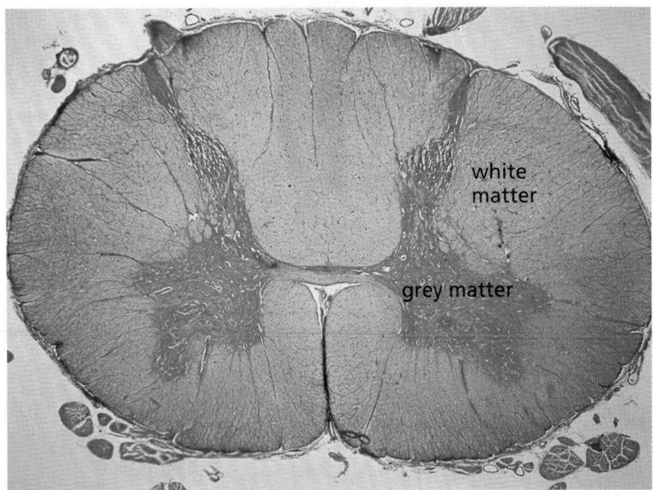

Figure 19.11 Section through spinal cord (× 7). The light area is the white matter, consisting largely of nerve fibres running to and from the brain. The darker central area is the grey matter, consisting largely of nerve cell bodies

All the cell bodies, apart from those in the dorsal root ganglia, are concentrated in the central region called the **grey matter**. The **white matter** consists of nerve fibres. Some of these will be passing from the grey matter to the spinal nerves and others will be running along the spinal cord connecting the spinal nerve fibres to the brain. The spinal cord is thus concerned with (a) reflex actions involving body structures below the neck, (b) conducting sensory impulses from the skin and muscles to the brain and (c) carrying motor impulses from the brain to the muscles of the trunk and limbs.

The brain

The brain may be thought of as the expanded front end of the spinal cord. Certain areas are greatly enlarged to deal with all the information arriving from the ears, eyes, tongue, nose and semicircular canals. A simplified diagram of the main regions of the brain as seen in vertical section is given in Figure 19.12d.

The **medulla** is involved in the breathing rhythm, heart rate, swallowing and vasoconstriction of the arterioles. These basically involve reflex actions but all are influenced by nerve impulses from other parts of the brain, e.g. the hypothalamus and the cerebrum. For example, the breathing rhythm can be altered voluntarily for speaking and singing.

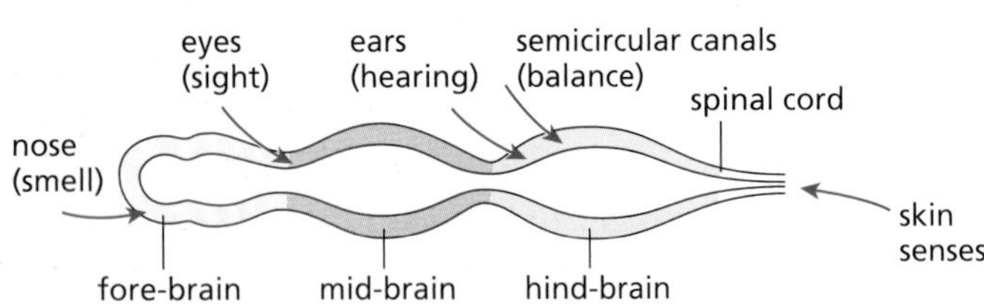

(a) The front of the spinal cord develops three bulges: the fore-, mid- and hind-brain. Each region receives impulses mainly from sense organs in the head.

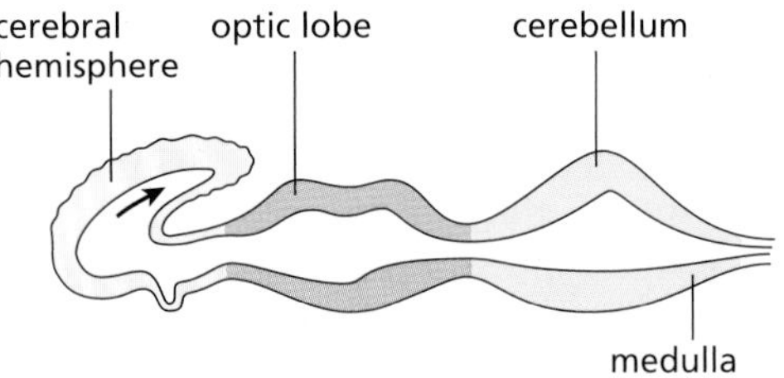

(b) The roofs of the fore-, mid- and hind-brain become thicker and form the cerebral hemispheres, optic lobes and cerebellum. The floor of the hind-brain thickens to form the medulla.

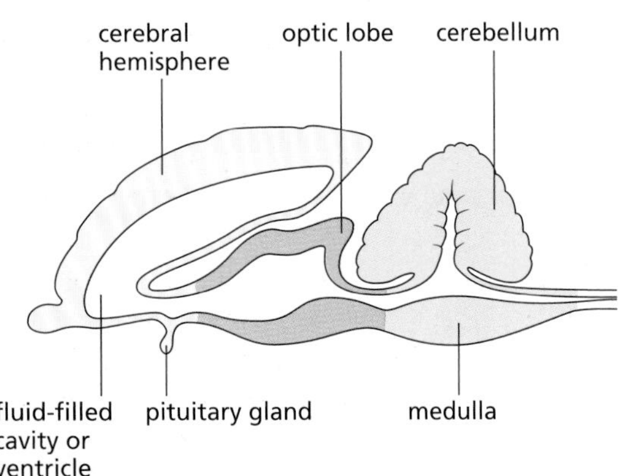

(c) A rabbit's brain would look something like this in vertical section.

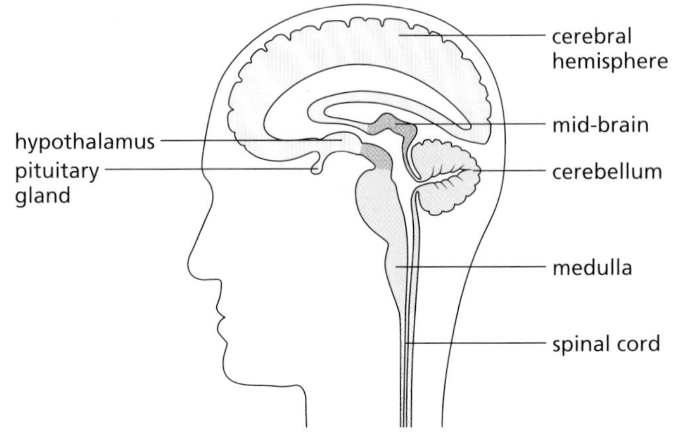

(d) The same regions are present in a human brain, but because of our upright position, the brain is bent through 90°.

Figure 19.12 Development of the brain of a mammal (vertical sections)

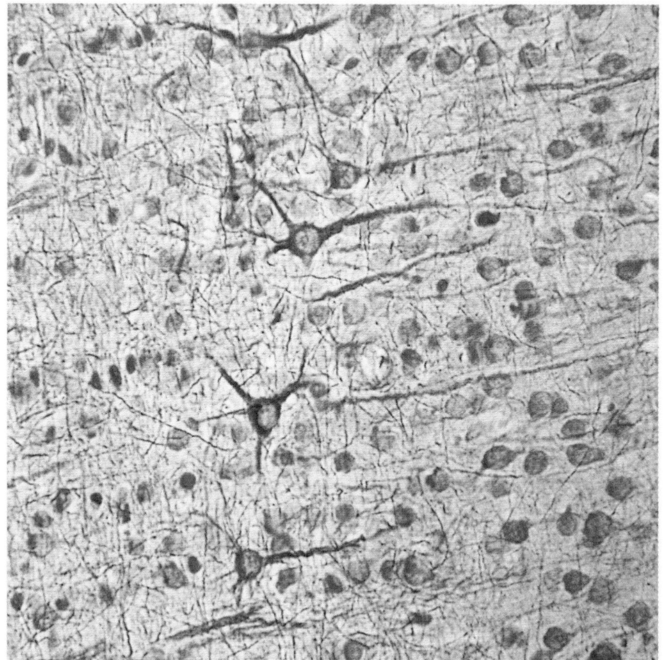

Figure 19.13 Multi-polar neurones in the cerebral cortex (×350). The cell bodies and their branching fibres have been darkly stained to make them show up

The **cerebellum** is principally involved in maintaining balance and co-ordinating movement. It receives sensory impulses from stretch receptors in the muscles and from the semicircular canals and utriculus (p. 162). It sends motor impulses to the muscles. However, it is in the cerebrum that the pattern of movement is 'decided' and relayed to the cerebellum, which makes sure that the systems work effectively together.

The **cerebrum** is the largest part of the brain and consists of two **cerebral hemispheres**. These are highly developed in mammals (Figure 19.12c), especially humans, and are the regions concerned with intelligence, memory, reasoning ability, acquired skills and consciousness.

In the cerebral hemispheres, there is an outer layer of cells, the **cortex**, with hundreds of thousands of multi-polar neurones (Figure 19.3c, p. 164) making possible an enormous number of synapse connections between the dendrites (Figure 19.13).

The cerebral hemispheres are basically the 'command centre' of the brain and, to a large extent, the output from the other regions depends on the 'instructions' from the cerebrum.

Functions of the brain

To sum up:

- The brain receives impulses from all the sensory organs of the body.
- As a result of these sensory impulses, it sends off motor impulses to the glands and muscles, causing them to function accordingly.
- In its association centres it correlates the various stimuli from the different sense organs and the memory.
- The association centres and motor areas co-ordinate bodily activities so that the mechanisms and chemical reactions of the body work efficiently together.
- It 'stores' information so that behaviour can be modified according to past experience.

Questions

1 Would you expect synapses to occur in grey matter or in white matter? Explain your answer.

2 Look at Figure 19.2, p. 164. If the spinal cord were damaged at a point about one-third of the way up the vertebral column, what effect would you expect this to have on the bodily functions?

3 **a** With which senses are the fore-, mid- and hind-brain mainly concerned?
b Which part of the brain seems to be mainly concerned with keeping the basic body functions going?

4 Describe the biological events involved when you hear a sound and turn your head towards it.

The endocrine system

Co-ordination by the nervous system is usually rapid and precise. Nerve impulses, travelling at up to 100 metres per second, are delivered to specific parts of the body and produce an almost immediate response. A different kind of co-ordination is brought about by the endocrine system. This system depends on chemicals, called **hormones**, which are released from special glands, called **endocrine glands**, into the bloodstream. The hormones circulate round the body in the blood and eventually reach certain organs, called **target organs**. Hormones speed up or slow down or alter the activity of those organs. After being secreted, hormones do not remain permanently in the blood but are changed by the liver into inactive compounds and excreted by the kidneys.

Unlike the digestive glands, the endocrine glands do not deliver their secretions through ducts. For this reason, the endocrine glands are sometimes called the 'ductless glands'. The hormones are picked up directly from the glands by the blood circulation.

Responses of the body to hormones are much slower than responses to nerve impulses. They depend, in the first instance, on the speed of the circulatory system and then on the time it takes for the cells to change their chemical activities. Many hormones affect long-term changes such as growth rate, puberty and pregnancy. Nerve impulses often cause a response in a very limited area of the body, such as an eye-blink or a finger movement. Hormones often affect many organ systems at once.

Serious deficiencies or excesses of hormone production give rise to illnesses. Small differences in hormone activity between individuals probably contribute to differences of personality and temperament.

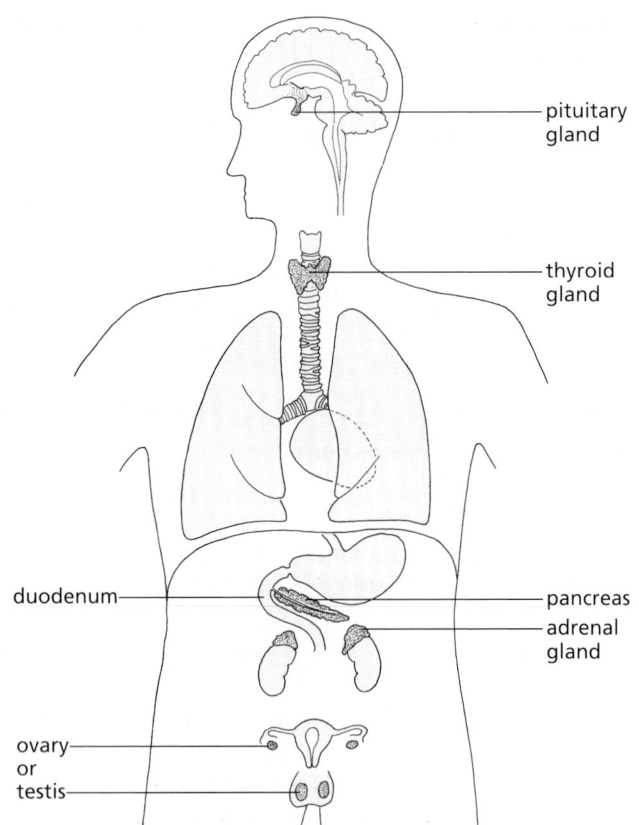

Figure 19.14 Position of endocrine glands in the body

The position of the endocrine glands in the body is shown in Figure 19.14. Notice that the pancreas and the reproductive organs have a dual function.

Thyroid gland

The thyroid gland is situated in the front part of the neck and lies in front of the windpipe. It produces a hormone called **thyroxine**, which is formed from an amino acid and iodine (p. 88). Thyroxine has a stimulatory effect on the metabolic rate of nearly all the body cells. It controls our level of activity, promotes normal skeletal growth and is essential for the normal development of the brain.

Adrenal glands

These glands are attached to the back of the abdominal cavity, one above each kidney. Each adrenal gland is made up of two distinct regions with different functions. There is an outer layer called the **adrenal cortex** and an inner zone called the **adrenal medulla**. The medulla receives nerves from the brain and produces the hormone **adrenaline**. The cortex has no nerve supply and produces a number of hormones called **corticosteroids**. The corticosteroids help to control the metabolism of carbohydrates, fats, proteins, salts and water.

Adrenaline, from the medulla, has less important but more obvious effects on the body. In response to a stressful situation, nerve impulses are sent from the brain to the adrenal medulla, which releases adrenaline into the blood. As adrenaline circulates round the body it affects a great many organs, as shown in Table 19.1 below.

All these effects make us more able to react quickly and vigorously in dangerous situations that might require us to run away or put up a struggle. However, in many stressful situations, such as taking examinations or giving a public performance, vigorous activity is not called for. So the extra adrenaline in our bodies just makes us feel tense and anxious. You will recognize the sensations described in column four of Table 19.1 as characteristic of fear and anxiety.

Most of the systems affected by adrenaline are also controlled by a section of the nervous system called the **sympathetic nervous system** which affects internal organs other than voluntary muscles. It is difficult to tell whether adrenaline or the sympathetic nervous system is mainly responsible for the stress reaction.

Adrenaline is quickly converted by the liver to a less active compound which is excreted by the kidneys. All hormones are similarly altered and excreted, some within minutes, others within days. Thus their effects are not long-lasting. The long-term hormones, such as thyroxine, are secreted continuously to maintain a steady level.

Table 19.1 Responses to adrenaline

Target organ	Effects of adrenaline	Biological advantage	Effect or sensation
Heart	beats faster	sends more glucose and oxygen to the muscles	thumping heart
Breathing centre of the brain	faster and deeper breathing	increased oxygenation of the blood; rapid removal of carbon dioxide	panting
Arterioles of the skin	constricts them (see p. 138)	less blood going to the skin means more is available to the muscles	person goes paler
Arterioles of the digestive system	constricts them	less blood for the digestive system allows more to reach the muscles	dry mouth
Muscles of alimentary canal	relax	peristalsis and digestion slow down; more energy available for action	'hollow' feeling in stomach
Muscles of body	tenses them	ready for immediate action	tense feeling; shivering
Liver	conversion of glycogen to glucose	glucose available in blood for energy production	no sensation
Fat depots	conversion of fats to fatty acids	fatty acids available in blood, for muscle contraction	

The pancreas

The pancreas is a digestive gland which secretes enzymes into the duodenum through the pancreatic duct (p. 100). It is also an endocrine (ductless) gland. Most of the pancreas cells produce digestive enzymes but some of them produce hormones. The hormone-producing cells are arranged in small isolated groups called **islets** (Figure 19.15) and secrete their hormones directly into the bloodstream. One of the hormones is called **glucagon** and the other is **insulin**.

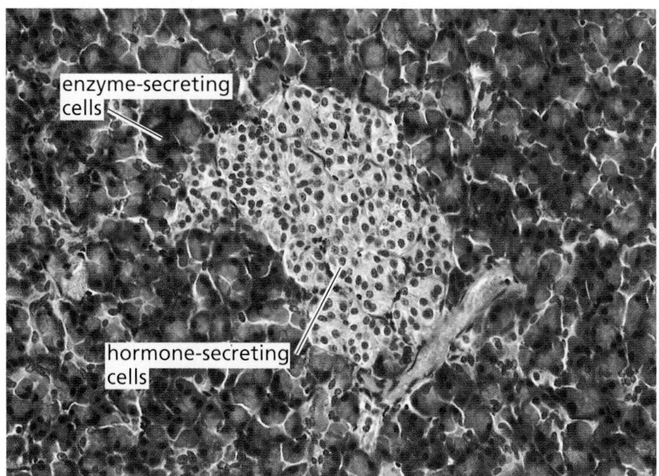

Figure 19.15 Section of pancreas tissue showing an islet (×250)

If the level of sugar in the blood falls, the islets release glucagon into the bloodstream. Glucagon acts on the cells in the liver and causes them to convert some of their stored glycogen into glucose and so restore the blood sugar level (p. 104).

Insulin has the opposite effect to glucagon. If the concentration of blood sugar increases (e.g. after a meal rich in carbohydrate), insulin is released from the islet cells. When the insulin reaches the liver it stimulates the liver cells to take up glucose from the blood and store it as glycogen.

Insulin has many other effects; it increases the uptake of glucose in all cells; it promotes the conversion of carbohydrates to fats and slows down the conversion of protein to carbohydrate.

All these changes have the effect of regulating the level of glucose in the blood to within narrow limits—a very important example of homeostasis (p. 135).

If anything goes wrong with the production or function of insulin, the person will show the symptoms of **diabetes**.

Diabetes

This may result from a failure of the islet cells to produce sufficient insulin or from the reduced ability of the body cells to use it. Two forms of diabetes are recognized, juvenile-onset and adult-onset diabetes.

Juvenile-onset diabetes is the less common form. It mainly affects young people and results from the islets producing too little insulin. There is a slight inherited tendency towards the disease, but it may be triggered by some event, possibly a virus infection, which causes the body's immune system to attack the islet cells which produce insulin. It is therefore classed as an **autoimmune** disease. The outcome is that the patient's blood is deficient in insulin and he or she needs regular injections of the hormone in order to control blood sugar level and so lead a normal life. This form of the disease is, therefore, sometimes called 'insulin-dependent' diabetes.

Adult-onset diabetes usually affects people after the age of 40. The level of insulin in their blood is often not particularly low but it seems that their bodies are unable to use the insulin properly. This condition can be controlled by careful regulation of the diet and does not usually require insulin injections.

In both forms of diabetes, the patient is unable to regulate the level of glucose in the blood. It may rise to such a high level that it is excreted in the urine or fall so low that the brain cells cannot work properly and the person goes into a coma.

All diabetics need a carefully regulated diet to keep the blood sugar within reasonable limits.

Reproductive organs

These produce hormones as well as gametes (sperms and ova) and their effects have been described on p. 148.

The hormones from the ovary, **oestrogen** and **progesterone**, both prepare the uterus for the implantation of the embryo, by making its lining thicker and increasing its blood supply.

The hormones **testosterone** (from the testes) and oestrogen (from the ovaries) play a part in the development of the secondary sexual characteristics (p. 148).

During pregnancy, the placenta produces a hormone which has effects similar to those of progesterone.

Pituitary gland

This gland is attached to the base of the brain (Figure 19.12d). It produces many hormones. One of these (**antidiuretic hormone**, ADH) acts on the kidneys and regulates the amount of water reabsorbed in the kidney tubules (p. 134). Another pituitary hormone (**growth hormone**) affects the growth rate of the body as a whole and the skeleton in particular. Several of the pituitary hormones act on the other endocrine glands and stimulate them to produce their own hormones. For example, the pituitary releases into the blood a **follicle-stimulating hormone** (FSH) which, when it reaches the ovaries, makes one of the follicles start to mature and to produce oestrogen. **Luteinizing hormone** (LH) is also produced from the pituitary and, together with FSH, induces ovulation.

A **thyroid-stimulating hormone** (TSH) acts on the thyroid gland and makes it produce thyroxine.

Homeostasis and feedback

Homeostasis

The endocrine system is important for maintaining the composition of the body fluids ('Homeostasis', p. 135).

A rise in blood sugar after a meal stimulates the pancreas to produce insulin. The insulin causes the liver to remove the extra glucose from the blood and store it as glycogen (p. 104). This helps to keep the concentration of blood sugar within narrow limits.

The brain monitors the concentration of the blood passing through it. If the concentration is too high, the pituitary gland releases ADH (antidiuretic hormone). When this reaches the kidneys (the target organs) it causes them to reabsorb more water from the blood passing through them (p. 134). If the blood is too dilute, production of ADH is suppressed and less water is absorbed in the kidneys. Thus ADH helps to maintain the amount of water in the blood at a fairly constant level.

Feedback

Some of the endocrine glands are themselves controlled by hormones. For example, pituitary hormones such as LH (luteinizing hormone) affect the endocrine functions of the ovaries. In some cases, the output of hormones is regulated by a process of **negative feedback**. (See also negative feedback in temperature control, p. 139.)

The feedback between the pituitary and the ovaries produces fluctuations which cause the menstrual cycle (Figure 19.16). When the level of oestrogen in the blood rises, it affects the pituitary gland, suppressing its production of FSH. A low level of FSH in the blood reaching the ovary will cause the ovary to slow down its production of oestrogen. With less oestrogen in the blood, the pituitary is able to resume its production of FSH which, in turn, makes the ovary start to produce oestrogen again. This cycle of events takes about a month and is the basis of the monthly menstrual cycle (p. 148).

The oestrogen and progesterone in the female contraceptive pill act on the pituitary and suppress the production of FSH. If there is not enough FSH, none of the follicles in the ovary will grow to maturity and so no ovum will be released.

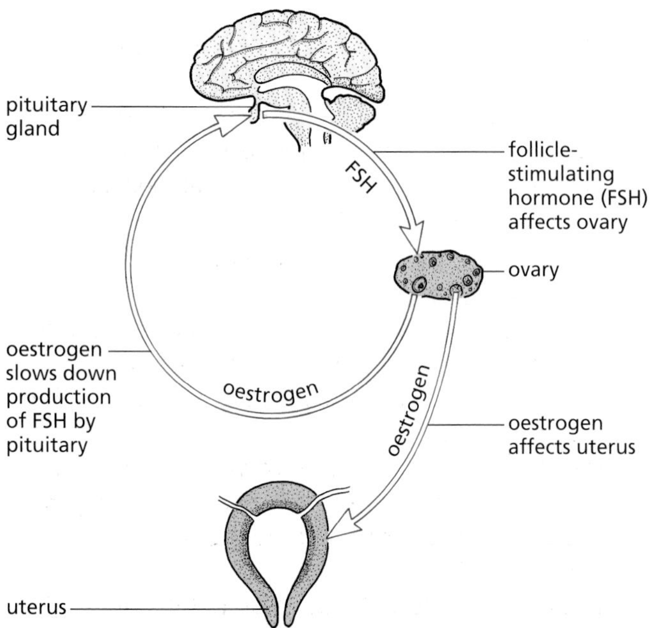

Figure 19.16 Feedback

Questions

1 Study Table 19.2 and give one example for each point of comparison.

2 The pancreas has a dual function in producing digestive enzymes as well as hormones. Which other endocrine glands have a dual function and what are their other functions? (See also p. 148.)

3 What are the effects on body functions of
a too much insulin,
b too little insulin?

4 Why do you think urine tests are carried out to see if a woman is pregnant?

Table 19.2 Endocrine and nervous control compared

Endocrine	Nervous
transmission of chemicals	transmission of electrical impulses
transmission via blood	transmission in nerves
slow transmission	rapid transmission
hormones dispersed throughout body	impulse sent directly to target organ
long-term effects	short-lived effects

Therapeutic and industrial uses of hormones

In humans, hormones can be used to treat various conditions. A pituitary growth hormone has been used successfully to promote growth in abnormally small children; oestrogen is used to prevent the thinning of bones (osteoporosis) in women after menopause (hormone replacement therapy or HRT).

LH and FSH help to cure certain cases of infertility; progesterone and oestrogen form the basis of the contraceptive pill. Injection of insulin is effective in treating juvenile-onset diabetes.

All these forms of hormone therapy have their drawbacks (side effects). HRT may increase the risk of cancer; LH may lead to multiple births. Some treatments, such as insulin injection, have been tried, tested and refined over many years; others are still in the early stages of development and may eventually be superseded by safer methods of treatment.

In agriculture, hormones may be used to increase productivity. Anabolic steroids (p. 20) are given to beef cattle to increase their meat production; bovine somatotrophin (BST) given to dairy cows can increase their milk yields by up to 25 per cent. There is still a good deal of controversy over the use of these hormones and BST is not licensed for use in Britain. It is claimed that the hormone makes cows more susceptible to disease and there are worries about the levels of the hormone or its residues in the milk.

The American Food and Drug Administration stated in February 1994 that milk produced with the aid of BST was indistinguishable from milk produced without it. However, the West does not really need more milk so the use of the hormone is related to profitability (for large-scale dairy farms and drug companies) rather than to food production.

Performance-enhancing hormones

In the last 20 years or so, some atheletes and sport persons have made use of drugs to boost their performance. Some of these drugs are synthetic forms of hormones.

Anabolic steroids are synthetic derivatives of testosterone. They affect protein metabolism, increasing muscle development and reducing body fat. Athletic performance is thus enhanced. There are serious long-term effects of taking anabolic steroids. The list is a long one but the main effects are sterility, masculinization in women, and liver and kidney malfunction.

Human growth hormone also increases muscle and reduces fat. Its side-effects are not fully known but it is a banned substance.

Erythropoietin (**EPO**), a hormone produced mainly in the kidneys, stimulates the production of red blood cells by the bone marrow. The additional red cells enhance the carriage of oxygen to the tissues and improve performance, particularly in endurance sports requiring stamina. The danger is that increasing the numbers of red cells in the circulation makes the blood more viscous. This, in turn, increases the risk of strokes and heart attacks.

Because these drugs enhance performance beyond what could be achieved by normal training, they are deemed unfair and banned by most sports organizations (Figure 19.17). Anabolic steroids are universally banned but different sports regulatory bodies have different rules for other substances.

The products of the steroid hormones can be detected in the urine and this is the basis of most tests for banned substances. Without these regulations, sport would become a competition between synthetic chemical substances rather than between individuals and teams.

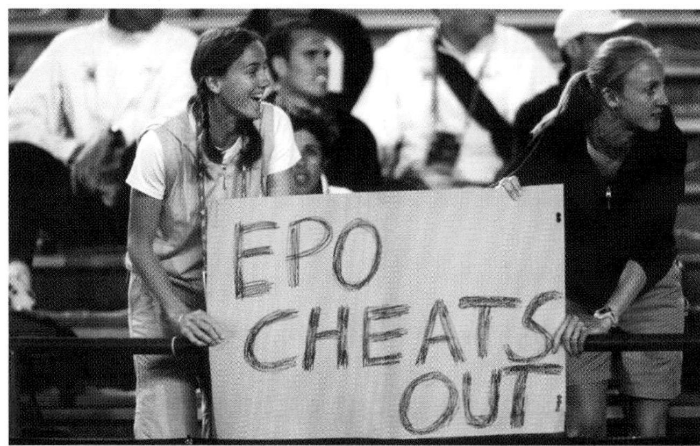

Figure 19.17 Athletes protest at the use of EPO

Checklist

- The body systems are made to work efficiently together by the nervous system and the endocrine system.

Nervous system

- The nervous system consists of the brain, the spinal cord and the nerves.
- The nerves consist of bundles of nerve fibres.
- Each nerve fibre is a thin filament which grows out of a nerve cell body.
- The nerve cell bodies are mostly in the brain and spinal cord.
- Nerve fibres carry electrical impulses from sense organs to the brain or from the brain to muscles and glands.
- A reflex is an automatic nervous reaction that cannot be consciously controlled.
- A reflex arc is the nervous pathway which carries the impulses causing a reflex action.
- The simplest reflex involves a sensory nerve cell and a motor nerve cell, connected by synapses in the spinal cord.
- The brain and spinal cord contain millions of nerve cells.
- The millions of possible connections between the nerve cells in the brain allow complicated actions, learning, memory and intelligence.

Endocrine system

- The thyroid, adrenals and pituitary are all endocrine glands.
- The testes, ovaries and pancreas are also endocrine glands in addition to their other functions.
- The endocrine glands release hormones into the blood system.
- When the hormones reach certain organs they change the rate or kind of activity of the organ.
- Too much or too little of a hormone can cause a metabolic disorder.
- Under-production of insulin or inability to use it causes diabetes.

20 Personal health

Diet
Dental health
Dental decay. Gum disease. Fluoridation of water supplies.
Smoking
Exercise
Mood-influencing drugs
Tolerance and dependence. Stimulants; depressants. Alcohol. Hallucinogens. Narcotics.
Solvent misuse

Health is not merely the absence of disease but a state of physical and mental well-being. However, you are not usually aware of your state of health until you are unwell. Therefore you may need to make a conscious effort to maintain good health.

The general rules for health are:

Do eat a balanced diet
take regular exercise
develop a positive attitude to life

Don't smoke
drink too much
misuse drugs.

If you feel this recipe for health is 'boring', ask yourself whether being overweight, having bronchitis, craving drugs and alcohol, and looking forward to premature heart failure are likely to make you happy and content.

You may know middle-aged or elderly people who claim to have smoked, got drunk, over-eaten and taken no exercise all their lives, and still remain healthy. You will also have heard of fit people who exercise regularly and do not smoke but die suddenly from a heart attack.

These stories prove nothing. We inherit part of our 'constitution' from our parents. So some people are more able than others to resist disease and to ill-use their bodies. You can do nothing about your genes, but you can do a lot to make the best of the constitution you have inherited.

You can find physiological reasons for some of the 'dos and don'ts' listed in the preceding chapters (for diet see pp. 90–1, smoking pp. 127–9 and exercise p. 156). This chapter will refer to them only briefly. In some cases, such as smoking and drug abuse, there is overwhelming evidence to support the advice. In others, such as diet and exercise, the evidence is less strong, but the advice given represents most expert opinion.

Diet

To remain healthy, you must eat enough food to meet all your energy requirements. It must contain protein and fat to make new cells. You also need vitamins and salts (see pp. 87–9).

A healthy diet should contain only a little salt and refined sugar (if any) and a low proportion of fat. The diet also needs to include plenty of vegetable fibre (Figure 20.1). A low fat intake may help reduce the chances of arterial and heart disease (p. 120). A high fibre intake probably helps to prevent diseases of the large intestine.

In some people, over-eating can lead to obesity with its attendant problems (such as high blood pressure and diabetes).

A healthy digestive system also depends on regular meals eaten in a relaxed atmosphere. Hasty snacks and irregular heavy meals can lead to digestive disorders.

The health aspects of diet are discussed more fully on pp. 90–2.

Figure 20.1 Some components of a healthy diet

Dental health

Sugary food and the neglect of oral hygiene can lead to toothache, gum disease and, ultimately, the loss of the teeth and the need to wear dentures.

Dental decay (dental caries)

Decay begins when small holes (cavities) appear in the enamel. The cavities are caused by bacteria on the tooth surface. The bacteria produce acids which dissolve the calcium salts in the tooth enamel. The enamel and dentine are dissolved away in patches, forming cavities. The cavities reduce the distance between the outside of the tooth and the nerve endings. The acids produced by the bacteria irritate the nerve endings and cause toothache. If the cavity is not cleaned and filled by a dentist, the bacteria will get into the pulp cavity and cause a painful abscess at the root. Often, the only way to treat this is to have the tooth pulled out.

Although some people's teeth are more resistant to decay than others, it seems that it is the presence of refined sugar (sucrose) that contributes to decay.

Western diets contain a good deal of refined sugar and children suck sweets between one meal and the next. The high level of dental decay in Western society is thought to be caused mainly by keeping sugar in the mouth for long periods of time.

The graph in Figure 20.2a shows how the pH in the mouth falls (i.e. becomes more acid) when a single sweet is sucked. The pH below which the enamel is attacked is called the **critical pH** (between 5.5 and 6). In this case, the enamel is under acid attack for about 10 minutes.

The graph in Figure 20.2b shows the effect of sucking sweets at the rate of four an hour. In this case the teeth are exposed to acid attack almost continually.

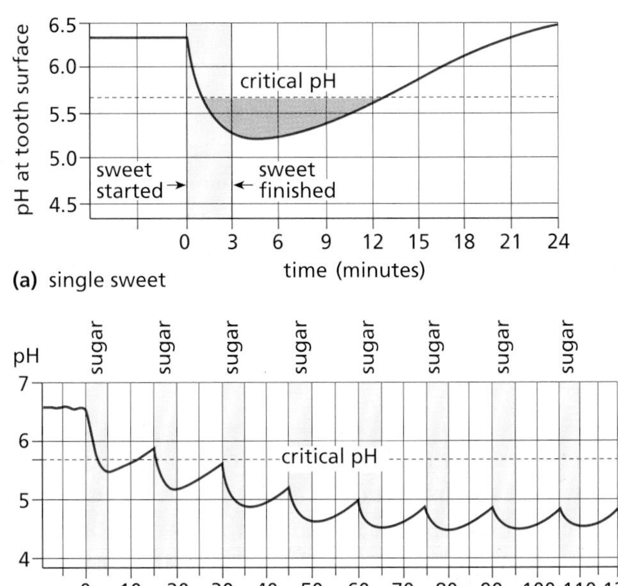

Figure 20.2 pH in the mouth when sweets are sucked

The best way to prevent tooth decay, therefore, is to avoid eating sugar at frequent intervals either in the form of sweets or in sweet drinks such as orange squash or cola drinks.

It is advisable also to visit the dentist every 6 months or so for a 'check-up' so that any caries or gum disease can be treated at an early stage.

Brushing the teeth is very important in the prevention of gum disease. It may not be so effective in preventing caries although the use of fluoride toothpaste does help to reduce the bacterial population on the teeth and to increase their resistance to decay (see below).

Gum disease (periodontal disease)

There is usually a layer of saliva and mucus over the teeth. This layer contains bacteria which live on the food residues in the mouth, building up a coating on the teeth, called **plaque**. If the plaque is not removed, mineral salts of calcium and magnesium are deposited on it, forming a hard layer of 'tartar' or **calculus**. If the bacterial plaque which forms on teeth is not removed regularly, it spreads down the tooth into the narrow gap between the gum and enamel. Here it causes inflammation, called **gingivitis**, which leads to redness and bleeding of the gums and to bad breath. It also causes the gums to recede and expose the cement. If gingivitis is not treated, it progresses to **periodontitis**; the fibres holding the tooth in the jaw are destroyed, so the tooth becomes loose and falls out or has to be pulled out.

There is evidence that cleaning the teeth does help to prevent gum disease. It is best to clean the teeth about twice a day using a toothbrush. No one method of cleaning has proved to be any better than any other but the cleaning should attempt to remove all the plaque from the narrow crevice between the gums and the teeth.

Drawing a waxed thread ('dental floss') between the teeth helps to remove plaque in these regions.

Fluoridation of water supplies

It has been known for many years that the presence of fluoride ions in drinking water reduces the incidence of dental decay by 30 per cent or more, particularly in children. The fluoride salts, which reach the teeth in the blood supply, are built into the enamel of the teeth and increase resistance to bacterial attack. Fluoride reaching the surface of the teeth has a similar effect and also reduces the bacterial population in the plaque.

To benefit from this effect, local authorities are permitted to add sodium fluoride to drinking water at a concentration of about 1 part per million (1 ppm). In regions where this has been done there has been a reducion in dental caries of up to 50 per cent. In the USA, 60 per cent of the population now receives fluoridated water. In Britain the figure is about 10 per cent.

After 50 years of intensive studies, there is no scientific evidence that fluoride at 1 ppm is harmful. Claims of increases in bone cancer, or bone fragility and many other hazards, have been shown to be unfounded. The one disadvantage is the small possibility of dental fluorosis in which the teeth develop pearly white patches. This occurs when the fluoride intake exceeds the 1 ppm level. For example, it may happen if children swallow fluoride toothpaste in regions where the water already has adequate fluoride. In the few cases where fluorosis occurs, the condition can often be observed only by expert dental inspection.

Opponents of fluoridation feel that it is a form of 'mass medication', a health measure forced on the population to achieve medical benefits rather than just making the water safer to drink, as is the case with chlorination (p. 340). They would say that parents should be free to decide whether to supplement their children's fluoride intake by using fluoride toothpaste or fluoride tablets. They would point out also that, by adopting good dietary and dental practices, caries can be reduced without resort to fluoridation

Smoking

There is a long list of diseases associated with smoking, including lung cancer, bronchitis, emphysema, arterial disease, stomach ulcers and bladder cancer. This does not mean that all smokers will develop these diseases. But the chances of their doing so are far higher than they are for non-smokers.

See pp. 127–9 for more detail.

Exercise

Exercise increases stamina, improves flexibility, makes muscles stronger and more efficient, and helps to keep your weight down (Figure 20.3). Exercise may also reduce your chance of a premature coronary heart attack, though the evidence is not clear-cut.

Figure 20.3 Exercise. Regular exercise contributes to good health

Most people agree that exercise makes you 'feel good'. But to enjoy the benefits, you need to take exercise throughout your life.

Exercise is discussed in more detail on p. 156.

Questions

1 List the features of your diet over the last 2 days which might be considered to make it
 a healthy,
 b unhealthy.

2 What are the most important things to do to avoid dental decay and gum disease?

Mood-influencing drugs

Any substance used in medicine to help our bodies fight illness or disease is called a **drug**. One group of drugs helps to control pain and ease feelings of distress. These are the mood-influencing drugs.

Sensations of excitement, well-being or depression are the products of our nervous and endocrine systems. The mood-influencing drugs mimic the effects of our hormones or neurotransmitters (p. 165) and produce corresponding states of mind.

If these drugs are used wisely and under medical supervision, they can be very helpful. A person who feels depressed to the point of wanting to commit suicide may be able to lead a normal life with the aid of an antidepressant drug which removes the feeling of depression. However, if drugs are used for trivial reasons, to produce feelings of excitement or calm, they may be extremely dangerous. This is because they can cause **tolerance** and **dependence**.

Tolerance

This means that if the substance is taken over a long period, the dosage has to keep increasing in order to have the same effect. People who take sleeping pills containing barbiturates may need to increase the dose from one to two or three tablets in order to get to sleep. People who drink alcohol in order to relieve anxiety may find that they have to keep drinking more and more before they feel relaxed. If the dosage continues to increase it will become so large that it causes death.

Dependence

This is the term used to describe the condition in which the user cannot do without the substance.

Sometimes a distinction is made between emotional and physical dependence, though the distinction is not always clear. A person with emotional dependence may feel a craving for the substance, may be bad-tempered, anxious or depressed without it, and may commit crimes in order to obtain it. Cigarette smoking is one example of emotional dependence.

Physical dependence involves the same experiences. But there are also physical symptoms, called **withdrawal symptoms**, when the substance is withheld. These may be nausea, vomiting, diarrhoea, muscular pain, uncontrollable shaking and hallucinations. Physical dependence is sometimes called **addiction**.

Not everyone who takes a mood-influencing drug develops tolerance or becomes dependent on it. Millions of people can take alcoholic drinks in moderation with no obvious physical or mental damage. Those who become dependent cannot drink in moderation. Their bodies seem to develop a need for permanently high levels of alcohol and dependent people (**alcoholics**) get withdrawal symptoms if they do not drink.

Physical and emotional dependence are very distressing states. Getting hold of the substance becomes the centre of the addicts' lives, and they lose interest in their personal appearance, their jobs and their families. Because the substances they need cannot be obtained legally or because they need the money to buy them, they turn to crime. Cures are slow, difficult and usually unpleasant.

There is no way of telling in advance which person will become dependent and which will not. Dependence is much more likely with some drugs than with others, however. These are, therefore, prescribed with great caution. Experimenting with drugs for the sake of emotional excitement is extremely unwise. Some people may become dependent on almost any substance which gives them a conscious sensation.

Some of the mood-influencing drugs will now be considered.

Stimulants

The **caffeine** in coffee, tea and cocoa is a mild stimulant and makes you more wakeful. In normal use there is no build-up of tolerance or dependence.

Amphetamines reduce fatigue and increase alertness but they also reduce accuracy and give a false sense of confidence. If taken to improve athletic performance they can cause dangerously high blood pressure. Also they are quite useless for helping examination candidates because although they increase confidence, they also reduce accuracy. Use of amphetamines can lead to tolerance and addiction.

MDMA (Ecstasy) is related to the amphetamines. It produces feelings of well-being and sociability and raises blood pressure and heart rate. Its use may be followed by a period of depression. It is thought to have caused about 70 deaths in the last 10 years, probably from overheating (heat stroke) and dehydration. Its long-term effects are not known for certain but it does temporarily damage the nerve fibres of certain brain cells.

Cocaine gives a temporary feeling of excitement but this is soon followed by depression and listlessness. It is very addictive and prolonged use constricts the arteries and causes mental disorders.

Depressants

Depressants, e.g. sedatives, act on the central nervous system to decrease emotional tension and anxiety. Different types of sedative probably affect different areas of the brain but they all lead to relaxation and, in sufficient doses, to sleep or anaesthesia. In excessive doses, they suppress the breathing centre of the brain and cause death.

Barbiturates

Barbiturates, such as phenobarbitone, are powerful sedatives, but in some cases they can be harmful or lead to addiction. As sedatives and sleeping pills they have largely been replaced by the safer benzodiazepine tranquillizers.

Tranquillizers

Some tranquillizers have been extremely valuable in treating severe mental illnesses such as schizophrenia and mania. Many thousands of mentally ill patients have been able to leave hospital and live normal lives as a result of using the tranquillizing drug **chlorpromazine**.

Nowadays, tranquillizers such as **diazepam** (Valium) and **nitrazepam** (Mogadon) are being prescribed in their millions for the relief of anxiety and tension. Some people think that these drugs are being used merely to escape the stresses of everyday life that could be overcome by a little more will-power and determination. Others think that there is no reason why people should suffer the distress of acute anxiety when drugs are available for its relief. On the other hand, some degree of anxiety is probably needed for mental and physical activity. These activities are unlikely to be very effective in people who tranquillize themselves every time a problem crops up. In some cases, use of certain tranquillizers has led to addiction.

Alcohol

The alcohol in wines, beer and spirits is a depressant of the central nervous system. Small amounts give a sense of well-being, with a release from anxiety. However, this is accompanied by a fall-off in performance in any activity requiring skill. It also gives a misleading sense of confidence in spite of the fact that one's judgement is clouded. The drunken driver usually thinks he or she is driving extremely well.

Even a small amount of alcohol in the blood increases our reaction time (the interval between receiving a stimulus and making a response). In some people, the reaction time is doubled even when the alcohol in the blood is well below the legal limit laid down for car drivers (Figure 20.4, overleaf). This can make a big difference in the time needed for a driver to apply the brakes after seeing a hazard such as a child running into the road.

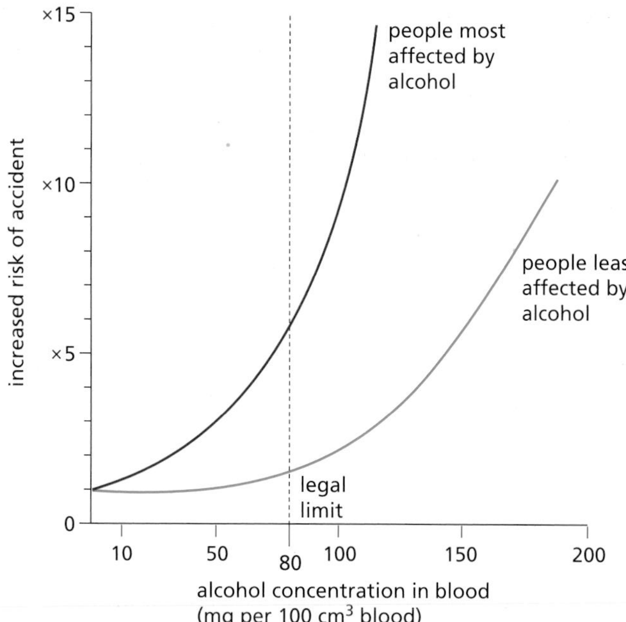

Figure 20.4 Increased risk of accidents after drinking alcohol. People vary in their reactions to alcohol. Body weight, for example, makes a difference

Alcohol causes vasodilation in the skin, giving a sensation of warmth but in fact leading to a greater loss of body heat (see p. 138). A concentration of 500 mg of alcohol in 100 cm^3 of blood results in unconsciousness. More than this will cause death because it stops the breathing centre in the brain. Ninety per cent of alcohol taken in is **detoxified** in the liver. That is, it is oxidized to carbon dioxide and water. Only 10 per cent is excreted by the kidneys. On average, the liver can oxidize about 75 mg alcohol per 1 kg body weight per hour. This rate varies considerably from one individual to the next but it indicates that it would take about 3 hours to oxidize the alcohol in a pint of beer or a glass of wine. If the alcohol intake exceeds this rate of oxidation, the level of alcohol in the blood builds up to toxic proportions; that is, it leads to **intoxication**.

Some people build up a tolerance to alcohol and this may lead to both emotional and physical dependence (alcoholism). High doses of alcohol can cause the liver cells to form too many fat droplets, leading to the disease called **cirrhosis**. A cirrhotic liver is less able to stop poisonous substances in the intestinal blood from reaching the general circulation (p. 105).

Pregnancy

Alcohol can cross the placenta and damage the fetus. Pregnant women who take as little as one alcoholic drink a day are at risk of having babies with lower than average birth weights. These underweight babies are more likely to become ill.

Heavy drinking during pregnancy can lead to deformed babies. All levels of drinking are thought to increase the risk of miscarriage.

Behaviour

Alcohol reduces inhibitions, because it depresses that part of the brain which causes shyness. This may be considered an advantage in 'breaking the ice' at parties. But it can also lead to irresponsible behaviour such as vandalism and aggression.

Moderate drinking

A moderate intake of alcoholic drink seems to do little physiological harm (except in pregnant women). But what is a 'moderate' intake?

A variety of drinks which all contain the same amount of alcohol is shown in Figure 20.5. Beer is a fairly dilute form of alcohol. Whisky, however, is about 40 per cent alcohol. Even so, half a pint of beer contains the same amount of alcohol as a single whisky. This amount of alcohol can be called a 'unit'. It is the number of units of alcohol, not the type of drink, which has a physiological effect on the body. The Health Development Agency recommends upper limits of 21–28 units for men and 14–21 units for women over a 1-week period.

Figure 20.5 Alcohol content of drinks. All these drinks contain the same amount of alcohol (1 unit). Although the alcohol is more dilute in the beer than in the whisky, it has the same effect on the body

Hallucinogens

Cannabis

Cannabis and other extracts of Indian hemp are chewed or smoked to produce a sense of well-being, detachment and sometimes hallucinations. There does not seem to be much evidence of tolerance or emotional and physical dependence. But unstable individuals looking for more 'exciting' experiences are thought to be likely to move on from cannabis to the 'hard drugs' such as morphine and heroin.

It is difficult to conduct research into the effects of an illegal drug. Nevertheless, evidence from animal studies and human surveys is accumulating and shows that cannabis-smoking can have harmful effects on the lungs, central nervous system, immune system and reproductive function. A particular worry is the long time that the products persist in the body (days rather than hours) so that the user may be unaware of the impairment of performance, days after the last smoke.

Animal and human studies show that sperm count and sexual drive are diminished, the immune system is depressed, and learning ability and memory are impaired. These effects are in addition to the lung damage which is known to occur.

As in the case of tobacco, it may take many years to establish the full extent of the harmful side-effects of cannabis use.

Narcotics

Morphine, codeine and heroin are narcotics made from opium. Morphine and heroin relieve severe pain and produce short-lived feelings of well-being and freedom from anxiety. They can both lead to tolerance and physical dependence within weeks, and so they are prescribed with caution.

The illegal use of heroin has terrible effects on the unfortunate addict. The overwhelming dependence on the drug leads many addicts into prostitution and crime in order to obtain the money to buy it. There are severe withdrawal symptoms and a 'cure' is a long and often unsuccessful process.

Additional hazards are that blood poisoning, hepatitis and AIDS may result from the use of unsterilized needles when injecting the drug.

Codeine is a less effective analgesic than morphine, but does not lead to dependence so easily. It is still addictive if used in large enough doses.

Solvent misuse

Solvent misuse or 'glue-sniffing' is the inhaling of vapours from various organic solvents (not only glue) in order to become intoxicated. The vapours produce effects similar to drunkenness, but they are more intense and last for only a few minutes. They include dizziness and stupor, followed by a loss of co-ordination and control and eventually by unconsciousness.

The short-term after-effects are headache, nausea, vomiting and, in some cases, convulsions. After-effects which last for several weeks include runny nose, bloodshot eyes, an acne-like rash round the mouth, irritability, lethargy and depression. In the long term, the liver and kidneys may be damaged.

The number of deaths resulting from solvent misuse has been increasing steadily in the last few years. It is estimated that 46 people died in the United Kingdom in 1981, and 152 in 1990, since when it has declined to 70 (in 1998).* Death may result from (1) accidents while the person is intoxicated and out of control, (2) suffocation by the plastic bag used for inhaling, (3) choking on vomit while unconscious and (4) toxic effects of the solvent.

Solvent misusers are mostly adolescents of 14 or 15 years old or younger. They may experiment with 'glue-sniffing' as a form of rebellion against authority, to 'keep in' with their friends, to relieve boredom or for 'kicks' (risk-taking for the sake of a thrill). These reasons can, of course, be motives for other forms of drug misuse and destructive behaviour.

It is not clear whether solvent misuse leads to tolerance and dependence. Most 'glue-sniffers' do not continue once they are old enough to buy alcoholic drinks. But they are likely to drink too much, and to be tempted to move on to other drugs.

Questions

1 What is the difference between
   a becoming tolerant of a drug and
   b becoming dependent on a drug?
   Which of these do you think is meant by being 'hooked' on a drug?

2 Why are amphetamine stimulants unsuitable for improving performance in
   a athletics,
   b examinations?

3 Why should drinking alcohol make you 'feel' warm, but cause you to lose heat?

4 Why is it dangerous to take alcoholic drinks before driving?

5 If morphine and heroin make addicts 'feel good', why can't they keep taking steady low doses to stay in this 'happy' state?

Checklist

- A good diet and regular exercise contribute to good health.
- Smoking and excessive drinking contribute to ill-health.
- Mood-influencing drugs may be useful for treating certain illnesses but are dangerous if used for other purposes.
- Tolerance means that the body needs more and more of a particular drug to produce the same effect.
- Dependence means that a person cannot do without a particular drug.
- Withdrawal symptoms are unpleasant physical effects experienced by an addict when the drug is not taken.
- Alcohol is a depressant drug which slows down reaction time and reduces inhibitions.
- Alcohol in a pregnant woman's blood can damage her fetus.
- Heroin and morphine are strongly addictive drugs.
- Solvent sniffing produces intoxication and has caused an increasing number of deaths.

* Data from St George's Hospital Medical School, UK

Human physiology

Examination questions

Do not write on this page. Where necessary copy drawings, tables or sentences.

1 **a** The food we eat contains nutrients.
This box shows a nutrient and the chemical test for it.

Name of nutrient	Chemical test
fat	alcohol emulsion

Look at boxes **A** to **E**.

A	simple sugar	biuret
B	protein	biuret
C	starch	Benedict's solution
D	starch	iodine solution
E	simple sugar	Benedict's solution

The nutrient and the chemical test are correctly matched in only **three** of the boxes. Write down the letters of these three boxes. (3)

b Describe how you would do the chemical test using Benedict's solution. (3)

c (i) The material that jeans are made from contains starch. A student spilt some amylase solution on her jeans.
What type of substance is amylase? (1)

(ii) If the amylase solution was not washed off her jeans it would digest the starch.
What would the starch be turned into? (1)

(OCR)

2 Charles investigates the digestion of fats by the enzyme lipase in a test tube. He finds that when lipase digests fats, it changes the pH of the solution.

a Explain how the digestion of fats changes the pH of the solution. (2)

Charles sets up three test tubes containing various liquids.

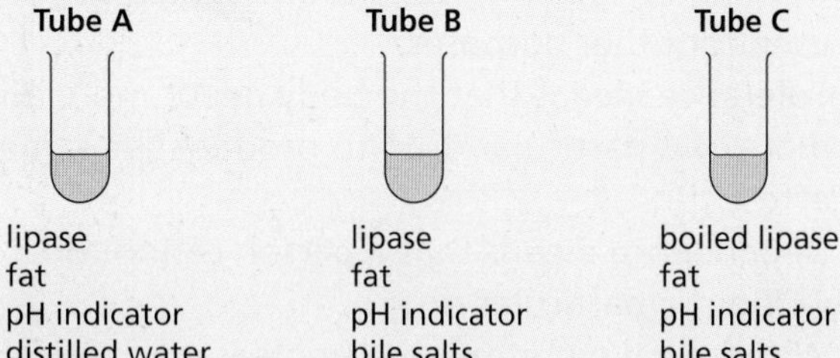

He times how long it takes for the indicator to change colour after the lipase is added.
The results are shown in the table.

Tube	Time taken in minutes
A	5
B	1
C	no change after 30 minutes

b Explain why the indicator in tube B changes colour much faster than the indicator in tube A. (3)

(OCR)

3 The drawing shows some of the structures in a human knee joint.

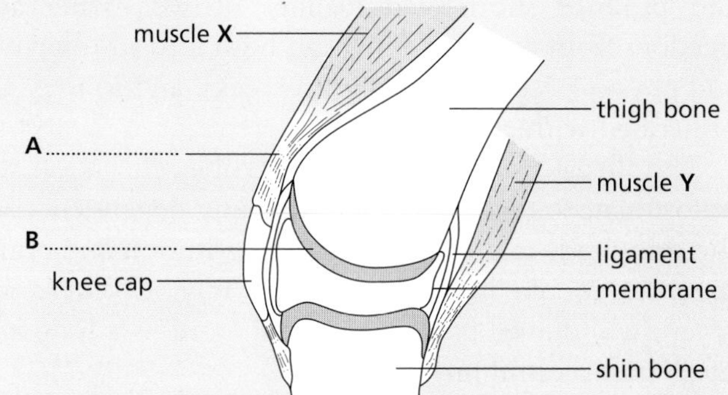

a Label the parts A and B on the drawing. (2)

b Explain how muscles X and Y straighten the leg at the knee joint. (2)

(AQA)

4 The drawing shows the front view of an eye in bright light.

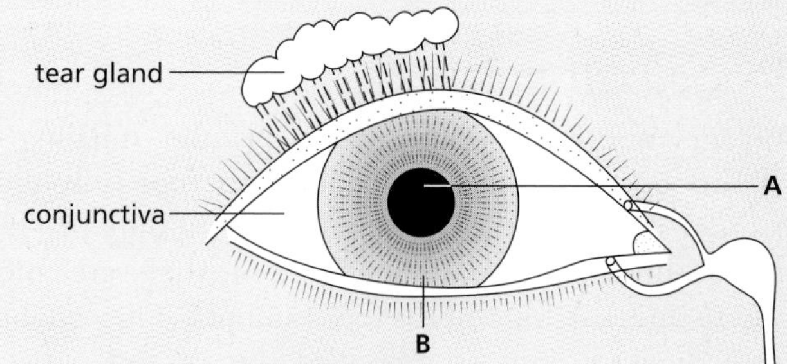

a Name the parts A and B. (2)

b Give **one** function of
(i) the tears, (ii) the conjunctiva. (2)

c Describe and explain how the appearance of the eye differs in dim light. (2)

(CCEA)

5 Use a word or phrase from the box to complete each sentence.
The first one has been done for you.

increases **decreases** **stays the same**

After injecting with a used needle, the chance of getting hepatitis *increases.*
After taking an antibiotic, the number of disease-causing micro-organisms in the body ______.
After taking heroin, the amount of pain ______ .
For regular smokers, the chance of getting lung cancer ______.
When a person is healthy, the number of white cells ______. (4)

(Edexcel)

Genetics and heredity

21 Cell division, chromosomes and genes

Heredity and genetics
Explanation of terms.

Chromosomes and mitosis
Mitosis: the movement of chromosomes at cell division. Function of chromosomes: genes on the chromosomes control the cell's physiology and structure. Number of chromosomes: a fixed number for each species.

Gamete production and chromosomes
Meiosis: the chromosomes are shared between the gametes. Meiosis and mitosis compared.

Genes
The structure of the gene. The role of DNA. The genetic code. Replication of DNA. Mutations.

Practical work
Observing chromosomes in plant cells.

Heredity and genetics

We often talk about people inheriting certain characteristics: 'John has inherited his father's curly hair', or 'Mary has inherited her mother's blue eyes'. We expect tall parents to have tall children. The inheritance of such characteristics is called **heredity** and the branch of biology which studies how heredity works is called **genetics**.

Genetics also tries to forecast what sort of offspring are likely to be produced when plants or animals reproduce sexually. What will be the eye colour of children whose mother has blue eyes and whose father has brown eyes? Will a mating between a black mouse and a white mouse produce grey mice, black-and-white mice or some black and some white mice?

To understand the method of inheritance, we need to look once again at the process of sexual reproduction and fertilization. In sexual reproduction, a new organism starts life as a single cell called a **zygote** (p. 140). This means that you started from a single cell. Although you were supplied with oxygen and food in the uterus, all your tissues and organs were produced by cell division from this one cell. So, the 'instructions' that dictated which cells were to become liver, or muscle, or bone must all have been present in this first cell. The 'instructions' which decided that you should be tall or short, dark or fair, male or female must also have been present in the zygote.

To understand how these 'instructions' are passed from cell to cell, we need to look in more detail at what happens when the zygote divides and produces an organism consisting of thousands of cells. This type of cell division is called **mitosis**. It does not take place only in a zygote but occurs in all growing tissues.

Question

1 **a** What are gametes? What are the male and female gametes of
(i) plants and
(ii) animals called, and where are they produced?
b What happens at fertilization?
c What is a zygote and what does it develop into?

See pp.71 and 140.

Chromosomes and mitosis

Mitosis

When a cell is not dividing, there is not much detailed structure to be seen in the nucleus even if it is treated with special dyes called stains. Just before cell division, however, a number of long, thread-like structures appear in the nucleus and show up very clearly when the nucleus is stained (Figures 21.1a and 21.2). These thread-like structures are called **chromosomes**. Although they are present in the nucleus all the time, they show up clearly only at cell division because at this time they get shorter and thicker.

Each chromosome is seen to be made up of two parallel strands, called **chromatids**. When the nucleus divides into two, one chromatid from each chromosome goes into each daughter nucleus. The chromatids in each nucleus now become chromosomes and later they will make copies of themselves ready for the next cell division. The process of copying is called **replication** because each chromosome makes a replica (an exact copy) of itself. As Figure 21.1 is a simplified diagram of mitosis, only two chromosomes are shown, but there are always more than this. Human cells contain 46 chromosomes.

Mitosis will be taking place in any part of a plant or animal which is producing new cells for growth or replacement. Bone marrow produces new blood cells by mitosis; the epidermal cells of the skin are replaced by mitotic divisions in the basal layer; new epithelial cells lining the alimentary canal are produced by mitosis; growth of muscle or bone in animals, and root, leaf, stem or fruit in plants, results from mitotic cell divisions.

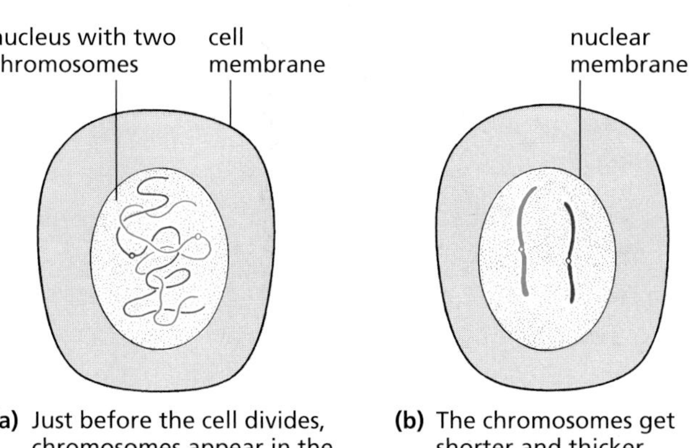

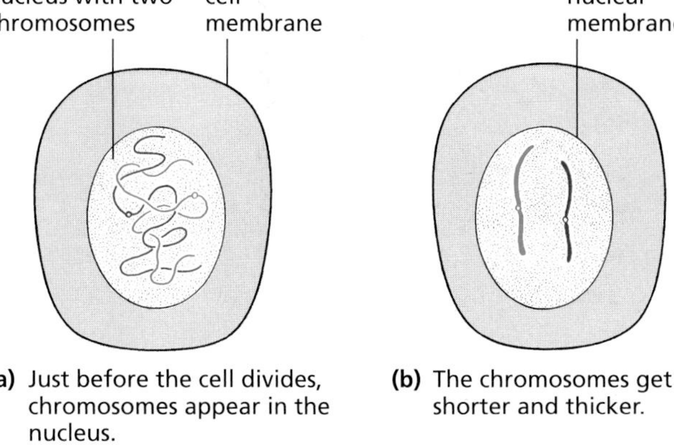

(a) Just before the cell divides, chromosomes appear in the nucleus.

(b) The chromosomes get shorter and thicker.

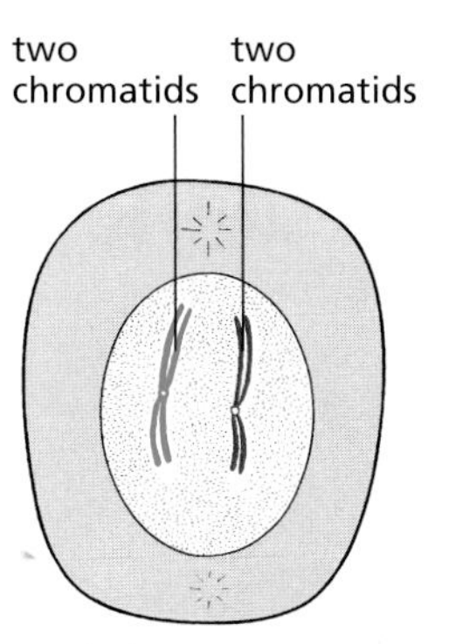

(c) Each chromosome is now seen to consist of two chromatids.

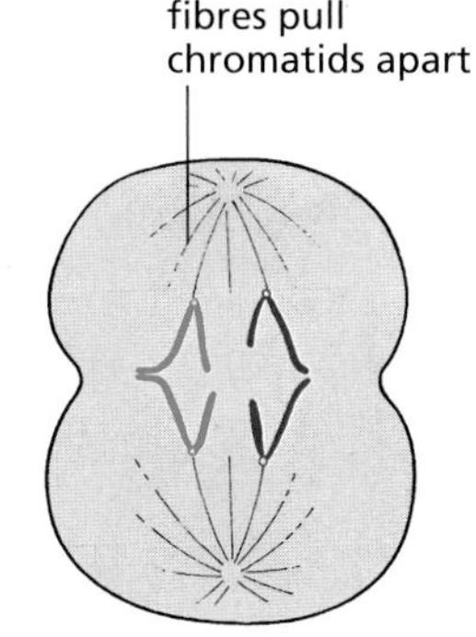

(d) The nuclear membrane disappears and the chromatids are pulled apart to opposite ends of the cell.

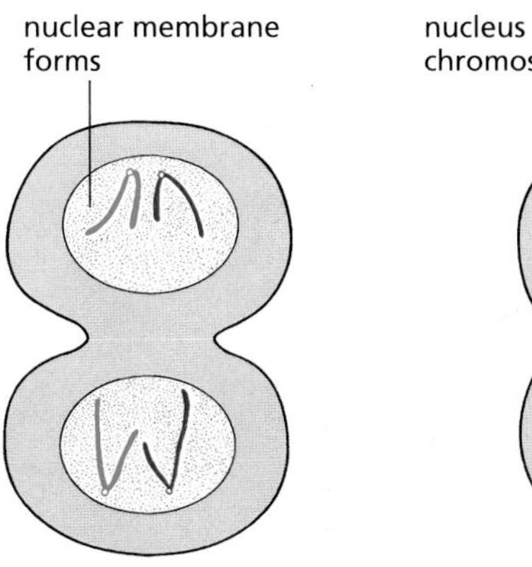

(e) A nuclear membrane forms round each set of chromatids, and the cell starts to divide.

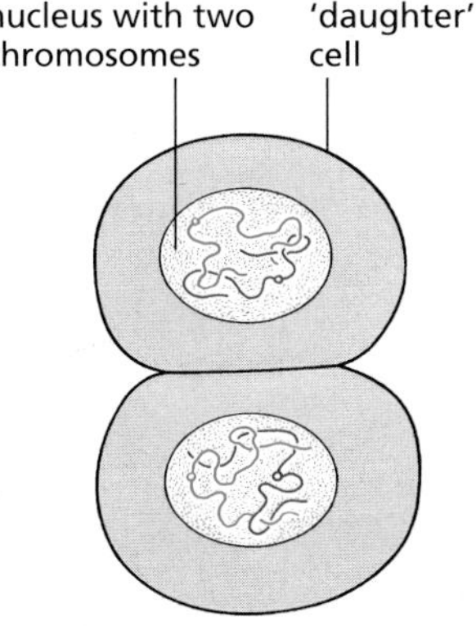

(f) Cell division completed, giving two 'daughter' cells, each containing the same number of chromosomes as the parent cell.

Figure 21.1 Mitosis. Only one pair of chromosomes is shown. Three of the stages described here are shown in Figure 21.2

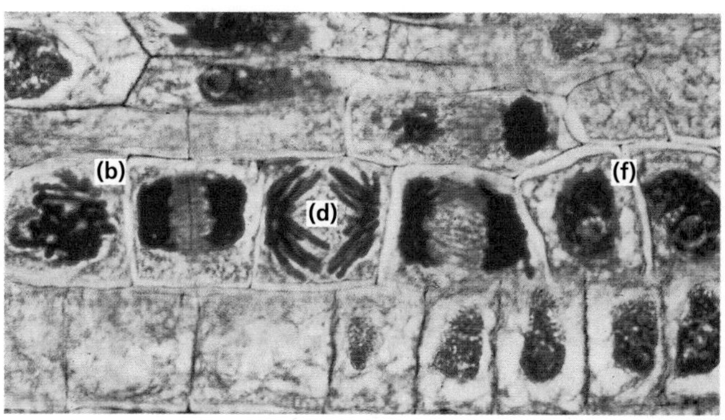

Figure 21.2 Mitosis in a root tip (× 500). The letters refer to the stages described in Figure 21.1. (The tissue has been squashed to separate the cells.)

An exception to this occurs in the final stages of gamete production in the reproductive organs of plants and animals. The cell divisions which give rise to gametes are not mitotic but meiotic, as explained on p. 185.

Cells which are not involved in the production of gametes are called **somatic cells**. Mitosis takes place only in somatic cells.

Questions

1 In the nucleus of a human cell just before cell division, how many chromatids will there be?

2 Why can chromosomes not be seen when a cell is not dividing?

3 Look at Figure 15.1 on p. 137. Where would you expect mitosis to be occurring most often?

4 In which human tissues would you expect mitosis to be going on in
a a 5-year-old child, **b** an adult?

The function of chromosomes

When a cell is not dividing, its chromosomes become very long and thin. Along the length of the chromosome is a series of chemical structures called **genes** (Figure 21.3, overleaf). The chemical which forms the genes is called DNA (which is short for deoxyribonucleic acid, p. 186). Each gene controls some part of the chemistry of the cell. It is these genes which provide the 'instructions' mentioned at the beginning of the chapter. For example, one gene may 'instruct' the cell to make the pigment which is formed in the iris of brown eyes. On one chromosome there will be a gene which causes the cells of the stomach to make the enzyme pepsin. When the chromosome replicates, it builds up an exact replica of itself, gene by gene (Figure 21.4, overleaf). When the chromatids separate at mitosis, each cell will receive a full set of genes. In this way, the chemical instructions in the zygote are passed on to all cells of the body. All the chromosomes, all the genes and, therefore, all the 'instructions' are faithfully reproduced by mitosis and passed on complete to all the cells.

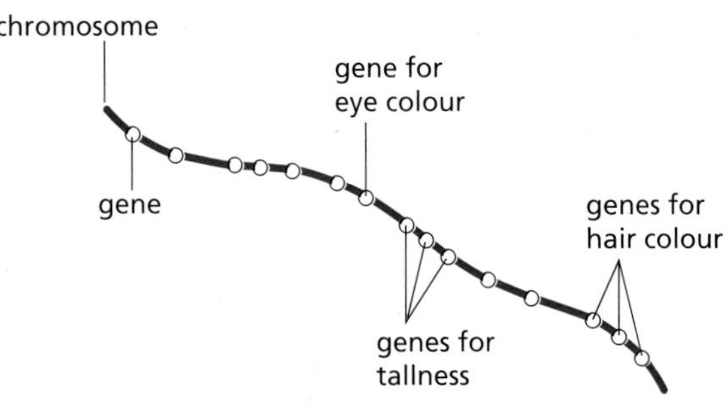

Figure 21.3 Relationship between chromosomes and genes. The drawing does not represent real genes or a real chromosome. There are probably thousands of genes on a chromosome

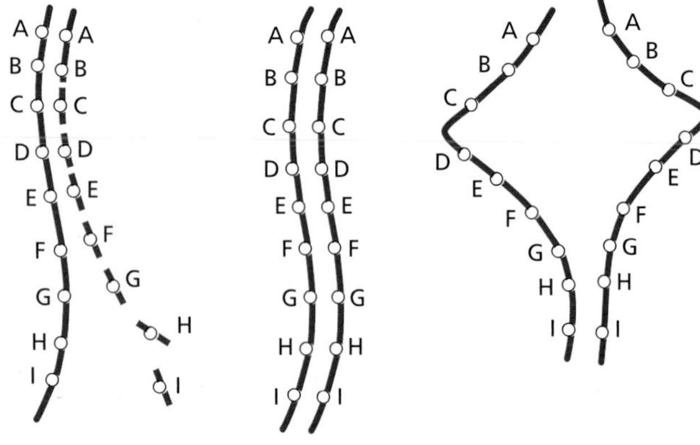

(a) A chromosome builds up a replica of itself.

(b) When the cell divides, the original and the replica are called chromatids.

(c) Mitosis separates the chromatids. Each new cell gets a full set of genes.

Figure 21.4 Replication. (A, B, C, etc. represent genes.)

Which of the 'instructions' are used depends on where a cell finally ends up. The gene which causes brown eyes will have no effect in a stomach cell and the gene for making pepsin will not function in the cells of the eye. So a gene's chemical instructions are carried out only in the correct situation.

The genes which produce a specific effect in a cell (or whole organism) are said to be **expressed**. In the stomach lining, the gene for pepsin is expressed. The gene for melanin (the pigment in brown eyes) is not expressed.

Number of chromosomes

- There is a fixed number of chromosomes in each species. Human body cells each contain 46 chromosomes, mouse cells contain 40, and garden pea cells 14 (see also Figure 21.5).
- The number of chromosomes in a species is the same in all of its body cells. There are 46 chromosomes in each of your liver cells, in every nerve cell, skin cell and so on.
- The chromosomes have different shapes and sizes and can be recognized by a trained observer.
- The chromosomes are always in pairs (Figure 21.5), e.g. two long ones, two short ones, two medium ones. This is because when the zygote is formed, one of each pair comes from the male gamete and one from the female gamete. Your 46 chromosomes consist of 23 from your mother and 23 from your father.
- The number of chromosomes in each body cell of a plant or animal is called the **diploid number**. Because the chromosomes are in pairs, it is always an even number.

The chromosomes of each pair are called **homologous** chromosomes. In Figure 21.7b, the two long chromosomes form one homologous pair and the two short chromosomes form another.

kangaroo (12)

human (46)

domestic fowl (36)

fruit fly (8)

Figure 21.5 Chromosomes of different species. Note that the chromosomes are always in pairs

Questions

1 How many chromosomes would there be in the nucleus of
  a a human muscle cell,
  b a mouse kidney cell,
  c a human skin cell that has just been produced by mitosis?

2 What is the diploid number in humans?

Gamete production and chromosomes

The genes on the chromosomes carry the 'instructions' which turn a single-cell zygote into a bird, or a rabbit or an oak tree. The zygote is formed at fertilization, when a male gamete fuses with a female gamete. Each gamete brings a set of chromosomes to the zygote. The gametes, therefore, must each contain only half the diploid number of chromosomes, otherwise the chromosome number would double each time an organism reproduced sexually. Each human sperm cell contains 23 chromosomes and each human ovum has 23 chromosomes. When the sperm and ovum fuse at fertilization (p. 140), the diploid number of 46 (23 + 23) chromosomes is produced (Figure 21.6).

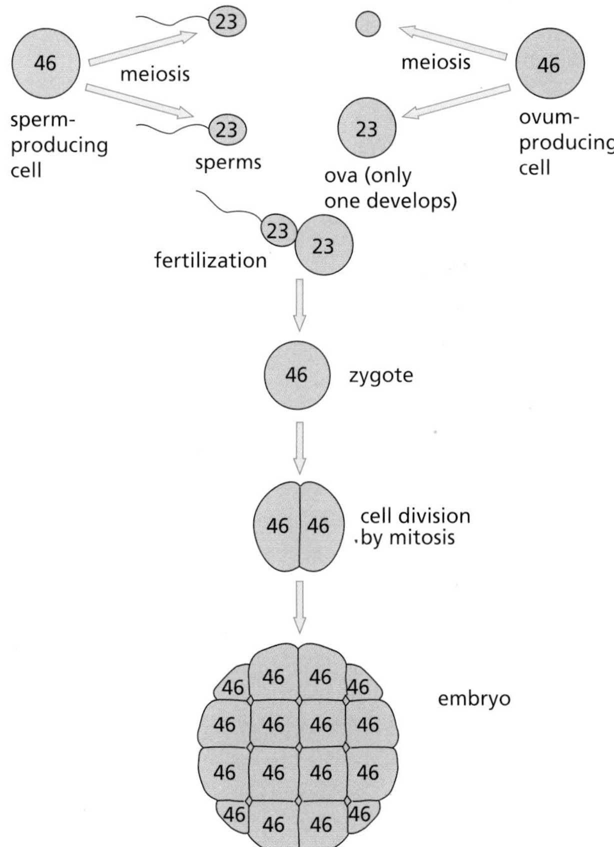

Figure 21.6 Chromosomes in gamete production and fertilization

The process of cell division which gives rise to gametes is different from mitosis because it results in the cells containing only half the diploid number of chromosomes. This number is called the **haploid number** and the process of cell division which gives rise to gametes is called **meiosis**.

Meiosis takes place only in reproductive organs.

Meiosis

In a cell which is going to divide and produce gametes, the diploid number of chromosomes shorten and thicken as in mitosis. The pairs of homologous chromosomes, e.g. the two long ones and the two short ones in Figure 21.7b, lie alongside each other and, when the nucleus divides for the first time, it is the chromosomes and not the chromatids which are separated. This results in only half the total number of chromosomes going to each daughter cell. In Figure 21.7(c) the diploid number of four chromosomes is being reduced to two chromosomes prior to the first cell division.

By now (Figure 21.7d), each chromosome is seen to consist of two chromatids and there is a second division of the nucleus (Figure 21.7e) which separates the chromatids into four distinct nuclei (Figure 21.7f).

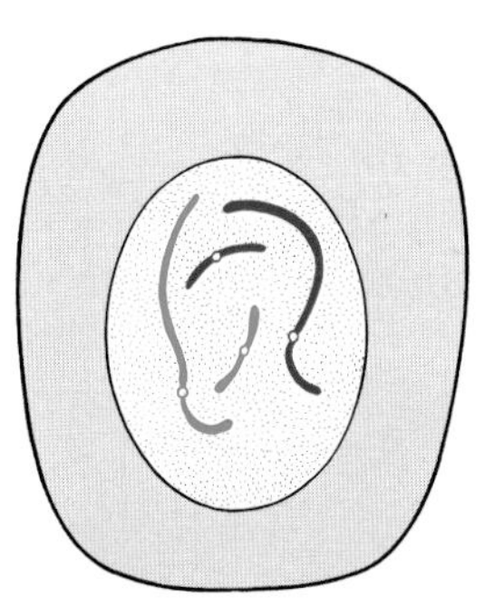

(a) The chromosomes appear. Those in red are from the organism's mother; the blue ones are from the father.

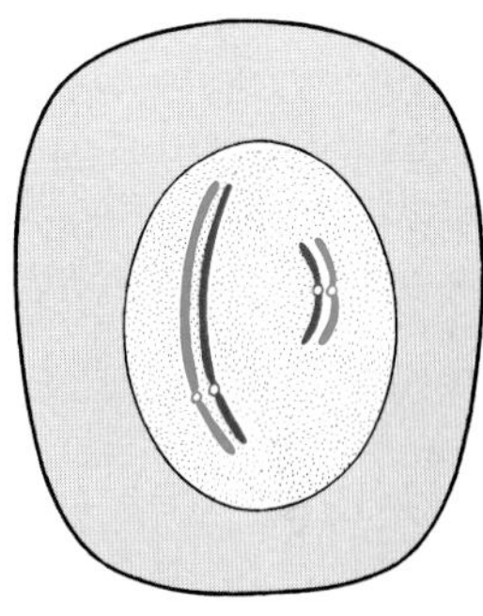

(b) Homologous chromosomes lie alongside each other.

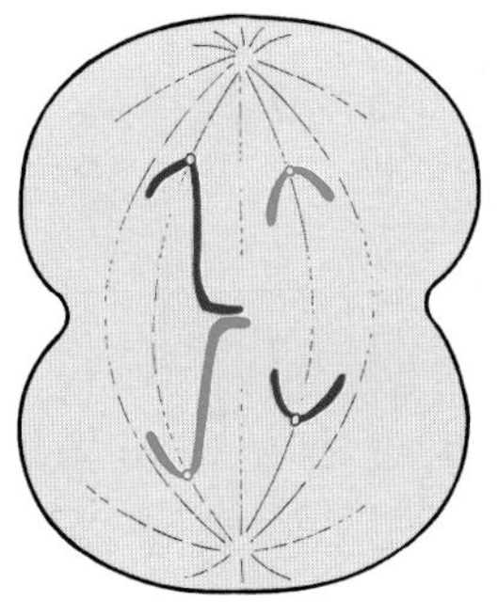

(c) The nuclear membrane disappears and corresponding chromosomes move apart to opposite ends of the cell.

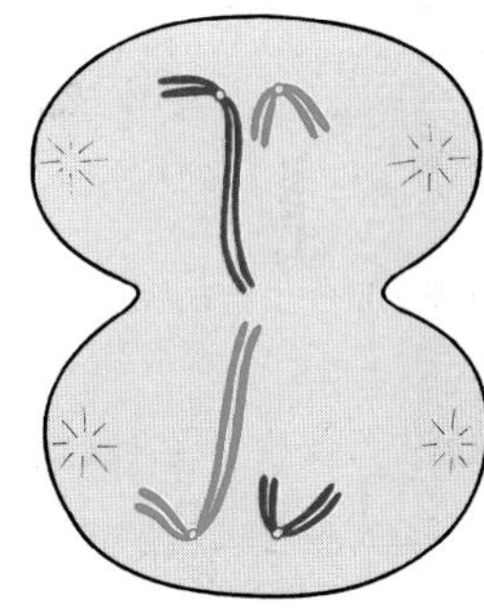

(d) By now each chromosome has become two chromatids.

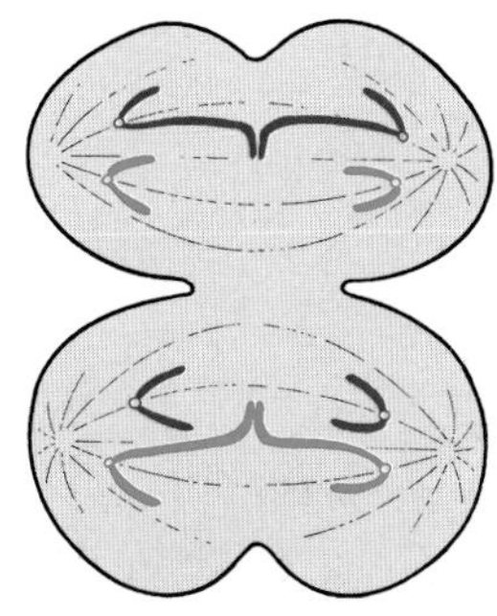

(e) A second division takes place to separate the chromatids.

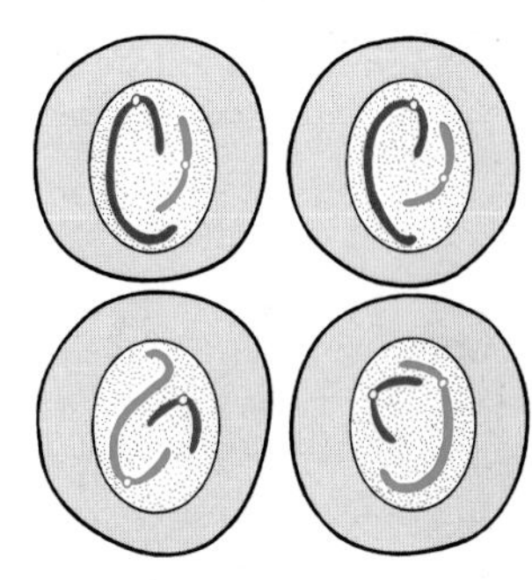

(f) Four gametes are formed. Each contains only half the original number of chromosomes.

Figure 21.7 Meiosis

This gives rise to four gametes, each with the haploid number of chromosomes. In the anther of a plant (p. 68), four haploid pollen grains are produced when a pollen mother cell divides by meiosis (Figure 21.8). In the testis of an animal, meiosis of each sperm-producing cell forms four sperms. In the cells of the ovule of a flowering plant or the ovary of a mammal, meiosis gives rise to only one mature female gamete. Four gametes may be produced initially, but only one of them turns into an egg cell which can be fertilized.

As a result of meiosis and fertilization, the maternal and paternal chromosomes meet in different combinations in the zygotes (p. 201). Consequently, the offspring will differ from their parents and from each other in a variety of ways.

Asexually produced organisms (p. 79) show no such variation because they are produced by mitosis and all their cells are identical to those of their single parent.

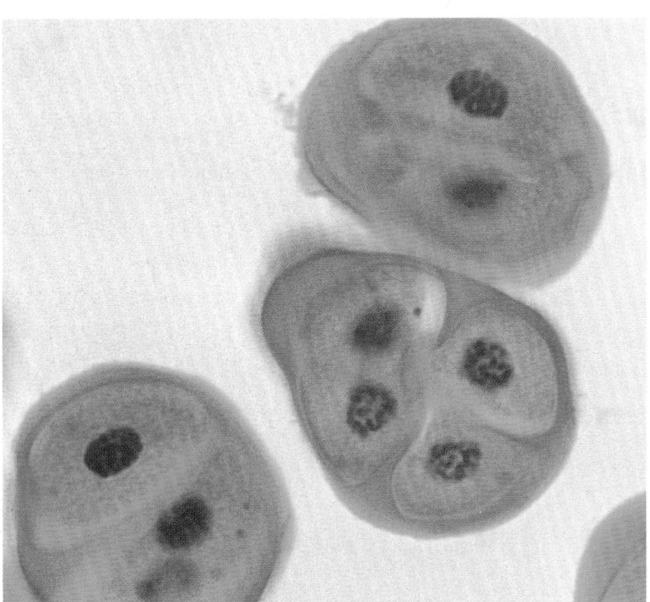

Figure 21.8 Meiosis in an anther (× 1000). The last division of meiosis in the anther of a flower produces four pollen grains

Questions

1 What is the haploid number for
a a human, **b** a fruit fly?

2 Which of the following cells would be haploid and which diploid: white blood cell, male cell in pollen grain, guard cell, root hair, ovum, sperm, skin cell, egg cell in ovule?

3 Where in the body of
a a human male,
b a human female and
c a flowering plant, would you expect meiosis to be taking place?

4 How many chromosomes would be present in
a a mouse sperm cell,
b a mouse ovum?

5 Why are organisms which are produced by asexual reproduction identical to each other?

Genes

The structure of the gene

The diagram of genes and a chromosome given in Figure 21.3 (p. 184) is greatly over-simplified. Chromosomes consist of a protein framework, with a long DNA molecule coiled round the framework in a complicated way (Figure 21.9). It is the DNA part of the chromosome which controls the inherited characters, and it is sections of the DNA molecule which constitute the genes.

A DNA molecule is a long chain of nucleotides. A **nucleotide** is a 5-carbon sugar molecule joined to a phosphate group ($—PO_3$) and an organic base (Figure 21.10). In DNA the sugar is deoxyribose, and the organic base is either **adenine** (A), **thymine** (T), **cytosine** (C) or **guanine** (G). The nucleotides are joined by their phosphate groups to form a long chain, often thousands of nucleotides long. The phosphate and sugar molecules are the same all the way down the chain but the bases may be any one of the four listed above (Figure 21.11).

Mitosis and meiosis compared

Mitosis	Meiosis
Occurs during cell division of somatic cells.	Occurs in the final stages of cell division leading to production of gametes.
A full set of chromosomes is passed on to each daughter cell. This is the diploid number of chromosomes.	Only half the chromosomes are passed on to the daughter cells, i.e. the haploid number of chromosomes.
The chromosomes and genes in each daughter cell are identical.	The homologous chromosomes and their genes are randomly assorted between the gametes. (See p. 201 for a fuller explanation of this.)
If new organisms are produced by mitosis in asexual reproduction (e.g. bulbs, p. 80) they will all resemble each other and their parents. They are said to form a 'clone'.	New organisms produced by meiosis in sexual reproduction will show variations from each other and from their parents.

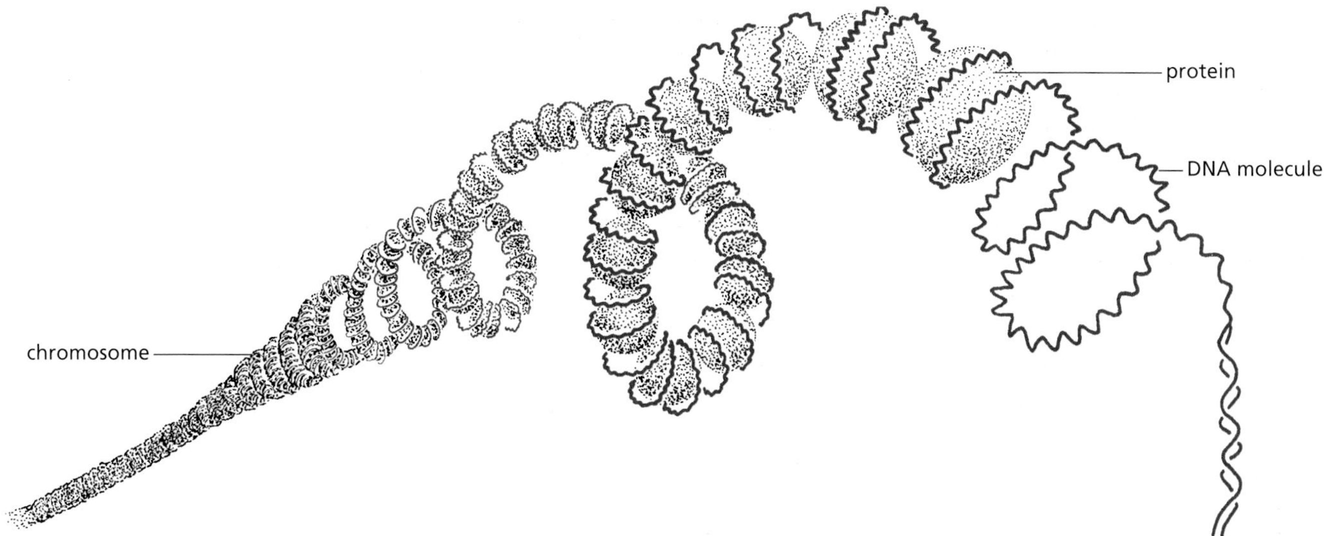

Figure 21.9 Simplified model of chromosome structure. This is a '1974' model, which has been superseded by something much more complicated

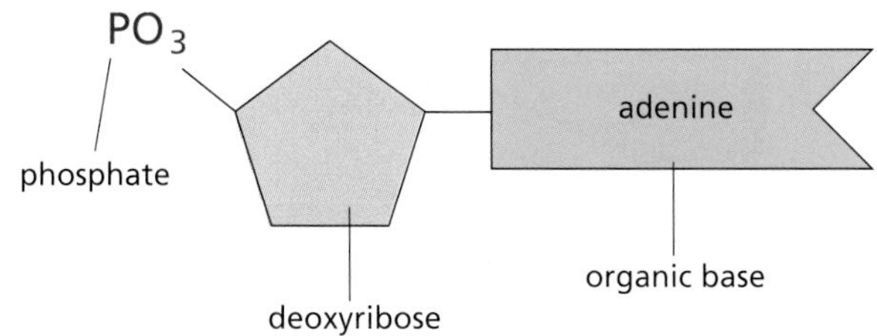

Figure 21.10 A nucleotide (adenosine monophosphate)

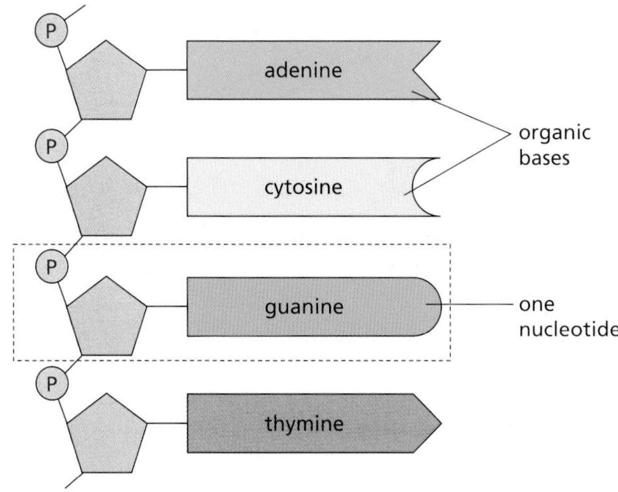

Figure 21.11 Part of a DNA molecule with four nucleotides

The sequence of bases down the length of the DNA molecule forms a code which instructs the cell to make particular proteins. Proteins are made from amino acids linked together (p.11). The type and sequence of the amino acids joined together will determine the kind of protein formed. For example, one protein molecule may start with the sequence *alanine–glycine–glycine* .... A different protein may start *glycine–serine–alanine* ....

It is the sequence of bases in the DNA molecule that decides which amino acids are used and in which order they are joined. Each group of three bases stands for one amino acid, e.g. the triplet of bases *cytosine–guanine–adenine* (CGA) specifies the amino acid *alanine*, the base triplet *cytosine–adenine–thymine* (CAT) specifies the amino acid *valine*, and the triplet *cytosine–cytosine–alanine* (CCA) stands for *glycine*. The tripeptide *valine–glycine–alanine* is specified by the DNA code CAT–CCA–CGA (Figure 21.12).

A gene, then, is a sequence of triplets of the four bases, which specifies an entire protein. Insulin is a small protein with only 51 amino acids. A sequence of 153 (i.e. 3 × 51) bases in the DNA molecule would constitute the gene which makes an islet cell in the pancreas produce insulin. Most proteins are much larger than this and most genes contain a thousand or more bases.

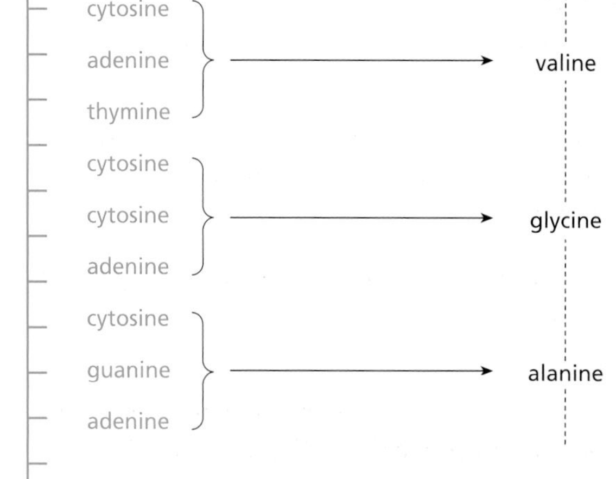

Figure 21.12 The genetic code (triplet code)

The chemical reactions which take place in a cell determine what sort of a cell it is and what its functions are. These chemical reactions are, in turn, controlled by enzymes. Enzymes are proteins. It follows, therefore, that the **genetic code** of DNA, in determining which proteins, particularly enzymes, are produced in a cell, also determines the cell's structure and function. In this way, the genes also determine the structure and function of the whole organism.

Replication of DNA

The DNA in a chromosome consists of two chains of nucleotides held together by chemical bonds between the bases. The size of the molecules ensures that adenine always pairs with thymine and cytosine pairs with guanine. The double strand is twisted to form a helix (like a twisted rope ladder with the base pairs representing the rungs) (Figures 21.13 and 21.14).

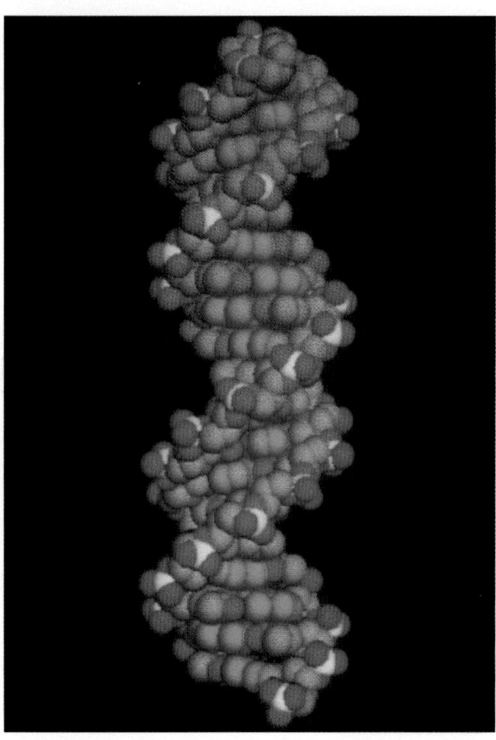

Figure 21.13 Model of the structure of DNA

Before cell division can occur, the DNA of the chromosome has to replicate, i.e. make replicas of itself. To do this, enzymes make the double strands of DNA unwind and separate into two single strands, rather like undoing a zip (Figure 21.15a).

Nucleotides are brought to the 'unzipped' DNA and joined to the exposed bases with the aid of enzymes (Figure 21.15b). The adenine of an arriving nucleotide always joins to the thymine of the DNA; the cytosine of a nucleotide always joins to the guanine of the DNA. Similarly, the thymine or guanine on a nucleotide will join to the adenine or cytosine, respectively, of the DNA. The pairing is always A with T, and C with G. Thus an exposed sequence of C–C–A–T–T–G–C–A on the single strand of DNA would build up a corresponding sequence of G, G, T, A, A, C, G, T nucleotides.

The new nucleotides join up to form a chain attached to the exposed strand. This happens all the way along each DNA strand. Since this is happening in both strands of DNA, the **double helix** is replicated and the full set of genetic instructions is passed to both daughter cells at cell division (Figure 21.15c).

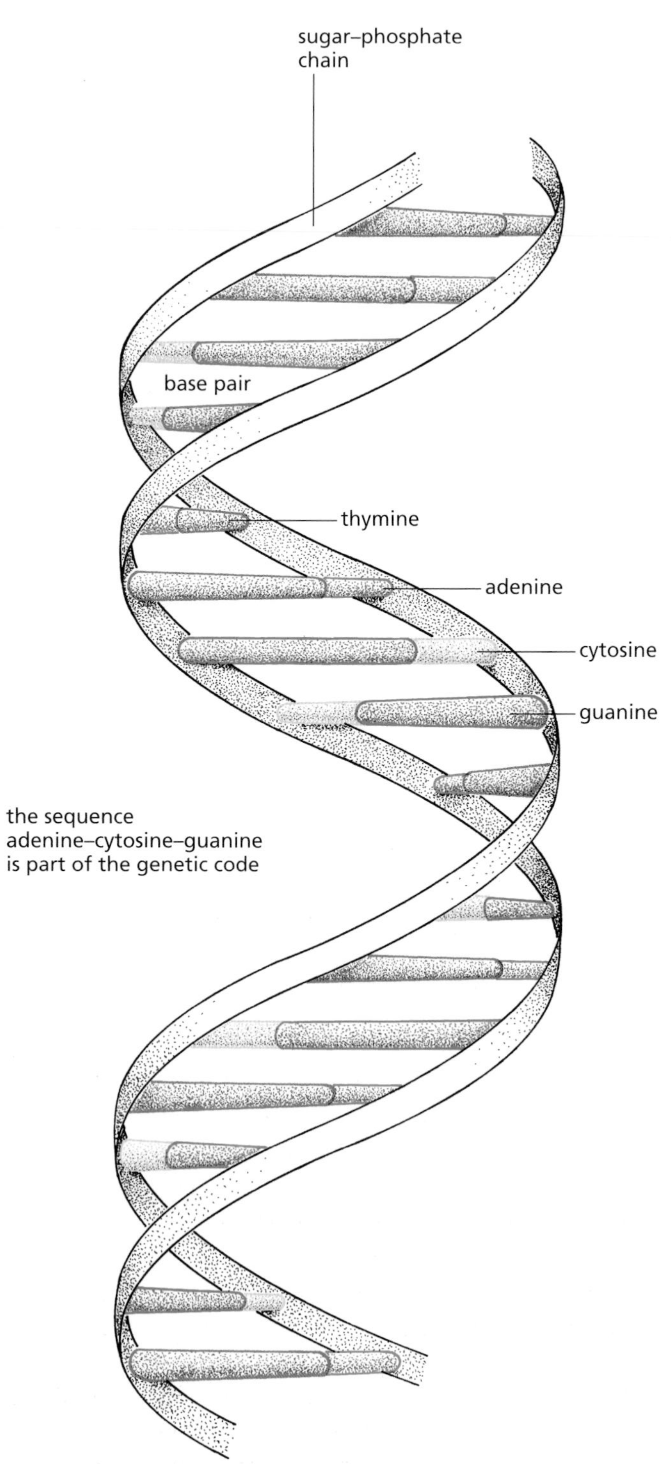

Figure 21.14 The drawing shows schematically part of a DNA molecule

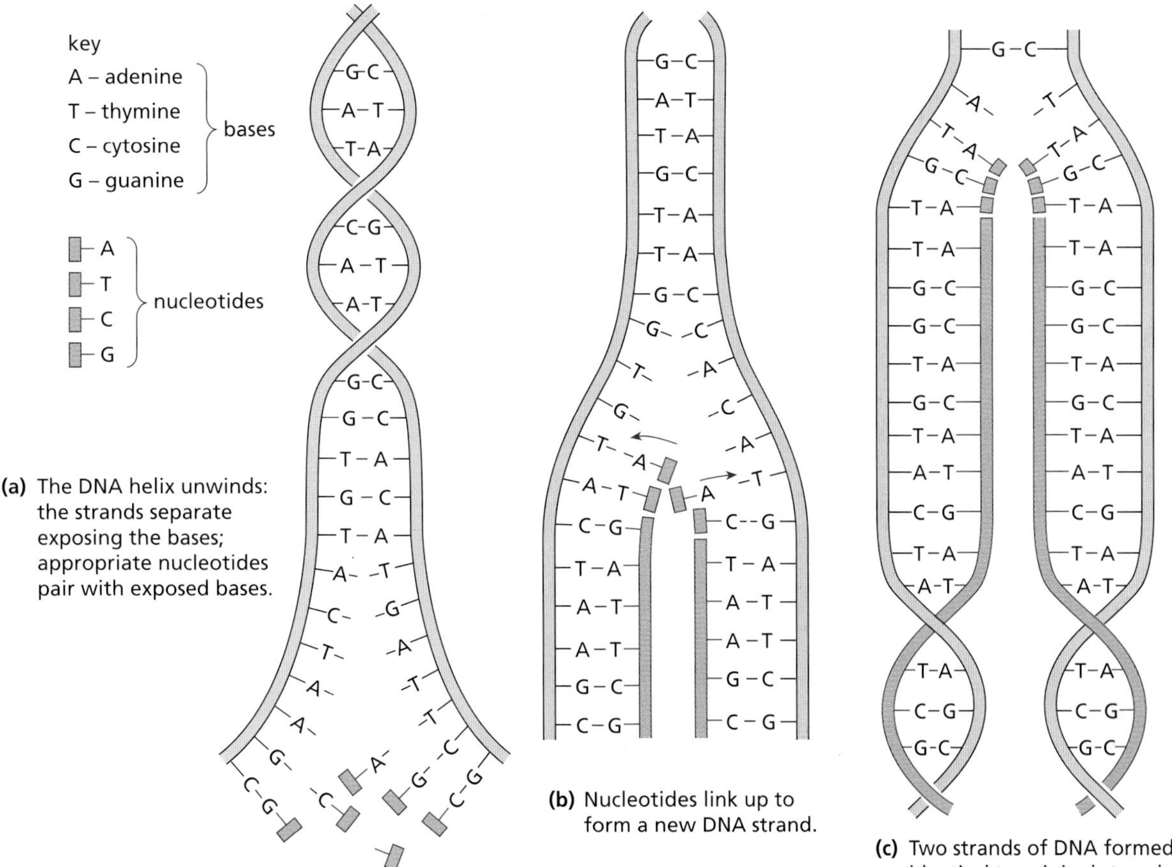

(a) The DNA helix unwinds: the strands separate exposing the bases; appropriate nucleotides pair with exposed bases.

(b) Nucleotides link up to form a new DNA strand.

(c) Two strands of DNA formed, identical to original strand.

Figure 21.15 Replication of DNA

Mutations

A mutation is a spontaneous change in a gene or a chromosome. In a gene mutation it may be that one or more genes are not replicated correctly. A chromosome mutation may result from damage to or loss of part of a chromosome during mitosis or meiosis, or even the gain of an extra chromosome as in Down's syndrome (see below).

An abrupt change in a gene or chromosome is likely to result in a defective enzyme and will usually disrupt the complex reactions in the cells. Most mutations, therefore, are harmful to the organism.

Surprisingly, only about 3 per cent of human DNA consists of genes. The rest consists of repeated sequences of nucleotides that do not code for proteins. This is sometimes called 'junk DNA', but that term only means that we do not know its function. If mutations occur in these non-coding sequences they are unlikely to have any effect on the organism and are, therefore, described as 'neutral'.

Rarely a gene or chromosome mutation produces a beneficial effect and this may contribute to the success of the organism (p. 203).

If a mutation occurs in a gamete, it will affect all the cells of the individual which develops from the zygote. Thus the whole organism will be affected. If the mutation occurs in a somatic cell (body cell), it will affect only those cells produced, by mitosis, from the affected cell.

Thus, a mutation in a gamete may result in a genetic disorder, e.g. haemophilia (p. 197) or cystic fibrosis. Mutations in somatic cells may give rise to cancers by promoting uncontrolled cell division in the affected tissue. For example, skin cancer results from uncontrolled cell division in the basal layer.

A mutation may be as small as the substitution of one organic base for another in the DNA molecule, or as large as the breakage, loss or gain of a chromosome.

A disease called **sickle-cell anaemia** (p. 204) results from a defective haemoglobin molecule which causes the red blood cells to distort when subjected to a low oxygen concentration. The defective haemoglobin molecule differs from normal haemoglobin by only one amino acid, i.e. *valine* replaces *glutamic acid*. This could be the result of faulty replication at meiosis. When the relevant parental chromosome replicated at gamete formation, the DNA could have produced the triplet –CAT– (which specifies *valine*) instead of –CTT– (which specifies *glutamic acid*).

An inherited form of mental and physical retardation, known as **Down's syndrome**, results from a chromosome mutation. During the meiosis which produces an ovum, one of the chromosomes fails to separate from its homologous partner, a process known as **non-disjunction**. As a result, the ovum carries 24 chromosomes instead of 23, and the resulting zygote has 47 instead of the normal 46 chromosomes.

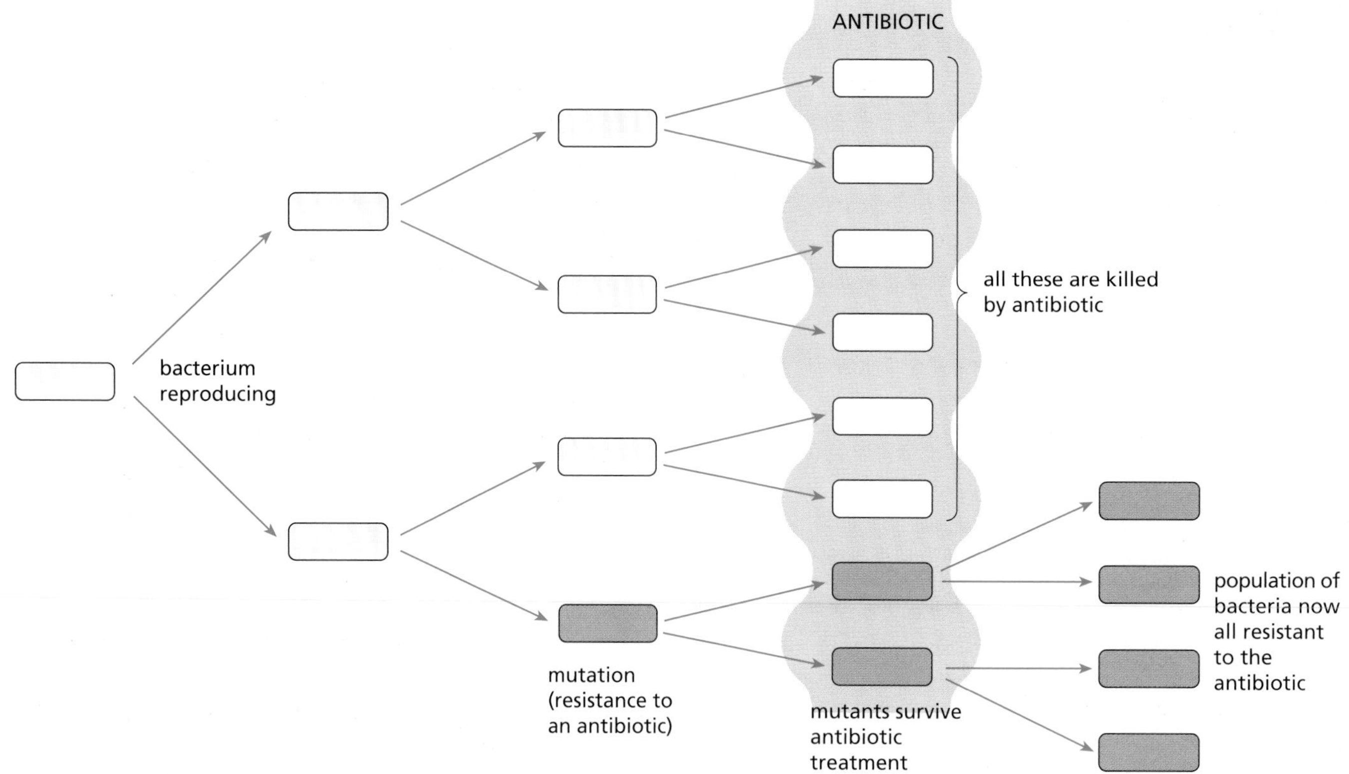

Figure 21.16 Mutation in bacteria can lead to drug resistance

Mutations in bacteria often produce resistance to drugs. Bacterial cells reproduce very rapidly, perhaps as often as once every 20 minutes. Thus a mutation, even if it occurs only rarely, is likely to appear in a large population of bacteria. If a population of bacteria, containing one or two drug-resistant mutants, is subjected to that particular drug, the non-resistant bacteria will be killed, but the drug-resistant mutants survive (Figure 21.16). Mutant genes are inherited in the same way as normal genes, so when the surviving mutant bacteria reproduce, all their offspring will be resistant to the drug.

Mutations are comparatively rare events; perhaps only one in every 100 000 replications results in a mutation. Nevertheless they do occur naturally all the time. (There are bound to be some mutants in the 500 million sperms produced in a human ejaculate.)

Exposure to **mutagens**, namely certain chemicals and radiation, is known to increase the rate of mutation. Some of the substances in tobacco smoke are mutagens which can cause cancer.

Ionizing radiation from X-rays and radioactive compounds, and ultraviolet radiation from sunlight, can both increase the mutation rate. It is uncertain whether there is a minimum dose of radiation below which there is negligible risk. It is possible that repeated exposure to low doses of radiation is as harmful as one exposure to a high dose. It has become clear in recent years that, in light-skinned people, unprotected exposure to the ultraviolet radiation from the sun can cause a form of skin cancer.

Generally speaking, however, exposure to natural and medical sources of radiation carries less risk than smoking cigarettes or driving a car but it is sensible to keep exposure to a minimum.

Questions

1. What peptide is specified by the DNA sequence CGA–CGA–CAT–CCA–CAT?
2. A mutation in the DNA sequence in question 1 produces a valine in place of the glycine. What change in the genetic code could have produced this result?
3. State briefly the connection between genes, enzymes and cell structure.
4. Why is it particularly important to prevent radiation from reaching the reproductive organs?

Practical work

Squash preparation of chromosomes using acetic orcein

Material *Allium cepa* (onion) root tips. Support onions over beakers or jars of water. Keep the onions in darkness for several days until the roots growing into the water are 2–3 cm long. Cut off about 5 mm of the root tips, place them in a watch glass and:

1 Cover them with 9 drops acetic orcein and 1 drop molar hydrochloric acid.
2 Heat the watch glass gently over a very small Bunsen flame till the steam rises from the stain, but do not boil.
3 Leave the watch glass covered for at least 5 minutes.
4 Place one of the root tips on a clean slide, cover with 45 per cent ethanoic (acetic) acid and cut away all but the terminal 1 mm.
5 Cover this root tip with a clean cover-slip and make a squash preparation as described below.

Making the squash preparation Squash the softened, stained root tips by lightly tapping on the cover-slip with a pencil: hold the pencil vertically and let it slip through the fingers to strike the cover-slip (Figure 21.17). The root tip will spread out as a pink mass on the slide; the cells will separate and the nuclei, many of them with chromosomes in various stages of mitosis (because the root tip is a region of rapid cell division), can be seen under the high power of the microscope (×400).

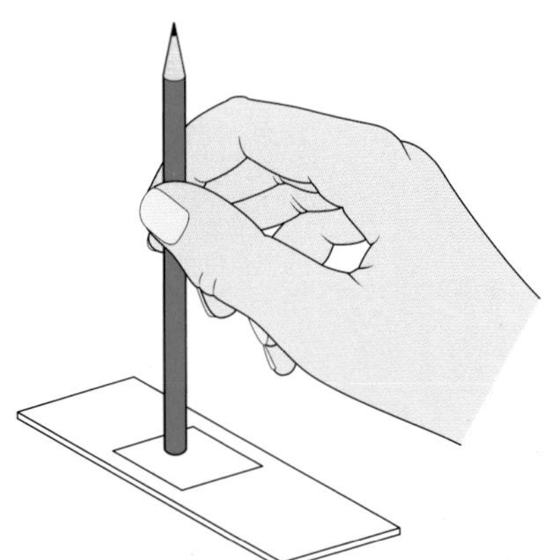

Figure 21.17 Tap the cover-slip gently to squash the tissue

Checklist

- In the nuclei of all cells there are thread-like structures called 'chromosomes'.
- The chromosomes are in pairs; one of each pair comes from the male and one from the female parent.
- On these chromosomes are carried the genes.
- The genes control the chemical reactions in the cells and, as a result, determine what kind of organism is produced.
- Each species of plant or animal has a fixed number of chromosomes in its cells.
- When cells divide by mitosis, the chromosomes and genes are copied exactly and each new cell gets a full set.
- At meiosis, only one chromosome of each pair goes into the gamete.
- The DNA molecule is coiled along the length of the chromosome.
- A DNA molecule is made up of a double chain of nucleotides in the form of a helix.
- The nucleotide bases in the helix pair up adenine–thymine (A–T) and cytosine–guanine (C–G).
- Triplets of bases control production of the specific amino acids which make up a protein.
- Genes consist of particular lengths of DNA.
- Most genes control the type of enzyme that a cell will make.
- At replication, DNA strands separate and build up new chains on the exposed bases.
- A mutation is a spontaneous change in a gene or chromosome. Most mutations produce harmful effects.

22 Heredity

Patterns of inheritance
Single-factor inheritance: dominant and recessive genes. Breeding true: homozygous and heterozygous individuals. Genotype and phenotype. Alleles. The three to one ratio: breeding experiments with mice. The recessive test-cross: testing for a heterozygote. Codominance and incomplete dominance. Sickle-cell anaemia. Determination of sex: X and Y chromosomes.

Hereditary diseases
Albinism, haemophilia, cystic fibrosis, Huntington's disease. Genetic counselling and screening.

Patterns of inheritance

A knowledge of mitosis and meiosis allows us to explain, at least to some extent, how heredity works. The gene in a mother's body cells which causes her to have brown eyes may be present on one of the chromosomes in each ovum she produces. If the father's sperm cell contains a gene for brown eyes on the corresponding chromosome, the zygote will receive a gene for brown eyes from each parent. These genes will be reproduced by mitosis in all the embryo's body cells and when the embryo's eyes develop, the genes will make the cells of the iris produce brown pigment (melanin) and the child will have brown eyes.

In a similar way, the child may receive genes for curly hair and Figure 22.1 shows this happening. But it does not, of course, show all the other chromosomes with thousands of genes for producing the enzymes, making different types of cell and all the other processes which control the development of the organism.

Single-factor inheritance

Because it is impossible to follow the inheritance of the thousands of characteristics controlled by genes, it is usual to start with the study of a single gene which controls one characteristic. We have used eye colour as an example so far. Probably more than one gene pair is involved, but the simplified example will serve our purpose. It has already been explained how a gene for brown eyes from each parent results in the child having brown eyes. Suppose, however, that the mother has blue eyes and the father brown eyes. The child might receive a gene for blue eyes from its mother and a gene for brown eyes from its father (Figure 22.2). If this happens, the child will, in fact, have brown eyes. The gene for brown eyes is said to be **dominant** to the gene for blue eyes. Although the gene for blue eyes is present in all the child's cells, it is not expressed. It is said to be **recessive** to brown.

Eye colour is a useful 'model' for explaining inheritance but it is not wholly reliable because 'blue' eyes vary in colour and sometimes contain small amounts of brown pigment.

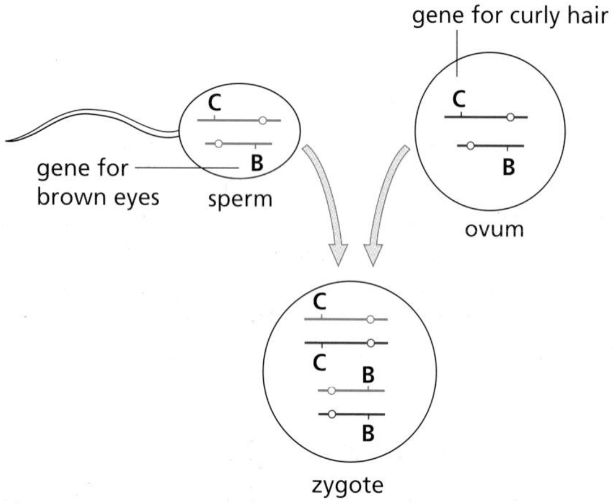

Figure 22.1 Fertilization. Fertilization restores the diploid number of chromosomes and combines the genes from the mother and father

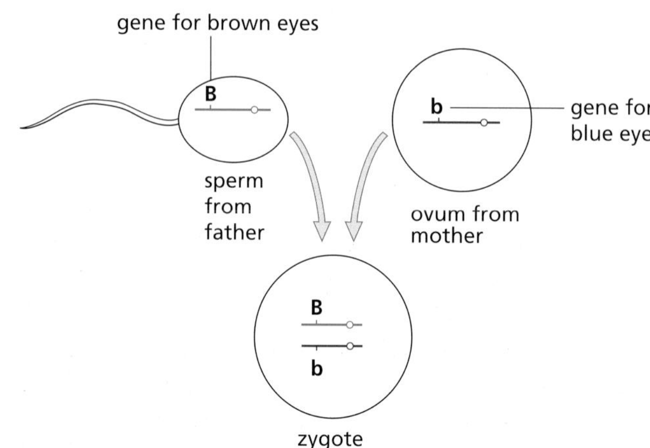

Figure 22.2 Combination of genes in the zygote (only one chromosome is shown). The zygote has both genes for eye colour; the child will have brown eyes

This example illustrates the following important points:

- There is a pair of genes for each characteristic, one gene from each parent.
- Although the gene pairs control the same characteristic, e.g. eye colour, they may have different effects. One tries to produce blue eyes, the other tries to produce brown eyes.
- Often one gene is dominant over the other.
- The genes of each pair are on corresponding chromosomes and occupy corresponding positions. For example, in Figure 22.1 the genes for eye colour are shown in the corresponding position on the two short chromosomes and the genes for hair curliness are in corresponding positions on the two long chromosomes. In diagrams and explanations of heredity:
  - genes are represented by letters;
  - genes controlling the same characteristic are given the same letter; and
  - the dominant gene is given the capital letter.

For example, in rabbits, the dominant gene for black fur is labelled **B**. The recessive gene for white fur is labelled **b** to show that it corresponds to **B** for black fur. If it were labelled **w**, we would not see any connection between **B** and **w**. **B** and **b** are obvious partners. In the same way **L** could represent the gene for long fur and **l** the gene for short fur.

Questions

1. Some plants occur in one of two sizes, tall or dwarf. This characteristic is controlled by one pair of genes. Tallness is dominant to shortness. Choose suitable letters for the gene pair.
2. Why are there two genes controlling one characteristic? Do the two genes affect the characteristic in the same way as each other?
3. The gene for red hair is recessive to the gene for black hair. What colour hair will a person have if he inherits a gene for red hair from his mother and a gene for black hair from his father?

Breeding true

A white rabbit must have both the recessive genes **b** and **b**. If it had **B** and **b,** the dominant gene for black (**B**) would override the gene for white (**b**) and produce a black rabbit. A black rabbit, on the other hand, could be either **BB** or **Bb** and, by just looking at the rabbit, you could not tell the difference. When the male black rabbit **BB** produces sperms by meiosis, each one of the pair of chromosomes carrying the **B** genes will end up in different sperm cells. Since the genes are the same, all the sperms will have the **B** gene for black fur (Figure 22.3a).

The black rabbit **BB** is called a true-breeding black and is said to be **homozygous** for black coat colour ('homo-' means 'the same'). If this rabbit mates with another black (**BB**) rabbit, all the babies will be black because all will receive a dominant gene for black fur. When all the offspring have the same characteristic as the parents, this is called 'breeding true' for this characteristic.

When the **Bb** black rabbit produces gametes by meiosis, the chromosomes with the **B** genes and the chromosomes with the **b** genes will end up in different gametes. So 50 per cent of the sperm cells will carry **B** genes and 50 per cent will carry **b** genes (Figure 22.3b). Similarly, in the female, 50 per cent of the ova will have a **B** gene and 50 per cent will have a **b** gene. If a **b** sperm fertilizes a **b** ovum, the offspring, with two **b** genes (**bb**), will be white. The black **Bb** rabbits are not true-breeding because they may produce some white babies as well as black ones. The **Bb** rabbits are called **heterozygous** ('hetero-' means 'different').

The black **BB** rabbits are homozygous dominant.

The white **bb** rabbits are homozygous recessive.

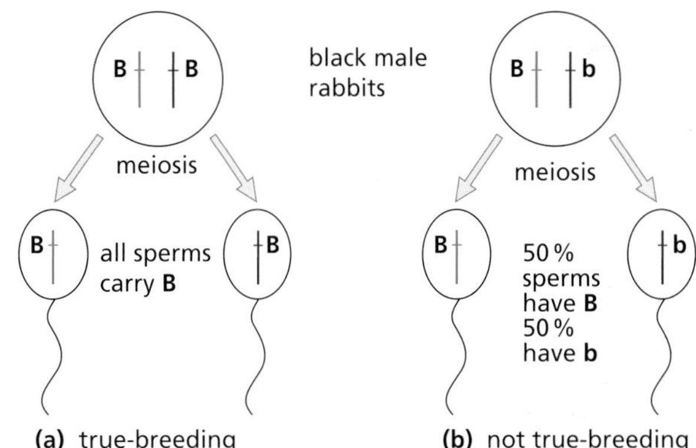

Figure 22.3 Breeding true

Questions

4. a Read question 3 again. Choose letters for the genes for red hair and black hair and write down the gene combination for having red hair.
   b Would you expect a red-haired couple to breed true?
   c Could a black-haired couple have a red-haired baby?
5. Use the words 'homozygous', 'heterozygous', 'dominant' and 'recessive' (where suitable) to describe the following gene combinations: **Aa, AA, aa.**
6. A plant has two varieties, one with red petals and one with white petals. When these two varieties are cross-pollinated, all the offspring have red petals. Which gene is dominant? Choose suitable letters to represent the two genes.

Genotype and phenotype

The two kinds of black rabbit **BB** and **Bb** are said to have the same **phenotype**. This is because their coat colours look exactly the same. However, because they have different gene pairs for coat colour they are said to have different **genotypes**, i.e. different combinations of genes. One genotype is **BB** and the other is **Bb**.

You and your brother might both be brown-eyed phenotypes but your genotype could be **BB** and his could be **Bb**. You would be homozygous dominant for brown eyes; he would be heterozygous for eye colour.

Alleles

The genes which occupy corresponding positions on homologous chromosomes and control the same characteristic are called **allelomorphic genes** or **alleles**. The word 'allelomorph ' means 'alternative form'. The genes **B** and **b** are alternative forms of a gene for eye colour. **B** and **b** are alleles.

There are often more than two alleles of a gene. The human ABO blood groups (p. 119) are controlled by three alleles, $\mathbf{I}^A$, $\mathbf{I}^B$ and **i**, though only two of these can be present in one genotype.

The three to one ratio

The result of a mating between a true-breeding (homozygous) black mouse (**BB**), and a true-breeding (homozygous) brown mouse (**bb**) is shown in Figure 22.4a. The illustration is greatly simplified because it shows only one pair of the 20 pairs of mouse chromosomes and only one pair of alleles on the chromosomes.

Because black is dominant to brown, all the offspring from this mating will be black phenotypes, because they all receive the dominant allele for black fur from the father. Their genotypes, however, will be **Bb** because they all receive the recessive **b** allele from the mother. They are heterozygous for coat colour. The offspring resulting from this first mating are called the $\mathbf{F_1}$ **generation**.

Figure 22.4b (opposite) shows what happens when these heterozygous, F_1 black mice are mated together to produce what is called the $\mathbf{F_2}$ **generation**. Each sperm or ovum produced by meiosis can contain only one of the alleles for coat colour, either **B** or **b**. So there are two kinds of sperm cell, one kind with the **B** allele and one kind with the **b** allele. There are also two kinds of ovum, with either **B** or **b** alleles. When fertilization occurs, there is no way of telling whether a **b** or a **B** sperm will fertilize a **B** or a **b** ovum, so we have to look at all the possible combinations as follows:

- A **b** sperm fertilizes a **B** ovum. Result: **bB** zygote.
- A **b** sperm fertilizes a **b** ovum. Result: **bb** zygote.
- A **B** sperm fertilizes a **B** ovum. Result: **BB** zygote.
- A **B** sperm fertilizes a **b** ovum. Result: **Bb** zygote.

There is no difference between **bB** and **Bb**, so there are three possible genotypes in the offspring – **BB**, **Bb** and **bb**. There are only two phenotypes – black (**BB** or **Bb**) and brown (**bb**). So, according to the laws of chance, we would expect three black baby mice and one brown. Mice usually have more than four offspring and what we really expect is that the **ratio** (proportion) of black to brown will be close to 3:1.

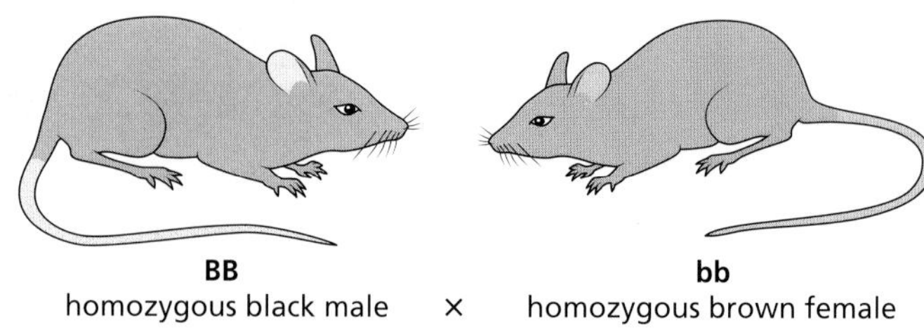

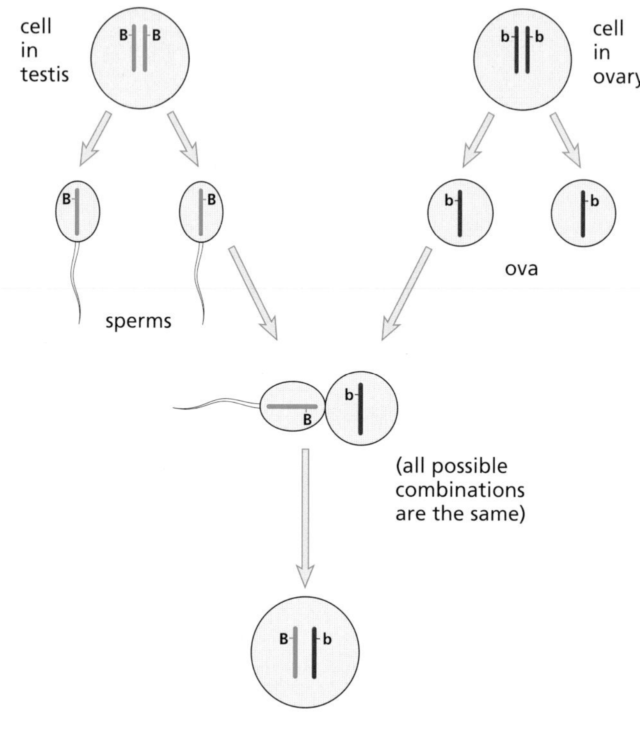

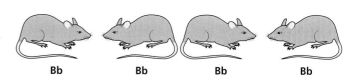

Bb Bb Bb Bb

(a) all the F_1 generation are heterozygous black

Figure 22.4 Inheritance of coat colour in mice

If the mouse had 13 babies, you might expect nine black and four brown, or eight black and five brown. Even if she had 16 babies you would not expect to find exactly 12 black and four brown because whether a **B** or **b** sperm fertilizes a **B** or **b** ovum is a matter of chance. If you spun ten coins, you would not expect to get exactly five heads and five tails. You would not be surprised at six heads and four tails or even seven heads and three tails. In the same way, we would not be surprised at 14 black and two brown mice in a litter of 16.

To decide whether there really is a 3:1 ratio, we need a lot of results. These may come either from breeding the same pair of mice together for a year or so to produce many litters, or from mating 20 black and 20 brown mice, crossing the offspring and adding up the number of black and brown babies in the F_2 families (see also Figure 22.5).

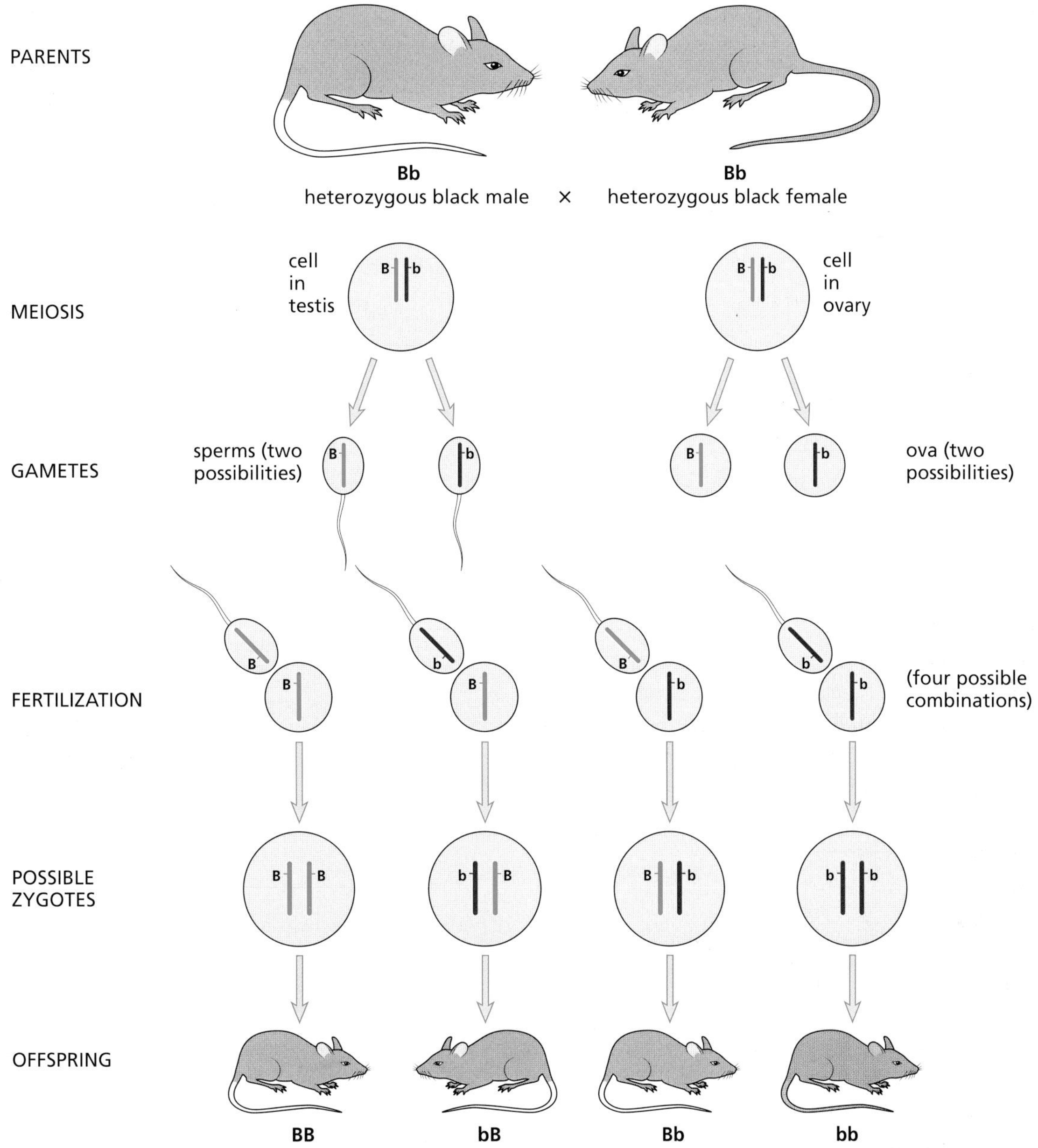

PARENTS

Bb heterozygous black male × **Bb** heterozygous black female

MEIOSIS

cell in testis B b

cell in ovary B b

GAMETES

sperms (two possibilities) B b

B b ova (two possibilities)

FERTILIZATION

B B b B B b b b (four possible combinations)

POSSIBLE ZYGOTES

B B b B B b b b

OFFSPRING

BB **bB** **Bb** **bb**

(b) the probable ratio of coat colours in the F_2 generation is 3 black : 1 brown

Figure 22.5 F_2 hybrids in maize. In the two left-hand cobs, the grain colour phenotypes appear in a 3:1 ratio (try counting single rows). What was the colour of the parental grains for each of these cobs?

Questions

1 Look at Figure 22.4a. Why is there no possibility of getting a **BB** or a **bb** combination in the offspring?

2 In Figure 22.4b what proportion of the F_2 black mice are true-breeding?

3 Two black guinea-pigs are mated together on several occasions and their offspring are invariably black. However, when their black offspring are mated with white guinea-pigs, half of the matings result in all black litters and the other half produce litters containing equal numbers of black and white babies. From these results, deduce the genotypes of the parents and explain the results of the various matings, assuming that colour in this case is determined by a single pair of alleles.

The recessive test-cross (back-cross)

A black mouse could have either the **BB** or the **Bb** genotype. One way to find out which, is to cross the black mouse with a known homozygous recessive mouse, **bb**. The **bb** mouse will produce gametes with only the recessive **b** allele. A black homozygote, **BB**, will produce only **B** gametes. Thus, if the black mouse is **BB**, all the offspring from the cross will be black heterozygotes, **Bb**.

Half the gametes from a black, **Bb**, mouse would carry the **B** allele and half would have the **b** allele. So, if the black mouse is **Bb**, half of the offspring from the cross will, on average, be brown homozygotes, **bb**, and half will be black heterozygotes, **Bb**.

The term 'back-cross' refers to the fact that, in effect, the black, mystery mouse is being crossed with the same genotype as its brown grandparent, the **bb** mouse in Figure 22.4a. Mouse ethics and speed of reproduction make the use of the actual grandparent quite feasible!

Question

1 Two black rabbits thought to be homozygous for coat colour were mated and produced a litter which contained all black babies. The F_2, however, resulted in some white babies, which meant that one of the grandparents was heterozygous for coat colour. How would you find out which parent was heterozygous?

Codominance and incomplete dominance

Codominance

If both genes of an allelomorphic pair produce their effects in an individual (i.e. neither allele is dominant to the other) the alleles are said to be codominant.

The inheritance of the human ABO blood groups provides an example of codominance. On p. 119 it was explained that, in the ABO system, there are four phenotypic blood groups, A, B, AB and O. The alleles for groups A and B are codominant. If a person inherits alleles for group A and group B, his or her red cells will carry both antigen A and antigen B.

However, the alleles for groups A and B are both completely dominant to the allele for group O. (Group O people have neither A nor B antigens on their red cells.)

Table 22.1 shows the genotypes and phenotypes for the ABO blood groups. (Note that the allele for group O is sometimes represented as **i** and sometimes as $\mathbf{I^o}$.)

Table 22.1 The ABO blood groups

Genotype	Blood group (phenotype)
I^AI^A or I^Ai	A
I^BI^B or I^Bi	B
I^AI^B	AB
ii	O

Since the genes for groups A and B are dominant to that for group O, a group A person could have the genotype $\mathbf{I^AI^A}$ or $\mathbf{I^Ai}$. Similarly a group B person could be $\mathbf{I^BI^B}$ or $\mathbf{I^Bi}$. There are no alternative genotypes for groups AB and O.

Incomplete dominance

This term is sometimes taken to mean the same as 'codominance' but, strictly, it applies to a case where the effect of the recessive allele is not completely masked by the dominant allele.

An example occurs with sickle-cell anaemia (p. 189). If a person inherits both recessive genes ($\mathbf{Hb^SHb^S}$) for sickle-cell haemoglobin, then he or she will show manifest signs of the disease, i.e. distortion of the red cells leading to severe bouts of anaemia.

A heterozygote ($\mathbf{Hb^AHb^S}$), however, will have a condition called '**sickle-cell trait**'. Although there may be mild symptoms of anaemia the condition is not serious or life-threatening. In this case, the normal haemoglobin allele ($\mathbf{Hb^A}$) is not completely dominant over the recessive ($\mathbf{Hb^S}$) allele.

Questions

1 What are the possible blood groups likely to be inherited by children born to a group A mother and a group B father? Explain your reasoning.

2 A woman of blood group A claims that a man of blood group AB is the father of her child. A blood test reveals that the child's blood group is O. Is it possible that the woman's claim is correct? Could the father have been a group B man? Explain your reasoning.

3 A red cow has a pair of alleles for red hairs. A white bull has a pair of alleles for white hairs. If a red cow and a white bull are mated, the offspring are all 'roan', i.e. they have red and white hairs equally distributed over their body.
- **a** Is this an example of codominance or incomplete dominance?
- **b** What coat colours would you expect among the offspring of a mating between two roan cattle?

Determination of sex

Whether you are a male or female depends on one particular pair of chromosomes called the 'sex chromosomes'. In females, the two sex chromosomes, called the X chromosomes, are the same size as each other. In males, the two sex chromosomes are of different sizes. One corresponds to the female sex chromosomes and is called the X chromosome. The other is smaller and is called the Y chromosome. So the female genotype is XX and the male genotype is XY.

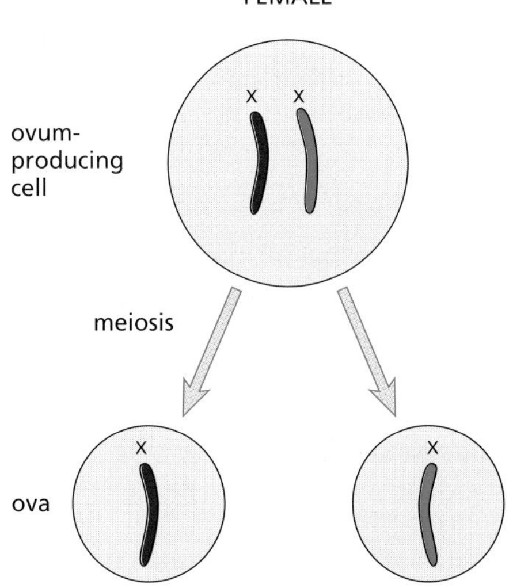

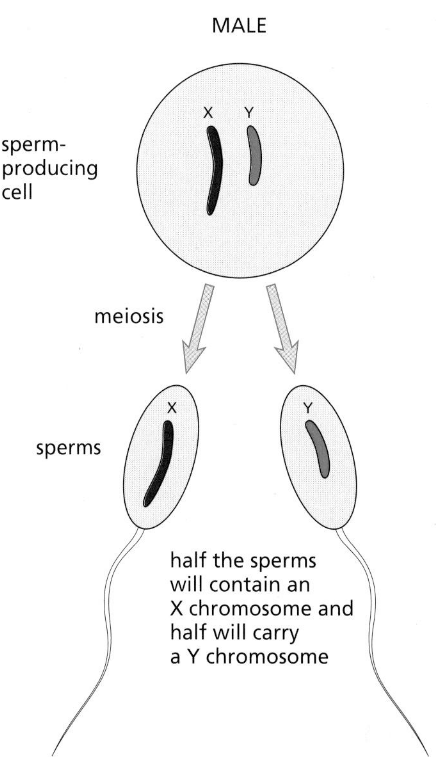

Figure 22.6 Determination of sex. Note that:
(i) only the X and Y chromosomes are shown
(ii) the Y chromosome is not smaller than the X in all organisms
(iii) details of meiosis have been omitted
(iv) in fact, four gametes are produced in each case, but two are sufficient to show the distribution of X and Y chromosomes

When meiosis takes place in the female's ovary, each ovum receives one of the X chromosomes, so all the ova are the same for this. Meiosis in the male's testes results in 50 per cent of the sperms getting an X chromosome and 50 per cent getting a Y chromosome (Figure 22.6). If an X sperm fertilizes the ovum, the zygote will be XX and will grow into a girl. If a Y sperm fertilizes the ovum, the zygote will be XY and will develop into a boy. There is an equal chance of an X or Y chromosome fertilizing an ovum, so the numbers of girl and boy babies are more or less the same.

Question

1 A married couple has four girl children but no boys. This does not mean that the husband produces only X sperms. Explain why not.

Hereditary diseases

Most of the characteristics we think of as important are controlled by more than one pair of alleles, and can be influenced by our environment. Your height, for example, is probably determined by several allelomorphic pairs. It will also depend on whether you get enough food while you are growing.

The most striking single-factor characteristics are those associated with inherited defects such as sickle-cell anaemia, haemophilia, cystic fibrosis and colour blindness.

These four diseases are caused by recessive genes. It follows that people who develop the disease must have received both recessive alleles, one from each of their parents. This, in turn, means that both parents must be heterozygous for the gene. They are described as 'carriers' because they do not exhibit characteristics of the disease although they carry the recessive gene. If the father, for example, had been homozygous normal, he would have passed the dominant gene to all his children and none would be affected.

A few genetic diseases are caused by dominant alleles, which means that even heterozygotes will develop the disease.

Albinism

An albino lacks a gene for producing the pigment melanin (p. 137). As a result the iris of the eye is pink, the hair is white or very pale yellow, and the skin is unpigmented and easily damaged by sunlight. The albinism allele is recessive to the pigment-producing allele.

Haemophilia

Haemophilia is a genetic disease in which blood clots very slowly. The blood of people with the disease lacks one of the plasma proteins, called factor VIII, which plays a part in clotting. The production of factor VIII is controlled by a single gene. A person who lacks this gene will be a haemophiliac. Quite minor cuts tend to bleed for a long time and internal bleeding may occur.

Although haemophilia is a genetically controlled disease, symptoms vary from mild to severe in different individuals. The condition can be largely controlled by daily or weekly injections of factor VIII.

Cystic fibrosis

Cystic fibrosis is the commonest inherited disease in white people, affecting about one in every 2500 children. The disease occurs when a person inherits two copies of a recessive allele which controls the production of a protein in cell membranes. The protein normally controls the passage of chloride ions across the membrane but is defective in homozygous recessive individuals. As a consequence, the epithelial cells lining the respiratory passages and digestive glands produce a thick, sticky mucus which makes the person very susceptible to infections and interferes with digestion.

Trials are taking place, with genetically engineered genes, which may alleviate the condition (p. 215).

Sickle-cell anaemia

(See 'Incomplete dominance', p. 196.)

Huntington's disease

Unlike the disease described above, Huntington's disease is caused by a dominant allele and can therefore be passed on from the one parent who has the disorder. However, the gene is not expressed until the carrier reaches middle age. Thus, although persons carrying the gene will eventually develop the disease, they may have passed the gene on to 50 per cent of their children before they were aware of the condition.

The disease is rare, affecting about 1 person in 10 000.

It is characterized by a progressive deterioration in mental and physical faculties.

Because the allele is dominant, and because the phenotypic symptoms do not appear until after reproductive age, members of families in which the disease has occurred may choose not to have children.

Genetic screening and counselling

Screening

Tests can now be carried out to predict, with varying degrees of certainty, whether a person is carrying genes for a heritable disorder. These tests can be made on the potential parents, on babies or even on the fetus in the uterus.

For example, an inherited disease called **phenylketonuria (PKU)** is caused by a recessive gene. Children who are homozygous for PKU have such a high level of the amino acid, phenylalanine, in their blood that their brains are damaged and they become mentally handicapped. However, by using a diet which is free from phenylalanine in the first 6 years, the damage can be avoided. In Britain, all babies are routinely screened for PKU by taking a blood sample immediately after birth (Figure 22.7).

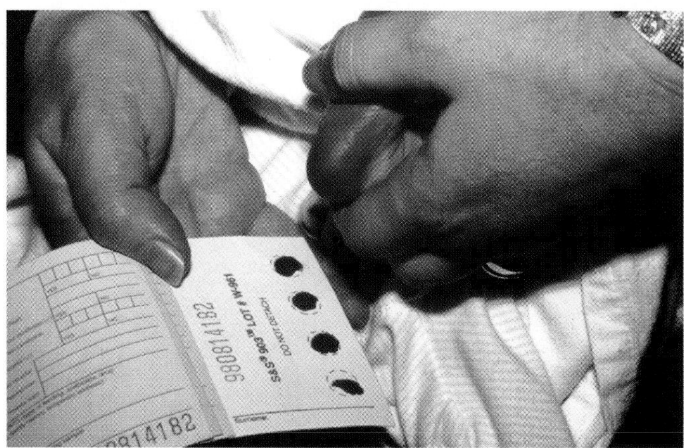

Figure 22.7 A blood sample is taken from the heel of a newborn baby to screen for PKU

A screening test for Down's syndrome (p. 189) is carried out by taking samples of amniotic fluid from a pregnant woman (by amniocentesis, Figure 22.8 and p. 147). This fluid will contain cells shed from the fetus. These cells can be examined under the microscope to see if they contain the extra chromosome indicative of Down's syndrome. In this case, the parents may have to decide whether or not to terminate the pregnancy.

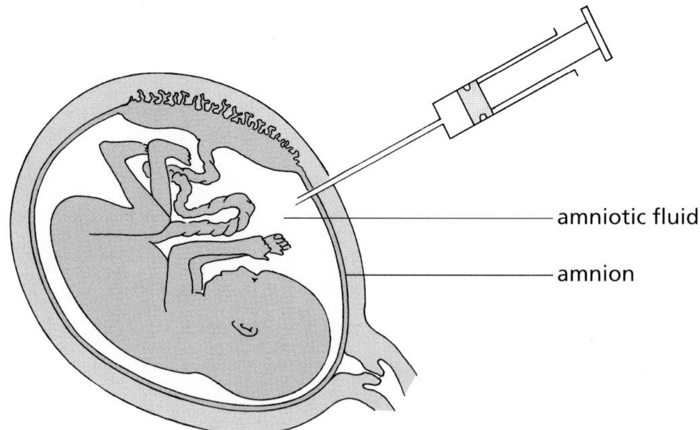

Figure 22.8 Screening by amniocentesis

Potential parents can also be screened to see if either or both are carrying recessive genes for a hereditary defect. Abnormal levels of sodium chloride in sweat may indicate (though not with 100 per cent certainty) that the individual is a carrier (i.e. heterozygous) for the cystic fibrosis gene. Carriers of the blood disease, sickle-cell anaemia (p. 196) can be detected by abnormalities in the size or shape of some of their red cells.

In many instances, it is now possible to detect the presence of a harmful gene in a DNA sample from a person's cells. This test may be used as a first resort or to confirm a less reliable screening test, e.g. for cystic fibrosis.

Who should be screened? Potential parents with a family history of a genetic disease are obvious candidates to see if they too carry the defective genes.

Routine screening of babies for PKU has the obvious advantage that, once recognized, the condition can be effectively treated by a special diet.

Although the gene for Huntington's disease can be detected by screening before the appearance of symptoms, the condition is incurable. Passing on the gene may be avoided but the affected person has to come to terms with the certainty of developing the disease.

There are also fears that widespread genetic screening for more and more conditions might lead to undue anxiety and also discrimination by employers and insurance companies.

You can now appreciate that genetic screening introduces many ethical problems. It would be very undesirable to use it, for example, to try to select for IQ, sex, physique or any natural attributes of an offspring.

Counselling

If normal parents have a child affected by an inherited disorder caused by a recessive gene, they can be sure that they are both heterozygous (**Nn**) for the gene.

Thus there is a 1 in 4 chance that any one of their later children will be affected too (see p. 194). If the condition is serious, the parents may decide that the risk of having another affected child is too high and therefore they will have no more children.

If one member of a couple has a genetic disorder, the chances that they will have an affected child can be worked out. For an albino man (**nn**) marrying a normal woman, this is done as follows:

1 About 1 person in 70 carries the recessive allele for albinism.
2 Thus there is a 1 in 70 chance that the woman is a carrier (heterozygous for albinism, **Nn**).
3 The chance of an affected child being born to an affected man (**nn**) and a heterozygous woman (**Nn**) is 1 in 2.
4 Combining probabilities 2 and 3 gives a chance of $\frac{1}{2} \times \frac{1}{70} = \frac{1}{140}$ that any one child will be an albino.

The parents may decide that this risk is low, and that it is worth starting a family.

If one of the parents obviously has the genetic defect, as in the case of albinism, there is clearly a case for seeking advice from a counsellor. If neither parent shows any symptoms and there is no family history, the presence of a faulty gene may not be recognized until the first child is born.

If there is a family history of a hereditary disease, or some other reason to suspect that an offspring might be affected, then genetic screening may be advised.

Question

1 A woman has the sickle-cell *trait* (p. 196). What are the chances of her children inheriting
(i) the sickle-cell trait
(ii) sickle-cell disease if she marries
a a normal man
b a man with sickle-cell trait
c a man with sickle-cell disease?

Checklist

- In breeding experiments, the effect of only one or two genes (out of thousands) is studied, e.g. colour of fur in rabbits or mice.
- The genes are in pairs (allelomorphic pairs), because the chromosomes are in pairs.
- Although each pair of alleles controls the same characteristic, the alleles do not necessarily have the same effect. For example, of a pair of alleles controlling fur colour, one may try to produce black fur and the other may try to produce white fur.
- Usually, one allele is dominant over the other, e.g. the allele (**B**) for black fur is dominant over the allele (**b**) for white fur.
- This means that a rabbit with the alleles **Bb** will be black even though it has a gene for white fur.
- Although **BB** rabbits and **Bb** rabbits are both black, only the **BB** rabbits will breed true.
- **Bb** black rabbits mated together are likely to have some white babies.
- The expectation is that, on average, there will be one white baby rabbit to every three blacks.
- Meiosis is the kind of cell division that leads to production of gametes.
- Only one of each chromosome pair goes into a gamete.
- A **Bb** rabbit would produce two kinds of gametes for coat colour; 50 per cent of the gametes would have the **B** gene and 50 per cent would have the **b** gene.
- In some cases, neither one of a pair of alleles is fully dominant over the other. This may be called incomplete dominance or codominance.
- Sex, in mammals, is determined by the X and Y chromosomes. Males are XY; females are XX.
- Most hereditary diseases result from a person inheriting two recessive alleles for the disease.
- Genetic counselling tries to assess the chances of having an affected child.
- Genetic screening can detect the presence of a defective gene or chromosome in some cases.

23 Variation, selection and evolution

Variation
New gene combinations and mutations. Meiosis and new gene combinations. Discontinuous variations: distinct differences. Continuous variation: graduated differences. Interaction of genes and environment.

Natural selection
The selection of more efficient varieties. Peppered moth: selection of the different forms. Sickle-cell anaemia: selection in different environments. Artificial selection for improved breeds.

The theory of evolution
First life; increasing complexity; vertebrate evolution.

Evidence for evolution
Fossil record. Evidence from anatomy. Extinction.

Variation

The term 'variation' refers to observable differences within a species. All domestic cats belong to the same species, i.e. they can all interbreed, but there are many variations of size, coat colour, eye colour, fur length, etc. Those variations which can be inherited are determined by genes. They are genetic or heritable variations.

There are also variations which are not heritable, but determined by factors in the environment. A kitten which gets insufficient food will not grow to the same size as its litter mates. A cat with a skin disease may have bald patches in its coat. These conditions are not heritable. They are caused by environmental effects. Similarly, a fair-skinned person may be able to change the colour of his or her skin by exposing it to the sun, so getting a tan. The tan is an **acquired characteristic**. You cannot inherit a suntan. Black skin, on the other hand, is an **inherited characteristic**.

Many features in plants and animals are a mixture of acquired and inherited characteristics (Figure 23.1). For example, some fair-skinned people never go brown in the sun, they only become sunburned. They have not inherited the genes for producing the extra brown pigment in their skin. A fair-skinned person with the genes for producing pigment will only go brown if he exposes himself to sunlight. So his tan is a result of both inherited and acquired characteristics.

Heritable variation may be the result of mutations (p. 189), or new combinations of genes in the zygote.

Figure 23.1 Acquired characteristics. These apples have all been picked from different parts of the same tree. All the apples have similar genotypes, so the differences in size must have been caused by environmental effects

New combinations of genes

If a grey cat with long fur is mated with a black cat with short fur, the kittens will all be black with short fur. If these offspring are mated together, in due course the litters may include four varieties: black–short, black–long, grey–short and grey–long. Two of these are different from either of the parents. (See 'Meiosis and new combinations of characteristics' on p. 201.)

Mutations

Many of the coat variations mentioned above may have arisen, in the first place, as mutations in a wild stock of cats. A recent variant produced by a mutation is the 'rex' variety, in which the coat has curly hairs.

Many of our high-yielding crop plants have arisen as a result of mutations in which the whole chromosome set has been doubled.

Meiosis and new combinations of characteristics

On p. 185 it was explained that, during meiosis, homologous chromosomes pair up and then, at the first nuclear division, separate again.

One of the homologous chromosomes comes from the male parent and the other from the female parent. The alleles for a particular characteristic occupy identical positions on the homologous chromosomes but they do not necessarily control the characteristic in the same way.

The allele for eye colour will be in the same position on the maternal and paternal chromosome but, in one case, it may be the allele for brown eyes and in the other case, for blue eyes. Separation of homologous chromosomes at meiosis means that the alleles for blue and brown eyes will end up in different gametes (Figure 23.2).

On a second pair of homologous chromosomes there may be allelomorphic genes for hair curliness (**C** = curly; **c** = straight). These chromosomes and their alleles will also be separated at the first division of meiosis.

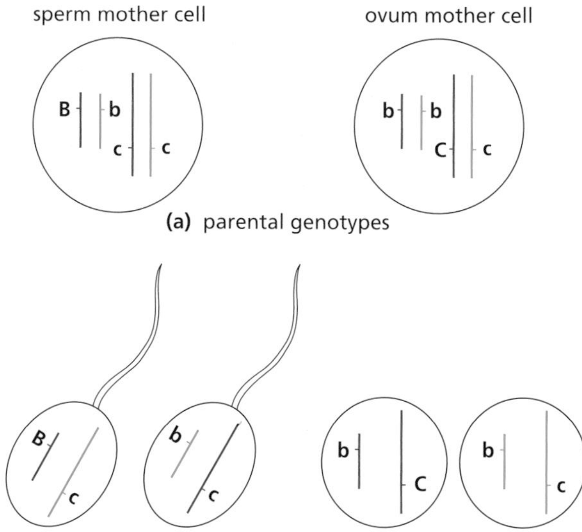

(a) parental genotypes

(b) possible gene combinations in gametes

ova / sperms	bC	bc
Bc	(1) Bb Cc	(2) Bb cc
bc	(3) bb Cc	(4) bb cc

(c) possible combinations of genes in the children

Figure 23.2 New combinations of characteristics

Suppose a father has brown eyes and straight hair, and a mother has blue eyes and curly hair. Also suppose that the father is heterozygous for eye colour (**Bb**) and the mother is heterozygous for hair curliness (**Cc**) (Figure 23.2a).

In the mother's ovary, the **b** and **b** alleles will be separated at the first division of meiosis and so will the **C** and **c** alleles. The **b** allele could finish up in the same gamete as either the **C** or **c** allele. So, the genotype of the ovum could be either **bC** or **bc** (Figure 23.2b). Similarly, meiosis in the father's testes will produce equal numbers of **Bc** and **bc** gametes.

At fertilization, it is a matter of chance which of the two types of sperm fertilizes which of the two types of ovum. (Although usually only one ovum is released, there is a 50:50 chance of its being **bC** or **bc**.)

The grid (**Punnett square**), in Figure 23.2c shows the possible genotypes of the children in the family. Offspring (2) and (3) would have the same combination of characteristics as their parents: (2) has brown eyes and straight hair (father's phenotype), and (3) has blue eyes and curly hair (mother's phenotype). Offspring (1) and (4), however, would have different combinations of these two characteristics; these combinations are not present in either parent, namely, (1) brown eyes and curly hair, (4) blue eyes and straight hair.

The separation of parental chromosomes at meiosis and their recombination at fertilization has thus introduced the possibility of new combinations of characteristics. It occurs because the homologous chromosomes derived from one parent do not all go into the same gamete, but move independently of each other.

This recombination of characteristics as a result of meiosis is important for plant and animal breeding programmes as described on p. 211. It is also important as a source of variation for natural selection to act on, as described on p. 203.

Meiosis takes place only at gamete formation and this is an essential feature of sexual reproduction. It can be claimed, therefore, that one of the important biological advantages of sexual reproduction is the production of new varieties which might be more successful than existing varieties.

Crossing over

At an early stage in meiosis, when homologous chromosomes are lying alongside each other, the chromatids break at various points. They may then rejoin with their original chromatid or with a chromatid from their homologous partner. As a result, portions of the maternal chromatids become attached to paternal chromatids and vice versa (Figure 23.3). Thus there will be new combinations of genes on these chromatids when they separate and produce gametes. This is a further source of genetic variation which results from meiosis.

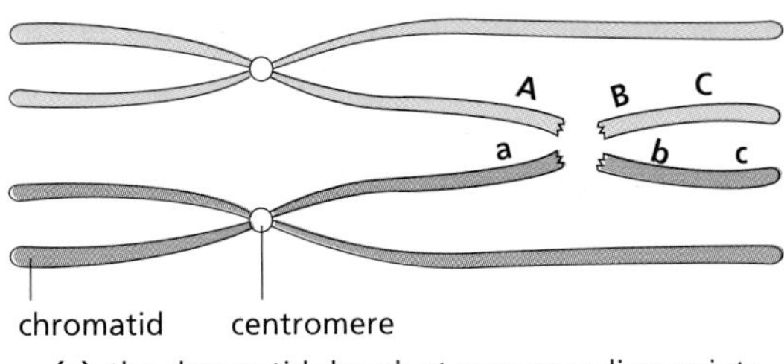

(a) the chromatids break at corresponding points

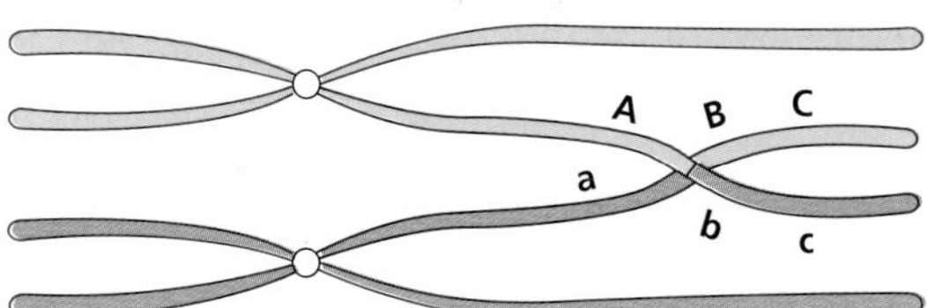

(b) they rejoin but to their opposite partner

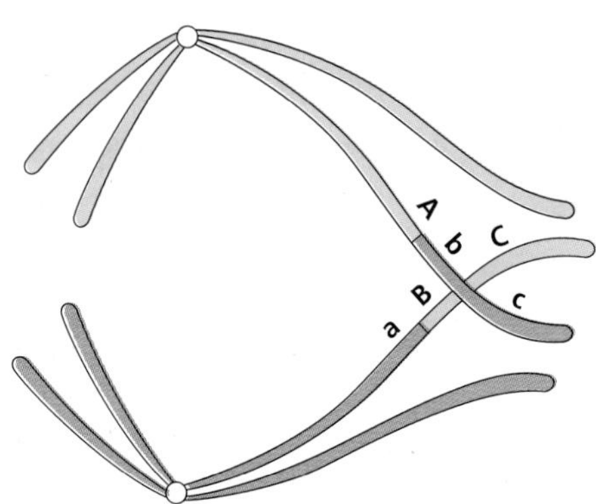

(c) when they separate they carry new gene combinations into the gametes

Figure 23.3 Crossing over

Questions

1 Which of the following do you think are
a mainly inherited characteristics,
b mainly acquired characteristics or
c a more or less equal mixture:
manual skills, facial features, body build, language, athleticism, ability to talk?

2 a Bearing in mind the role of the sex chromosomes, suggest two other new variations which might occur among the children in Figure 23.2.
b In Figure 23.2, if the mother had been homozygous for hair curliness and the father had been homozygous for eye colour, is there a possibility of new combinations of those characters in the offspring?

3 What new combinations of characters are possible as a result of crossing a tall plant with yellow seeds **(TtYy)** with a dwarf plant with green seeds **(ttyy)**?

4 What are the environmental effects which might have caused the variation in apple size in Figure 23.1?

5 What new combination of genes has arisen from the crossing over in Figure 23.3?

Discontinuous variations

These are variations under the control of a single pair of alleles or a small number of genes. The variations take the form of distinct, alternative phenotypes with no intermediates (Figure 23.4). The mice in Figure 22.4 on p. 194 are either black or brown; there are no intermediates. You are either male or female. Apart from a small number of abnormalities, sex is inherited in a discontinuous way.

Discontinuous variations cannot usually be altered by the environment. You cannot change your blood group or eye colour by altering your diet. A genetic dwarf cannot grow taller by eating more food.

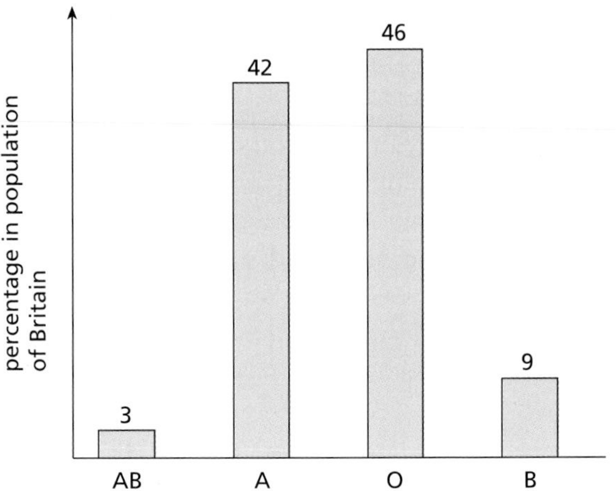

Figure 23.4 Discontinuous variation. Frequencies of ABO blood groups in Britain. The figures could not be adjusted to fit a smooth curve because there are no intermediates

Continuous variation

There are no distinct categories of height; people are not either tall or short. There are all possible intermediates between very short and very tall. This is a case of continuous variation (Figure 23.5).

Continuously variable characteristics are usually controlled by several pairs of alleles. There might be five pairs of alleles for height – (**Hh**), (**Tt**), (**Ll**), (**Ee**) and (**Gg**) – each dominant allele adding 4 cm to your height. If you inherited all ten dominant genes (**HH, TT**, etc.) you could be 40 cm taller than a person who inherited all ten recessive genes (**hh**, **tt**, etc.).

The actual number of genes which control height, intelligence, and even the colour of hair and skin, is not known.

Continuously variable characteristics are greatly influenced by the environment. A person may inherit genes for tallness and yet not get enough food to grow tall. A plant may have the genes for large fruits but not get enough water, minerals or sunlight to produce large fruits. Continuous variations in human populations, such as height, physique and intelligence, are always the result of interaction between the genotype and the environment.

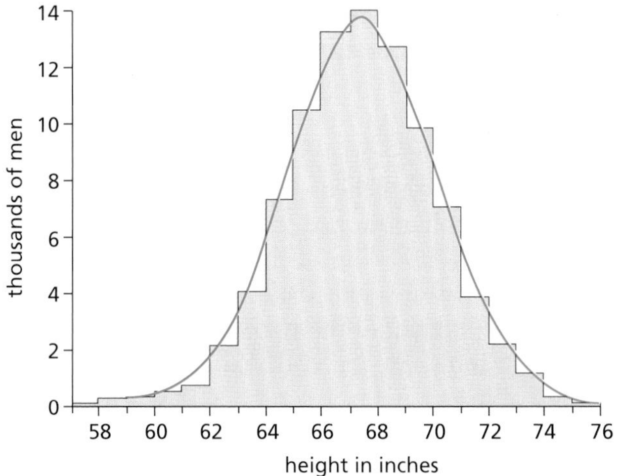

Figure 23.5 Continuous variation. Heights of 90 000 recruits in 1939. The apparent 'steps' in the distribution are the result of arbitrarily chosen categories, differing in height by one inch. But heights do not differ by exactly one inch. If measurements could be made accurately to the nearest millimetre, there would be a smooth curve like the one shown in colour

There are many characteristics which are difficult to classify as either wholly continuous or discontinuous variations. Human eye colour has already been mentioned. People can be classified roughly as having blue eyes or brown eyes, but there are also categories described as grey, hazel or green eyes. Probably there is a small number of genes for eye colour and a dominant gene for brown eyes which overrides all the others when it is present. Similarly, red hair is a discontinuous variation but it is masked by genes for other colours and there is a continuous range of hair colour from blond to black.

Natural selection

Theories of evolution have been put forward in various forms for hundreds of years. In 1858, Charles Darwin and Alfred Russel Wallace published a theory of evolution by natural selection which is still an acceptable theory today (see pp. 350–1).

The theory of natural selection suggests that:

- Individuals within a species are all slightly different from each other (Figure 23.6). These differences are called **variations**.
- If the climate or food supply changes, some of these variations may be better able to survive than others. A variety of animal that could eat the leaves of shrubs as well as grass would be more likely to survive a drought than one which fed only on grass.
- If one variety lives longer than others, it is also likely to leave behind more offspring. A mouse that lives for 12 months may have ten litters of five babies (50 in all). A mouse that lives for 6 months may have only five litters of five babies (25 in all).
- If some of the offspring inherit the variation that helped the parent survive better, they too will live longer and have more offspring.
- In time, this particular variety will outnumber and finally replace the original variety.

This is sometimes called 'the survival of the fittest'. However, 'fitness', in this case, does not mean good health but implies that the organism is well fitted to the conditions in which it lives.

Thomas Malthus, in 1798, suggested that the increase in the size of the human population would outstrip the rate of food production. He predicted that the number of people would eventually be regulated by famine, disease and war. When Darwin read the Malthus essay, he applied its principles to other populations of living organisms.

He observed that animals and plants produce vastly more offspring than can possibly survive to maturity and he reasoned that, therefore, there must be a 'struggle for survival'.

For example, if a pair of rabbits had eight offspring which grew up and formed four pairs, eventually having eight offspring per pair, in four generations the number of rabbits stemming from the original pair would be 512 (i.e. $2 \rightarrow 8 \rightarrow 32 \rightarrow 128 \rightarrow 512$). The population of rabbits, however, remains more or less constant. Many of the offspring in each generation must, therefore, have failed to survive to reproductive age.

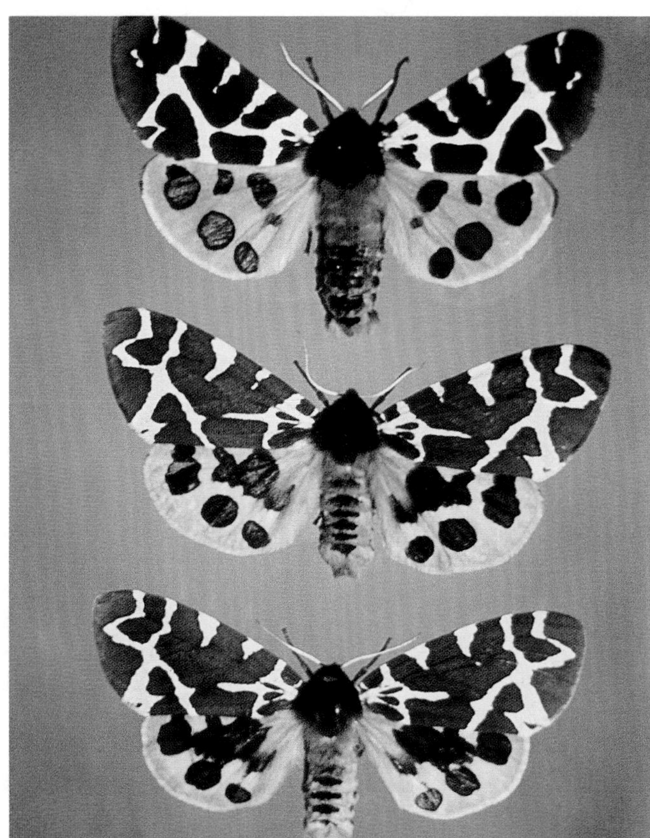

Figure 23.6 Variation. The garden tiger moths in this picture are all from the same family. There is a lot of variation in the pattern on the wings

Competition and selection

There will be **competition** between members of the rabbit population for food, burrows and mates. If food is scarce, space is short and the number of potential mates limited, then only the healthiest, most vigorous, most fertile and otherwise well-adapted rabbits will survive and breed.

The competition does not necessarily involve direct conflict. The best adapted rabbits may be able to run faster from predators, digest their food more efficiently, have larger litters or grow coats which camouflage them better or more effectively reduce heat losses. These rabbits will survive longer and leave more offspring. If the offspring inherit the advantageous characteristics of their parents, they may give rise to a new race of faster, different coloured, thicker furred and more fertile rabbits which gradually replace the original, less well-adapted varieties. The new variations are said to have **survival value**.

This is natural selection; the better adapted varieties are 'selected' by the pressures of the environment (**selection pressures**).

For natural selection to be effective, the variations have to be heritable. Variations which are not heritable are of no value in natural selection. Training may give athletes more efficient muscles, but this characteristic will not be passed on to their children.

Evolution

Most biologists believe that natural selection, among other processes, contributes to the evolution of new species and that the great variety of living organisms on the Earth is the product of millions of years of evolution, involving natural selection (p. 205).

The peppered moth

A possible example of natural selection is provided by a species of moth called the peppered moth. The common form is speckled but there is also a variety which is black. The black variety was rare in 1850, but by 1895 in the Manchester area its numbers had risen to 98 per cent of the population of peppered moths. Observation showed that the light variety was concealed better than the dark variety when they rested on tree-trunks covered with lichens (Figure 23.7). In the Manchester area, pollution had caused the death of the lichens and the darkening of the tree-trunks with soot. In this industrial area the dark variety was the better camouflaged (hidden) of the two and was not picked off so often by birds. So the dark variety survived better, left more offspring and nearly replaced the light form.

The selection pressure, in this case, was presumed to be mainly predation by birds. The adaptive variation which produced the selective advantage was the dark colour.

Although this is an attractive and plausible hypothesis of how natural selection could occur, some of the evidence does not support the hypothesis or has been called into question.

For example, the moths settle most frequently on the under-side of branches rather than conspicuously on tree trunks, as in Figure 23.7. Also, in several unpolluted areas the dark form is quite abundant, for example 80 per cent in East Anglia. Research is continuing in order to test the hypothesis.

Sickle-cell anaemia

This condition has already been mentioned on p. 196. A person with sickle-cell disease has inherited both recessive alleles (**Hb^{S}Hb^S**) for defective haemoglobin. The distortion and destruction of the red cells which occurs in low oxygen concentrations leads to bouts of severe anaemia (Figure 23.8). In many African countries, sufferers have a reduced chance of reaching reproductive age and having a family. There is thus a selection pressure which tends to remove the homozygous recessives from the population. In such a case, you might expect the harmful **Hb^S** allele to be selected out of the population altogether. However, the heterozy-

(a) **(b)**

(a) and **(b)** show the two forms against the background on which they are best concealed

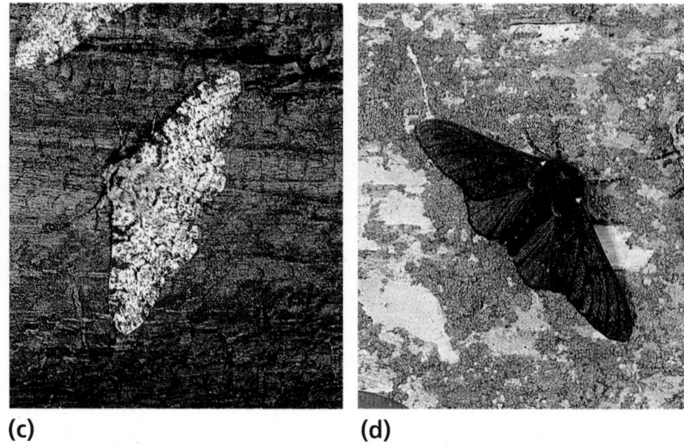

(c) **(d)**

(c) and **(d)** show the forms on alien backgrounds

Figure 23.7 Selection for varieties of the peppered moth

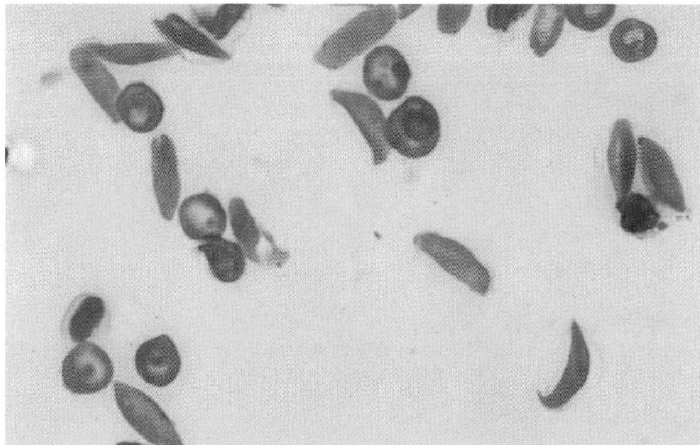
Figure 23.8 Sickle-cell anaemia (×800). At low oxygen concentration the red cells become distorted

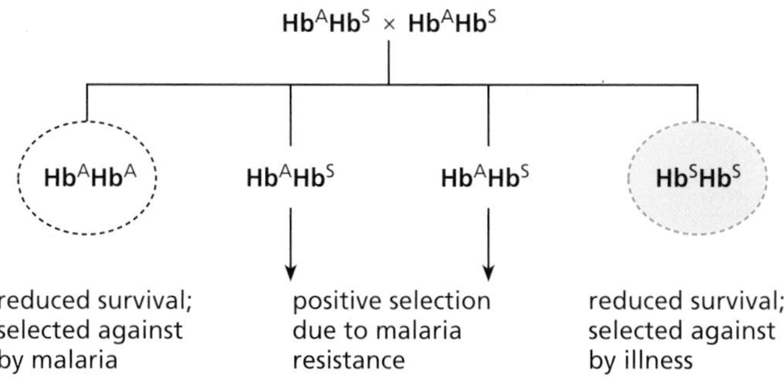

Figure 23.9 Selection in sickle-cell disease

gotes ($\mathbf{Hb^AHb^S}$) have virtually no symptoms of anaemia but do have the advantage that they are more resistant to malaria than the homozygotes $\mathbf{Hb^AHb^A}$.

The selection pressure of malaria, therefore, favours the heterozygotes over the homozygotes and the potentially harmful $\mathbf{Hb^S}$ allele is kept in the population (Figure 23.9).

When Africans migrate to countries where malaria does not occur, the selective advantage of the $\mathbf{Hb^S}$ allele is lost and the frequency of this allele in the population diminishes.

Artificial selection

Human communities practise a form of selection when they breed plants and animals for specific characteristics. The many varieties of dog that you see today have been produced by selecting individuals with short legs, curly hair, long ears, etc. One of the puppies in a litter might vary from the others by having longer ears. This individual, when mature, is allowed to breed. From the offspring, another long-eared variant is selected for the next breeding stock, and so on, until the desired or 'fashionable' ear length is established in a true-breeding population (Figure 23.10).

More important are the breeding programmes to improve agricultural livestock or crop plants. Animal-breeders will select cows for their high milk yield, sheep for their wool quality, and pigs for their long backs (more bacon rashers). Plant-breeders will select varieties for their high yield and resistance to fungus diseases (Figure 23.11). (See also p. 211.)

Figure 23.10 Selective breeding. Three varieties of dog produced by artificial selection over many years

Figure 23.11 Selective breeding in tomatoes. Different breeding programmes have selected genes for fruit size, colour and shape. Similar processes have given rise to most of our cultivated plants and domesticated animals

Questions

1. What features of a bird's appearance and behaviour do you think might help it compete for a mate?
2. What selection pressures do you think might be operating on the plants in a lawn?

The theory of evolution

On p. 203 it was pointed out that Darwin's theory of natural selection in 1858 provided a good explanation for the process of evolution, but theories of evolution were in existence long before that.

Aristotle (384–322 BC) suggested a classification system which implied a progression from simple to complex creatures. However, the spread of Judaeo-Christian religion effectively deterred people from putting forward explanations that differed from the biblical story of Creation until the 18th and 19th centuries.

Jean Baptiste Lamarck (1744–1829) was a professor of zoology in Paris. He was perhaps the first person to suggest openly that species could change. His explanation, however, depended on the selection of characteristics acquired during the lifetime of the organism rather than selection of characteristics that arose genetically by chance (p. 349).

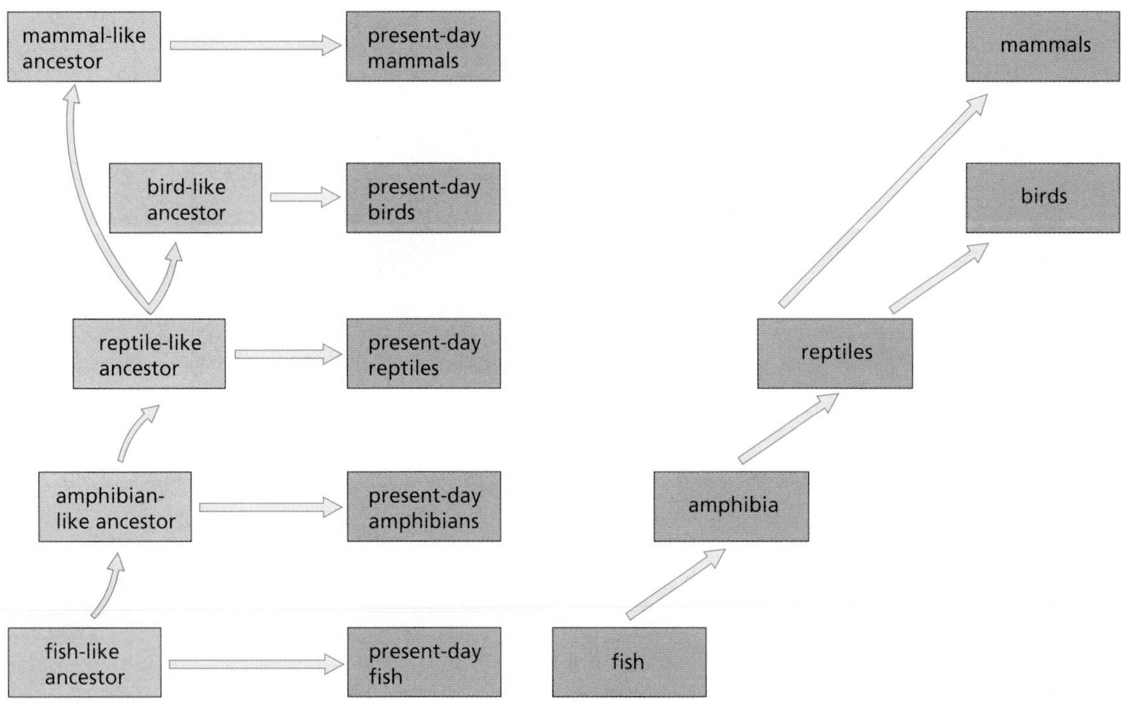

(a) 'Correct' representation of vertebrate evolution.

(b) 'Incorrect' representation of vertebrate evolution. It seems to suggest that present-day fish gave rise to present-day amphibia, and so on.

Figure 23.12 The theory of evolution. This says that fish and amphibia, for example, share a common ancestor. It does *not* say that fish give rise to amphibia. However, we do not know for sure what the common ancestors were like

The first life

The theory, as it stands at present, suggests that at one time there were no living organisms of any kind on the Earth, possibly because it was too hot. As the Earth cooled, the theory supposes that conditions were just right for certain kinds of chemical reaction to take place in the water. These chemical reactions might have produced compounds like amino acids and enzymes, which could make other reactions take place. If these chemicals somehow came together inside a membrane, they would form the first single-celled creatures, such as bacteria, which could feed, grow and reproduce. There is a great deal of argument among biologists about how this could have happened or whether it could have happened at all. It is possible to make amino acids (p. 11) in laboratory experiments in the conditions suggested, but it is impossible to be confident that these really were the conditions on the Earth millions of years ago.

Organisms became more numerous and more complicated

The first single-celled creatures might have given rise to many-celled creatures if the cells stuck together after cell division. The many-celled creatures could have started as a simple ball of cells. Some cells might have become specialized to carry food, conduct nerve impulses, or contract to produce movement. As a result, the creatures would become more complicated and different forms would arise (Figure 23.13). Some could stick to rocks and produce tentacles like sea anemones. Some might swim in the surface waters like shrimps or burrow in the sand like worms.

Organisms came on to the land

If evolution has occurred, it most likely started in the water. Gradually some plants and animals became able to survive on land for longer and longer periods. Until green plants developed, there was probably no oxygen in the air. The plants' photosynthesis produced oxygen and this would have made it possible for animals to evolve.

Evolution of vertebrates

Some biologists think that most evolutionary change takes place as a result of a series of very small changes ('gradualism'). Others claim that there were long periods with no change and then short periods of rapid change ('punctuated equilibrium'). In either case it would take millions of years for a fish-like creature to turn into a frog-like creature. Over a period of about 400 million years it is thought that some fish-like creatures developed legs and lungs and so became amphibia, like newts. Some amphibian ancestors could have developed scales and the ability to lay eggs on land and so become reptiles. One group of ancient reptiles might have given rise to the birds by developing feathers and wings and another group, by developing fur and producing milk, could have become the mammals.

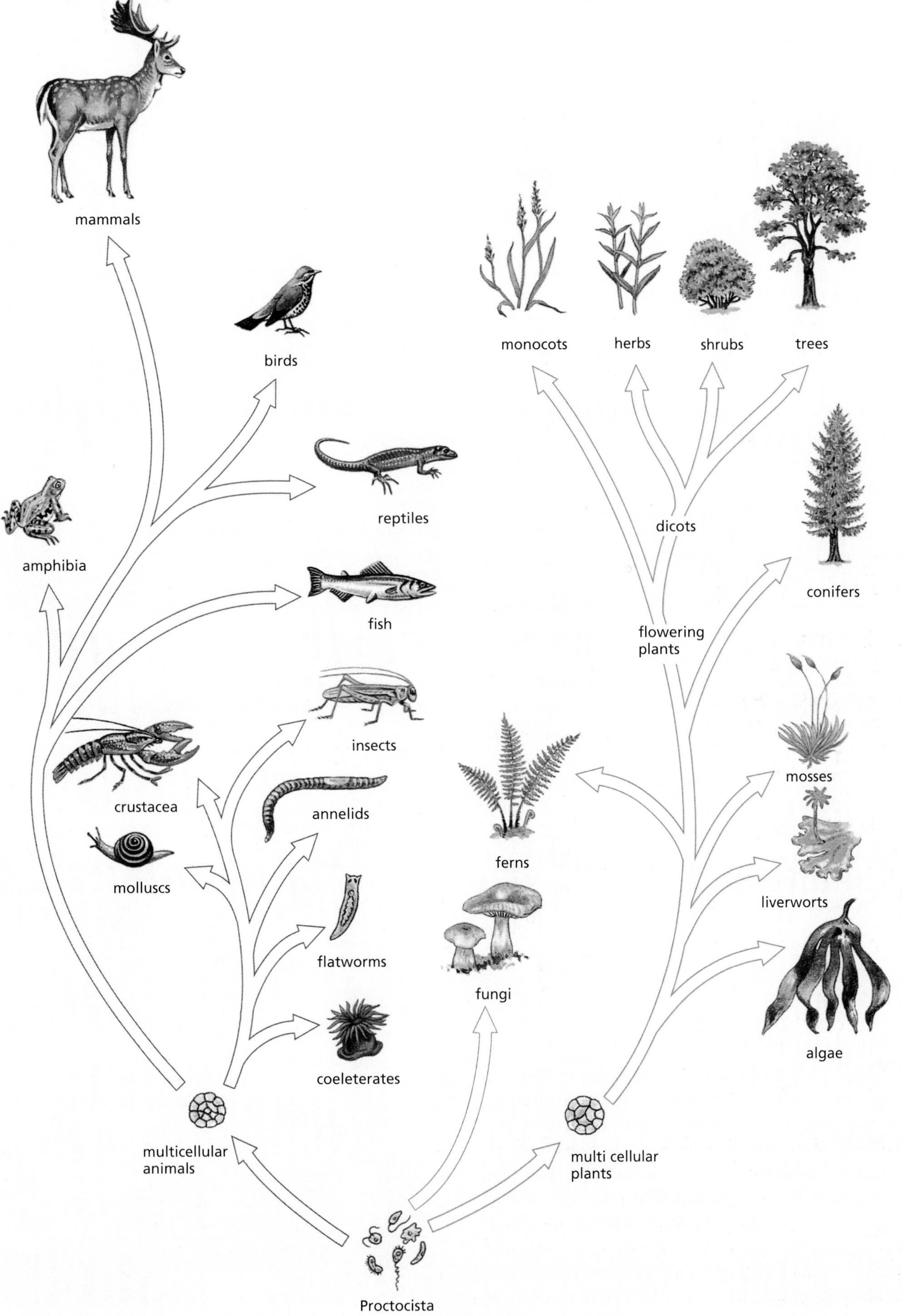

Figure 23.13 Organic evolution. Only a few of the main types of organism are shown

The theory of evolution does not suggest that frogs turned into lizards or that lizards turned into birds. Frogs and lizards are themselves thought to be the products of millions of years of evolution from ancestral amphibia and reptiles. It is supposed that some of the ancestral amphibia gave rise to two main groups of descendants. One of these groups developed into primitive reptiles and the other continued evolving to become frogs and toads (Figure 23.12).

Evidence for evolution

Fossils

Fossils are the remains of animals and plants preserved in various ways and Figure 23.14 shows how a skeleton of the fish might become embedded in mud or sand which was settling down on the bottom of a lake. After a few million years, with the pressure of more layers building up on top of it, the mud or sand becomes rock. Massive earth movements may raise the rocks above the water and when they form cliffs or are quarried, the preserved skeletons can be found and chipped out of the rock.

This is not the only method of fossilization. Any environment that prevents rapid decay may produce fossils. Insects have been found trapped in amber which is formed from resin exuding from trees; pollen grains are preserved in the anaerobic conditions of bogs and their fossil remains are found in peat; whole animals have been found frozen in ancient ice.

It is usually the hard tissues which are preserved, e.g. bones, shells, wood, seeds, etc. These decay more slowly than the softer tissues and, in many cases, the organic matter is entirely replaced by minerals which do not decay.

The skeletons, shells, tree-trunks and other fossil remains give us some idea of the animals and plants that were living millions of years ago. The deepest layers of rock are likely to contain the oldest fossils (Figure 23.14). So by studying the fossil record, the theory of evolution can be put to the test.

The extent of the fossil remains of the five vertebrate classes in rock layers of increasing depth is shown in Figure 23.15. The width of the coloured bands represents the range of fossil remains found. For example, 100 million years ago, it looks as if there were many more varieties of reptile than there are today. Also you will notice that fossils of mammals do not appear at all in rocks which are older than about 300 million years.

The fossil record appears to support the idea that fish-like creatures could have given rise to mammals because, 300 million years ago, there were plenty of fish but no mammals. Most of the organisms preserved in the fossil record appear to be different from the organisms we know today. They are obviously related but different in important ways. For example, Figure 23.16 (overleaf) shows a fish that lived 350 million years ago. It is clearly a fish, but quite different from any fish living today.

So the fossil record shows that plants and animals have changed from one form to another over a long period – or does it?

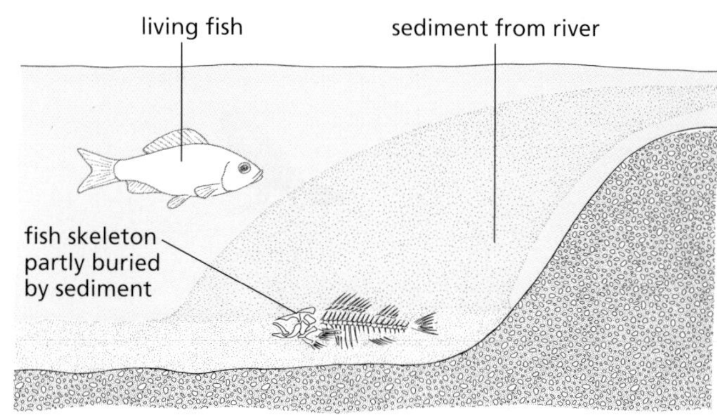

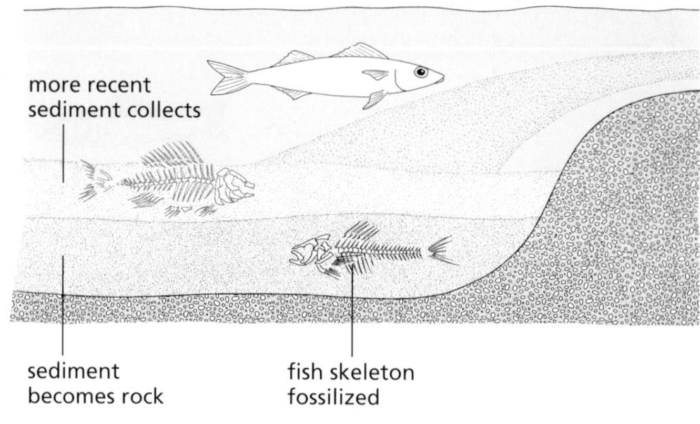

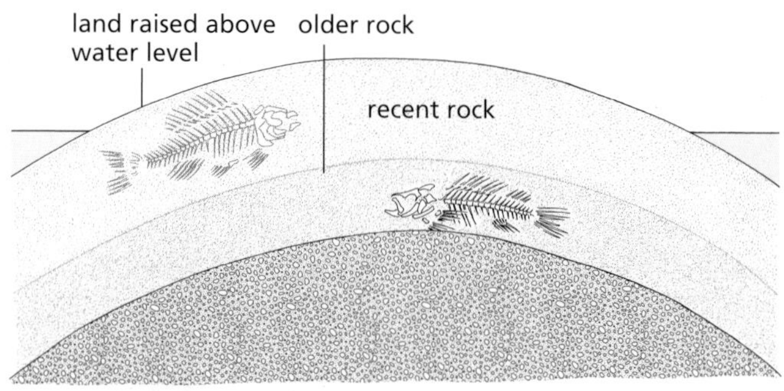

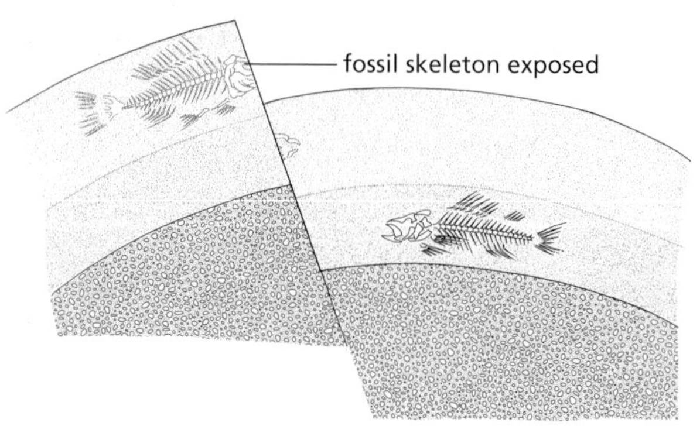

Figure 23.14 Formation of fossils

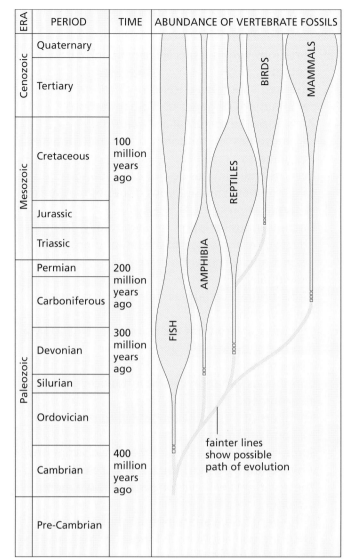

ERA	PERIOD	TIME	ABUNDANCE OF VERTEBRATE FOSSILS
Cenozoic	Quaternary		
	Tertiary		
Mesozoic	Cretaceous	100 million years ago	
	Jurassic		
	Triassic		
Paleozoic	Permian	200 million years ago	
	Carboniferous		
	Devonian	300 million years ago	
	Silurian		
	Ordovician		
	Cambrian	400 million years ago	
	Pre-Cambrian		

FISH
AMPHIBIA
REPTILES
BIRDS
MAMMALS
fainter lines show possible path of evolution

Figure 23.15 Abundance of vertebrate fossils found in rocks

Could it be that 350 million years ago there were all the fish, reptiles, amphibia, birds and mammals that we know today plus all those we know as fossils? Perhaps there were so few mammals at that time that we have not yet found any in the rocks of that period. In such a case, all that the fossil record would show is that some creatures have become more abundant, some have become scarce and some have died out altogether.

This seems a very improbable interpretation of the fossil evidence and fails to offer any scientific explanation of the origin of life or the diversity of organisms.

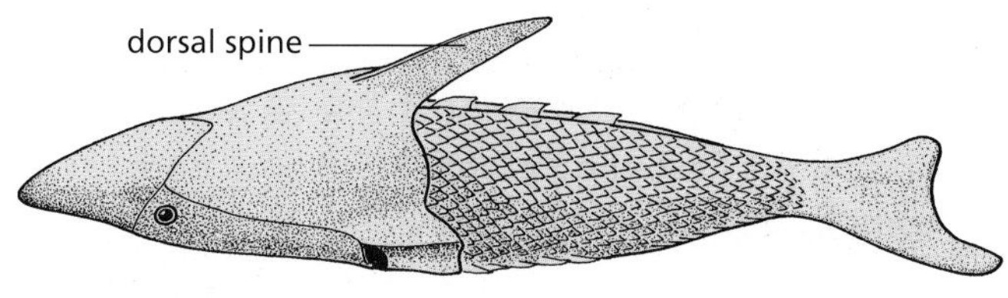

Figure 23.16 An extinct 'armour-plated' fish of 350 million years ago (reconstructed from fossil remains)

Evidence from anatomy

The skeletons of the front limb of five types of vertebrate are shown in Figure 23.17. Although the limbs have different functions such as grasping, flying, running and swimming, the arrangement and number of the bones is almost the same in all five. There is a single top bone (humerus), with a ball and socket joint at one end and a hinge joint at the other. It makes a joint with two other bones (radius and ulna) which join to a group of small wrist bones. The limb skeleton ends with five groups of bones (hand and fingers), although some of these groups are missing in the bird.

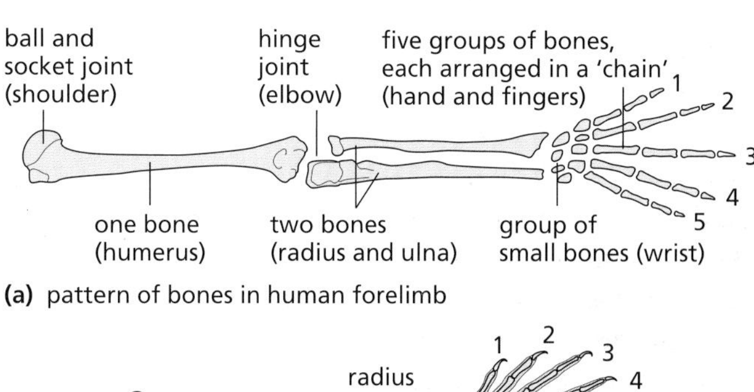

(a) pattern of bones in human forelimb

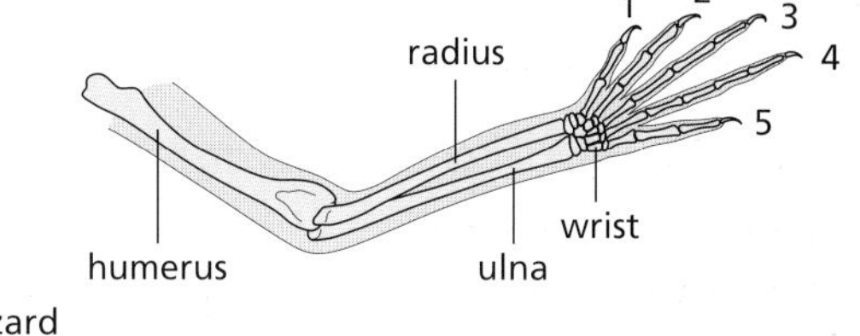

(b) lizard

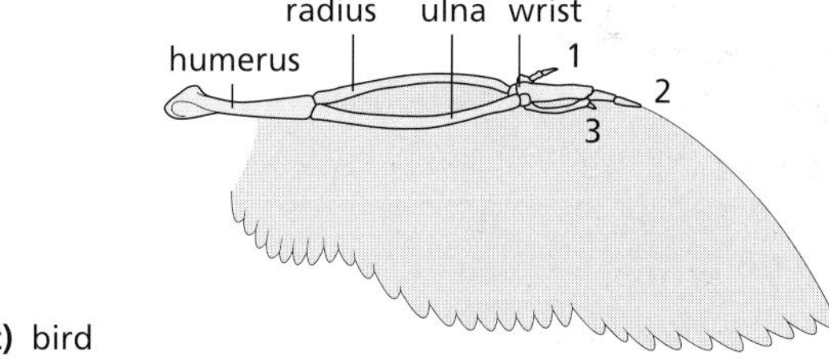

(c) bird

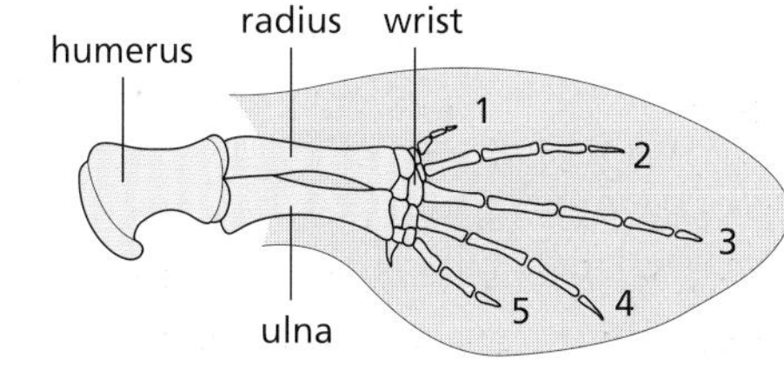

(d) whale

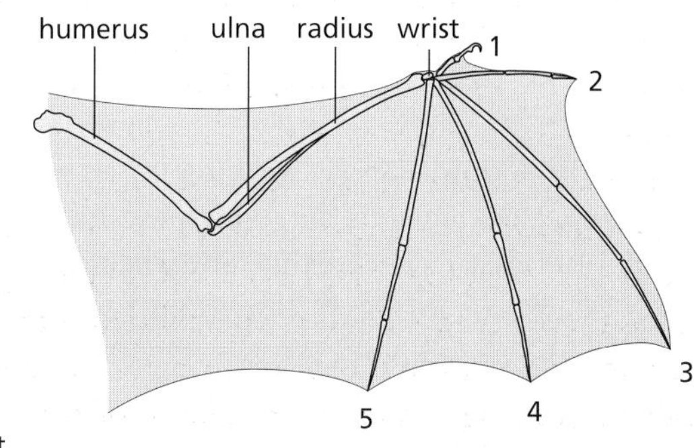

(e) bat

Figure 23.17 Skeletons of five vertebrate limbs

The argument for evolution says that, if these animals are not related, it seems very odd that such a similar limb skeleton should be used to do such different things as flying, running and swimming. If, on the other hand, all the animals came from the same ancestor, the ancestral skeleton could have changed by small stages in different ways in each group. So we would expect to find that the basic pattern of bones was the same in all these animals. There are many other examples of this kind of evidence among the vertebrate animals.

Questions

1 According to Figure 23.15 when were the amphibia most abundant?

2 What features of Figure 23.16 suggest that the fossil animal is related to present-day fishes?

3 In Figure 23.13 why do you think that mammals and birds are drawn at the top of the diagram and worms and insects lower down?

Extinction

From the fossil record, it is obvious that a great many organisms have become extinct. It is not usually possible to be sure of the causes of extinction in the distant past. Perhaps there were drastic changes in climate, outbreaks of disease or an invasion of more successful competitors. There have been many periods of mass extinction, one of which included the decline of the dinosaurs 65 million years ago. One theory suggests that the impact of a giant asteroid produced such a vast dust cloud that the Earth became dark and cold. As a result, plants and the animals which fed on them died out.

More recent extinctions usually result from human activities. The dodo, a flightless bird on the island of Mauritius, was driven to extinction in 1680 by hunting and the introduction of pigs. The Polynesians are thought to have exterminated half the population of birds on the islands which they colonized in the 4th and 5th centuries.

The expanding human population is hunting some animals to the verge of extinction and destroying their habitats by cutting down forests for timber or agriculture. The tiger and the giant panda are obvious examples.

The World Conservation Monitoring Centre estimates that over 300 species have become extinct since 1600 and Friends of the Earth claim that one species per day is being lost for ever. The exact figures may be disputed but the message is clear.

Checklist

- Variations within a species may be inherited or acquired.
- Inherited variations arise from different combinations of genes or from mutations.
- At meiosis the maternal and paternal chromosomes are randomly distributed between the gametes.
- Because the gametes do not carry identical sets of genes, new combinations of genes may arise at fertilization.
- Discontinuous variation results, usually, from the effects of a single pair of alleles, and produces distinct and consistent differences between individuals.
- Discontinuous variations cannot be changed by the environment.
- Continuous variations are usually controlled by a number of genes affecting the same characteristic.
- Continuous variation can be influenced by the environment.
- Members of a species compete with each other for food and mates.
- Some members of a species may have variations which enable them to compete more effectively.
- These variants will live longer and leave more offspring.
- If the beneficial variations are inherited, the offspring will also survive longer.
- The new varieties may gradually replace the older varieties.
- Natural selection involves the elimination of less well-adapted varieties by environmental pressures.
- Artificial selection is used to improve commercially useful plants and animals.
- The theory of evolution tries to explain how present-day plants and animals came into existence.
- It supposes that simple living organisms, such as bacteria, were formed from non-living matter.
- These simple organisms changed over hundreds of millions of years to become many-celled and more complicated.
- They also became more varied, giving rise to many different groups of plants and animals.
- The evidence from the fossil record supports the theory of evolution because it shows animals and plants becoming more varied and more complicated as time goes on.

24 Applied genetics

Selective breeding
Cross-breeding. Selection.

Genetic engineering
Restriction enzymes, plasmids and recombinant DNA.

Applications of genetic engineering
Enzymes, hormones and vaccines. Gene therapy. GM crops and food.

Cloning
Asexual reproduction. Cloned animals. Stem cells.

The principles of genetic fingerprinting
Separation of restriction fragments.

The human genome project
Objectives. Mapping. Sequencing. Applications.

A knowledge of genetics can be applied in a number of ways, some of which have already been described. For example, artificial selection is mentioned on p. 205; genetic screening and counselling are introduced on p. 198.

Plant- and animal-breeding programmes may be achieved by conventional crossing of varieties followed by selection of those offspring with the desired characteristics. Similar results can be achieved by genetic engineering.

Aspects of human health can be addressed by screening and counselling but also, in the future, by gene therapy.

Selective breeding

Cross-breeding

It is possible for biologists to use their knowledge of genetics to produce new varieties of plants and animals. For example, suppose one variety of wheat produces a lot of grain but is not resistant to a fungus disease. Another variety is resistant to the disease but has only a poor yield of grain. If these two varieties are cross-pollinated (Figure 24.1), the F_1 offspring should be disease-resistant and give a good yield of grain (assuming that the useful characteristics are controlled by dominant genes).

R represents a dominant allele for resistance to disease, and **r** is the recessive allele for poor resistance. **H** is a dominant allele for high yield and **h** is the recessive allele for low yield. The high-yield/low-resistance variety (**HHrr**) is crossed with the low-yield/high-resistance variety (**hhRR**). Each pollen grain from the **HHrr** plant will contain one **H** and one **r** allele (**Hr**). Each ovule from the **hhRR** plant will contain an **h** and an **R** allele (**hR**). The seeds will, therefore, all be **HhRr**. The plants which grow from these seeds will have dominant alleles for both high yield and good disease resistance.

The offspring from crossing two varieties are called **hybrids**. If the F_1 hybrids from this cross bred true, they could give a new variety of disease-resisting, high-yielding wheat. However, as you learned on p. 194, the F_1 generation from a cross does not necessarily breed true. The F_2 generation of wheat may contain:

- high-yield, disease-resistant
- low-yield, disease-prone
- low-yield, disease-resistant } parental
- high-yield, disease-prone } types

This would not give such a successful crop as the F_1 plants.

With some commercial crops, the increased yield from the F_1 seed makes it worthwhile for the seedsman to make the cross and sell the seed to the growers. The hybrid corn (maize) grown in America is one example. The F_1 hybrid gives nearly twice the yield of the standard varieties.

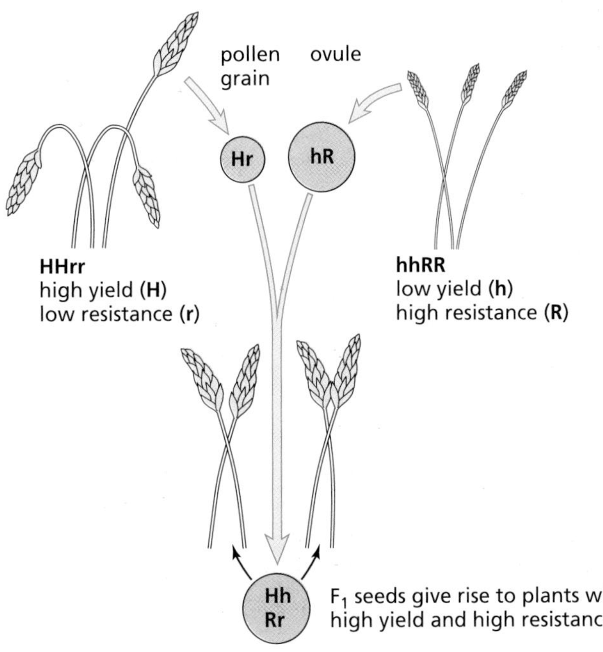

Figure 24.1 Combining useful characteristics

Figure 24.2 The genetics of bread wheat. A primitive wheat **(a)** was crossed with a wild grass **(b)** to produce a better-yielding hybrid wheat **(c)**. The hybrid wheat **(c)** was crossed with another wild grass **(d)** to produce one of the varieties of wheat **(e)** which is used for making flour and bread

In other cases it is possible to work out a cross-breeding programme to produce a hybrid which breeds true (Figure 24.2). If, instead of the **HhRr** in Figure 24.1, an **HHRR** could be produced, it would breed true.

Selection

An important part of any breeding programme is the selection of the desired varieties. The largest fruit on a tomato plant might be picked and its seeds planted next year. In the next generation, once again only seeds from the largest tomatoes are planted. Eventually it is possible to produce a true-breeding variety of tomato plant which forms large fruits. Figure 23.11(p. 205) shows the result of such selective breeding.

The same technique can be used for selecting other desirable qualities, such as flavour and disease resistance.

Similar principles can be applied to farm animals. Desirable characteristics, such as high milk yield and resistance to disease, may be combined. Stock-breeders will select calves from cows which give large quantities of milk. These calves will be used as breeding stock to build a herd of high yielders. A characteristic such as milk yield is probably under the control of many genes. At each stage of selective breeding the farmer, in effect, is keeping the beneficial genes and discarding the less useful genes from his or her animals.

Selective breeding in farm stock can be slow and expensive because the animals often have small numbers of offspring and breed only once a year.

By producing new combinations of genes, selective breeding achieves the same objectives as genetic engineering but it takes much longer and is less predictable.

In selective breeding, the transfer of genes takes place between individuals of the same or closely related species. Genetic engineering involves transfer between unrelated species.

Selective breeding and genetic engineering both endeavour to produce new and beneficial combinations of genes. Selective breeding, however, is much slower and less precise than genetic engineering. On the other hand, cross-breeding techniques have been around for a very long time and are widely accepted.

One of the drawbacks of selective breeding is that the whole set of genes is transferred. As well as the desirable genes, there may be genes which, in a homozygous condition, would be harmful. It is known that artificial selection repeated over a large number of generations tends to reduce the fitness (p. 203) of the new variety.

A long-term disadvantage of selective breeding is the loss of variability. By eliminating all the offspring who do not bear the desired characteristics, many genes are lost from the population. At some future date, when new combinations of genes are sought, some of the potentially useful ones may no longer be available.

In attempting to introduce, in plants, characteristics such as salt tolerance or resistance to disease or drought, the geneticist goes back to wild varieties, as shown in Figure 24.2. However, with the current rate of extinction, this source of genetic material is diminishing.

In the natural world, reduction of variability could lead to local extinction if the population was unable to adapt, by natural selection, to changing conditions.

Questions

1 Suggest some good characteristics that an animal-breeder might try to combine in sheep by mating different varieties together.

2 A variety of barley has a good ear of seed but has a long stalk and is easily blown over. Another variety has a short, sturdy stalk but a poor ear of seed.
Suggest a breeding programme to obtain and select a new variety which combined both of the useful characteristics.
Choose letters to represent the genes and show the genotypes of the parent plants and their offspring.

3 A farmer plants a crop of F_1 maize and is pleased with the yield. Hybrid seed is expensive so he saves seed from his own crop and plants it the following year.
Explain why he might be disappointed with the result.

Genetic engineering

Genetic engineering involves the transfer of genes from one organism to (usually) an unrelated species.

To understand the principles of genetic engineering you need to know something about **bacteria** (p. 283) and **restriction enzymes**.

Bacteria are microscopic single-celled organisms with cytoplasm, cell membranes and cell walls, but without a proper nucleus. Genetic control in a bacterium is exercised by a double strand of deoxyribonucleic acid (DNA) in the form of a circle, but not enclosed in a nuclear membrane. This circular DNA strand carries the genes that control bacterial metabolism.

In addition, there are present in the cytoplasm a number of small, circular pieces of DNA called **plasmids**. The plasmids often carry genes which give the bacterium resistance to particular antibiotics such as tetracycline and ampicillin.

Restriction enzymes are produced by bacteria. They 'cut' DNA molecules at specific sites, e.g. between the A and the T in the sequence GAA–TTC. Restriction enzymes can be extracted from bacteria and purified. By using a selected restriction enzyme, DNA molecules extracted from different organisms can be cut at predictable sites and made to produce lengths of DNA which contain specific genes.

DNA from human cells can be extracted and restriction enzymes used to 'cut' out a sequence of DNA which includes a gene, e.g. the gene for production of insulin (Figure 24.3). Plasmids are extracted from bacteria and 'cut open' with the same restriction enzyme. If the human DNA is then added to a suspension of the plasmids, some of the human DNA will attach to some of the plasmids, which will then close up again, given suitable enzymes. The DNA in these plasmids is called **recombinant DNA**.

The bacteria can be induced to take up the plasmids and, by ingenious culture methods using antibiotics, it is possible to select the bacteria which contain the recombinant DNA. The human DNA in the plasmids continues to produce the same protein as it did in the human cells. In the example mentioned, this would be the protein, insulin (p. 171). The plasmids are said to be the **vectors** which carry the human DNA into the bacteria and the technique is sometimes called **gene-splicing**.

Given suitable nutrient solutions, bacteria multiply rapidly and produce vast numbers of offspring. The bacteria reproduce by mitosis and so each daughter bacterium will contain the same DNA and the same plasmids as the parent. The offspring form a **clone** and the insulin gene is said to be **cloned** by this method.

The bacteria are cultured in special vessels called **fermenters** (p. 330) and the insulin which they produce can be extracted from the culture medium and purified for use in treating diabetes (p. 171).

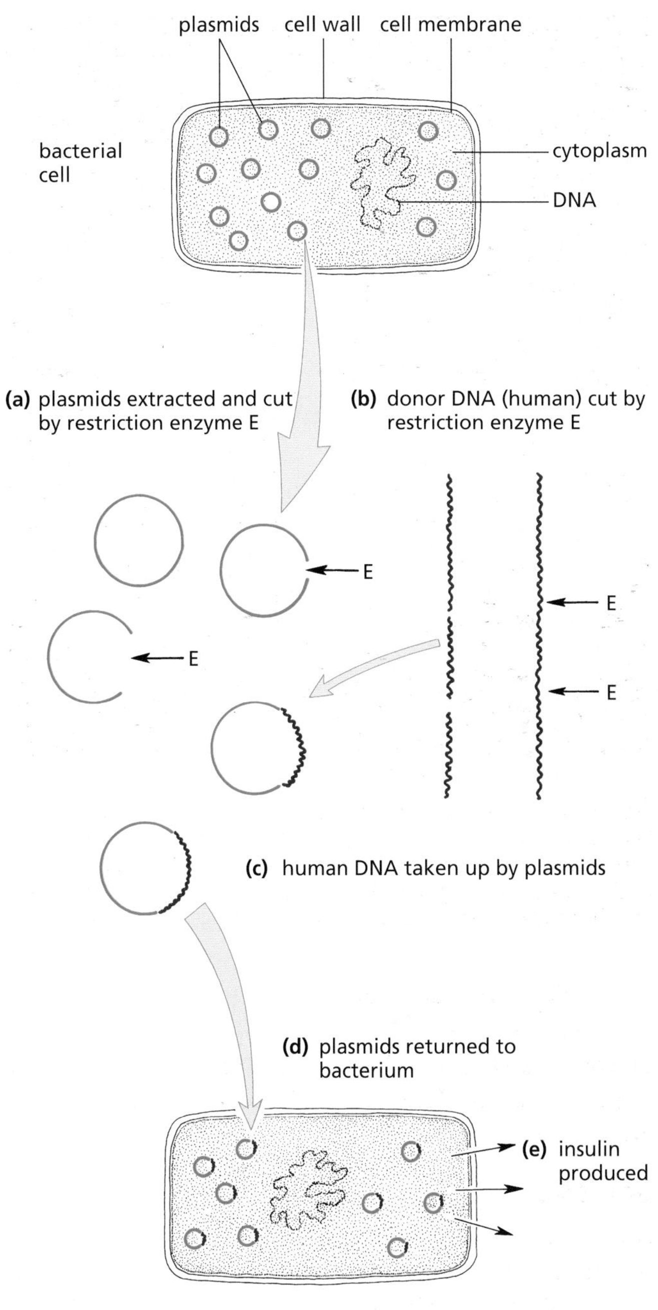

Figure 24.3 The principles of genetic engineering

This is only one type of genetic engineering. The vector may be a virus rather than a plasmid; the DNA may be inserted directly, without a vector; the donor DNA may be synthesized from nucleotides rather than extracted from cells; yeast may be used instead of bacteria. The outcome, however, is the same. DNA from one species is inserted into a different species and made to produce its normal proteins (Figure 24.3).

In the example shown in Figure 24.3, the gene product, insulin, is harvested (p. 330) and used to treat diabetes. In other cases, genes are inserted into organisms to promote changes which may be beneficial. Bacteria or viruses are used as vectors to deliver the genes. For example, a bacterium is used to deliver a gene for herbicide resistance in crop plants.

Applications of genetic engineering

It is possible to list only a few examples of the progress of genetic engineering, a progress becoming more rapid every day. Some products, such as insulin and chymosin, are in full scale production. A few genetically modified (**GM**) crops, e.g. maize and soya bean, are being grown on a large scale in the USA. Many other projects are still at the experimental stage, undergoing trials, awaiting approval by regulatory bodies or simply on a 'wish list'.

Enzymes, hormones and vaccines

Chymosin

This enzyme is naturally secreted by the gastric glands in the stomach lining. It coagulates the protein, casein, in milk as the first step in its digestion. Commercially, the enzyme is used in the manufacture of hard cheese. It precipitates the milk casein, forming a solid product that is further processed to make the cheese.

For a long time, the enzyme used for this process was extracted from calves' stomachs. When this supply became inadequate, other sources were found. Today, 90 per cent of hard cheese is made using chymosin produced by genetically modified micro-organisms, principally yeast. The GM chymosin is purer and more reliable than that from other sources.

Alpha-anti-trypsin

This is an enzyme inhibitor which blocks the action of protease enzymes in the lungs of people suffering from hereditary emphysema (p. 128). The gene for its production has been transferred to the milk-producing cells in the mammary glands of sheep. These sheep secrete the anti-enzyme in their milk, from which it can be extracted and purified. Clinical trials are planned.

Animals and plants which have been genetically engineered to produce substances that are not part of their normal metabolism are called **transgenic** (Figure 24.4).

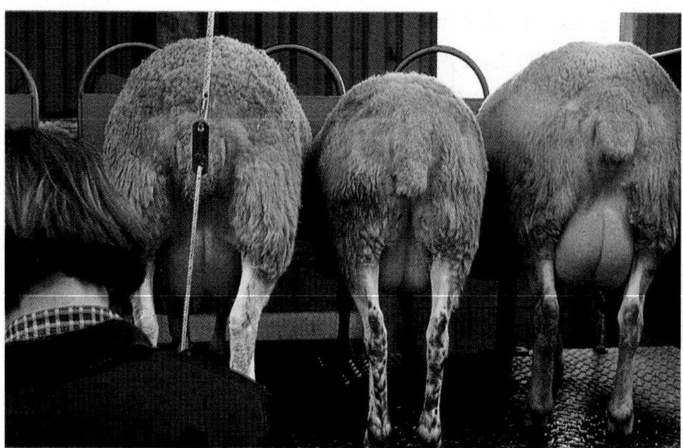

Figure 24.4 Transgenic sheep. The three sheep sharing the milking platform are all daughters of a transgenic ram and all three produce alpha-anti-trypsin in their milk

An anti-blood-clotting agent used in heart surgery has been produced in the milk of transgenic goats. This product is in the final stages of clinical trials.

There are at least five different transgenic species, engineered to produce specific proteins in their milk, including an anti-cancer antibody.

Hepatitis B vaccine

The gene for the protein coat (p. 285) of the hepatitis virus is inserted into yeast cells. When these are cultured, they produce a protein which acts as an antigen (a vaccine, p. 118) and promotes the production of antibodies to the disease.

Transgenic plants have been engineered to produce vaccines that can be taken effectively by mouth. These include vaccines against rabies and cholera. Several species of plant have been used including the banana, which is cheap and widespread in the tropics, can be eaten without cooking and does not produce seeds (Figure 24.5).

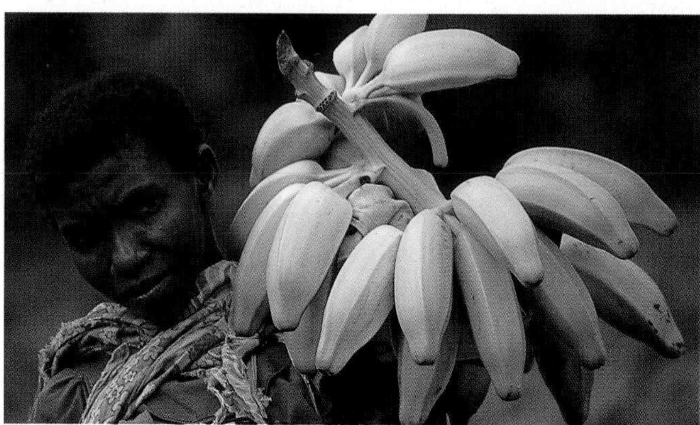

Figure 24.5 It is important to ensure that plants engineered to produce drugs and vaccines cannot find their way, by chance, into the human food chain. Strict control measures have to be applied

Insulin

This hormone can be produced by genetically modified bacteria and has been in use since 1982. The human insulin produced in this way (Figure 24.6) is purer than insulin prepared from pigs or cattle, which sometimes provokes allergic reactions owing to traces of 'foreign' protein. Another hormone produced from GM bacteria is human growth hormone.

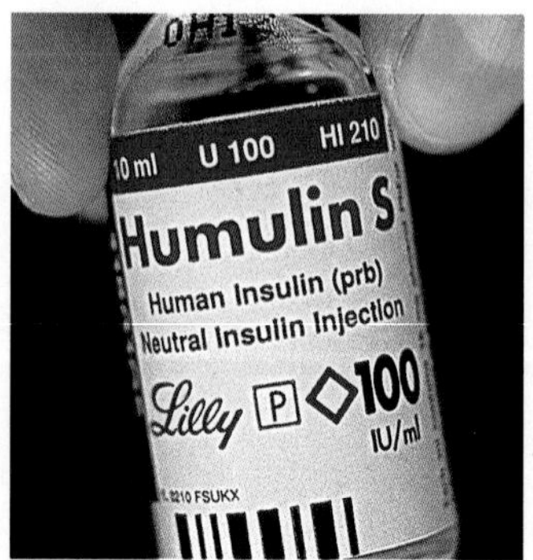

Figure 24.6 Human insulin prepared from genetically engineered bacteria. Though free from foreign proteins, it does not suit all patients

Gene therapy

The purpose of gene therapy is to effect a cure by using recombinant DNA, delivered by a vector or by direct injection. The therapy is directed principally towards hereditary diseases such as cystic fibrosis and sickle-cell anaemia (p. 198) or towards cancer.

Hereditary diseases result from the inheritance of a defective gene. If a normal gene could be introduced permanently into the affected cells the disease would be cured once and for all. The healthy gene is delivered either by viruses or **liposomes** or by direct injection. The virus vectors have to be disabled so that, although they can get into the affected cells, they cannot reproduce. Liposomes are microscopic droplets containing the gene and surrounded with a lipid membrane, which fuses with the cell's plasma membrane to introduce the genes.

Viruses are good at getting into cells but are too small to carry effective quantities of DNA. They can also cause inflammation. Liposomes can carry more DNA but are less efficient at penetrating cells.

Whichever delivery method is used, unless the recombinant DNA is incorporated into the cell's nuclear DNA, its effects are not permanent and the therapy has to be repeated.

Cystic fibrosis

Absence of a gene for a cell membrane protein leads to the production of a thick, sticky mucus in the airways and makes the patient very susceptible to infections (Figure 24.7). Trials are taking place in which the normal gene is sprayed into the patient's lungs. The vector is either recombinant viruses or liposomes. The virus is the disabled common cold virus; although it cannot reproduce it can cause inflammation.

The introduced gene is incorporated into the lung epithelial cells but is effective for only a short period and the treatment has to be repeated regularly.

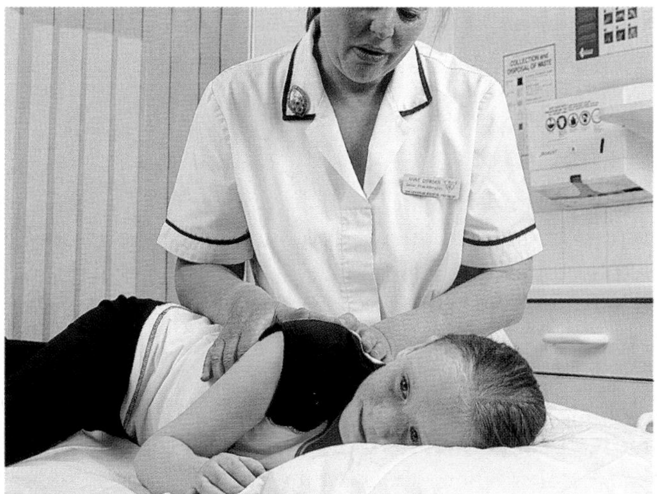

Figure 24.7 Physiotherapy for a young cystic fibrosis patient: regular treatment is required to move copious viscous mucus from the lung, to help breathing

Cancers

Researchers are working on, for example, lung cancer, breast cancer and skin cancer. One technique involves taking white blood cells from the patient and introducing a gene for **tumour necrosis factor** (TNF). The lymphocytes are returned to the patient and, on reaching the tumour, they release the TNF, which kills the cancer cells but does not harm normal cells. Clinical trials are taking place.

A similar technique has produced promising results with cancers of the kidney and prostate gland. A sample of tumour cells has a gene inserted which alters the antigens on the cell membrane and so programmes the patient's T cells (p. 118) to kill the rest of the tumour cells. Direct injection of similar DNA (in liposomes) into the tumour has had early successes with skin cancer.

GM crops

Genetic engineering has huge potential benefit in agriculture but, apart from a relatively small range of crop plants, most developments are in the experimental or trial stages. In the USA, 50 per cent of the soya bean crop and 30 per cent of the maize harvest consist of genetically modified plants which are resistant to herbicides and insect pests.

In the UK at the moment, GM crops are grown only on a trial basis.

Pest resistance

The bacterium, *Bacillus thuringiensis*, produces a toxin which kills caterpillars and other insect larvae. The toxin has been in use for some years as an insecticide. The gene for the toxin has been successfully introduced into some plant species using a bacterial vector. The plants produce the toxin and show increased resistance to attack by insect larvae. The gene is also passed on to the plant's offspring. Unfortunately there are signs that insects are developing immunity to the toxin.

Most American GM maize, apart from its herbicide-resistant gene, also carries a pesticide gene which reduces the damage caused by a stem-boring larva of a moth (Figure 24.8).

Figure 24.8 The maize stem borer can cause considerable losses by killing young plants

Herbicide resistance

Some of the safest and most effective herbicides are those, such as glyphosate, which kill any green plant but become harmless as soon as they reach the soil. These herbicides cannot be used on crops because they kill the crop plants as well as the weeds. A gene for an enzyme which breaks down glyphosate can be introduced into a plant cell culture (p. 81). This should lead to a reduced use of herbicides.

Modifying plant products

A gene introduced to oilseed rape and other oil-producing plants can change the nature of the oils they produce to make them more suitable for commercial processes, e.g. detergent production. This might be very important when stocks of petroleum run out. It could be a renewable source of oil which would not contribute to global warming (p. 245).

The tomatoes in Figure 24.9 have been modified to improve their keeping qualities.

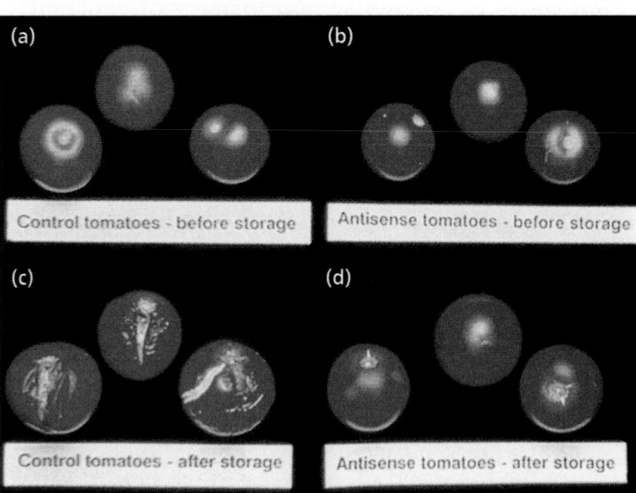

Figure 24.9 Genetically engineered tomatoes. In the engineered tomatoes (labelled antisense), biologists have deleted the gene which produces the enzyme which makes fruit go soft. **(a)** and **(c)** Control tomatoes before and after storage. **(b)** and **(d)** Genetically engineered tomatoes before and after storage

Other applications

One of the objections to GM crops is that, although they show increased yields, this has benefited only the farmers and the chemical companies in the developed world. So far, genetic engineering has done little to improve yields or quality of crops in the developing world, except perhaps in China. In fact, there are a great many trials in progress which hold out hopes of doing just that. Here are just a few.

Inadequate intake of iron is one of the major dietary deficiencies (p. 88), world wide. An enzyme in some plant roots enables them to extract more iron from the soil. The gene for this enzyme can be transferred to plants, such as rice, enabling them to extract iron from iron-deficient soils.

Over 100 million children in the world are deficient in vitamin A. This deficiency often leads to blindness. A gene for beta-carotene, a precursor of vitamin A (p. 88), can be inserted into plants to alleviate this widespread deficiency. This is not, of course, the only way to increase vitamin A availability but it could make a significant contribution.

Some acid soils contain levels of aluminium which reduce yields of maize by up to 80 per cent. About 40 per cent of soils in tropical and subtropical regions have this problem. A gene introduced into maize produces citrate, which binds the aluminium in the soil and releases phosphate ions. After 15 years of trials, the GM maize was made available to farmers, but pressure from environmental groups has blocked its adoption.

As a result of irrigation, much agricultural land has become salty and unproductive. Transferring a gene for salt tolerance from, say, mangrove plants to crop plants could bring these regions back into production.

If the gene, or genes, for nitrogen fixation (p. 229) from bacteria or leguminous plants could be introduced to cereal crops, yields could be increased without the need to add fertilizers.

Similarly, genes for drought resistance would make arid areas available for growing crops.

Genes coding for human vaccines have been introduced into plants (p. 214).

Possible hazards of GM crops

One of the possible harmful effects of planting GM crops is that their modified genes might get into wild plants. If a gene for herbicide resistance found its way, via pollination, into a 'weed' plant, this plant might become resistant to herbicides and so become a 'super weed'. The purpose of field trials is to assess the likelihood of this happening. Until it is established that this is a negligible risk, licences to grow GM crops will not be issued.

To prevent the transfer of pollen from GM plants, other genes can be introduced which stop the plant from producing pollen and induce the seeds and fruits to develop without fertilization. This is a process which occurs naturally in many cultivated and wild plants.

Apart from specific hazards, there is also a sense of unease about introducing genes from one species into a totally different species. This is something which does not happen 'in nature' and therefore long-term effects are not known. In conventional cross-breeding, the genes transferred come from the same, or a closely related, species. However, in cross-breeding the whole raft of genes is transferred and this has sometimes had bad results when genes other than the target genes have combined to produce harmful products. Genetic engineering offers the advantage of transferring only those genes which are required.

The differences between the genetic make-up of different organisms is not as great as we tend to think. Plants and animals share 60 per cent of their genes and humans have 50 per cent of their genes in common with fruit flies. Not all genetic engineering involves transfer of 'alien' genes. In some cases it is the plant's own genes which are modified to improve its success in the field.

At least some of the protests against GM crops may be ill-judged (Figure 24.10).

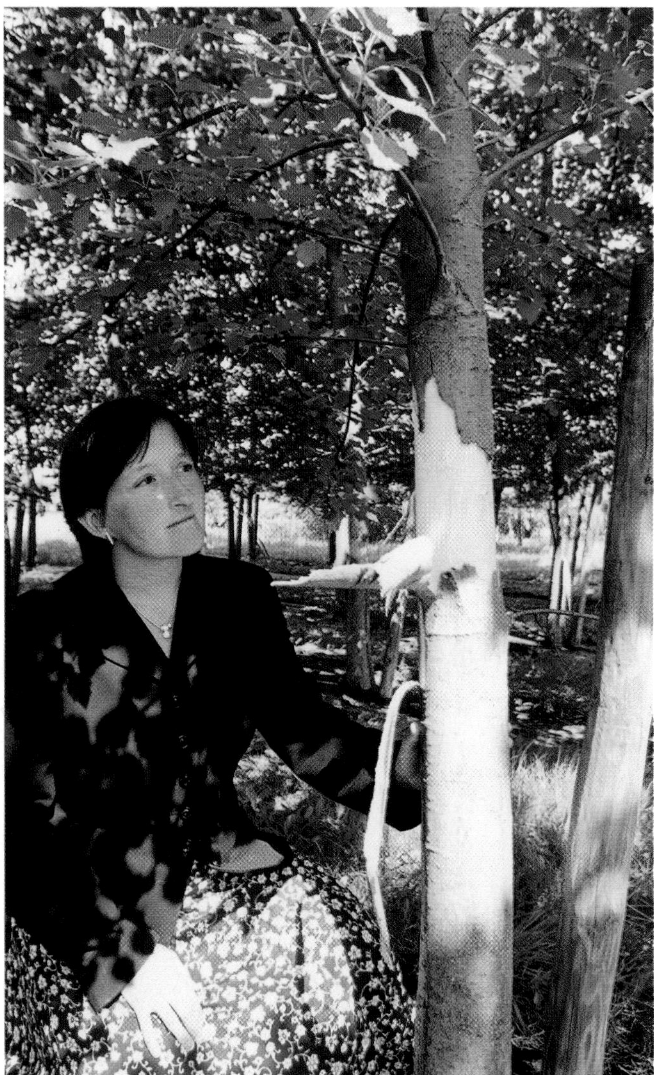

Figure 24.10 Ill-judged protest. These vandalized poplars carried a gene that softened the cell walls, reducing the need for environmentally damaging chemicals used in paper making. They were also all female plants so no pollen could have been produced

GM food

This is food prepared from GM crops. Most genetic modifications are aimed at increasing yields rather than changing the quality of food. However, it is possible to improve the protein, mineral or vitamin content of food and the keeping qualities of some products (Figure 24.9).

Possible hazards of GM food

One of the worries is that the vectors for delivering recombinant DNA contain genes for antibiotic resistance. The antibiotic-resistant properties are used to select only those vectors which have taken up the new DNA. If, in the intestine, the DNA managed to get into potentially harmful bacteria, it might make them resistant to antibiotic drugs.

Although there is no evidence to suggest this happens in experimental animals, the main biotech companies are trying to find methods of selecting vectors without using antibiotics.

Another concern is that GM food could contain pesticide residues or substances which cause allergies (allergens). However, it has to be said that all GM products are rigorously tested for toxins and allergens over many years, far more so than any products from conventional cross-breeding. The GM products have to be passed by a series of regulatory and advisory bodies before they are released on to the market. In fact only a handful of GM foods are available. One of these is soya, which is included, in one form or another, in 60 per cent of processed foods.

Cloning

Cloning refers, in effect, to any form of asexual reproduction. Since meiosis and recombination (p. 201) are not involved, all the offspring, whether cells or organisms, are genetically identical and form a **clone** (Figure 24.11). Natural propagation by bulbs, corms and rhizomes (p. 79) produces clones of identical plants. Gardeners clone plants artificially by propagating from cuttings or by grafting.

Tissue culture (pp. 9 and 81) is a form of cloning. Cells in culture divide mitotically and each cell receives the diploid set of chromosomes.

Bacteria reproduce by cell division (p. 284) and produce clones (Figure 24.11, overleaf). In genetic engineering, recombinant DNA is cloned by inserting it into bacteria which reproduce themselves and the introduced DNA.

It was pointed out on p. 184 that all cells (other than gametes) in an organism carry a full set of chromosomes and genes and, therefore, contain all the 'instructions' for producing a complete organism. On p. 81 it was explained how plant cells can be cultured to produce a complete plant.

In recent years this has been done with mammalian cells but in a rather different way. One of the first experiments was with sheep. The nucleus (haploid) from a sheep's ovum was removed. Then the nucleus (diploid) from a cell taken from the sheep's mammary gland was inserted into the 'empty' ovum. This ovum was implanted into the uterus (Figure 24.12, overleaf). It took 277 trials before a lamb, 'Dolly', was born

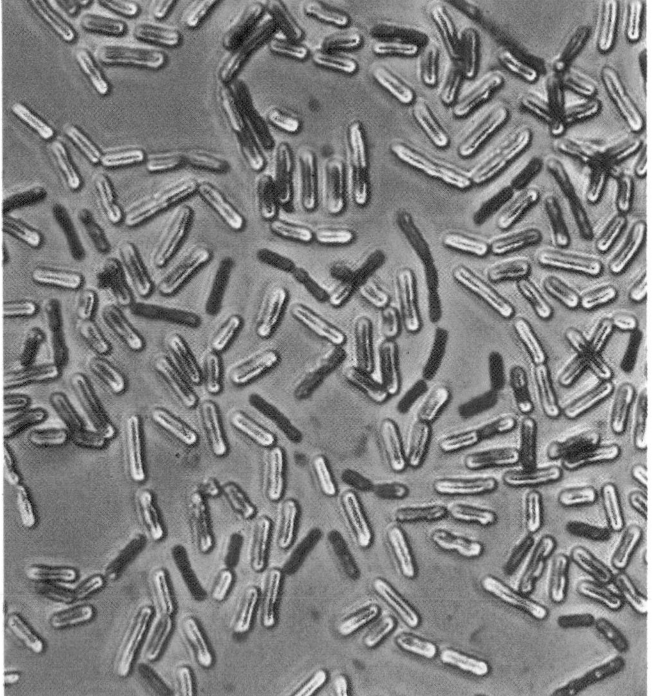

Figure 24.11 Asexual reproduction in bacteria produces a clone of identical cells (×1250)

using this technique. This lamb was a clone from the sheep which donated the original mammary gland cell. All its cells contained the diploid set of chromosomes derived exclusively from the donor mother, with which it was identical.

Since Dolly was born, in 1996, pigs and cows have been cloned. This raises the possibility of eventually being able to clone domestic animals with favourable characteristics and without the risk of losing some of the beneficial genes or introducing unwanted combinations as a result of standard crossing techniques.

Stem cells

Recent developments in tissue culture have involved stem cells. Stem cells are those cells in the body which have retained their power of division (p. 5). Examples are the basal cells of the skin (p. 137), which keep dividing to make new skin cells, and cells in the red bone marrow which constantly divide to produce the whole range of blood cells (p. 108).

In normal circumstances this type of stem cell can produce only one type of tissue, epidermis, blood, muscle, nerves, etc. Even so, culture of these stem cells could lead to effective therapies by introducing healthy stem cells into the body, to take over the function of diseased or defective cells.

Cells taken from early embryos can be induced to develop into almost any kind of cell, but there are ethical objections to using human embryos for this purpose. However, it has recently been shown that, given the right conditions, brain stem cells can become muscle or blood cells, and liver cells have been cultured from blood stem cells.

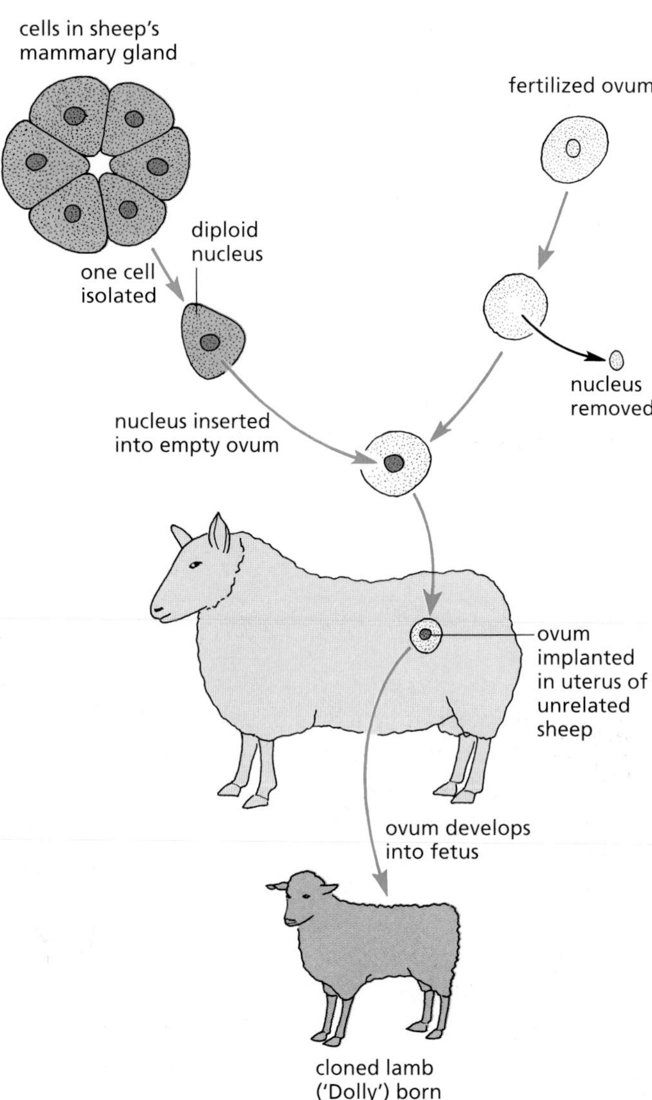

Figure 24.12 Cloning 'Dolly'. The technique varies in different laboratories

The principles of genetic fingerprinting

It has been pointed out (p. 189) that only 3 per cent of human DNA consists of genes coding for proteins. Some of the remaining 97 per cent includes sequences of apparently functionless DNA. Some of these consist of repeated sequences of, in many cases, two to four base pairs, for example:

–CATG–CATG–CATG–CATG–

The number of repeats is very variable between different individuals. But since these regions do not code for proteins, such variations do not produce observable effects, though they can be revealed in the laboratory. In closely related individuals the length of the repeats is very similar. The more unrelated people are, the more will be the variation in length of the repeated sequences.

Samples of DNA are extracted and treated with restriction enzymes (p. 213). These restriction enzymes are chosen so that they cut at only the beginning and the end of the repeat sequence:

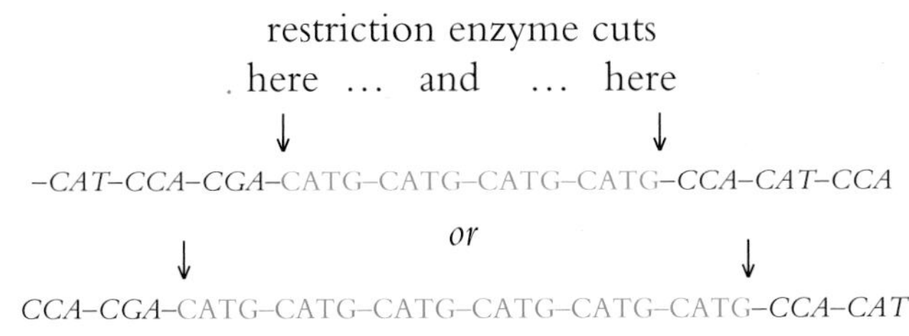

producing **restriction fragments** of differing lengths:

CATG–CATG–GATG–GATG

or

CATG–CATG–CATG–CATG–CATG–CATG

After treatment with the restriction enzyme, the DNA samples are placed at one end of a gelatinous sheet and an electric current is applied. The restriction fragments carry a small electric charge and move through the gel under the influence of the electric field. The shorter fragments, however, move faster and further than the longer fragments, so their final positions on the gel sheet are quite different.

The DNA fragments are, at first, invisible on the gel sheet but, by a series of operations, their positions are revealed. Figure 24.13 shows the result of a separation of restriction fragments from a child, her mother and two men both of which claim to be the father. The child will have inherited one restriction fragment from her mother and a different one from her father. If you look in column 2, you can see that one of the child's restriction fragments occupies the same position as the mother's, and the other matches one from man 'A', who must therefore be the father. This example shows the result of investigating a single restriction fragment.

In other cases, in criminal investigations for example, many more restriction sites are used. Figure 24.14 shows the result of one such analysis. The left-hand column shows the positions of 20 or so restriction fragments from the DNA of a victim of an assault. The second column shows the DNA profile from a sample of blood taken from the scene of the crime. The three columns on the right are DNA profiles of three suspects. You can see that the profile of suspect number 1 matches the sample taken from the scene of the crime. Although this is powerful evidence, you can probably imagine the problems confronting the prosecuting counsel in explaining it to the jury.

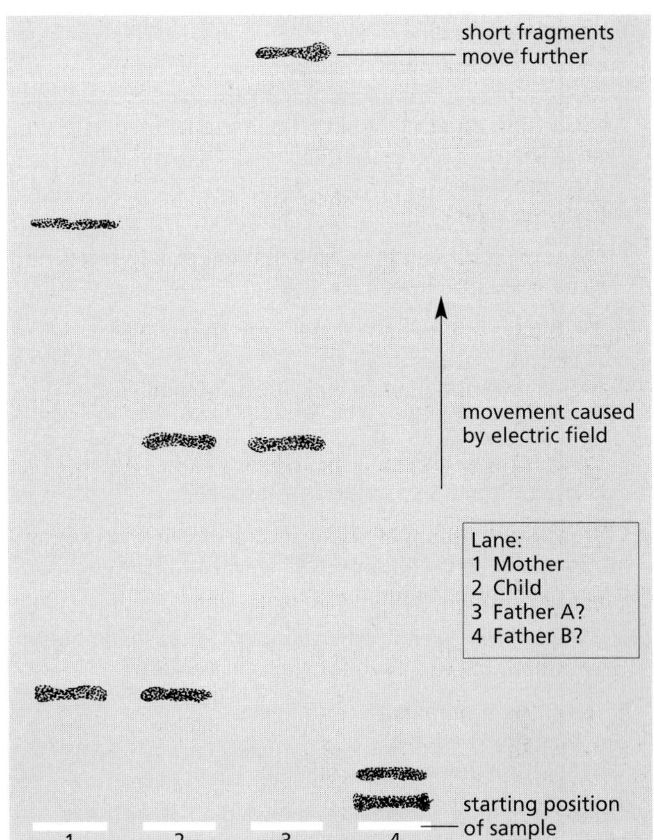

Figure 24.13 Genetic fingerprinting in a paternity case. One of the child's restriction fragments matches one of its mother's, and one matches that of supposed father 'A'

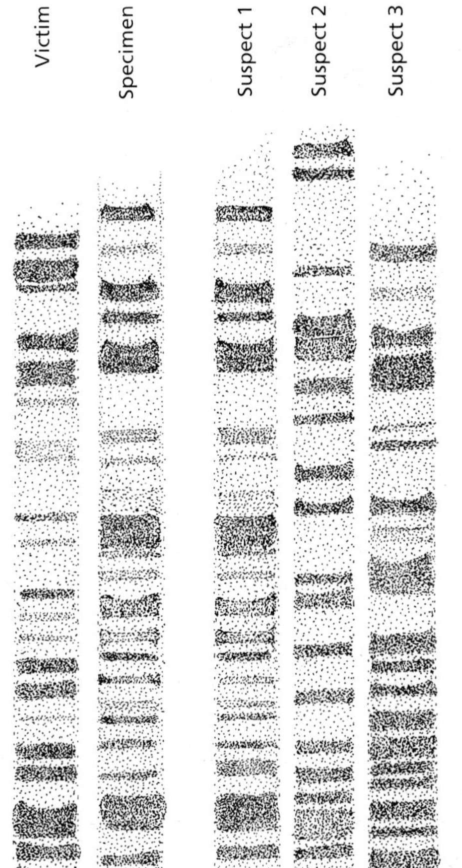

Figure 24.14 Forensic DNA profiling. More than 20 of suspect 1's fragments match those in samples taken from the scene

Questions

1 On p. 15 it was explained that enzymes act on specific substrates. What is the substrate for a restriction enzyme
 a in general,
 b in particular?

2 Name three different vectors that might be used in genetic engineering.

3 What advantages have genetically engineered
 a insulin,
 b chymotrypsin over the traditionally produced substances?

4 What is the principle in the use of genetic engineering in the treatment of hereditary diseases?

5 What is the potential disadvantage of using viruses as vectors for recombinant DNA? How can this disadvantage be minimized?

6 Can you think of any ethical objections to sheep being used to produce a drug for treating humans?

7 State two examples of
 a potential benefits,
 b potential hazards of GM crops.

8 In what major respect does a cloned animal differ from one produced by sexual reproduction?

9 How do stem cells differ from any other cell? Would you regard a zygote as a stem cell? Give your reasons.

10 a What is a restriction fragment?
 b In what way do corresponding restriction fragments differ between unrelated individuals?
 c How is this difference revealed?

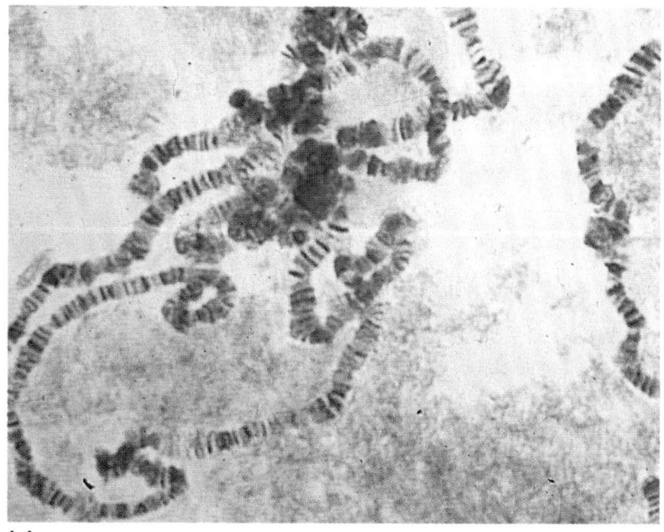

(a)

if bands in this region are missing, the fly has no colour in its eyes (normally red)

if this band is missing, there is an irregularity in the wing

loss of this band leads to a change in texture of the eye surface

one or more of the bands in this region controls the normal development of bristles

(b)

Figure 24.15 Gene mapping. **(a)** 'Giant chromosomes' of the fruit fly (×250). By using special stains, bands appear on the chromosomes. These are not necessarily the genes but can be used as markers for the position of genes. **(b)** In certain mutant flies, bands may be missing and corresponding changes are seen in the phenotype

The human genome project

A **genome** is the total genetic content of any cell in an organism. It consists of all the genes on all the chromosomes. It is thought that the **human genome** consists of up to 40 000 genes distributed between 46 chromosomes. The genome also includes all the DNA on the non-coding sections.

The **human genome project** aims to:

- locate the position of all the genes on each chromosome; this involves **mapping**,
- work out the entire sequence of bases for the whole genome; this is called **sequencing**.

Mapping

It has long been possible to recognize, in chromosomes, regions which have a specific function. Figure 24.15a shows stained bands in a 'giant' chromosome from the salivary gland of a fruit fly. Figure 24.15b shows that, if certain bands are missing, there is a mutation in the fruit fly, and corresponding changes are seen in the phenotype. This shows that the bands include, at least, the gene or genes controlling a specific feature.

Cross-breeding experiments also enable geneticists to identify positions of genes on chromosomes.

Modern biochemical techniques allow many more regions of chromosomes and their DNA to be identified.

The recognizable regions are not necessarily genes. After all, only about 3 per cent of the human genome consists of genes. The recognizable (mapped) regions may represent genes (Figure 24.16), stretches of DNA which turn genes on and off, or even lengths of DNA which have no known function.

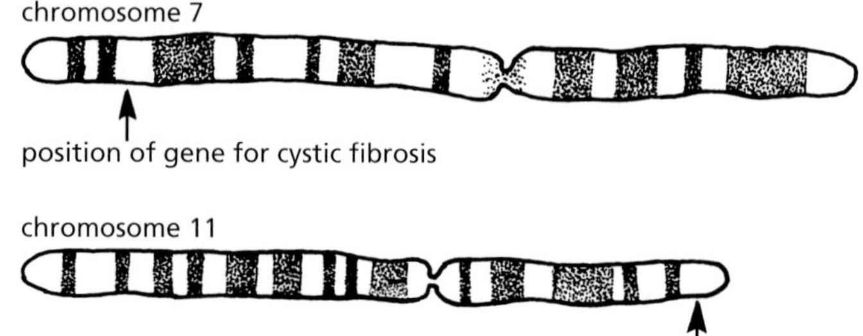

Figure 24.16 Mapped genes for diseases on chromosomes 7 and 11

Sequencing

Although mapping gives an overall picture of the genome, it does not reveal the sequence of bases in the DNA of these regions. A street map might lead you accurately to a position in the town but will not tell you the names of the houses in the street. Sequencing aims to reveal the sequence of nucleotides in the whole genome. Currently 85 per cent of the human genome has been sequenced with a reasonable degree of certainty.

The techniques for sequencing DNA have become progressively more sophisticated so that much of the analysis can be automated. It is even possible to analyse DNA by putting it on specially prepared computer chips. As a result, the complete genome should be known by the time this book is published.

Knowing the complete gene sequence is not the end of the story, however. Researchers will still need to find out (for example) which sequences actually code for proteins, what these proteins are and what is their role in the cell's metabolism.

Applications

It is hoped that a knowledge of the human genome will lead to:

- identification of defective genes and hence the opportunity to offer early treatment,
- identification of genes which confer a susceptibility to certain diseases and so enable individuals to take preventive measures,
- prediction of the proteins that the genes produce, giving an opportunity to design appropriate drugs to enhance or inhibit the activities of these proteins.

These outcomes, desirable though they are, are pretty distant. Not everybody thinks that the millions of dollars spent on the project is worthwhile or could not have been better spent on more conventional forms of health care.

There is also some unease about the other uses to which the knowledge could be put, for example in the realms of health or life insurance. If it can be established that you are carrying genes for susceptibility to high blood pressure, will you be able to obtain such insurance? (See also, 'Genetic counselling and screening' p. 198.)

Checklist

- A knowledge of genetics enables breeders to produce new varieties of plants or animals.
- Cross-breeding two varieties together enables beneficial genes to be brought together in a new variety.
- Genetic engineering involves the transfer of genes between unrelated species, or the alteration of genes in one species.
- Plasmids, viruses and liposomes are vectors used to deliver the genes.
- Genetic engineering is used in the production of enzymes, hormones and drugs.
- Crop plants can be genetically modified to resist insect pests and herbicides.
- There is concern that the genes introduced into crop plants might spread to wild plants.
- A clone refers to the offspring from asexual reproduction. They will be genetically identical.
- Animals have been cloned by fusing a body cell or nucleus with a fertilized ovum whose nucleus has been removed.
- Many stretches of DNA consist of repeated sequences of nucleotides which do not code for enzymes or proteins.
- Variation in the length of these stretches is the basis for genetic fingerprinting.
- The human genome project aims to map the position of genes on the chromosomes and discover the sequence of bases in the DNA.

Genetics and heredity

Examination questions

Do not write on this page. Where necessary copy drawings, tables or sentences.

1 Cystic fibrosis is an inherited condition caused by mutation.

a What is mutation? (2)

b Give **one** cause of mutations. (1)

c Why is cystic fibrosis described as an inherited condition? (1)

(CCEA)

2 Some people suffer from a disease called diabetes. They need to take regular injections of the hormone insulin. Insulin can be produced by genetic engineering. The diagram shows four stages in this process, but they are **not** in the correct order.

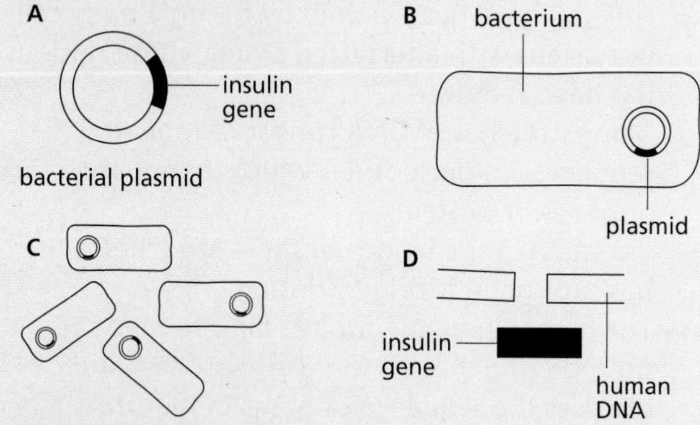

a Put the stages into their correct order, and describe what is happening at each stage. (7)

b (i) Genes can sometimes mutate. State one possible cause of these mutations. (1)

(ii) State two effects that a mutation might have on the structure of a gene. (2)

(OCR)

3 Hair colour in mice is controlled by a gene with two alleles. A homozygous black-haired mouse was bred with a homozygous brown-haired mouse. All the offspring were black-haired.

a (i) Explain what is meant by the terms *homozygous* and *recessive*. (2)

(ii) Which is the dominant hair colour in mice? (1)

b One of the heterozygous black-haired offspring was bred with a homozygous brown-haired mouse.

(i) Using the symbols B and b to represent the two alleles, draw a genetic diagram to show the outcome of this cross. (4)

(ii) State the ratio of the phenotypes of the offspring. (1)

(IGCSE)

4 PKU (phenylketonuria) is an inherited disease. The allele (n) for the disease is recessive to the normal allele (N). The diagram shows how PKU was inherited in a family.

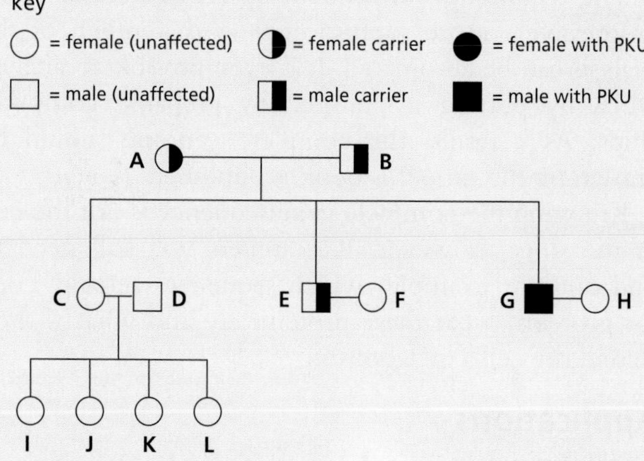

a Give the genotypes of each individual in the table below.

Individual	Genotype
B	
J	

(2)

b How many of the children of A and B are homozygous? (1)

c If G and H have a child, what is the probability that it will have PKU? (1)

d C and D have four children, all of whom are female. What is the probability that their next child will be female? (1)

(Edexcel)

5 The sentences describe how a gene can be transferred from one organism to another. Use words from the list to complete the sentences.

adrenaline bacterium chromosome cytoplasm hormone insulin ligase plasmid protease restriction enzyme

a The gene for the production of the ______ insulin was cut from the DNA of a human ______ using a restriction enzyme.

b A plasmid removed from the bacterium was cut using the same ______.

c The insulin gene was inserted into a circle of DNA called a ______.

d This circle of DNA was then returned to the ______ and identical copies of this circle of DNA were produced.

e Microbes were produced in large numbers then used in a fermenter to make perfect human ______ for diabetics. (7)

(Edexcel)

Organisms and their environment

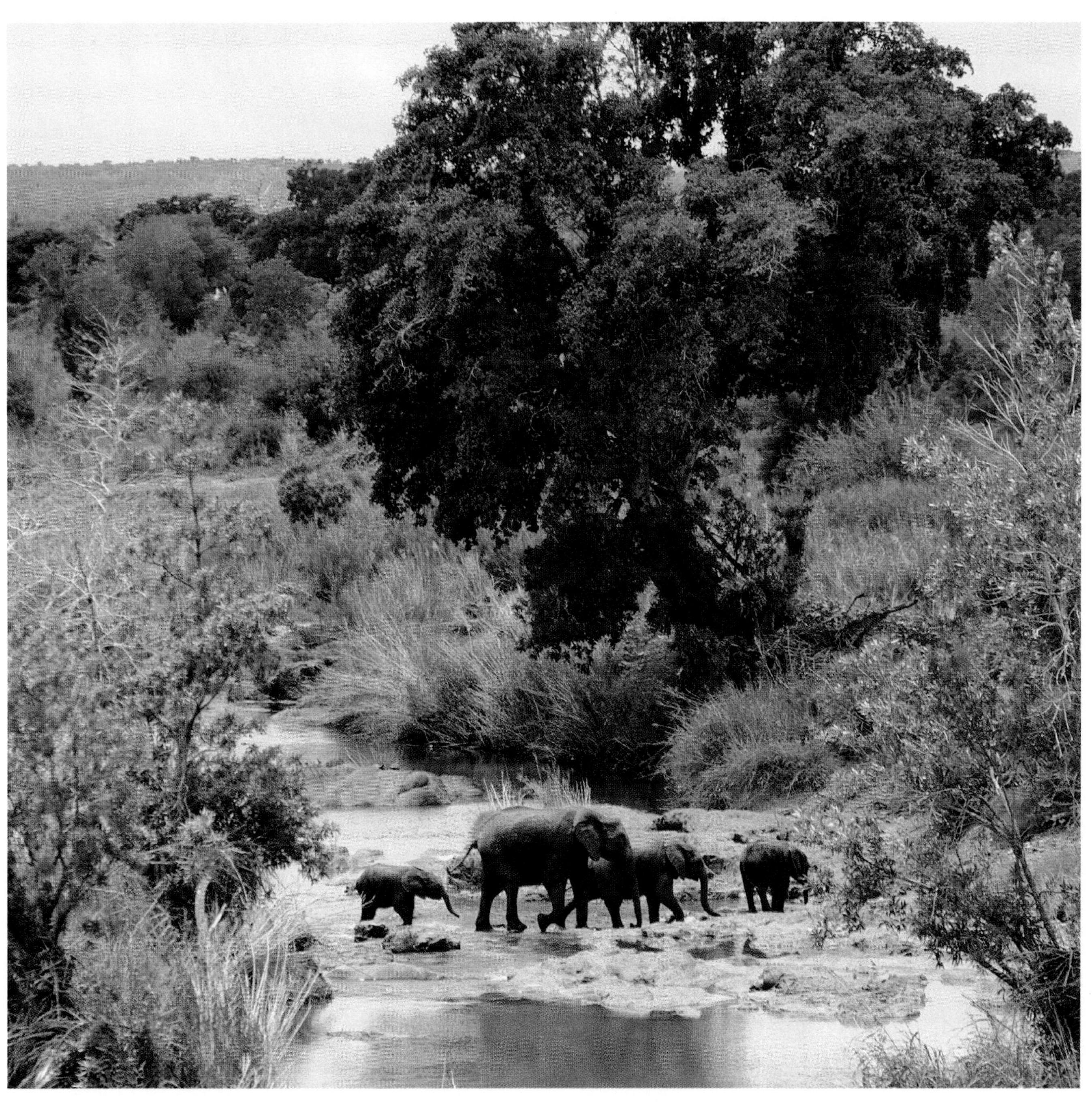

25 The interdependence of living organisms

Food chains and food webs
Pyramids of numbers and biomass, recycling.
The carbon cycle
The nitrogen cycle
The water cycle
Agriculture
Decay
Energy flow in an ecosystem

'Interdependence' means the way in which living organisms depend on each other in order to remain alive, grow and reproduce. For example, bees depend for their food on pollen and nectar from flowers. Flowers depend on bees for pollination (p. 69). Bees and flowers are, therefore, interdependent.

Food chains and food webs

One important way in which organisms depend on each other is for their food. Many animals, such as rabbits, feed on plants. Such animals are called **herbivores**. Animals called **carnivores** eat other animals. A **predator** is a carnivore which kills and eats other animals. A fox is a predator which preys on rabbits. **Scavengers** are carnivores which eat the dead remains of animals killed by predators. These are not hard and fast definitions. Predators will sometimes scavenge for their food and scavengers may occasionally kill living animals.

Food chains

Basically, all animals depend on plants for their food. Foxes may eat rabbits, but rabbits feed on grass. A hawk eats a lizard, the lizard has just eaten a grasshopper but the grasshopper was feeding on a grass blade. This relationship is called a **food chain** (Figure 25.1).

Figure 25.1 A food chain. The caterpillar eats the leaf; the blue tit eats the caterpillar but may fall prey to the kestrel

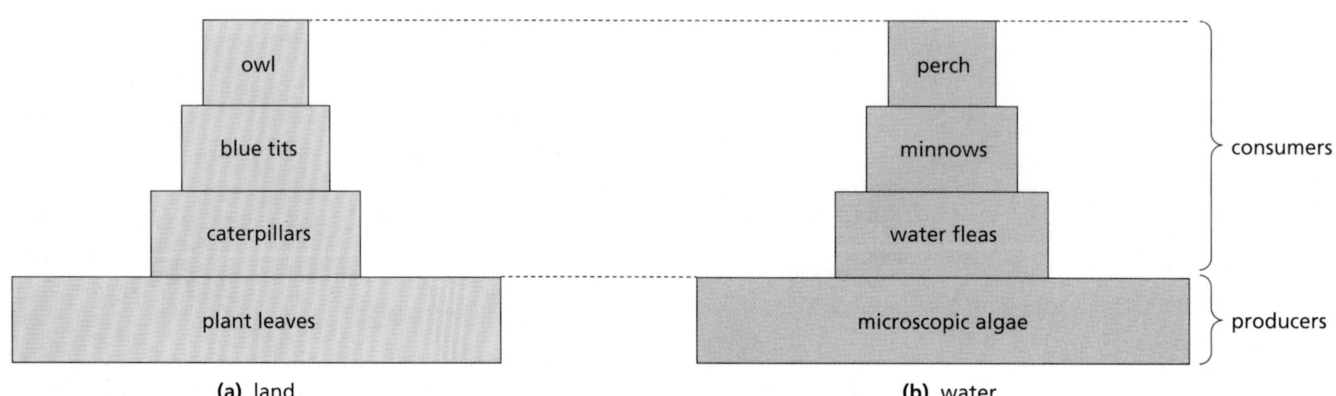

Figure 25.2 Examples of food pyramids (pyramids of numbers)

(a) phytoplankton (× 100) These microscopic algae form the basis of a food pyramid in the water

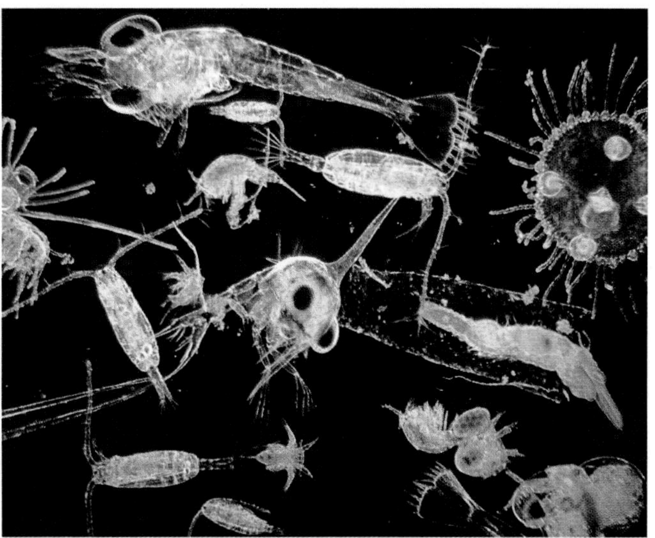

(b) zooplankton (× 20) These crustacea will eat microscopic algae

Figure 25.3 Plankton. The microscopic organisms which live in the surface waters of the sea or fresh water are called collectively, plankton. The single-celled algae (see p. 279) are the phytoplankton. They are surrounded by water, salts and dissolved carbon dioxide. Their chloroplasts absorb sunlight and use its energy for making food by photosynthesis. The phytoplankton is eaten by small animals in the zooplankton, mainly crustacea (see p. 274). Small fish will eat the crustacea

The organisms at the beginning of a food chain are usually very numerous while the animals at the end of the chain are often large and few in number. The **food pyramids** in Figure 25.2 show this relationship. There will be millions of microscopic, single-celled algae in a pond (Figure 25.3a). These will be eaten by the larger but less numerous water fleas and other crustacea (Figure 25.3b), which in turn will become the food of small fish, like minnow and stickleback. The hundreds of small fish may be able to provide enough food for only four or five large carnivores, like pike or perch.

The organisms at the base of the food pyramids in Figure 25.2 are plants. Plants produce food from carbon dioxide, water and salts (see 'Photosynthesis', p. 35), and are, therefore, called **producers**. The animals which eat the plants are called **primary consumers**, e.g. grasshoppers. Animals which prey on the plant-eaters are called **secondary consumers**, e.g. shrews, and these may be eaten by **tertiary consumers**, e.g. weasels or kestrels (Figure 25.4).

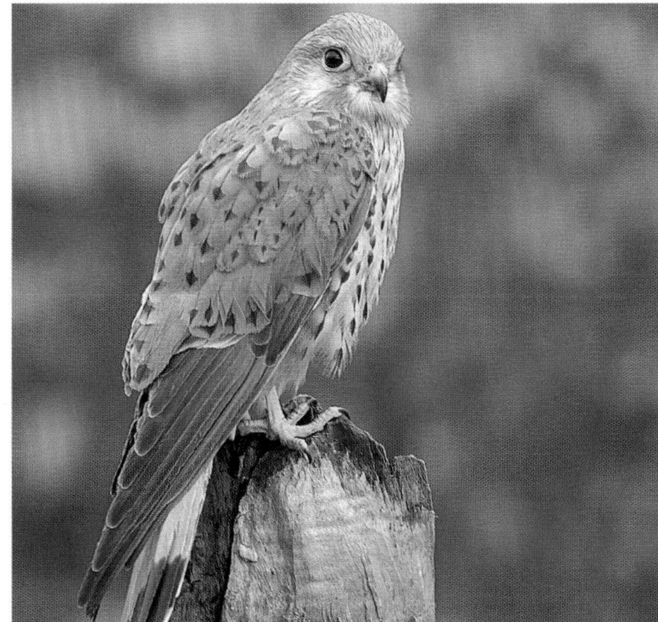

Figure 25.4 The kestrel, a secondary or tertiary consumer

The position of an organism in a food pyramid is sometimes called its **trophic level**.

Pyramids of numbers and biomass

The width of the bands in Figure 25.2 is meant to represent the relative number of organisms at each trophic level. So, the diagrams are sometimes called **pyramids of numbers**.

However, you can probably think of situations where a pyramid of numbers would not show the same effect. For example, a single sycamore tree may provide food for thousands of greenfly. One oak tree may feed hundreds of caterpillars. In these cases the pyramid of numbers is upside-down.

The way round this problem is to consider not the single tree, but the mass of the leaves that it produces in the growing season, and the mass of the insects which can live on them. **Biomass** is the term used when the mass of living organisms is being considered, and pyramids of biomass can be constructed as in Figure 25.16, p. 233.

An alternative is to calculate the energy available in a year's supply of leaves and compare this with the energy needed to maintain the population of insects which feed on the leaves. This would produce a **pyramid of energy**, with the producers at the bottom having the greatest amount of energy. Each successive trophic level would show a reduced amount of energy.

Later in the chapter, the recycling of matter is discussed. The elements which make up living organisms are recycled, i.e. they are used over and over again. This is not the case with energy, which flows from producers to consumers and is eventually lost to the atmosphere as heat. (See 'Energy flow', p. 231.)

Food webs

Food chains are not really as straightforward as described above, because most animals eat more than one type of food. A fox, for example, does not feed entirely on rabbits but takes beetles, rats and voles in its diet. To show these relationships more accurately, a **food web** can be drawn up (Figure 25.5).

The food webs for land, sea and fresh water, or for ponds, rivers and streams, will all be different. Food webs will also change with the seasons when the food supply changes.

If some event interferes with a food web, all the organisms in it are affected in some way. For example, if the rabbits in Figure 25.5 were to die out, the foxes, owls and stoats would eat more beetles and rats. Something like this happened in 1954 when the disease myxomatosis wiped out nearly all the rabbits in England. Foxes ate more voles, beetles and blackberries, and attacks on lambs and chickens increased. Even the vegetation was affected because the tree seedlings which the rabbits used to nibble off were able to grow. As a result, woody scrubland started to develop on what had been grassy downs. A similar effect is shown in Figure 25.6.

Dependence on sunlight

If you take the idea of food chains one step further you will see that all living organisms depend on sunlight and photosynthesis (p. 35). Green plants make their food by photosynthesis, which needs sunlight. Since all animals depend, in the end, on plants for their food, they therefore depend indirectly on sunlight. A few examples of our own dependence on photosynthesis are given below.

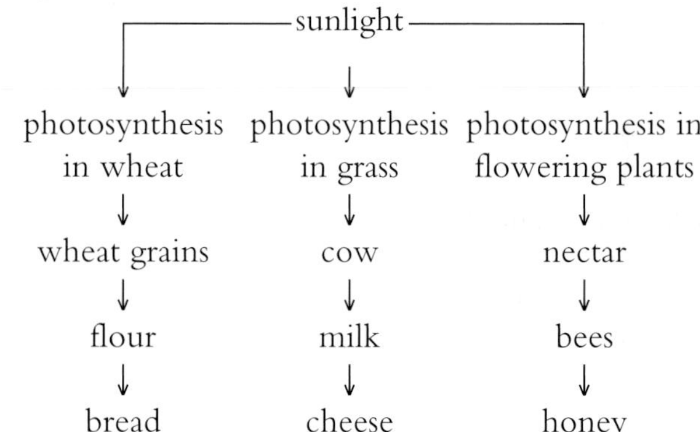

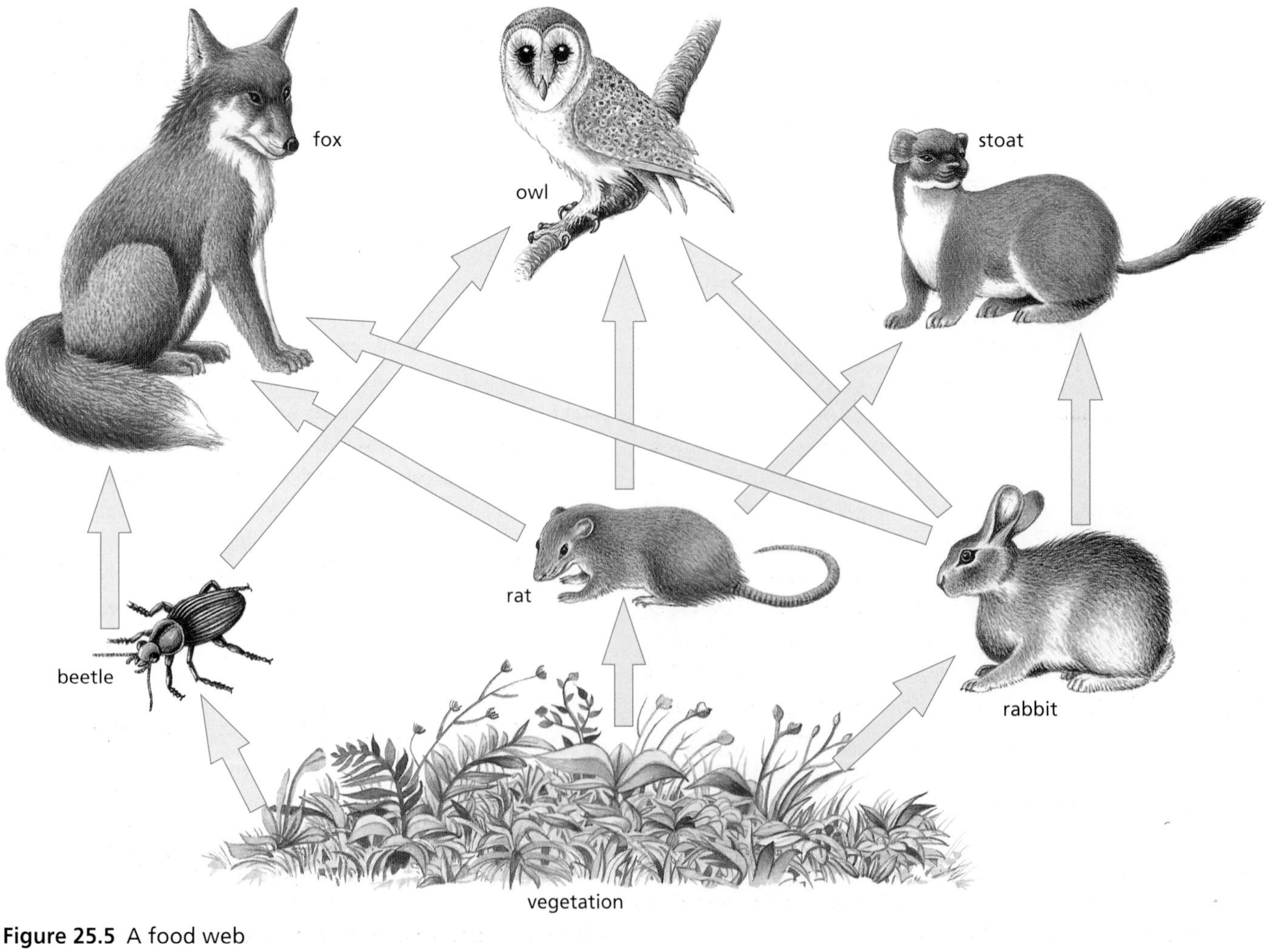

Figure 25.5 A food web

(a) Sheep have eaten any seedlings which grew under the trees

(b) Ten years later, the fence has kept the sheep off and the tree seedlings have grown

Figure 25.6 Effect of grazing

Nearly all the energy released on the Earth can be traced back to sunlight. Coal comes from tree-like plants, buried millions of years ago. These plants absorbed sunlight for their photosynthesis when they were alive. Petroleum was formed, also millions of years ago, probably from the partly decayed bodies of microscopic algae which lived in the sea. These, too, had absorbed sunlight for photosynthesis.

Today it is possible to use mirrors and solar panels to collect energy from the sun directly, but the best way, so far, of trapping and storing energy from sunlight is to grow plants and make use of their products, such as starch, sugar, oil, alcohol and wood, for food or as energy sources. For example, sugar from sugar-cane can be fermented (p. 20) to alcohol, and used as a motor fuel instead of petrol.

Recycling

There are a number of organisms which have not been fitted into the food webs or food chains described so far. Among these are the **saprotrophs**. Saprotrophs do not obtain their food by photosynthesis, nor do they kill and eat living animals or plants. Instead they feed on dead and decaying matter such as dead leaves in the soil or rotting tree-trunks (Figure 25.7). The most numerous examples are the fungi, such as mushrooms, toadstools or moulds, and the bacteria, particularly those which live in the soil. They produce extracellular enzymes (p. 15) which digest the decaying matter and then they absorb the soluble products back into their cells. In so doing, they remove the dead remains of plants and animals which would otherwise collect on the Earth's surface. They also break these remains down into substances which can be used by other organisms. Some bacteria, for example, break down the protein of dead plants and animals and release nitrates which are taken up by plant roots and there built into new amino acids and proteins (p. 11). This use and re-use of materials in the living world is called **recycling**.

Figure 25.7 Saprotrophs. These toadstools are getting their food from the rotting tree stump

The general idea of recycling is illustrated in Figure 25.8. The green plants are the producers, and the animals which eat the plants and each other are the consumers. The bacteria and fungi, especially those in the soil, are called the **decomposers** because they break down the dead remains and release the chemicals for the plants to use again. Two examples of recycling, one for carbon and one for nitrogen, are described next.

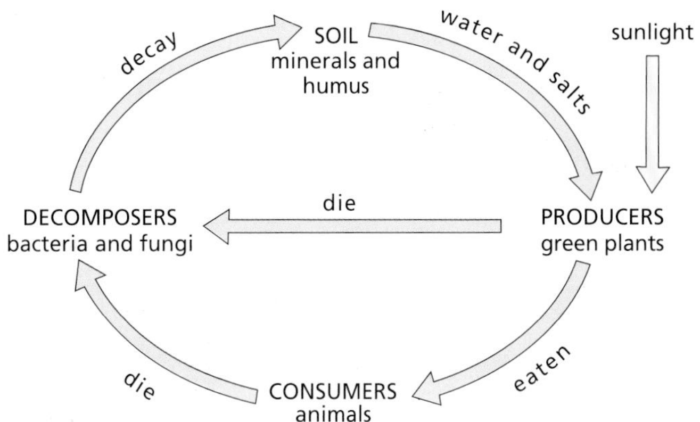

Figure 25.8 Recycling in an ecosystem

Questions

1 Try to construct a simple food web using the following: sparrow, fox, wheat seeds, cat, kestrel, mouse.

2 Describe briefly all the possible ways in which the following might depend on each other: grass, earthworm, blackbird, oak tree, soil.

3 Explain how the following foodstuffs are produced as a result of photosynthesis: wine, butter, eggs, beans.

4 An electric motor, a car engine and a race horse can all produce energy. Show how this energy could come, originally, from sunlight. What forms of energy on the Earth are *not* derived from sunlight?

5 How do you think evidence is obtained in order to place animals such as a fox and a pigeon in a food web?

6 When humans colonized islands they often introduced their domestic animals such as goats or cats. This usually had a devastating effect on the natural food webs. Suggest reasons for this.

The carbon cycle

Carbon is an element which occurs in all the compounds which make up living organisms. Plants get their carbon from carbon dioxide in the atmosphere and animals get their carbon from plants. The carbon cycle, therefore, is mainly concerned with what happens to carbon dioxide (Figure 25.9).

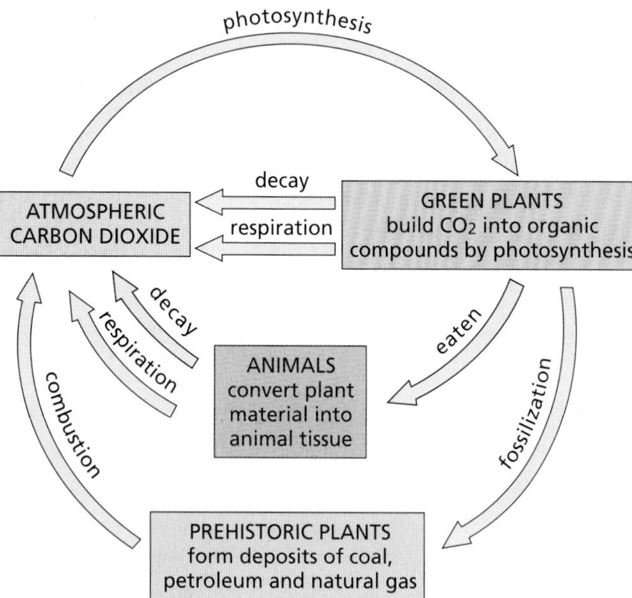

Figure 25.9 The carbon cycle

Removal of carbon dioxide from the atmosphere

Photosynthesis

Green plants remove carbon dioxide from the atmosphere as a result of their photosynthesis. The carbon of the carbon dioxide is built first into a carbohydrate such as sugar. Some of this is changed into starch or the cellulose of cell walls, and the proteins, pigments and other compounds of a plant. When the plants are eaten by animals, the organic plant material is digested, absorbed and built into the compounds making up the animals' tissues. Thus the carbon atoms from the plant become part of the animal.

Addition of carbon dioxide to the atmosphere

Respiration

Plants and animals obtain energy by oxidizing carbohydrates in their cells to carbon dioxide and water (p. 19). The carbon dioxide and water are excreted and so the carbon dioxide returns once again to the atmosphere.

Decay

The organic matter of dead animals and plants is used by saprotrophs, especially bacteria and fungi, as a source of energy. These micro-organisms decompose the plant and animal remains and turn the carbon compounds into carbon dioxide.

Combustion (burning)

When carbon-containing fuels such as wood, coal, petroleum and natural gas are burned, the carbon is oxidized to carbon dioxide ($C + O_2 \rightarrow CO_2$). The hydrocarbon fuels, such as coal and petroleum, come from ancient plants which have only partly decomposed over the millions of years since they were buried.

So, an atom of carbon which today is in a molecule of carbon dioxide in the air may tomorrow be in a molecule of cellulose in the cell wall of a blade of grass. When the grass is eaten by a cow, the carbon atom may become part of a glucose molecule in the cow's bloodstream. When the glucose molecule is used for respiration, the carbon atom will be breathed out into the air once again as carbon dioxide.

The same kind of cycling applies to nearly all the elements of the Earth. No new matter is created, but it is repeatedly rearranged. A great proportion of the atoms of which you are composed will, at one time, have been part of other organisms.

Questions

1 a Why do living organisms need a supply of carbon?
  b Give three examples of carbon-containing compounds which occur in living organisms (see pp. 11–12).
  c Where do
    (i) animals,
    (ii) plants,
    get their carbon from?

2 Write three chemical equations
  a to illustrate that respiration produces carbon dioxide (see p. 19),
  b to show that burning produces carbon dioxide,
  c to show that photosynthesis uses up carbon dioxide (see p. 35).

3 Outline the events that might happen to a carbon atom in a molecule of carbon dioxide which entered the stoma in the leaf of a potato plant, and became part of a starch molecule in a potato tuber which was then eaten by a man. Finally the carbon atom is breathed out again in a molecule of carbon dioxide.

4 Construct a diagram, on the lines of the carbon cycle (Figure 25.9), to show the cycling process for hydrogen (starting from the water used in photosynthesis).

The nitrogen cycle

When a plant or animal dies, its tissues decompose (p. 231), partly as a result of the action of saprotrophic bacteria. One of the important products of this decay is **ammonia** (NH_3, a compound of nitrogen), which is washed into the soil (Figure 25.10).

The excretory products of animals contain nitrogenous waste products such as ammonia, urea and uric acid (p. 133). The organic matter in their droppings is also decomposed by soil bacteria.

Processes which add nitrates to soil

Nitrifying bacteria

These are bacteria living in the soil which use the ammonia from excretory products and decaying organisms as a source of energy (as we use glucose in respiration). In the process of getting energy from ammonia, the bacteria produce **nitrates**.

- The 'nitrite' bacteria oxidize ammonium compounds to nitrites ($NH_4^- \rightarrow NO_2^-$).
- 'Nitrate' bacteria oxidize nitrites to nitrates ($NO_2^- \rightarrow NO_3^-$).

Although plant roots can take up ammonia in the form of its compounds, they take up nitrates more readily, so the nitrifying bacteria increase the fertility of the soil by making nitrates available to the plants.

Nitrogen-fixing bacteria

This is a special group of nitrifying bacteria which can absorb nitrogen as a gas from the air spaces in the soil, and build it into compounds of ammonia. Nitrogen gas cannot itself be used by plants. When it has been made into a compound of ammonia, however, it can easily be changed to nitrates by other nitrifying bacteria. The process of building the gas, nitrogen, into compounds of ammonia is called **nitrogen fixation**. Some of the nitrogen-fixing bacteria live freely in the soil. Others live in the roots of **leguminous plants** (peas, beans, clover), where they cause swellings called **root nodules** (Figure 32.11, p. 297). These leguminous plants are able to thrive in soils where nitrates are scarce, because the nitrogen-fixing bacteria in their nodules make compounds of nitrogen available for them. Leguminous plants are also included in crop rotations (p. 231) to increase the nitrate content of the soil.

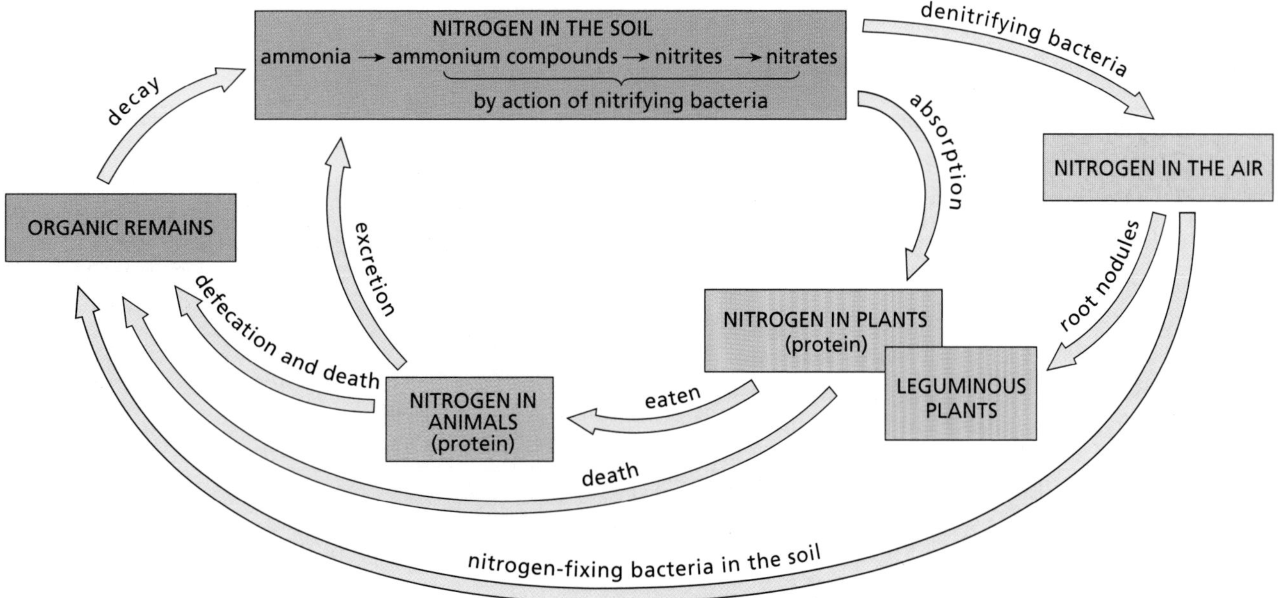

Figure 25.10 The nitrogen cycle

Lightning

The high temperature of lightning discharge causes some of the nitrogen and oxygen in the air to combine and form oxides of nitrogen. These dissolve in the rain and are washed into the soil as weak acids, where they form nitrates. Although several million tonnes of nitrate may reach the Earth's surface in this way each year, this forms only a small fraction of the total nitrogen being recycled.

Processes which remove nitrates from the soil

Uptake by plants

Plant roots absorb nitrates from the soil and combine them with carbohydrates to make proteins (p. 43).

Leaching

Nitrates are very soluble (i.e. dissolve easily in water), and as rain water passes through the soil it dissolves the nitrates and carries them away in the run-off or to deeper layers of the soil. This is called **leaching**.

Denitrifying bacteria

These are bacteria which obtain their energy by breaking down nitrates to nitrogen gas which then escapes from the soil into the atmosphere.

These processes are summed up in Figure 25.10.

All the elements which make up living organisms are recycled, not just carbon and nitrogen. It would be possible to trace out cycles for hydrogen, oxygen, phosphorus, sulphur, iron, etc.

Questions

1 On a lawn growing on nitrate-deficient soil, the patches of clover often stand out as dark green and healthy against a background of pale green grass. Suggest a reason for this contrast. (See also p. 44.)

2 Very briefly explain the difference between nitrifying, nitrogen-fixing and denitrifying bacteria.

The water cycle

The 'water cycle' is somewhat different from the others because only a tiny proportion of the water which is recycled passes through living organisms.

Animals lose water by evaporation (p. 138), defecation (p. 102), urination (p. 133) and exhalation (p. 126). They gain water from their food and drink. Plants take up water from the soil (p. 63) and lose it by transpiration (p. 59). Millions of tonnes of water are transpired, but only a tiny fraction of this has taken part in the reactions of respiration (p. 19) or photosynthesis (p. 35).

The great proportion of water is recycled without the intervention of animals or plants. The sun shining and the wind blowing over the oceans evaporate water from their vast, exposed surfaces. The water vapour produced in this way enters the atmosphere and eventually forms clouds. The clouds release their water in the form of rain or snow (precipitation). The rain collects in streams, rivers and lakes and ultimately finds its way back to the oceans. The human population diverts some of this water for drinking, washing, cooking, irrigation, hydro-electric schemes and other industrial purposes, before allowing it to return to the sea.

Agriculture

In a natural community of plants and animals, the processes which remove and replace mineral elements in the soil are in balance. In agriculture, most of the crop is usually removed so that there is little or no organic matter for nitrifying bacteria to act on. In a farm with animals, the animal manure, mixed with straw, is ploughed back into the soil or spread on the pasture. The manure thus replaces the nitrates and other minerals removed by the crop. It also gives the soil a good structure and improves its water-holding properties.

When animal manure is not available in large enough quantities, artificial fertilizers are used. These are mineral salts made on an industrial scale. Examples are ammonium sulphate (for nitrogen and sulphur), ammonium nitrate (for nitrogen) and compound NPK fertilizer for nitrogen, phosphorus and potassium (see p. 44). These are spread on the soil in carefully calculated amounts to provide the minerals, particularly nitrogen, phosphorus and potassium, that the plants need. These artificial fertilizers increase the yield of crops from agricultural land, but they do little to maintain a good soil structure because they contain no organic matter (Figures 25.11 and 25.12).

Figure 25.11 Experimental plots of wheat. The rectangular plots have been treated with different fertilizers

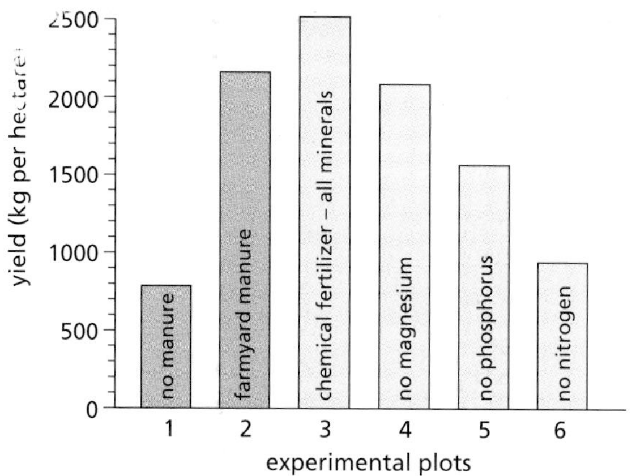

Figure 25.12 Average yearly wheat yields from 1852 to 1925, Broadbalk field, Rothamsted Experimental Station. Plot 1 received no manure or chemical fertilizer for 73 years. Plot 2 received an annual application of farmyard manure. Plot 3 received chemical fertilizer with all necessary minerals. Plots 4 to 6 received chemical fertilizer lacking one element

Crop rotation

Different crops make differing demands on the soil; potatoes and tomatoes use much potassium, for example. By changing the crop grown from year to year, no single group of minerals is continuously removed from the soil. Leguminous crops such as clover and beans may help to replace the nitrogen content of the soil because their root nodules contain nitrogen-fixing bacteria. The nitrates are released into the soil when the plant dies and decays.

The use of artificial fertilizers has made crop rotation, at least for the reasons above, largely unnecessary. However, turning arable land over to grass for a year or two does improve the structure of the soil, its drainage and other properties. Rotation also reduces the chances of infectious diseases that can enter the crop through the soil. For example, repeated crops of potatoes in the same field will increase the population of the fungus causing the disease 'potato blight'. If potatoes are not planted for a few years, the crop will suffer less from this disease when potatoes are grown again.

Questions

1 To judge from Figure 25.12, which mineral element seems to have the most pronounced effect on the yield of wheat? Explain your answer.

2 Draw up two columns headed A and B. In A list the processes which add nitrates to the soil and in B list those which remove nitrates from the soil. How might the activities of humans alter the normal balance between these two processes?

3 Suggest
  a some advantages,
  b some disadvantages of
  (i) organic manures (e.g. compost or farmyard manure) and
  (ii) chemical fertilizers.

Decay

A crucial factor in recycling is the process of decay. If it were not for decay, essential materials would not be released from dead organisms. When an organism dies, the enzymes in its cells, freed from normal controls, start to digest its own tissues (auto-digestion). Soon, scavengers appear on the scene and eat much of the remains; blowfly larvae devour carcases, earthworms consume dead leaves.

Finally the decomposers, fungi and bacteria (collectively called **micro-organisms**), arrive and invade the remaining tissues (Figure 25.13). These saprotrophs secrete extracellular enzymes (p. 15) into the tissues and reabsorb the liquid products of digestion (p. 227). When the micro-organisms themselves die, auto-digestion takes place, releasing the products such as nitrates, sulphates, phosphates, etc. into the soil or the surrounding water to be taken up again by the producers in the ecosystem.

Figure 25.13 Mould fungus growing on an over-ripe orange

The speed of decay depends on the abundance of micro-organisms, temperature, presence of water and, in many cases, oxygen. High temperatures speed up decay because they speed up respiration of the micro-organisms. Water is necessary for all living processes and oxygen is needed for aerobic respiration of the bacteria and fungi. Decay can take place in anaerobic conditions but it is slow and incomplete, as in the waterlogged conditions of peat bogs.

Energy flow in an ecosystem

An **ecosystem** is a community of living organisms and the habitat in which they live (p. 253). Most ecosystems are sustained by a constant input of energy from sunlight. This energy drives the chemistry of photosynthesis in the producers and passes to the other organisms through the food webs in the ecosystem. A pond is an ecosystem consisting of plants, animals, water, dissolved air, minerals and mud. The input of energy from the sun and a supply of water from rain is all that the pond community needs to maintain its existence. A forest is an ecosystem, so is an ocean. The whole of the Earth's surface may be considered as one vast ecosystem.

With the exception of atomic energy and tidal power, all the energy released on Earth is derived from sunlight. The energy released by animals comes, ultimately, from plants that they or their prey eat and the plants depend on sunlight for making their food (p. 226). The energy in organic fuels also comes ultimately from sunlight trapped by plants. Coal is formed from fossilized forests, and petroleum probably comes from the cells of ancient marine algae.

Use of sunlight

To try and estimate just how much life the Earth can support it is necessary to examine how efficiently the sun's energy is used. The amount of energy from the sun reaching the Earth's surface in 1 year ranges from 2 million to 8 million kilojoules per 1 m^2 ($2–8 \times 10^9 J\,m^{-2} yr^{-1}$) depending on the latitude. When this energy falls onto grassland, about 20 per cent is reflected by the vegetation, 39 per cent is used in evaporating water from the leaves (transpiration), 40 per cent warms up the plants, the soil and the air, leaving only about 1 per cent to be used in photosynthesis for making new organic matter in the leaves of the plants (Figure 25.14).

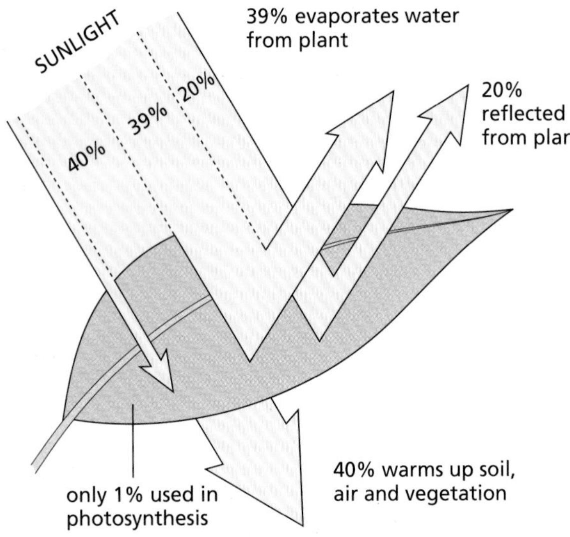

Figure 25.14 Absorption of sun's energy by plants

This figure of 1 per cent will vary with the type of vegetation being considered and with climatic factors such as availability of water and the soil temperature. Sugar-cane grown in ideal conditions can convert 3 per cent of the sun's energy into photosynthetic products; sugar-beet at the height of its growth has nearly a 9 per cent efficiency. Tropical forests and swamps are far more productive than grassland but it is difficult, and, in some cases undesirable, to harvest and utilize their products (p. 239).

In order to allow crop plants to approach their maximum efficiency they must be provided with sufficient water and mineral salts. This can be achieved by irrigation and the application of fertilizer.

Energy transfer between organisms

Having considered the energy conversion from sunlight to plant products the next step is to study the efficiency of transmission of energy from plant products to primary consumers. On land, primary consumers eat only a small proportion of the available vegetation. In a deciduous forest only about 2 per cent is eaten; in grazing land, 40 per cent of the grass may be eaten by cows. In open water, however, where the producers are microscopic plants (phytoplankton, see Figure 25.3a) and are swallowed whole by the primary consumers in the zooplankton (see Figure 25.3b), 90 per cent or more may be eaten. In the land communities, the parts of the vegetation not eaten by the primary consumers will eventually die and be used as a source of energy by the decomposers.

A cow is a primary consumer; over 60 per cent of the grass it eats passes through its alimentary canal (p. 97) without being digested. Another 30 per cent is used in the cow's respiration to provide energy for its movement and other life processes. Less than 10 per cent of the plant material is converted into new animal tissue to contribute to growth (Figure 25.15). This figure will vary with the diet and the age of the animal. In a fully grown animal all the digested food will be used for energy and replacement and none will contribute to growth. Economically it is desirable to harvest the primary consumers before their rate of growth starts to fall off.

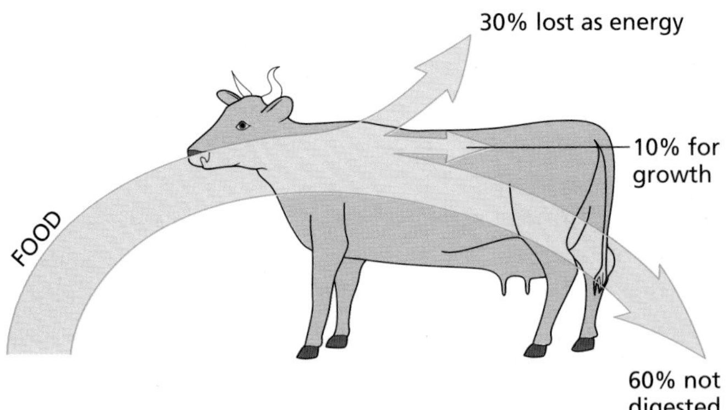

Figure 25.15 Energy transfer from plants to animals

The transfer of energy from primary to secondary consumers is probably more efficient since a greater proportion of the animal food is digested and absorbed than is the case with plant material. The transfer of energy at each stage in a food chain may be represented by classifying the organisms in a community as producers, or primary, secondary or tertiary consumers, and showing their relative masses in a pyramid such as the one shown in Figure 25.2 but on a more accurate scale. In Figure 25.16 the width of the horizontal bands is proportional to the masses (dry weight) of the organisms in a shallow pond.

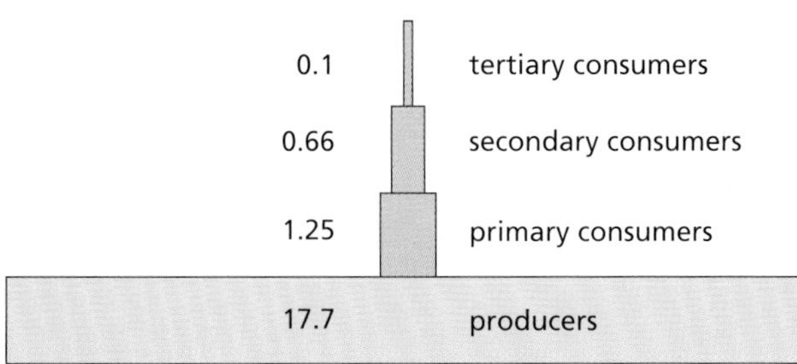

Figure 25.16 Biomass (dry weight) of living organisms in a shallow pond (grams per square metre)

Energy transfer in agriculture

In human communities, the use of plant products to feed animals which provide meat, eggs and dairy products is wasteful, because only 10 per cent of the plant material is converted to animal products. It is more economical to eat bread made from the wheat than to feed the wheat to hens and then eat the eggs and chicken meat. This is because eating the wheat as bread avoids using any part of its energy to keep the chickens alive and active. Energy losses can be reduced by keeping hens indoors in small cages where they lose little heat to the atmosphere and cannot use much energy in movement (Figure 25.17). The same principles can be applied in 'intensive' methods of rearing calves. However, many people feel that these methods are less than humane, and the saving of energy is far less than if the plant products were eaten directly by humans, as is the case in vegetarians.

Figure 25.17 Battery chickens. The hens are well fed but kept in crowded and cramped conditions with no opportunity to move about or scratch in the soil as they would normally do. Often the upper mandible of the beak is cut off to stop the birds pecking each other

Consideration of the energy flow of a modern agricultural system reveals other sources of inefficiency. To produce 1 tonne of nitrogenous fertilizer takes energy equivalent to burning 5 tonnes of coal. Calculations show that if the energy needed to produce the fertilizer is added to the energy used to produce a tractor and to power it, the energy derived from the food so produced is less than that expended in producing it.

Questions

1 It can be claimed that the sun's energy is used indirectly to produce a muscle contraction in your arm. Trace the steps in the transfer of energy which would justify this claim.

2 Discuss the advantages and disadvantages of human attempts to exploit a food chain nearer to its source, e.g. the plankton in Figure 25.3.

Checklist

- All animals depend, ultimately, on plants for their source of food.
- Since plants need sunlight to make their food, all organisms depend, ultimately, on sunlight for their energy.
- Plants are the producers in a food web; animals may be primary, secondary or tertiary consumers.
- The materials which make up living organisms are constantly recycled.
- Plants take up carbon dioxide during photosynthesis; all living organisms give out carbon dioxide during respiration; the burning of carbon-containing fuels produces carbon dioxide.
- The uptake of carbon dioxide by plants balances the production of carbon dioxide from respiration and combustion.
- Soil nitrates are derived naturally from the excretory products of animals and the dead remains of living organisms.
- Nitrifying bacteria turn these products into nitrates which are taken up by plants.
- Nitrogen-fixing bacteria can make nitrogenous compounds from gaseous nitrogen.
- An ecosystem is a self-contained community of organisms.
- Only about 1 per cent of the sun's energy which reaches the Earth is trapped by plants during photosynthesis.
- At each step in a food chain, only a small proportion of the food is used for growth. The rest is used for energy to keep the organism alive.

26 The human impact on the environment

The human impact on natural communities
Hunting, fishing, agriculture, pesticides, eutrophication.

Humans and forests
Erosion, flooding, climate, biodiversity.

Agriculture and soil
Erosion, pesticides.

Water pollution
Sewage, eutrophication, chemical pollution.

Air pollution
Smoke, particulates, sulphur dioxide, oxides of nitrogen, smog, carbon monoxide, chlorofluorocarbons. Control of air pollution. The greenhouse effect and global warming.

A few thousand years ago, most of the humans on the Earth probably obtained their food by gathering leaves, fruits or roots and by hunting animals. The population was probably limited by the amount of food that could be collected in this way.

Human faeces, urine and dead bodies were left on or in the soil and so played a part in the nitrogen cycle (p. 229). Life may have been short, and many babies may have died from starvation or illness, but humans fitted into the food web and nitrogen cycle like any other animal.

Once agriculture had been developed, it was possible to support much larger populations and the balance between humans and their environment was upset.

Population increase in the last 300 years has had three main effects on the environment.

1 Intensification of agriculture

Forests and woodland are cut down and the soil is ploughed up in order to grow more food. This destroys important wildlife habitats and may affect the climate.

Tropical rainforest is being cut down at the rate of 111 400 square kilometres per year. Since 1950, between 30 and 50 per cent of British deciduous woodlands have been felled to make way for farmland or conifer plantations.

The application of chemical fertilizers can cause deterioration of the soil structure and, in some cases, results in pollution of rivers and streams. Application of pesticides often kills beneficial creatures as well as pests.

2 Urbanization

The development of towns and cities makes less and less land available for wildlife. In addition, the crowding of growing populations into towns leads to problems of waste disposal. The sewage and domestic waste from a town of several thousand people can cause disease and pollution in the absence of effective means of disposal.

When fuels are burned for heating and transport, they produce gases which pollute the atmosphere.

3 Industrialization

In some cases, an increasing population is accompanied by an increase in manufacturing industries which produce gases and other waste products which damage the environment.

The effects of the human population on the environment are complicated and difficult to study. They are even more difficult to forecast. In their ignorance, humans have destroyed many plants and animals and great areas of natural vegetation. Unless we control our consumption of the Earth's resources, limit our own numbers and treat our environment with more care and understanding, we could make the Earth's surface impossible to live on and so cause our own extinction.

The account which follows mentions just some of the ways in which our activities damage the environment.

The human impact on natural communities

Hunting

Hunting animals for food has always been a human activity and a small human population using basic methods of capture had little effect on the numbers of prey species.

Today, some wild animals are still hunted to sustain small populations of indigenous people. Most hunting, however, is not for food but to obtain parts of animals that are valuable or believed to have curative properties. Rhino horns and tiger bones, for example, are prized ingredients of traditional medicines in some countries (Figure 26.1). Elephants are killed for the ivory in their tusks. This is used for making decorative objects which fetch a high price.

The result is that many animals have been hunted to the point of extinction and now survive only in isolated pockets of country or in special reserves where they can be protected.

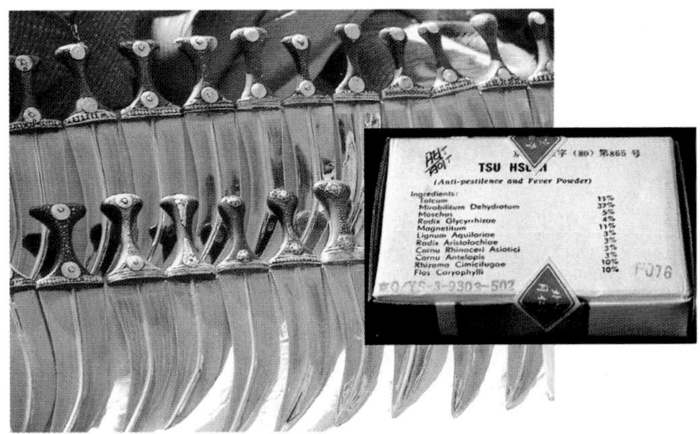

Figure 26.1 The rhinoceros is endangered because some people believe, mistakenly, that powdered rhino horn (Cornu Rhinoceri Asiatici) has medicinal properties, and others greatly prize rhino horn handles for their daggers

The World Wide Fund for Nature (WWF) believes that 15–20 per cent of all species on Earth will soon disappear if we do not change our patterns of consumption and destruction. This is a thousand times faster than the natural extinction rate.

Fishing

Small populations of humans, taking fish from lakes or oceans and using fairly basic methods of capture, had little effect on fish numbers. At present, however, commercial fishing has intensified to the point where some fish stocks are threatened or can no longer sustain fishing. In the past 100 years, fishing fleets have increased and the catching methods have become more sophisticated.

If the number of fish removed from a population exceeds the number of young fish reaching maturity, then the population will decline (Figure 26.2). At first, the catch size remains the same but it takes longer to catch it. Then the catch starts to contain a greater number of small fish so that the return per day at sea goes down even more. Eventually the stocks are so depleted that it is no longer economical to exploit them. The costs of the boats, the fuel and the wages of the crew exceed the value of the catch. Men are laid off, boats lie rusting in the harbour and the economy of the fishing community and those who depend on it is destroyed. Over-fishing has severely reduced stocks of many fish species: herring in the North Sea, halibut in the Pacific and anchovies off the Peruvian coast, for example. In 1965, 1.3 million tonnes of herring were caught in the North Sea. By 1977 the catch had diminished to 44 000 tonnes, i.e. about 3 per cent of the 1965 catch.

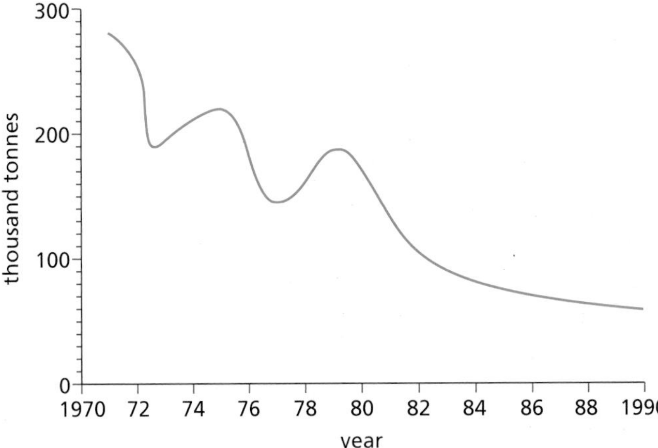

Figure 26.2 Landings of North Sea cod from 1970 to 1990

Similarly, whaling has reduced the population of many whale species to levels which give cause for concern. The blue whale's numbers have been reduced from about 2 000 000 to 6000 as a result of intensive hunting.

Agriculture

Monoculture

The whole point of crop farming is to remove a mixed population of trees, shrubs, wild flowers and grasses (Figure 26.3) and replace it with a dense population of only one species such as wheat or beans (Figure 26.4). When a crop of a single species is grown on the same land, year after year, it is called a **monoculture**.

Figure 26.3 Natural vegetation. Uncultivated land carries a wide variety of species

Figure 26.4 A monoculture. Only wheat is allowed to grow. All competing plants are destroyed

Figure 26.5 Weed control by herbicide spraying. A young wheat crop is sprayed with herbicide to suppress weeds

Figure 26.6 Effect of a herbicide spray. The crop has been sprayed except for a strip which the tractor driver missed

In a monoculture, every attempt is made to destroy organisms which feed on, compete with or infect the crop plant. So, the balanced life of a natural plant and animal community is displaced from farmland and left to survive only in small areas of woodland, heath or hedgerow. We have to decide on a balance between the amount of land to be used for agriculture, roads or building and the amount of land left alone in order to keep a rich variety of wildlife on the Earth's surface.

Pesticides

Monocultures, with their dense populations of single species and repeated planting, are very susceptible to attack by insects or the spread of fungus diseases. To combat these threats, pesticides are used. A pesticide is a chemical which destroys agricultural pests or competitors.

For a monoculture to be maintained, plants which compete with the crop plant for root space, soil minerals and sunlight are killed by chemicals called **herbicides** (Figures 26.5 and 26.6). The crop plants are protected against fungus diseases by spraying them with chemicals called **fungicides** (Figure 26.7). To destroy insects which eat and damage the plants, the crops are sprayed with **insecticides**.

Pesticide	*kills*
insecticide	insects
fungicide	parasitic fungi
herbicide	'weed' plants

The trouble with most pesticides is that they kill indiscriminately. Insecticides, for example, kill not only harmful insects but the harmless and beneficial ones, such as the bees, which pollinate flowering plants, and ladybirds, which eat aphids.

Figure 26.7 Control of fungus disease. The tree bearing the apples on the right has been sprayed with a fungicide. The apples on the unsprayed tree have developed apple scab

In about 1960, a group of chemicals, including **aldrin** and **dieldrin**, were used as insecticides to kill wireworms and other insect pests in the soil. Dieldrin was also used as a seed dressing. If seeds were dipped in the chemical before planting, it prevented certain insects from attacking the seedlings. This was thought to be better than spraying the soil with dieldrin which would have killed all the insects in the soil. Unfortunately pigeons, rooks, pheasants and partridges dug up and ate so much of the seed that the dieldrin poisoned them. Thousands of these birds were poisoned and, because they were part of a food web, birds of prey and foxes, which fed on them, were also killed. The use of dieldrin and aldrin was restricted in 1981 and banned in 1992.

One alternative to pesticides is the use of biological control though this also is not without its drawbacks unless it is thoroughly researched and tested.

Pesticides in the food chain

The concentration of insecticide often increases as it passes along a food chain (Figure 26.8). Clear Lake in California was sprayed with DDT to kill gnat larvae. The insecticide made only a weak solution of 0.015 parts per million (ppm) in the lake water. The microscopic plants and animals which fed in the lake water built up concentrations of about 5 ppm in their bodies. The small fish which fed on the microscopic animals had 10 ppm. The small fish were eaten by larger fish, which in turn were eaten by birds called grebes. The grebes were found to have 1600 ppm of DDT in their body fat and this high concentration killed large numbers of them.

A similar build-up of pesticides can occur in food chains on land. In the 1950s in the USA, DDT was sprayed on to elm trees to try and control the beetle which spread Dutch elm disease. The fallen leaves, contaminated with DDT, were eaten by earthworms. Because each worm ate many leaves, the DDT concentration in their bodies was increased ten times. When birds ate a large number of worms, the concentration of DDT in the birds' bodies reached lethal proportions and there was a 30–90 per cent mortality among robins and other song birds in the cities.

Even if DDT did not kill the birds, it caused them to lay eggs with thin shells. The eggs broke easily and fewer chicks were raised. In Britain, the numbers of peregrine falcons and sparrow hawks declined drastically between 1955 and 1965. These birds are at the top of a food web and so accumulate very high doses of the pesticides which are present in their prey, such as pigeons. After the use of DDT was restricted, the population of peregrines and sparrow hawks started to recover.

These new insecticides had been thoroughly tested in the laboratory to show that they were harmless to humans and other animals when used in low concentrations. It had not been foreseen that the insecticides would become more and more concentrated as they passed along the food chain.

Insecticides like this are called persistent because they last a long time without breaking down. This makes them good insecticides but they also persist for a long time in the soil, in rivers, lakes and the bodies of animals, including humans. This is a serious disadvantage.

Pesticides in food

Pesticides have to be poisonous in order to kill the target pests. In high doses they are also poisonous to humans. Many items of our food contain small amounts of residual pesticides. Some of these are suspected of causing cancer and other disorders but whether they do so in the very low doses we ingest is not certain. Some scientists think the levels are so low as to be negligible.

All pesticides have to be approved by a governmental regulatory body and there are legal maximum residue limits for 62 of them. New European Union (EU) rules are being drawn up which will probably extend the list and lower the maximum residue limits, but hundreds of pesticides are not covered by these regulations. In 1990–91 food samples tested in Britain had residues in 29 per cent of fruit and vegetables, 32 per cent of cereals, 48 per cent of potatoes and 55 per cent of milk, though only 1 per cent of all the samples had pesticides above the maximum residue limit.

Peeling apples and potatoes removes most of the surface pesticides but there is not much you can do to reduce any residues on the inside. Cooking seems to have variable effects, depending on the particular residue.

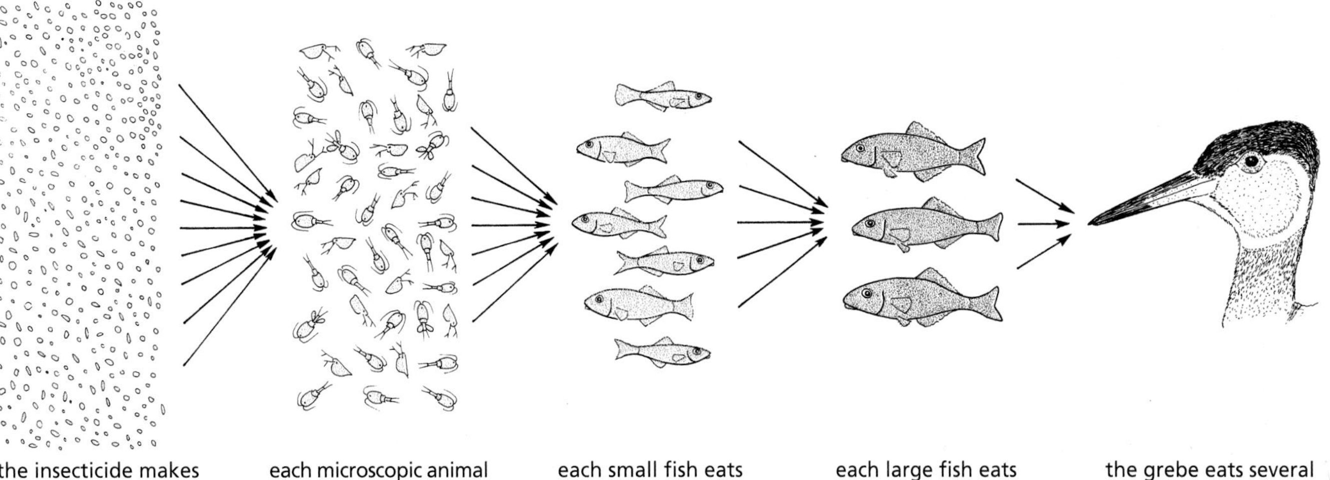

the insecticide makes only a weak solution in the water, but the microscopic plants take up the DDT
each microscopic animal eats many microscopic plants
each small fish eats many microscopic animals
each large fish eats several small fish
the grebe eats several large fish

Figure 26.8 Pesticides may become more concentrated as they move along a food chain. The intensity of colour represents the concentration of DDT

Eutrophication

On p. 43 it was explained that plants need a supply of nitrates for making their proteins, and a source of phosphates for many chemical reactions in their cells. The rate at which plants grow is often limited by how much nitrate and phosphate they can obtain. In recent years, the amount of nitrate and phosphate in our rivers and lakes has been greatly increased. This leads to an accelerated process of **eutrophication**.

Eutrophication is the enrichment of natural waters with nutrients which allow the water to support an increasing amount of plant life. This process takes place naturally in many inland waters but usually very slowly. The excessive enrichment which results from human activities leads to an overgrowth of microscopic algae (Figure 26.9).

Figure 26.9 Growth of algae in a lake. Abundant nitrate and phosphate from treated sewage and from farmland make this growth possible

Figure 26.11 Fish killed by pollution. The water may look clear but is so short of oxygen that the fish have died from suffocation

These aquatic algae are at the bottom of the food chain. The extra nitrates and phosphates from the processes listed below enable them to increase so rapidly that they cannot be kept in check by the microscopic animals which normally eat them. So they die and fall to the bottom of the river or lake. Here, their bodies are broken down by bacteria. The bacteria need oxygen to carry out this breakdown and the oxygen is taken from the water (Figure 26.10). So much oxygen is taken that the water becomes deoxygenated and can no longer support animal life. Fish and other organisms die from suffocation (Figure 26.11).

The following processes are the main causes of eutrophication.

Discharge of treated sewage

In a sewage treatment plant, human waste is broken down by bacteria (pp. 331–2) and made harmless, but the breakdown products include phosphates and nitrates. When the water from the sewage treatment is discharged into rivers it contains large quantities of phosphate and nitrate which allow the microscopic plant life to grow very rapidly (Figure 26.9).

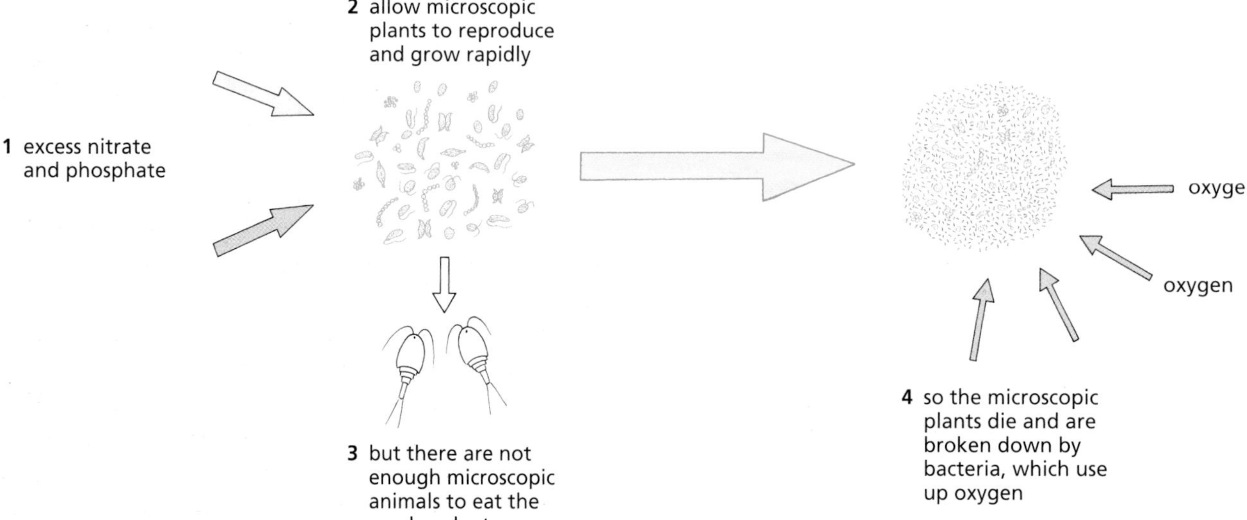
1 excess nitrate and phosphate
2 allow microscopic plants to reproduce and grow rapidly
3 but there are not enough microscopic animals to eat the surplus plants
4 so the microscopic plants die and are broken down by bacteria, which use up oxygen
oxygen
oxygen

Figure 26.10 Effects of eutrophication

Use of detergents

Some detergents contain a lot of phosphate. This is not removed by sewage treatment and is discharged into rivers. The large amount of phosphate encourages growth of microscopic plants (algae).

Arable farming

Since the Second World War, more and more grassland has been ploughed up in order to grow arable crops such as wheat and barley. When soil is exposed in this way, the bacteria, aided by the extra oxygen and water, produce soluble nitrates which are washed into streams and rivers where they promote the growth of algae. If the nitrates reach underground water stores they may increase the nitrate in drinking water to levels considered 'unsafe' for babies.

Some people think that it is excessive use of artificial fertilizers which causes this pollution but there is not much evidence for this.

'Factory farming'

Chickens, calves and pigs are often reared in large sheds instead of in open fields. Their urine and faeces are washed out of the sheds with water forming 'slurry'. If this slurry gets into streams and rivers it supplies an excess of nitrates and phosphates for the microscopic algae.

The degree of pollution of river water is often measured by its **biochemical oxygen demand (BOD)**. This is the amount of oxygen used up by a sample of water in a fixed period of time. The higher the BOD, the more polluted the water is likely to be.

It is possible to reduce eutrophication by using:

- detergents with less phosphates;
- agricultural fertilizers that do not dissolve so easily;
- animal wastes on the land instead of letting them reach rivers.

Questions

1 What are the disadvantages of catching increasingly small fish in order to maintain the total catch?

2 Give three examples of monocultures. What are
  a the advantages and
  b the disadvantages of monocultures?

3 What are
  a the benefits
  b the hazards of using pesticides?

4 DDT is a fat-soluble compound, so it was often stored in the fat depots of the birds which ingested it. The harmful effects were often unnoticed until a spell of bad weather occurred. Suggest reasons for this.

5 Explain briefly why too much nitrate could lead to too little oxygen in river water.

Humans and forests

Forests have a profound effect on climate, water supply and soil maintenance. They have been described as environmental buffers. For example, they intercept heavy rainfall and release the water steadily and slowly to the soil beneath and to the streams and rivers that start in or flow through them. The tree roots hold the soil in place.

At present, we are destroying forests, particularly tropical forests, at a prodigious rate (*a*) for their timber, (*b*) to make way for agriculture, roads (Figure 26.12) and settlements, and (*c*) for firewood. At the current rate of destruction, it is estimated that all tropical rainforests will have disappeared in the next 85 years.

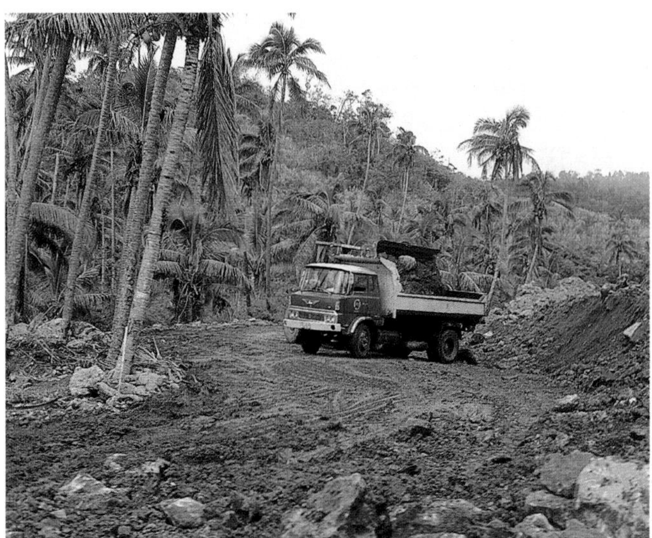

Figure 26.12 Cutting a road through a tropical rainforest. The road not only destroys the natural vegetation, it also opens up the forest to further exploitation

Removal of forests allows soil erosion, silting up of lakes and rivers, floods and the loss for ever of thousands of species of animals and plants.

Trees can grow on hillsides even when the soil layer is quite thin. When the trees are cut down and the soil is ploughed, there is less protection from the wind and rain. Heavy rainfall washes the soil off the hillsides into the rivers. The hillsides are left bare and useless and the rivers become choked up with mud and silt which can cause floods (Figures 26.13 and 26.16). For example, Argentina spends 10 million dollars a year on dredging silt from the River Plate estuary to keep the port of Buenos Aires open to shipping. It has been found that 80 per cent of this sediment comes from a deforested and overgrazed region 1800 km upstream which represents only 4 per cent of the river's total catchment area. Similar sedimentation has halved the lives of reservoirs, hydroelectric schemes and irrigation programmes. The disastrous floods in India and Bangladesh in recent years may be attributed largely to deforestation.

Figure 26.13 Soil erosion. Removal of forest trees from steeply sloping ground has allowed the rain to wash away the topsoil

The soil of tropical forests is usually very poor in nutrients. Most of the organic matter is in the leafy canopy of the tree tops. For a year or two after felling and burning, the forest soil yields good crops but the nutrients are soon depleted and the soil eroded. The agricultural benefit from cutting down forests is very short-lived, and the forest does not recover even if the impoverished land is abandoned.

Forests and climate

About half the rain which falls in tropical forests comes from the transpiration of the trees themselves. The clouds which form from this transpired water help to reflect sunlight and so keep the region relatively cool and humid. When areas of forest are cleared, this source of rain is removed, cloud cover is reduced and the local climate changes quite dramatically. The temperature range from day to night is more extreme and the rainfall diminishes.

In North Eastern Brazil, for example, an area which was once rainforest is now an arid wasteland. If more than 60 per cent of a forest is cleared, it may cause irreversible changes in the climate of the whole region. This could turn the region into an unproductive desert.

Forests and biodiversity

One of the most characteristic features of tropical forests is the enormous diversity of species they contain. In Britain, a forest or wood may consist of only one or two species of tree such as oak, ash, beech or pine. In tropical forests there are many more species and they are widely dispersed throughout the habitat. It follows that there is also a wide diversity of animals which live in such habitats. In fact, it has been estimated that half of the world's 10 million species live in tropical forests.

Destruction of tropical forest, therefore, destroys a large number of different species, driving many of them to the verge of extinction, and also drives out the indigenous populations of humans. In addition, we may be depriving ourselves of many valuable sources of chemical compounds which the plants and animals produce. The US National Cancer Institute has identified 3000 plants which have products active against cancer cells and 70 per cent of them come from the rainforest (Figure 26.14).

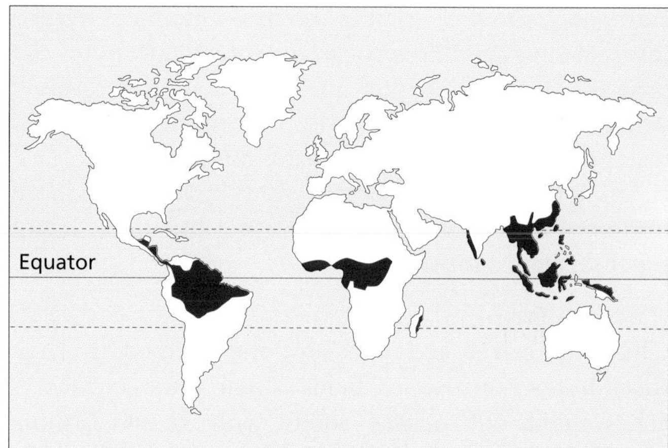

Figure 26.14 The world's rainforests

Agriculture and the soil

Soil erosion

Bad methods of agriculture lead to soil erosion. This means that the soil is blown away by the wind (Figure 26.15), or washed away by rain water. Erosion may occur for a number of reasons (Figure 26.16).

Deforestation

The soil cover on steep slopes is usually fairly thin but can support the growth of trees. If the forests are cut down to make way for agriculture, the soil is no longer protected by a leafy canopy from the driving rain. Consequently, some of the soil is washed away eventually reaching streams and rivers (Figure 26.13).

Figure 26.15 Topsoil blowing in the wind. A dry, sandy soil can easily be eroded by the wind

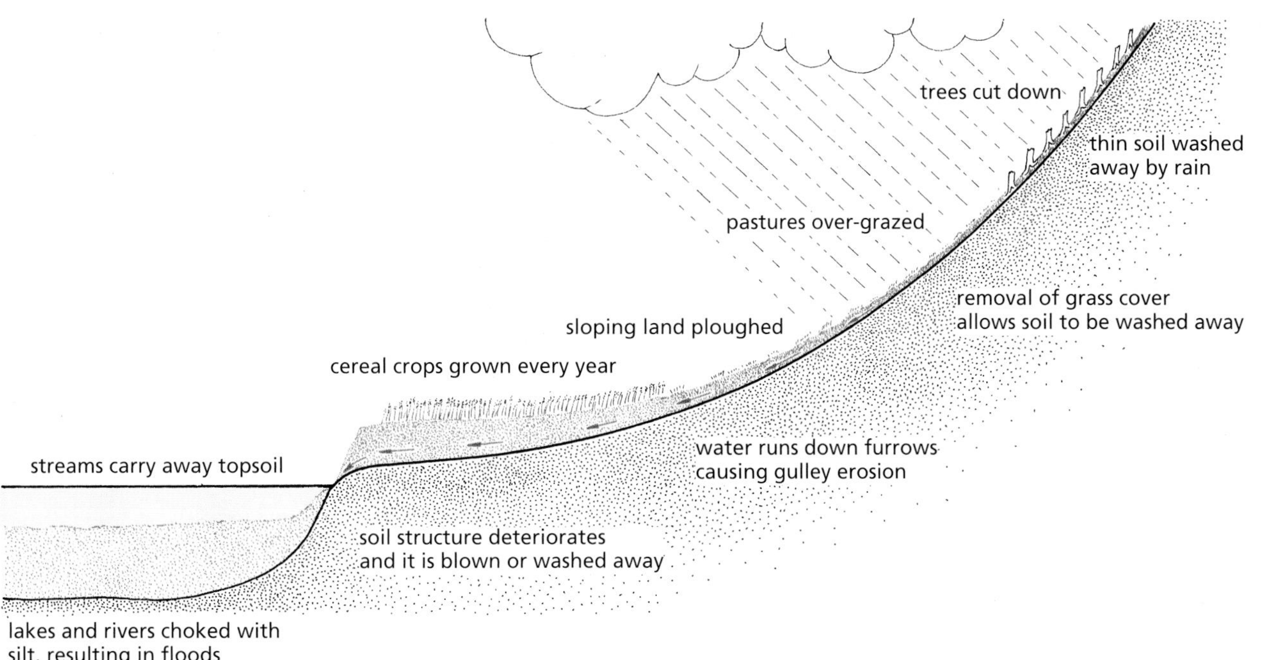

Figure 26.16 The causes of soil erosion

Bad farming methods

If land is ploughed year after year and treated only with chemical fertilizers, the soil's structure may be destroyed and it becomes dry and sandy. In strong winds it can be blown away as dust (Figure 26.15), leading to the formation of 'dust bowls', as in central USA in the 1930s, and even to deserts.

Overgrazing

If too many animals are kept on a pasture, they eat the grass down almost to the roots, and their hooves trample the surface soil into a hard layer. As a result, the rain water will not penetrate the soil and so it runs off the surface, carrying the soil with it.

Questions

1 What pressures lead to destruction of tropical forest? Give three important reasons for trying to preserve tropical forests.

2 In what ways might trees protect the soil on a hillside from being washed away by the rain?

3 If a farmer ploughs a steeply sloping field, in what direction should the furrows run to help cut down soil erosion?

4 What is the possible connection between
  a cutting down trees on hillsides and flooding in the valleys, and
  b clear-felling (logging) in tropical forests and local climate change?

5 The graph in Figure 26.17 shows the change in the numbers of mites and springtails in the soil after treating it with an insecticide. Mites eat springtails. Suggest an explanation for the changes in numbers over the 16-month period.

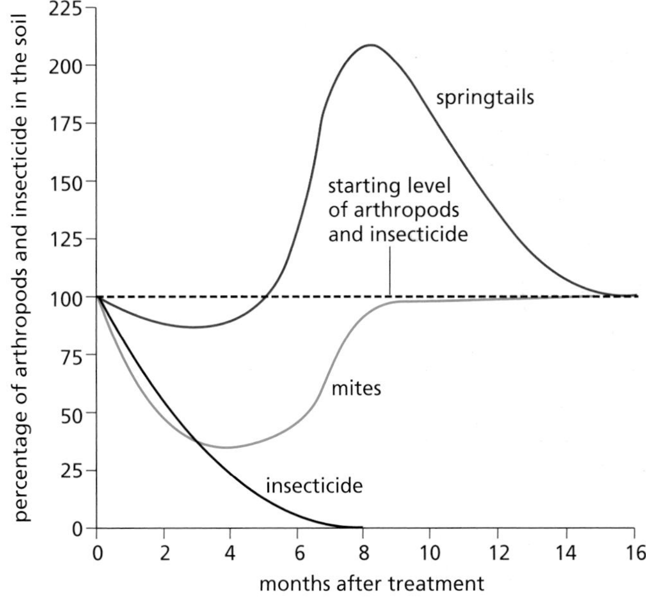

Figure 26.17 The effect of insecticide on some soil organisms

Use of pesticides

On p. 236 it was explained that most pesticides are non-selective. That is, they kill beneficial organisms as well as harmful ones.

When insecticides get into the soil, they kill the insect pests but they also kill other organisms. The effects of this on the soil's fertility are not very clear. An insecticide called **aldrin** was found to reduce the number of species of soil animals in a pasture to half the original number. Ploughing up a pasture also reduces the number of species to the same extent, so the harm done by the insecticide is not obvious.

Water pollution

Human activity sometimes pollutes streams, rivers (Figure 26.18), lakes and even coastal waters. This affects the living organisms in the water and sometimes poisons humans or infects them with disease.

Figure 26.18 River pollution. The river is badly polluted by the effluent from a paper mill

Sewage

Diseases like typhoid and cholera are caused by certain bacteria when they get into the human intestine. The faeces passed by people suffering from these diseases will contain the harmful bacteria. If the bacteria get into drinking water they may spread the disease to hundreds of other people. For this reason, among others, untreated sewage must not be emptied into rivers. It is treated at the sewage works so that all the solids are removed and the water discharged into rivers is free from harmful bacteria and poisonous chemicals (but see 'Eutrophication' on p. 238).

Eutrophication

When nitrates and phosphates from farmland and sewage escape into water they cause excessive growth of microscopic green plants. This may result in a serious oxygen shortage in the water as explained on p. 238.

Chemical pollution

Many industrial processes produce poisonous waste products. Electroplating, for example, produces waste containing copper and cyanide. If these chemicals are released into rivers they poison the animals and plants and could poison humans who drink the water. It is estimated that the River Trent receives 850 tonnes of zinc, 4000 tonnes of nickel and 300 tonnes of copper each year from industrial processes.

In 1971, 45 people in Minamata Bay in Japan died and 120 were seriously ill as a result of mercury poisoning. It was found that a factory had been discharging a compound of mercury into the bay as part of its waste. Although the mercury concentration in the sea was very low, its concentration was increased as it passed through the food chain (see p. 237). By the time it reached the people of Minamata Bay, in the fish and other sea food which formed a large part of their diet, it was concentrated enough to cause brain damage, deformity and death.

High levels of mercury have also been detected in the Baltic Sea and in the Great Lakes of North America.

Oil pollution of the sea has become a familiar event. In 1989, a tanker called the *Exxon Valdez* ran on to Bligh Reef in Prince William Sound, Alaska, and 11 million gallons of crude oil spilled into the sea. Around 400 000 sea birds were killed by the oil (Figure 26.19) and the populations of killer whales, sea otters and harbour seals among others, were badly affected. The hot water high pressure hosing techniques and chemicals used to clean up the shoreline killed many more birds and sea creatures living on the coast. Since 1989, there have continued to be major spillages of crude oil from tankers and off-shore oil wells.

Figure 26.19 Oil pollution. Oiled sea birds like these long-tailed ducks cannot fly to reach their feeding grounds. They also poison themselves by trying to clean the oil from their feathers

Questions

1 What are the possible dangers of dumping and burying poisonous chemicals on the land?

2 Before most water leaves the waterworks, it is exposed for some time to the poisonous gas, chlorine. What do you think is the point of this?

3 If the concentration of mercury in Minamata Bay was very low, why did it cause such serious illness in humans?

Air pollution

Some factories and all motor vehicles release poisonous substances into the air. Factories produce smoke and sulphur dioxide; cars produce lead compounds, carbon monoxide and the oxides of nitrogen which lead to smog (Figure 26.23, p. 244).

Smoke

This consists mainly of tiny particles of carbon and tar which come from burning coal either in power stations or in the home. The tarry drops contain chemicals which may cause cancer. When the carbon particles settle, they blacken buildings and damage the leaves of trees. Smoke in the atmosphere cuts down the amount of sunlight reaching the ground. For example, since the Clean Air Act of 1956, London has received 70 per cent more sunshine in December.

Particulates

Although smoke has been largely eliminated from our towns, vehicle exhaust gases (particularly from diesels), contain microscopic particles coated with hydrocarbons. The particles may be referred to as PM10s or PM2.5s because their diameters are less than 10 or 2.5 micrometres (μm) respectively. The particles are thought to be a cause of about 10 000 deaths per year, particularly of people already suffering from chronic lung diseases such as emphysema and bronchitis.

Sulphur dioxide and oxides of nitrogen

Coal and oil contain sulphur. When these fuels are burned, they release sulphur dioxide (SO_2) into the air (Figure 26.20). Although the tall chimneys of factories (Figure 26.21) send smoke and sulphur dioxide high into the air, the sulphur dioxide dissolves in rain water and forms an acid. When this acid falls on buildings, it slowly dissolves the limestone and mortar. When it falls on plants, it reduces their growth and damages their leaves.

Figure 26.21 Air pollution by industry. Tall chimneys keep pollution away from the immediate surroundings but the atmosphere is still polluted

This form of pollution has been going on for many years and is getting worse. In North America, Scandinavia and Scotland, forests are being destroyed (Figure 26.22) and fish are dying in lakes, at least partly as a result of **'acid rain'**.

Figure 26.22 Effects of acid rain on conifers in the Black Forest, Germany

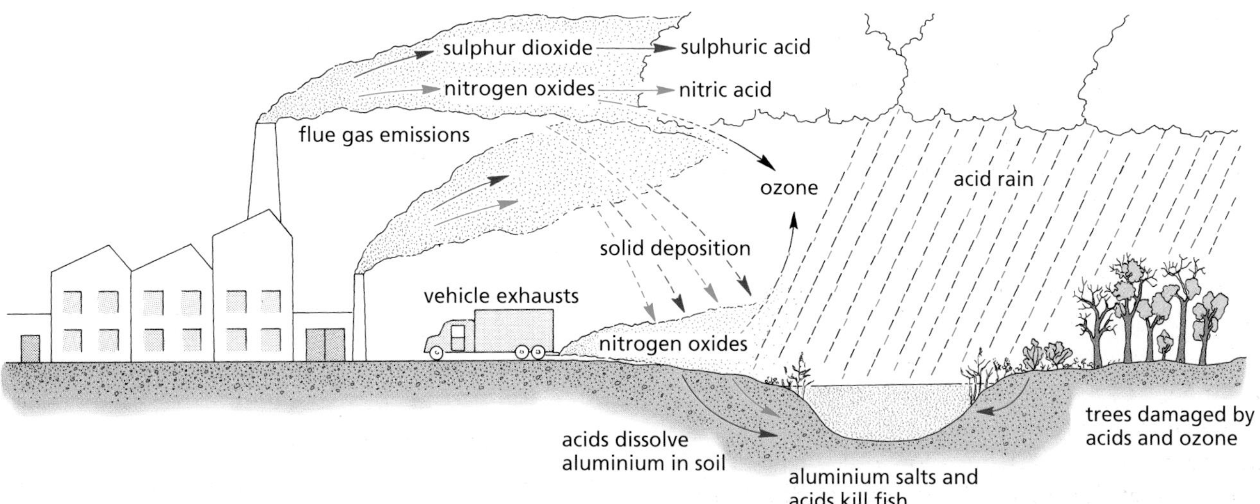

Figure 26.20 Unnaturally acid rain in Britain. The pollution comes from British factories, power stations, homes and vehicles. Most emissions start as dry gases and are converted slowly to dilute sulphuric and nitric acids

Oxides of nitrogen from power stations and vehicle exhausts also contribute to atmospheric pollution and acid rain. The nitrogen oxides dissolve in rain drops and form nitric acid.

Oxides of nitrogen also take part in reactions with other atmospheric pollutants and produce ozone. It may be the ozone and the nitrogen oxides which are largely responsible for the damage observed in forests.

One effect of acid rain is that it dissolves out the aluminium salts in the soil. These salts eventually reach toxic levels in streams and lakes.

There is still some argument about the source of the acid gases which produce acid rain. For example, a large proportion of the sulphur dioxide in the atmosphere comes from the natural activities of certain marine algae. These microscopic 'plants' produce the gas, dimethyl-sulphide, which is oxidized to sulphur dioxide in the air.

Nevertheless, there is considerable circumstantial evidence that industrial activities in Britain, America and Central and Eastern Europe add large amounts of extra sulphur dioxide and nitrogen oxides to the atmosphere.

Smog

This is a thin fog which occurs in cities in certain climatic conditions (Figure 26.23). Smog is irritating to the eyes and lungs and also damages plants. It is produced when sunlight and ozone (O_3) in the atmosphere act on the oxides of nitrogen and unburnt hydrocarbons released from vehicle exhausts. This type of smog is called 'photochemical smog' to distinguish it from the smoke plus fog that used to afflict British cities.

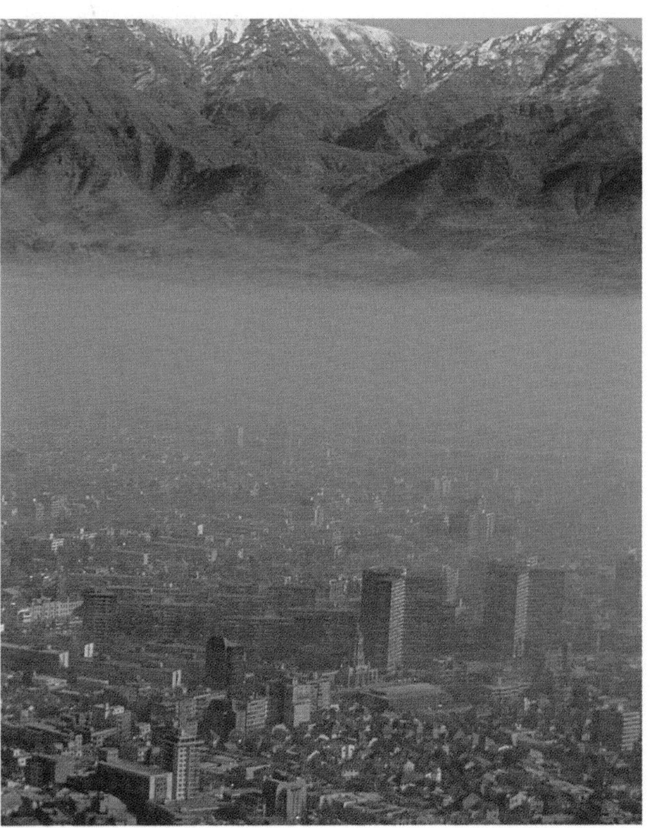

Figure 26.23 Photochemical 'smog' over a city

Carbon monoxide (CO)

This gas is also a product of combustion in the engines of cars and trucks. When inhaled, carbon monoxide combines with haemoglobin in the blood to form a fairly stable compound, carboxyhaemoglobin. The formation of carboxyhaemoglobin reduces the oxygen-carrying capacity of the blood and this can be harmful, particularly in people with heart disease or anaemia.

A smoker is likely to inhale far more carbon monoxide from cigarettes than from the atmosphere. Nevertheless, the carbon monoxide levels produced by heavy traffic in towns can be harmful.

Chlorofluorocarbons (CFCs)

These are gases which readily liquefy when compressed. This makes them useful as refrigerants, propellants in aerosol cans and in plastic foams. Chlorofluorocarbons are very stable and accumulate in the atmosphere, where they react with ozone (O_3).

Ozone is present throughout the atmosphere but reaches a peak at about 25 km, where it forms what is called the **'ozone layer'**. This layer filters out much of the ultraviolet radiation in sunlight.

The chlorine from CFCs reacts with ozone and reduces its concentration in the ozone layer. As a result, more ultraviolet (UV) radiation reaches the Earth's surface. Higher levels of UV radiation can lead to an increased incidence of skin cancer. It can also affect crops, damage marine plankton and even distort weather patterns.

The reactions involved are very complex. There are also natural processes which destroy or generate ozone.

Control of air pollution

The Clean Air Acts of 1956 and 1968

These Acts designated certain city areas as 'smokeless zones'. The use of coal for domestic heating was prohibited and factories were not allowed to emit black smoke. This was effective in abolishing dense fogs in cities but did not stop the discharge of sulphur dioxide and nitrogen oxides in the country as a whole.

Reduction of acid gases

The concern over the damaging effects of acid rain has led many countries to press for regulations to reduce emissions of sulphur dioxide and nitrogen oxides.

Reduction of sulphur dioxide can be achieved either by fitting desulphurization plants to power stations or by changing the fuel or the way it is burnt. In 1986, Britain decided to fit desulphurization plants to three of its major power stations, but also agreed to a United Nations protocol to reduce sulphur dioxide emissions to 50 per cent of 1980 levels by the year 2000, and to 20 per cent by 2010. This will be achieved largely by changing from coal-fired to gas-fired power stations.

Reduction of vehicle emissions

Oxides of nitrogen come, almost equally, from industry and from motor vehicles (Figure 26.20). Flue gases from industry can be treated to remove most of the nitrogen oxides. Vehicles can have **catalytic converters** fitted to their exhaust systems. These converters remove most of the nitrogen oxides, carbon monoxide and unburned hydrocarbons. They add £200–600 to the cost of a car and will work only if lead-free petrol is used, because lead blocks the action of the catalyst.

Another solution is to redesign car engines to burn petrol at lower temperatures (**'lean burn'** engines). These emit less nitrogen oxide but just as much carbon monoxide and hydrocarbons as normal engines.

In the long term, it may be possible to use fuels such as alcohol or hydrogen which do not produce so many pollutants.

The European Union has set limits on exhaust emissions. From 1989, new cars over 2 litres had to have catalytic converters and from 1993 smaller cars had to fit them as well.

Regulations introduced in 1995 should cut emissions of particulates by 75 per cent and nitrogen oxides by 50 per cent.

These reductions will have less effect if the volume of traffic continues to increase. Significant reduction of pollutants is more likely if the number of vehicles is stabilized and road freight is reduced.

Protecting the ozone layer

The appearance of 'ozone holes' in the Antarctic and Arctic, and the thinning of the ozone layer elsewhere, spurred countries to get together and agree to reduce the production and use of CFCs (p. 244) and other ozone-damaging chemicals.

1987 saw the first Montreal protocol which set targets for the reduction and phasing out of these chemicals. In 1990, nearly 100 countries, including Britain, agreed to the next stage of the Montreal protocol which committed them to reduce production of CFCs by 85 per cent in 1994 and phase them out completely by 2000.

Although production has been phased out in Western countries, global production has still risen by 5 per cent; this increase is largely due to China. China and the developing countries will now phase out production and use of CFCs and similar chemicals by 2100.

The 'greenhouse effect' and global warming

The Earth's surface receives and absorbs radiant heat from the Sun. It re-radiates some of this heat back into space. The Sun's radiation is mainly in the form of short-wavelength energy and penetrates our atmosphere easily. The energy radiated back from the Earth is in the form of long wavelengths (infrared or IR), much of which is absorbed by the atmosphere. The atmosphere acts like the glass in a greenhouse. It lets in light and heat from the Sun but reduces the amount of heat which escapes (Figure 26.24).

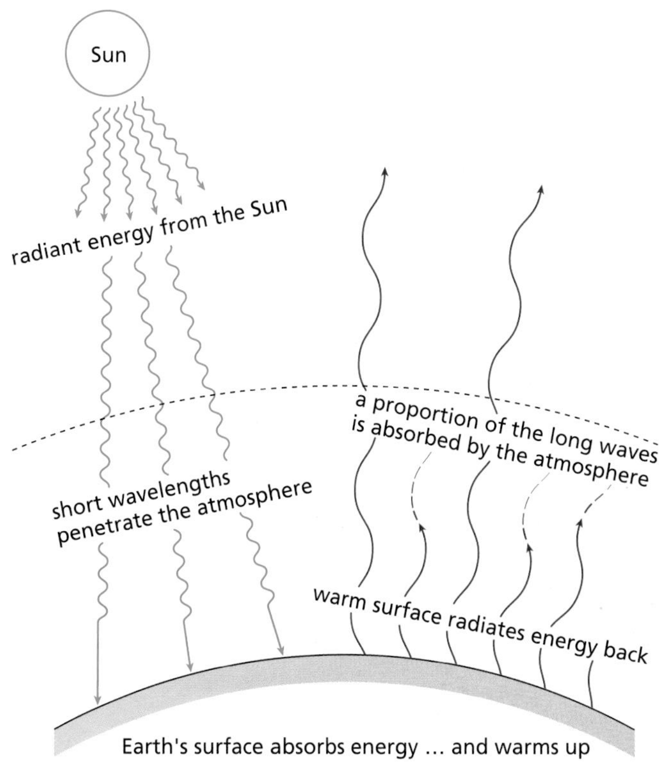
Sun
radiant energy from the Sun
a proportion of the long waves is absorbed by the atmosphere
short wavelengths penetrate the atmosphere
warm surface radiates energy back
Earth's surface absorbs energy ... and warms up

Figure 26.24 The 'greenhouse effect'

If it were not for this 'greenhouse effect' of the atmosphere, the Earth's surface would probably be at −18 °C. The 'greenhouse effect', therefore, is entirely natural and desirable.

Not all the atmospheric gases are equally effective at absorbing IR radiation. Oxygen and nitrogen, for example, absorb little or none. The gases which absorb most IR radiation, in order of maximum absorption, are water vapour, carbon dioxide (CO_2), methane and atmospheric pollutants such as oxides of nitrogen and CFCs. Apart from water vapour, these gases are in very low concentrations in the atmosphere, but some of them are strong absorbers of IR radiation. It is assumed that if the concentration of any of these gases were to increase, the greenhouse effect would be enhanced and the Earth would get warmer.

In recent years, attention has focused principally on CO_2. If you look back at the 'carbon cycle' on p. 228, you will see that the natural processes of photosynthesis, respiration and decay would be expected to keep the CO_2 concentration at a steady level. However, since the Industrial Revolution, we have been burning the 'fossil fuels' derived from coal and petroleum and releasing extra CO_2 into the atmosphere. As a result, the concentration of CO_2 has increased from 0.029 to 0.035 per cent since 1860. It is likely to go on increasing as we burn more and more fossil fuel.

Although it is not possible to prove beyond all reasonable doubt that production of CO_2 and other 'greenhouse gases' is causing a rise in the Earth's temperature, i.e. global warming, the majority of scientists and climatologists agree that it is happening now and will get worse unless we take drastic action to reduce the output of these gases.

Predictions of the effects of global warming depend on computer models. But these depend on very complex and uncertain interaction of variables.

Changes in climate might increase cloud cover and this might reduce the heat reaching the Earth from the Sun. Oceanic plankton absorb a great deal of CO_2. Will the rate of absorption increase or will a warmer ocean absorb less of the gas? An increase in CO_2 should, theoretically, result in increased rates of photosynthesis, bringing the system back into balance.

None of these possibilities is known for certain. The worst scenario is that the climate and rainfall distribution will change, and disrupt the present pattern of world agriculture; the oceans will expand and the polar ice-caps will melt causing a rise in sea level; extremes of weather may produce droughts and food shortages.

An average of temperature records from around the world suggests that, since 1880, there has been a rise of 0.5–0.7 °C, most of it in the last 15 years (Figure 26.25), but this is too short a period from which to draw firm conclusions about long-term trends. If the warming trend continues, however, it could produce a rise in sea level of between 0.2 and 1.5 metres in the next 50–100 years.

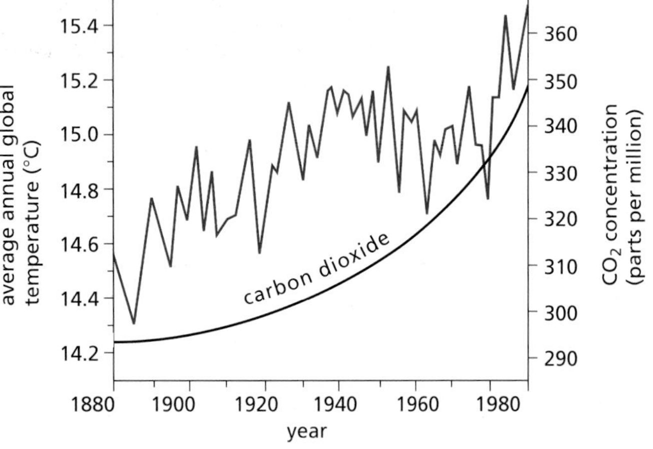

Figure 26.25 Annual average global temperatures since 1880

The first Kyoto Conference (Japan) in 1997 set targets for the industrialized countries to reduce CO_2 emissions by 2010. Europe, as a whole, has agreed to cuts of 8 per cent, though this average allows some countries to increase their emissions.

Britain plans to reduce emissions by 20 per cent of 1990 levels by 2010 but we really need an overall cut of 60 per cent to halt the progress of global warming. The big industrialized countries who contribute 80 per cent of the greenhouse gases, particularly the USA, are opposed to measures which might interfere with their industries, claiming that global warming is not a proven fact. The Kyoto Conference of 2000 failed to reach any agreement.

The precautionary principle suggests that, even if global warming is not taking place, our supplies of fossil fuels will eventually run out and we need to develop alternative sources of energy now.

Questions

1 To what extent do tall chimneys on factories reduce atmospheric pollution?

2 What are thought to be the main causes of 'acid rain'?

3 Why are carbon dioxide and methane called 'greenhouse gases'?

Checklist

- The plants and animals in a food web are so interdependent that even a small change in the numbers of one group has a far-reaching effect on all the others.
- Hunting activities and farming upset the natural balance between other living organisms.
- Pesticides kill insects, weeds and fungi that could destroy our crops.
- Pesticides help to increase agricultural production but they kill other organisms as well as pests.
- A pesticide or pollutant which starts off at a low, safe level can become dangerously concentrated as it passes along a food chain.
- Eutrophication of lakes and rivers results in the excessive growth of algae followed by an oxygen shortage when the algae die and decay.
- Soil erosion results from removal of trees from sloping land, use of only chemical fertilizers on ploughed land and putting too many animals on pasture land.
- The conversion of tropical forest to agricultural land usually results in failure because forest soils are poor in nutrients.
- Removal of forests can lead to erosion, silting-up of lakes and rivers and to flooding.
- We pollute our lakes and rivers with industrial waste and sewage effluent.
- We pollute the sea with crude oil and factory wastes.
- We pollute the air with smoke, sulphur dioxide and nitrogen oxides from factories, and carbon monoxide and nitrogen oxides from motor vehicles.
- The acid rain resulting from air pollution leads to poisoning of lakes and possibly destruction of trees.
- The extra carbon dioxide from fossil fuels might lead to global warming.

27 Conservation

Extinction

Conservation of species
Captive breeding and reintroductions.

Conservation of genes

Conservation of habitats
International initiatives. Conservation in Britain. Farmland. ESAs.

Conservation of non-renewable resources
Recycling.

Extinction

In the course of evolution, species become extinct. After all, the fossil remains of plants and animals represent organisms that became extinct hundreds of thousands of years ago. There have been periods of mass extinction, such as that which wiped out the dinosaurs during the Cretaceous era, 65 million years ago.

The 'background' extinction rate for, say, birds might be one species in 100–1000 years. Today, as a result of human activity, the rate of extinction has gone up by at least ten times and possibly as much as 1000 times. Some estimates suggest that the world is losing one species every day and within 20 years at least 25 per cent of all forms of wildlife could become extinct. Reliable evidence for these figures is hard to obtain, however.

A classic example is the colonization of the Pacific islands by the Polynesians. They killed and ate the larger bird species, and introduced pigs and rats which ate the eggs and young of ground-nesting species. Their goats and cattle destroyed plant species. Of about 1000 plant species, 85 per cent has been lost since they were first discovered.

This may be an extreme example but the same sort of changes are happening all over the world. For example, only about 6000 tigers survive in the wild. This is a mere 5 per cent of their number in 1900 (Figure 27.1).

Humans have accelerated the rate of extinction by killing the organisms, introducing alien species and destroying habitats. Apart from the fact that we have no right to wipe out species for ever, the chances are that we will deprive ourselves not only of the beauty and diversity of species but also of potential sources of valuable products such as drugs. Many of our present-day drugs are derived from plants (e.g. quinine and aspirin) and there may be many more sources as yet undiscovered. We are also likely to deprive the world of genetic resources (see below).

Figure 27.1 In 100 years the tiger population has fallen from 120 000 to 6000

Conservation of species

Species can be conserved by passing laws which make killing or collecting them an offence, by international agreements on global bans or trading restrictions, and by conserving habitats (Figure 27.2, overleaf).

In Britain, it is an offence to capture or kill almost all species of wild birds or to take eggs from their nests; wild flowers in their natural habitats may not be uprooted; badgers, otters and bats are just three of the protected species of mammal (Figure 27.3, overleaf).

CITES (Convention on International Trade in Endangered Species) gives protection to about 1500 animals and thousands of plants by persuading governments to restrict or ban trade in endangered species or their products, e.g. snake skins or rhino horns. There are about 70 countries which are party to the Convention.

Figure 27.2 Trying to stop the trade in endangered species. A customs official checks an illegal cargo impounded at an Indian customs post

Figure 27.3 Badger. One of a number of species protected by law

The **WWF** (World Wide Fund for Nature) operates on a global scale and is represented in 25 countries. The WWF raises money for conservation projects in all parts of the world, but with particular emphasis on endangered species and habitats.

The **IWC** (International Whaling Commission) was set up to try and avoid the extinction of whales as a result of uncontrolled whaling, and has about 40 members.

The IWC allocates quotas of whales that the member countries may catch but, having no powers to enforce its decisions, cannot prevent countries from exceeding their quotas.

In 1985, the IWC declared a moratorium (i.e. a complete ban) on all whaling, which was reaffirmed in 2000 despite opposition from Japan and Norway. Japan continues to catch whales 'for scientific purposes'.

Captive breeding and reintroductions

Provided a species has not become totally extinct, it may be possible to boost its numbers by breeding in captivity and releasing the animals back into the environment. In Britain, modest success has been achieved with otters (Figure 27.4). It is important (a) that the animals do not become dependent on humans for food and (b) that there are suitable habitats left for them to recolonize.

Sea eagles, red kites (Figure 27.5) and ospreys have been introduced from areas where they are plentiful to areas they had become extinct.

Figure 27.4 The otter has been bred successfully in captivity and released

Figure 27.5 Red kites from Spain and Sweden have been reintroduced to Britain

Conservation of genes

On p. 212 it was explained that crossing a wild grass with a strain of wheat produced an improved variety. This is only one example of many successful attempts to improve yield, drought resistance and disease resistance in food plants. Some 25 000 plant species are threatened with extinction at the moment. This could result in a devastating loss of hereditary material (see 'Genes', p. 186) and a reduction of about 10 per cent in the genes available for crop improvement. 'Gene banks' have been set up to preserve a wide range of plants, but these banks are vulnerable to accidents, disease and human error. The only secure way of preserving the full range of genes is to keep the plants growing in their natural environments.

Questions

1 What kinds of human activity can lead to the extinction of a species?

2 How do the roles of CITES and WWF differ? In what respects might their activities overlap?

3 How might the loss of a species affect
  a our health (indirectly) and
  b the prospect of developing new varieties of crop plants resistant to drought?

Conservation of habitats

If animals and plants are to be conserved it is vital that their habitats are conserved also.

Habitats are many and varied: from vast areas of tropical forest to the village pond, and including such diverse habitats as wetlands, peat bogs, coral reefs, mangrove swamps, lakes and rivers, to list but a few.

International initiatives

In the last 30 years it has been recognized that conservation of major habitats needed international agreements on strategies. In 1992, the Convention on Biological Diversity was opened for signature at the 'Earth Summit' Conference in Rio, and 168 countries signed it. The Convention aims to preserve biological diversity ('biodiversity').

Biodiversity encompasses the whole range of species in the world. The Convention will try to share the costs and benefits between developed and developing countries, promote 'sustainable development' and support local initiatives.

'Sustainable development' implies that industry and agriculture should use natural resources sparingly (p. 251) and avoid damaging natural habitats and the organisms in them.

The Earth Summit meeting addressed problems of population, global warming, pollution, etc., as well as biodiversity.

There are several voluntary organizations which work for world-wide conservation, e.g. WWF (p. 248), Friends of the Earth and Greenpeace.

Habitat conservation in Britain

English Nature, the **Countryside Council for Wales** and **Scottish Natural Heritage** were formed from the Nature Conservancy Council (NCC). They are regulatory bodies committed to establish, manage and maintain nature reserves, protect threatened habitats and conduct research into matters relevant to conservation.

The NCC established 195 nature reserves (Figure 27.6) but, in addition, had responsibility for notifying planning authorities of **Areas of Special Scientific Interest (ASSIs)**, also known as Sites of Special Scientific Interest (SSSIs). These are privately owned lands which include important habitats or rare species (Figure 27.7). English Nature and other conservation bodies establish management agreements with the owners so that the sites are not damaged by felling trees, ploughing land or draining fens (Figure 27.8).

Figure 27.6 An English Nature National Nature Reserve at Bridgewater Bay in Somerset. The mudflats and saltmarsh attract large numbers of wintering wildfowl

Figure 27.7 Area of Special Scientific Interest. This heathland in Surrey is protected by a management agreement with the landowner

Figure 27.8 The Royal Society for the Protection of Birds (RSPB) maintains this wet grassland by Loch Leven, Scotland, for nesting redshanks, snipe, lapwings and ducks

There are now about 5000 ASSIs, and the Countryside and Rights of Way Act of 2000 has strengthened the rules governing the maintenance of ASSIs.

There are several other, non-governmental organizations which have set up reserves and which help to conserve wildlife and habitats. The Nature Conservation Trust Reserve has about 1400 reserves, the Royal Society for the Protection of Birds (RSPB) has 150, the Woodland Trust has 102 and there are about 160 other reserves managed by other organizations.

The National Parks Commission has set up ten National Parks covering some 9 per cent of England and Wales, e.g. Dartmoor, Snowdonia and the Lake District. Although the land is privately owned, the Park Authorities are responsible for protecting the landscape and wildlife, and for planning public recreation such as walking, climbing or gliding.

The European Commission's **Habitats Directive** of 1994 requires member states to designate **Special Areas of Conservation (SACs)** to protect some of the most seriously threatened habitats and species throughout Europe. The UK has submitted a list of 340 sites, though many of these are already protected areas, such as ASSIs.

Desirable though ASSIs, National Parks and SACs are, they represent only relatively small, isolated areas of land. Birds can move freely from one area to another, but plants and small animals are confined to an isolated habitat so are subject to risks which they cannot escape. If more farmland were managed in a way 'friendly' to wildlife, these risks could be reduced.

Farmland

Farmland is not a natural habitat but, at one time, hedgerows, hay meadows and stubble fields were important habitats for plants and animals. Hay meadows and hedgerows supported a wide range of wild plants as well as providing feeding and nesting sites for birds and animals.

Figure 27.9 Destruction of a hedgerow. Permission now has to be sought from the local authority before this can happen

Figure 27.10 Grass for silage. There is no variety of plant life and, therefore, an impoverished population of insects and other animals

Figure 27.11 Traditional hay meadow. The variety of wild flowers will attract butterflies and other insects

Intensive agriculture has destroyed many of these habitats; hedges have been grubbed out (Figure 27.9) to make fields larger, a monoculture of silage grasses (Figure 27.10) has replaced the mixed population of a hay meadow (Figure 27.11) and planting of winter wheat has denied animals access to stubble fields in autumn. As a result, populations of butterflies, flowers and birds such as skylarks, grey partridges, corn buntings and tree sparrows have crashed.

Recent legislation now prohibits the removal of hedgerows without approval from the local authority but the only hedges protected in this way are those deemed to be 'important' because of species diversity or historical significance.

The **Farming and Wildlife Advisory Group** can advise farmers how to manage their land in ways which encourage wildlife. This includes, for example, leaving strips of uncultivated land round the margins of fields or planting new hedgerows. Even strips of wild grasses and flowers between fields significantly increase the population of beneficial insects.

ESAs

Certain areas of farmland have been designated as Environmental Sensitive Areas (ESAs), and farmers are paid a subsidy for managing their land in ways that conserve the environment.

Questions

1 What do you understand by
 a biodiversity and
 b sustainable development?

2 What is the difference between an ASSI and a nature reserve?

Conservation of non-renewable resources

Coal, oil, natural gas and minerals (including metallic ores) cannot be replaced once their sources have been totally depleted. Estimates of how long these stocks will last are unreliable but in some cases, e.g. lead and tin, they are less than 100 years.

By the time that fossil fuels run out, we will have to have alternative sources of energy. Even the uranium used in nuclear reactors is a finite resource and will, one day, run out.

The alternative sources of energy available to us are hydro-electric, nuclear, wind and wave power, wood and other plant products. The first two are well established; the others are either in the experimental stages, making only a small contribution, or are more expensive (at present) than fossil fuels (Figure 27.12).

Figure 27.12 Wind generators in the USA. On otherwise unproductive land or offshore, these generators make an increasing input to the electricity supply

Plant products are **renewable resources** and include alcohol distilled from fermented sugar (from sugar cane), which can replace or supplement petrol (Figure 27.13), and sunflower oil, which can replace diesel fuel, and wood from fast-growing trees. In addition, plant and animal waste material can be decomposed anaerobically in fermenters to produce **biogas** (Figure 27.14), which consists largely of methane.

Figure 27.13 An alcohol-powered car in Brazil

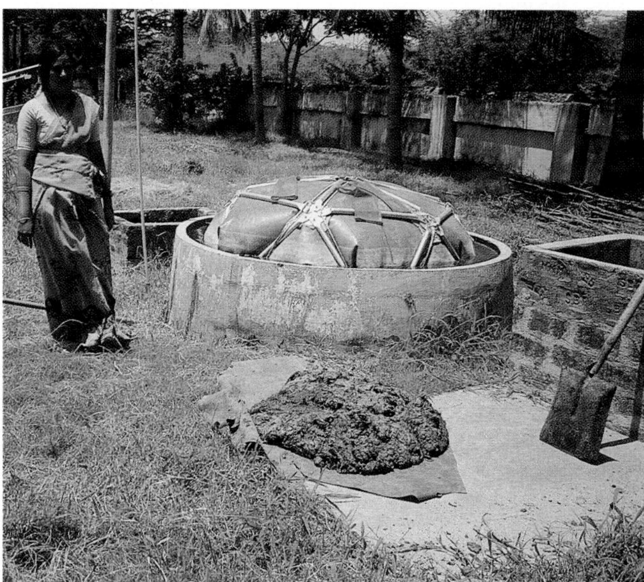

Figure 27.14 Animal and plant waste can be fermented to produce biogas

Chemicals for industry or drugs, currently derived from petroleum, will have to be made from plant products.

When **non-renewable resources** run out they will have to be replaced by recycling or by using man-made materials derived from plant products. Already some bacteria have been genetically engineered to produce substances which can be converted to plastics.

Recycling

As minerals and other resources become scarcer, they also become more expensive. It then pays to use them more than once. The recycling of materials may also reduce the amount of energy used in manufacturing. In turn this helps to conserve fuels and reduce pollution.

For example, producing aluminium alloys from scrap uses only 5 per cent of the energy that would be needed to make them from aluminium ores. In 2000, Britain recycled over 35 000 tonnes of aluminium.

We also recycle 56 per cent of the lead used in Britain. This seems quite good until you realize that it also means that 44 per cent of this poisonous substance enters the environment.

Manufacturing glass bottles uses about three times more energy than if they were collected, sorted, cleaned and re-used. Recycling the glass from bottles does not save energy but does reduce the demand for sand used in glass manufacture. In 1997, 425 000 tonnes of glass were recycled. Polythene waste is now also recycled (Figure 27.15).

Waste paper can be pulped and used again, mainly for making paper and cardboard. Newspapers are de-inked and used again for newsprint. One tonne of waste paper is equivalent to perhaps 17 trees. (Paper is made from wood-pulp.) So collecting waste paper may help to cut our import bill for timber and spare a few more hectares of moorland from the spread of commercial forestry.

Figure 27.15 Recycling polythene. Polythene waste is recycled for industrial use

Questions

1 Explain why some renewable energy sources depend on photosynthesis.

2 In what ways does the recycling of materials help to save energy and conserve the environment?

3 Explain why some of the alternative and renewable energy sources are less likely to cause pollution than coal and oil.

Checklist

- Although extinction is a natural phenomenon, human activities are causing a great increase in the rates of extinction.
- Conservation of species requires international agreements and regulations.
- These regulations may prohibit killing or collecting species and prevent trade in them or their products.
- Loss of a plant species deprives us of (a) a possible source of genes and (b) a possible source of chemicals for drugs.
- Conserving a species by captive breeding is of little use unless its habitat is also conserved.
- The Earth Summit Conference tried to achieve international agreement on measures to conserve wildlife and habitats, and reduce pollution.
- National Parks, nature reserves, ASSIs and SACs all try to preserve habitats but they cover only a small proportion of the country and exist as isolated communities.
- Intensive farming has resulted in habitat deterioration and reduction of wildlife.
- Incentives exist for farming in a way that is friendly to wildlife.
- When supplies of fossil fuels run out or become too expensive, we will need to develop alternative sources of energy.
- Raw materials, such as metal ores, will one day run out.
- Recycling metals, paper, glass and polythene helps to conserve these materials and save energy.

28 Ecosystems

Definitions
Environment, habitat, population, community, ecosystem, competition.

Factors which affect communities
Biotic and abiotic factors.

Natural and artificial ecosystems

Adaptation
Camels and polar bears.

Definitions

Ecology is the study of living organisms in relation to their natural environment, as distinct from in the laboratory. This does not mean that laboratory work is ruled out, but its object is always to explain how the organism survives, how it relates to other organisms and why it is successful in its particular environment.

The following are some of the terms used in any discussion of ecology.

Environment

This means everything in the surroundings of an organism that could possibly influence it. The environment of a tadpole consists of water. The temperature of the water will influence the tadpole's rate of growth and activity. The watery environment contains plants and animals on which the tadpole will feed, but it also contains fish and insects which may eat the tadpole. The water contains dissolved oxygen which the tadpole breathes by means of its gills. The water, the oxygen, the food and the predators are all part of the tadpole's environment.

Habitat

A habitat is where an organism lives, i.e. where it obtains its food and shelter, and where it reproduces. The habitat of a limpet is a rocky shore. The environment includes air, sea water and sunlight but the habitat is the shore. The habitat of the tapeworm is the intestine of a mammal. Its environment, however, is the warm digested food and digestive juices of its host. The habitat of an aphid may be a bean plant, but its environment will include sun, wind, rain, ladybirds, ants and bacteria.

Population

In biology, this term always refers to a single species. A biologist might refer to the population of sparrows in a farmyard or the population of carp in a lake. In each case this would mean the total numbers of sparrows or the total numbers of carp in the stated area.

Community

A community is made up of all the plants and animals living in a habitat. In the soil there is a community of organisms which includes earthworms, springtails and other insects, mites, fungi and bacteria. In a lake, the animal community will include fish, insects, crustacea, molluscs and protozoa.

The plant community will consist of rooted plants with submerged leaves, rooted plants with floating leaves, reed-like plants growing at the lake margin, plants floating freely on the surface, filamentous algae and single-celled algae (p. 279) in the surface waters.

Ecosystem

The community of organisms in a habitat, plus the non-living part of the environment (air, water, soil, light, etc.) make up an ecosystem. A lake is an ecosystem which consists of the plant and animal communities mentioned above, and the water, minerals, dissolved oxygen, soil and sunlight on which they depend. An ecosystem is self-supporting (Figure 28.1).

individuals of the same species } = POPULATION
+
populations of other species } = COMMUNITY

non-living part of environment
+
COMMUNITY } = ECOSYSTEM

In a woodland ecosystem, the plants absorb light and rain water for photosynthesis, the animals feed on the plants and on each other. The dead remains of animals and plants, acted upon by fungi and bacteria, return nutrients to the soil.

Lakes and ponds are clear examples of ecosystems. Sunlight, water and minerals allow the plants to grow and support animal life. The recycling of materials from the dead organisms maintains the supply of nutrients.

So, a *population* of carp forms part of the animal *community* living in a *habitat* called a lake. The communities in this habitat, together with their watery *environment*, make up a self-supporting *ecosystem*.

Figure 28.1 The 'Ecosphere'. The 5-inch globe contains sea water, bacteria, algae, snails and a few Pacific shrimps. Given a source of light it is a self-supporting system and survives for several years (at least). The shrimps live for up to 7 years but few reproduce

A carp is a *secondary consumer* at the top of a *food chain*, where it is in *competition* with other species of fish for food and with other carp for food and mates.

The whole of that part of the Earth's surface which contains living organisms (called the **biosphere**), may be regarded as one vast ecosystem.

No new material (in significant amounts) enters the Earth's ecosystem from space and there is no significant loss of materials. The whole system depends on a constant input of energy from the sun (p. 35) and recycling of the chemical elements (p. 227).

Distribution in an ecosystem

All ecosystems contain producers, consumers and decomposers as described on p. 227. The organisms are not distributed uniformly throughout the ecosystem but occupy habitats that suit their way of life.

For example, fish may range freely within an aquatic ecosystem but most of them will have preferred habitats in which they feed and spend most of their time. Plaice, sole and flounders feed on molluscs and worms on the sea floor, whereas herring and mackerel feed on plankton in the surface waters. In a pond, the snails do not range much beyond the plants where they feed. On a rocky coast, limpets and barnacles can withstand exposure between the tides and colonize the rocks. Sea anemones, on the other hand, are restricted mainly to the rocky pools left at low tide.

Competition

Living organisms compete with each other for resources such as food, light, rooting space and breeding partners. As explained on p. 203, all organisms produce more offspring than can possibly survive, so competition is unavoidable.

Interspecific competition occurs between different species. On p. 260 there is a description of competition between two species of *Paramecium* leading to a decline in population of one of them. Wildlife programmes on television will have shown how lions, cheetahs and hyenas compete for the same carcass.

Figure 28.2 Reduced interspecific competition. The giraffe and antelope do not compete for food, though both are herbivores

Animals occupying the same habitat often have adaptations that reduce competition between them. Giraffes and antelopes are both herbivores of the African savannah ecosystem, but giraffes, with their long necks, browse on tree leaves while antelopes graze the grasses and other plants growing on the ground (Figure 28.2).

In a plant community, interspecific competition is for light, water and soil nutrients. A plant that grows taller and more quickly will get more sunlight for photosynthesis and so reduce the amount of light reaching the shorter plants near to it. A plant with a root system that goes deeply into the soil will be able to withstand drought better than shallow-rooted plants. Some plant root systems (for example, rhododendrons) even secrete chemicals from their roots, so inhibiting the growth of competing species.

Intraspecific competition takes place between members of the same species competing for food, territory, nesting sites and mates (Figure 28.3). Variations between individuals may result in one being a more successful competitor than the other. This is the basis of 'natural selection' (p. 203).

In animals, physical conflict is often avoided by each member taking up a territory that is defended by a threat display or sounds, such as bird song. It is at the boundaries of the territory that most confrontations occur. Many species feed or breed in herds or colonies. Gannets, for example, form colonies where territory may be only the pecking distance between nests (see Figure 29.3, p. 260).

Figure 28.3 Intraspecific competition in red deer. The 'pushing contest' will determine who has access to the females

Factors which affect communities

A great many factors contribute to the stability or decline of a community (Figure 28.4). They can be grouped under two headings.

Biotic factors are those which involve the activities of living organisms – competition, predation and parasitism, for example.

Abiotic factors are those which do not involve living organisms – e.g. rainfall, temperature range, prevailing wind or soil quality. In an aquatic ecosystem, the amount of oxygen dissolved in the water will affect the animals living there (see 'Eutrophication', p. 238).

Human factors include activities which alter the ecosystem for farming, forestry, building and all activities which add pollutants to the atmosphere and water (see Chapter 26).

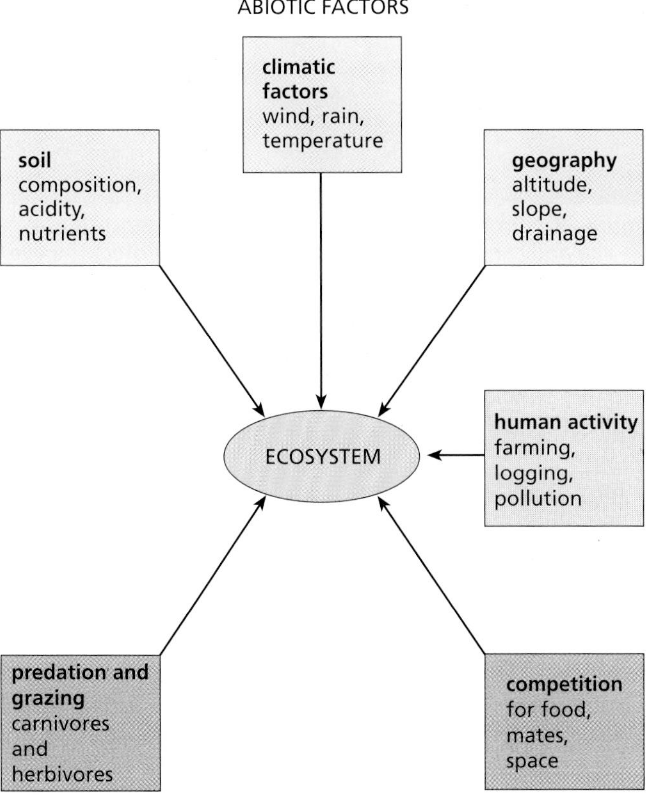

Figure 28.4 Some examples of factors which affect communities and ecosystems

Questions

1 What communities might be present in an area of woodland?

2 a What is the habitat of an earthworm?
  b What makes up the environment of the earthworm?

3 Name some of the producers, consumers and decomposers that might be present in a grassland ecosystem.

4 Plants make their food by photosynthesis, so in what ways can they be said to 'compete for food'?

Natural and artificial ecosystems

Natural ecosystems

Many examples of these have already been described. They may be ponds, rivers, lakes, woods, sand dunes, rocky shores, bogs and swamps to name only a few.

They are each self-supporting, needing only sunlight, air, water and mineral nutrients to survive. They usually contain a wide range of interdependent plants and animals and need no input from humans.

Artificial ecosystems

To a large extent these are farms, market gardens, nurseries, orchards and domestic gardens or parks. On arable farms, the plants are deliberately introduced and grow in dense stands of a single species. They are usually specifically bred for their food value or commercial value to humans. Such plants would be unlikely to survive for long in a natural environment.

Attempts are made to abolish food chains and food webs in the artificial ecosystem. Competing plants are excluded with the aid of herbicides; fungi are kept at bay with fungicides and insects are eliminated, as far as possible, with insecticides. Animals, other than insects, usually find the ecosystem inhospitable. Such ecosystems could not be maintained without constant human intervention.

The soil in these situations is, to some extent, still a natural ecosystem but it is modified by constant ploughing and addition of fertilizers.

Ecologically, artificial ecosystems are not very exciting but they are essential to our survival. We would starve without them. Unless carefully managed, however, they may do long-term damage to their own and other environments. The potential harm from agricultural practices is outlined on pp. 235–8.

Figure 28.5 An orchard is an example of an artifical ecosystem

Adaptation

When biologists say that a plant or animal is *adapted* to its habitat they usually mean that, in the course of evolution (p. 203), changes have occurred in the organism which make it more successful in exploiting its habitat, e.g. animals finding and digesting food, selecting nest sites or hiding places, or plants exploiting limited mineral resources or tolerating salinity or drought. It is tempting to assume that because we find a plant or animal in a particular habitat it must be adapted to its habitat. There is some logic in this; if an organism was not adapted to its habitat, presumably it would be eliminated by natural selection. However, it is best to look for positive evidence of adaptation.

Sometimes, just by looking at an organism and comparing it with related species, it is possible to make reasoned guesses about adaptation. For example, there seems little doubt that the long, hair-fringed hind legs of a water beetle are adaptations to locomotion in water when compared with the corresponding legs of a land-living relative (Figure 28.6).

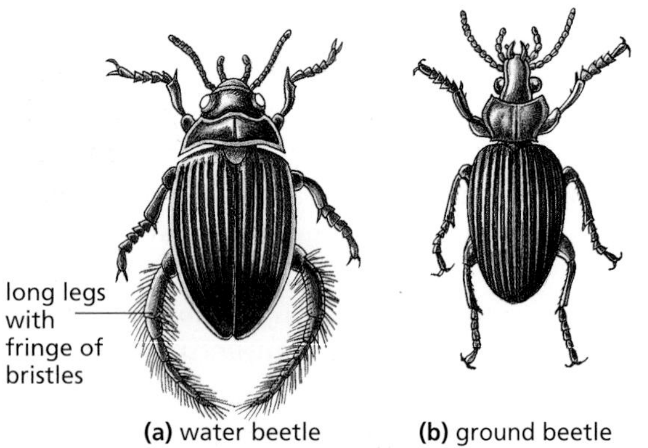

Figure 28.6 Adaptation to locomotion in water and on land

Similarly, in Figure 23.17 on p. 209 it seems reasonable to suppose that, compared with the generalized mammalian limb, the forelimbs of whales are adapted for locomotion in water.

By studying animals which live in extreme habitats, it is possible to suggest ways in which they might be adapted to these habitats especially if the obervations are supported by physiological evidence.

The **camel** is adapted to survive in a hot, dry and sandy environment. Adaptive physical features are the closable nostrils and long eyelashes, which help keep out wind-blown sand (Figure 28.7). The feet are broad and splay out under pressure, so reducing the tendency to sink into the sand. The thick fur insulates the body against heat gain in the intense sunlight.

Physiologically, the camel is able to survive without water for 6–8 days. Its stomach has a large water-holding capacity, though it drinks to replace water lost by evaporation rather than in anticipation of water deprivation.

The body temperature of a 'thirsty' camel rises to as much as 40 °C during the day and falls to about 35 °C at night. The elevated daytime temperature reduces the heat gradient between the body and the surroundings, so less heat is absorbed. A camel is able to tolerate water loss equivalent to 25 per cent of its body weight, compared with humans for whom a 12 per cent loss may be fatal. The blood volume and concentration are maintained by withdrawing water from the body tissues.

Figure 28.7 Protection against blown sand. The nostrils are slit-like and can be closed. The long eyelashes protect the eye

The nasal passages are lined with mucus. During exhalation, the dry mucus absorbs water vapour. During inhalation the now moist mucus adds water vapour to the inhaled air. In this way, water is conserved.

The role of the camel's humps in water conservation is more complex. The humps contain fat and are therefore an important reserve of energy-giving food. However, when the fat is metabolized during respiration, carbon dioxide and water (metabolic water) are produced. The water enters the blood circulation and would normally be lost by evaporation from the lungs but the water-conserving nasal mucus will trap at least a proportion of it.

The **polar bear** lives in the Arctic, spending much of its time on snow and ice. Several physical features contribute to its adaptation to this cold environment.

It is a very large bear (Figure 28.8), which means that the ratio of its surface area to its volume is relatively small (see p. 298). The relatively small surface area means that the polar bear loses proportionately less heat than its more southerly relatives. Also its ears are small, another feature which reduces heat loss (Figure 28.9).

It has a thick coat with long loosely packed coarse hairs (guard hairs) and a denser layer of shorter woolly hairs forming an insulating layer. The long hairs are oily and water-repellant and enable the bear to shake off water when it emerges from a spell of swimming.

Figure 28.8 The polar bear and the sun bear (from SE Asia). The smaller surface area/volume ratio in the polar bear helps conserve heat

Figure 28.9 The heavy coat and small ears also help the polar bear to reduce heat losses

The principal thermal insulation comes from a 10 cm layer of fat (blubber) beneath the skin. The thermal conductivity of fat is little different from any other tissue but it has a limited blood supply. This means that very little warm blood circulates close to the skin surface.

The hollow hairs of the white fur are thought to transmit the sun's heat to the black skin below. Black is an efficient colour for absorbing heat. The white colour is also probably an effective camouflage when hunting its prey, mainly seals.

A specific adaptation to walking on snow and ice is the heat-exchange arrangement in the limbs. The arteries supplying the feet run very close to the veins returning blood to the heart. Heat from the arteries is transferred to the veins before the blood reaches the feet (Figure 28.10). So, little heat is lost from the feet but their temperature is maintained above freezing point, preventing frost-bite.

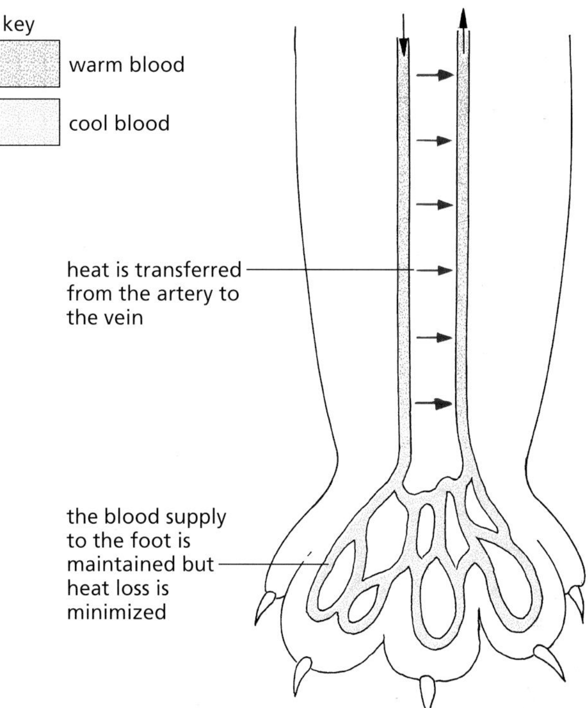
key
warm blood
cool blood
heat is transferred from the artery to the vein
the blood supply to the foot is maintained but heat loss is minimized

Figure 28.10 The heat-exchange mechanism in the polar bear's limb

The polar bear breeds in winter when temperatures fall well below zero. However, the pregnant female excavates a den in the snow in which to give birth and rear her two cubs. In this way the cubs are protected from the extreme cold.

The female remains in the den for about 140 days, suckling her young on the rich milk which is formed from her fat reserves.

Adaptations of plants to arid conditions

This is discussed on pp. 61 and 62.

Questions

1 In an organically grown field of wheat what intraspecific and interspecific competition might there be?

2 Judging from the external features of the beetles in Figure 28.6, which of these features might be adaptations to their respective habitats or way of life?

3 A bird's wings cannot really be described as an *adaptation* to flight because without wings there could be no flight. From the information on pp. 313–14, suggest features which could be regarded as adaptations to flight.

4 List the ways in which a camel
   a reduces its loss of water,
   b is able to withstand desiccation.

5 List the physical features of a polar bear that help to reduce heat loss.

Checklist

- A habitat is where an organism lives, feeds and breeds.
- A community is all the organisms living in a habitat.
- An ecosystem is a self-supporting community plus the physical features of its environment.
- A population is the number of a given species in a defined habitat.
- There is competition within and between species for food, light, space and mates.
- Artifical ecosystems such as arable farmland consist of monocultures. Competitors to the crop plants are eliminated.
- A community or ecosystem may be affected by biotic and abiotic factors.
- Biotic factors are those involving other living organisms, e.g. predators. Abiotic factors involve non-living systems, e.g. rainfall.
- Adaptations are changes in an organism, in the course of evolution, which help to make it more successful in its habitat.
- The camel is adapted to desert conditions by conserving water and reducing heat gain.
- The polar bear is adapted to Arctic conditions by having a small surface area to volume ratio and effective thermal insulation.

29 Populations

Population growth and fluctuation
Limits to population growth.

Human population
Growth, birth and death rates, world population.

Population changes

On p. 253 it was explained that a biological population is defined as the total number of individuals of any one species in a particular habitat. Such a population will not necessarily be evenly spread throughout the habitat, nor will its numbers remain steady. The population will also be made up of a wide variety of individuals: adults (male and female), juveniles, larvae, eggs or seeds, for example. In studying populations, these variables often have to be simplified.

Population growth

In the simplest case, where a single species is allowed to grow in laboratory conditions, the population develops more or less as shown in Figure 29.1.

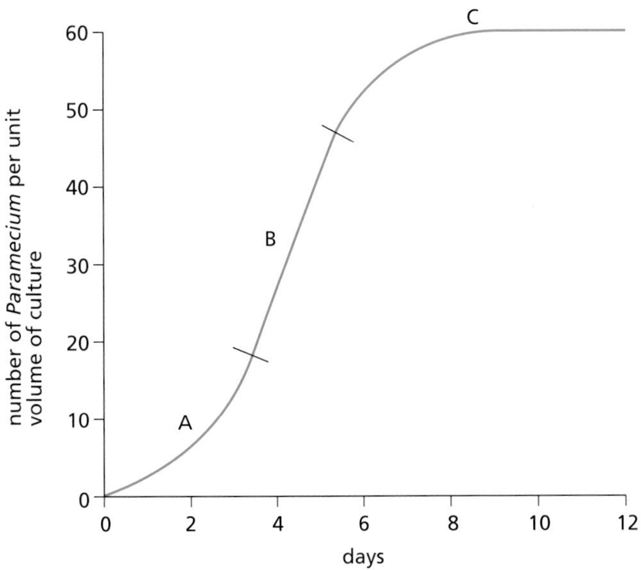

Figure 29.1 The sigmoid curve (*Paramecium caudatum*). This is the characteristic growth pattern of a population when food is abundant at first and there are no other factors limiting growth and reproduction

The population might be of yeast cells (p. 288) growing in a sugar solution, flour beetles in wholemeal flour or weevils in a grain store. The experiment illustrated by Figure 29.1 uses a single-celled organism called *Paramecium* (see p. 268), which reproduces by dividing into two (binary fission). The **sigmoid** (S-shaped) form of the graph can be explained as follows:

A The population increase is exponential or logarithmic, i.e. it does not increase 2–4–6–8, etc. but 2–4–8–16–32, etc. One *Paramecium* divides into two, the two offspring each divide, producing 4, the 4 divide into 8 cells, and so on. In other words, the population doubles at each generation. In ten generations it would reach 1024. When a population of four organisms doubles, it is not likely to strain the resources of the habitat, but when a population of 1024 doubles, there is likely to be considerable competition for food and space.

B The population continues to grow but at a steady rate. This may be because the food resources are limiting the rate of growth and reproduction; or the effect of crowding may itself reduce the reproduction rate. Also, some of the mature organisms may be dying.

C At this point the population ceases to grow. The **reproduction rate** equals the **mortality rate** (death rate). The number of offspring produced will still be greater than the number of adults which die, but fewer of these offspring will live long enough to reproduce.

After this stage, the population may start to decline. This can happen because the food supply is insufficient, waste products contaminate the habitat or disease spreads through the dense population.

Questions

1 In Figure 29.1, how many days does it take for the mortality rate to equal the replacement rate?

2 From the graph in Figure 29.1, what is the approximate increase in the population of *Paramecium*
 a between day 0 and day 2,
 b between day 2 and day 4, and
 c between day 8 and day 10?

Limits to population growth

The sigmoid curve is a very simplified model of population growth. Few organisms occupy a habitat on their own, and the conditions in a natural habitat will be changing all the time. The steady state of the population in part C of the sigmoid curve is rarely reached in nature. In fact, the population is unlikely to reach its maximum theoretical level because of the many factors limiting its growth. These are called **limiting factors**.

Competition

If, in the laboratory, two species of *Paramecium* (*P. aurelia* and *P. caudatum*) are placed in an aquarium tank, the population growth of *P. aurelia* follows the sigmoid curve but the population of *P. caudatum* soon declines to zero because *P. aurelia* takes up food more rapidly than *P. caudatum* (Figure 29.2).

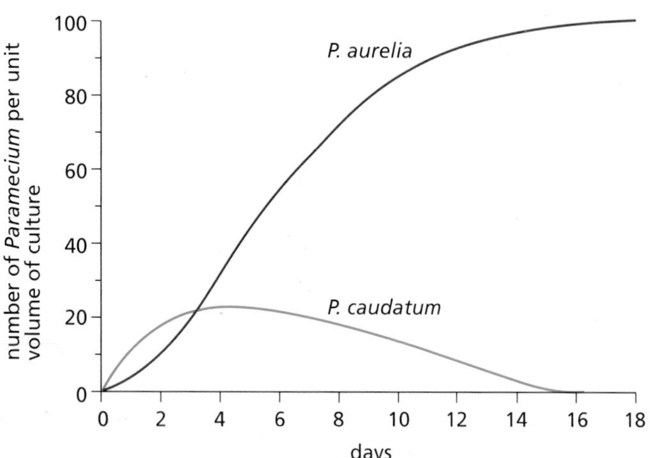

Figure 29.2 The effect of competition. *Paramecium aurelia* and *P. caudatum* eat the same food but *P. aurelia* can capture and ingest it faster than *P. caudatum*

This example of competition for food is only one of many factors in a natural environment which will limit a population or cause it to change.

Abiotic and biotic limiting factors

Plant populations will be affected by **abiotic** (non-biological) factors such as rainfall, temperature and light intensity. The population of small annual plants may be greatly reduced by a period of drought; a severe winter can affect the numbers of more hardy perennial plants. **Biotic** (biological) factors affecting plants include their leaves being eaten by browsing and grazing animals or by caterpillars and other insects, and the spread of fungus diseases.

Animal populations, too, will be limited by abiotic factors such as seasonal changes. A cold winter can severely reduce the populations of small birds. However, animal populations are also greatly affected by biotic factors such as the availability of food, competition for nest sites (Figure 29.3), predation (i.e. being eaten by other animals), parasitism and diseases. (See also p. 255.)

The size of an animal population will also be affected by the numbers of animals entering from other localities (immigration) or leaving the population (emigration).

In a natural environment, it is rarely possible to say whether the fluctuations observed in a population are mainly due to one particular factor because there are so many factors at work. In some cases, however, the key factors can be identified as mainly responsible for limiting the population.

Figure 29.3 Colony of nesting gannets. Availability of suitable nest sites is one of the factors which limits the population

Predator–prey relationship

A classical example of predator–prey relationships comes from an analysis of the fluctuating populations of lynxes and snowshoe hares in Canada. The figures are derived from the numbers of skins sold by trappers to the Hudson's Bay Company between 1845 and 1945.

The lynx preys on the snowshoe hare, and the most likely explanation of the graph in Figure 29.4 is that an increase in the hare population allowed the predators to increase. Eventually the increasing numbers of lynxes caused a reduction in the hare population.

However, seasonal or other changes affecting one or both of the animals could not be ruled out. (See also p. 255.)

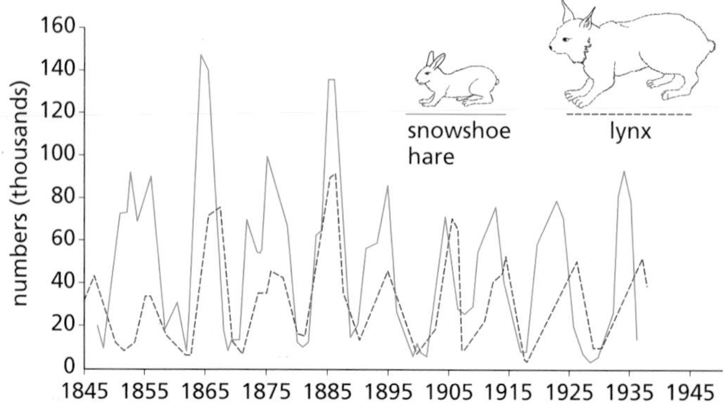

Figure 29.4 Prey–predator relationships: fluctuations in the numbers of pelts received by the Hudson's Bay Company for lynx (predator) and snowshoe hare (prey) over a 100-year period

Questions

1 In section B of the graph in Figure 29.1, what is the approximate reproduction rate of *Paramecium* (i.e. the number of new individuals per day)?

2 In 1937, 2 male and 6 female pheasants were introduced to an island off the NW coast of America. There were no other pheasants and no natural predators. The population for the next 6 years increased as follows:

1937 – 24	1940 – 563
1938 – 65	1941 – 1122
1939 – 253	1942 – 1611

Plot a graph of these figures and say whether it corresponds to any part of the sigmoid curve.

3 In Figure 29.2, which part of the curve approximately represents the exponential growth of the *P. aurelia* population? Give the answer in days.

4 What forms of competition might limit the population of sticklebacks in a pond?

5 Suggest
a some abiotic factors,
b some biotic factors
that might prevent an increase in the population of sparrows in a farmyard.

Human population

In AD1000, the world population was probably about 300 million. In the early 19th century it rose to 1000 million (1 billion), and by 1984 it had reached 4.7 billion. In 2000 it reached about 6 billion and might stabilize at 10 billion by 2100. The graph in Figure 29.5 shows that the greatest population surge has taken place in the last 300 years.

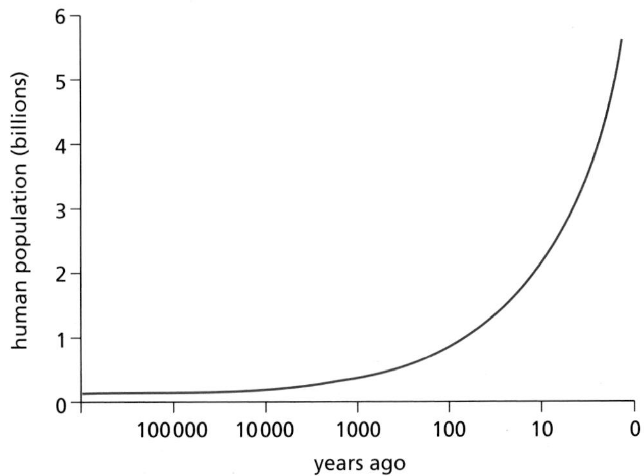

Figure 29.5 World population growth. The time scale (horizontal axis) is logarithmic. The right-hand space (0–10) represents only 10 years, but the left-hand space (100 000–1 million) represents 900 000 years. The greatest population growth has taken place in the last 300 years

Population growth

About 20 years ago, the human population was increasing at the rate of 2 per cent a year. This may not sound very much. But it means that the world population was doubling every 35 years. This doubles the demand for food, water, space and other resources. Recently, the growth rate has slowed to 1.3 per cent. But it is not the same everywhere. Nigeria's population is growing by 2.9 per cent each year, but Western Europe's grows at only 0.1 per cent.

Traditionally, it is assumed that population growth is limited by famine, disease or war. These factors are affecting local populations in some parts of the world today but they are unlikely to have a limiting effect on the rate of overall population growth.

Diseases such as malaria (spread by mosquitoes), and sleeping sickness (spread by tsetse flies) have for many years limited the spread of people into areas where these insects carry the infections.

Diseases such as bubonic plague and influenza have checked population growth from time to time, and the current AIDS epidemic in sub-Saharan Africa is having significant effects on population growth and life expectancy.

Factors affecting population growth

If a population is to grow, the birth rate must be higher than the death rate. Suppose a population of 1000 people produces 100 babies each year but only 50 people die each year. This means that 50 new individuals are added to the population each year and the population will double in 20 years (or less if the new individuals start reproducing at 16) (Figure 29.6).

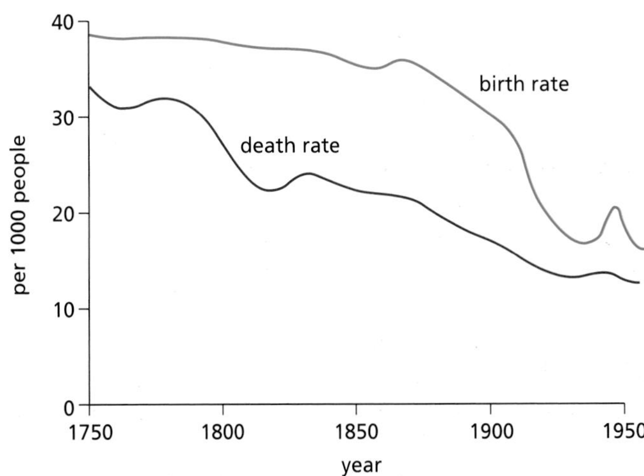

Figure 29.6 Birth and death rates in England and Wales from 1750 to 1950. Although the birth rate fell during this period, so did the death rate. As a result, the population continued to grow. Note the 'baby boom' after the Second World War. (Used by permission of Carolina Biological Supply Company.)

One of the factors affecting population growth is **infant mortality**, i.e. the death rate for children less than 1 year old. Populations in the developing world are growing, not because of an increase in the number of babies born per family, but because more babies are surviving to reach reproductive age. Infant mortality is falling and more people are living longer. That is, **life expectancy** is increasing.

Increase in life expectancy

The life expectancy is the average age to which a newborn baby can be expected to live. In Europe between 1830 and 1900 the life expectancy was 40–50 years. Between 1900 and 1950 it rose to 65 and now stands at 73–74 years. In sub-Saharan Africa, life expectancy was rising to 58 years until the AIDS epidemic reduced it to about 45 years.

These figures are averages. They do not mean, for example, that everyone in the developing world will live to the age of 58. In the developing world, 40 per cent of the deaths are of children younger than 5 years and only 25–30 per cent are deaths of people over 60. In Europe, only 5–20 per cent of deaths are those of children below the age of 5, but 70–80 per cent are of people over 60.

An increase in the number of people over the age of 60 does not change the rate of population growth much, because these people are past child-bearing age. On the other hand, if the death rate among children falls and the extra children survive to reproduce, the population will continue to grow. This is the main reason for the rapid population growth in the developing world since 1950.

Causes of the reduction in death rate

The causes are not always easy to identify and vary from one community to the next. In 19th-century Europe, agricultural development and economic expansion led to improvements in nutrition, housing and sanitation, and to clean water supplies. These improvements reduced the incidence of infectious diseases in the general population. And better-fed children could resist these infections when they did meet them. The drop in deaths from infectious diseases probably accounted for three-quarters of the total fall in deaths.

The social changes probably affected the population growth more than did the discovery of new drugs or improved medical techniques. Because of these techniques, particularly immunization, diphtheria, tuberculosis and polio are now rare (Figure 29.7), and by 1977 smallpox had been wiped out by the World Health Organization's vaccination campaign.

In the developing world, sanitation, clean water supplies and nutrition are improving slowly. The surge in the population since 1950 is likely to be at least 50 per cent due to modern drugs, vaccines and insecticides.

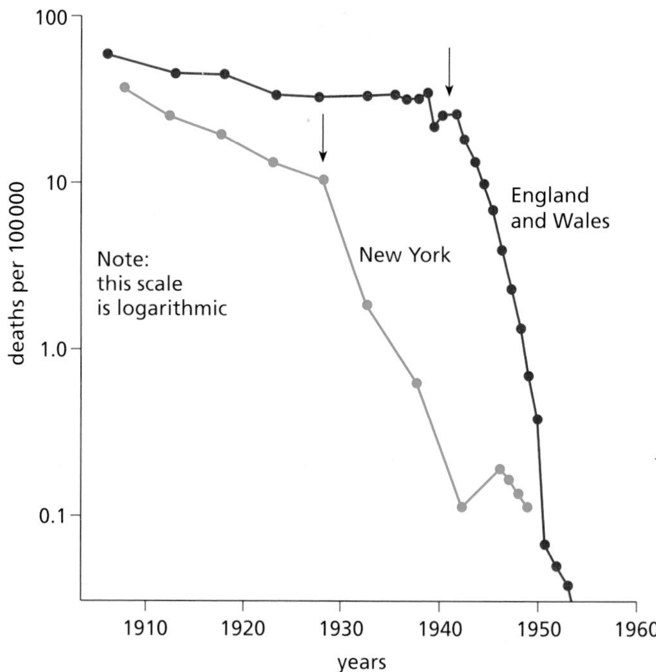

Figure 29.7 Fall in death rate from diphtheria as a result of immunization. The arrows show when 50 per cent or more of children were vaccinated. Note that the rate was already falling but was greatly increased by immunization.

Stability and growth

Up to 300 years ago, the world population was relatively stable. Fertility (the birth rate) was high and so was the mortality rate (death rate). Probably less than half the children born lived to have children of their own. Many died in their first year (infant mortality), and many mothers died during childbirth.

No one saw any point in reducing the birth rate. If you had a lot of children, you had more help on your land and a better chance that some of them would live long enough to care for you in your old age.

In the past 300 years, the mortality rate has fallen but the birth rate has not gone down to the same extent. As a result the population has expanded rapidly.

In 18th-century Europe, the **fertility rate** was about 5. This means that, on average, each woman would have five children. When the death rate fell, the fertility rate lagged behind so that the population increased. However, the fertility rate has now fallen to somewhere between 1.4 and 2.6, and the European population is more or less stable.

A fall in the fertility rate means that young people will form a smaller proportion of the population. There will also be an increasing proportion of old people for the younger generation to look after. In Britain it is estimated that, between 1981 and 1991, the number of people aged 75–84 increased by 16 per cent. The number of those over 85 increased by about 46 per cent (Figure 29.8).

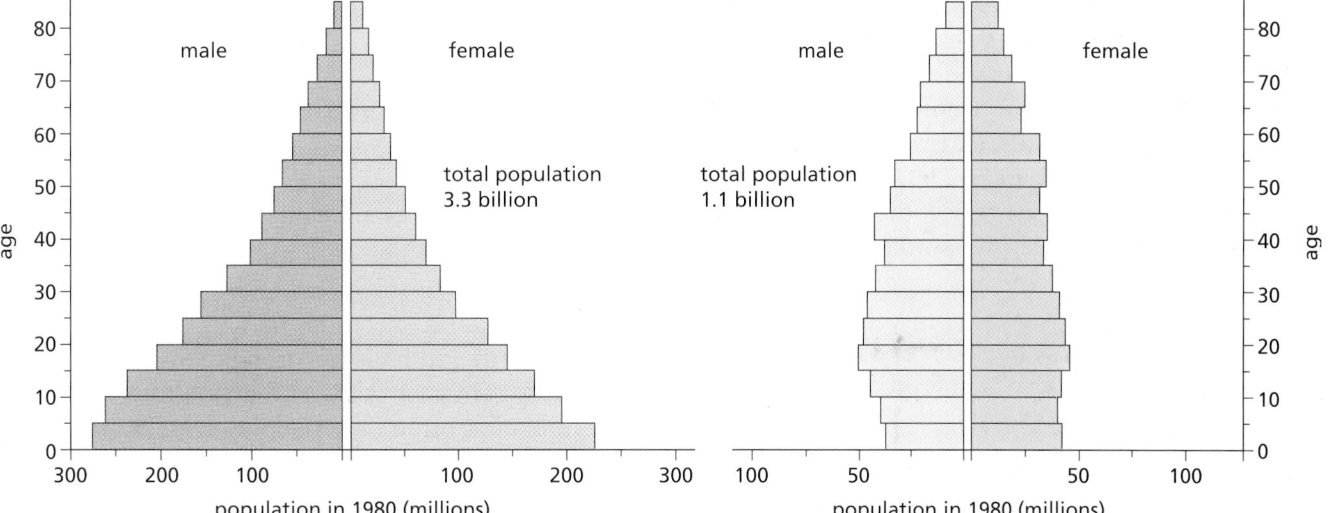

(a) The developing regions. The tapering pattern is characteristic of a population with a high birth rate and low average life expectancy. The bulk of the population is under 25.

(b) The developed regions. The almost rectangular pattern is characteristic of an industrialized society, with a steady birth rate and a life expectancy of about 70. (The horizontal scale is not the same as in **a**.)

Figure 29.8 Age distribution of population in 1980

In the developing world, the fertility rate has dropped from about 6.2 to 3.0. This is still higher than the mortality rate. An average fertility rate of 2.1 is necessary to keep the population stable.

As a community grows wealthier, the birth rate goes down. There are believed to be four reasons.

Longer and better education

Marriage is postponed and a better-educated couple will have learned about methods of family limitation.

Better living conditions

Once people realize that half their offspring are not going to die from disease or malnutrition, family sizes fall.

Agriculture and cities

Modern agriculture is no longer labour intensive. Farmers do not need large families to help out on the land. City dwellers do not depend on their offspring to help raise crops or herd animals.

Application of family planning methods

Either natural methods of birth control or the use of contraceptives is much more common.

It takes many years for social improvements to produce a fall in birth rate. Some countries are trying to speed up the process by encouraging couples to limit their family size (Figure 29.9), or by penalizing families who have too many children.

Meanwhile the population goes on growing. The United Nations expect that the birth rate and death rate will not be in balance until the year 2100. By that time the world population may have reached 10 billion, assuming that the world supply of food will be able to feed this population.

In the past few decades, the world has produced enough food to feed, in theory, all the extra people. But the extra food and the extra people are not always in the same place. As a result, 72 per cent of the world's population has a diet which lacks energy, as well as other nutrients.

Every year between 1965 and 1975, food production in the developed nations rose by 2.8 per cent, while the population rose by 0.7 per cent. In the developing nations during the same period, food production rose by only 1.5 per cent each year, while the annual population rise was 2.4 per cent.

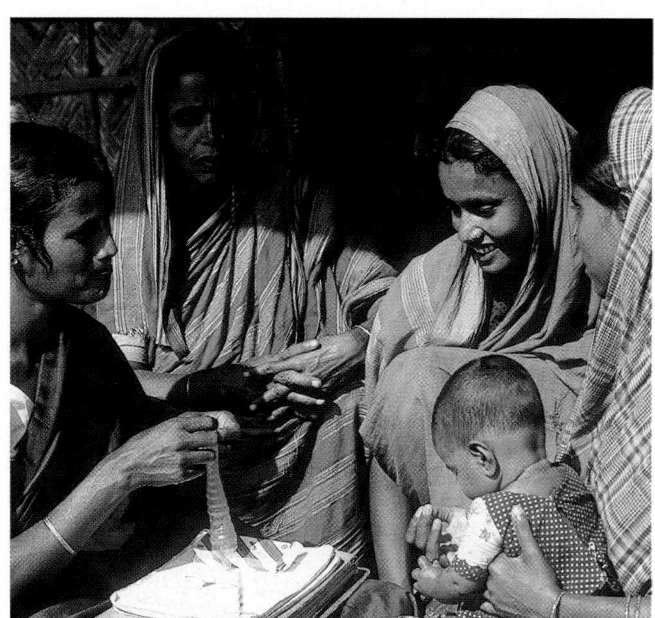

Figure 29.9 Family planning. A health worker in Bangladesh explains the use of a condom

The Western world can produce more food than its people can consume. Meanwhile people in the drier regions of Africa face famine due to drought and population pressure on the environment. Even if the food could be taken to the developing world, people there are often too poor to buy it. Ideally, each region needs to grow more food or reduce its population until the community is self-supporting. Some countries grow tobacco, cotton, tea and coffee (cash crops) in order to obtain foreign currency for imports from the Western world. This is fine, so long as they can also feed their people. But when food is scarce, people cannot live on the cash crops.

Population pressures

More people, more agriculture and more industrialization will put still more pressure on the environment unless we are very watchful (see Chapters 26 and 27). If we damage the ozone layer (p. 244), increase atmospheric carbon dioxide (p. 245), release radioactive products or allow farmland to erode, we may meet with additional limits to population growth.

Questions

1 Look at the graph in Figure 29.6, p. 261.
  a When did the post-war 'baby-boom' occur?
  b What was the growth rate of the population in 1800?

2 Which of the following causes of death are likely to have most effect on the growth rate of a population: smallpox, tuberculosis, heart disease, polio, strokes, measles?
  Give reasons for your answer.

3 Suggest some reasons why the birth rate tends to fall as a country becomes wealthier.

4 Give examples of the kind of demands that an increasing population makes on the environment. In what ways can these demands lead to environmental damage?

5 If there are 12 000 live births in a population of 400 000 in 1 year, what is the birth rate?

6 Try to explain why, on average, couples need to have just over two children if the population is to remain stable.

7 Study Figure 29.8 (p. 263) and then comment on
  a the relative number of boy and girl babies,
  b the relative number of men and women of reproductive age (20–40), and
  c the relative numbers of the over-70s.

8 In Figure 29.7 (p. 262) what might be the reasons for the fall in death rate from diphtheria even before 50 per cent immunization was achieved?

Checklist

- A population is the total number of individuals of a species in a habitat.
- The growth of plant populations is limited by competition for, e.g. light, water, minerals, rooting space and the abundance of herbivores.
- The growth of animal populations is limited by competition for food, water, breeding space and the abundance of predators.
- The world population is growing at the rate of 1.7 per cent each year. At this rate, the population more than doubles every 50 years.
- The rate of increase is slowing down and the population may stabilize at 10 billion by the year 2100.
- A population grows when the birth rate exceeds the death rate, provided the offspring live to reproduce.
- In the developed countries, the birth rate and the death rate are now about the same.
- In the developing countries, the birth rate exceeds the death rate and their populations are growing. This is not because more babies are born, but because more of them survive.
- The increased survival rate may be due to improved social conditions, such as clean water, efficient sewage disposal, better nutrition and better housing.
- It is also the result of vaccination, new drugs and improved medical services.
- As a population becomes more wealthy, its birth rate tends to fall.
- Food production in the developed countries has increased faster than the population growth.
- Food production in the developing countries has not kept pace with population growth.

Organisms and their environment

Examination questions

Do not write on this page. Where necessary copy drawings, tables or sentences.

1 The diagram below represents the carbon cycle.

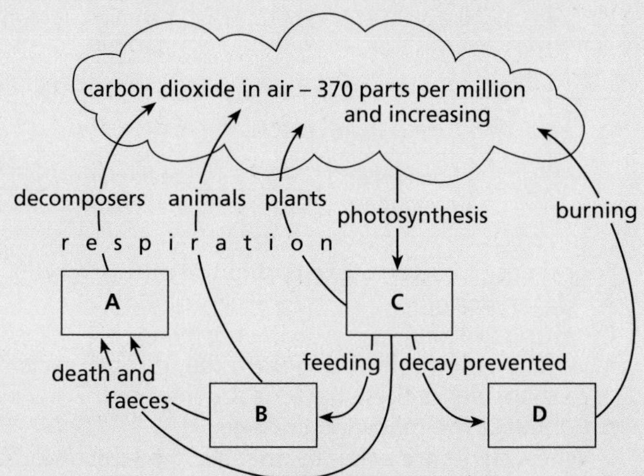

Fill in the last column in the table below to show which statement matches with which box, A to D, on the diagram. (3)

Statements	Box
Carbon in organic compounds in plants *matches with*	
Carbon in fossil fuels e.g. coal, oil, natural gas *matches with*	
Carbon in organic compounds of animals *matches with*	

(WJEC)

2 The graph shows nitrate concentration in a river flowing through a farm in Wales, over 50 years.

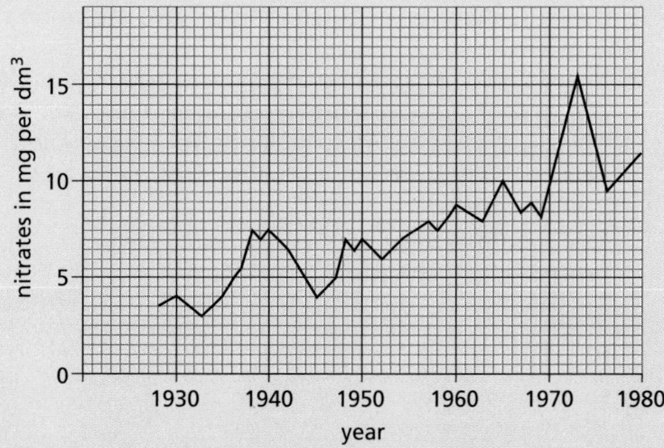

a What was the **difference** in concentration between the highest and lowest records? (1)

b Explain fully the increase in nitrate concentration **in the river** over the 50 years. (4)

c Explain how an increase in nitrate concentration in the river could lead to the death of the fish. (5)

(WJEC)

3 The figure shows a food web which includes some organisms in the African grasslands.

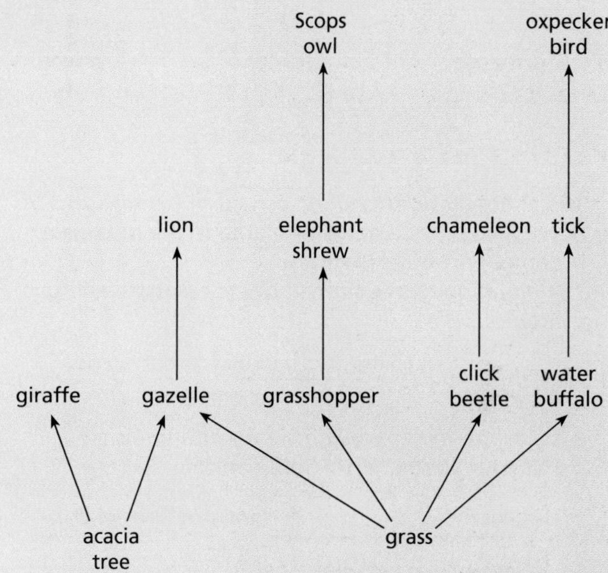

a (i) Draw a food chain consisting of **four** organisms. The organisms must be part of the food web. (2)

(ii) Using examples from the food web, explain the difference between producers and consumers. (4)

b When weather conditions are favourable the grasshopper population can suddenly increase enormously.
Predict and explain the effect this might have on the (i) Scops owl population, (ii) water buffalo population, (iii) giraffe population. (7)

(IGCSE)

4 Modern farming methods keep production of a cereal crop at high levels. Some organisms affect the cereal crop, reducing the yield. Here are examples:

Organism	Effect
Fungus	Uses crop's carbohydrates
Couch grass	Competes with the crop for resources
Virus	Attacks leaves, which turn brown
Greenfly (aphids)	Spreads viruses

a Explain how the farmer could increase the yield if:
(i) all four organisms were present
(ii) none of the organisms were present. (4)

b Explain how a natural ecosystem in a pond near his farm might be damaged by the effects of the methods used to improve the yield of the crop. (4)

c Suggest **two** ways in which competition from couch grass may lead to a reduced yield. (2)

d Viruses cause the leaves to turn brown. Explain why this reduces crop yield. (1)

(WJEC)

5 The diagram shows a food web with organisms of types A, B, C, D and E. Numbers on arrows show the energy available to these organisms in kJ per m^2 per year. Numbers in brackets show the energy that becomes part of the biomass of the organisms in kJ per m^2 per year.

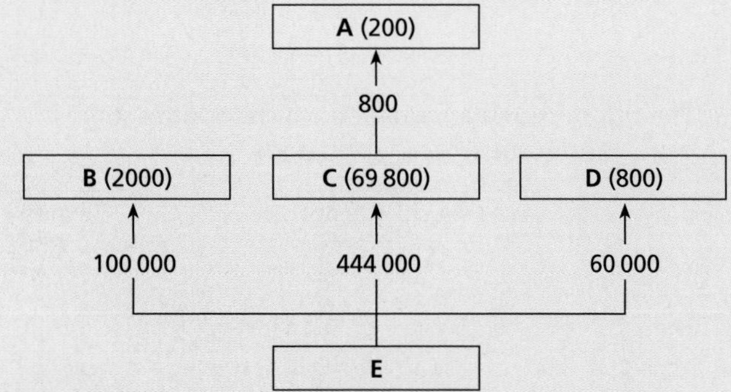

The energy efficiency of an organism is a measure of how much of the energy available to the organism becomes part of its biomass.
The equation below shows how to calculate energy efficiency.

$$\text{energy efficiency} = \frac{\text{energy that becomes part of biomass}}{\text{energy available}} \times 100\%$$

a Calculate the energy efficiency of organism B. Put your answer in the table below.

Organism	Energy efficiency (%)
A	25.0
B	
C	15.7
D	1.3

(1)

b Suggest **three** reasons why organism D has a low energy efficiency. (3)

(Edexcel)

6 The graph shows the numbers of bacteria over 30 hours, in a flask of liquid.

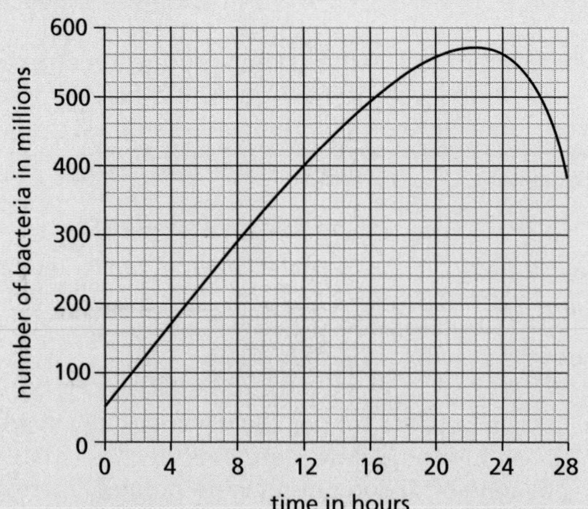

a What evidence is there that the liquid contained food? (1)

b At which times did the flask contain 400 million bacteria? (2)

c Give **one** reason why the numbers of bacteria decreased after 24 hours. (1)

(Edexcel)

7 The diagram shows the nitrogen cycle.

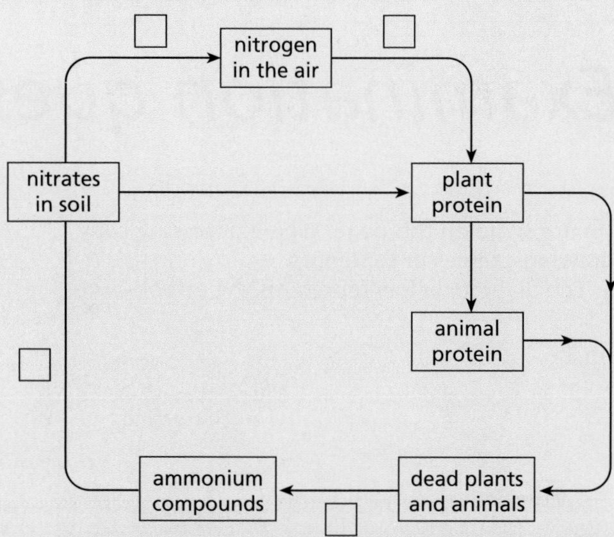

Four types of bacteria take part in the nitrogen cycle.

A – decomposing B – denitrifying
C – nitrogen fixing D – nitrifying

a Write **one** letter in each box on the nitrogen cycle to show where these bacteria are involved. (4)

b Complete the following passage:
When plants are eaten by animals, the large insoluble molecules of plant protein are ______ into small soluble molecules called ______.
This process is catalysed by ______ released from the ______. The small soluble molecules are ______ into the blood and used to make animal protein. (5)

(Edexcel)

8 Humans have damaged the environment by deforestation.

a Explain what is meant by the term deforestation. (1)

b Suggest how deforestation affects the number of species living in this area. (1)

c Describe **two** other ways the environment is changed by deforestation. (2)

(CCEA)

9 Algae are microscopic plants often found in water. They produce food by photosynthesis. Researchers plan to grow large numbers of algae to help solve the world's energy crisis.

a Write a letter suggesting what the researchers could do in order to grow large numbers of the algae. (6)

b The algae can be used to make petrol. This would reduce the need to obtain fuel by destroying the world's forests.
Suggest **three** advantages of reducing the destruction of the world's forests. (3)

(Edexcel)

Diversity of organisms

30 Classification

Kingdoms
Monera, Protoctista, Fungi, Plants, Animals and their subdivisions.
Species
Binomial nomenclature.
Animal kingdom
Plant kingdom
Keys for identification

You do not need to be a biologist to realize that there are millions of different organisms living on the Earth, but it takes a biologist to sort them into a meaningful order, i.e. to **classify** them.

There are many possible ways of classifying organisms. You could group all aquatic organisms together or put all black and white creatures into the same group. However, these do not make very meaningful groups; a seaweed and a porpoise are both aquatic organisms, a magpie and a zebra are both black and white but neither of these pairs has much in common apart from being living organisms and the latter two being animals. These would be **artificial systems** of classification.

A biologist looks for a **natural system** of classification using important features which are shared by as large a group as possible. In some cases it is easy. Birds all have wings, beaks and feathers; there is rarely any doubt about whether a creature is a bird or not. In other cases it is not so easy. As a result, biologists change their ideas from time to time about how living things should be grouped. New groupings are suggested and old ones abandoned.

Kingdoms

The largest group of organisms recognized by biologists is the kingdom. But how many kingdoms should there be? Most biologists used to favour the adoption of two kingdoms, namely **Plants** and **Animals**. This, however, caused problems in trying to classify fungi, bacteria and single-celled organisms which do not fit obviously into either kingdom. A scheme now favoured by many biologists is the Whittaker 5-kingdom scheme comprising the **Monera**, **Protoctista**, **Fungi**, **Plants** and **Animals**.

Viruses are not living organisms (p. 285) and are therefore not included in this scheme.

Kingdom Monera

These are the bacteria (p. 283) and the blue-green algae. They consist of single cells but differ from other single-celled organisms because their chromosomes are not organized into a nucleus.

Kingdom Protoctista

These are single-celled (unicellular) organisms which have their chromosomes enclosed in a nuclear membrane to form a nucleus.

Some of them, e.g. *Euglena* (Figure 30.1), possess chloroplasts and make their food by photosynthesis. These protoctista are often referred to as unicellular 'plants' or **protophyta**. Organisms such as *Amoeba* and *Paramecium* (Figure 30.1) take in and digest solid food and thus resemble animals in their feeding. They may be called unicellular 'animals' or **protozoa**.

Amoeba (Figure 30.1) is a protozoan which moves by a flowing movement of its cytoplasm. It feeds by picking up bacteria and other microscopic organisms as it goes. *Vorticella* has a contractile stalk and feeds by creating a current of water with its cilia (p. 6). The current brings particles of food to the cell. *Euglena* and *Chlamydomonas* have chloroplasts in their cells and feed, like plants, by photosynthesis.

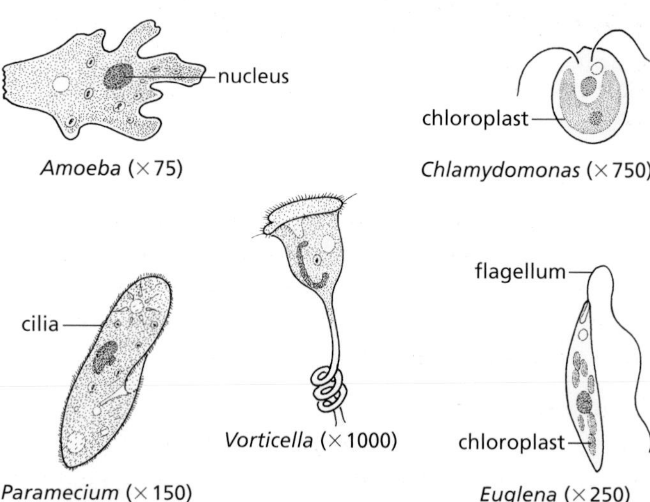

Figure 30.1 Protoctista. *Chlamydomonas* and *Euglena* have chloroplasts and can photosynthesize. The others are protozoa and ingest solid food

Kingdom Fungi

Most fungi are made up of thread-like hyphae (p. 286), rather than cells, and there are many nuclei distributed throughout the cytoplasm in their hyphae.

The fungi include fairly familiar organisms such as mushrooms, toadstools, puffballs, and the bracket fungi that grow on tree-trunks. There are also the less obvious, but very important, mould fungi which grow on stale bread, cheese, fruit or other food. Many of the mould fungi live in the soil or in dead wood (Figure 30.2). The yeasts are single-celled fungi similar to the moulds in some respects.

Some fungal species are parasites and live in other organisms, particularly plants, where they cause diseases which can affect crop plants (Figure 30.3).

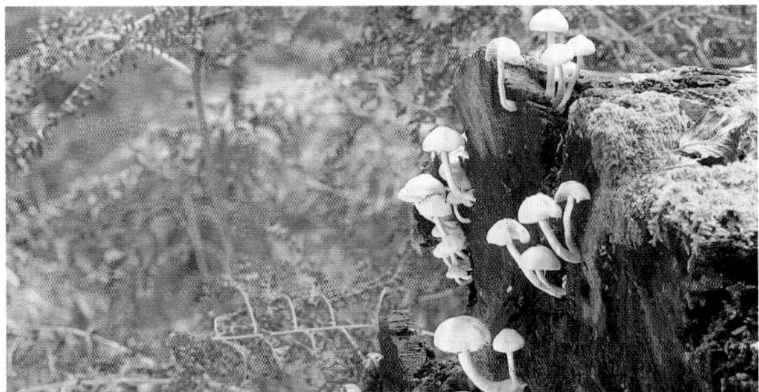

Figure 30.2 Toadstools. They are getting their food from the rotting tree stump

Figure 30.3 Mildew on wheat. Most of the hyphae are inside the leaves, digesting the cells, but some grow out and produce the powdery spores seen here

Kingdom Plants

These are made up of many cells – they are multicellular. Plant cells have an outside wall made of cellulose. Many of the cells in plant leaves and stems contain chloroplasts with photosynthetic pigments, e.g. chlorophyll. Plants make their food by photosynthesis.

Kingdom Animals

Animals are multicellular organisms whose cells have no cell walls or chloroplasts. Most animals ingest solid food and digest it internally.

It is still not easy to fit all organisms into this scheme. For example, many protoctista with chlorophyll (the protophyta) show important resemblances to some members of the algae, but the algae are classified into the plant kingdom. The viruses are not included in the scheme because, in many respects, viruses are not independent living organisms.

This kind of problem will always occur when we try to devise rigid classificatory schemes with distinct boundaries between groups. The process of evolution would hardly be expected to result in a tidy scheme of classification for biologists to use.

An outline classification of plants and animals is given below and illustrated in Figures 30.4–30.7 (pp. 270–3). Bacteria and fungi are described more fully on pp. 283–9.

Question

1 Which kingdoms contain organisms with
a many cells, **b** nuclei in their cells,
c cell walls, **d** hyphae, **e** chloroplasts?

Species

The smallest natural group of organisms is the species. Robins, blackbirds and sparrows are three different species of bird. Apart from small variations, members of a species are almost identical in their anatomy, physiology and behaviour.

Members of a species also often resemble each other very closely in appearance, unless humans have taken a hand in the breeding programmes. All dogs belong to the same species but there are wide variations in the appearance of different breeds (p. 205).

Organisms belonging to the same species can successfully breed together. A spaniel and a labrador retriever may look very different but they have no problems in breeding together.

Closely related species are grouped into a **genus** (plural **genera**). For example, stoats, weasels and polecats are grouped into the genus *Mustela*.

Binomial nomenclature

Species must be named in such a way that the name is recognized all over the world.

The 'cuckoo flower' and the 'lady's smock' are two common names for the same wild plant. If you are not aware that these are alternative names this could lead to confusion. If the botanical name, *Cardamine pratensis*, is used, however, there is no chance of error. The Latin form of the name allows it to be used in all the countries of the world irrespective of language barriers.

Binomial means 'two names'; the first name gives the genus and the second gives the species. For example, the stoat and weasel are both in the genus *Mustela* but they are different species; the stoat is *Mustela erminea* and the weasel is *Mustela nivalis*.

The name of the genus (generic name) is always given a capital letter and the name of the species (specific name) always starts with a small letter.

Frequently, the specific name is descriptive, e.g. *hirsutum* = hairy, *aquatilis* = living in water, *bulbosus* = having a bulb.

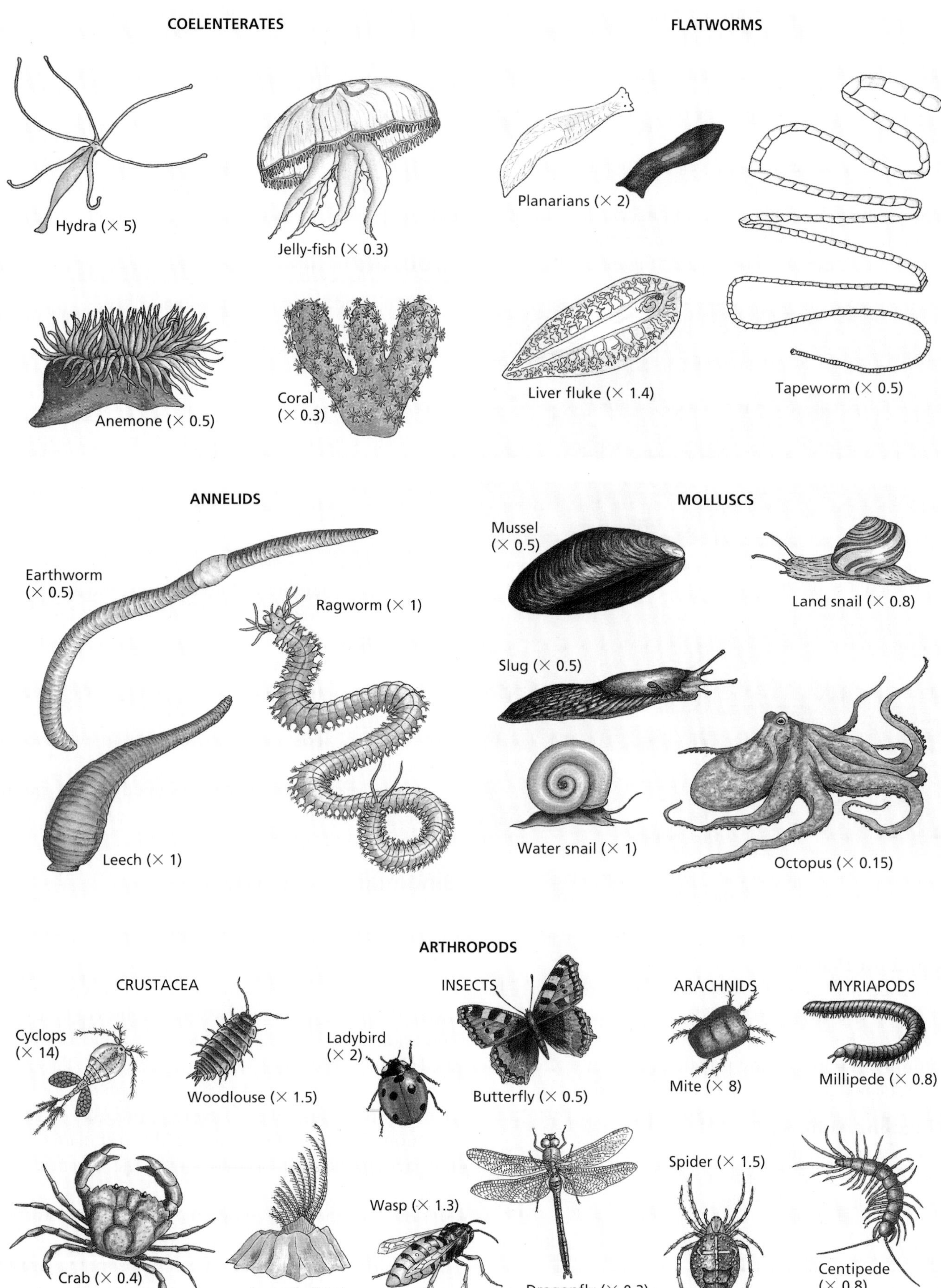

Figure 30.4 The animal kingdom; examples of five invertebrate groups (phyla)

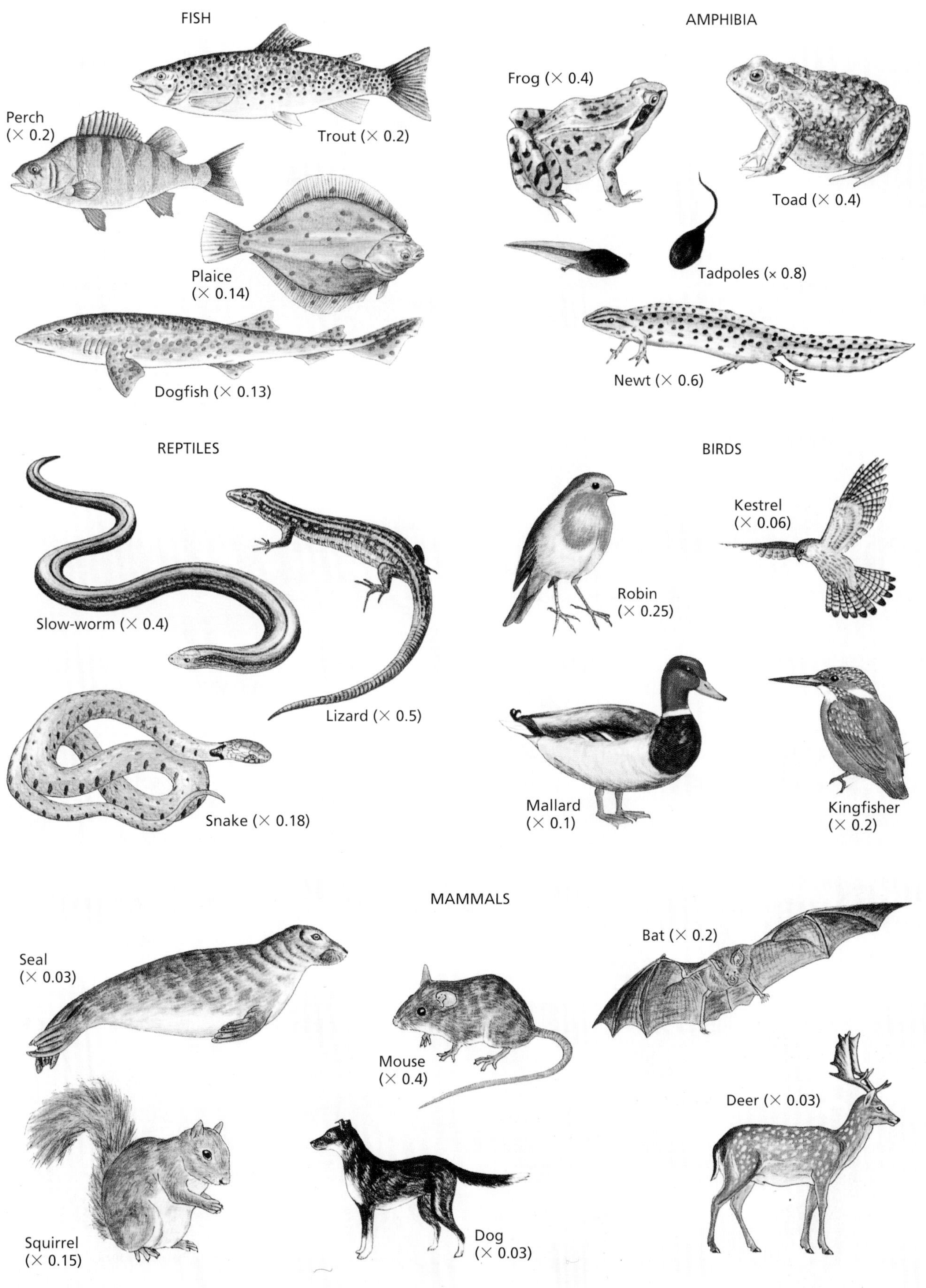

Figure 30.5 The animal kingdom; the vertebrate classes

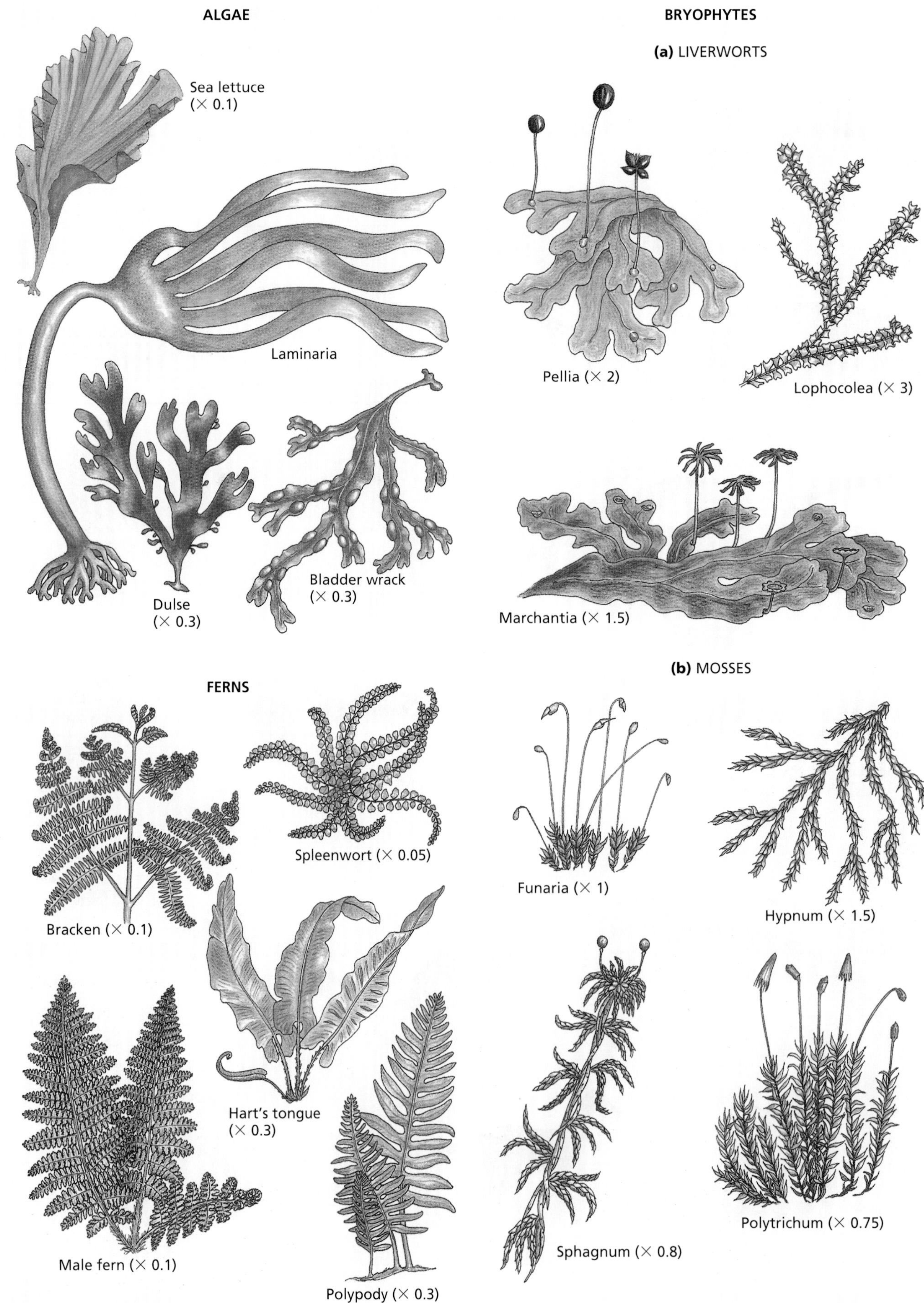

Figure 30.6 The plant kingdom; plants that do not bear seeds

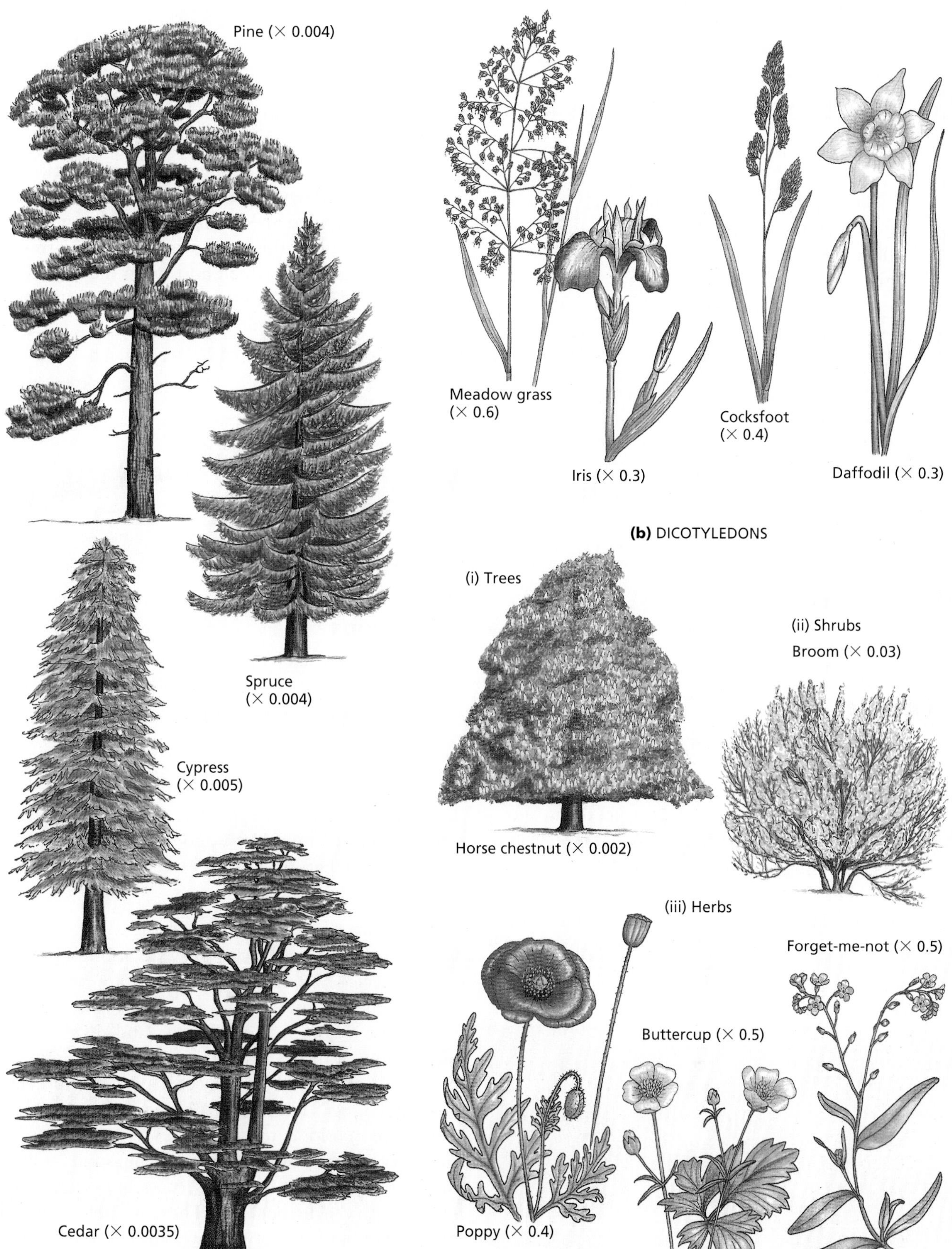

Figure 30.7 The plant kingdom; seed-bearing plants

Animal kingdom

(Only 8 groups out of 23 are listed here.)

* **Coelenterates** (sea anemones, jellyfish)
Flatworms
Nematode worms
Annelids (segmented worms)
Arthropods
CLASS
Crustacea (crabs, shrimps, water fleas)
Insects
Arachnids (spiders and mites)
Myriapods (centipedes and millipedes)
Molluscs (snails, slugs, mussels, octopuses)
Echinoderms (starfish, sea urchins)

Vertebrates
CLASS
Fish
Amphibia (frogs, toads, newts)
Reptiles (lizards, snakes, turtles)
Birds
Mammals
(Only 4 subgroups out of about 26 are listed.)
Insectivores
Carnivores
Rodents
Primates

*All the organisms which do not have a vertebral column are often referred to as invertebrates. Invertebrates are not a natural group, but the term is convenient to use.

Nematodes

These are often called 'roundworms'. Their bodies, which are not divided into segments, are circular in cross-section and pointed at both ends (Figure 30.8). There are thousands of different species ranging from microscopic to 10 cm or more, but their simple body plan is virtually universal within the phylum.

They are widely distributed, free-living in the soil or as parasites of plants and animals. 'Worms' living in the intestine of humans or other vertebrates are usually nematodes. In the tropics thay can cause river blindness and elephantiasis.

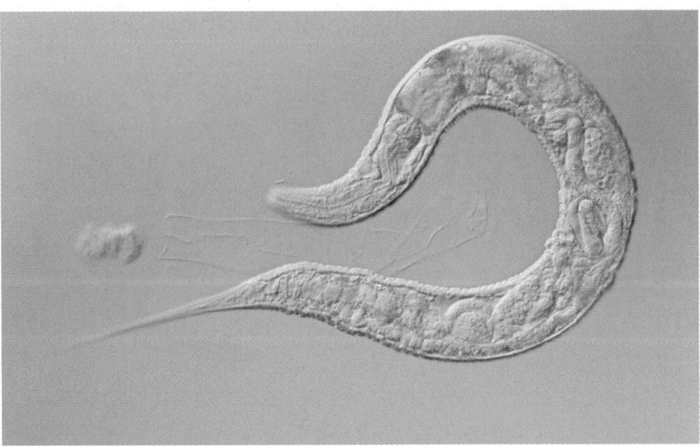

Figure 30.8 A nematode (*Caenorhabditis elegans*) (×130)

Annelids

Annelids are worms (see Figure 30.4 on p. 270). Most of them have elongated, cylindrical bodies which are divided into segments. All the segments have identical sets of organs, though those at the front end may have specialized structures. Some organs, e.g. the alimentary canal, the nerve cord and the main blood vessels, run the whole length of the body.

In most annelids, each segment carries a number of bristles, called **chaetae,** which help in locomotion.

Earthworms are annelids (Figure 30.9), but there are many more annelid species living in fresh water and the sea. Lugworms, bristle-worms and ragworms, for example, are annelids which burrow in the sand on the sea shore. *Tubifex* is a freshwater annelid living in the mud at the bottom of ponds.

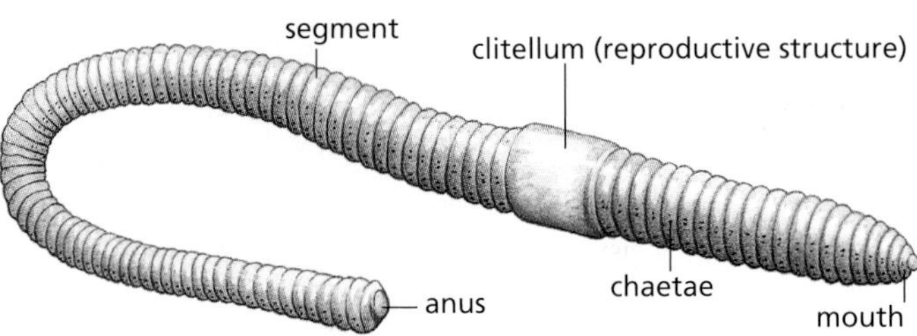

Figure 30.9 Earthworm

Arthropods

The arthropods include the crustacea, insects, centipedes and spiders (Figure 30.4 on p. 270). The name arthropod means jointed limbs, and this is a feature common to them all. They also have a hard, firm external skeleton, called a **cuticle**, which encloses their bodies. Their bodies are segmented and, between the segments, there are flexible joints which permit movement. In most arthropods, the segments are grouped together to form distinct regions, e.g. the head, thorax and abdomen.

Crustacea

Marine crustacea are crabs, prawns, lobsters, shrimps and barnacles. Freshwater crustacea are water fleas, *Cyclops*, the freshwater shrimp (*Gammarus*) and the water louse (*Asellus*). Woodlice are land-dwelling crustacea. Some of these crustacea are illustrated on p. 270.

Like all arthropods, crustacea have an exoskeleton and jointed legs. They also have two pairs of antennae which are sensitive to touch and to chemicals, and they have **compound eyes**. Compound eyes are made up of tens or hundreds of separate lenses with light-sensitive cells beneath. They are able to form a crude image and are very sensitive to movement.

Typically, crustacea have a pair of jointed limbs on each segment of the body, but those on the head segments are modified to form antennae or specialized mouth parts for feeding (Figure 30.10).

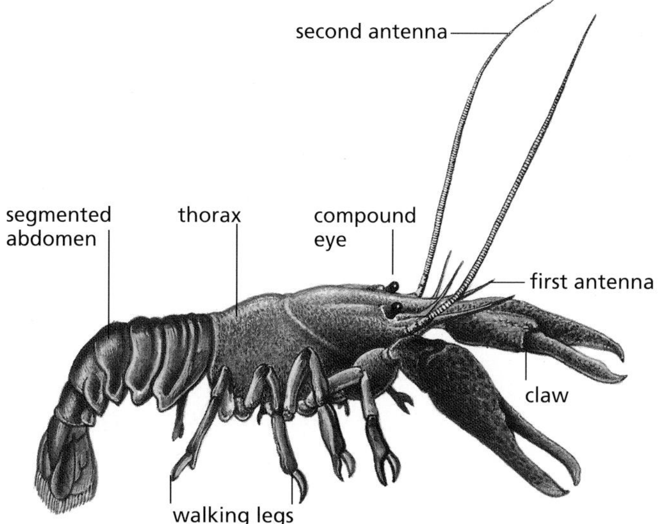

Figure 30.10 Lobster (×0.2)

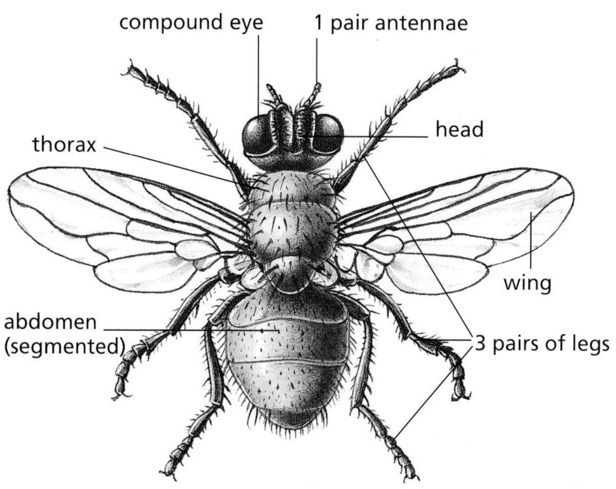

Figure 30.11 External features of an insect (greenbottle, ×5). Flies, midges and mosquitoes have only one pair of wings

Insects

The insects form a very large class of arthropods. Bees, butterflies, mosquitoes, houseflies, earwigs, greenfly and beetles are just a few of the subgroups in this class.

Insects have segmented bodies with a firm exoskeleton, three pairs of jointed legs, compound eyes and, typically, two pairs of wings. The segments are grouped into distinct head, thorax and abdomen regions (Figure 30.11).

Insects differ from crustacea in having wings, only one pair of antennae and only three pairs of legs. There are no limbs on the abdominal segments.

The insects have very successfully colonized the land. One reason for their success is the relative impermeability of their cuticles, which prevents desiccation even in very hot, dry climates.

Arachnids

These are the spiders, scorpions, mites and ticks. Their bodies are divided into two regions, the cephalothorax and abdomen (Figure 30.12). They have four pairs of limbs on the cephalothorax, two pedipalps and two chelicerae. The pedipalps are used in reproduction; the chelicerae pierce the prey and paralyse it with a poison secreted by a gland at their base. There are usually several pairs of simple eyes.

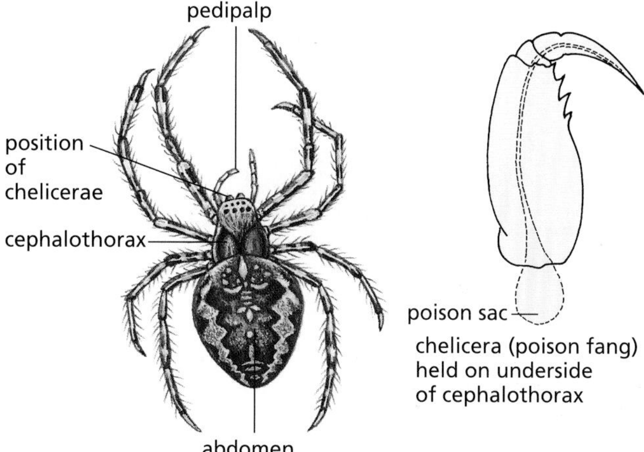

Figure 30.12 Spider (×2.5)

Myriapods

These are millipedes and centipedes. They have a head and a segmented body which is not obviously divided into thorax and abdomen. There is a pair of legs on each body segment but in the millipede the abdominal segments are fused in pairs and it looks as if it has two pairs of legs per segment (Figure 30.13).

As the myriapod grows, additional segments are formed. The myriapods have one pair of antennae and simple eyes. Centipedes are carnivorous; millipedes feed on vegetable matter.

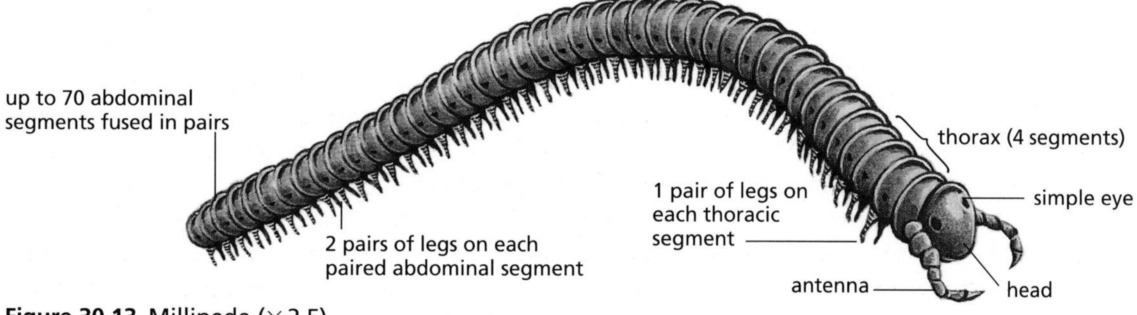

Figure 30.13 Millipede (×2.5)

Molluscs

Molluscs include snails, whelks, slugs, mussels, oysters and (perhaps surprisingly) squids and octopuses (see Figure 30.4 on p. 270).

Many of the molluscs have a shell. In snails, the shell is usually a coiled, tubular structure. In mussels and clams (the bivalves), the shell consists of two halves which can be partially open or tightly closed. In squids the shell is a plate-like structure enclosed in the body. In other molluscs, the shell is reduced or absent, e.g. slugs, octopuses.

All molluscs have a muscular **foot**. In the snails and slugs it forms a flattened structure which protrudes from the shell during locomotion (Figure 30.14). In bivalves, the foot can protrude from between the halves of the shell and burrow in the sand (e.g. cockles). In the squids and octopuses, the foot has become the array of tentacles.

Slugs and snails breathe air by means of a simple 'lung'. The bivalves, squids and octopuses have gills.

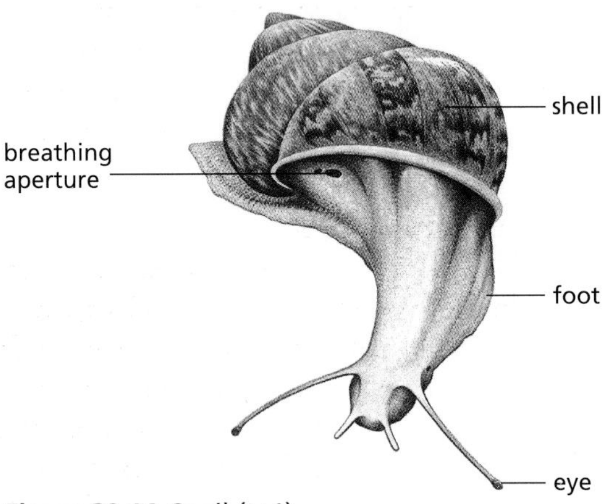

Figure 30.14 Snail (× 1)

Vertebrates

Vertebrates are animals which have a vertebral column. The vertebral column is sometimes called the spinal column or just the spine and consists of a chain of cylindrical bones (vertebrae) joined end to end (p. 152).

Each vertebra carries an arch of bone on its dorsal (upper) surface. This arch protects the spinal cord (p. 166), which runs most of the length of the vertebral column. The front end of the spinal cord is expanded to form a brain which is enclosed and protected by the skull.

The skull carries a pair of jaws which, in most vertebrates, have rows of teeth.

The five classes of vertebrates are fish, amphibia, reptiles, birds and mammals.

Body temperature

Fish, amphibia and reptiles are often referred to as 'cold-blooded'. This is a misleading term. A fish in a tropical lagoon, or a lizard basking in the sun, will have warm blood. The point is that these animals have a variable body temperature which, to some extent, depends on the temperature of their surroundings. Reptiles, for example, may control their temperature by moving into sunlight or retreating into shade but there is no internal regulatory mechanism.

So-called 'warm-blooded' animals, for the most part, have a body temperature higher than that of their surroundings. The main difference, however, is that these temperatures are kept more or less constant despite any variations in external temperature. There are internal regulatory mechanisms (p. 138), which keep the body temperature within narrow limits.

It is better to use the terms **poikilothermic** (variable temperature) and **homoiothermic** (constant temperature).

The advantage of homoiothermy is that an animal's activity is not dependent on the surrounding temperature. A lizard may become sluggish if the surrounding temperature falls. This could be a disadvantage if the lizard is being pursued by a homoiothermic predator whose speed and reactions are not affected by low temperatures.

Questions

1 What position do
   a earthworms,
   b snails, occupy in their respective food webs?

2 An earthworm can withdraw into its burrow; a snail can retract into its shell. What is the likely benefit of this behaviour to these two organisms?

3 Why do you think poikilothermic animals are slowed down by low temperatures (see p. 15)?

Fish

Fish are poikilothermic vertebrates. Many of them have a smooth, streamlined shape which offers minimal resistance to the water through which they move (Figure 30.15). Their bodies are covered with overlapping scales and they have fins which play a part in movement (p. 312).

Fish breathe by means of filamentous gills which are protected by a bony plate, the gill cover (p. 300).

Cartilaginous fish such as dogfish, sharks and rays differ from the 'bony' fish described above in having spiny scales and separate gill openings.

Fish reproduce sexually but fertilization usually takes place externally, i.e. the female lays eggs and the male sheds sperms on them after they are laid.

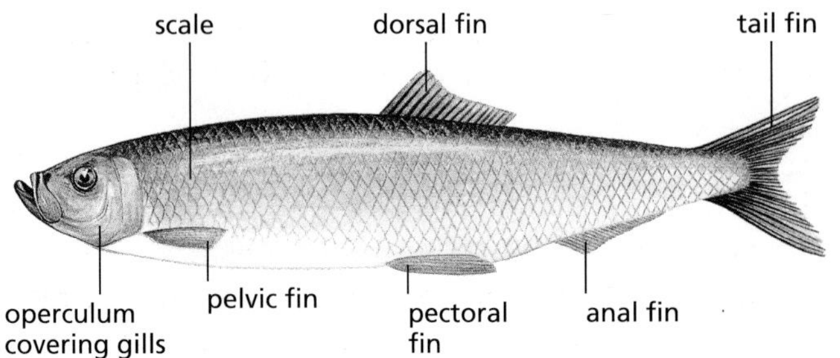

Figure 30.15 Herring (*Clupea*) (× 0.3)

Amphibia

Amphibia are poikilothermic vertebrates with four limbs and no scales. The class includes frogs, toads and newts. The name, amphibian, means double life and refers to the fact that the organism spends part of its life in water and part on the land. In fact, most frogs, toads and newts spend much of their time on the land, in moist situations, and return to ponds, etc. only to lay eggs.

The external features of the common frog are shown in Figure 30.16: Figure 30.5 on p. 271 shows the toad and the newt.

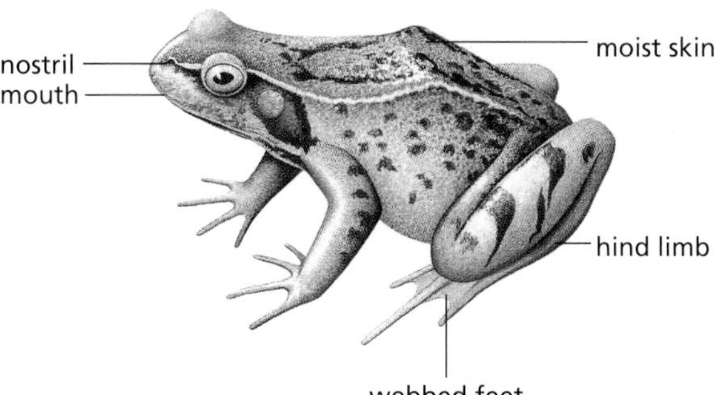

Figure 30.16 *Rana* (× 0.75)

The toad's skin is drier than that of the frog and it has glands which can exude an unpleasant-tasting chemical which discourages predators. Newts differ from frogs and toads in having a tail. All three groups are carnivorous.

Amphibians have four limbs. In frogs and toads, the hind feet have a web of skin between the toes. This offers a large surface area to thrust against the water when the animal is swimming. Newts swim by a wriggling, fish-like movement of their bodies and make less use of their limbs for swimming.

Amphibia have moist skins with a good supply of capillaries which can exchange oxygen and carbon dioxide with the air or water. They also have lungs which can be inflated by a kind of swallowing action. They do not have a diaphragm or ribs.

Frogs and toads migrate to ponds where the males and females pair up. The male climbs on the female's back and grips firmly with his front legs (Figure 34.4, p. 302). When the female lays eggs, the male simultaneously releases sperms over them. Fertilization, therefore, is external even though the frogs are in close contact for the event.

Reptiles

Reptiles are land-living vertebrates. Their skins are dry and the outer layer of epidermis forms a pattern of scales. This dry, scaly skin resists water loss. Also the eggs of most species have a tough, parchment-like shell. Reptiles, therefore, are not restricted to damp habitats, nor do they need water in which to breed.

The reptiles are poikilothermic but they can regulate their temperature to some extent. They do this by basking in the sun until their bodies warm up.

When reptiles warm up, they can move about rapidly in pursuit of insects and other prey.

The reptiles include lizards, snakes, turtles, tortoises and crocodiles. In Britain we have only three species of lizard and three species of snake (Figure 30.17 and Figure 30.5 on p. 271).

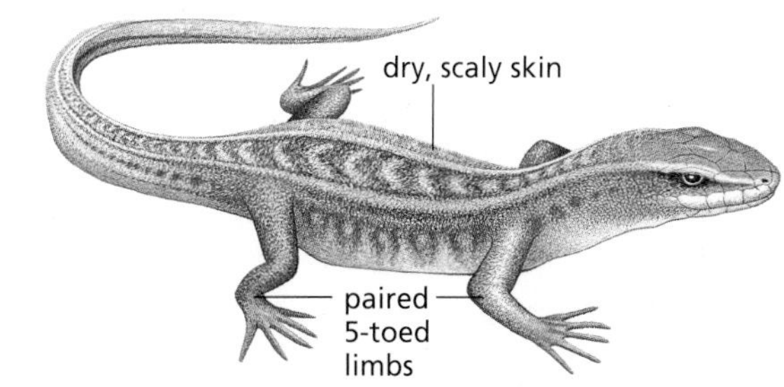

Figure 30.17 *Lacerta* (× 1.5)

Apart from the snakes, reptiles have four limbs, each with five toes. Some species of snake still retain the vestiges of limbs and girdles.

Male and female reptiles mate, and sperms are passed into the female's body. The eggs are, therefore, fertilized internally before being laid. In some species, the female retains the eggs in the body until they are ready to hatch.

Birds

Birds are homoiothermic vertebrates.

The vertebral column in the neck is flexible but the rest of the vertebrae are fused to form a rigid structure. This is probably an adaptation to flight, as the powerful wing muscles need a rigid air-frame to work against.

The epidermis over most of the body produces a covering of feathers but, on the legs and toes, the epidermis forms scales. The feathers are of several kinds. The fluffy down feathers form an insulating layer close to the skin; the contour feathers cover the body and give the bird its shape and coloration; the large quill feathers on the wing are essential for flight.

Birds have four limbs, but the forelimbs are modified to form wings. The feet have four toes with claws which help the bird to perch, scratch for seeds or capture prey, according to the species.

The upper and lower jaws are extended to form a beak which is used for feeding in various ways (Figure 30.18).

In birds, fertilization is internal and the female lays hard-shelled eggs in a nest where she incubates them.

Figure 30.18 The main features of a bird

Mammals

Mammals are homoiothermic vertebrates with four limbs. They differ from birds in having hair rather than feathers. Unlike the other vertebrates they have a diaphragm which plays a part in breathing (p. 124). They also have mammary glands and suckle their young on milk.

A sample of mammals is shown in Figure 30.5 on p. 271 and Figure 30.19 illustrates some of the mammalian features.

Figure 30.19 Mammalian features. The furry coat, the external ear pinnae and the facial whiskers (vibrissae) are visible mammalian features in this gerbil

Humans are mammals and most of the physiology described in Section 3 applies to all mammals.

Mammals give birth to fully formed young instead of laying eggs. The eggs are fertilized internally (see p. 143) and undergo a period of development in the uterus.

The young may be blind and helpless at first, e.g. dogs and cats, or they may be able to stand up and move about soon after birth, e.g. sheep and cows. In either case, the youngster's first food is the milk which it sucks from the mother's teats. The milk is made in the mammary glands and contains all the nutriments that the offspring need for the first few weeks or months, depending on the species.

As the youngsters get older, they start to feed on the same food as the parents. In the case of carnivores, the parents bring the food to the young until they are able to fend for themselves.

Questions

1 Which vertebrate classes
   a are warm-blooded,
   b have four legs,
   c lay eggs,
   d have internal fertilization and
   e have some degree of parental care?

2 Draw up a key (see p. 281) which a biologist could use to place an animal in its correct group.

Plant kingdom

DIVISION

Red algae
Brown algae
Green algae
} seaweeds and filamentous forms; mostly aquatic

Bryophytes (no specialized conducting tissue)

CLASS
Liverworts
Mosses

Vascular plants (well-developed xylem and phloem)

CLASS
Ferns
[1] { Conifers (seeds not enclosed in fruits)
Flowering plants (seeds enclosed in fruits)
SUBCLASS
Monocotyledons[2] (grasses, lilies)
Dicotyledons (trees, shrubs, herbaceous plants)
FAMILY
e.g. Ranunculaceae (one of about 70 families)
GENUS
e.g. *Ranunculus*
SPECIES
e.g. *Ranunculus bulbosus* (bulbous buttercup)

[1] Sometimes called, collectively, 'seed-bearing plants'.
[2] The monocotyledons (monocots for short), are flowering plants which have only one cotyledon in their seeds (p. 74). Most, but not all, monocots also have long, narrow leaves (e.g. grasses, daffodils, bluebells) with parallel leaf veins (Figure 30.20).
The dicotyledons (dicots for short), have two cotyledons in their seeds (p. 74). Their leaves are usually broad and the leaf veins form a branching network.

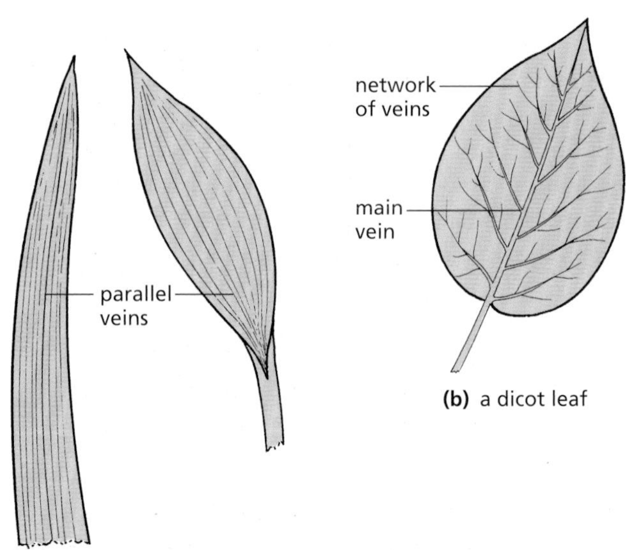

Figure 30.20 Leaf types in flowering plants

Algae

The algae are plants which range from very simple filaments, consisting of a single row of identical cells, to large seaweeds. The Protoctista with chloroplasts (the protophyta) may also, with good reason, be classified as algae.

Most members of the algae have cellulose cell walls and chloroplasts. The chloroplasts contain chlorophyll, but other pigments may be present in brown and red algae.

The algae, for the most part, grow in water and have little need of specialized supporting or water-conducting tissue. Compared with vascular plants, therefore, their cells and tissues are relatively unspecialized. The aquatic algae cannot survive prolonged desiccation when out of water.

In open water, it is the single-celled algae which form the basis of the food web. These algae, particularly the forms called **diatoms**, are abundant in the surface waters of ponds, lakes and the sea. Usually they are invisible to the naked eye, but occasionally, in a pond or neglected swimming-pool, the unicellular plants are so numerous that the water looks green.

Bryophytes

The bryophytes include the liverworts and mosses.

Liverworts

Liverworts are small, flat, green plants, usually growing close together in very moist, shady places such as in the mouths of caves or on a stream bank near the water-line. They look like small, overlapping strips of dark green leaf stuck to the rock or the bank (Figure 30.21).

Figure 30.21 A colony of liverworts. The small black knobs on the ends of the green stalks are the spore capsules

Liverworts have no stems or roots but single-celled **rhizoids** (like root hairs, p. 56) grow from their lower surface, anchor them in the ground and absorb water.

Mosses

Mosses are simple but successful land plants. Each plant is quite small, on average about 1–5 cm long, but they usually grow in dense tufts, which probably helps to support them and to conserve moisture.

Each plant has a slender stem with numerous small leaves arising from it (Figure 30.22). Sometimes the stem is horizontal and has many branches. Most moss leaves are only one cell thick, so there is no specialization of cells like there is in the leaves of flowering plants. In the stem, the innermost cells conduct water and food, but apart from being longer than other cells, they are not particularly specialized in their structure. The outer cells have thick walls and probably give the stem strength.

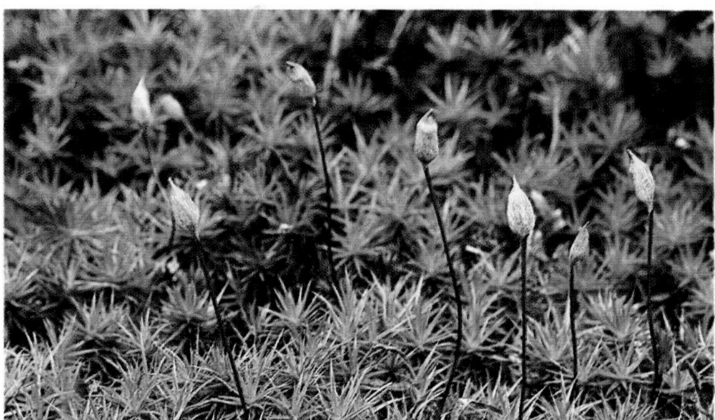

Figure 30.22 Moss plants growing in a wood. Some of the plants have produced spore capsules

As in liverworts, there are no proper roots. The rhizoids may consist of several cells and grow from the base of the stem into the soil.

Four different species of moss are illustrated in Figure 30.6 on p. 272.

Mosses and liverworts reproduce sexually but do not form seeds. The male gametes have flagella and swim in the film of water on the surface of the plant in order to reach the female gamete. When the gametes fuse the zygote grows into a spore capsule on a long stalk. The spore capsules eventually break open and release thousands of single-celled **spores**. Each of these spores can grow into a new moss or liverwort plant.

The rhizoids of liverworts and mosses penetrate only into the top few millimetres of the soil. The plants lack a waxy cuticle and so are unable to resist desiccation. The male gametes have to reach the female gamete by swimming in water. These characteristics restrict most bryophytes to damp, shady habitats.

Ferns

Ferns are land plants, usually much larger than mosses, with more highly developed structures. Their stems, leaves and roots are very similar to those of the flowering plants.

The stem is usually entirely below ground and takes the form of a **rhizome** (p. 79). In bracken, the rhizome grows horizontally below ground, sending up leaves at intervals. The roots which grow from the rhizome are called adventitious roots. This is the name given to any roots which grow directly from the stem rather than from other roots.

The stem and leaves have sieve tubes and water-conducting cells similar to those in the xylem and phloem of a flowering plant (p. 54). For this reason, the ferns and seed-bearing plants are sometimes referred to as vascular plants, because they all have vascular bundles or vascular tissue. Ferns also have multicellular roots with vascular tissue.

The leaves of ferns vary from one species to another (Figure 30.23 and Figure 30.6 on p. 272), but they are all much larger than those of mosses and are several cells thick. Most of them have an upper and lower epidermis, a layer of palisade cells and a spongy mesophyll similar to the leaves of a flowering plant.

Figure 30.23 Young fern leaf. Ferns do not form buds like those of the flowering plants. The midrib and leaflets of the young leaf are tightly coiled and unwind as it grows

Ferns produce gametes but no seeds. The zygote gives rise to the fern plant, which then produces single-celled spores from numerous **sporangia** (spore capsules) on its leaves. The sporangia are formed on the lower side of the leaf but their position depends on the species of fern. The sporangia are usually arranged in compact groups (Figure 30.24).

Figure 30.24 Polypody fern. Each brown patch on the underside of the leaf is made up of many sporangia

Conifers

Conifers are trees or shrubs often with needle-like leaves (Figure 30.25). They reproduce by seeds rather than spores but the seeds are formed in cones and not in flowers and are not enclosed in an ovary.

Figure 30.25 Pine needles and male cones. The needle-like shape of the leaves helps to reduce the rate of transpiration (p. 59) in windy conditions

There are male and female cones on the same plant. The male cones produce pollen which is light and powdery and falls, or is blown, on to the female cones (Figure 30.26). Familiar conifers are the pine, larch, spruce and cypress.

Figure 30.26 Larch cones. The green cones are male and are producing pollen. The pink female cones face upwards and will produce the seeds

Flowering plants

Flowering plants reproduce by seeds which are formed in flowers. The seeds are enclosed in an ovary. The structure, physiology and life cycle of flowering plants are described on pp. 50–82.

Questions

1 Try to classify the following organisms: beetle, sparrow, weasel, gorilla, pine tree, buttercup, moss.
e.g. Butterfly: Kingdom, animal; Group, arthropod; Class, insect.

2 The white deadnettle is *Lamium album*; the red deadnettle is *Lamium purpureum*. Would you expect these two plants to cross-pollinate successfully?

3 If a heath fire destroys all the above-ground vegetation, the bracken will still grow well next season. Suggest why this should be so.

4 Which groups of plants are sometimes called vascular plants and why are they so called?

Keys for identification

Once you know the main characteristics of a group, it is possible to draw up a systematic plan for identifying an unfamiliar organism. One such plan is shown in Figure 30.27.

An alternative form of key is the **dichotomous key**. Dichotomous means two branches, so you are confronted with two possibilities at each stage.

DICHOTOMOUS KEY FOR VERTEBRATE CLASSES

1	Poikilothermic	2
	Homoiothermic	4
2	Has fins but no limbs	**Fish**
	Has 4 limbs	3
3	Has no scales on body	**Amphibian**
	Has scales	**Reptile**
4	Has feathers	**Bird**
	Has fur	**Mammal**

Above is an example of a dichotomous key that could be used to place an unknown vertebrate in the correct class. Item 1 gives you a choice between two alternatives. If the animal is poikilothermic you move to item 2 and make a further choice. If it is homoiothermic you move to item 4 for your next choice.

The same technique may be used for assigning an organism to its class, genus or species. However, the important features may not always be easy to see and you have to make use of less fundamental characteristics.

Figure 30.28 is a key for identifying some of the possible invertebrates to be found in a compost heap. Of course, you do not need a key to identify these familiar animals but it does show you how a key can be constructed.

INHABITANTS OF A COMPOST HEAP

1	Has legs	2
	No legs	5
2	More than 6 legs	3
	6 legs	4
3	Short, flattened grey body	**Woodlouse**
	Long brown/yellow body	**Centipede**
4	Pincers on last segment	**Earwig**
	Hard wing covers	**Beetle**
5	Body segmented	**Earthworm**
	Body not segmented	6
6	Has a shell	**Snail**
	No shell	**Slug**

Figure 30.28 Key for some invertebrates in a compost heap

Questions

1 Figure 30.1 on p. 268 shows some protoctista. Using only the features shown in the drawings, construct a dichotomous key that could be used to identify these organisms.

2 Construct a dichotomous key that would lead an observer to discriminate between the following plants: daffodil, snowdrop, lupin, couch grass, gorse. (There is more than one way.)
Try to explain why this is an 'artificial' key rather than a 'natural' key.

3 Revise pp. 278–81 and try to construct a dichotomous key to sort the plant kingdom (excluding algae) into its classes and, where possible, its subclasses.

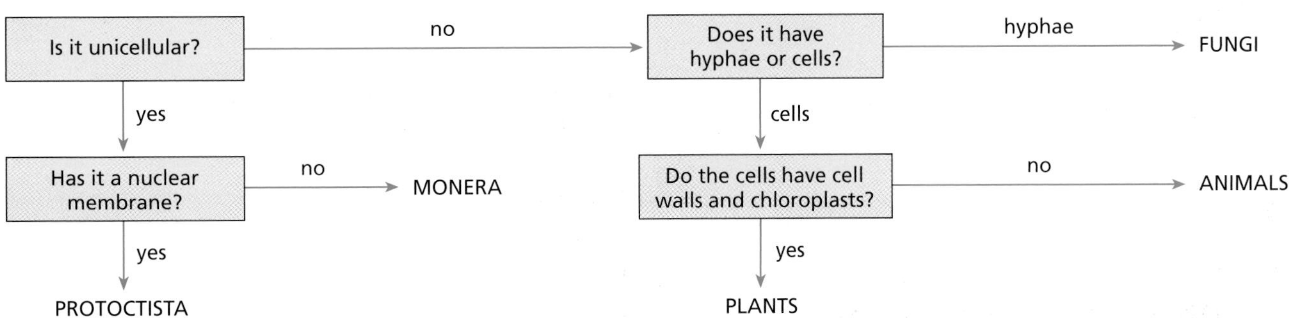

Figure 30.27 Identification plan

Checklist

- BACTERIA are microscopic organisms; they have no proper nucleus.
- PROTOCTISTA are single-celled organisms containing a nucleus.
- FUNGI are made up of thread-like hyphae. They reproduce by spores.
- PLANTS make their food by photosynthesis.
  - **a** Algae are simple plants with few special organs. Most of them live in water.
  - **b** Liverworts and mosses are land plants. Mosses have stems and leaves.
  - **c** Ferns have well-developed stems, leaves and roots. They reproduce by spores.
  - **d** Seed-bearing plants reproduce by seeds.
    - i Conifers have no flowers; their seeds are not enclosed in an ovary.
    - ii Flowering plants have flowers; their seeds are in an ovary which forms a fruit. Monocots have one cotyledon in the seed; dicots have two cotyledons in the seed.
- ANIMALS get their food by eating plants or other animals.
  - **a** Coelenterates live mostly in the sea. They have tentacles and sting cells.
  - **b** Flatworms have flat bodies; some have cilia. They live in fresh water or are parasitic.
  - **c** Annelids have tubular bodies divided into segments.
  - **d** Molluscs have a creeping or holding 'foot' and many have a shell.
  - **e** Arthropods have a hard exoskeleton and jointed legs.
    - i Crustacea mostly live in water and have more than three pairs of legs.
    - ii Insects mostly live on land and have wings and three pairs of legs.
    - iii Arachnids have four pairs of legs and poisonous mouth parts.
    - iv Myriapods have many pairs of legs.
  - **f** Vertebrates have a spinal column and skull.
    - i Fish have gills, fins and scales.
    - ii Amphibia can breathe in air or in water.
    - iii Reptiles are land animals; they lay eggs with leathery shells.
    - iv Birds have feathers, beaks and wings; they are 'warm-blooded'.
    - v Mammals have fur, and suckle their young; the young develop inside the mother.

31 Micro-organisms

Bacteria
Structure. Physiology. Reproduction.

Viruses
Structure. Multiplication.

Fungi
Structure. Nutrition. Reproduction.
Penicillium. Yeast.

Culturing micro-organisms
Liquid and solid media; aseptic techniques.

Practical work
Culturing bacteria, effect of antibiotics, growing mould fungi, starch digestion.

The term 'micro-organism' includes viruses, bacteria, protoctista and some fungi and algae. Most of these organisms are, indeed, microscopic.

Bacteria and protoctista are single-celled organisms. Fungi and algae are multicellular. (That means, their bodies are made up of many cells.) Viruses do not have a cellular structure.

'Micro-organisms' is a convenient term by which to refer to a wide variety of fairly simple organisms. But the word is not used in classifications, in the way we use words like 'mammal' or 'vertebrate'.

Bacteria

Bacterial structure

Bacteria (singular = bacterium) are very small organisms consisting of single cells rarely more than 0.01 mm in length. They can be seen only with the higher powers of the microscope.

Their cell walls are made, not of cellulose, but of a complex mixture of proteins, sugars and lipids. Some bacteria have a **slime capsule** outside their cell wall. Inside the cell wall is the cytoplasm, which may contain granules of glycogen, lipid and other food reserves (Figure 31.1).

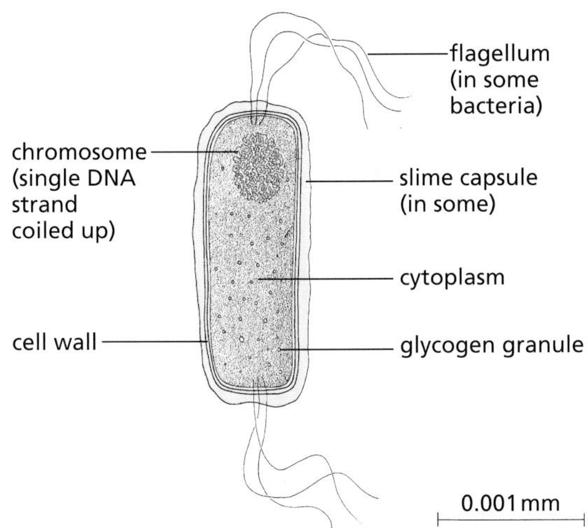

Figure 31.1 Generalized diagram of a bacterium

Each bacterial cell contains a single chromosome, consisting of a circular strand of DNA (p. 186). The chromosome is not enclosed in a nuclear membrane but is coiled up to occupy part of the cell (Figure 31.2).

Individual bacteria may be spherical, rod-shaped or spiral and some have filaments, called **flagella**, projecting from them. The flagella can flick, and so move the bacterial cell about.

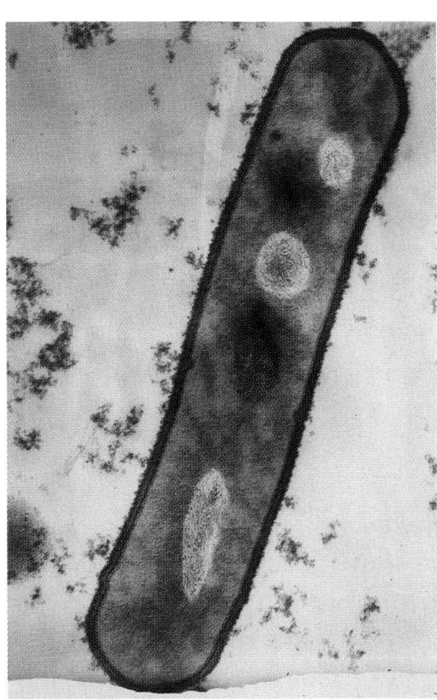

Figure 31.2 Longitudinal section through a bacterium (×27 000). The light areas are coiled DNA strands. There are three of them because the bacterium is about to divide twice (see Figure 31.3)

Bacterial physiology

Nutrition

There are a few species of bacteria which contain a photosynthetic pigment like chlorophyll, and can build up their food by photosynthesis. Most bacteria, however, live in or on their food. They produce and release enzymes which digest the food outside the cell. The liquid products of digestion are then absorbed back into the bacterial cell.

Respiration

The bacteria which need oxygen for their respiration are called **aerobic bacteria**. Those which do not need oxygen for respiration are called **anaerobic bacteria**. The bacteria used in the filter beds of sewage plants (p. 331) are aerobic, but those used to digest sewage sludge and produce methane are anaerobic.

Reproduction

Bacteria reproduce by cell division or **fission**. Any bacterial cell can divide into two and each daughter cell becomes an independent bacterium (Figure 31.3). In some cases, this cell division can take place every 20 minutes so that, in a very short time, a large colony of bacteria can be produced. This is one reason why a small number of bacteria can seriously contaminate our food products.

This kind of reproduction, without the formation of gametes, is called **asexual reproduction** (p. 303).

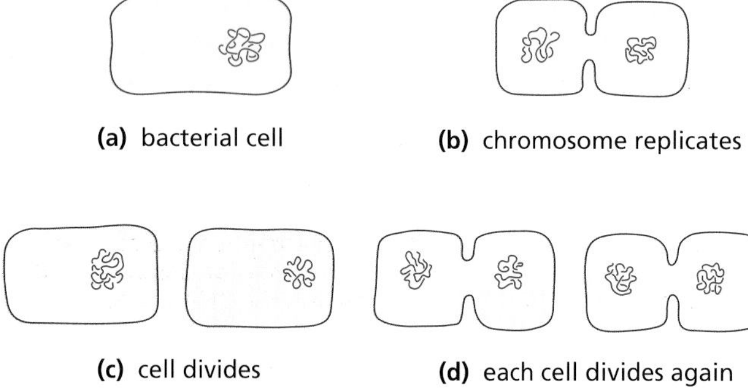

Figure 31.3 Bacterium reproducing. This is asexual reproduction by cell division (see 'Mitosis', p. 182)

Effect of heat

Bacteria, like any other living organisms, are killed by high temperatures. The process of cooking destroys any bacteria in food, provided high enough temperatures are used. If drinking water is boiled, any bacteria present are killed.

However, some bacteria can produce spores which are resistant to heat. When the cooked food or boiled water cools down, the spores germinate to produce new colonies of bacteria, particularly if the food is left in a warm place for many hours. For this reason, cooked food should be eaten at once or immediately refrigerated. (Refrigeration slows down bacterial growth and reproduction.) After refrigeration, food should not merely be warmed up but either eaten cold or heated to a temperature high enough to kill any bacteria that have grown, i.e. to 90 °C or more (p. 93).

Useful and harmful bacteria

When people talk about bacteria, they are usually thinking about those which cause disease or spoil our food. In fact, only a tiny minority of bacteria are harmful. Most of them are harmless or extremely useful.

Bacteria which feed saprotrophically (p. 293) bring about decay. They secrete enzymes into dead organic matter and liquefy it. This may be a nuisance if the organic matter is our food but, in most cases, it consists of the excreta and dead bodies of organisms. If it were not for the activities of the decay bacteria (and fungi), we should be buried in ever-increasing layers of dead vegetation and animal bodies.

The decay bacteria also release essential elements from the dead remains. For example, proteins are broken down to ammonia and the ammonia is turned into nitrates by nitrifying bacteria (p. 229). The nitrates are taken up from the soil by plants, which use them to build up their proteins. In a similar way, sulphur, phosphorus, iron, magnesium and all the elements essential to living organisms are recycled in the course of bacterial decomposition.

Humans exploit bacterial physiology in the course of **biotechnology** (p. 326).

The bacteria which cause disease are parasites (p. 293). They live in the cells of plants or animals and feed on the cytoplasm. Parasitic organisms which cause disease are called **pathogens** (Figure 31.4). The organism in which they live and reproduce is called the **host**.

Pathogenic bacteria (p. 333) may cause diseases because of the damage they do to the host's cells, but most bacteria also produce poisonous waste products called **toxins**. The toxin produced by the *Clostridium* bacteria (which cause tetanus) is so poisonous that as little as 0.00023 g is fatal.

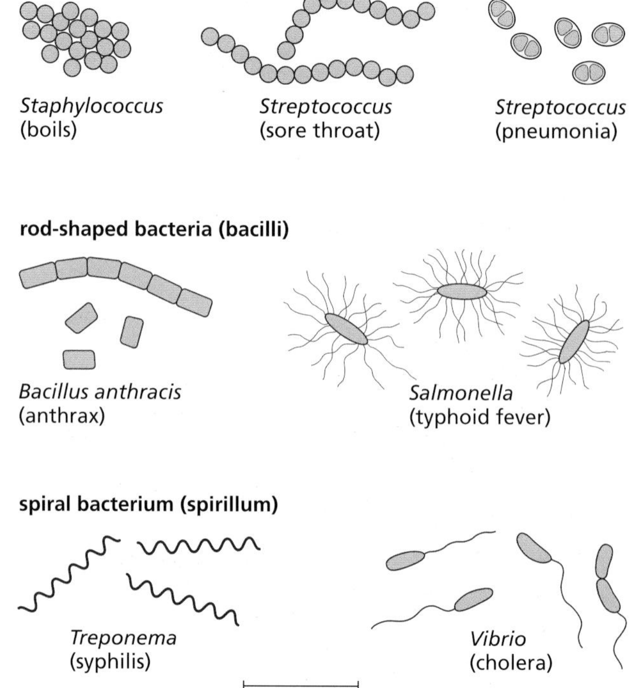

Figure 31.4 Some pathogenic bacteria

Viruses

Most viruses are very much smaller than bacteria and can be seen only with the electron microscope at magnifications of about × 30 000.

Virus structure

There are many different types of virus and they vary in their shape and structure. All viruses, however, have a central core of RNA or DNA (p. 186) surrounded by a protein coat. Viruses have no nucleus, cytoplasm, cell organelles, or cell membrane, though some forms have a membrane outside their protein coats.

Virus particles, therefore, are not cells. They do not feed, respire, excrete or grow and it is debatable whether they can be classed as living organisms. Viruses do reproduce, but only inside the cells of living organisms, using materials provided by the host cell.

A generalized virus particle is shown in Figure 31.5. The nucleic acid core is a coiled single strand of RNA. The coat is made up of regularly packed protein units called **capsomeres** each containing many protein molecules. The protein coat is called a **capsid**. Outside the capsid, in the influenza and some other viruses, is an envelope which is probably derived from the cell membrane of the host cell (Figure 31.6).

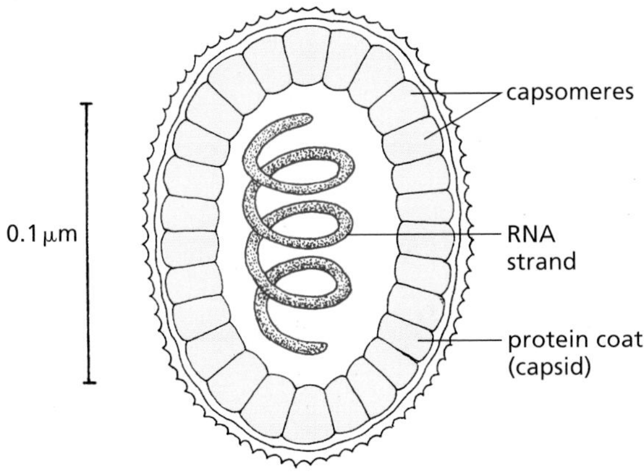

Figure 31.5 Generalized structure of a virus

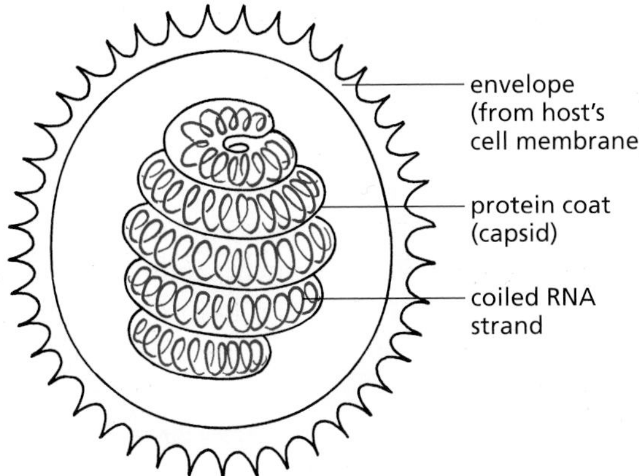

Figure 31.6 Structure of the influenza virus

Multiplication of viruses

Viruses can survive outside the host cell, but in order to reproduce they must penetrate into a living cell. How they do this, in many cases, is not known for certain. In most instances, the virus particle first sticks to the cell membrane. It may then 'inject' its DNA or RNA into the cell's cytoplasm or the whole virus may be taken in by a kind of endocytosis (p. 27).

Once inside the host cell, the virus is 'uncoated', i.e. its capsid is dispersed, exposing its DNA or RNA. The DNA or RNA then takes over the host cell's physiology. It arrests the normal syntheses in the cell and makes the cell produce new viral DNA or RNA and new capsomeres. The nucleic acid and the capsomeres are assembled in the cell to make new virus particles which escape from the cell (Figure 31.7).

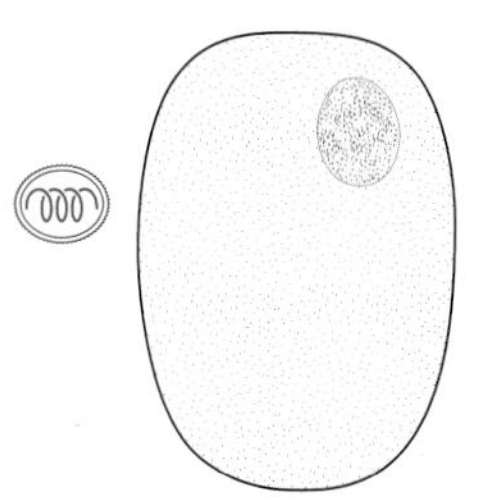

(a) The virus sticks to the membrane of a suitable host cell.

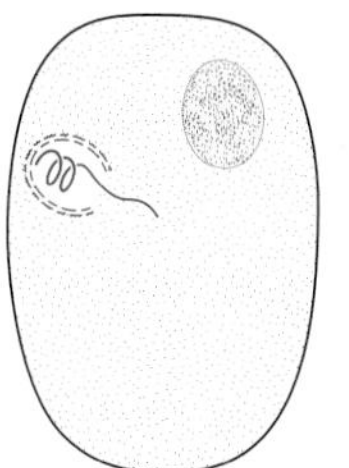

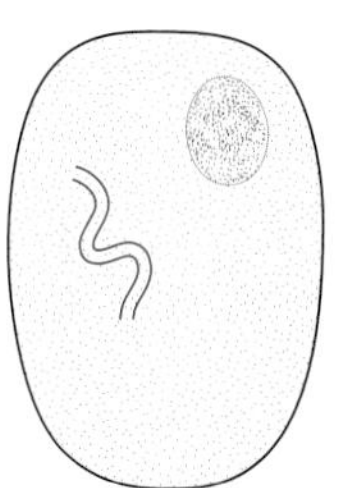

(b) The virus enters the cell, the protein coat breaks down and releases the DNA or RNA.

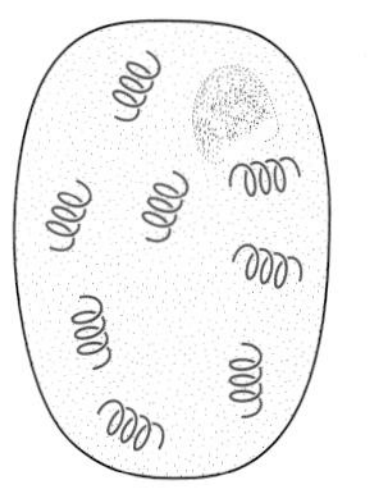

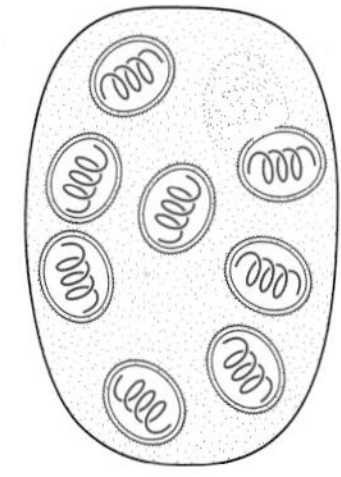

(c) The viral DNA or RNA replicates and directs the host cell to make new protein coats.

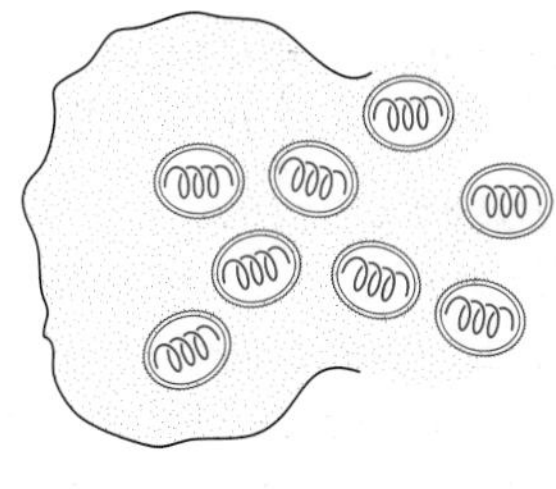

(d) The new viruses escape from the host cell.

Figure 31.7 Reproduction of a virus

The cell may be destroyed in this process or the viruses may escape, wrapping themselves in pieces of the host's cell membrane as they do so. These activities give rise to the signs and symptoms of disease.

Many viruses cause diseases in plants and animals. Human virus diseases include the common cold, poliomyelitis, measles, mumps, chickenpox, herpes, rubella, influenza and AIDS (p. 335).

Tobacco mosaic virus affects tomato plants as well as tobacco, but the virus which causes the striking patterns (called 'breaks') in tulip petals does not seem to harm the plant.

Viruses may be used as vectors to 'deliver' recombinant DNA in genetic engineering (p. 213).

Questions

1 Which of the following structures are present in both bacterial cells and plant cells: cytoplasm, cellulose, DNA, cell wall, nucleus, chromosome, vacuole, glycogen?

2 If five bacteria landed in some food and reproduced at the maximum possible rate, what would be the population of bacteria after 4 hours?

3 a Why is a virus particle not considered to be a cell?
b Why are viruses not easy to classify as living organisms?

4 How does the reproduction of a virus differ from that of a bacterium?

Fungi

The kingdom of the fungi includes fairly familiar organisms such as mushrooms, toadstools, puffballs, and the bracket fungi that grow on tree-trunks. There are also the less obvious, but very important, mould fungi which grow on stale bread, cheese, fruit or other food. Many of the mould fungi live in the soil or in dead wood. The yeasts are single-celled fungi similar to the moulds in some respects.

Some fungal species are parasites and live in other organisms, particularly plants, where they cause diseases which can affect crop plants.

Structure

Many fungi are not made up of cells but of microscopic threads called **hyphae**. The branching hyphae spread through the material on which the fungus is growing, and absorb food from it. The network of hyphae that grow over or through the food material is called the **mycelium** (Figure 31.8).

Mushrooms and toadstools are the reproductive structures, 'fruiting bodies', of an extensive mycelium that spreads through the soil or the dead wood on which the fungus is growing.

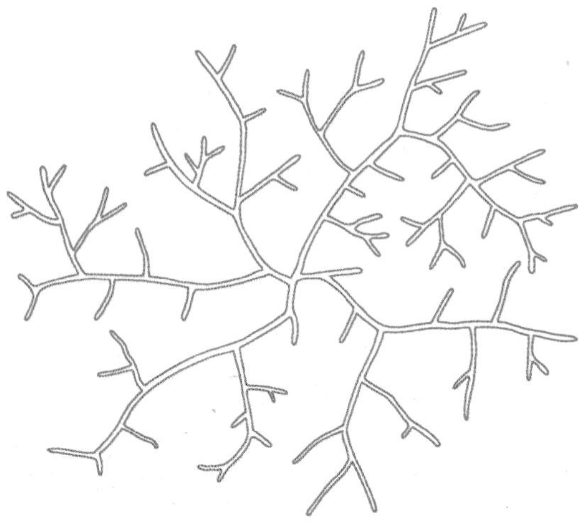

Figure 31.8 The branching hyphae form a mycelium

The hyphae are like microscopic tubes lined with cytoplasm. In the centre of the older hyphae there is a vacuole, and the cytoplasm contains organelles and inclusions (p. 3). The inclusions may be lipid droplets or granules of glycogen but, unlike plants, there are no chloroplasts or starch grains (Figure 31.9).

The hyphal wall may contain cellulose or chitin or both, according to the species. Chitin is similar to cellulose but the chitin molecule contains nitrogen atoms.

In some species of fungi, there are incomplete cross-walls dividing the hyphae into cell-like regions, but the cytoplasm is free to flow through large pores in these walls. In the species which do have cross-walls (septa), there may be one, two or more nuclei in each compartment. In the species without septa, the nuclei are distributed throughout the cytoplasm.

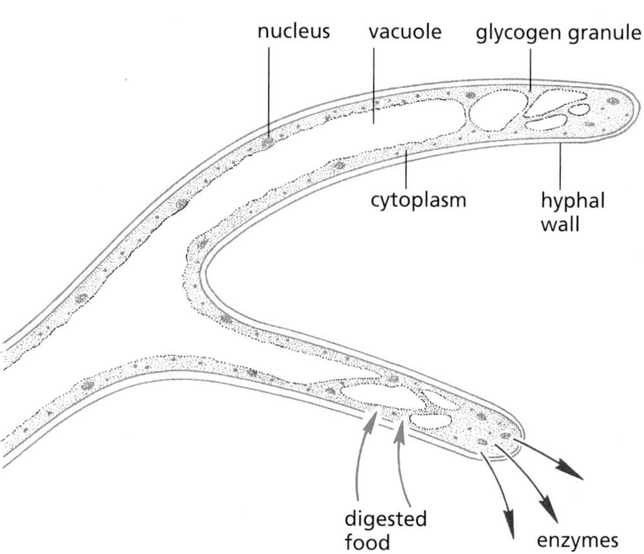

Figure 31.9 Structure of fungal hyphae

Nutrition

Saprotrophic fungi

Most fungi are saprotrophs (p. 227) living on dead organic matter (Figure 31.10).

The hyphae secrete enzymes into the organic material and digest it to liquid products. The digested products are then absorbed back into the hyphae and used for energy or for the production of new cytoplasm or hyphal walls.

The type of enzymes produced by the hyphae will depend on the species of fungus and the material it is growing on. Those species which produce a **cellulase**, i.e. an enzyme which digests cellulose, are important in helping to break down plant remains (Figure 31.11). Saprotrophic fungi and bacteria are the 'decomposers' in most food webs (p. 226), and are largely responsible for the recycling of essential nutrients in any ecosystem (p. 227).

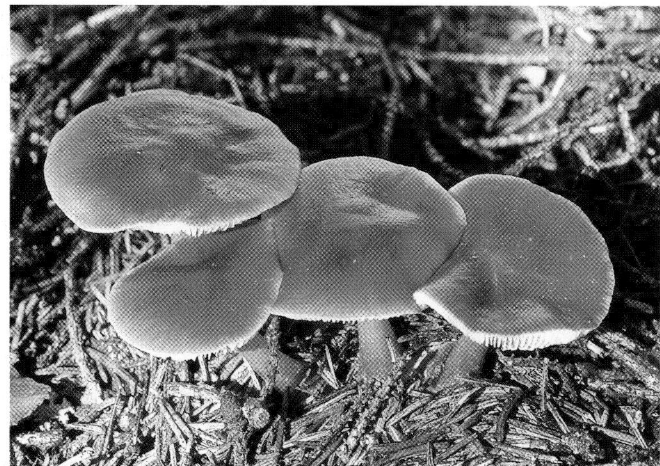

Figure 31.10 Toadstools growing on a fallen tree. The toadstools are the reproductive structures which produce spores. The feeding hyphae are inside the tree, digesting the wood

Figure 31.11 Saprotrophic bacteria and fungi have started to rot some of the apples. The saprotrophs gained access via wasp damage

Parasitic fungi

The hyphae of parasitic fungi penetrate the tissues of their host plant and digest the cells and their contents. If the mycelium spreads extensively through the host, it usually causes the death of the plant. The bracket fungus shown in Figure 31.12 is the fruiting body of a mycelium that is spreading through the tree and will eventually kill it.

Fungus diseases such as blight, mildews or rusts (Figure 30.3, p. 269) are responsible for causing considerable losses to arable farmers, and there is a constant search for new varieties of crop plants which are resistant to fungus disease, and for new chemicals (fungicides) to kill parasitic fungi without harming the host.

A few parasitic fungi cause diseases in animals, including humans. One group of these fungi cause tinea or ringworm. The fungus grows in the epidermis of the skin and causes irritation and inflammation. One form of tinea is athlete's foot, in which the skin between the toes becomes infected.

Tinea is very easily spread by contact with infected towels or clothing, but can usually be cured quickly with a fungicidal ointment.

Figure 31.12 A parasitic fungus. The 'brackets' are the reproductive structures. The mycelium in the trunk will eventually kill the tree

Reproduction

Fungi reproduce by releasing microscopic, single-celled **spores**. You can see a cloud of these spores if you squeeze a mature puffball (Figure 31.13).

The spores are budded off from the tips of special hyphae. Each spore contains a little cytoplasm and one or more nuclei, depending on the species. The spores are dispersed in air currents or by other methods. When a spore lands on suitable organic matter or on a new host, it germinates to produce a mycelium.

The spores are produced from a single mycelium without the involvement of gametes or any sexual process, and so this is an example of **asexual reproduction** (p. 303).

Most fungi, however, do have a sexual process in their life cycle and this may precede the production of asexual spores.

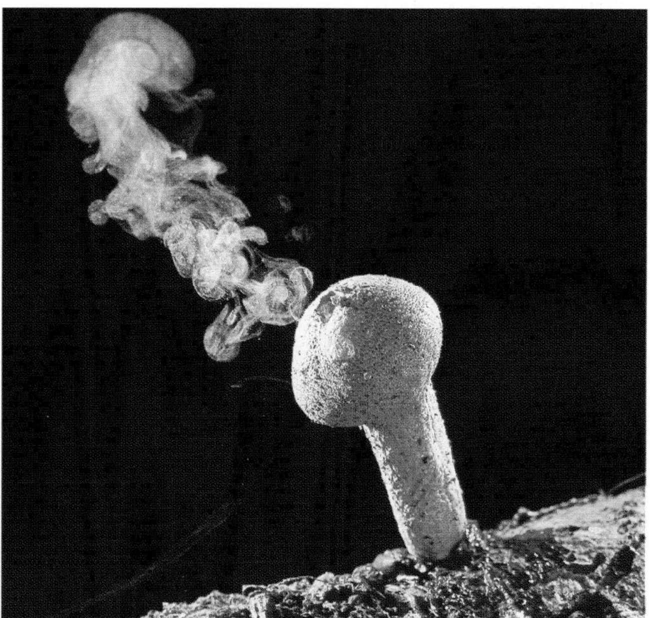

Figure 31.13 Puffball dispersing spores. When a rain drop hits the ripe puffball, a cloud of spores is ejected

Penicillium

Penicillium is a genus of mould fungi that grow on decaying vegetable matter, damp leather and citrus fruits. The mycelium grows over the food, digesting it and absorbing nutrients. Vertical hyphae grow from the mycelium and, at their tips, produce chains of spores (Figures 31.14 and 31.15). These give the colony a blue-green colour and a powdery appearance (Figure 25.13, p. 231).

The spores are dispersed by air currents and, if they reach a suitable substrate, grow into a new mycelium.

There are many species of *Penicillium*. *P. notatum* was the species which led to the discovery of penicillin when Alexander Fleming, in 1928, observed that the mould suppressed bacterial growth in one of his culture plates (p. 353). *P. camemberti* and *P. roqueforti* are used in the preparation of cheeses. *P. griseofulvin* is used to produce an antibiotic effective against fungal skin diseases such as athlete's foot.

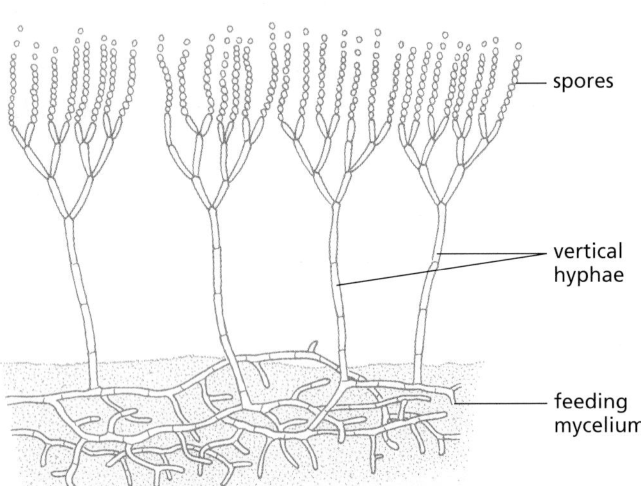

Figure 31.14 *Penicillium sp*

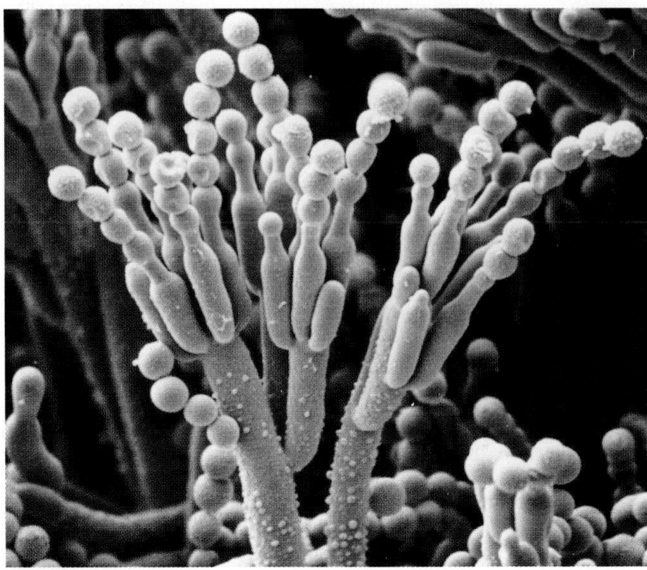

Figure 31.15 Scanning electron micrograph of *Penicillium* spores

Questions

1. When something goes mouldy what is actually happening to it?
2. Suggest why bread, wood and leather may go mouldy, while glass and plastic do not.
3. Suggest why toadstools may be found growing in very dark areas of woodland where green plants cannot flourish.

Yeast

The yeasts are a rather unusual family of fungi. Only a few of the several species can form true hyphae. The majority of them consist of separate, spherical cells, which can be seen only under the microscope. They live in situations where sugar is likely to be available, e.g. in the nectar of flowers or on the surface of fruits.

The thin cell wall encloses the cytoplasm, which contains a nucleus and a vacuole. In the cytoplasm are granules of glycogen and other food reserves (Figure 31.16a).

The cells reproduce by budding. An outgrowth from the cell appears, enlarges and is finally cut off as an independent cell. When budding occurs rapidly, the individuals do not separate at once and, as a result, small groups of attached cells may sometimes be seen (Figure 31.16b).

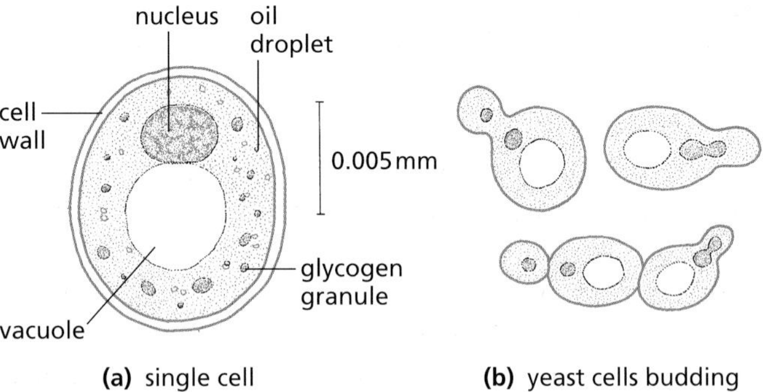

Figure 31.16 Yeast

Fermentation

Yeasts are of economic importance in promoting alcoholic fermentation. Yeast cells contain many enzymes, some of which can break down sugar into carbon dioxide and alcohol. This chemical change provides energy for the yeast cells to use in their vital processes.

$$C_6H_{12}O_6 \rightarrow 2CO_2 + \underset{\text{alcohol}}{2C_2H_5OH} + 118\ \text{kJ}$$

Alcoholic fermentation is a form of anaerobic respiration (p. 20) but, if the yeast is supplied with carbohydrates other than sugar, the yeast needs oxygen to convert these substances to sugar first.

In brewing, barley is allowed to germinate. During germination, the barley grains convert their starch reserves into maltose (p. 13). The germinating barley is then killed by heat and the sugars are dissolved out with water. Yeast is added to this solution and ferments it to carbon dioxide and alcohol.

In making beer, hops are added to give the brew a bitter flavour and the liquid is sealed into casks or bottles so that the carbon dioxide is under pressure. When the pressure is released, the dissolved carbon dioxide escapes from the liquid giving it a 'fizz'.

In making spirits such as whisky, the fermentation is allowed to go on longer and the alcohol is distilled off. Wine is made by extracting the juice from fruit, usually grapes, and allowing the yeasts, which live on the surface of the fruit, to ferment the sugar to alcohol.

If too much oxygen is admitted, or fungi and bacteria are allowed to enter, the alcohol may be oxidized to ethanoic (acetic) acid, i.e. vinegar.

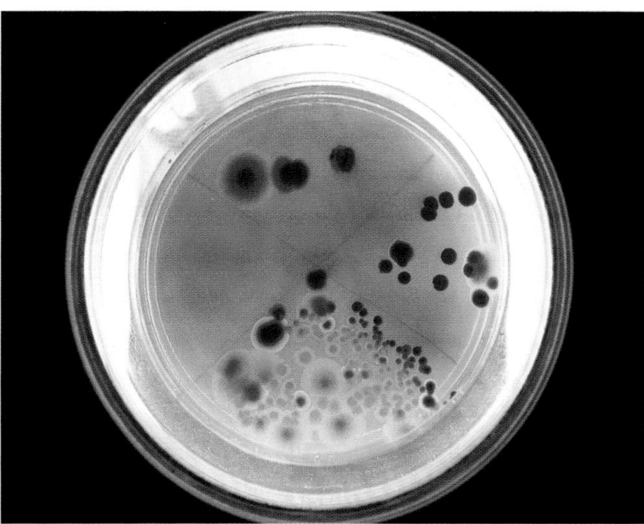

Figure 31.17 Micro-organisms from the air. The nutrient agar was exposed to the air for a few minutes. Subsequently, these bacterial colonies grew. Each colony has developed from a single micro-organism

Culturing micro-organisms

Most micro-organisms will grow under controlled conditions in the laboratory. By carefully controlling the composition of the culture medium, its temperature and pH it is possible to encourage the growth of some organisms while suppressing others. For example, bacteria grow readily on an alkaline medium; fungi prefer an acid medium. Most micro-organisms need a culture medium which contains an organic carbon source such as glucose, and some need an organic source of nitrogen (e.g. amino acids) as well. Others will grow if supplied with potassium nitrate as their sole source of nitrogen.

In genetic engineering, culture media containing antibiotics are used to isolate bacteria carrying recombinant DNA (p. 213).

The culture medium can either be a liquid or a solid. If it is a solid, it is based on a jelly made from **agar**. Agar is a jelly derived from seaweed. Bacteria cannot grow on it unless nutrients are added. This makes it possible to be selective as described above.

When fungi and bacteria grow on agar, they reproduce rapidly and form visible colonies (Figure 31.17). If the culture plate was inoculated with only a few organisms, each colony will have grow from a single individual. By subculturing from one of these colonies, it is possible to produce a pure strain of the micro-organism.

If bacteria are cultured in a liquid medium, their rapid reproduction soon causes the liquid to become cloudy. If this culture is repeatedly diluted, the concentration of bacteria can be reduced to a point where pouring a little of it on to an agar plate will produce isolated colonies, each derived from a single bacterium. If the dilution is known, the population of bacteria in the original colony can be calculated.

Viruses reproduce only in living cells so they cannot be cultured on non-living media. One technique is to inoculate them into living hens' eggs (Figure 31.18); another is to use tissue cultures of mammalian cells (p. 9).

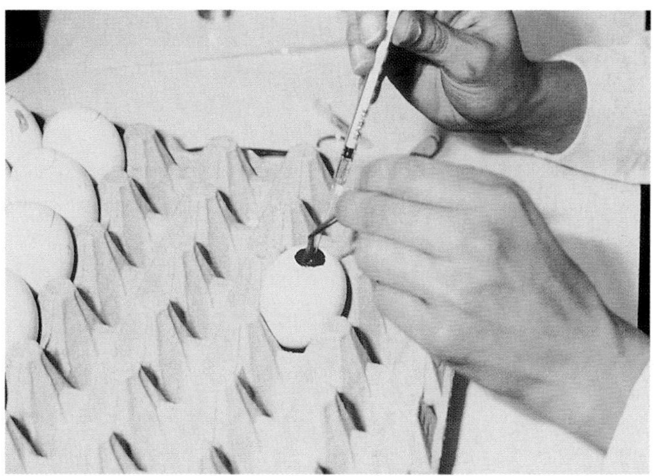

Figure 31.18 Culturing viruses in hens' eggs

Aseptic techniques

Fungal spores and bacteria are everywhere, in the air, on surfaces, on clothing and the skin. If you want to culture one particular species of micro-organism, it is important to ensure that all other species are excluded. This means using **aseptic techniques**. The glassware and the culture media must be sterilized, and precautions must be taken to avoid allowing bacteria from the air or from the experimenter to contaminate the experiments.

The glassware and media are sterilized by heating them to temperatures which destroy all bacteria and their spores. Contamination is avoided by handling the apparatus in a way which reduces the chances of unwanted bacteria entering the experiments.

Practical work

1 Culturing bacteria

The glassware must be sterilized by super-heated steam in an autoclave or pressure cooker for 15 minutes at 1 kg/cm^2 so that the unwanted bacteria on the glassware are destroyed.

1.5 g agar is stirred into 100 cm^3 of hot distilled water and 1 g beef extract, 0.2 g yeast extract, 1 g peptone and 0.5 g sodium chloride are added. The mixture is sterilized in an autoclave and poured into sterile petri dishes which are covered at once and allowed to cool.

Precautions Culture methods can give rise to very dense colonies of potentially harmful bacteria. **All bacteria should be treated as if they were harmful.** When the petri dishes have been inoculated with the bacterial samples, the lids should be sealed in place with adhesive tape and not removed until the plates have been sterilized again at the end of the experiment. Any colonies which appear must be examined with the lids still on the dishes.

Inoculating the plates If a little cooked potato or other vegetable matter is allowed to rot for a few days in water, bacteria will arrive and grow in the liquid. A drop of this liquid is picked up in a sterile wire loop[1] and streaked lightly across the surface of the agar jelly, lifting the lid of the dish just far enough to admit the wire loop, but not enough to let bacteria from the air fall on to the agar (Figure 31.19). The lid is then sealed on the dish with adhesive tape, and the dish is kept, upside-down[2], in an incubator (no higher than 25 °C) or similar warm place for about 2 days. A 'control' dish is left unopened.

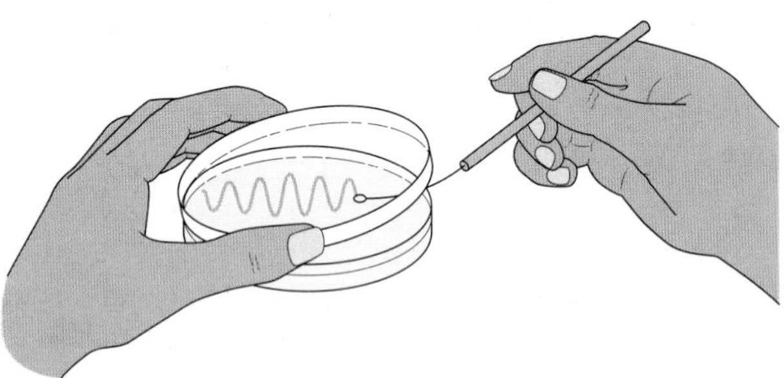

Figure 31.19 Lift the lid as little as possible

Bacterial colonies should grow on the streaked dish, following the path taken by the loop (Figure 31.20). There should be no growth of bacteria in the control dish, showing that the bacteria in the experiment came from the inoculating liquid and not from the glassware or the medium.

All the dishes, still sealed, are then sterilized once more in the autoclave before they are washed up.

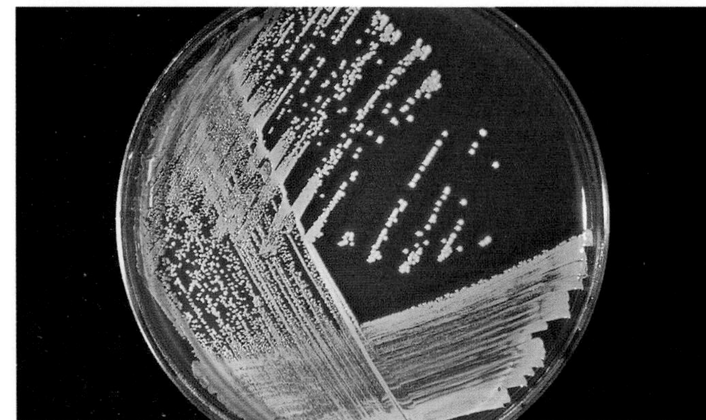

Figure 31.20 A 'streak plate'. Notice how the bacterial colonies have grown where the plate was streaked

2 The effect of antibiotics

The culture medium is made as before. When it is cool, but before it sets, a few drops of a pure bacteria culture[3] are added and thoroughly mixed in with the medium. The medium is then poured into sterile petri dishes as before and allowed to set.

Using sterile forceps, some discs containing antibiotics are placed on the surface of the agar and the lids sealed on. The dishes are then incubated at about 25 °C for 24 hours.

The bacteria, which are dispersed in the agar, will grow and make the medium look cloudy, but in regions surrounding some of the antibiotic discs, bacterial growth will have been suppressed and the agar will look clear (Figure 31.21).

If the bacteria are able to grow right up to the edge of a disc, it means that the antibiotic in that disc is unable to suppress the growth of the species of bacteria used in the trial.

1 The wire loop is sterilized by heating it to redness in a Bunsen flame.
2 If water condenses in the petri dish, it will fall on the lid and not spread over the surface of the agar.
3 e.g. *Escherichia coli* or *Staphylococcus albus* from a reputable supplier.

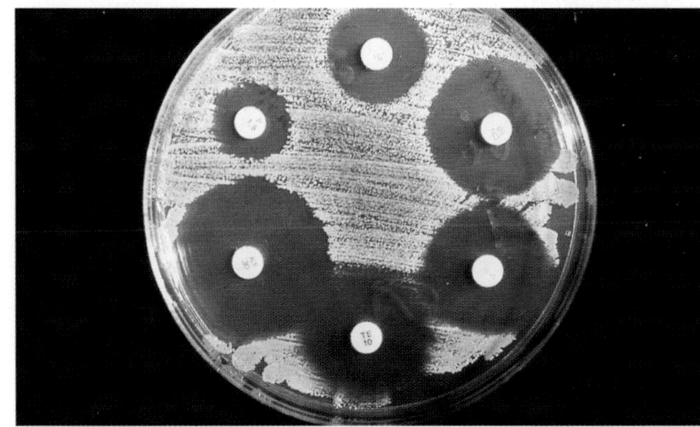

Figure 31.21 Testing antibiotics. Antibiotics have diffused out from the discs and suppressed the growth of bacteria

Question

1 Suggest reasons for the variation in extent of the clear areas round the disc in Figure 31.21.

3 Growing mould fungi

Moisten a piece of stale bread, place it in a petri dish base and leave it exposed to the air. After a day, moisten it again if it has dried out, and then cover it with a beaker or small jar. In the humid atmosphere inside the jar, colonies of mould fungi should grow on the bread within a few days.

The experiment can be duplicated by using different food material, e.g. pieces of banana or cooked apple.

4 Culturing fungi

Mould fungi or even small toadstools can be cultured on dishes of agar as described for bacteria on p. 290. Follow the same techniques as described for bacteria, but use vegetable agar in place of nutrient agar. The vegetable agar is made simply by dissolving 2 g agar and 25 cm^3 tomato juice in 100 cm^3 water. The vegetable agar and petri dishes are sterilized in an autoclave. The cool, but still liquid agar is poured into the sterile dishes and allowed to set.

The dishes are then exposed to the air or inoculated with fungus spores or hyphae using a sterile wire loop as shown in Figure 31.19. The fungi may be selected from one of the colonies that grew in Experiment 3.

The dishes are incubated for 24 hours at about 35 °C or left at room temperature for 2 or 3 days.

5 Starch digestion by a mould fungus

Prepare starch-agar by dissolving 0.3 g starch powder and 1 g agar in 100 cm^3 water. If the experiment can be examined 1 or 2 days after setting up, there is no need to sterilize the agar or the glassware.

Pour the agar, when cool, into petri dishes, replace the lids and allow the medium to set.

Inoculate the agar with fungal spores or hyphae, by making one or two strokes across it with a sterile wire loop that has been dipped into a mature fungal colony, as described for Experiment 1. Replace the lid and leave the dish for 1 or 2 days.

After this time, the fungus should be seen growing along the lines of the streaks.

Remove the lid, pour iodine solution into the dish to cover the agar and leave for a minute or two. Finally wash away the iodine with water.

Iodine solution will turn starch-agar blue. So, any areas which remain clear have no starch in them and it is reasonable to assume that the fungus has secreted an enzyme into the agar and digested the starch.

A control should be set up at the same time as the experiment, by streaking a starch-agar plate with a wire loop that has been heated to redness in a Bunsen flame. This plate should also be tested with iodine solution.

For experiments on yeast, see p. 23.

Checklist

- Bacteria are single cells; they have a cell wall, cytoplasm and a single chromosome.
- Bacteria produce enzymes which digest the surrounding medium.
- Bacteria reproduce by cell division.
- Bacteria are killed by heat but their spores survive.
- Most bacteria are saprotrophs and help to bring about decay.
- Bacterial decay releases essential substances for recycling.
- Viruses are smaller than bacteria and cannot, strictly, be classed as living organisms.
- Each virus particle consists of a DNA or RNA core enclosed in a protein coat.
- Viruses can reproduce only inside a living cell.
- Viruses take over the host cell's physiology and make it produce new virus particles.
- Viruses and bacteria can cause disease.
- Tetanus, pneumonia, typhoid and syphilis are bacterial diseases.
- Influenza, colds, herpes and AIDS are viral diseases.
- Fungi are formed from thread-like hyphae rather than cells.
- The branching hyphae produce a network called a mycelium.
- The hyphae have walls, cytoplasm, nuclei, vacuoles, organelles and inclusions.
- Fungi secrete enzymes into their food and absorb the digested products.
- Saprotrophic fungi digest dead organic matter.
- Parasitic fungi digest living tissues.
- The saprotrophic bacteria are important as decomposers; they release essential nutrients from dead organic matter.
- Parasitic fungi cause plant diseases some of which affect our crops.
- Fungi reproduce asexually by releasing spores which can grow into new mycelia.
- Yeasts are single-celled fungi which are important in brewing, baking and other forms of biotechnology.
- Yeasts respire anaerobically, causing a fermentation process which produces CO_2 and alcohol.

Characteristics of living organisms

All living organisms, whether they are single-celled, many-celled, plants or animals, do the following things:

Feed They may take in solid food as animals do, or digest it first and absorb it later like fungi do, or build it up for themselves like plants do (p. 293).

Breathe They take in oxygen and give out carbon dioxide. This exchange of gases takes place between the organism and the air or between the organism and water. The oxygen is used for respiration (p. 298).

Respire They break down food to obtain energy. Most organisms need oxygen for this (p. 19).

Excrete Respiration and other chemical changes in the cells produce waste products such as carbon dioxide. Living organisms expel these substances from their bodies in various ways (p. 131).

Grow Bacteria and single-celled creatures increase in size. Many-celled organisms increase the numbers of cells in their bodies, become more complicated and change their shape as well as increasing in size (p. 306).

Reproduce Single-celled organisms and bacteria may simply keep dividing into two. Many-celled plants and animals may reproduce sexually or asexually (p. 301).

Respond The whole animal or parts of plants respond to stimuli (p. 317).

Move Most single-celled creatures and animals move about as a whole. Fungi and plants may make movements with parts of their bodies (p. 311).

The next five chapters describe some of these characteristics in different living things.

32 Feeding

Types of nutrition
Autotrophic and heterotrophic.
Saprotrophs
Parasites
Animals
Filter-feeders, insects, mammals, carnivores, herbivores and omnivores.
Symbiosis and mutualism

All living organisms need food. Some of this food is used to make new tissues for growth and replacement but most of it is used to provide energy.

Energy is obtained from food by breaking it down chemically to carbon dioxide and water (see p. 19).

Types of nutrition

There are two principle methods of obtaining food, called **autotrophic** and **heterotrophic** nutrition. Autotrophic organisms (**autotrophs**) build up all the organic molecules they need from simple inorganic substances. Plants are autotrophs. They build up large, complex molecules (carbohydrates, lipids and proteins) from small inorganic molecules (carbon dioxide, water and mineral salts) by photosynthesis. In any ecosystem, it is the autotrophs which are the producers (p. 225).

Heterotrophic organisms (**heterotrophs**) use ready-made organic compounds as their food source. These organic compounds will have been made, originally, by autotrophs (mainly plants). The heterotrophs digest the organic compounds to simpler substances and absorb the products into their bodies. Animals, fungi and some bacteria and protoctista are heterotrophs.

Animals take in food, in the form of complex organic molecules (carbohydrates, lipids and proteins, p. 11) and digest them to simpler compounds (e.g. glucose and amino acids, p. 98) which can be absorbed. Fungi and heterotrophic bacteria also digest their food source but do so by a process of external digestion.

In all plants and other autotrophs the process of photosynthesis is similar. The heterotrophs have a variety of methods of obtaining food.

Saprotrophs

These are organisms which feed on dead and decaying matter. Often they release digestive enzymes into their food and absorb the soluble products back into their bodies. Bacteria and fungi are examples of saprotrophs which digest dead wood, rotting vegetation or the humus in the soil (see p. 227).

Parasites

A parasite is an organism which derives its food from another organism, called the **host**, while the host is still alive. **Ectoparasites** live, for most of their life cycle, on the surface of their host: **endoparasites** live inside the host.

A flea is an ectoparasite which lives in the fur or feathers of mammals or birds and sucks blood from the skin. An aphid (greenfly) is a parasite on plants and sucks food from the veins in their leaves or stems. A tapeworm is an endoparasite living in the intestine of a vertebrate host and absorbing the host's digested food. In some cases the host may be weakened by the presence of the parasite, or its metabolism may be upset by the parasite's excretory products. In many cases, however, the host appears to suffer no serious disadvantage. Often, this will depend on how many parasites the host is carrying.

Animals

Animals feed on either plants or other animals. There is a variety of ways in which they obtain their food.

Filter-feeders

Many unrelated animals which live in water filter it and ingest the particles which they extract from it (Figure 32.1).

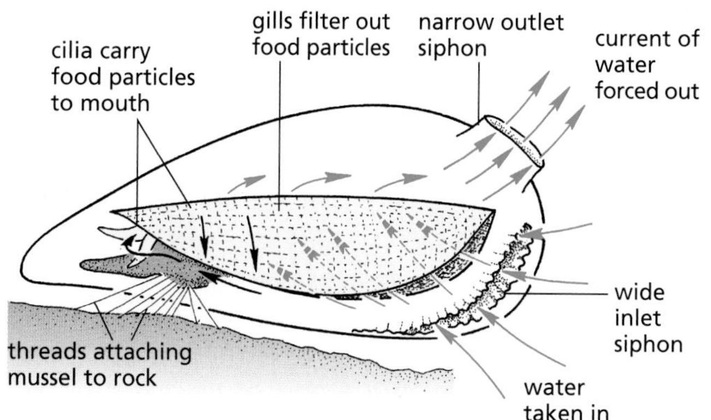

Figure 32.1 Filter-feeding in a mussel

Figure 32.2 Mussels filter-feeding. The water is going in through the opening fringed with 'tentacles' and coming out of the smooth-edged opening. If anything touches the tentacles, the mussel closes its shells at once. Some of the mussels have barnacles growing on them

The mussel (Figures 32.1 and 32.2) has large gills covered with cilia. The beat of the cilia draws a current of water in between the shells and forces it through the gills. The gills filter out any food particles and the filtered water is expelled through the siphon. The cilia on the gills also carry the trapped food particles forward to the mouth.

Mussels stay in one place and create a water current. Many animals swim, filtering the water as they go. As the herring swims along, it takes in water through its mouth and lets it out through the gill covers (see Figure 33.7 on p. 300). This flow of water is partly for gaseous exchange as described on p. 300, but the water is filtered by gill rakers (Figure 32.3). These form a bony grid between the gills which allows water to pass out but traps small shrimps and other crustacea swimming in the surface waters of the sea. Certain types of whale filter-feed in a similar way.

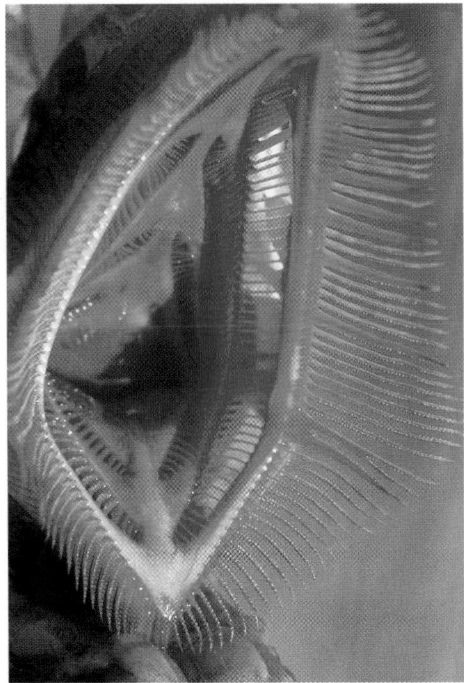

Figure 32.3 Gill rakers, dissected out of a herring. When water passes between the gill bars, food is trapped on them

Insects

Insects have a great variety of feeding methods. Grasshoppers, locusts and caterpillars have simple jaws with which they bite off pieces of leaf small enough to be ingested.

The butterfly's mouth parts form a long, fine tube which reaches into flowers and sucks the nectar (Figure 32.4).

The greenfly also has tubular mouth parts but these are used to pierce through leaves and stems of plants and suck food from the phloem (Figure 32.5).

Figure 32.4 Butterfly feeding. The mouth parts form a tube, like a fine drinking straw, which probes into the flower to suck nectar

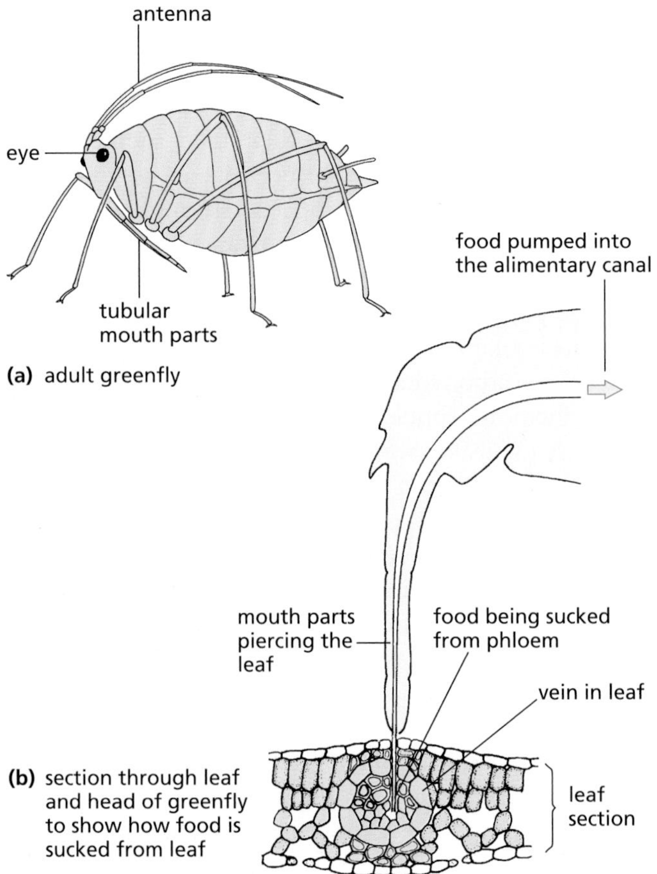

Figure 32.5 Feeding in the aphid

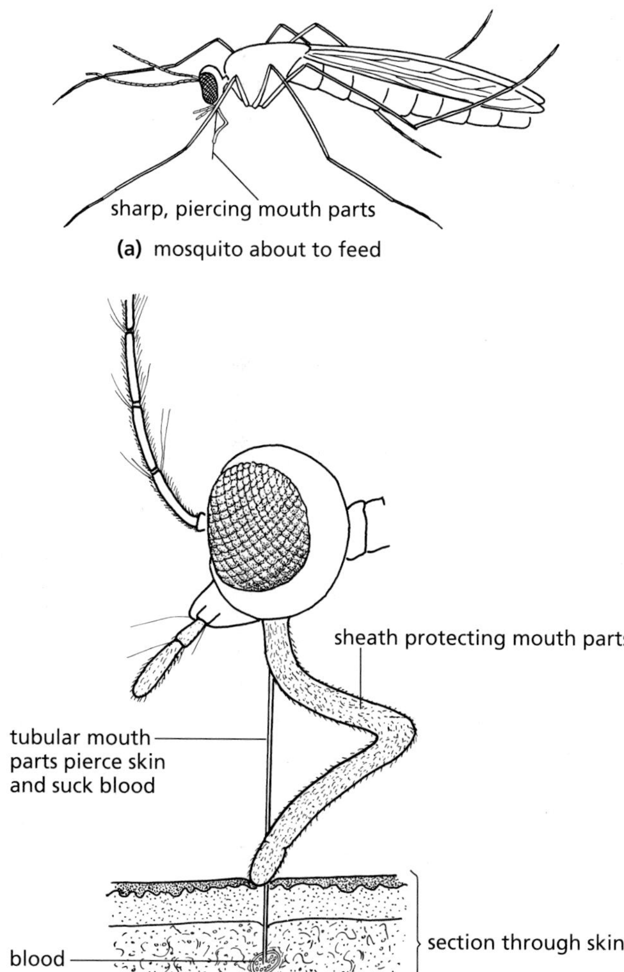

Figure 32.6 Feeding in the mosquito

The tubular mouthparts of the female mosquito are adapted for piercing the skin and sucking blood from the skin capillaries (Figure 32.6).

The feeding method of the housefly is described on p. 339.

Mammals

Mammals may be classed as **carnivores** (flesh-eaters), **herbivores** (plant-eaters) or **omnivores** (mixed diet).

Carnivores

A carnivore's teeth are adapted to catching and killing its prey, cutting flesh from the carcase, and cracking bones, so that the food is reduced to portions that can be swallowed. The position and shapes of its teeth are adapted to these varied functions, as shown in Figure 32.7.

Herbivores

A herbivore's teeth are adapted to the single function of collecting vegetation and grinding it up finely. Sheep and cows have no upper incisors but grip the grass between the lower incisors and the top gum. The canines are similar in shape and function to the incisors. There is a gap between the canines and pre-molars which allows the tongue to manipulate the food between the back teeth as they slide over each other and grind it up. The premolars and molars are similar to each other and throughout the life of the animal they grow as fast as they are worn down by their grinding action.

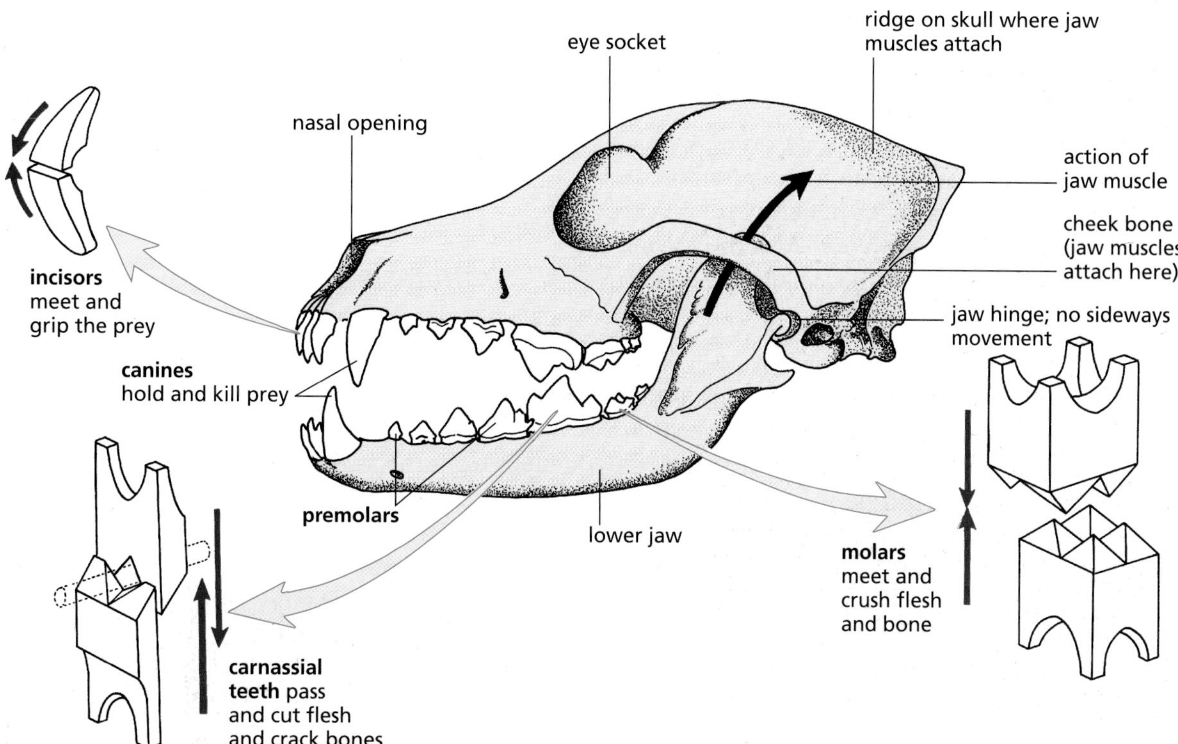

Figure 32.7 Dog's skull and action of teeth

In herbivores, the alimentary canal also is adapted to the diet of grass and leaves. Mammals cannot make an enzyme for digesting the cellulose of plant cell walls. They depend upon bacteria and protoctista living in their alimentary canals to do the digestion for them. These micro-organisms digest the cellulose slowly and, later on, the micro-organisms are digested by the herbivore. This is an example of mutualism (see below).

In cattle and sheep, there are several compartments to the stomach (Figure 32.8). The grass is chewed, swallowed and directed into the first compartment, called the **rumen**. In the rumen, the micro-organisms start to digest the grass. After a time the partly digested grass, called the **cud**, is sent back up the gullet to the mouth and chewed some more. When it is swallowed it goes, not to the rumen, but to the main stomach and so on its way through the rest of the alimentary canal.

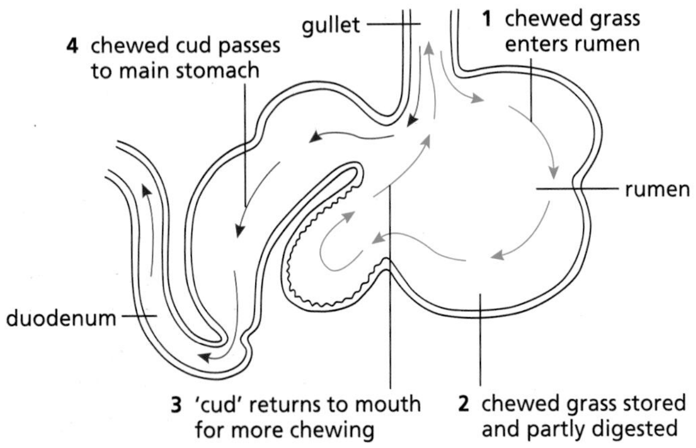

Figure 32.8 Stomach of sheep

Rabbits, at night, produce soft, moist faeces which they promptly eat. In this way, the grass is put through the digestive system twice. The dry, hard pellets produced during the day are not eaten.

Digestion of vegetation is a very slow process which is why the rabbit's double digestion is important. In addition, the alimentary canal of a herbivore is very much longer than that of a carnivore. Also the caecum and appendix, where much of the cellulose digestion takes place, are very much larger.

Omnivores

Pigs and humans are omnivores but humans cannot digest the cellulose in plant material. Human teeth are not used for catching, holding, killing and tearing up prey, and we cannot cope with bones. Thus, although we have incisors, canines, premolars and molars, they do not show such big variations in size and shape as the dog's. Figure 32.9 shows the position of teeth in the upper jaw and Figure 32.10 shows how they appear in both jaws when seen from the side.

Our top incisors pass in front of out bottom incisors and cut pieces off the food as when biting into an apple or taking a bite out of a piece of toast.

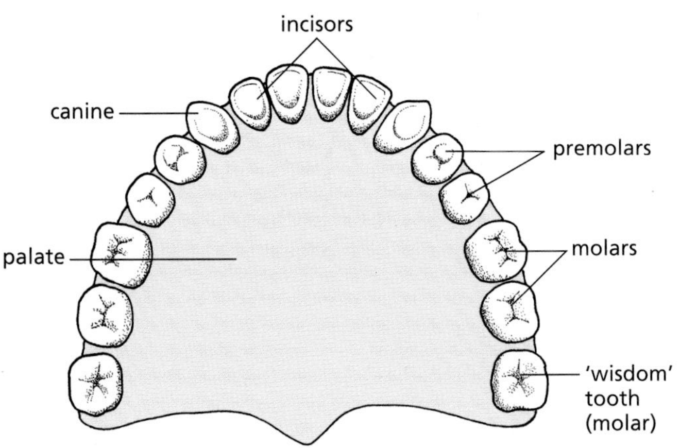

Figure 32.9 Teeth in human upper jaw

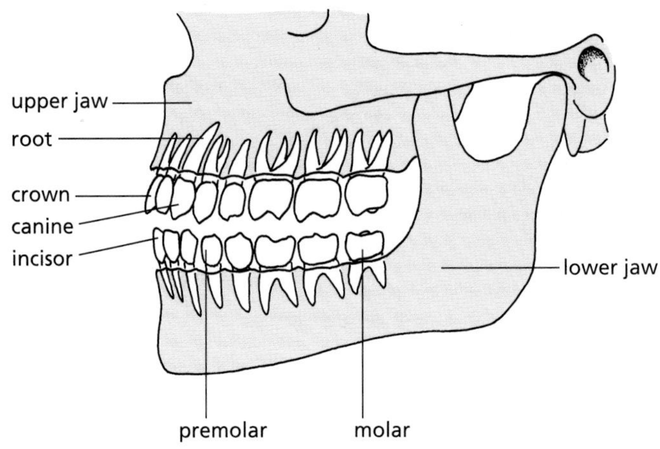

Figure 32.10 Human jaws and teeth

Our canines are more pointed than the incisors but are not much larger and they function like extra incisors.

Our premolars and molars are similar in shape and function. Their knobbly surfaces meet when the jaws are closed, and crush the food into small pieces. Small particles of food are easier to digest than large chunks.

Symbiosis and mutualism

The term 'symbiosis' strictly refers to two unrelated organisms living more or less permanently together, irrespective of whether one of them is a parasite. Mutualism implies that both organisms derive some benefit from the association.

For example, in the stomachs of cattle and sheep there live large numbers of bacteria. They cause no symptoms of illness and they are thought to be of value to the animal because they help to digest the cellulose in its food. The cow benefits from the relationship because it is better able to digest grass. The bacteria are thought to benefit by having an abundant supply of food, though they are themselves digested when the grass moves along the cow's digestive tract.

The nitrogen-fixing bacteria in root nodules (Figure 32.11) provide a further example of symbiosis. The plant benefits from the extra nitrates that the bacteria provide, while the bacteria are protected in the plant's cells and can also use the sugars made by the plant's photosynthesis.

It is often difficult to produce good evidence that one or both organisms are getting some benefit or suffering damage from these relationships. Consequently, it is not always easy to make a distinction between parasitism and mutualism. The argument is about words rather than biology. It is far more important to try and find out what the relationship between organisms really is rather than to argue about what word to apply to the relationship.

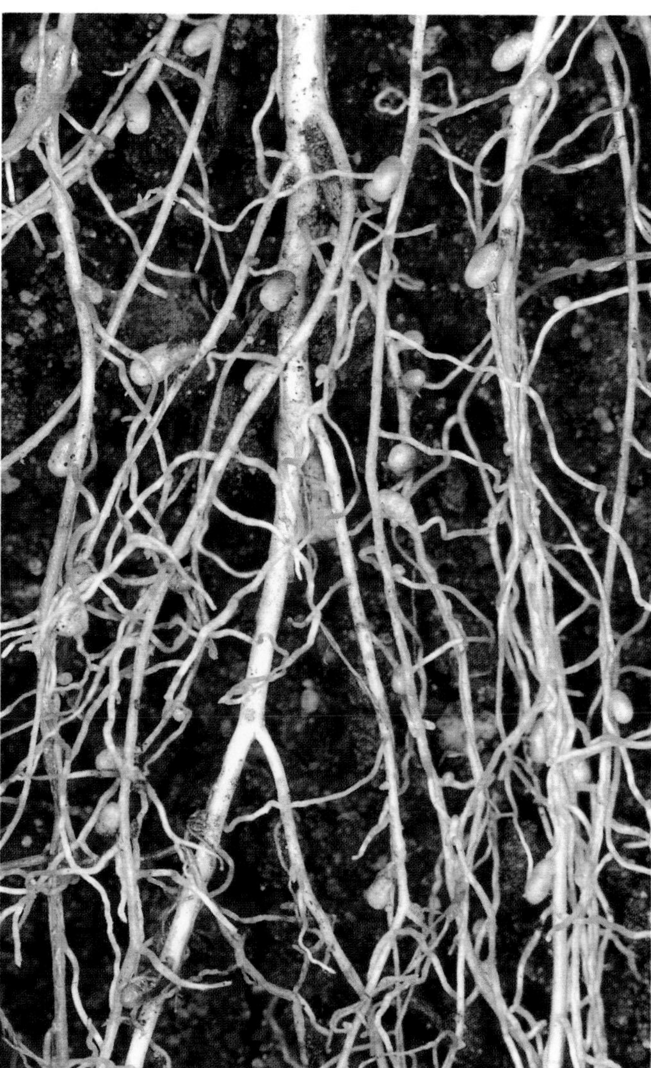

Figure 32.11 Root nodules of white clover – a leguminous plant

Questions

1 What type of nutrition is characteristic of
- **a** an apple tree,
- **b** a toadstool,
- **c** a human,
- **d** a mosquito,
- **e** a *Streptococcus*?

2 Apart from carbon dioxide, water and salts, what do plants need in order to make their food?

3 In animals, where does digestion and absorption take place?

Checklist

- All living organisms need to obtain food to make new tissue and to provide energy.
- All green plants make their food by photosynthesis.
- Saprophytes feed on decaying organic matter.
- Parasites live in or on another organism and obtain food from it.
- Animals eat plants or other animals. They ingest, digest and absorb their food.
- There is a great variety of ways in which animals obtain their food, such as filter-feeding, chasing and catching prey or chewing up vegetation.
- The teeth and alimentary canals of animals are adapted to deal with the kind of food they eat.
- Mutualism applies to two unrelated organisms living together, both deriving benefit from this association.

33 Breathing

Gaseous exchange in flowering plants

Surface area and volume
Diffusion or special respiratory organs, surface area/volume ratio.

Earthworm and frog
Breathing through the skin.

Fish
Breathing by gills.

In order to get energy from their food, animals and plants have to break it down to carbon dioxide and water by the chemical process called respiration (p. 19). One form of respiration, aerobic respiration, uses oxygen for this process. The oxygen is obtained by gaseous exchange (p. 126) with the air or water around the organism. The same process gets rid of carbon dioxide produced by respiration.

In some animals, gaseous exchange is helped by a process of ventilation (p. 124) but many plants and animals depend on diffusion alone.

Very small organisms, like bacteria and protoctista, have no need of ventilation. The distances over which the oxygen and carbon dioxide have to travel in these creatures are so small that diffusion is fast enough to meet their needs (Figure 33.1).

Larger animals need special organs for gaseous exchange and some method of ventilating these organs.

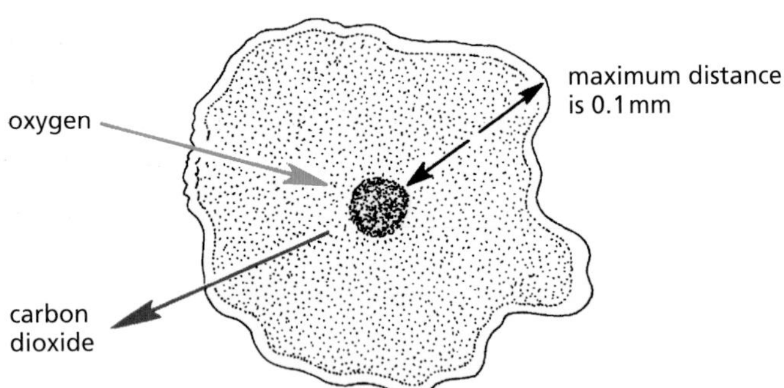

Figure 33.1 The distance is so small in *Amoeba* that diffusion is a rapid enough process for the cell's needs

Gaseous exchange in flowering plants

Plants do not have special breathing organs. Because they do not move about, like animals, they do not use up oxygen or produce carbon dioxide very rapidly. Their leaves present a very large surface area to the air, and diffusion of oxygen and carbon dioxide through the stomata into the intercellular spaces (p. 53) is fast enough for the respiration going on in their cells. Because most plant leaves are thin, the distances for diffusion are also very short (see also pp. 41 and 53).

Plant stems exchange gases with the air through their stomata or structures called lenticels. Roots use oxygen dissolved in the soil water which they absorb.

Surface area and volume

The ratio of surface area to volume in a small organism is greater than the ratio in a large organism.

As a crude example, imagine an organism that has a volume of 8 cm^3 and a surface area of 24 cm^2 (Figure 33.2a). The ratio of surface area to volume is 24/8 = 3. A smaller organism, having half the volume (4 cm^3) has a surface area of 16 cm^2. Its surface area to volume ratio is 16/4 = 4 (Figure 33.2b).

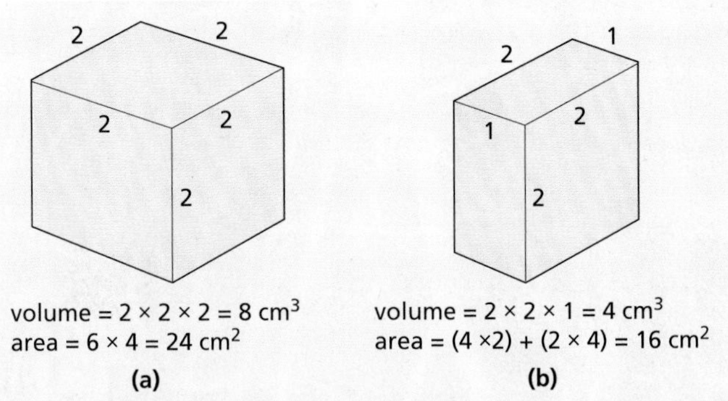

Figure 33.2 Surface area and volume

The demand for oxygen depends on the volume of active cytoplasm but the supply of oxygen depends on the exposed surface area. Consequently smaller organisms have the advantage of a relatively large absorbing surface to meet the oxygen requirements of a comparatively small volume of respiring tissue.

In larger organisms, not only is the ratio of surface area to volume low, but also the skin is not very permeable to gases and there is a large distance between the skin and the internal organs. Absorption through the skin plus diffusion through the tissues would not be rapid enough to meet the demands of the living cells.

In these organisms, a blood circulatory system largely replaces diffusion as the means of delivering oxygen to the tissues.

Earthworm and frog

Some comparatively large organisms still rely on their skin as a means of absorbing oxygen. Earthworms and frogs are examples, though the frog has lungs which can supplement the supply of oxygen when necessary (Figures 33.3 and 33.4). In both cases, the thin, moist skin is supplied with a network of capillaries which absorb oxygen and deliver it, via the circulatory system, to the rest of the body (Figure 33.5). Earthworms are fairly inactive animals, so their oxygen requirements are easily met by the relatively small absorbing surface of the skin.

Figure 33.3 Earthworm in its burrow. The thin, moist skin of the earthworm makes it vulnerable to water loss. It spends most of the day in its burrow where evaporation is at a minimum

Figure 33.4 Frog. The skin can exchange gases with either the air or water. Frogs spend most of their life on land but their moist skins restrict them to damp, shady habitats

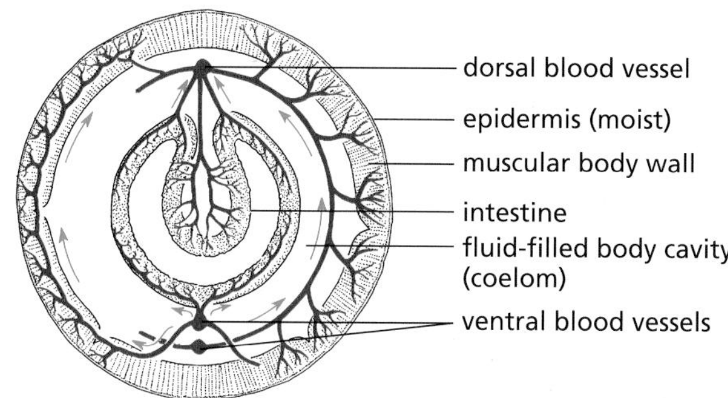

(a) section through a worm to show blood supply to the skin and intestine

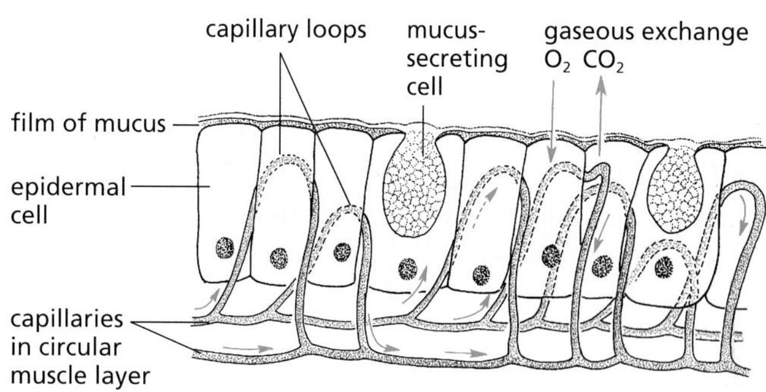

(b) high power section through the skin to show capillary network

Figure 33.5 Gaseous exchange in an earthworm

Active animals have special absorbing surfaces such as lungs or gills. These are internal organs but they still provide a very large absorbing surface. In the gills of a fish, this is achieved by a large number of branched filaments. In mammals, the surface area of the lungs is greatly increased by thousands of tiny air pockets (alveoli, p. 124).

All respiratory surfaces have a dense network of capillaries to absorb oxygen and get rid of carbon dioxide. Steep diffusion gradients (p. 27) for oxygen and carbon dioxide are maintained by ventilation mechanisms which constantly exchange the air or water in contact with the respiratory surface.

Fish

Although the water molecule (H_2O) contains oxygen, this cannot be used by aquatic organisms. The oxygen they breathe comes from the atmosphere and is dissolved in the water.

Fish absorb dissolved oxygen from the water by means of gills. The herring has four gills on each side of the head underneath a bony gill cover (Figure 30.15 on p. 276). Each gill consists of a curved bar with gill filaments sticking out from one side and gill rakers projecting forwards from the other side (Figure 33.6, overleaf). The gill filaments are branched structures with blood capillaries running in them. The branched filaments present a very large surface to the water for absorbing oxygen into the blood and getting rid of carbon dioxide.

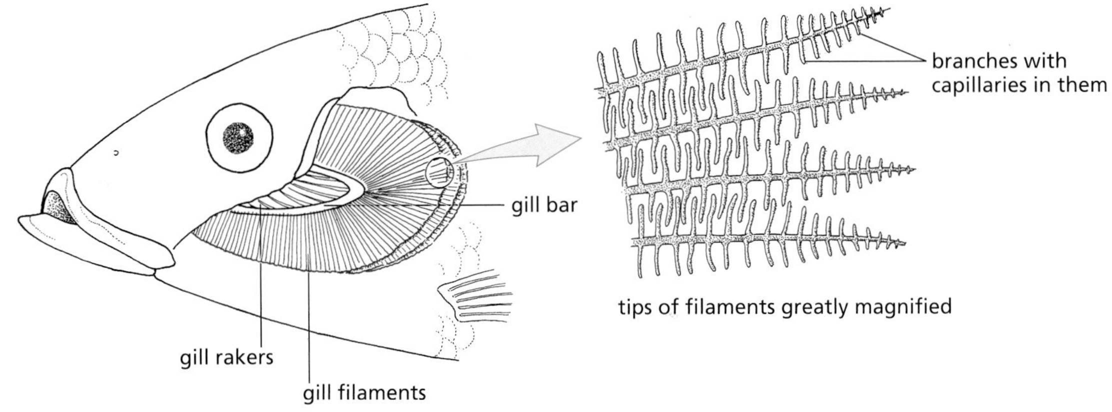

Figure 33.6 Herring gills. (The gill cover has been cut away to show the gills.)

By movements of the floor of the mouth and the gill covers, a fish takes water in through the mouth, and forces it between the gills and out through a gap between the gill cover and its body (Figure 33.7). This method of ventilation changes the water in contact with the gills, so bringing fresh supplies of oxygen and carrying away carbon dioxide.

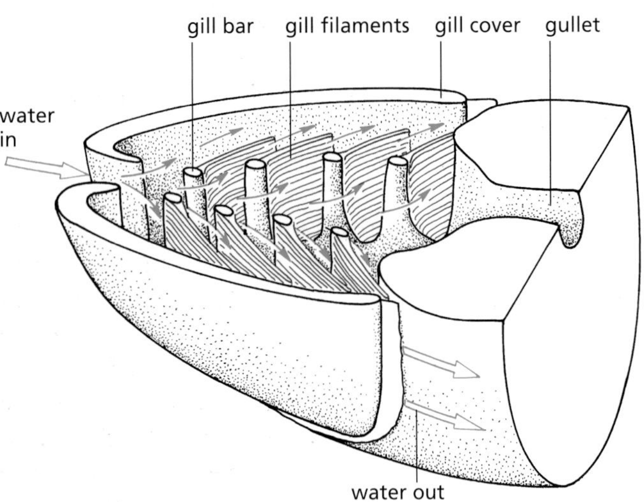

Figure 33.7 Diagram to show ventilation of gills

The lungs of mammals and gills of fish both show the basic requirements of an efficient structure for gaseous exchange. These structures need a large surface area. The gills achieve this by having hundreds of filaments, with many branches on each filament. In the lungs, it is the millions of alveoli (p. 124) which give a large surface area.

In both gills and lungs, there is a dense network of capillaries which enable rapid exchange of oxygen and carbon dioxide between the blood and the water or air. The capillaries are separated from the water or air by only a thin epithelium (single layer of cells), so that diffusion of the gases is as rapid as possible.

Questions

1. What would be the surface area to volume ratio in an organism half the size of the one in Figure 33.2b, p. 298?
2. What would be the maximum diffusion distance for oxygen in the bacterium in Figure 31.1 on p. 283?
3. In frogs, gaseous exchange takes place continually through the skin. In what conditions would you expect a frog to use its lungs?
4. Why do large, active animals need special breathing organs and a circulatory system?
5. In what way might the method of ventilation of a fish's gills be considered more efficient than the ventilation of lungs? (See 'Residual volume' on p. 125.)
6. Ventilation of gills and lungs requires muscular activity. Where does this muscular activity take place in
   a mammals,
   b fish?

Checklist

- Diffusion is rapid enough to supply microscopic organisms with the oxygen they need.
- Plants exchange carbon dioxide and oxygen with the air by diffusion through their stems, roots and leaves.
- The ratio of surface area to volume decreases with increasing size of an organism.
- Respiratory organs and blood circulation overcome the disadvantage of this low ratio in large animals.
- All breathing organs have a large surface, thin epithelium and many capillaries.
- Ventilation maintains a favourable diffusion gradient at a respiratory surface.
- Frogs and earthworms breathe through their skin.
- Fish force water over their gills, which absorb the dissolved oxygen.

34 Reproduction

Sexual reproduction
Plants and animals, external and internal fertilization.

Asexual reproduction
Animals, fungi, flowering plants.

Sexual and asexual reproduction compared

No organism can live for ever, but part of it lives on in its offspring. Offspring are produced by the process of reproduction. This process may be **sexual** or **asexual** (see below), but in either case it results in the continuation of the species.

Sexual reproduction

The following statements apply equally to plants and animals. Sexual reproduction involves the production of sex cells. These sex cells are called **gametes** and they are made in reproductive organs. The process of cell division which produces the gametes is called **meiosis** and is described on p. 185. In sexual reproduction, the male and female gametes come together and **fuse**, that is, their cytoplasm and nuclei join together to form a single cell called a **zygote**. The zygote then grows into a new individual (Figure 16.1 on p. 140).

Sexual reproduction in flowering plants

The male gamete is a cell in the pollen grain. The female gamete is an egg cell in the ovule. The process which brings the male gamete within reach of the female gamete (i.e. from stamen to stigma) is called **pollination**. The pollen grain grows a microscopic tube which carries the male gamete the last few millimetres to reach the female gamete for fertilization. The zygote then grows to form the seed. These processes are all described in more detail on pp. 67–72.

Sexual reproduction in animals

The male gamete is the sperm; the female gamete is the ovum. The release of the sperms and ova is co-ordinated in such a way that they come close to each other. This may involve behaviour patterns or mating or both. The sperms then swim to the ovum and fertilize it. The zygote grows into an embryo. Sexual reproduction in humans is described more fully on pp. 140–51.

In both plants and animals, the male gamete is microscopic and mobile (i.e. can move from one place to another). The sperms swim to the ovum; the pollen cell moves down the pollen tube (Figure 34.1). The female gametes are always larger than the male gametes and are not mobile. Pollination in the seed-bearing plants and mating in most animals bring the male and female gametes close together. Fertilization is the fusion of the gametes, and as a result the zygote undergoes rapid cell division to produce an embryo.

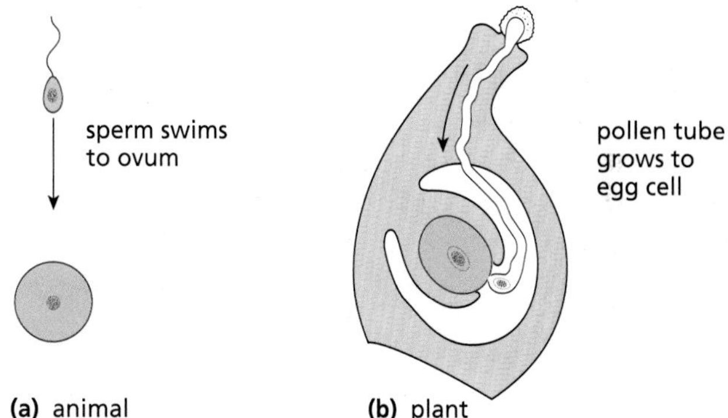

Figure 34.1 The male gamete is small and mobile; the female gamete is larger

External fertilization

In many animals which live or breed in water, the female sheds her unfertilized eggs into the water and the male releases sperms over them. The eggs are fertilized outside the body of the female and so the process is described as external fertilization. For fertilization to be successful, the eggs and sperms must be released at the same time and close to each other. This is usually achieved by a behaviour pattern in which the male and female are first attracted towards each other and then stimulate each other to produce gametes.

The male stickleback, in the breeding season, develops a red belly and blue eyes (Figure 34.2). These colour changes seem to keep other males at bay. The male digs a small hollow at the bottom of the pond and roofs it over with pieces of vegetation. When a mature female enters the male's territory, she is attracted by his bright colours. The male's next behaviour is triggered by the sight of her abdomen, swollen with eggs. He swims towards her and then down to the nest. She follows him and enters the nest (Figures 34.2 and 34.3, overleaf). The male taps the female's tail with his snout and this stimulates her to lay eggs. When she leaves the nest, the male enters and sheds sperms on the eggs. The nest and the courtship behaviour ensure that eggs and sperms are shed in the same place and at the same time.

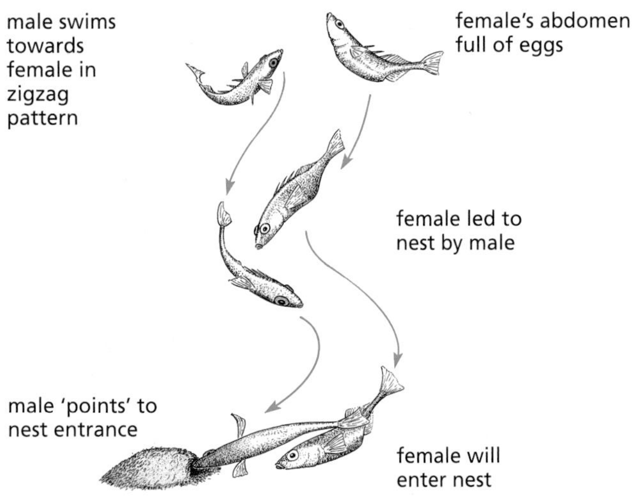
male swims towards female in zigzag pattern
female's abdomen full of eggs
female led to nest by male
male 'points' to nest entrance
female will enter nest

Figure 34.2 Courtship in the three-spined stickleback

Figure 34.3 Sticklebacks at the nest. The male's movements in the nest entrance induce the female to enter and lay eggs. Notice her abdomen swollen with eggs

Frogs and toads, in the spring, migrate to ponds. Here the male climbs on the back of the female and is carried about by her till she lays her eggs (Figure 34.4). When the male feels the eggs being laid, he releases his sperms which fertilize the eggs. This is external fertilization, but the male's behaviour in remaining attached to the female makes sure that the sperms are released at the same time as the eggs.

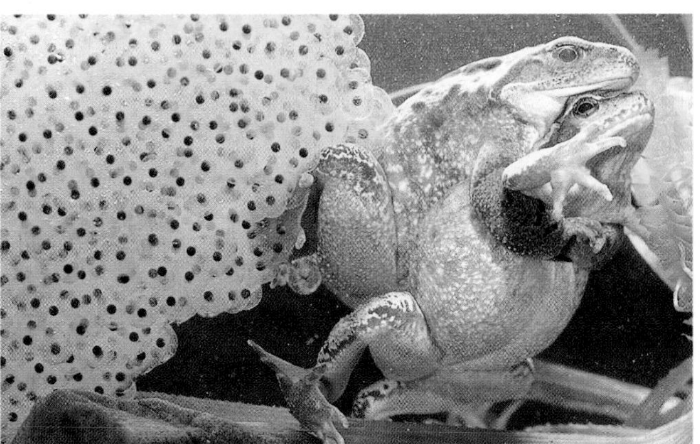

Figure 34.4 Frogs pairing. The male clings to the female's back and releases his sperm as she lays the eggs

Many animals, having laid their eggs, take no further part in their development. Others, birds and mammals for example, protect and feed their young till they can fend for themselves. This is called **parental care**.

Where there is little parental care, there is usually a large number of eggs (Figure 34.5). Because the parents do not protect the eggs and young or bring them food, many offspring will die. The large number of eggs, however, ensures that some offspring will survive to maturity. The parental care shown by birds and mammals gives the young a better chance of survival and they have small numbers of offspring.

Figure 34.5 Large numbers of offspring. One method of survival is to produce a large number of eggs, as in this egg-rope of the perch. Many of the young fish will die but enough will survive to continue the species

Internal fertilization

In reptiles, birds and mammals, the eggs are fertilized by the male placing sperms inside the body of the female. This is internal fertilization and it requires a behaviour pattern which brings male and female together at the right time so that sperms are released when the eggs are mature. The first step is **pair formation**, which ensures that the male and female stay together at least for the breeding period. There is usually a courtship pattern which brings the pair together in the first place, stimulates them to mate, and keeps them together while the young are reared (Figure 34.6). In birds, the male often feeds the female as part of the courtship ritual.

The reproductive organs have to develop in such a way that internal fertilization can occur. This usually means that the male animal has developed a penis which can be inserted into the female's reproductive organs in order to deposit the sperms.

Figure 34.6 Gannets at nest site. The beak 'fencing' ritual helps to maintain the pair bond

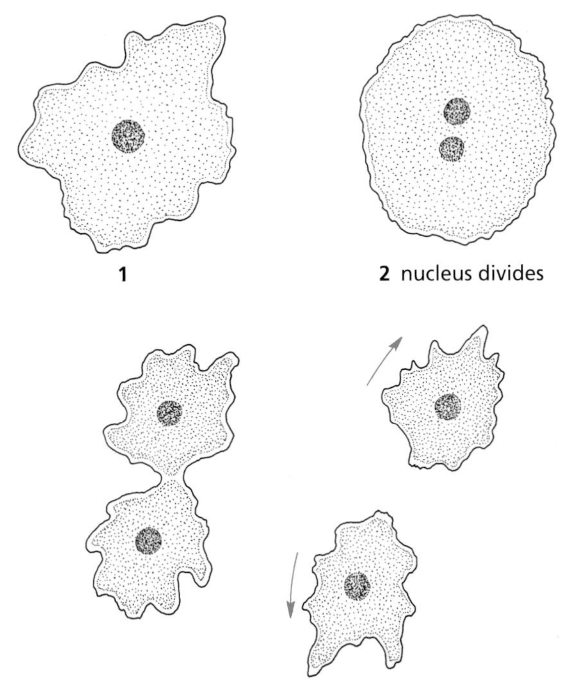

Figure 34.7 Asexual reproduction in *Amoeba*. This is called 'binary fission'. See 'Bacteria' (p. 284) and 'Yeast' (p. 288) for other examples of binary fission

Internal fertilization is essential for animals living on land. External fertilization here is impossible because the eggs and sperms would dry up when exposed to the air. In reptiles and birds, the egg is prevented from drying up by having a shell, so the sperms have to get to the egg before the shell is put on it. This means that internal fertilization is necessary. In mammals, the egg develops inside the body of the female after internal fertilization.

Other land-dwelling animals such as insects, spiders and snails rely on methods of internal fertilization for their reproduction.

Asexual reproduction

Asexual means 'without sex' and this method of reproduction does not involve gametes. In the single-celled protoctista or in bacteria, the cell simply divides into two and each new cell becomes an independent organism (Figure 34.7 and Figure 31.3 on p. 284).

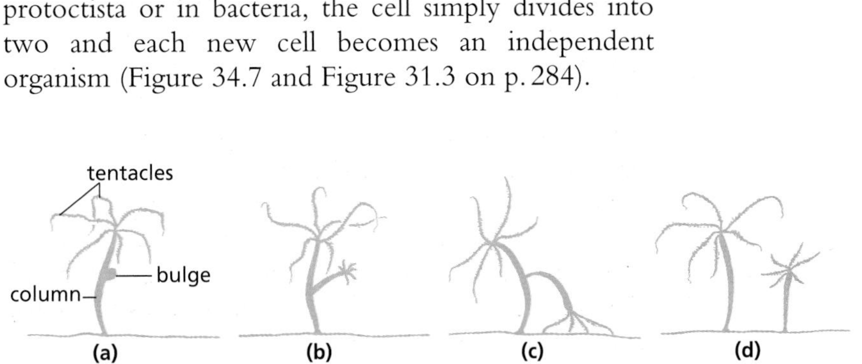

Figure 34.8 Asexual reproduction in *Hydra*
(a) a group of cells on the column start dividing rapidly and produce a bulge
(b) the bulge develops tentacles
(c) the daughter *Hydra* pulls itself off the parent
(d) the daughter becomes an independent animal

In more complex organisms, part of the body may grow and develop into a separate individual. For example, a small piece of stem planted in the soil may form roots and grow into a complete plant.

Asexual reproduction in animals

Some species of invertebrate animals are able to reproduce asexually.

Hydra is a small animal, 5–10 mm long, which lives in ponds attached to pondweed. It traps small animals with its tentacles, swallows and digests them. *Hydra* reproduces sexually by releasing its male and female gametes into the water but it also has an asexual method which is shown in Figure 34.8.

(e) *Hydra* with bud

Questions

1 Draw up a table with three columns as shown below. In the first column write
- a male reproductive organs,
- b female reproductive organs,
- c male gamete,
- d female gamete,
- e place where fertilization occurs,
- f zygote grows into.

Now complete the other two columns.

		flowering plants	mammals
a	male reproductive organs		
b	female reproductive organs		
c	male gamete etc.		

2 Which of the following animals would you expect to have external and which internal fertilization: butterfly, mussel, trout, sparrow, earthworm? In each case give the reason for your decision.

3 Why do you think courtship behaviour is necessary for the success of both internal and external fertilization?

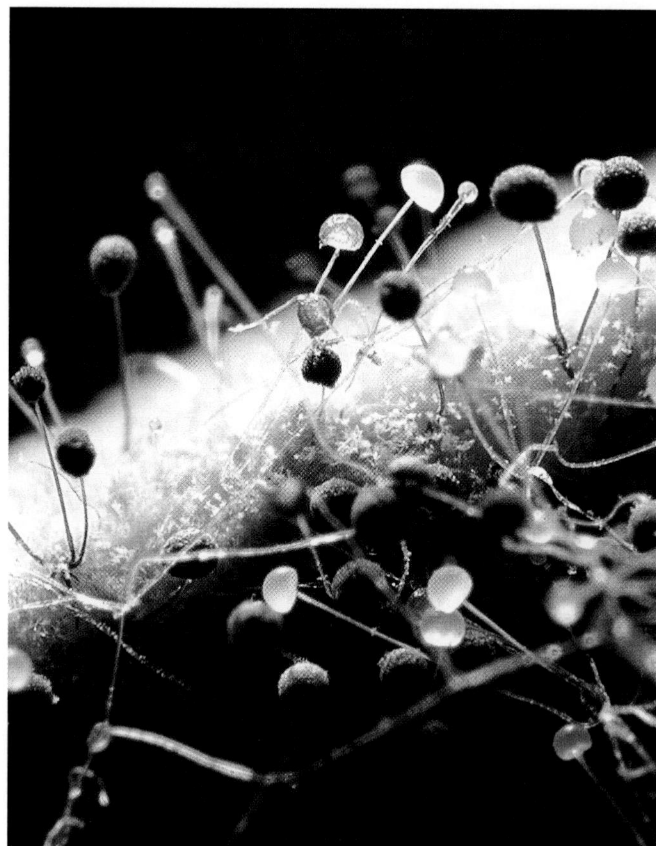

Figure 34.9 Asexual reproduction in *Mucor*. The black spheres are sporangia which have not yet discharged their spores (× 160)

Asexual reproduction in fungi

Fungi have sexual and asexual methods of reproduction. In the asexual method they produce single-celled, haploid spores. These are dispersed, often by air currents and, if they reach a suitable situation, they grow new hyphae which develop into a mycelium.

Penicillium and *Mucor* are examples of mould fungi which grow on decaying food or vegetable matter. *Mucor* feeds, grows and reproduces in a similar way to *Penicillium* (as described on p. 288) but the spores are produced in a slightly different way. Instead of chains of spores at the tips of the vertical hyphae, *Mucor* forms spherical sporangia, each containing hundreds of spores (Figure 34.9). These are dispersed on the feet of insects or by the splashes of rain drops.

The gills on the underside of a mushroom or toadstool (Figure 25.7 on p. 227) produce spores. Puffballs release clouds of spores (Figure 31.13 on p. 287).

Asexual reproduction in flowering plants (vegetative propagation)

Although all flowering plants reproduce sexually (that is why they have flowers), many of them also have asexual methods.

In most cases a lateral bud on the stem, instead of producing a flower or a leafy shoot, gives rise to a complete plant which eventually becomes independent of its parent.

Several of these asexual methods (also called 'vegetative propagation') are described on pp. 79–81. An unusual method is shown in Figure 34.10.

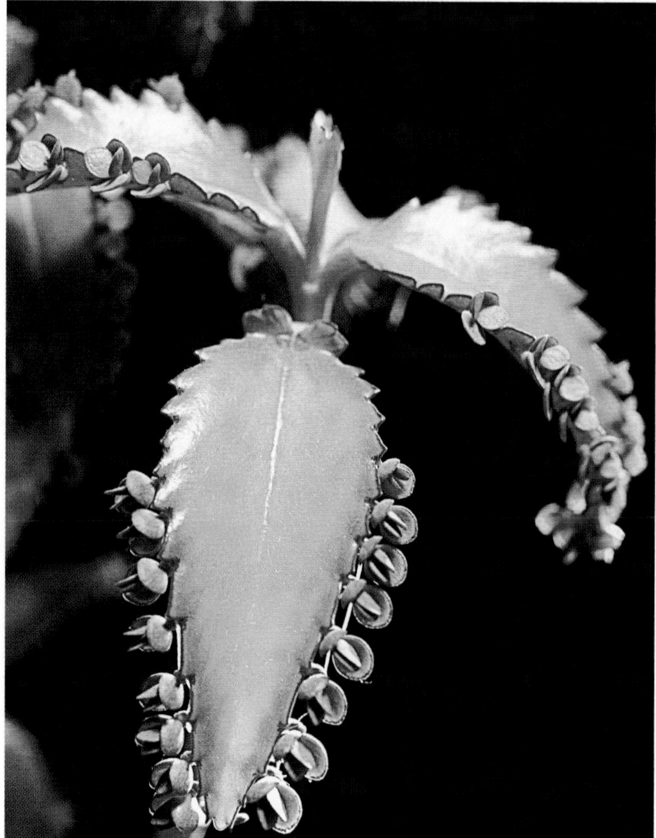

Figure 34.10 Bryophyllum. The plantlets are produced from the leaf margin. When they fall to the soil below, they grow into independent plants

Sexual and asexual reproduction compared

The relative advantages of asexual and sexual reproduction are discussed on p. 82 in the context of flowering plants. However, the points made there are equally applicable to most forms of sexual and asexual reproduction.

In asexual reproduction, the offspring are usually produced by mitosis (p. 182), mostly in diploid cells. In sexual reproduction the offspring result from the fusion of two haploid cells (gametes) produced by meiosis.

The outcome is that the offspring in asexual reproduction inherit identical sets of genes and there is no variation between them and their parent. This has the advantage of preserving the 'good' characteristics of a successful species from generation to generation. The disadvantage is that there is no variability for natural selection (p. 203) to act on in the process of evolution.

The new combinations of genes which result from meiosis and fertilization (p. 201) produce variety among the offspring. This has the disadvantage that some combinations will produce less successful individuals. On the other hand, there are likely to be some more successful combinations that have greater survival value or produce individuals which can thrive in new or changing environments.

In agriculture and horticulture, asexual reproduction is exploited to preserve desirable qualities in crops; sexual reproduction is exploited to produce new varieties of animals and plants by cross-breeding.

Questions

1 a In what ways does asexual reproduction in *Mucor* differ from asexual reproduction in flowering plants?
  b How does a spore differ from a seed? (See p. 74 for seed structure.)

2 A gardener finds a new and attractive plant produced as a result of a chance mutation (p. 189). Should she attempt to produce more of the same plant by self-pollination or by vegetative propagation? Explain your reasoning.

3 A farmer wants to breed pigs with less fat and longer backs (more lean rashers). Discuss some of the factors he will have to consider in deciding whether this will be cost-effective.

4 Which of the following do not play a part in asexual reproduction: mitosis, gametes, meiosis, cell division, chromosomes, zygote?

5 Revise pp. 79–81 and then say how we exploit the process of asexual reproduction in plants.

6 A fish may lay hundreds of eggs; a bird may lay only five or six. Despite this difference in the number of eggs the numbers of birds and fish do not change much. Suggest reasons for this.

Checklist

- Sexual reproduction involves the male and female sex cells (gametes) joining together.
- The male gamete is small and mobile. The female gamete is larger and not often mobile.
- The male gamete of an animal is a sperm. The male gamete of a flowering plant is the pollen nucleus.
- The female gamete of an animal is an ovum. The female gamete of a flowering plant is an egg cell in an ovule.
- Fish and frogs have external fertilization. The sperms are placed on the eggs after they are laid.
- Reptiles, birds and mammals have internal fertilization. The sperms are placed in the female's body to fertilize the eggs.
- Asexual reproduction does not involve gametes.
- Fungi can reproduce asexually by single-celled spores.
- Many flowering plants reproduce asexually by vegetative propagation.
- Sexual reproduction can produce new varieties of plants and animals.
- Asexual reproduction keeps the characteristics of the organism the same from one generation to the next.

35 Growth and development

Growth
Cell division and enlargement, change in shape.
Measurement of growth. Development.

Metamorphosis in insects
Complete and incomplete metamorphosis.

Control of growth

Practical work

Growth

Most living organisms start their lives as a single cell (zygote, p. 140), too small to be seen with the naked eye. This cell divides many times (by mitosis, p. 182) to produce an organism made up of thousands or millions of cells. Not only does this cell division eventually increase the size of the organism, it also makes it far more complicated than it was to start with. The new cells become specialized and form tissues and organs (p. 6). So growth involves an increase in size and mass, and also an increase in complexity.

Cell division on its own does not always produce growth. The fertilized frog's egg in Figure 35.1a divides into hundreds of cells but the embryo in Figure 35.1d is no bigger than the single cell from which it came. It consists of hundreds of tinier cells. As the cells continue to divide, the tissues they produce fold or roll up to form organs. They tuck in at the front to form a mouth and alimentary canal. Along the top, they roll up to form a tube which will later become the spinal cord and brain.

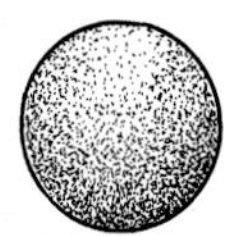
(a) fertilized egg (one cell)

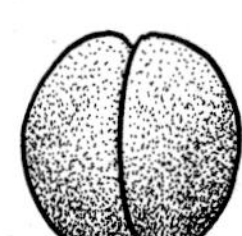
(b) cell divides into two

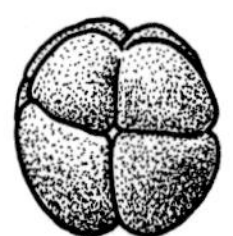
(c) eight cells

(d) many cells

Figure 35.1 Cell division in a fertilized frog's egg. At this stage there is a vast increase in cell numbers but no increase in the size of the embryo

In the early stages of growth, all the cells are able to divide. In the later stages, the dividing cells are restricted to certain regions, e.g. root tips (p. 56), buds, basal layer of the skin (p. 137). Specialized cells lose their ability to divide (but see 'Tissue culture' (p. 81) and 'stem cells' (p. 218)).

In plants, there are special groups of rapidly dividing cells in the tip of the root or the shoot. Some of these cells, when they stop dividing, start to increase in size because their vacuoles enlarge and stretch the cell walls (see Figure 35.9, p. 309). As a result of hundreds of these cells all getting larger, the root grows down into the soil and the shoot grows upwards.

In most organisms, growth is a gradual process, though the rate of growth may vary at different ages. Figure 35.2 shows that your increase in weight is rapid for the first 3 years and slows down in the next 2 years. Between the ages of 10 and 16 (adolescence), there is an increase in the growth rate, but after the age of 20 you will probably not grow at all.

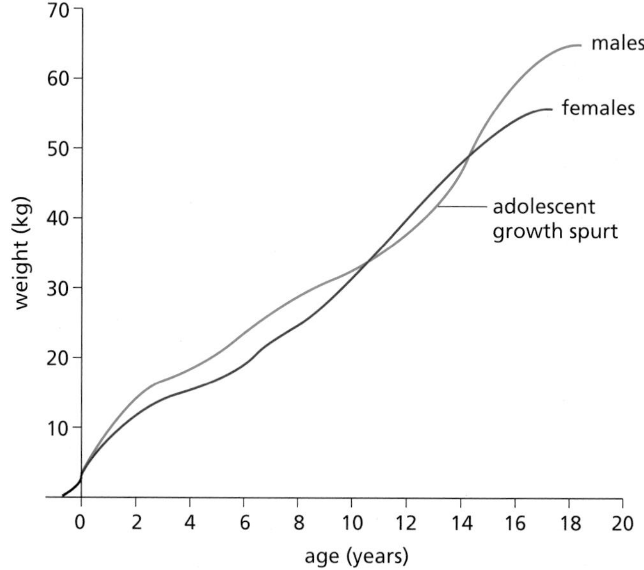

Figure 35.2 Weight increase in humans

Some fish go on growing all their lives but the growth rate gets slower as they get older.

In the arthropods (p. 274) growth in size takes place in spurts. The hard shell, called the **exoskeleton** or **cuticle**, covering the body of a crustacean like a crab, cannot expand. The crab has to get out of this cuticle before it can grow in size. So the cuticle is partly dissolved and softened from inside, then it splits open, the crab crawls out and allows its body to expand. When the body has expanded, the new cuticle forms and hardens on the outside, and the crab stays this size until the next time it sheds its cuticle.

Although an insect's cuticle is much thinner than the crab's it cannot be extended. So insects grow by a series of moults. Figure 35.3 shows a dragonfly that has emerged from its last larval cuticle.

Figure 35.3 The final moult in a dragonfly. The brown object is the larval cuticle from which the adult insect has emerged

Measurement of growth

To investigate growth, it is possible to measure increase in length, mass, volume or area. The apparent pattern of growth revealed by these measurements will differ according to which method is used. For example, when a zygote undergoes cell division it does not increase in size or mass; a relatively large single cell simply divides to form a number of smaller cells (Figure 35.1).

If growth is defined as increase in the total amount of cytoplasm, then the only reliable indicator of growth is the increase of **dry mass** (p. 21). An increase of living mass may be the result of a temporary intake of water by osmosis, which is not necessarily a feature of growth.

However, measuring dry mass involves killing the organism and heating it in an oven at 110 °C until it loses no more weight. This is often neither desirable nor feasible, so one of the other measurements has to be used, but with regard to its possible limitations.

Development

Growth involves an increase in size and complexity, but it also results in a change in shape. A tadpole is a very different shape from an adult frog. Also, growth does not take place uniformly in all parts of an organism. In young humans, the head grows relatively little compared with the limbs (Figure 35.4).

If our bones were to grow simply by adding cells to the outside, they would become very thick and short (Figure 35.5a). Figure 35.5b shows that they grow more rapidly at the ends than they do at the sides. Also, in the process of growth, some parts actually have to be dissolved away and remodelled.

When a seed germinates to produce a plant, the embryo completely changes its shape and proportions as shown in Figure 8.18 on p. 75.

In some organisms, the change in shape at different stages of growth is so extreme that the young form of the creature seems to be totally unlike the adult form. The caterpillar growing into a butterfly or the tadpole into a frog involves drastic changes like this. The process is called **metamorphosis**.

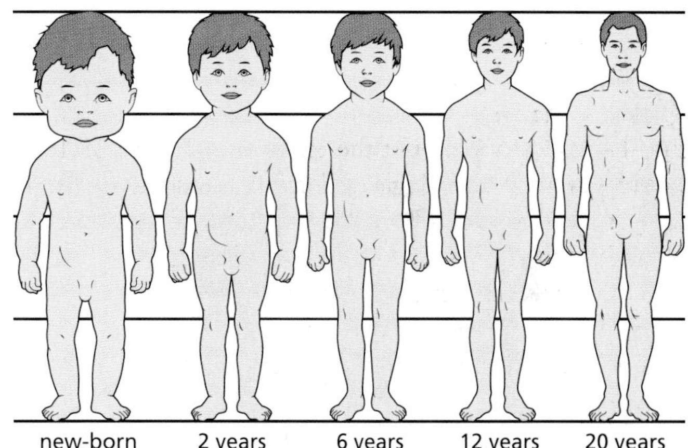

Figure 35.4 Human growth. All the figures are drawn to the same height to show how the body proportions change with age

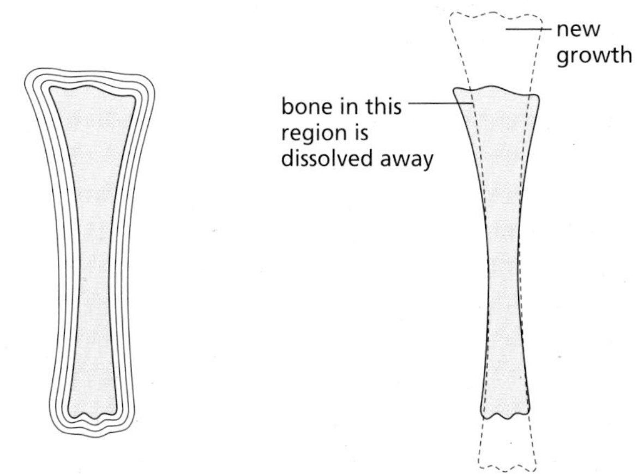

Figure 35.5 Growth of limb bones

Metamorphosis in insects

All insects lay eggs. The eggs hatch either into larvae like maggots or caterpillars (e.g. butterflies, bees, houseflies) or into **nymphs**, which are miniature, immature forms of the adult (e.g. dragonflies, cockroaches). At each moult the nymphs become more like the adult. This is called **incomplete metamorphosis**. With caterpillars and maggots, the early moults result only in an increase in size, but the final moult, after a resting stage, brings about a dramatic change from the larval to the adult form. This is called **complete metamophosis**.

The **large white butterfly** ('cabbage white') lays its eggs on the underside of brassica leaves, (e.g. cabbage, broccoli). The larva which emerges from the egg (Figure 35.6) is a caterpillar, with a cylindrical, segmented body, three pairs of short legs on the first segments and four pairs of grasping 'prolegs' on segments 6–9 (Figure 35.7a). The prolegs have fine hooks which grip the food plant firmly. The caterpillar is wingless and has biting jaws on its head. It feeds non-stop on the brassica leaf, using its jaws to bite off chunks of leaf, and grows rapidly, moulting several times.

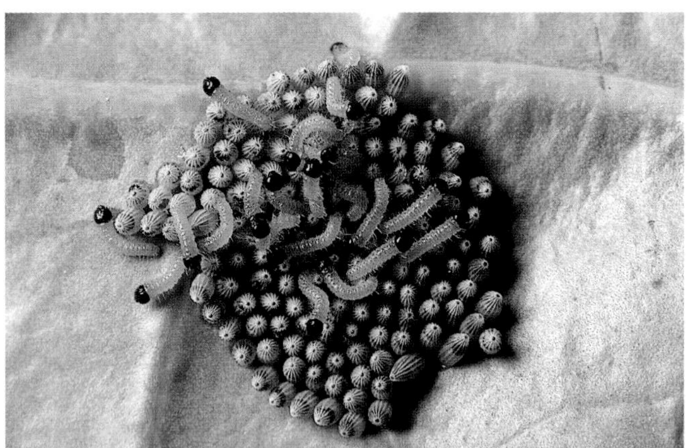

Figure 35.6 Caterpillars (large white butterfly) hatching

Eventually it stops feeding and enters a resting stage called a **pupa** (Figure 35.7b). In the pupa, the structures of the larva are digested away and replaced by adult structures such as wings, long jointed legs and tubular mouth parts.

After 2 or 3 weeks, the pupal cuticle splits; the adult butterfly pulls itself out and expands its wings by pumping blood into them (Figures 35.7c and 35.8). The adult does not grow or moult its cuticle.

In many insects, the nymphs and larvae exploit completely different habitats and food sources. The caterpillar lives on its food plant and uses external jaws to bite off pieces of leaf. The butterfly ranges widely and feeds on nectar, which it sucks from flowers using a long tubular proboscis (Figure 32.4, p. 294). The adults and larvae are, therefore, not in competition.

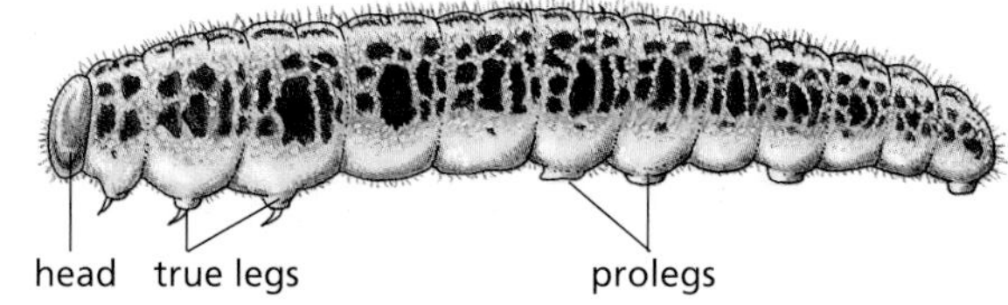

(a) Caterpillar

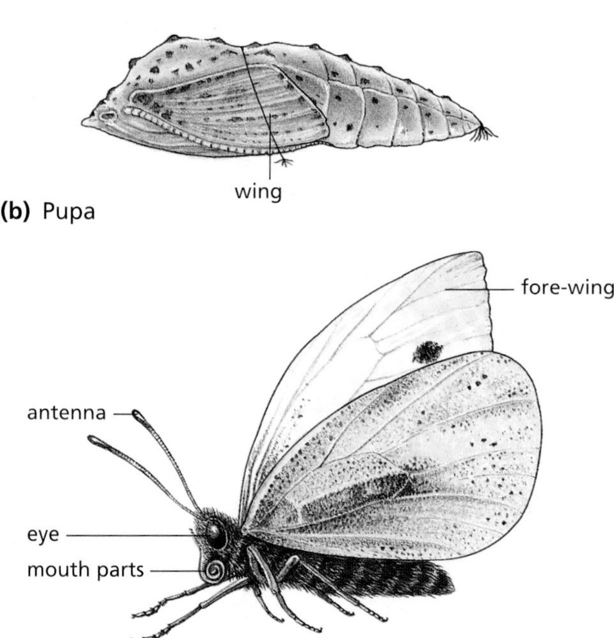

(b) Pupa

(c) Adult

Figure 35.7 Metamorphosis in the large white butterfly

Figure 35.8 Small tortoiseshell butterfly emerging from pupa. Note the coiled proboscis (in two sections) and the unexpanded wings

Control of growth

In animals and plants, the growth rate and extent of growth are controlled by chemicals: **hormones** in animals and **growth substances** in plants. Additionally, growth may be limited in animals by the availability of food, and in plants by light, water and minerals.

In humans, growth is regulated by a growth hormone, **somatotrophin**, secreted by the pituitary gland (p. 171). Its main effects are to promote protein synthesis, cell division and cell enlargement.

There are many different **growth substances** ('plant hormones') in plants. They are similar in some ways to animal hormones because they are produced in specific regions of the plant and transported to 'target' organs such as roots, shoots and buds. The sites of production are not specialized organs, as in animals, but regions of actively dividing cells such as the tips of shoots and roots.

One of the growth substances is **auxin**. Chemically it is indoleacetic acid (IAA). It is produced in the tips of actively growing roots and shoots and carried by active transport (p. 28) to the regions of extension where it promotes cell enlargement (Figure 35.9).

The responses made by shoots and roots to light and gravity are influenced by growth substances (p. 319) and so is the production of whole plants from tissue culture (p. 81).

Growth substances also control seed germination, bud burst, leaf fall, initiation of lateral roots and many other processes.

Questions

1 The following figures show the growth of an insect over 200 days.
a Plot a graph of the figures with the days on the horizontal axis.
b Say whether you think the graph is sigmoid.

Mass in mg	10	20	40	120	240	400	640	940	950	960
Time in days	20	40	60	80	100	120	140	160	180	200

2 Some of the cells in Figure 35.1d will become skin, some will become nerves and muscles. What kind of changes must take place in one of the cells for it to become a nerve cell?

3 According to the graph in Figure 35.2, what is the average increase in weight
a from the age of 1 to 5 years,
b from 5 to 10 years?

4 What are
a the advantages,
b the disadvantages, of using dry mass as a measure of growth?

5 In Figure 35.4, what is the ratio of head size to body length
a at 2 years,
b at 20 years?

6 Study Figures 35.6 and 35.7. What changes have taken place in changing from a caterpillar to a butterfly?

7 In a shoot, why does the region of most rapid growth occur at some distance from the shoot tip?

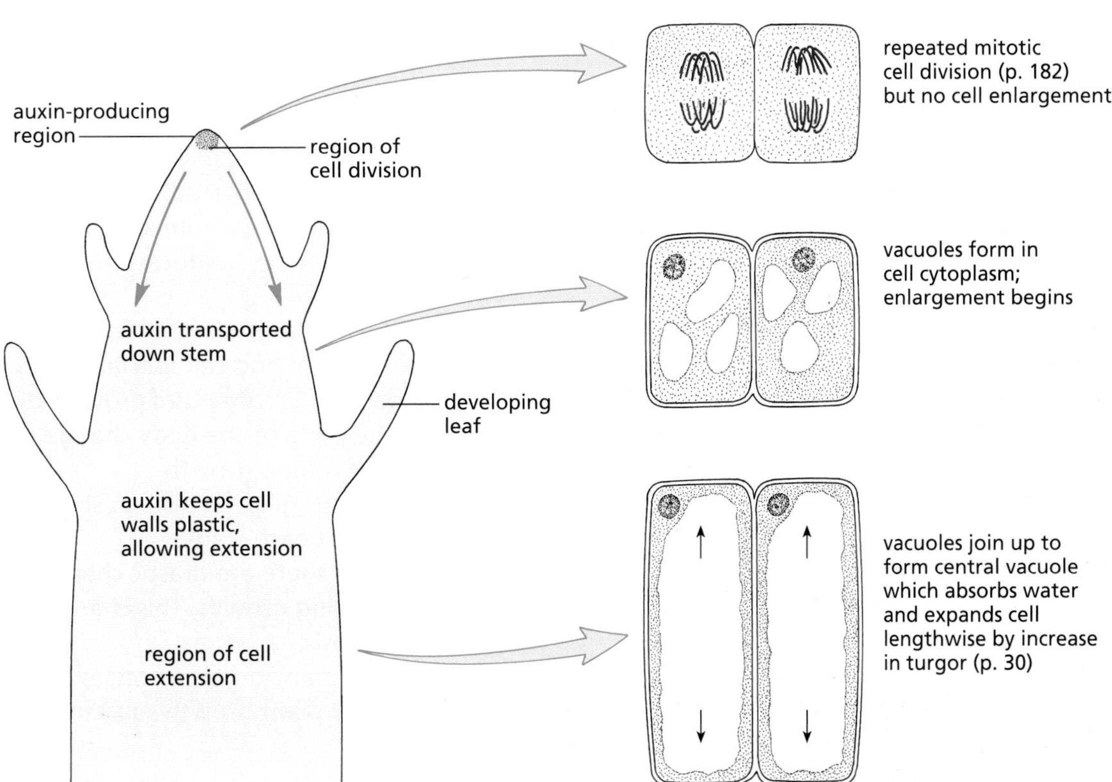

Figure 35.9 Extension growth at shoot tip

Use of plant growth substances

Chemicals can be manufactured which closely resemble natural growth substances and may be employed to control various aspects of growth and development of crop plants.

An artificial auxin sprayed on to tomato flowers will induce all of them to produce fruit, whether or not they have been pollinated. Another growth substance, sprayed on to fruit trees, prevents early fruit fall and enables all the fruit to be harvested at the same time.

The weedkiller, 2,4-D, is very similar to one of the auxins. When sprayed on a lawn, it affects the broad-leaved weeds (e.g. daisies and dandelions) but not the grasses. (It is called a 'selective weedkiller'.) Among other effects, it distorts the weeds' growth and speeds up their rate of respiration to the extent that they exhaust their food reserves and die.

Gardeners and horticulturists often dip cuttings into a powder ('rooting hormone') containing a growth substance which initiates the production of roots from the stem (p. 81).

In the production of plants by tissue culture (p. 81) the correct balance of growth substances is necessary to promote root and shoot formation from the undifferentiated tissue.

Practical work

Use of hormone rooting powder

Cut ten shoots, about 15 cm long, from the new growth of an untrimmed *Lonicera* hedge in June–August. (*Lonicera nitida* is bush honeysuckle; cuttings from flowering currant and laurel also work well but take longer than 2 weeks.) Remove the leaves from the lower 5 cm of the stems and, with five of the cuttings, dip the exposed stem first into water and then into hormone rooting powder. Tap off the excess powder and push the stems into potting compost in a flowerpot labelled 'A'. Dip the other five stems into water only and push them into compost in a flowerpot labelled 'B'. **Wash your hands to remove any traces of the powder.**

Leave both pots in the same conditions of light and temperature and keep the compost moist for 2 weeks. After this time, remove the cuttings and wash the compost off the stem bases.

Result Most of the cuttings will have developed roots from the part of the stem in the compost, but the cuttings treated with hormone rooting powder will have many more roots (Figure 35.10). Some of the untreated cuttings may have failed to develop any roots at all.

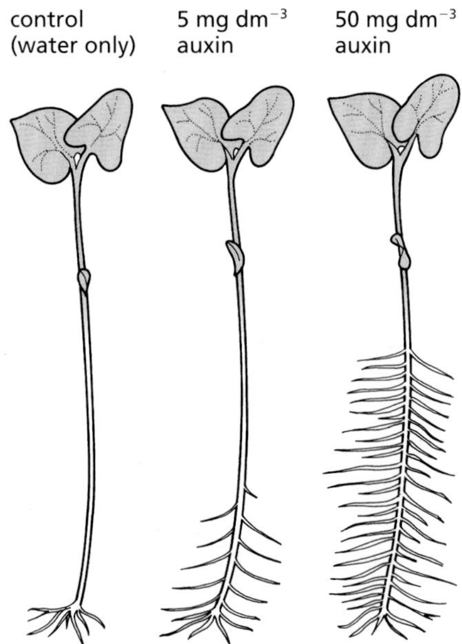

Figure 35.10 Stimulation of root initiation in bean cuttings using an auxin

Interpretation The hormone rooting powder contains an auxin. This chemical has promoted the formation of roots in the stem cuttings.

Note The rooting powder may also contain a fungicide, which makes interpretation of the results less certain.

Checklist

- When animals and plants grow, they increase their size and weight, and their structures become more complicated.
- Most organisms grow from a single cell, by cell division, to become creatures consisting of millions of cells.
- Growth takes place as a result of cell division, cell enlargement and cell specialization.
- Not all parts of an organism grow at the same rate, so the parts of the body change their proportions during growth.
- The arthropods grow in distinct stages, shedding their cuticle at each stage.
- In the insects there are drastic changes in body structure during growth. This change is called metamorphosis.
- Growth and many other processes in plants are controlled by plant growth substances.

36 Movement and locomotion

Micro-organisms
Invertebrates
Arthropods
Vertebrates
Fish, birds and mammals.
Flowering plants

Locomotion refers to the movement of an organism from place to place. **Movement** refers to a change in position of any part of an organism's body, but does not necessarily involve locomotion. Chewing, breathing, the heart beat and blinking are all examples of movement in humans.

Most animals and protoctista (p. 268) exhibit locomotion. Even animals such as mussels, barnacles and sea anemones, which remain in one place throughout their lives, have larval stages which are freely mobile.

Mature plants do not exhibit locomotion but the male gametes of algae and bryophytes swim freely during the process of fertilization. Parts of plants do show movement, however. Examples are the tropisms described on p. 317, the movements of insectivorous plants and the opening and closing of petals shown on p. 316.

Locomotion gives organisms the advantage of being able to seek food, escape predators, search for mates or move to more congenial conditions.

Micro-organisms

Many protozoans (p. 268) propel themselves through water by means of structures called **cilia** or **flagella**. Cilia are fine cytoplasmic 'hairs' which cover the surface membranes of some protozoans, such as *Paramecium* (Figure 36.1a). The cilia flick rhythmically to drive the organism through the water. Flagella are longer than cilia and occur singly or in pairs, e.g. *Euglena* (Figure 36.1b). Flagella flick, whirl or ripple in a way that pulls the protozoan through the water.

Some bacteria move by means of flagella (Figure 31.1, p. 283), but their structure is different from those of protozoans.

Many male gametes are motile, travelling through water or body fluids to reach the female gamete. Algae, liverworts, mosses and ferns all have motile male gametes. The 'tail' sections of vertebrate sperms act as flagella (Figure 16.1, p. 140).

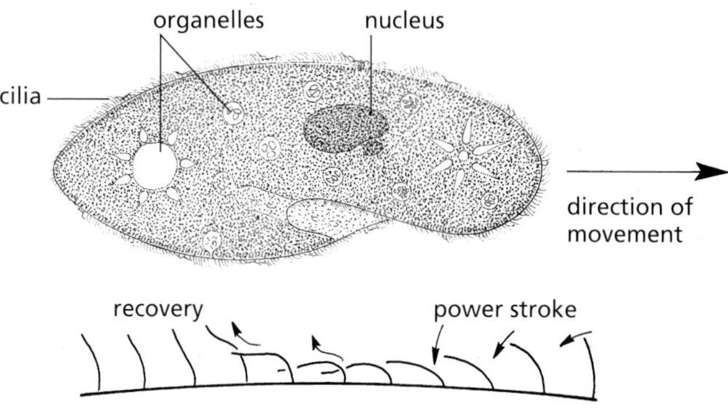

(a) ***Paramecium*** (×250)

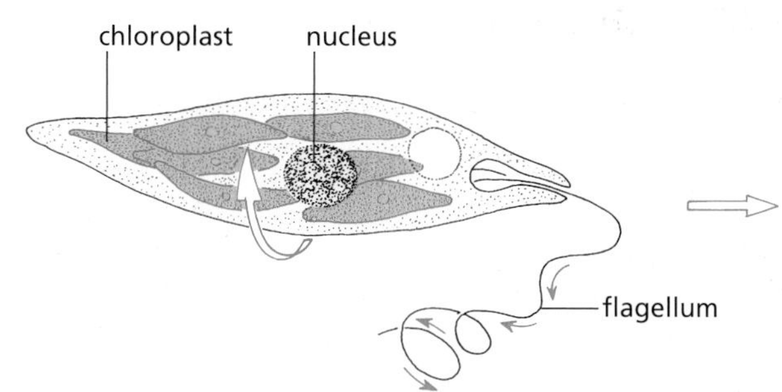

(b) ***Euglena*** (×400)

Figure 36.1 Locomotion in *Paramecium* and *Euglena*

Invertebrates

There is an enormous variety of methods of locomotion in invertebrate animals. Arthropods have jointed legs for walking, leaping and running, and insects have wings for flying.

Slugs and snails have a foot with which they creep over the soil or over plants. Muscular ripples pass down the length of the base of the foot and propel the mollusc forward, usually over a coating of mucus which lubricates the track.

The way in which the limbs of arthropods propel them during walking is shown in Figure 36.2 (overleaf).

The cuticle of arthropods (p. 274) forms an **exoskeleton**. The muscles which produce movement are attached to the *inside* of the exoskeleton (Figure 36.2a).

Compare this with the muscle attachment in vertebrates (e.g. Figure 36.6), which have an **endoskeleton**.

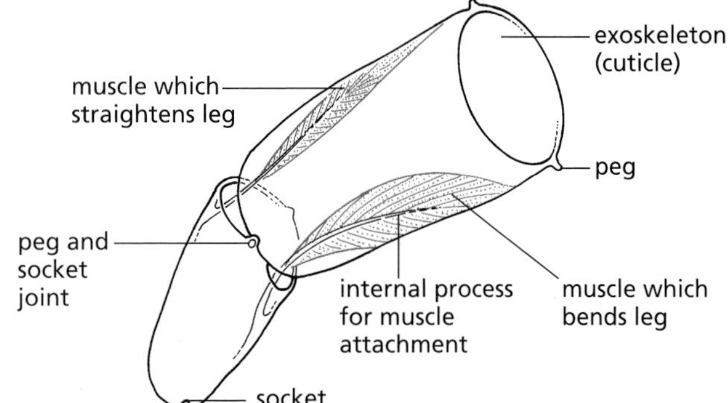

(a) muscle attachment in arthropod limb

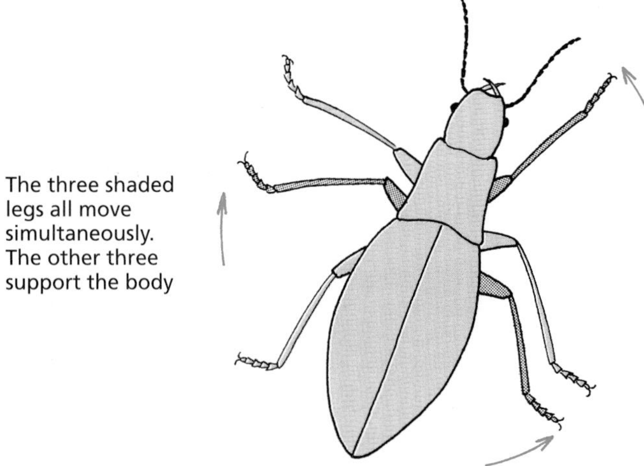

(b) walking pattern in a beetle

Figure 36.2 Arthropod muscle attachment and movement

Vertebrates

Vertebrates move by making their muscles contract and pull on the bones of their skeletons. This process is described more fully on p. 154.

Fish

The external features of a bony fish are shown in Figure 30.15 on p. 276. Locomotion is achieved by lateral contractions of the whole of the muscular body with a final thrust by the tail (Figure 36.3a). The muscles on opposite sides of the body contract alternately to send the tail first to one side and then to the other (Figure 36.3b). As the tail moves, it pushes sideways and backwards on the water. The sideways movements to left and right cancel each other out, and the backwards thrust of the tail on the water drives the fish forward.

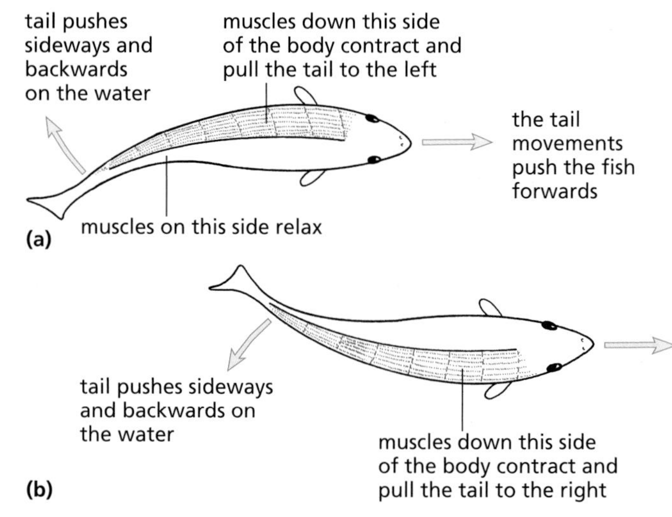

Figure 36.3 Locomotion in bony fish

The tail fin contributes to the forward propulsion but the other fins, in most fish, serve mainly to steer and stabilize (Figure 36.4). The dorsal and ventral fins are called **median fins**; the pectoral and pelvic fins are the **paired fins**.

The streamlined shape of the fish offers the least possible resistance to forward movement through the water.

The fish may, therefore, be said to be adapted to movement in water by having a streamlined shape, powerful body muscles to move its tail and fins to help it steer.

Many fish also have an air bladder in their bodies which makes them buoyant, so that they don't sink down every time they stop swimming.

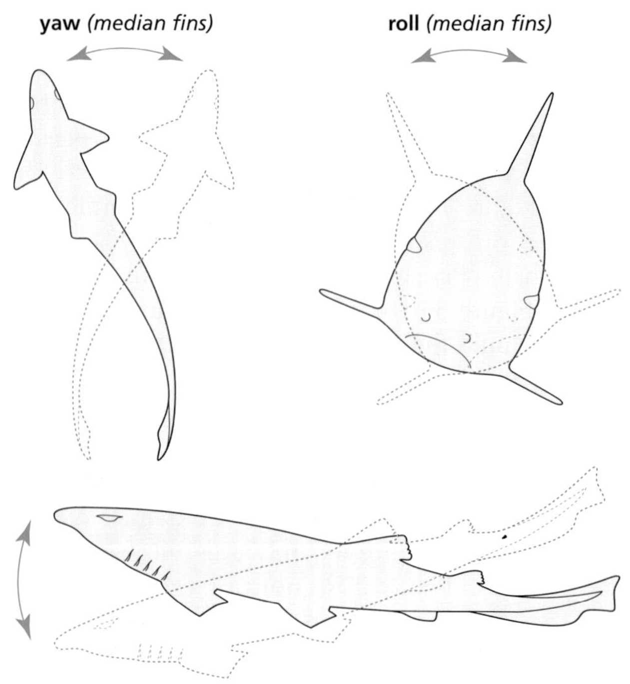

Figure 36.4 Movements controlled by fins

Birds

In mammals, the muscles for locomotion act mainly on the hind limbs. In birds, they act on the forelimbs which are the wings. One pair of muscles pulls the wings down and an antagonistic pair pulls the wings up.

Figure 36.5 shows the skeleton of a bird and you will see that the breast bone has a deep keel projecting down. This provides a large surface for the attachment of the powerful flight muscles that act on the wing. The **coracoid** bone acts as a strut between the breast bone and the spine so that when the flight muscles contract, they pull the wing bone down rather than pull the breast bone up.

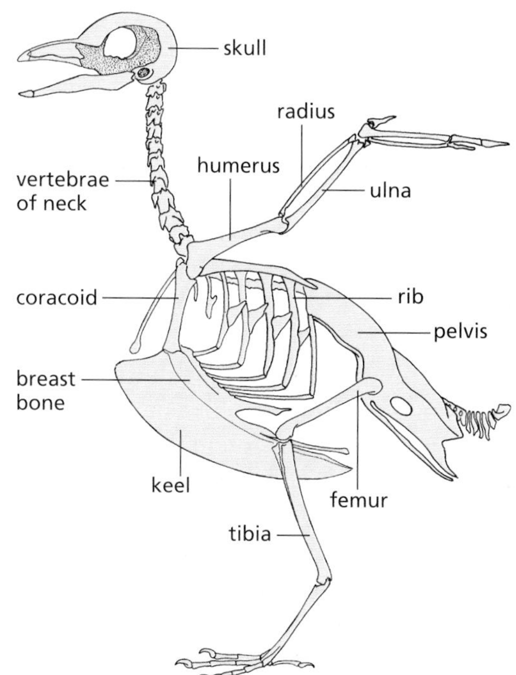

Figure 36.5 Bird's skeleton

Figure 36.6 shows this part of the bird's skeleton as seen from the front. You can see how the flight muscles are attached to the breast bone at one end and the humerus at the other.

When the large flight muscles contract (Figure 36.6a), they pull the wings down. The small flight muscles, when they contract (Figure 36.6b), pull the wings up because their tendons run over a groove in each coracoid, like a pulley, and attach to the upper side of the humerus. So the large and small flight muscles are antagonistic to each other. When they contract alternately, they pull the wing up and down.

During the upstroke, the wing is bent at the wrist and so it does not offer much air resistance (Figure 36.6b). In the downstroke the wing is spread out and pushes downwards on the air. The air resistance causes an upthrust on the wing and so lifts the bird up.

Gliding flight

In gliding flight a bird's wings act as aerofoils (i.e. like aeroplane wings). Air flowing over the curved upper surface moves faster than it does over the lower surface. This effectively produces a high pressure on the lower surface and low pressure on the upper surface and so provides a lifting force (Figure 36.7).

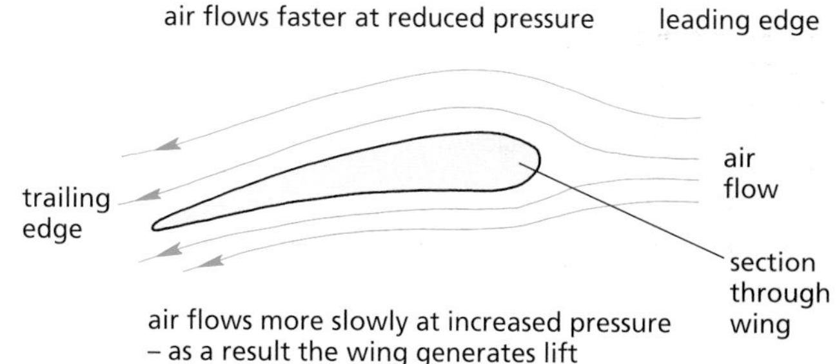

Figure 36.7 The wings as an aerofoil

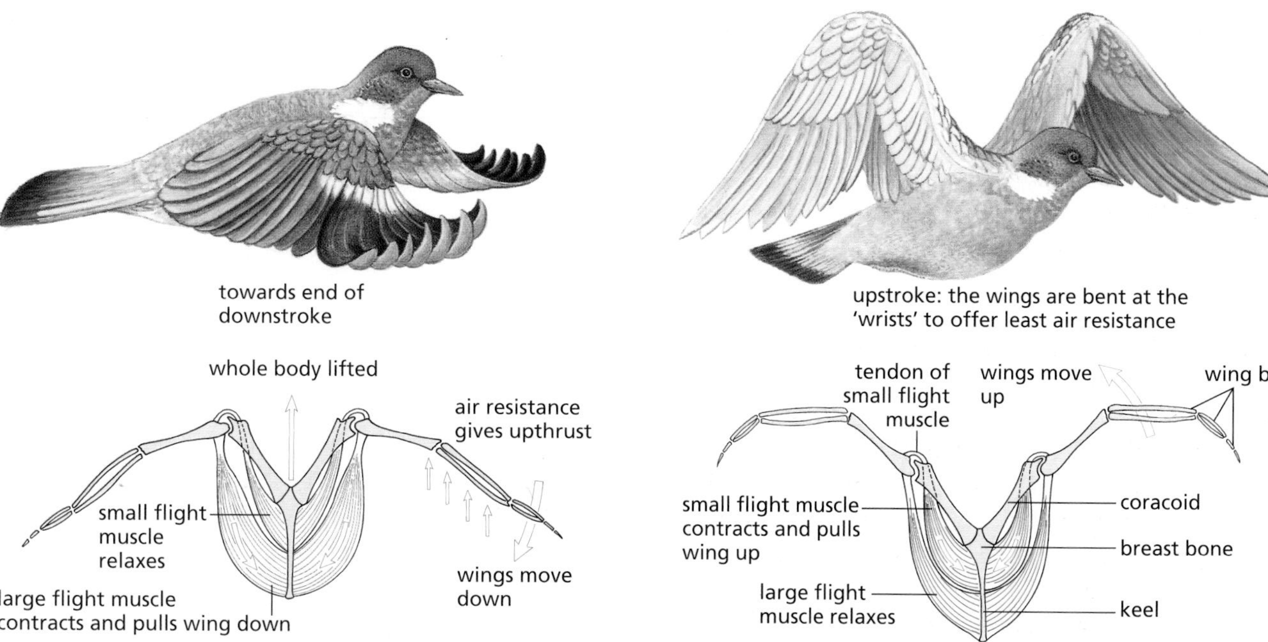

Figure 36.6 Front view of bird's skeleton to show how the muscles work the wings

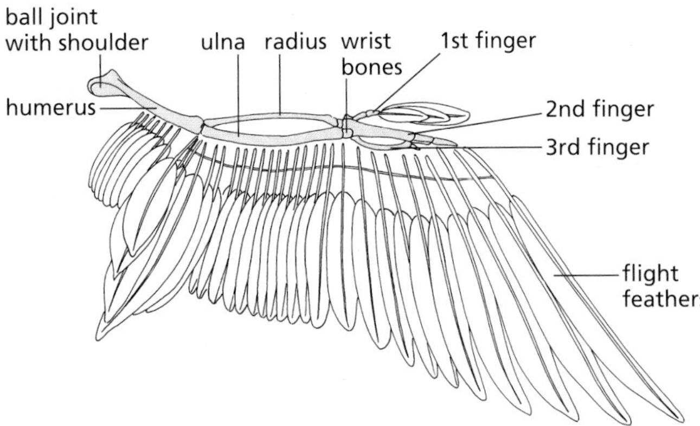

Figure 36.8 Skeleton and flight feathers of a bird's wing

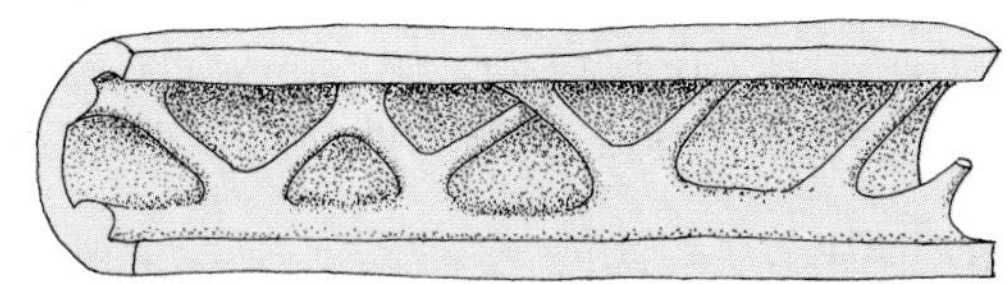

Figure 36.9 Section through wing bone. The bones are hollow and light but internally strutted for strength

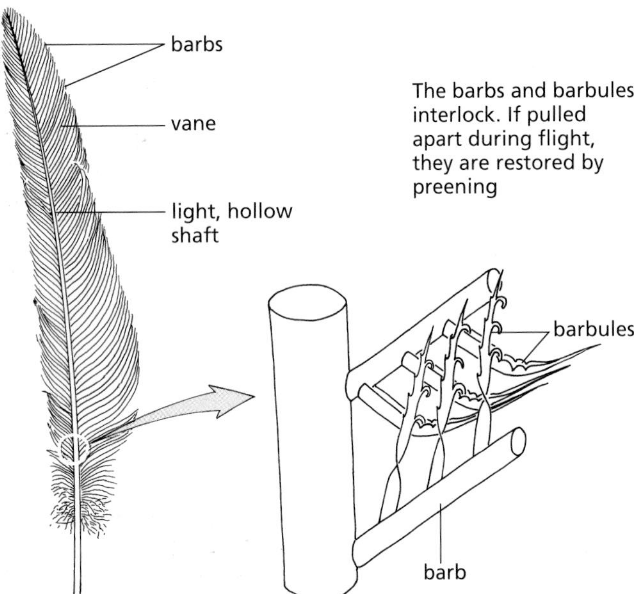

Figure 36.10 The flight feathers are light and air-resistant but easily repaired

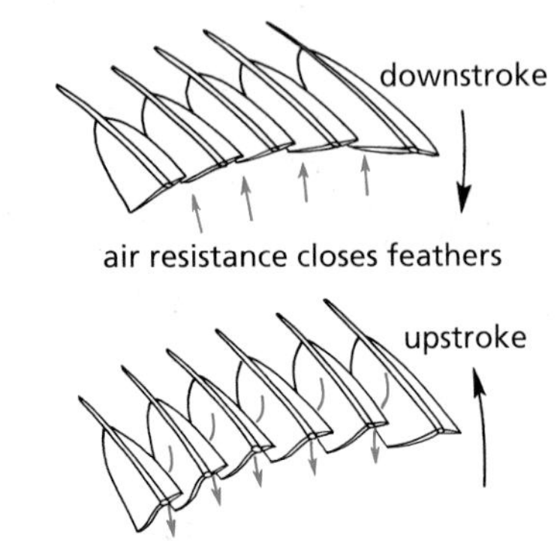

Figure 36.11 Action of wing feathers during flight. The overlap makes the downstroke effective

With wings outstretched, a bird moves forward and loses height. The lift produced by the air flow over its wings keeps the bird airborne. Most birds glide at some stages in their flight pattern, especially when landing. On upcurrents of air near cliff faces or updraughts caused by thermals, some birds can remain airborne for long periods without flapping their wings.

Adaptations for flight

The skeleton, limbs and musculature of a bird are modified from the basic vertebrate pattern in the following ways:

- Its front legs are modified to form wings (Figure 36.8).
- It has powerful flight muscles (Figure 36.6).
- The deep keel provides a large surface for attachment of these muscles (Figure 36.5).
- Its limb bones are hollow and light (Figure 36.9).
- Its skeleton forms a rigid framework to stand up to the contractions of the flight muscles.
- Its flight feathers are light but provide a large air-resistant surface (Figures 36.10 and 11).

Mammals

Locomotion in most terrestrial mammals is by walking or running in which the four limbs move, usually, in a diagonal pattern, i.e. right forelimb and left hind limb moving at the same time, followed by left forelimb and right hind limb. As the pace quickens, the hind leg is lifted before the diagonal forelimb reaches the ground.

The way the muscles act on the limb bones to produce movement is described for humans on p. 315 and, in general, is applicable to most walking and running mammals.

Mammals' limbs are often adapted to their way of life or to the terrain over which they move, e.g. running, burrowing, climbing. Figure 23.17 on p. 209 shows how some mammalian limbs are adapted to their use in locomotion.

Questions

1. In vertebrates, what systems are directly involved in producing co-ordinated movement?
2. In Figure 36.2b (p. 312), how do you think the beetle completes its walking movement?
3. In fish, how do the functions of the tail fin, median fins and paired fins differ?
4. In the hind limbs of terrestrial vertebrates, the extensor muscles are usually larger than the flexor muscles. Why do you think this is so?
5. How do birds achieve 'lift'
   a in flapping flight,
   b in gliding flight?

Humans

Locomotion in humans is brought about by the limb muscles contracting and relaxing in an orderly (i.e. co-ordinated) manner. The photograph in Figure 36.12 shows a sprinter at the start of a race, and Figure 36.13 shows how some of his leg muscles are acting on the bones to thrust him forward. When muscle A contracts, it pulls the femur backwards. Contraction of muscle B straightens the leg at the knee. Muscle C contracts and pulls the foot down at the ankle. When these three muscles contract at the same time, the leg is pulled back and straightened and the foot is extended, pushing the foot downwards and backwards against the ground. If the ground is firm, the straightening of the leg pushes upwards against the pelvic girdle, which in turn pushes the vertebral column and so lifts the whole body upwards and forwards.

Figure 36.12 Leg muscles used in running (compare with Figure 36.13)

While muscles A, B and C are contracting to extend the leg, their antagonistic muscles are kept in a state of relaxation. At the end of the extension movement, muscles A, B and C relax, and their antagonistic partners contract to flex the leg.

Flowering plants

Plants do not show locomotion; they remain in one place for their whole life. They do make movements, however, of parts of their bodies. These movements, such as tropisms (p. 317), are also seen in the opening up of flowers, folding up of leaves at night or climbing plants twining round supports. When these movements are photographed by 'time-lapse' photography and shown speeded up, they reveal plants as active organisms, even though they do not move about as a whole.

Comparatively rapid movements are the 'sleep movements' made by some plants at night or when light intensity falls, e.g.

- the closing up of the petals of crocuses and tulips (Figure 36.14, overleaf)
- the closing of the outer florets of daisies
- the folding of the leaflets of wood sorrel.

In some cases, the movement is brought about by changes in growth rate. In other cases it is changes in the turgor (p. 30) of groups of cells which bring about the movement.

It is not entirely clear what the advantages of 'sleep movements' are.

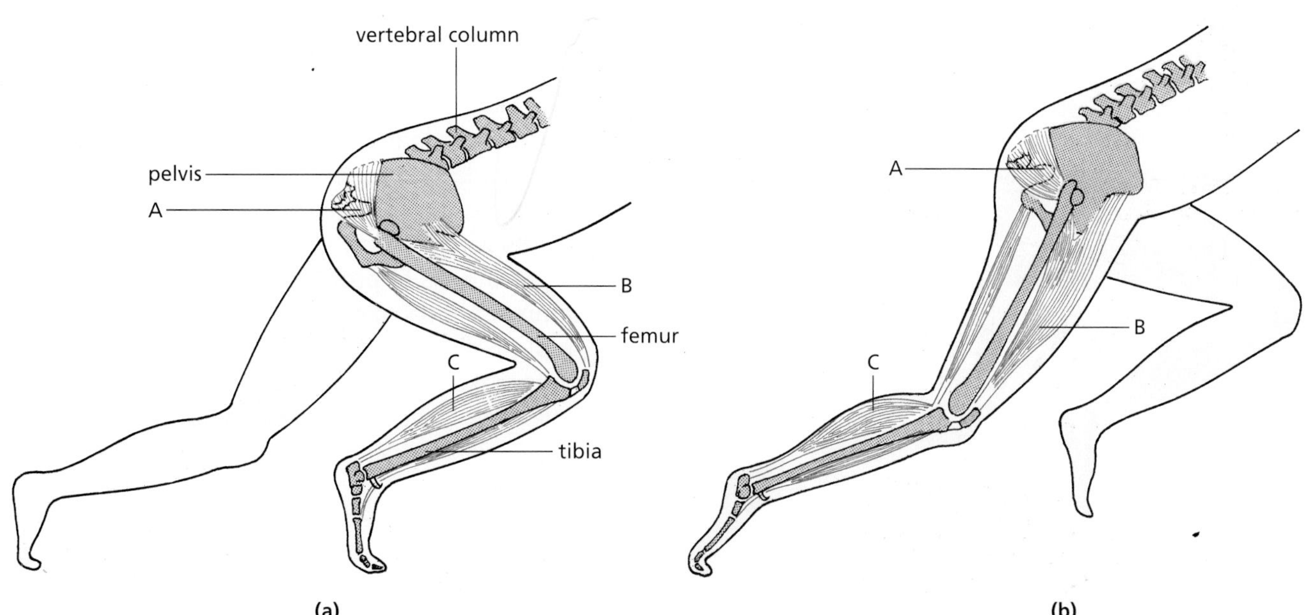

Figure 36.13 Action of leg muscles. There are many other muscles which are not shown here. A, B and C are all contracting to straighten the leg and foot

Some insectivorous plants make easily visible movements. The sticky tentacles on the leaves of the sundew curl gradually round a trapped insect. The Venus flytrap closes very rapidly when triggered by an insect walking between the leaves (Figure 36.15); the leaves of the 'sensitive plant' (*Mimosa pudica*) collapse and droop at once if they are touched (Figure 36.16).

(a) Early Morning

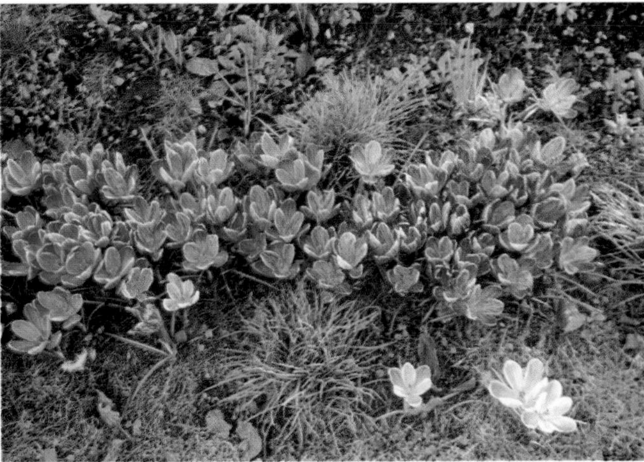

(b) Two hours later

Figure 36.14 'Sleep movements' of crocuses

Figure 36.15 Venus flytrap with trapped blowfly, which will eventually be digested

(a) Plant in unstimulated condition

(b) Two seconds after plant was touched by a pencil

Figure 36.16 The sensitive plant, *Mimosa pudica*

Questions

1. How do plants survive despite being unable to move about?
2. Rhizomes and runners both enable plants to reach more distant areas, but what aspect of some plant life cycles covers the greatest distance?
3. The advantages of 'sleep movements' may not be clear, but what might they be?
4. Speculate on the possible advantages to the 'sensitive plant' of its rapid response to touch.

Checklist

- Nearly all living things make movements with parts of their body.
- Protozoans and animals move their whole bodies around (locomotion).
- Some protozoans use cilia and flagella to move about.
- Arthropods have exoskeletons with muscles attached to the inside of the cuticle.
- A fish is adapted by its shape and muscle arrangement to swim in water.
- A bird's skeleton and flight muscles are adapted in special ways to produce flight.
- Mammals' limbs may be specialized for running, jumping, digging or climbing.
- Plants make growth movements (tropisms) and move their leaves and flower petals.

37 Sensitivity

Plants
Tropisms, growth substances.
Practical work
Experiments on tropisms.

Animals
Non-directional and directional responses. Sense organs.

Sensitivity is the ability of living organisms to respond to stimuli. A **stimulus** is a change in the external or internal environment of an organism.

Plants

Although plants do not respond by moving their whole bodies, parts of them do respond to stimuli. Some of these responses are described as tropic responses or **tropisms**.

Tropisms

Tropisms are growth movements related to directional stimuli, e.g. a shoot will grow towards a source of light but away from the direction of gravity. Growth movements of this kind are usually in response to the *direction* of light, or gravity. Responses to light are called **phototropisms**; responses to gravity are **geotropisms** (or **gravitropisms**).

If the plant organ responds by growing towards the stimulus, the response is said to be 'positive'. If the response is growth away from the stimulus, it is said to be 'negative'. For example, if a plant is placed horizontally, its stem will change its direction and grow upwards, away from gravity (Figure 37.1).

Figure 37.1 Negative geotropism. The tomato plant has been left on its side for 24 hours

The shoot is **negatively geotropic**. The roots, however, will change their direction of growth to grow vertically downwards towards the pull of gravity (Experiment 1). Roots, therefore, are **positively geotropic**.

Phototropism and geotropism are best illustrated by some simple controlled experiments. Seedlings are good material for experiments on sensitivity because their growing roots and shoots respond readily to the stimuli of light and gravity.

Practical work

Experiments on tropisms

1 Geotropism in pea radicles

Soak about 20 peas in water for a day and then let them germinate in a roll of moist blotting-paper (see Figure 8.22, p. 77). After 3 days, choose 12 seedlings with straight radicles and pin six of these to the turntable of a clinostat so that the radicles are horizontal. Pin another six seedlings to a cork that will fit in a wide-mouthed jar. Leave the jar on its side. A **clinostat** is a clockwork or electric turntable which rotates the seedlings slowly about four times an hour. Although gravity is pulling sideways on their roots, it will pull equally on all sides as they rotate.

Place the jar and the clinostat in the same conditions of lighting or leave them in darkness for 2 days.

Result The radicles in the clinostat will continue to grow horizontally but those in the jar will have changed their direction of growth, to grow vertically downwards (Figure 37.2).

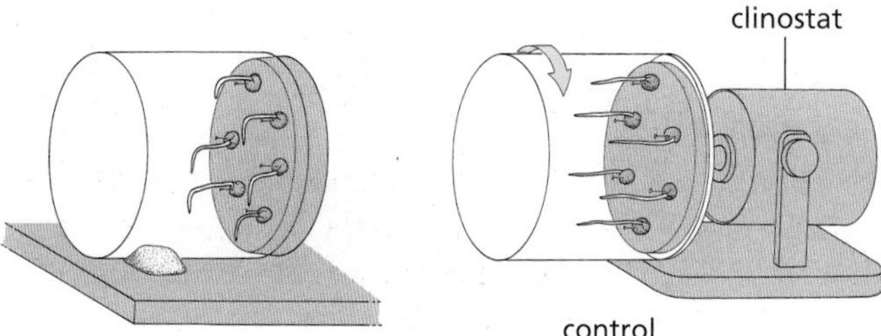

Figure 37.2 Geotropism in roots – results

Interpretation The stationary radicles have responded to the stimulus of one-sided gravity by growing towards it. The radicles are positively geotropic.

The radicles in the clinostat are the controls. Rotation of the clinostat has allowed gravity to act on all sides equally and there is no one-sided stimulus, even though the radicles were horizontal.

2 Phototropism in shoots

Select two potted seedlings, e.g. sunflower or runner bean, of similar size and water them both. Place one of them under a cardboard box with a window cut in one side so that light reaches the shoot from one direction only (Figure 37.3). Place the other plant in an identical situation but on a clinostat. This will rotate the plant about four times per hour and expose each side of the shoot equally to the source of light. This is the control.

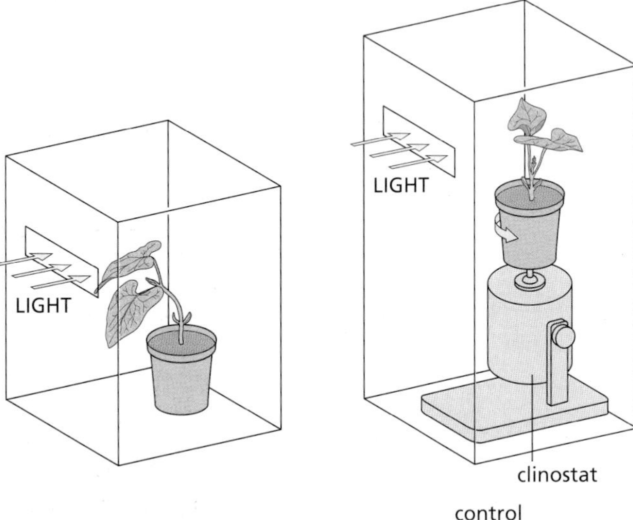

Figure 37.3 Phototropism in a shoot

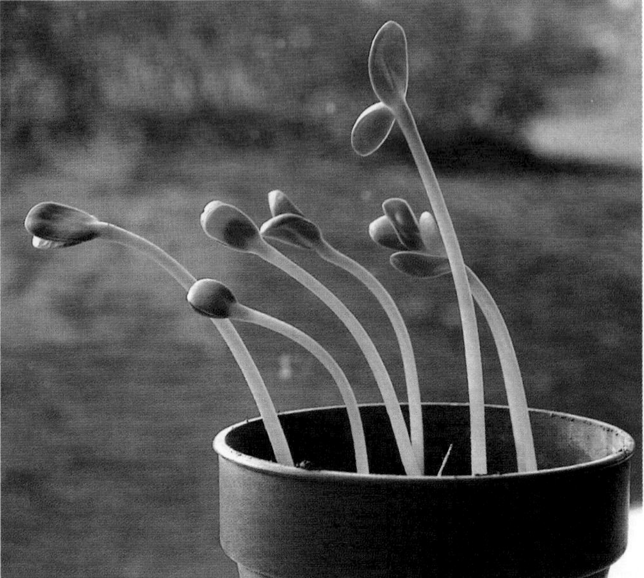

Figure 37.4 Positive phototropism. The sunflower seedlings have received one-sided lighting for a day

Result After 1 or 2 days, the two plants are removed from the boxes and compared. It will be found that the stem of the plant with one-sided illumination has changed its direction of growth and is growing towards the light (Figure 37.4). The control shoot has continued to grow vertically.

Interpretation The results suggest that the young shoot has responded to one-sided lighting by growing towards the light. The shoot is said to be positively phototropic because it grows towards the direction of the stimulus.

However, the results of an experiment with a single plant cannot be used to draw conclusions which apply to green plants as a whole. The experiment described here is more of an illustration than a critical investigation. To investigate phototropisms thoroughly, a large number of plants from a wide variety of species would have to be used.

Advantages of tropic responses

Positive phototropism of shoots

By growing towards the source of light, a shoot brings its leaves into the best situation for photosynthesis. Similarly, the flowers are brought into an exposed position where they are most likely to be seen and pollinated by flying insects.

Negative geotropism in shoots

Shoots which are negatively geotropic grow vertically. This lifts the leaves and flowers above the ground and helps the plant to compete for light and carbon dioxide. The flowers are brought into an advantageous position for insect or wind pollination. Seed dispersal may be more effective from fruits on a long, vertical stem. However, these advantages are a product of a tall shoot rather than negative geotropism.

Stems which form rhizomes (p. 79) are not negatively geotropic; they grow horizontally below the ground, though the shoots which grow up from them are negatively geotropic.

Branches from upright stems are not negatively geotropic; they grow at 90 degrees or, usually, at a more acute angle to the directional pull of gravity. The lower branches of a potato plant must be partially **positively** geotropic when they grow down into the soil and produce potato tubers (p. 80).

Positive geotropism in roots

By growing towards gravity, roots penetrate the soil which is their means of anchorage and their source of water and mineral salts. Lateral roots are not positively geotropic; they grow at right angles or slightly downwards from the main root. This **diageotropism** of lateral roots enables a large volume of soil to be exploited and helps to anchor the plants securely.

Hydrotropism

It may be that some plants have roots which respond positively to a water gradient in the soil by changing their direction of growth towards the wetter region. More often, the side of the whole root system in the wetter region simply grows faster than the side in the drier region. This is a growth response but not a tropism.

Questions

1 a To what directional stimuli do
  (i) roots,
  (ii) shoots respond?
 b Name the plant organs which are
  (i) positively phototropic,
  (ii) positively geotropic,
  (iii) negatively geotropic.

2 Why is it incorrect to say
 a 'Plants grow towards the light',
 b 'If a root is placed horizontally, it will bend towards gravity'?

3 Explain why a clinostat is used for the controls in tropism experiments.

4 Look at Figure 37.1. What will the shoot look like in 24 hours after the pot has been stood upright again? (Just draw the outline of the stem.)

5 Why does a bean seed still germinate successfully even if you plant it 'upside-down'?

6 What do you think might happen if a potted plant were placed on its side and the shoot illuminated from below (i.e. light and gravity are acting from the same direction)?

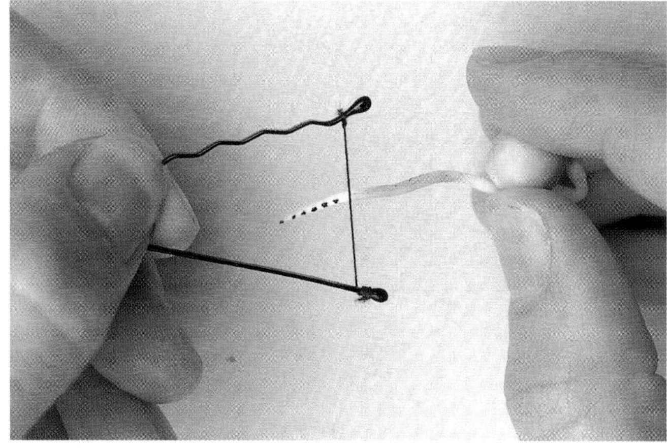

Figure 37.5 Marking a root. A piece of cotton is held by the hairpin and dipped into black ink

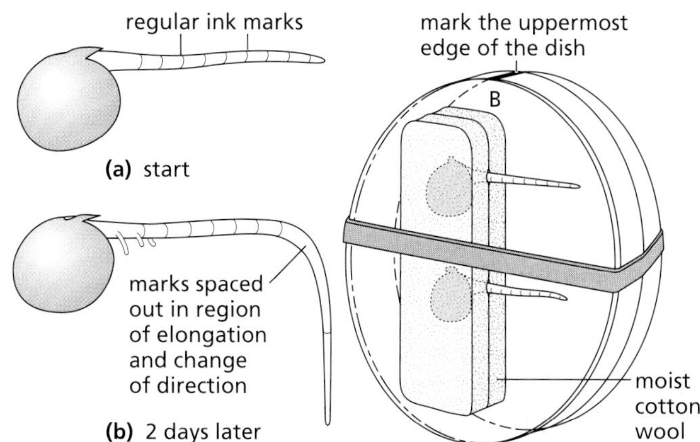

Figure 37.6 Region of response in radicles. Result of Experiment 3 on the B seedlings

Practical work

3 Region of response

Grow pea seedlings as described for maize on p. 77 and select four with straight radicles about 25 mm long. Mark all the radicles with lines about 1 mm apart (Figures 37.5 and 37.6a). Use four strips of moist cotton wool to wedge two seedlings in each of two Petri dishes (Figure 37.6). Leave the dishes on their sides for 2 days, one (A) with the radicles vertical and the other (B) with the radicles horizontal.

Result The ink marks will be more widely spaced in the region of greatest extension (Figure 37.6b). By comparing the seedlings in the two dishes, it can be seen that the region of curvature in the B seedlings corresponds to the region of extension in the A seedlings.

Interpretation The response to the stimulus of one-sided gravity takes place in the region of extension. It does not necessarily mean that this is also the region which detects the stimulus.

Plant growth substances and tropisms

On p. 309 it was explained that growth substances, e.g. auxin, are produced by the tips of roots and shoots and can stimulate or, in some cases, inhibit extension growth. Tropic responses could be explained if the one-sided stimuli produced a corresponding one-sided distribution of growth substance.

In the case of positive geotropism in roots there is evidence that, in a horizontal root, more growth substance accumulates on the lower side. In this case the growth substance is presumed to inhibit extension growth, so that the root tip curves downwards (Figure 37.7, overleaf).

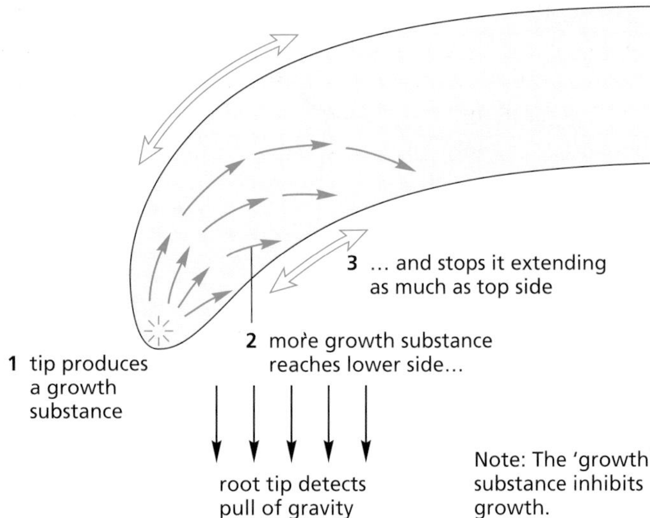
3 … and stops it extending as much as top side
2 more growth substance reaches lower side…
1 tip produces a growth substance
root tip detects pull of gravity
Note: The 'growth' substance inhibits growth.

Figure 37.7 Possible explanation of positive geotropism in roots

In the case of phototropism, it is assumed that the distribution of growth substance causes reduced extension on the illuminated side and/or increased extension on the non-illuminated side. There is not much supporting evidence for this, however.

Questions

1 In Figure 37.8 the two sets of pea seedlings were sown at the same time, but the pot on the left was kept under a light-proof box. From the evidence in the picture,
  a what effects does light appear to have on growing seedlings and
  b how might this explain positive phototropism?

2 It is suggested that it is the very tip of the radicle which detects the one-sided pull of gravity even though it is the region of extension which responds. How could you modify Experiment 3 to test this hypothesis? What snags might you encounter in interpreting your results? (See p. 57 for root growth.)

Figure 37.8 Effect of light on shoots (see Question 1)

Animals

Obvious responses by animals are such things as moving towards food or away from danger, but among the protoctista and invertebrate animals there are many examples of non-directional responses; the animal moves about at random until it escapes from an unpleasant stimulus.

Non-directional responses (taxes)

If *Paramecium*, observed on a microscope slide, swims into an obstacle in its path, the contact acts as a stimulus. The *Paramecium* responds by reversing the beat of its cilia (p. 311) and moving backwards, away from the obstacle (Figure 37.9). Then *Paramecium* turns through a small angle and moves forward again. The direction of turning is not related to the direction of the obstacle; it could be a turn to the left or to the right. If the *Paramecium* bumps into the obstacle again on its forward path, it simply repeats the 'reverse–turn–forward' pattern. By a series of random turns, it will sooner or later get past the obstacle. In its normal environment, *Paramecium's* reactions are probably more varied than this simple response.

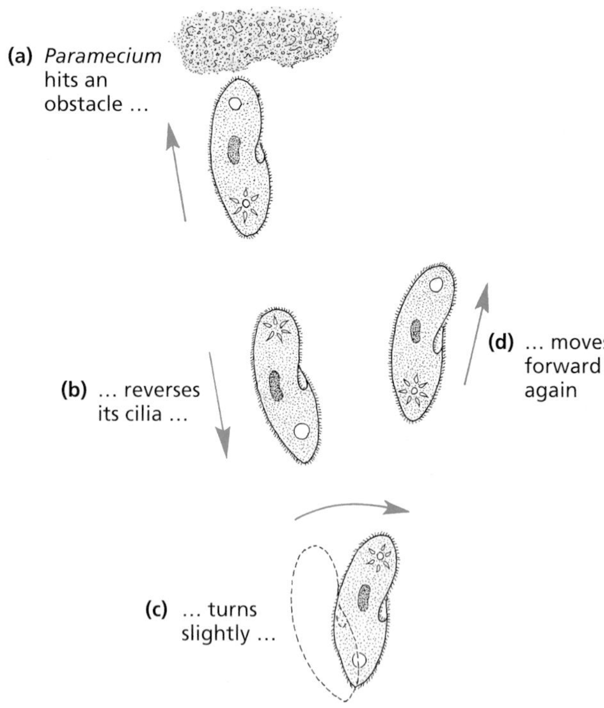
(a) Paramecium hits an obstacle …
(b) … reverses its cilia …
(c) … turns slightly …
(d) … moves forward again

Figure 37.9 Non-directional response in *Paramecium*

Figure 37.10 shows a choice chamber. The air on the left side will be moister (more humid) because of the water in the lower compartment. The air on the right will be dryer because the silica gel absorbs water vapour. Some woodlice are kept in a dry container for 3 hours. If ten of these woodlice are then placed in the top compartment of the choice chamber, they will move about at random. On the dry side, they will move rapidly but on the moist side they will move slowly and, sooner or later, they will stop.

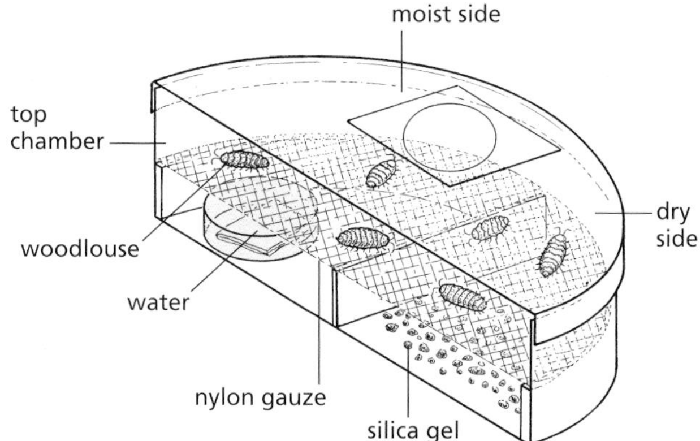

Figure 37.10 A choice chamber (only one half shown)

It looks as if they have moved deliberately towards the moist side but they have not. Their final positions result from the fact that the stimulus of dry air makes them move rapidly in any direction. When a movement happens to bring them by chance into the moist air, they stop moving.

If you do an experiment like this, you must take certain precautions if your results are to be reliable.

1 You must make sure that there is no stimulus other than moisture (humidity) which could affect the woodlice. If light was coming from the right side it could be said that the woodlice were responding to the light by moving away from it, rather than responding to the humidity. The light would have to come from above and not from one side. Alternatively you could repeat the experiment several times with light coming from each side in turn to show it made no difference to the results, or you could cover the choice chamber with a box so that it was in total darkness and lift the box at 1-minute intervals to see where the woodlice were.
2 You must use at least ten woodlice. If you had only five and they ended up with two on the dry side and three on the moist side, this is just as likely to happen by chance. For the results to be significant, the woodlice must end up with at least nine on the moist side and one or none on the dry side. Even 8:2 would not be accepted as very much better than chance.
3 You must do the experiment several times. The more often you repeat the experiment, and the more often you get a 9:1 or 10:0 distribution of woodlice, the more confident you can be that the animals are responding to humidity.

Directional responses

In these responses, the direction of the response is related to the direction of the stimulus. Tropisms are directional responses made by plant organs. A shoot is positively phototropic; it grows towards the source of light. The most obvious responses of animals are directional: towards food, towards a mate, away from an enemy.

A simple directional response can be shown with blowfly larvae (maggots). In Figure 37.11 the maggot is moving away from the light source at A. Then light A is switched off and light B is switched on. The maggots change direction and move away from light B.

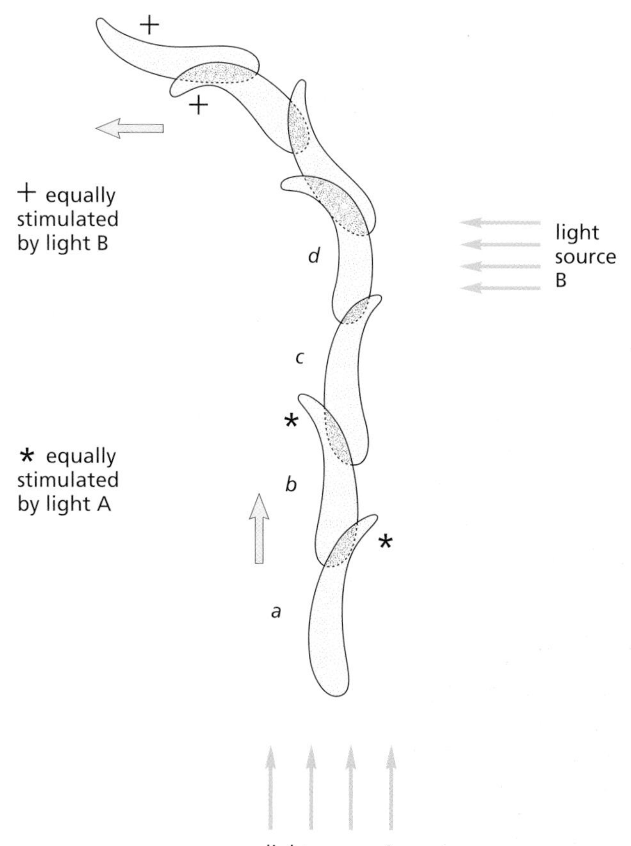

Figure 37.11 Directional response of blowfly larva

Close study of their movement shows that as they wriggle forwards, the maggots swing their head end from side to side. There is a group of cells in the head which are sensitive to light. If these cells are stimulated by light coming from the right, they send off nerve impulses to the maggot's body which make it swing its head more violently to the left.

When light B is switched on, the maggot at position *d* responds by swinging its head further to the left and so altering course. When the maggot is at positions *a*, *b* or *c*, light A affects both sides of its head equally as it swings from side to side. So the maggot continues to move away from the light.

Responses contribute to survival

Woodlice are unable to control their water losses as efficiently as insects do (most crustacea are aquatic). Their response to a humidity gradient keeps them in areas where their water losses are minimal, e.g. under stones, in rotting wood or in the soil.

Blowflies lay their eggs in the dead bodies of animals. The eggs hatch into larvae (maggots) which burrow through the dead animal, digesting and eating its flesh. The response of moving away from light occurs in fully grown maggots and makes them move downwards from the dead body and burrow into the soil. Here they turn into pupae protected by the soil from changes of temperature, from drying out and from being eaten by birds.

Sense organs

Before an animal can respond to a stimulus, it must be able to detect it. This detection is done by sensory organs. Sometimes these are sensory cells all over the bodies of the animals. Some of these cells detect changes in light, some detect changes in temperature and some respond to chemicals or touch. In the vertebrates, the skin contains some sensory cells of this kind (p. 158), but there are also special sense organs where the sensitive cells are packed closely together in one place and the stimulus is directed on to them. In our eyes (p. 159) the retina consists of thousands of closely packed cells, all able to respond to light. The stimulus of light is directed on to the sensitive cells by the cornea and lens.

The cochlea (p. 162) contains thousands of cells which detect vibrations. The ear drum and ossicles direct the stimulus on to the sensitive cells.

Figures 37.12–37.17 show parts of the sensory equipment of various animals. Although some parts of plants, such as the growing points, are more sensitive to stimuli than others, there are no special sense organs.

Figure 37.12 Common frog. The special sense organs of animals are concentrated on the head. The frog's circular ear drum is seen behind and below the eye, and its nostril in front of the eye. The nostrils can be closed when the frog is under the water

Figure 37.13 Hare. The long ears help to pick up and locate sound vibrations. The eyes at the side of the head give the hare good vision all around

Figure 37.14 Grass snake; sense of smell. The flicking tongue picks up chemicals in the air and carries them to a sense organ in the roof of the mouth which 'tastes' them

Figure 37.15 Long-eared bat. The bat gives out high-pitched sounds which are reflected back, from its prey and from obstacles, to its ears and sensitive patches on its face. By timing these echoes the bat can judge its distance from the obstacle or prey

Figure 37.16 Tawny owl. The owl's eyes point forwards and help it to judge distances accurately and so capture prey. The large size of the eyes helps to pick up what little light there is at night time

Figure 37.17 Gerbils. The whiskers on the face affect nerve endings in the skin. The slightest movement of the whiskers, even those caused by air movements, will cause nerve impulses to be sent to the brain

Questions

1 What directional responses do wasps appear to make
   a when they pester you at a picnic
   b when they try to escape from a room?

2 In the choice chamber experiment, what differences, other than humidity and light, might affect the final distribution of the woodlice?

3 What would be a good control experiment (see page 23) for the choice chamber experiment to show that it was humidity and not some other stimulus that affected the woodlice? What results would you expect in the control experiment?

4 How would you design an experiment to see if woodlice responded to the stimulus of light?

5 Why do you think that most animals have their main sense organs on their heads?

Checklist

- Living organisms respond to stimuli. They are said to be sensitive.
- Stimuli are such things as touch, heat, cold, and light acting on the organism.
- Animals have sensory organs for detecting these stimuli. Plants are sensitive but do not have specialized sense organs.
- The roots and shoots of plants may respond to the stimuli of light or gravity.
- A response related to the direction of the stimulus is a tropism.
- Phototropism is a growth response to the direction of light.
- Geotropism is a growth response to the direction of gravity.
- Growth towards the direction of the stimulus is called 'positive'; growth away from the stimulus is called 'negative'.
- Tropic responses bring shoots and roots into the most favourable positions for their life-supporting functions.
- In animals, non-directional responses are made at random until favourable conditions are reached.
- A non-directional response is called a taxis (Pl. taxes).
- A directional response is related to the direction of the stimulus.
- Higher animals have their sensory cells concentrated into specialized sensory organs.

Diversity of organisms
Examination questions

Do not write on this page. Where necessary copy drawings, tables or sentences.

1 The diagram shows the results of an experiment to find the effect of light on the growth of seedlings.

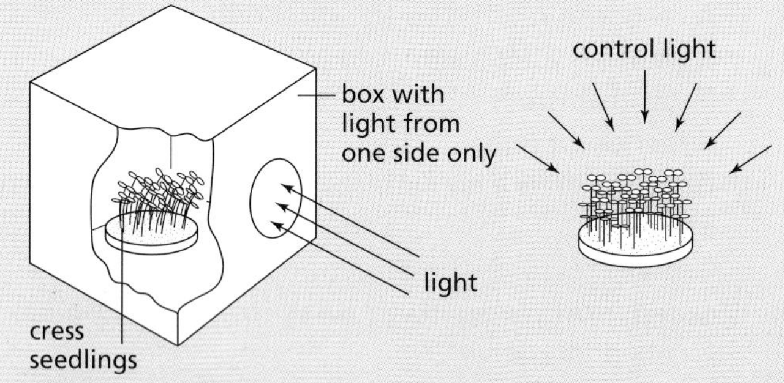

a Describe the results of the experiment. (2)

b Name the type of substance that controls growth in plants. (1)

c Complete the sentence about growth in plant roots. Plant roots grow towards ______ and in the direction of the force of ______. (2)

(AQA)

2 The drawing shows an aardvark.

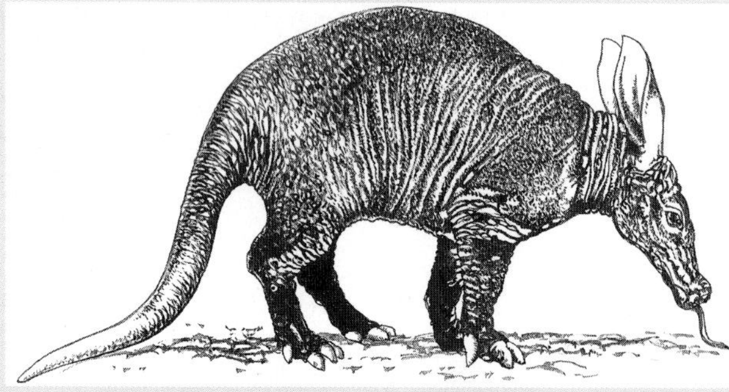

a The aardvark has receptors which are sensitive to changes in the environment. Name **one** part of the aardvark's body where there are:
(i) receptors which are sensitive to light;
(ii) receptors which help it keep its balance;
(iii) receptors which are sensitive to chemicals. (3)

b The aardvark feeds mainly at night. It is hunted by several predators. Use information from the drawing to give one feature of the aardvark which is an adaptation for sensing predators at night. (1)

(AQA)

3 This question is about insect life cycles.

a (i) The diagram below shows stages in the life cycle of a blowfly. Write down the name of each stage. One has been done for you. (2)

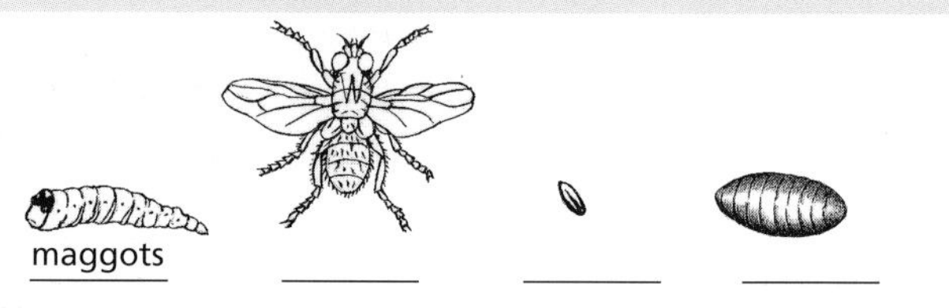

(ii) The life cycle of the blowfly shows metamorphosis. Explain what **metamorphosis** means. Give an example from the life cycle of the blowfly. (2)
(iii) The different stages in the life cycle of the blowfly eat different foods. Explain how this helps the blowfly survive. (3)

b Blowfly maggots feed on dead and decaying matter. In 1500 it was found that placing live maggots of blowflies in infected wounds often helped the healing process.
In 1917, during the First World War, soldiers had wounds which were naturally infected with maggots. The maggots helped to heal their wounds more quickly.In 1930 doctors used maggots to help heal the wounds of some children with bone disease.
Suggest why the wounds healed more quickly when infected with maggots. (2)

(OCR)

4 The key shows some of the features of chordates.*

a Name **one** feature common to all chordates. (1)

b Identify chordate groups A, B, C and D. (4)

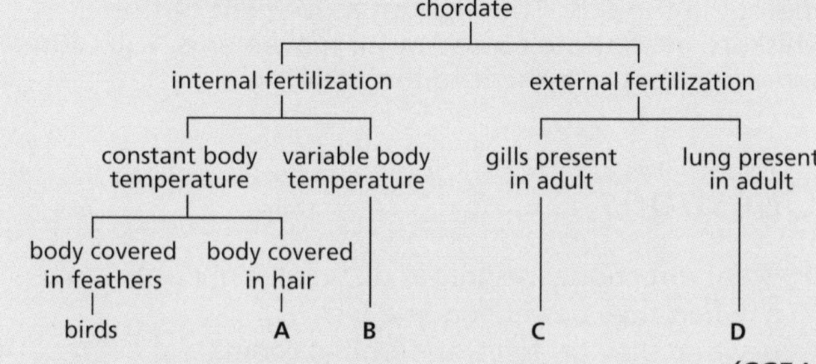

(CCEA)

* The term 'vertebrate' is used in this book.

5 Complete the table below. (6)

(CCEA)

Classification group	Vascular tissue	Stem, root and leaves present	Method of reproduction	Example
Algae	No	No	Spores	
Bryophytes		No	Spores	Moss
Pteridophytes	Yes		Spores	Fern
Angiosperms	Yes	Yes		
	No	No	Spores	Toadstool

Micro-organisms and humans

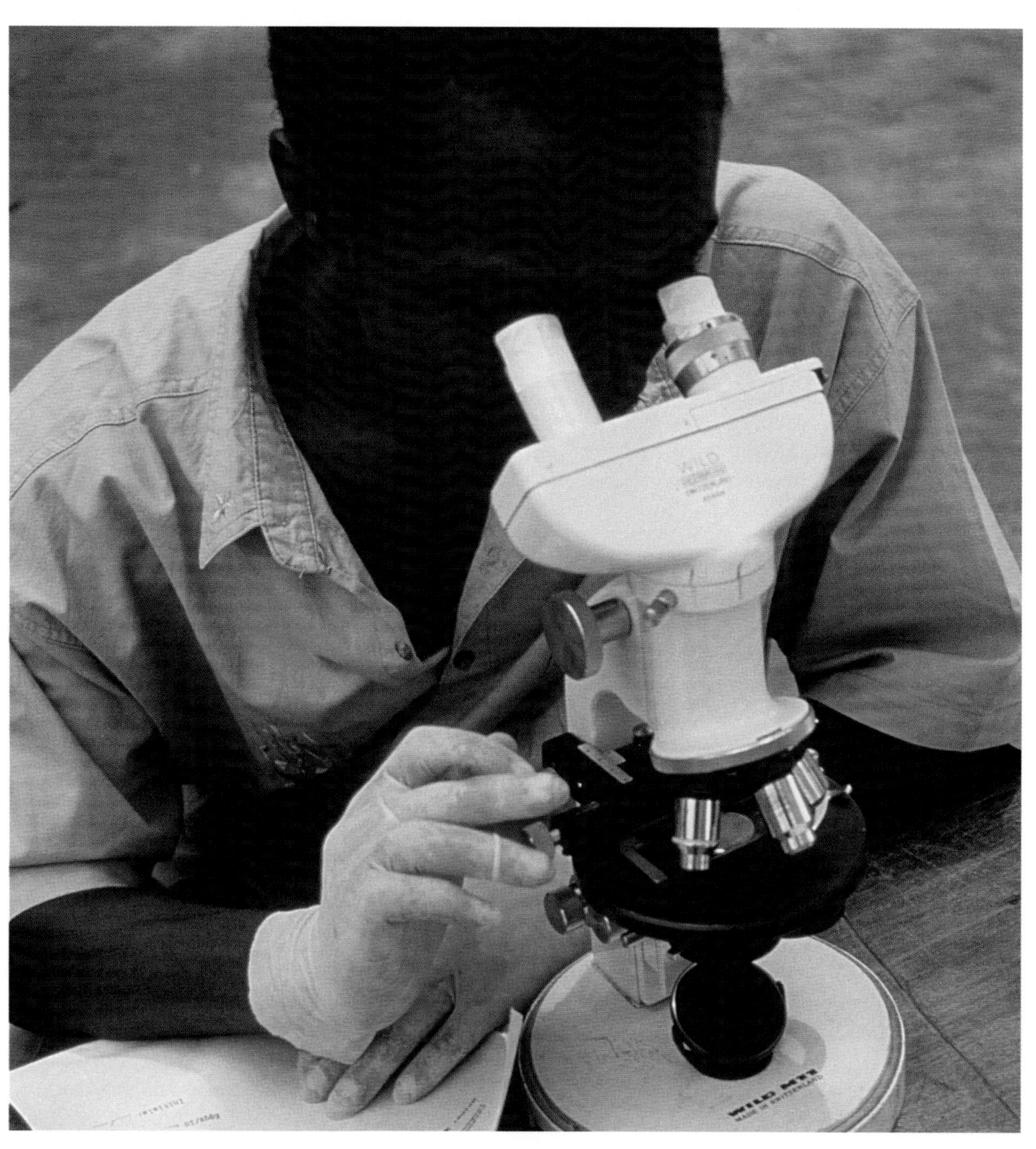

38 Biotechnology

Fermentation products
Yoghurt, cheese, beer, wine, vinegar, bread, soy sauce, biofuels.

Antibiotics
Penicillin. Commercial production.

Single-cell protein
Mycoprotein.

Enzymes

Sewage treatment
Activated sludge process. Sludge digestion.

Biotechnology can be defined as the application of biological organisms, systems or processes to manufacturing and service industries.

Although biotechnology is 'hot news', we have been making use of it for hundreds of years. Wine-making, the brewing of beer, the baking of bread and the production of cheese all depend on fermentation processes brought about by yeasts, other fungi and bacteria, or enzymes from these organisms.

Antibiotics, such as penicillin, are produced by mould fungi or bacteria. The production of industrial chemicals such as citric acid or lactic acid needs bacteria or fungi to bring about essential chemical changes.

Sewage disposal (p. 331) depends on bacteria in the filter beds to form the basis of the food chain which purifies the effluent.

Biotechnology is not concerned solely with the use of micro-organisms. Cell cultures and enzymes also feature in modern developments. In this chapter, however, there is space to consider only a representative sample of biotechnological processes which use micro-organisms.

The processes are described only in outline and relate to the basic, traditional steps. In modern processes, genetically engineered organisms, industrially produced enzymes and a variety of different substrates, nutrients and additives may be involved.

Fermentation products

The term 'fermentation' does not apply only to alcoholic fermentation but to a wide range of reactions, brought about by enzymes or micro-organisms. On p. 20 the anaerobic respiration of glucose to alcohol or lactic acid was described as a form of fermentation.

Micro-organisms which bring about fermentation are using the chemical reaction to produce the energy which they need for their living processes. The reactions which are useful in fermentation biotechnology are mostly those that produce incompletely oxidized compounds. A reaction that goes all the way to carbon dioxide and water is not much use in this context.

The micro-organisms are encouraged to grow and multiply by providing nutrients such as glucose, with added salts and, possibly, vitamins. Oxygen or air is bubbled through the culture if the reaction is aerobic, or excluded if the process is anaerobic. An optimum pH and temperature are maintained for the species of microbe being cultured.

Yoghurt

Yoghurt and cheese have been made by herdsmen for thousands of years. Both products result from the fermentation of milk by bacteria. Initially these bacteria came from natural sources, e.g. from the animals themselves but, since there would be a mixture of different bacteria in different conditions, the products were often unpredictable.

Today, the micro-organisms and the conditions are carefully controlled so that the wanted product is assured.

Milk from cows, sheep or goats may be used to prepare yoghurt. On a commercial scale the fat and protein content of the milk is adjusted and the milk is homogenized. This is a mechanical process which breaks up the fat droplets and prevents them from separating out. The milk is then pasteurized (p. 93).

Pasteurization destroys pathogenic (potentially harmful) bacteria and other micro-organisms that might cause unwanted changes.

The pasteurized milk is then fermented by adding a 'starter culture' of bacteria, usually a mixture of *Streptococcus thermophilus* and *Lactobacillus bulgaricus*. These bacteria act on the milk sugar, **lactose**, and convert it to lactic acid which, in turn coagulates the milk protein, **casein**, to produce the thick creamy consistency of yoghurt. The fermentation works best at a temperature of 46 °C and, when complete, the yoghurt is cooled to 5 °C to stop the bacterial processes. The lactic acid gives the yoghurt its slightly sour taste.

Throughout the process, the materials and containers must be kept in a sterile condition, i.e. free from any micro-organisms that might compete with the *Streptococcus* and *Lactobacillus* and produce unwanted substances.

Cheese

The basic processes for cheese production are the same as for yoghurt. The composition of the milk is adjusted; it is homogenized and pasteurized. A variety of bacteria cultures may be used to ferment the milk, depending on the type of cheese to be produced, but they are usually species of *Streptococcus* or *Lactobacillus* and they work best at about 40 °C. The process differs from yoghurt production, at this stage, in that a mixture of enzymes, called **rennet**, is added. Rennet contains the enzyme **chymosin**, which coagulates milk casein and forms the semi-solid 'curds'. Chymosin can be obtained from calves' stomachs but is now mainly produced from genetically engineered yeast which carries the chymosin gene.

The liquid 'whey' is drained from the curds which are partially dried and compressed (Figure 38.1). The degree of drying and the method of ripening depends on the variety of cheese required. In principle, the bacterial enzymes act on the proteins and fats in the curd, partly digesting them to amino acids and fatty acids which give the cheese its flavour and aroma.

For varieties of 'blue cheese', *Penicillium* spores are added at the fermentation stage (p. 288).

Figure 38.1 Cheese-making on a small scale. The casein has formed the curds which are being scooped out from the liquid whey.

Beer

As with yoghurt and cheese, the production of beer and wine dates back thousands of years. The fermentation in this case is brought about by yeast, not bacteria.

Commercial techniques have refined the traditional methods so that the process is controlled from start to finish.

Beer is made from barley. The barley grains are soaked in water for 2 days at 15–20 °C and then allowed to germinate for 4 days. During germination, the starch-digesting enzymes in the seeds are activated. The germinating seeds are then dried at temperatures which kill the seeds but do not denature the enzymes. This process is called 'malting' and the product is 'barley malt'.

The dried grains are crushed (milled) and mixed with water at about 65 °C and extra starch from wheat or rice is added. The barley enzymes now digest the starch to maltose and glucose. The sugary solution (called 'wort') is filtered and boiled with hops which suppress the growth of bacteria and give the beer its bitter flavour.

Yeast is added to the wort. The most commonly used species of yeast are *Saccharomyces cerevisiae* and *S. carlsbergensis*. Enzymes in the yeast convert maltose to glucose. In the early stages of fermentation (Figure 38.2), yeast respires aerobically, oxidizing the glucose to carbon dioxide and water and dividing rapidly. As the oxygen supply diminishes, anaerobic respiration takes over and the glucose is oxidized only as far as alcohol and carbon dioxide (p. 20).

Figure 38.2 Fermenting beer. The wort is being fermented in a large vat

Fermentation takes 5–14 days, depending on the type of beer, and results in the conversion of all the sugar to alcohol, giving a 3–5 per cent solution. After this, the dead yeast, proteins and hop resins are allowed to settle out, helped by 'clarification agents'. Beer for kegs, bottles or cans is pasteurized and its carbon dioxide level is adjusted to give the necessary 'sparkle' and fizzy taste. 'Real ale' is not pasteurized; the residual living yeast continues fermenting and producing carbon dioxide in the barrel.

The flavour of the final product depends on a multitude of substances, as well as alcohol and carbon dioxide, produced in the course of fermentation.

Wine

Grape juice is the basis for wine-making, and yeast is the micro-organism that causes the fermentation.

The grapes are crushed and the extracted juice is treated with sulphur dioxide to kill bacteria and naturally occurring yeasts. The sugar content and pH are adjusted before adding a starter culture of yeast, usually a variety of *Saccharomyces cerevisiae*. The natural yeasts on the grape skins would bring about fermentation but the outcome would be unpredictable.

The juice is aerated at first; the yeast respires aerobically and its population increases rapidly. When aeration stops, the yeast ferments the grape sugar (glucose) anaerobically to alcohol and carbon dioxide. The alcohol content rises to 10–15 per cent, a concentration which eventually kills the yeast. Low temperatures prolong the fermentation and increase the alcohol content.

The wine is transferred to barrels and any particles are allowed to settle, with the aid of clearing agents. This may be repeated several times until the wine is clear, then it is bottled.

The variety of wine depends on the type of grape used and the details of fermentation and storage. Red wine comes from 'black' grapes; the skins are left in the grape juice for a time and the alcohol dissolves out the red pigment. White wines can be prepared from black or white grapes but the skins are removed after crushing. The taste and flavour ('bouquet') of wine depends on the organic acids and many aromatic organic compounds present in the grape juice and produced during fermentation.

Vinegar

Acetobacter is a genus of rod-shaped bacteria widely distributed in nature and especially in rotting plant remains. Different species of *Acetobacter* can oxidize a variety of organic compounds to produce organic acids. Some of these species are exploited in the production of vinegar from beer, cider or wine (*vin aigre* (Fr.) = sour wine).

Malt vinegar is prepared from crushed malted barley which is mixed with hot water and fermented with yeast to produce a rough kind of beer called **gyle**.

The process is much the same as in brewing beer but there is no need to filter or sterilize the liquid, which is allowed to settle for 3–4 months.

The gyle is then inoculated with *Acetobacter* and pumped into large vats which contain two layers of birch twigs. The gyle is pumped over the twigs and the *Acetobacter* grows on the large surface they provide.

With air supply and temperature carefully controlled, most of the ethanol (alcohol) in the gyle is converted aerobically to ethanoic (acetic) acid.

$$\underset{\text{ethanol}}{C_2H_5OH} + \underset{\text{oxygen}}{O_2} \rightarrow \underset{\text{ethanoic acid}}{CH_3COOH} + \underset{\text{water}}{H_2O} + \text{energy}$$

The product is allowed to mature for up to 6 months before filtering and bottling.

There are more traditional and more commercial ways of making vinegar, but this is the principle behind them all.

If wine, beer or cider is left exposed to the air for any length of time, *Acetobacter* will arrive and oxidize the alcohol to ethanoic acid making the beverage sour and undrinkable.

Bread

Yeast is the micro-organism used in bread-making but the only fermentation product needed is carbon dioxide. The carbon dioxide makes bubbles in the bread dough. These bubbles make the bread 'light' in texture.

Flour, water, salt, oil and yeast are mixed to make a dough. Yeast has no enzymes for digesting the starch in flour but the addition of water activates the amylases already present in flour and these digest some of the starch to sugar. With highly refined white flour, it may be necessary to add sugar to the dough. The yeast then ferments the sugar to alcohol and carbon dioxide.

A protein called **gluten** gives the dough a sticky, plastic texture which holds the bubbles of gas. The dough is repeatedly folded and stretched ('kneaded') either by hand, in the home, or mechanically in the bakery. The dough is then left for an hour or two at a temperature of about 27 °C while the yeast does its work. The accumulating carbon dioxide bubbles make the dough rise to about double its volume (Figure 38.3). The dough may then be kneaded again or put straight into baking tins and into an oven at about 200 °C. This temperature makes the bubbles expand more, kills the yeast and evaporates the small quantities of alcohol before the dough turns into bread.

Figure 38.3 Carbon dioxide produced by the yeast has caused the dough to rise

Soy sauce

This is a food flavouring used chiefly with oriental and Asian dishes. It has been in use for, perhaps, over 2000 years after its discovery in China.

Today it is made from a mixture of wheat and soya beans crushed with water. The fungus *Aspergillus oryzae* is introduced to digest the starches to sugars. The mixture is then mixed with salt water and fermented using lactic acid bacteria (*Lactobacillus*) and yeast. The fermentation takes months and produces a red–brown mixture with a wide variety of chemicals which give the sauce its distinctive taste and flavour. Finally, it is filtered, pasteurized and bottled.

An alternative process, using hydrochloric acid rather than fermentation, is quicker but produces an inferior sauce.

Biofuels

On p. 251, it was pointed out that ethanol (alcohol), produced from fermented sugar or surplus grain, could replace, or at least supplement, petrol.

Brazil, Zimbabwe and the USA produce ethanol as a renewable source of energy for the motor car. Since 1990, 30 per cent of new cars in Brazil can use ethanol and many more use a mixture of petrol and ethanol. As well as being a renewable resource, ethanol produces less pollution than petrol does.

However, although the productivity of sugar cane has been greatly increased, and the stems and leaves (bagasse) of the cane are burnt to drive the distilleries, the price of oil has fallen sufficiently to make the production of ethanol unprofitable unless subsidized. The price of oil will inevitably increase as stocks run low, so the technology of ethanol production remains important.

Another biofuel, oil from rapeseed or sunflower seed, can with suitable treatment replace diesel fuel. It is less polluting than diesel but more expensive to produce.

Biogas production on a small scale (as in villages) is mentioned on p. 251 and on a larger scale on p. 332.

Antibiotics

When micro-organisms are used for the production of antibiotics, it is not their fermentation products which are wanted, but complex organic compounds, called **antibiotics**, that they synthesize.

Most of the antibiotics we use come from bacteria or fungi which live in the soil. The function of the antibiotics in this situation is not clear. One theory suggests that the chemicals help to suppress competition for limited food resources, but the evidence does not support this theory.

One of the most prolific sources of antibiotics is *Actinomycetes*. These are filamentous bacteria which resemble microscopic mould fungi. The actinomycete *Streptomyces* produces the antibiotic **streptomycin**.

Perhaps the best known antibiotic is **penicillin** which is produced by the mould fungus *Penicillium* (p. 352) and was discovered by Sir Alexander Fleming in 1928. Penicillin is still an important antibiotic but it is produced (Figure 38.4) by mutant forms of a different species of *Penicillium* from that studied by Fleming. The different mutant forms of the fungus produce different types of penicillin.

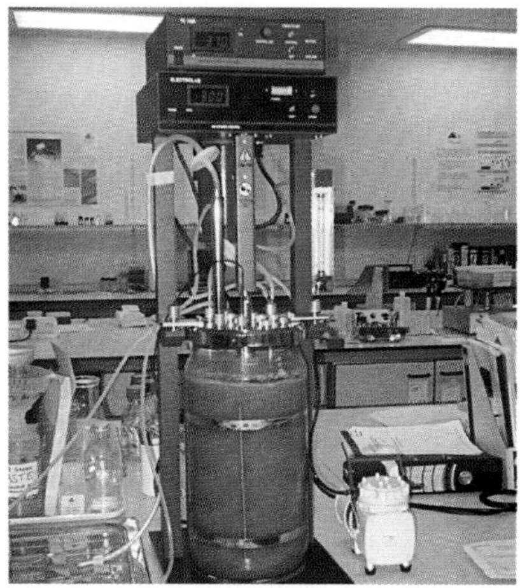

Figure 38.4 A laboratory fermenter for antibiotic production which will eventually be scaled up to 10 000-litre fermention vessels

The penicillin types are chemically altered in the laboratory to make them more effective and to 'tailor' them for use with different diseases. 'Ampicillin', 'methicillin' and 'oxacillin' are examples.

Antibiotics attack bacteria in a variety of ways. Some of them disrupt the production of the cell wall and so prevent the bacteria from reproducing, or even cause them to burst open; some interfere with protein synthesis and thus arrest bacterial growth. Those that stop bacteria from reproducing are said to be **bacteriostatic**; those that kill the bacteria are **bacteriocidal**.

Animal cells do not have cell walls, and the cell structures involved in protein production are different. Consequently, antibiotics do not damage human cells although they may produce some side-effects such as allergic reactions.

Commercial production

Antibiotics are produced in giant fermenting tanks, up to 100 000 litres in capacity. The tanks are filled with a nutrient solution. For penicillin production, the carbohydrate source is sugar, mainly lactose or 'corn-steep liquor' a by-product of the manufacture of cornflour and maize starch; it contains amino acids as well as sugars. Mineral salts are added, the pH is adjusted to between 5 and 6, the temperature is maintained at about 26 °C, air is blown through the liquid and it is stirred. The principles of industrial fermentation are shown in Figure 38.5. The nutrient liquid is seeded with a culture of the appropriate micro-organism which is allowed to grow for a day or two (Figure 38.4). Sterile conditions are essential. If 'foreign' bacteria or fungi get into the system they can completely disrupt the process. As the nutrient supply diminishes, the micro-organisms begin to secrete their antibiotics into the medium.

The nutrient fluid containing the antibiotic is filtered off and the antibiotic extracted by crystallization or other methods.

Single-cell protein

The object of this exercise is to produce micro-organisms in bulk to use as human food or animal feed. Unicellular algae, fungi and bacteria have been used.

The reasoning behind such projects was expressed by an eminent biologist many years ago when he said, 'In 24 hours half a tonne of bullock will make 500 grams of protein; half a tonne of yeast will make 50 tonnes and needs only a few square yards to do it on.' Moreover, the feedstocks for micro-organisms can often be produced cheaply from the waste products of other commercial processes. The production can be precisely controlled and is not dependent on the weather.

In theory, such a prodigous rate of reproduction in micro-organisms could make up for a world shortage of protein. In fact, very few projects have become commercially viable. Not only have the products turned out to be more expensive than conventional crops but they are also colourless and tasteless. Furthermore, because of the high ratio of nucleus to cytoplasm, they contain toxic amounts of nucleic acids which have to be removed.

One product which has reached the market is produced from a filamentous fungus, *Fusarium graninearum*. This is not, of course, a 'single cell' but can be used to produce a **mycoprotein** known as 'Quorn'. The fungus is grown in a medium containing glucose and mineral salts at 30 °C using specially designed fermenters. The process used is called **continuous culture**. The medium is circulated round the fermenter and the fungus is filtered off at a steady rate and dried. The culture medium is topped up as necessary. This is cheaper and simpler than the **batch processes** described above and avoids repeated sterilization of the fermenter vessels.

The filamentous nature of the fungus gives the product a texture similar to meat but it has a bland taste. It can be flavoured with natural products and cooked in a similar way to meat.

Its attraction is that it is rich in protein, low in fat and has a good level of dietary fibre. This is a healthier combination than is found in most meat products.

Enzymes

Enzymes can be produced by commercial fermentation using readily available feedstocks such as corn-steep liquor or molasses. Fungi (e.g. *Aspergillus*) or bacteria (e.g. *Bacillus*) are two of the commonest organisms used to produce the enzymes.

These organisms are selected because they are non–pathogenic (p. 284) and do not produce antibiotics. The fermentation process is similar to that described for penicillin. If the enzymes are extracellular (p. 15) then the liquid feedstock is filtered from the organism and the enzyme is extracted (Figure 38.5). If the enzymes are

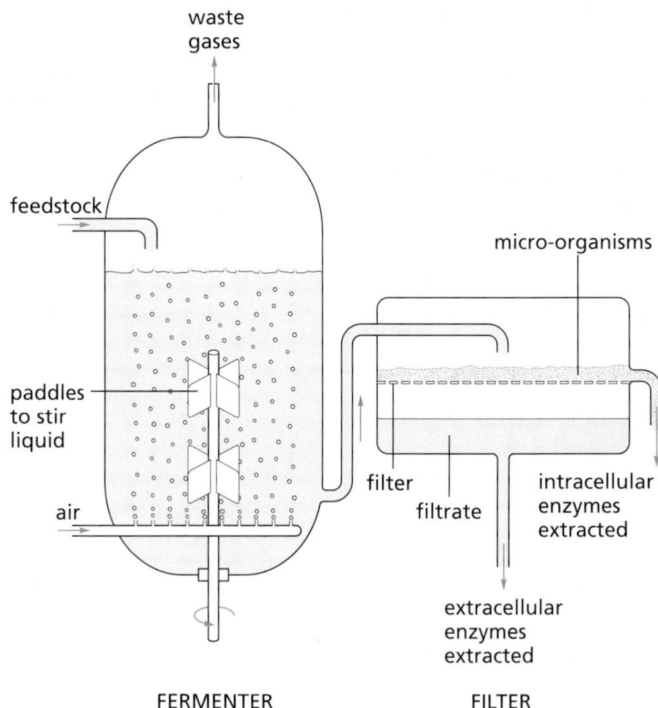

Figure 38.5 Principles of enzyme production from micro-organisms

intracellular, the micro-organisms have to be filtered from the feedstock. They are then crushed and the enzymes extracted with water or other solvents.

By the techniques of genetic engineering (p. 213), new genes can be introduced into the microbes to 'improve' the action of the enzymes coded for by the genes (e.g. making the enzymes more heat stable).

Genes from entirely different organisms can be introduced to make the microbes produce enzymes which are not in their normal repertoire. For example, the mammalian chymosin gene has been introduced into yeast cells (see 'Cheese' above).

One effective way of using enzymes is by 'immobilizing' them. The enzymes or the micro-organisms which produce them are held in or on beads or membranes of an insoluble and inert substance, e.g. plastic. The beads or membranes are packed into columns and the substrate is poured over them at the optimum rate. This method has the advantage that the enzyme is not lost every time the product is extracted. Immobilized enzymes also allow the process to take place in a continuous way rather than a batch at a time.

Some commercial uses of enzymes are listed below and on p. 18.

- *Proteases.* In washing powders for dissolving stains from, e.g. egg, milk and blood; removing hair from animal hides; cheese manufacture; tenderizing meat.
- *Lipases.* Flavour enhancer in cheese; in washing powders for removal of fatty stains.
- *Pectinases.* Clarification of fruit juices; maximizing juice extraction.
- *Amylases.* Production of glucose from starch.

Sewage treatment

Micro-organisms, mainly bacteria and protozoa, play an essential part in the treatment of sewage to make it harmless.

Sewage contains bacteria from the human intestine which can be harmful (p. 333). These bacteria must be destroyed in order to prevent the spread of intestinal diseases. Sewage also contains substances from household wastes (such as soap and detergent) and chemical from factories. These too must be removed before the sewage effluent is released into the rivers. Rain water from the streets is also combined with the sewage.

Inland towns have to make their sewage harmless in a sewage treatment plant before discharging the effluent into rivers. A sewage works removes solid and liquid waste from the sewage, so that the water leaving the works is safe to drink.

In a large town, the main method of sewage treatment is by the activated sludge process (Figures 38.6 and 38.7).

The activated sludge process

1 *Screening.*The sewage entering the sewage works is first 'screened'. That is, it is made to flow through a metal grid which removes the solids like rags, plastics, wood and so forth. The 'screenings' are raked off and disposed of – by incineration, for example.

2 *Grit.* The sewage next flows slowly through long channels. As it flows, grit and sand in it settle down to the bottom and are removed from time to time. The grit is washed and used for landfill.

3 *First settling tanks.* The liquid continues slowly through another series of tanks. Here about 40 per cent of the organic matter settles out as crude sludge. The rest of the organic matter is in the form of tiny suspended particles which pass, with the liquid, to the aeration tanks.

The semi-liquid sludge from the bottom of the tank is pumped to the sludge digestion plant (see p. 332).

4 *Aeration tanks.* Oxygen is added to the sewage liquid, either by stirring it or by bubbling compressed air though it. Aerobic bacteria and protozoa (p. 268) grow and reproduce rapidy in these conditions.

Figure 38.7 Sewage treatment – activated sludge method. In the foreground are the rectangular aeration tanks. Behind these are the circular settlement tanks. In the background are the sludge digesters

These micro-organisms clump the organic particles together. Enzymes from the bacteria digest the solids to soluble products which are absorbed by the bacteria and used for energy and growth

Dissolved substances in the sewage are used in the same way. Different bacteria turn urea into ammonia, ammonia into nitrates and nitrates into nitrogen gas. The bacteria derive energy from these chemical changes. The protozoa (Figure 38.8, overleaf) eat the bacteria.

In this way, the suspended solids and dissolved substances in sewage are converted to nitrogen, carbon dioxide (from respiration) and the cytoplasm of the bacteria and protozoa, leaving fairly pure water.

5 *Second settling tanks.* The micro-organisms settle out, forming a fine sludge which is returned to the aeration tanks to maintain the population of micro-organisms. This is the 'activated sludge' from which the process gets its name. The sewage stays in the aeration tanks for only 6–8 hours but the recycling of activated sludge allows the micro-organisms to act on it for 20–30 days.

6 When all the sludge has settled, the water is pure enough to discharge into a river and the sludge passes to a digester.

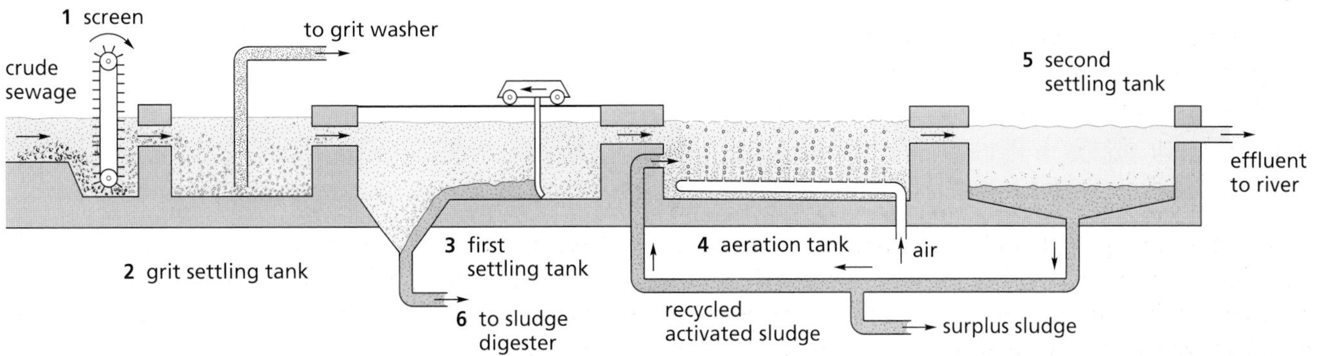

Figure 38.6 Sewage treatment – activated sludge process

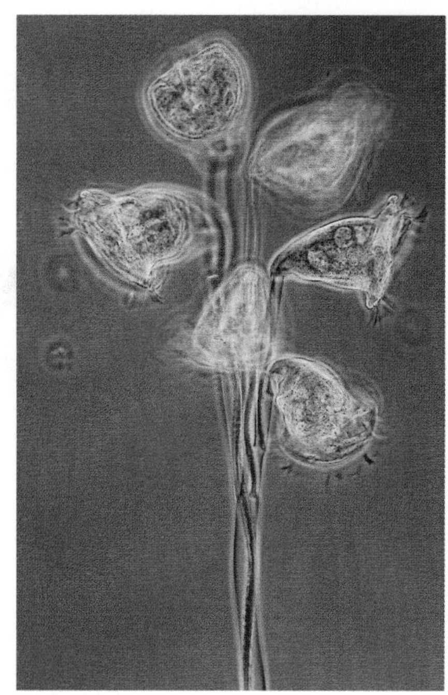

Figure 38.8 Protozoa in activated sludge (× 150). These single-celled organisms ingest bacteria in the liquid sewage

Sludge digestion

In this process, the principal role of the micro-organisms is to react with sewage sludge and convert it to harmless or useful substances. This is of great value in reducing potential sources of pollution.

The sludge is fed into large, enclosed tanks (anaerobic digesters) and heated initially to about 25 °C. Conditions soon become anaerobic and the wide range of bacteria in the waste material digest the organic compounds to fatty acids, monosaccharides and amino acids. These are acted on by other bacteria to produce organic acids and alcohols. Finally, **methanogenic** bacteria convert these to methane (CH_4) and water.

The methane is drawn off and used as an energy source to provide the heat needed to start the reactions and there is usually a substantial surplus for other purposes such as central heating or driving the compressors for activated sludge.

The advantages of the anaerobic biodigester are:

- the bulk of the sludge is greatly reduced;
- the high temperatures of anaerobic fermentation destroy potential pathogens so that the residue can be safely used as a soil conditioner or compost and the remaining liquid can be used or sold as a liquid fertilizer;
- the odours are greatly lessened (particularly valuable in intensive pig farms);
- dumping of sludge is reduced and the potential for water pollution with farmyard slurry is avoided;
- the methane ('biogas') is a valuable, relatively non-polluting source of energy.

The dried sludge from the digester can be used as a soil improver provided it is free from pathogenic bacteria and heavy metals. Otherwise, it is incinerated.

Biogas production is not confined to sludge. Many organic wastes, e.g. those from sugar factories, can be fermented anaerobically to produce biogas. In developing countries, biogas generators use animal dung to produce methane for whole villages (p. 251).

Questions

1. Name three micro-organisms used in industrial fermentation and state their fermentation products.
2. Apart from using their fermentation products, how else are micro-organisms exploited commercially?
3. What are the general conditions that need to be controlled for successful growth of micro-organisms on a large scale? Why is it necessary to ensure sterile conditions?
4. What are 'curds and whey'?
5. Why do you think bio-digesters need anaerobic conditions?
6. Yeast can digest sugar but not starch, so how can fermentation proceed in brewing of beer and bread-making where the original substrate is cereal starch?
7. Why does wine 'go off' if left exposed to the air?
8. How do modern penicillins differ from the one discovered in 1928?
9. What part do micro-organisms (bacteria and protozoa) play in sewage treatment?

Checklist

- Biotechnology is the application of living organisms, systems or processes in industry.
- Many biotechnological processes use micro-organisms (fungi and bacteria) to bring about the reactions.
- Most biotechnological processes are classed as 'fermentations'.
- Fermentation may be aerobic or anaerobic.
- The required product of biotechnology may be the organism itself (e.g. mycoprotein) or one of its products (e.g. alcohol).
- Yoghurt and cheese are made from fermented milk using bacteria.
- Beer and wine are produced from fermented sugars using yeast.
- Antibiotics are produced from bacteria and fungi.
- Enzymes from micro-organisms can be produced on an industrial scale and used in other biotechnology processes.
- Sterile conditions are essential in biotechnology to avoid contamination by unwanted microbes.
- Bacteria and protozoa play an essential role in the purification of sewage.

39 Disease: causes, transmission and control

Bacterial diseases
Salmonella food poisoning, gonorrhoea, syphilis.

Viral diseases
Common cold, influenza, AIDS.

Fungal parasites
Tinea.

Protozoan parasites
Malaria, amoebic dysentery.

Disease transmission
By air, water and food; vectors; contagion.

Methods of prevention
Natural barriers: prevention of transmission by air, water, food and contagion; water treatment; immunization; drug therapy.

Global travel

Non-transmissible diseases

These are diseases which are not infectious. In origin they may be genetic (e.g. cystic fibrosis, p. 198), nutritional (e.g. vitamin deficiency diseases, p. 89), degenerative (e.g. arthritis, heart attacks, strokes), or they may have environmental causes (e.g. silicosis resulting from exposure to mineral dust), or no external or clearly defined cause (e.g. diabetes and many cancers).

Transmissible diseases

These are the infectious diseases which can be passed from one person to another. They are caused by viruses, bacteria, fungi or protozoa and some of them are described below.

Bacterial diseases

Most bacteria are useful or harmless as explained on pp. 229 and 284. Even the bacteria which live and reproduce on our skin or in our breathing passages or intestines are usually harmless. In fact, they may keep more harmful competitors away.

Some bacteria in the human large intestine digest vegetable fibre in our food to form fatty acids, and others produce vitamin K. It is possible that we absorb and use these substances.

However, some species of parasitic bacteria cause disease when they invade our bodies. These are called **pathogenic bacteria**.

Species of the bacterium ***Streptococcus*** cause sore throats, blood poisoning and scarlet fever. Species of ***Clostridium*** bacteria cause tetanus and botulism. ***Staphylococcus*** species cause tonsillitis and boils. Tuberculosis (TB), cholera, typhoid, diphtheria, whooping cough, food poisoning, gonorrhoea and syphilis are all bacterial diseases.

The ill-effects of a bacterial disease are caused mainly by poisonous products produced either by the bacteria or by the cells which they invade. Bacterial poisons are called **toxins**. Toxins damage the cells in which the bacteria are growing. They also upset some of the systems in the body. This gives rise to a raised temperature, headache, tiredness and weakness, and sometimes to diarrhoea and vomiting.

Salmonella food poisoning

One of the commonest causes of food poisoning is the toxin produced by the bacteria *Salmonella typhimurium*, *S. enteritidis* or some varieties of *Escherichia coli* (*E. coli*). These bacteria live in the intestines of cattle, pigs, chickens and ducks without causing disease symptoms. Humans, however, may develop food poisoning if they drink milk or eat meat or eggs which are contaminated with *Salmonella* bacteria from the alimentary canal of an infected animal.

Intensive methods of animal rearing may contribute to a spread of infection unless care is taken to reduce the exposure of animals to infected faeces.

The symptoms of food poisoning are diarrhoea, vomiting and abdominal pain. They occur from 12 to 24 hours after eating the contaminated food. Although these symptoms are unpleasant, the disease is not usually serious and does not need treatment with drugs. Elderly people and very young children, however, may be made very ill by food poisoning.

The *Salmonella* bacteria are killed when meat is cooked or milk is pasteurized (p. 93). Infection is most likely if untreated milk is drunk, meat is not properly cooked, or cooked meat is contaminated with bacteria transferred from raw meat (Figure 39.1, overleaf). Frozen poultry must be thoroughly defrosted before cooking, otherwise the inside of the bird may not get hot enough during cooking to kill the *Salmonella*.

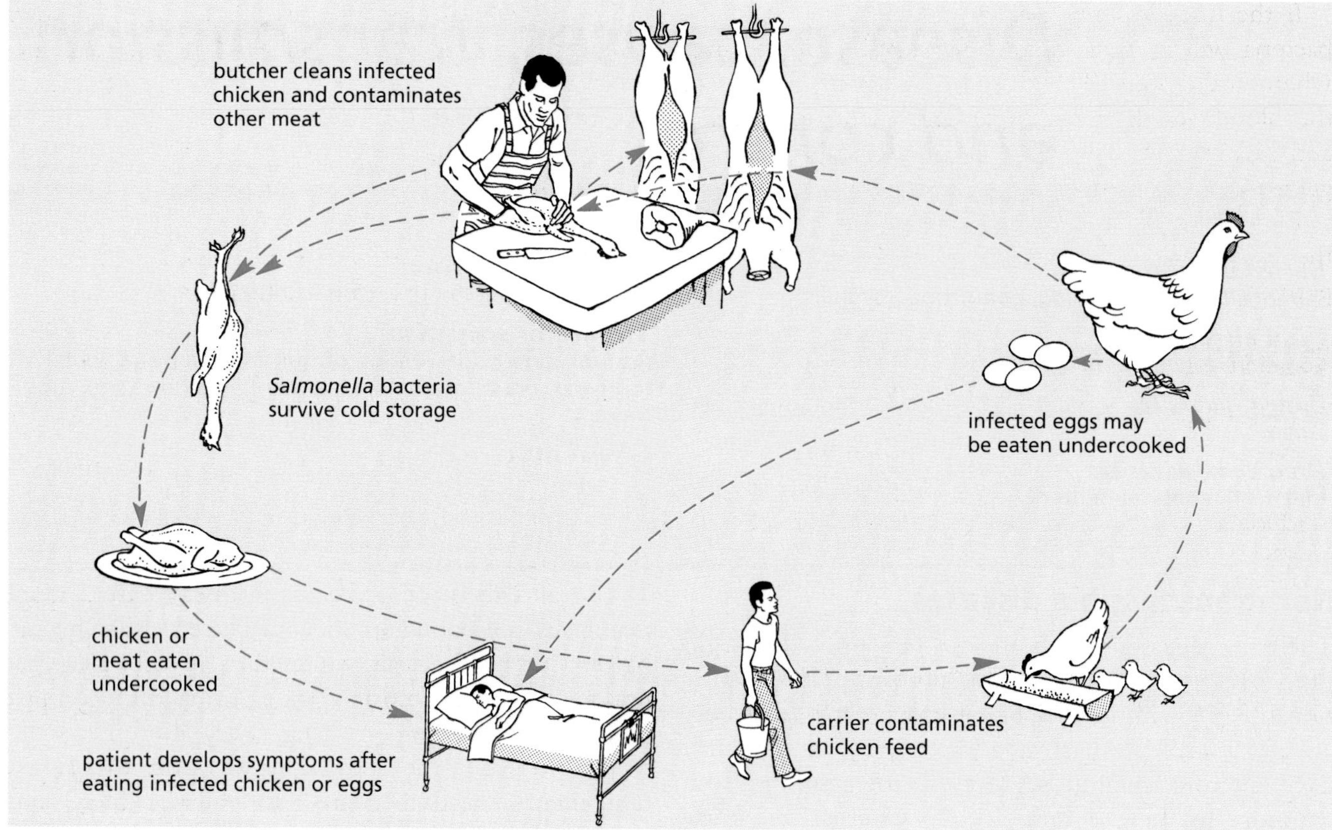

Figure 39.1 Transmission of *Salmonella* food poisoning

It follows that, to avoid the disease, all milk should be pasteurized and meat should be thoroughly cooked. People such as shop assistants and cooks should not handle cooked food at the same time as they handle raw meat. If they must do so, they should wash their hands thoroughly between the two activities.

The liquid which escapes when a frozen chicken is defrosted may contain *Salmonella* bacteria. The dishes and utensils used while the bird is defrosting must not be allowed to come into contact with any other food.

Uncooked meat or poultry should not be kept alongside any food which is likely to be eaten without cooking. Previously cooked meat should never be warmed up; the raised temperature accelerates the reproduction of any bacteria present. The meat should be eaten cold or cooked at a high temperature.

In the past few years there has been an increase in the outbreaks of *Salmonella* food poisoning in which the bacteria are resistant to antibiotics. Some scientists suspect that this results from the practice of feeding antibiotics to farm animals to increase their growth rate. This could allow populations of drug-resistant salmonellae to develop.

In the 1970s another genus of bacteria, *Campylobacter*, was identified as a cause of food poisoning. This bacterium causes acute abdominal pains and diarrhoea for about 24 hours. The sources of infection are thought to be undercooked meat, particularly 'burgers'.

Gonorrhoea

This is a sexually transmitted disease. That means that it is almost always caught by having sexual intercourse with an infected person. The disease is caused by a bacterium, *Neisseria gonorrhoea.*

The first symptoms in men are pain and a discharge of pus from the urethra. In women, there may be similar symptoms, or no symptoms at all.

In men, the disease leads to a blockage of the urethra and to sterility. A woman can pass the disease to her child during birth. The bacteria in the vagina invade the baby's eyes and cause blindness.

The disease can be cured with penicillin, but some strains of *Neisseria* have become resistant to this antibiotic. There is no immunity to gonorrhoea. Having the disease once does not prevent you catching it again.

Syphilis

Syphilis is also a sexually transmitted disease. It is caused by a bacterium called *Treponema pallidum.*

In the first stage of the disease, a lump or ulcer appears on the penis or the vulva, 1 week to 3 months after being infected. The ulcer usually heals without any treatment after about 6 weeks. By this time the bacteria have entered the body and may affect any tissue or organ. There may be a skin rash, a high temperature and swollen lymph nodes. But the symptoms are variable and the infected person may appear to be in good health for many years.

If the disease is not treated in the early stages, the bacteria will in time cause inflammation almost anywhere in the body. They can do permanent damage to the blood vessels, heart or brain, leading to paralysis and insanity. In a pregnant woman, the bacteria can get across the placenta and infect the fetus.

Penicillin will cure syphilis. But unless it is used in the early stages of the disease, the bacteria may do permanent damage.

Questions

1 Explain why rewarming cooked meat might lead to food poisoning.

2 In preparing for a reception, a cook defrosted a frozen chicken, placed it in the oven and then sliced some cooked ham which he put into the refrigerator. Then he washed his hands before preparing the salad.
After the reception several guests who ate the cold chicken and ham salad suffered from food poisoning.
  a How could they have become infected, in spite of the chicken being cooked at a high temperature and the ham being refrigerated?
  b How could the outbreak have been avoided?

3 Syphilis and gonorrhoea are sexually transmitted diseases. So how is it that babies can be infected?

Viral diseases

Viruses can reproduce only inside other cells, and so all viruses are parasitic. In humans they cause diseases such as colds, influenza, herpes, mumps, measles, chickenpox, rubella, hepatitis and AIDS. Some viruses can remain dormant (inactive) in certain body cells without immediately producing symptoms of disease.

Viruses do not produce toxins. The harm they cause is probably the result of the destruction of the cells they invade. When they destroy cells in, for example, the lining of the windpipe and bronchi, bacteria can invade the damaged tissues.

After entering the body and reproducing in a small group of cells, the viruses may be carried to other organs in the circulation.

Although there are many effective drugs against bacterial infections, drugs against viral infections are only just being developed. However, some viral infections do give long-term immunity, and many virus diseases can be prevented by immunization (p. 341).

The common cold

This is caused by a **rhinovirus**. It is spread by droplet infection and contact. The symptoms of the disease develop within 12–78 hours after infection and are very familiar: dry throat, watering of the eyes, production of watery mucus from the nose and swollen (congested) nasal membranes, making it difficult to breathe through the nose.

These symptoms last for a few days. But the damage done by the virus to the nose and throat membranes often allows *Streptococcus* bacteria to invade. This **secondary bacterial infection** may give rise to a sore throat, a cough and catarrh.

Although the body develops temporary immunity to the 'cold' virus, there are many different strains of rhinovirus, and other species of virus also cause colds. Immunity to one of these strains does not extend to others. This is why you can have one cold after another.

There is no cure for a cold. Antibiotics are ineffective against rhinoviruses, but they may be prescribed if an acute or persistent secondary bacterial infection appears.

Influenza ('flu)

Influenza is caused by a virus which exists in three strains, A, B and C. The virus attacks the lining of the throat and breathing passages, giving rise to inflammation of the trachea, bronchi and bronchioles. The patient will have a raised temperature, headache, dry cough and a mild sore throat and will feel generally 'rotten'. The symptoms subside in 2–4 days, but the damage to the respiratory linings may allow *Streptococci* to invade, causing a secondary bacterial infection.

There are no specific drugs. Aspirin helps to lower the temperature, and antibiotics may be used against any secondary infection.

A bout of infection confers immunity for several years. But the 'A' strain of the virus undergoes mutations (p. 189) very readily and immunity to one form is not effective against other mutants.

There are sometimes severe epidemics of influenza. If it is known which mutant is responsible, it is possible to prepare a vaccine which gives protection for a few months to people most at risk and to doctors and nurses dealing with the epidemic.

AIDS

The initials stand for Acquired Immune Deficiency Syndrome. (A 'syndrome' is a pattern of symptoms associated with a particular disease.) The virus which causes AIDS is the human immunodeficiency virus (**HIV**). It attacks certain kinds of lymphocyte (p. 108), and thus weakens the body's immune responses.

As a result, the patient has little or no resistance to a range of diseases that would not normally invade the body. Once the symptoms appear, the outlook for many of the patients is bleak. Death does not result from the virus itself, but from diseases such as pneumonia, blood disorders, skin cancer or damage to the nervous system which the body cannot resist.

After a person has been infected, years may pass before symptoms develop. So people may carry the virus yet not show any symptoms. They can still infect other people, however. It is not known for certain what proportion of HIV carriers will eventually develop AIDS: perhaps 30–50 per cent, or even more, may do so.

HIV is transmitted by direct infection of the blood. Drug users who share needles contaminated with infected blood run a high risk of the disease. It can also be transmitted sexually, both between men and women and, especially, between homosexual men who practise anal intercourse.

Haemophiliacs (p. 197) have also fallen victim to AIDS. Haemophiliacs have to inject themselves with a blood product which contains a clotting factor. Before the risks were recognized, infected carriers sometimes donated blood which was used to produce the clotting factor.

Babies born to HIV carriers may become infected with HIV, either in the uterus or during birth or from the mother's milk. The rate of infection varies from about 40 per cent in parts of Africa to 14 per cent in Europe. If the mother is given drug therapy during labour and the baby within 3 days, this method of transmission is reduced.

There is no evidence to suggest that the disease can be passed on by droplets (p. 338), by saliva or by normal everyday contact.

The disease has appeared comparatively recently and there are no fully effective drugs. One, 'zidovudine' does slow the progress of the disease but has harmful side-effects. Intensive research to find a vaccine and more effective drugs is going on.

There is a range of blood tests designed to detect HIV infection. These tests do not detect the virus but do indicate whether antibodies to the virus are in the blood. If HIV antibodies are present, the person is said to be **HIV positive**. The tests vary in their reliability and some are too expensive for widespread use. The American Food and Drug Administration claims a 99.8 per cent accuracy but this figure is disputed.

Questions

1. Antibiotics are ineffective against virus diseases. Why then are antibiotics sometimes given to people suffering from a virus infection such as influenza?
2. Blood donors now have their blood tested to see if it contains antibodies to AIDS. What do you think is the value of this test?
3. People living in closed, isolated communities, such as Arctic explorers, often get a cold or two at first and then have several months free from colds. However, when the supply ship arrives the colds start again. Try to explain this phenomenon.
4. Why do you think that HIV infection usually fails to provoke a strong enough immune response to combat the disease?

The measures needed to avoid catching AIDS are the same as those for any sexually transmitted disease (p. 341).

Fungal parasites

Tinea ('ringworm')

Several species of fungus give rise to the various forms of this disease. The fungus attacks the epidermis (p. 137) and produces a patch of inflamed tissue. On the skin the infected patch spreads outwards and heals in the centre, giving a ring-like appearance ('ringworm').

The different species of tinea fungi may live on the skin of humans or domestic animals, or in the soil. The region of the body affected will depend on the species of fungus.

One kind affects the scalp and causes circular bald patches. The hair usually grows again when the patient recovers from the disease.

The species of fungus which affect the feet usually cause cracks in the skin between the toes. This is known as 'athlete's foot'.

Tinea of the crutch is a fungus infection, occurring usually in males, which affects the inner part of the thighs on each side of the scrotum. It causes a spreading, inflamed area of skin with an itching or burning sensation.

All forms of the disease are very contagious. That means, they are spread by contact with an infected person or their personal property. Tinea of the scalp is spread by using infected hair-brushes, combs or pillows. Tinea of the crutch can be caught by using towels or bedclothes contaminated by the fungus or its spores, and 'athlete's foot' by wearing infected socks or shoes, or from the floors of showers and swimming pools.

When an infection is diagnosed, the clothing, bed linen, infected hair-brushes, combs or towels must be boiled to destroy the fungus. It is best, anyway, to avoid sharing these items as their owners may be carrying the infection without knowing or admitting it.

In young people, tinea infections often clear up without treatment. Where treatment is needed, a fungicide cream or dusting powder is applied to the affected areas of skin. Infected feet may be dipped in a solution of potassium permanganate (potassium manganate(VII)).

Candida

Candida albicans is a yeast-like fungus that usually lives harmlessly in the mouth or vagina (see 'Mutualism', p. 296). In infants and people weakened by illness it may become parasitic, invading the epithelium and causing white patches of damaged epithelial cells.

This condition is known as 'thrush' or candidiasis. It can be cured by mouthwashes or a drug called 'nystatin'.

Protozoan parasites

There are relatively few protozoa which parasitize humans, but those that do cause serious diseases. Two of these are malaria and amoebic dysentery.

Malaria

About 280 million people suffer from malaria in over 100 countries (Figure 39.2), and 2 million or more die each year from the disease.

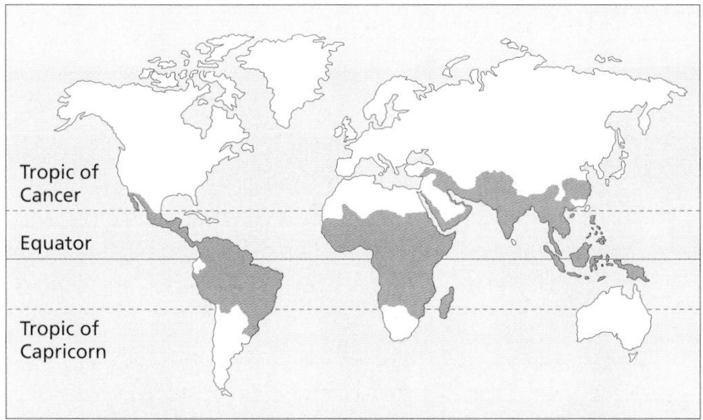

Figure 39.2 The world-wide distribution of malaria

The disease is caused by a protozoan parasite called *Plasmodium* which is transmitted from person to person by the bites of infected mosquitoes of the genus *Anopheles*. The mosquito is said to be the **vector** of the disease. When a mosquito 'bites' a human, it inserts its sharp, pointed mouth parts through the skin till they reach a capillary (p. 295). The mosquito then injects saliva which stops the blood from clotting. If the mosquito is infected, it will also inject hundreds of malarial parasites.

The parasites reach the liver via the circulation and burrow into the liver cells and reproduce. A week or two later, the daughter cells break out of the liver cells and invade the red blood cells. Here they reproduce rapidly and then escape from the original red cells to invade others (Figure 39.3).

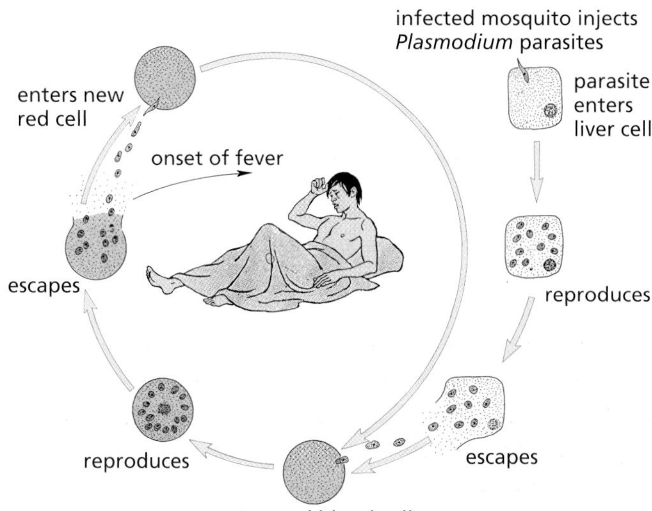

Figure 39.3 *Plasmodium*, the malarial parasite

The cycle of reproduction in the red cells takes 2 or 3 days (depending on the species of *Plasmodium*). Each time the daughter *Plasmodia* are released simultaneously from thousands of red cells the patient experiences the symptoms of malaria. These are chills accompanied by violent shivering followed by a fever and profuse sweating (Figure 39.4). With so many red cells being destroyed, the patient will also become anaemic (p. 88).

If a mosquito sucks blood from an infected person, it will take up the parasites in the red cells. The parasites reproduce in the mosquito and finally invade the salivary glands, ready to infect the next human.

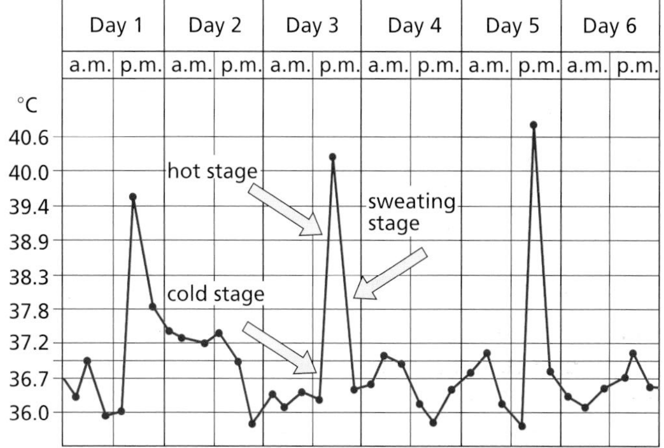

Figure 39.4 Temperature graph of a patient with one form of malaria

Control

There are drugs which kill the parasites in the bloodstream but they do not reach those in the liver. The parasites in the liver may emerge at any time and start the cycle again. If these drugs are taken by a healthy person before entering a malarious country, they kill any parasites as soon as they are injected. This is a protective or **prophylactic** use of the drug.

Unfortunately there are now many mutant forms of *Plasmodium* which have developed resistance to these drugs.

A great deal of work has been devoted to finding an effective vaccine, without much success. Trials are currently taking place of a vaccine which may offer at least partial protection against the disease.

The most far-reaching form of malarial control is based on the elimination of the mosquito. It is known that mosquitoes lay their eggs in stagnant water and that the larvae hatch, feed and grow in the water, but have to come to the surface to breathe air.

Spraying stagnant water with oil and insecticides suffocates or poisons the larvae and pupae. Spraying must include not only lakes and ponds but any accumulation of fresh water which mosquitoes can reach, e.g. drains, gutters, tanks, tin cans and old car tyres. By draining swamps and turning sluggish rivers into swifter streams, the breeding grounds of the mosquito are destroyed.

Spraying the walls of dwellings with chemicals like DDT was once very effective because the insecticide remained active for several months and the mosquito picked up a lethal dose merely by settling on the wall.

However, in at least 60 countries, many species of *Anopheles* have developed resistance to these insecticides and this method of control is now far less effective. The emphasis has changed back to the removal of the mosquito's breeding grounds or the destruction of the larvae and pupae.

Amoebic dysentery

Entamoeba histolytica is a species of small amoebae which normally live harmlessly in the human intestine, feeding on food particles or bacteria. In certain conditions, however, *Entamoeba* invades the lining of the intestine causing ulceration and bleeding, with pain, vomiting and diarrhoea: the symptoms of amoebic dysentery.

The diarrhoea and vomiting lead to a loss of water and salts from the body and if they persist for very long can cause **dehydration**. Dehydration, if untreated, can lead to kidney failure and death. The treatment for dehydration is to give the patient a carefully prepared mixture of water, salts and sugar. The intestine absorbs this solution more readily than water and it restores the volume and concentration of the body fluids. This simple, effective and inexpensive treatment is called **oral rehydration therapy** and has probably saved thousands of lives since it was first discovered. There are also drugs which attack *Entamoeba*.

The faeces of infected people contain *Entamoeba* amoebae which, if they reach food or drinking water, can infect other people. The disease is prevalent in tropical, sub-tropical and, to some extent, temperate countries and is associated with low standards of hygiene and sanitation.

Questions

1 a What are the two main lines of attack on malaria?
  b What is the connection between stagnant water and malaria?
  c What are the principal 'set-backs' in the battle against malaria?

2 In what ways might improved sanitation and hygiene help to reduce the spread of amoebic dysentery?

3 How might a medical officer try to control an outbreak of amoebic dysentery?

Disease transmission

Airborne, 'droplet' or aerosol infection

When we sneeze, cough, laugh, speak or just breathe out, we send a fine spray of liquid drops into the air. These droplets are so tiny that they remain floating in the air for a long time. They may be breathed in by other people or fall on to exposed food (Figure 39.5). If the droplets contain viruses or bacteria, they may cause disease when they are eaten with food or inhaled.

Figure 39.5 Droplet infection. The visible drops expelled by this sneeze will soon sink to the floor, but smaller droplets will remain suspended in the air

Virus diseases like colds, 'flu, measles and chickenpox are spread in this way. So are the bacteria (*Streptococci*) that cause sore throats. When the water in the droplets evaporates, the bacteria often die as they dry out. The viruses remain infectious, however, floating in the air for a long time.

In buses, trains, cinemas and discos, the air is warm and moist, and full of floating droplets. These are places where you are likely to pick up one of these infections.

Contamination of water

If disease bacteria get into water supplies used for drinking, hundreds of people can become infected. Diseases of the alimentary canal, like typhoid and cholera, are especially dangerous. Millions of bacteria infest the intestinal lining of a sick person.

Some of these bacteria will pass out with the faeces. If the faeces get into streams or rivers, the bacteria may be carried into reservoirs of water used for drinking. Even if faeces are left on the soil or buried, rain water may wash the bacteria into a nearby stream.

To prevent this method of infection, faeces must be made harmless, and drinking water purified.

Contamination of food

Contamination by people

The bacteria most likely to get into food are the ones which cause diseases of the alimentary canal such as typhoid and *Salmonella* food poisoning (p. 333). The bacteria are present in the faeces of infected people and may reach food from the unwashed hands of the sufferer.

People recovering from one of these diseases may feel quite well. But bacteria may still be present in their faeces. If they don't wash their hands thoroughly after going to the lavatory, they may have small numbers of bacteria on their fingers. If they then handle food, the bacteria may be transferred to the food. When this food is eaten by healthy people, the bacteria will multiply in their bodies and give them the disease.

People working in food shops, kitchens and food-processing factories could infect thousands of other people in this way if they were careless about their personal cleanliness.

Some forms of food poisoning result from poisons (toxins) that are produced by bacteria which get into food. Cooking kills the bacteria in the food, but does not destroy the toxins which cause the illness. Only one form of this kind of food poisoning, called **botulism**, is dangerous. It is also very rare.

Contamination by houseflies

Flies walk about on food. They place their mouth parts on it and pump saliva onto the food. Then they suck up the digested food as a liquid.

This would not matter much if flies fed only on clean food, but they also visit decaying food or human faeces. Here they may pick up bacteria on their feet or their mouth parts. They then alight on our food and the bacteria on their bodies are transferred to the food. Figure 39.6 shows the many ways in which this can happen.

Food poisoning (p. 333), amoebic dysentery (p. 338) and polio can be spread by houseflies.

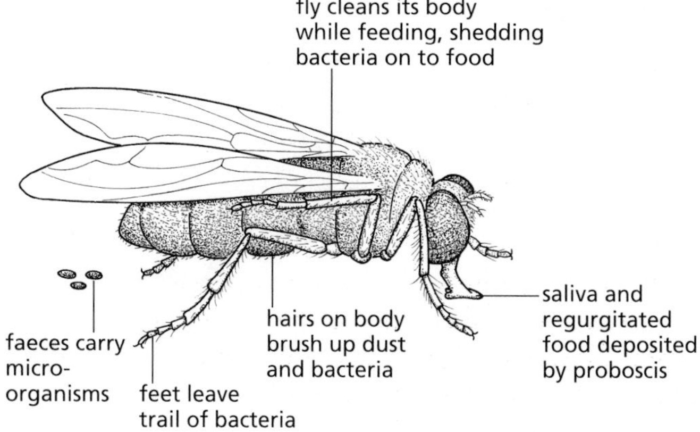
fly cleans its body while feeding, shedding bacteria on to food
saliva and regurgitated food deposited by proboscis
hairs on body brush up dust and bacteria
faeces carry micro-organisms
feet leave trail of bacteria

Figure 39.6 Transmission of bacteria by houseflies

Vectors

A vector is an animal, often an insect, which carries pathogens from one person to another. Some species of mosquito carry the protozoa which cause malaria (p. 337), and other species carry the viruses of yellow fever. Rat fleas carry the bacteria of bubonic plague. Houseflies spread polio, and a variety of intestinal diseases, such as food poisoning.

Contagion

A contagious disease can be spread by contact with an infected person, or with their clothing, bed linen or towels. The fungus disease tinea (described on p. 336) is very contagious.

Colds may also be spread by contact. The viruses are transferred from the hands of an infected person to the hands of healthy people, who then infect themselves by touching their eyes or nose.

Sexually transmitted diseases

Syphilis and gonorrhoea are spread almost exclusively by sexual intercourse with an infected person. AIDS too is spread mainly by sexual contact, though drug addicts who share contaminated needles are also at risk.

There is no easy way of recognizing an infected person. People who have many sexual partners, such as prostitutes, are the most likely to be infected.

Questions

1. Why should people who sell, handle and cook food be particularly careful about their personal hygiene?
2. Coughing or sneezing without covering the mouth and nose with a handkerchief is thought to be inconsiderate behaviour. Why is this?
3. Inhaling cigarette smoke can stop the action of cilia in the trachea and bronchi for about 20 minutes. Why should this increase a smoker's chance of catching a respiratory infection?
4. Although wasps often walk on our food, they are not suspected of spreading disease. Why do you think they are less harmful in this respect than houseflies?

Methods of preventing spread of disease

Natural barriers

Although many bacteria live on the surface of the skin, the outer layer of the epidermis (p. 137) seems to act as a barrier which stops them getting into the body. But if the skin is cut or damaged, the bacteria may get into the deeper tissues and cause infection.

Tears contain an enzyme called **lysozyme**. This dissolves the cell walls of some bacteria and so protects the eyes from infection.

The acid conditions in the stomach destroy most of the bacteria which may be taken in with food. The moist lining of the nasal passages traps many bacteria. So does the mucus produced by the lining of the trachea and bronchi. The ciliated cells of these organs carry the trapped bacteria away from the lungs.

When bacteria get through these barriers, the body has two more lines of defence – the white cells and the antibodies. The way these work is described on p. 117.

Airborne infection

There is no single way to prevent the spread of airborne pathogens, apart from programmes of immunization against individual diseases.

Children with infectious diseases such as rubella or chickenpox are usually kept away from school during their infectious period. However, it is probably better to catch these 'childhood' diseases when young and so develop immunity, because older people are more severely affected.

You might perhaps reduce your chances of catching a cold or 'flu by avoiding crowded, humid, poorly ventilated places. But you might not enjoy living in permanent isolation.

Similarly, you can hardly stay at home every time you have a cold. But you might infect fewer people if you cough and sneeze into a handkerchief, rather than spraying droplets over a wide area.

In an operating theatre, the surgeon and staff wear sterile gauze masks to trap any droplets that might otherwise infect the patient (Figure 39.7).

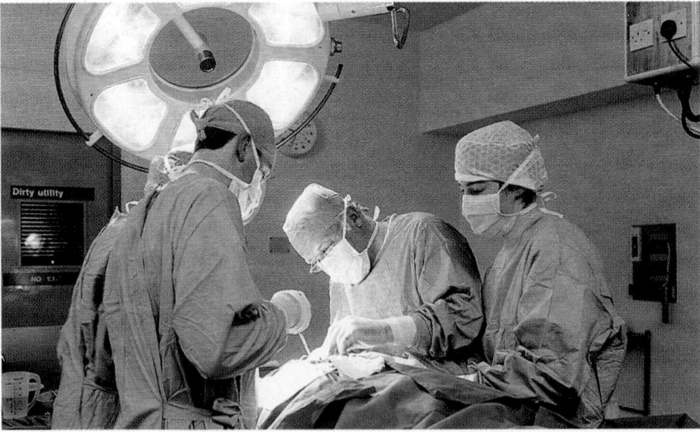

Figure 39.7 Prevention of infection in an operating theatre. Staff wear sterile masks and clothing in order to protect the patient from bacteria from their own bodies

Waterborne infection

On a small scale, simply boiling the water used for drinking will destroy any pathogens. On a large scale, water supplies are protected by (a) ensuring that untreated human sewage cannot reach them and (b) treating the water to make it safe.

Sewage treatment

This has been described on pp. 331–32.

Water treatment

The treatment needed to make water safe for drinking depends on the source of the water. Some sources, e.g. mountain streams, may be almost pure; others, e.g. sluggish rivers, may be contaminated.

The object of the treatment is to remove all micro-organisms which might cause disease.

This is done by filtration and chlorination. The water is passed through beds of sand in which harmless bacteria and protozoa are growing. These produce a gelatinous film, which acts as a fine filter and removes pathogens.

Finally, chlorine gas is added to the filtered water and remains in contact with it for long enough to kill any bacteria which have passed through the filter. How much chlorine is added and the length of the contact time both depend on how contaminated the water source is likely to be. Most of the chlorine disappears before the water reaches the consumers.

Ozone is to largely replace chlorine in the next few years.

The purified water is pumped to a high-level reservoir or water tower. These are enclosed to ensure that no pathogens can get into the water. The height of the reservoir provides the pressure needed to deliver the water to the consumer.

Food-borne infection

People

People who handle and prepare food need to be extremely careful about their personal hygiene. It is essential that they wash their hands before touching food, particularly after they have visited the lavatory (Figure 39.8). Hand-washing is also important after handling raw meat, particularly poultry (see p. 334).

Some people carry intestinal pathogens without showing any symptoms of disease. These people are called 'carriers'. Once identified, they should not be allowed to work in canteens or food-processing factories.

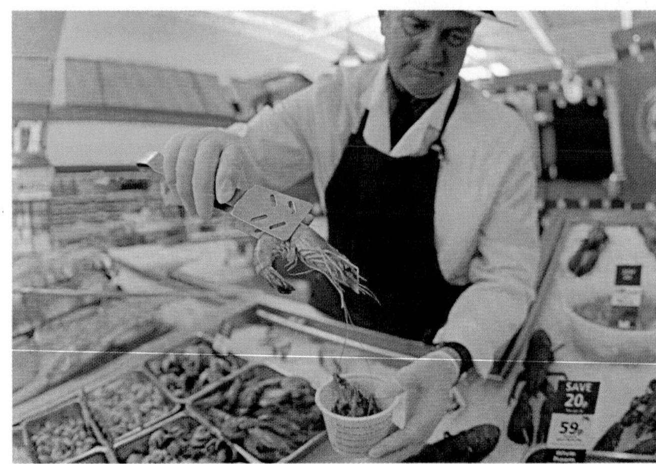

Figure 39.8 Hygienic handling of food. Shop assistants avoid handling meat and shellfish with their fingers by using disposable gloves and tongs

Houseflies

Houseflies must be prevented from carrying pathogens to food. There are four main ways to do this:

1 Keep all unwrapped food in fly-proof containers such as refrigerators or larders (Figure 39.9).
2 Enclose all food waste in fly-proof dustbins so that the flies cannot pick up bacteria.
3 Never leave human faeces where flies can reach them (if faeces cannot be flushed into the sewage system, they must be buried).
4 Destroy houseflies, wherever possible, in the places where they breed, such as rubbish tips and manure heaps.

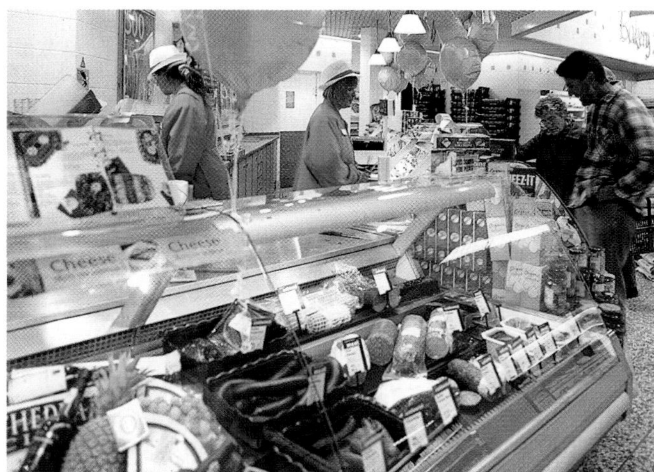

Figure 39.9 Protection of food on display. The glass barrier stops customers from touching the food and helps stop droplets from coughs and sneezes falling on the food

Processing

Cooking destroys any bacteria present in food. Refrigeration and freezing slow down or prevent bacterial reproduction. Dehydration, pickling, canning and irradiation are methods of preserving food. These processes destroy the bacteria which might cause disease. They also stop the food from going bad (see p. 93).

Contagion

Prevention of the spread of tinea is described on p. 336. The methods listed there apply equally well to any contagious disease other than those transmitted sexually.

Sexual transmission

The best way to avoid sexually transmitted diseases is to avoid having sexual intercourse with an infected person. However, the symptoms of the disease are often not obvious and it is difficult to recognize an infected individual. So the disease is avoided by not having sexual intercourse with a person who *might* have the disease. Such persons are (a) prostitutes who offer sexual intercourse for money, (b) people who are known to have had sexual relationships with many others ('slept around'), (c) casual acquaintances whose background and past sexual activities are not known.

These are good reasons, among many others for being faithful to one partner.

The risk of catching a sexually transmitted disease can be greatly reduced if the man uses a condom (p. 150). This acts as a barrier to bacteria or viruses.

If a person suspects that he or she has caught a sexually transmitted disease, treatment must be sought at once. Information about treatment can be obtained by phoning one of the numbers listed under 'Venereal Disease' or 'Health Information Service' in the telephone directory. Treatment is always confidential. The patients must, however, ensure that anyone they have had sexual contact with also gets treatment. There is no point in one partner being cured if the other is still infected.

Immunization

In the course of an infectious disease, the patient makes antibodies to the pathogen. The antibodies to some diseases remain in the blood for a long time. Others can be produced again very rapidly. So a pathogen which gets into the body for the second time is quickly destroyed. The person is said to be **immune** to that disease. Immunity may last for a few months or for many years, depending on the disease (see also p. 117).

Not all diseases produce immunity. There seems to be no immunity to syphilis or gonorrhoea.

You can acquire immunity to some diseases without actually catching them. One way is to be injected with a harmless form of the pathogen or its toxin (p. 333). This makes your body produce the antibody, without suffering the symptoms of the disease. Then if the real pathogens enter your body, you have the antibodies ready to attack them. You have been **immunized** against that disease.

The substance with which you are injected is called a **vaccine**. The vaccine against whooping cough consists of the killed bacteria. The dead bacteria cannot cause the disease, but they do stimulate the antibody reaction. The vaccines for diphtheria and tetanus are the toxins extracted from the bacteria and made harmless by heat or chemicals. The BCG vaccine against tuberculosis consists of a harmless form of bacteria closely related to the TB bacteria. The polio vaccine is a harmless form of the virus and can be taken by mouth.

You were probably immunized against diphtheria, polio, tetanus, whooping cough and measles during your first 5 years, and against tuberculosis between 10 and 13 years. Girls are usually immunized against rubella (Figure 39.10, overleaf). Babies may receive a combined inoculation against measles, mumps and rubella (MMR) at 18 months. This replaces the measles vaccination at 15 months and the later rubella inoculation.

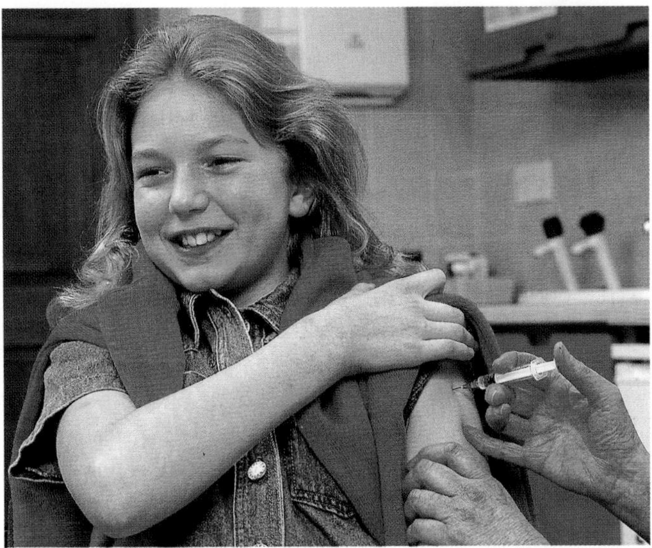

Figure 39.10 Immunization. The girl is being immunized against rubella (German measles)

There is a small risk of serious side-effects from vaccines, just as there is with all medicines. These risks are always far lower than the risk of catching the disease itself. For example, the measles vaccine carries a risk of 1 in 87 000 of causing encephalitis (inflammation of the brain). This is much less than the risk of getting encephalitis as a result of catching measles. Also, the vaccines themselves are becoming much safer, and the risk of side-effects is now almost nil.

Routine immunization not only protects the individual but also prevents the spread of infectious disease. Diseases like diphtheria and whooping cough were once common, and are now quite rare. This is the result of improved social conditions and routine immunization. Smallpox was completely wiped out throughout the world by a World Health Organization programme of immunization between 1959 and 1980.

Drug therapy

Drugs cure diseases rather than prevent them. But if a drug cures a disease before the infectious period is over, it might help to reduce the spread of that disease.

Although drugs play a valuable part in keeping us healthy, many people have come to rely on them for trivial ailments. Many diseases are self-limiting – the patient will get better quite quickly without using drugs. However, when some people consult a doctor, they may feel cheated if medicine of some kind or other is not prescribed. In addition, high pressure advertising gives the idea that headaches, coughs and colds, indigestion and other minor complaints can be relieved by taking patent medicines. In some cases this may be true. But many of these conditions will get better anyway.

There are fewer effective drugs against viruses than against bacteria. This is because the virus is so closely involved in its host cell's physiology that any chemical that harms the virus will also harm the cell.

Drugs which relieve the symptoms of a disease but do not cure it are called **analgesics**. They often take the form of pain relievers such as aspirin, ibuprofen or paracetamol.

Antibiotics

The ideal drug for curing disease would be a chemical that destroyed the pathogen without harming the tissues of the host. In practice, modern antibiotics such as penicillin come pretty close to this ideal for bacterial infections. Antibiotics cause the bacterial cell wall to break down by interfering with the chemical processes which maintain it. Since these processes do not occur in animal cells, the host tissues are unaffected. Nevertheless, many antibiotics cause some side-effects such as allergic reactions in some people.

Not all bacteria are killed by antibiotics. Some bacteria have a nasty habit of mutating to forms which are resistant to these drugs (p. 190). For this reason it is important not to use antibiotics in a diluted form, for too short a period or for trivial complaints. These practices lead to a build-up of a resistant population of bacteria. The drug resistance can be passed from harmless bacteria to pathogens.

Disinfectants and antiseptics

These compounds are not drugs but are used as antibacterial agents.

Disinfectants may be used to destroy bacteria on surfaces or in drains. They are poisonous compounds and would cause damage to living tissues. They may be used to remove bacteria from surfaces, e.g. a toilet seat, or to sterilize surgical instruments.

Antiseptics also act against bacteria but are safe to use on human tissues. They may be used to treat wounds but, unless carefully selected and used at the correct dilution, they can cause tissue damage and delay healing. A widely used antiseptic is 'chlorhexidine' (Figure 39.11). Used in solution it does little harm to the tissues up to 0.05 per cent.

Figure 39.11 Chlorohexidine is an antiseptic that can be safely used on human tissue in solution

Global travel

In the 18th and 19th centuries, explorers, traders and missionaries carried European disease to countries where the population had no natural immunity. It is thought that devastating epidemics of smallpox and measles in, for example, North American Indians and Australian aborigines resulted from contact with infected Europeans. Reliable evidence on these events of 200 or 300 years ago is not likely to be forthcoming.

Today, the ease with which we can travel around the world raises the possibility that travellers may catch a disease in a region where it is **endemic** and subsequently introduce it into a region where the incidence of disease is low or non-existent.

An 'endemic' disease is one which is constantly present in a population. Figure 39.2 shows areas in which malaria is endemic. Small numbers of travellers returning to Britain from such a region may have become infected during their stay. Fortunately, British mosquitoes do not transmit malaria, but global warming might change this.

If you plan to visit a country where an infectious disease is endemic, you are likely to be offered advice on vaccination. There is no vaccine against malaria but, if you are travelling to a malarious country, you will probably be advised to take a drug (e.g. chloroquine) which kills malarial parasites, starting a week or more before your departure, throughout your stay and for a few weeks after your return. Drugs such as this which help to *prevent* you getting a disease are called **prophylactics**.

Also, you may find your aircraft cabin being sprayed with insecticide to kill any malaria-carrying mosquitoes which might have entered.

If you visit a country where a disease, e.g. yellow fever, is endemic, you may be required to produce a certificate of vaccination (Figure 39.12) before being allowed into a country where the disease does not occur.

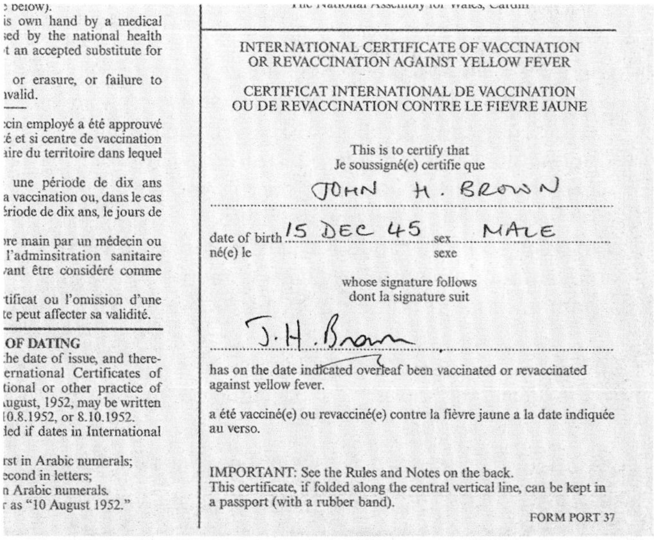

INTERNATIONAL CERTIFICATE OF VACCINATION OR REVACCINATION AGAINST YELLOW FEVER

CERTIFICAT INTERNATIONAL DE VACCINATION OU DE REVACCINATION CONTRE LE FIEVRE JAUNE

This is to certify that
Je soussigné(e) certifie que

JOHN H. BROWN

date of birth 15 DEC 45 sex MALE
né(e) le sexe

whose signature follows
dont la signature suit

J.H. Brown

has on the date indicated overleaf been vaccinated or revaccinated against yellow fever.

a été vacciné(e) ou revacciné(e) contre la fièvre jaune a la date indiquée au verso.

IMPORTANT: See the Rules and Notes on the back.
This certificate, if folded along the central vertical line, can be kept in a passport (with a rubber band).

FORM PORT 37

Figure 39.12 International certificate of vaccination

Questions

1. How might a harmful bacterium be destroyed or removed by the body if it arrived
   - **a** on the cornea,
   - **b** on the hand,
   - **c** in a bronchus,
   - **d** in the stomach?
2. After a disaster such as an earthquake, the survivors are urged to boil all drinking water. Why do you think this is so?
3. Revise pp. 117–18. Now explain why immunization against diphtheria does not protect you against polio as well.
4. Even if there have been no cases of diphtheria in a country for many years, children may still be immunized against it. What do you think is the point of this?
5. Why, do you think, has it not been possible to produce a vaccine against the common cold?
6. **a** What is the difference between HIV and AIDS?
   - **b** What does 'HIV positive' mean?
   - **c** Which groups of people are most at risk of HIV infection?

Checklist

- Transmissible diseases are infections caused by viruses, bacteria, fungi or protozoa.
- Non-transmissible diseases may be inherited, or acquired during the lifetime.
- Infectious diseases may be transmitted by air, water, food or contact.
- Sexually transmitted diseases are caught during sexual intercourse with an infected person.
- Waterborne diseases are controlled by sewage treatment and water purification.
- Food-borne diseases are controlled by (a) hygienic handling, (b) wrapping or covering, (c) preservation and processing, (d) control of houseflies.
- Sexually transmitted diseases can be prevented by avoiding sexual contact with infected people.
- A vaccine stimulates the blood system to produce antibodies against a disease, without causing the disease itself.
- The presence of antibodies in the blood, or the ability to produce them rapidly, gives immunity to a disease.
- Systematic immunization can protect whole populations.
- Antibiotics are very effective against disease bacteria. But bacteria can become resistant if the antibiotics are used unwisely.

Micro-organisms and humans

Examination questions

Do not write on this page. Where necessary copy drawings, tables or sentences.

1 In order to produce uncontaminated cultures of microbes, certain important steps must be taken.

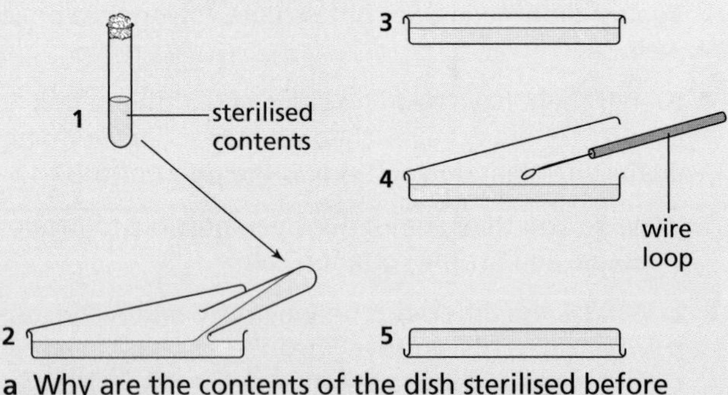

a Why are the contents of the dish sterilised before use? (1)

b Why is the lid of the dish never completely removed? (1)

c What is the wire loop used for? (1)

d How is the wire loop sterilised? (1)

e Why do we need to sterilise it? (1)

f How do we ensure that the lid of the dish remains tightly closed? (1)

(WJEC)

2 The drawing shows a section through a fermenter used to produce the antibiotic penicillin. The fermenter contains the mould *Penicillium*.

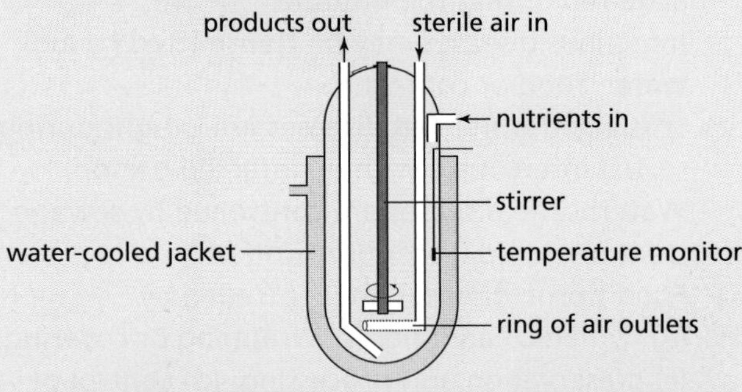

a Name **two** nutrients which moulds such as *Penicillium* need for growth. (2)

b Explain why air must be bubbled through the fermenter. (1)

c Explain why the water-cooled jacket is necessary. (1)

(AQA)

3 The table gives information about some diseases. Complete the table below. (4)

(Edexcel)

4 Most children start being vaccinated when they are 3 months old.

a How do new-born babies obtain some immunity to disease? (1)

b Children are no longer vaccinated against smallpox, as the disease has been eradicated.
Edward Jenner developed a vaccine against smallpox. Describe how he discovered his vaccine.
You will be given credit for the correct use of technical terms and for the correct use of spelling, punctuation and grammar. (4)

c Edward Jenner lived over 200 years ago.
Suggest why he would not be allowed to carry out his research today. (2)

d The statements show five stages in making and using vaccines.

A: vaccine is tested in a clinical trial

B: bacteria are grown and toxins collected

C: contact with bacteria causes memory cells to produce antibodies

D: toxins inactivated by heat or chemicals

E: vaccine used on population

Write the order of letters to show the correct sequence for making and using the vaccine. (3)

e Diseases caused by bacteria are sometimes treated with antibiotics. Explain why doctors only prescribe antibiotics for serious infections. (2)

(OCR)

5 a In the UK immunisation of children against polio was started in 1956. Children are given the vaccine, often on a sugar lump, by mouth. The vaccine contains a non-virulent form of the polio virus.
Explain as fully as you can how this vaccine prevents a child from getting polio. (6)

b In some countries polio is still common. Give **two** advantages of continuing with a polio immunisation programme in the UK. (2)

(AQA)

Disease	Type of organism causing the disease	How the organism is spread	Symptoms of the disease
Cholera	bacterium		diarrhoea and dehydration
	protozoan	by mosquito	shivering and fever
Influenza		by droplets in air	

Ideas and evidence

Charles Darwin (1809–82)

40 Development of biological ideas

Ideas about the circulatory system
Galen, Harvey.

Ideas about classification
Ray, Linnaeus.

Ideas about heredity
Mendel.

Ideas about evolution
Erasmus Darwin, Lamarck, Uniformitarianism vs catastrophism, Charles Darwin, Wallace; Neo-Darwinism, Galton.

Ideas about disease transmission and micro-organisms
Jenner, Pasteur, spontaneous generation, the germ theory.

Ideas about antibiotics
Fleming, Florey and Chain.

Ideas about DNA
Nature of the gene. Crick, Watson, Wilkins and Franklin.

In any account of the history of ideas, certain people are credited with a new discovery or a new theory. There will always be critics who claim that the discoveries were made earlier by a totally different person, often going back thousands of years to China or Ancient Egypt. Ideas and theories do not occur in a neat series, one idea refining the previous one until a 'final' theory is accepted. Ideas crop up in different places, with different people and at different times. No new theory emerges from a vacuum of ideas, but rather takes shape from a variety of sources.

So, it is not really worth disputing whether Harvey, Aristotle or Leonardo was the 'very first' person to discover the circulation of blood, but it was as a result of Harvey's observations and experiments that the idea of circulation was eventually accepted. Leonardo da Vinci made many biological observations which anticipated later discoveries, but he wrote them up in code so that they were not made generally available.

Ideas about the circulatory system

There must have been a knowledge of human internal anatomy thousands of years ago. This might have come, for example, from the practice of removing internal organs before the process of mummification in Ancient Egypt. However, there seems to have been little or no systematic study of human anatomy in the sense that the parts were named, described or illustrated.

Some of the earliest records of anatomical study come from the Greek physician, Galen.

Galen (AD130–200)

Galen dissected dogs, goats, monkeys and other animals and produced detailed and accurate records. He was not allowed to dissect human bodies, so his descriptions were often not applicable to human anatomy.

The anatomical knowledge was important but the functions of the various parts could only be guessed at. It was known that the veins contained blood but arteries at death are usually empty and it was assumed that they carried air or, more obscurely, 'animal spirit'. Galen observed the pulse but thought that it was caused by surges of blood into the veins.

William Harvey (1578–1657)

In the 15th and 16th centuries, vague ideas about the movement of blood began to emerge, but it was William Harvey, an English physician, who produced evidence to support the circulation theory.

Harvey's predecessors had made informed guesses, but Harvey conducted experiments to support his ideas. He noted that the valves in the heart would permit blood to pass in one direction only. So the notion that blood shunted back and forth was false. When he restricted the blood flow in an artery he observed that it bulged on the side nearest the heart, whereas a vein bulged on the side away from the heart.

Figure 40.1 shows a simple experiment that reveals the presence of valves in the veins and supports the idea of a one-way flow.

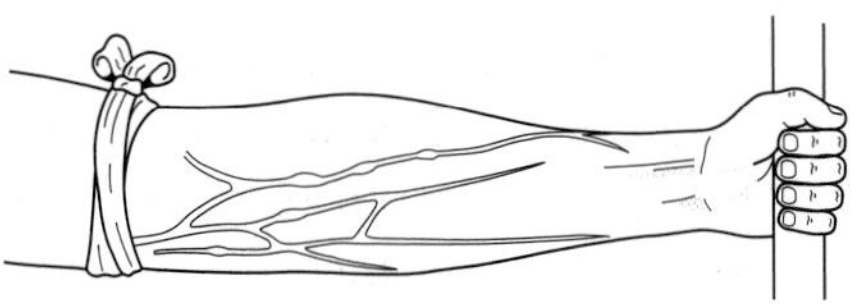

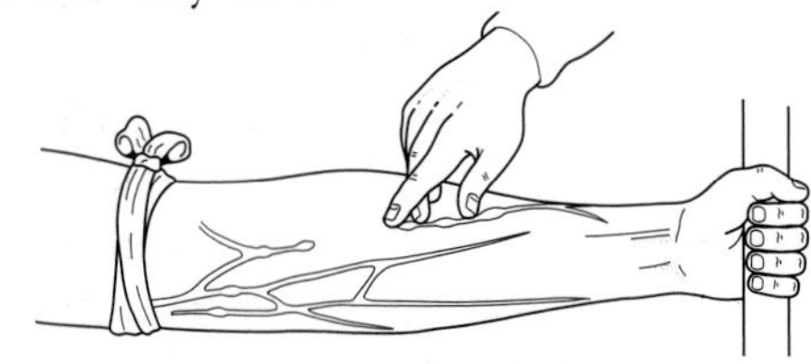

Figure 40.1 Harvey's demonstration of valves and one-way flow in a vein. The vein is compressed and the blood expelled by running a finger up the arm. The vein refills, but only as far as the valve. (Compare with Figure 12.19, p. 115.)

Harvey published his results in 1628. They were at first rejected and ridiculed, not because anyone tried his experiments or tested his observations, but simply because his conclusions contradicted the writings of Galen 1500 years ago.

By 1654, Harvey's theory of circulation was widely accepted but it was still not known how blood passed from the arteries to the veins. Harvey observed that arteries and veins branched and rebranched until the vessels were too small to be seen and suggested that the connection was made through these tiny vessels. This was confirmed after the microscope had been invented in 1660 and the vessels were called 'capillaries'.

The significance of this history is that, although it is reasonable to make an informed guess at the function of a structure or organ, it is only by testing these guesses by experiment that they can be supported or disproved. (See 'Hypothesis testing', p. 24.)

Ideas about classification

From the earliest days, humans must have given names to plants and animals they observed, particularly those that were useful as food or medicine. Over the years, there have been many attempts to sort plants and animals into related groups. Aristotle's 'Ladder of Nature' (Figure 40.2) organized about 500 animal species into broad categories.

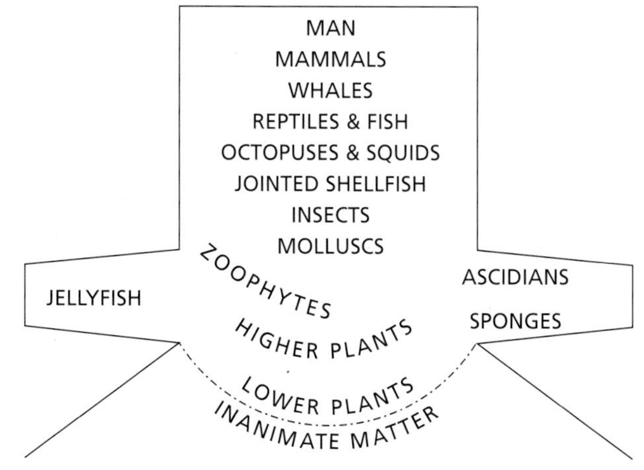

Figure 40.2 Aristotle's 'Ladder of Nature'

The 16th-century herbalists, such as John Gerard, divided the plant world into 'kindes' such as grasses, rushes, grains, irises and bulbs. Categories such as 'medicinal plants' and 'sweet-smelling plants', however, did not constitute a 'natural' classification based on structural features. The herbalists also gave the plants descriptive Latin names, e.g. *Anemone tenuifolia flore coccinea* ('the small-leaved scarlet anemone'). The first name shows a recognition of relationship to *Anemone nemorum flore pleno albo* ('the double white wood anemone'), for example. This method of naming was refined and popularized by Carl Linnaeus (see right).

John Ray (1625–1705)

Ray was the son of a blacksmith who eventually became a Fellow of the Royal Society. He travelled widely in Britain and Europe making collections of plants, animals and rocks.

In 1667 and 1682 he published a catalogue of British plants based on the structure of their flowers, seeds, fruits and roots. He was the first person to make a distinction between monocots and dicots (p. 278). Ray also published a classification of animals, based on hooves, toes and teeth. Ultimately he devised classificatory systems for plants, birds, mammals, fish and insects. In doing this, he brought order out of a chaos of names and systems.

At the same time he studied functions, adaptations and behaviour of organisms.

In 1691 he claimed that fossils were the mineralized remains of extinct creatures, possibly from a time when the Earth was supposedly covered by water. This was quite contrary to established (but varied) views on the significance of fossils. Some thought that the fossils grew and developed in the rocks, others supposed that God had put them there 'for his pleasure', and still others claimed that the Devil put them in the rocks to 'tempt, frighten or confuse'. A more plausible theory was that a huge flood had washed marine creatures on to the land.

Despite Ray's declaration, the modern idea of the significance of fossils was not generally accepted until Darwin's day (p. 350).

Carl Linnaeus (1707–78)

Linnaeus was a Swedish naturalist who initially graduated in medicine but became interested in plants. He travelled in Scandinavia, England and Eastern Europe, discovering and naming new plant species.

In 1735 he published his '*System Naturae*', which accurately described about 7700 plant species and classified them, largely on the basis of their reproductive structures (stamens, ovaries, etc., p. 67). He further grouped species into genera, genera into classes, and classes into orders. ('Phyla' came later.) He also classified over 4000 animals, but rather less successfully, into mammals, birds, insects and worms.

Linnaeus refined and popularized the binomial system of naming organisms, in which the first name represents the genus and the second name the species (p. 269). This system is still the official starting point for naming or revising the names of organisms.

Although the classificatory system must have suggested some idea of evolution, Linnaeus steadfastly rejected the theory and insisted that no species created by God had ever become extinct.

Ideas about heredity

Gregor Mendel (1822–84)

Mendel was an Augustinian monk in the town of Brünn (now Brno) in Czechoslovakia (now the Czech Republic). He studied maths and science at the University of Vienna in order to teach at a local school.

He was the first scientist to make a systematic study of patterns of inheritance which involved single characteristics. This he did by using varieties of the pea plant, *Pisum sativum*, which he grew in the monastery garden. He chose pea plants because they were self-pollinating (p. 69). Pollen from the anthers reached the stigma of the same flower even before the flower bud opened.

Mendel selected varieties of pea plant which bore distinctive and contrasting characteristics, such as green seeds vs yellow seeds, dwarf vs tall, round seeds vs wrinkled (Figure 40.3). He used only plants which bred true (p. 193).

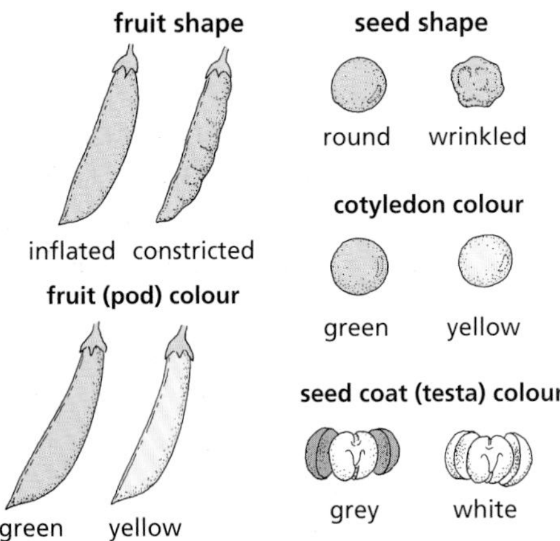

Figure 40.3 Some of the characteristics investigated by Mendel

He then crossed pairs of the contrasting varieties. To do this he had to open the flower buds, remove the stamens and use them to dust pollen on the stigmas of the contrasting variety. The offspring of this cross he called the 'first filial' generation, or F_1 (p. 194).

The first thing he noticed was that all the offspring of the F_1 cross showed the characteristic of only one of the parents. For example, tall plants crossed with dwarf plants produced only tall plants in the first generation.

Next he allowed the plants of the F_1 generation to self-pollinate and so produce a second filial generation, or F_2. Surprisingly, the dwarf characteristic that had, seemingly, disappeared in the F_1 reappeared in the F_2. This characteristic had not, in fact, been lost but merely concealed or suppressed in the F_1 to re-emerge in the F_2. Mendel called the repressed feature 'recessive' and the expressed feature 'dominant'.

Also, it must be noted, the plants were all either tall or dwarf; there were no intermediates, as might be expected if the characteristics blended.

Mendel noticed that pollen from tall plants, transferred to the stigmas of short plants, produced the same result as transferring pollen from short plants to the stigmas of tall plants. This meant that male and female gametes contributed equally to the observed characteristic.

When Mendel counted the number of contrasting offspring in the F_2, he found that they occurred in the ratio of 3 dominant to 1 recessive. For example, of 1064 F_2 plants from the tall × dwarf cross, 787 were tall and 277 dwarf, a ratio of 2.84 : 1. This F_2 ratio occurred in all Mendel's crosses, for example:

round vs wrinkled seeds	5474 : 1850 = 2.96 : 1
yellow vs green seeds	6022 : 2001 = 3.01 : 1
green vs yellow pods	428 : 152 = 2.82 : 1

Two-thirds of the dominant tall F_2 plants did not breed true when self-pollinated but produced the 3 : 1 ratio of tall : dwarf. They were therefore similar to the plants of the F_1 generation.

It is not clear whether Mendel speculated on how the characteristics were represented in the gametes or how they achieved their effects. At one point he wrote of 'the differentiating elements of the egg and pollen cells', but it is questionable whether he envisaged actual structures being responsible.

Similarly, when Mendel wrote 'exactly similar factors must be at work', he meant that there must be similar processes taking place. He does not use the term 'factor' to imply particles or any entities that control heritable characteristics.

His symbols **A**, **Ab** and **b** seem to be shorthand for the types of plants he studied: **A** = true-breeding dominant, **b** = true-breeding recessive and **Ab** = the non-true-breeding 'hybrid'. The letters represented the visible characteristics, whereas today they represent the genes responsible for producing the characteristic. For example, Mendel never refers to **AA** or **bb** so he probably did not appreciate that each characteristic is represented twice in the somatic cells but only once in the gametes.

When Mendel crossed plants, each carrying two contrasting characteristics, he found that the characteristics turned up in the offspring independently of each other. For example, in a cross between a tall plant with green seeds and a dwarf plant with yellow seeds, some of the offspring were tall with yellow seeds and some dwarf with green seeds.

So, Mendel's work was descriptive and mathematical rather than explanatory. He showed that certain characteristics were inherited in a predictable way, that the gametes were the vehicles, that these characteristics did not blend but retained their identity and could be inherited independently of each other. He also recognized dominant and recessive characteristics and, by 'hybridization', that in the presence of the dominant characteristic the recessive characteristic, though not expressed, did not 'disappear'.

Mendel published his results in 1866 in '*Transactions of the Brünn Natural History Society*', which, understandably, did not have a wide circulation. Only when Mendel's work was rediscovered in 1900 was the importance and significance of his findings appreciated.

Mendel's observations are sometimes summarized in the form of 'Mendel's laws', but Mendel did not formulate any laws and these are the product of modern knowledge of genetics.

- The first 'law' (the law of segregation) is expressed as 'of a pair of contrasted characters only one can be represented in the gamete'.
- The second 'law' (the law of independent assortment) is given as 'each of a pair of contrasting characters may be combined with either of another pair'.

On pp. 192–6 there is a modern interpretation of Mendelian inheritance.

Ideas about evolution

According to Darwin, at least 20 people in the past had put forward the idea of evolution. Aristotle (384–322BC) classified animals and arranged them in a series of increasing complexity, which was suggestive of an evolutionary series (Figure 40.2).

Linnaeus refined the classificatory system and popularized the binomial system of naming species (p. 269). Linnaeus, however, remained totally opposed to the idea of evolution. He claimed to have merely discovered the design of the Creator.

One of the obstacles to evolutionary theory was that Bishop Ussher in 1650 had calculated, from his interpretation of the Bible, that the Earth was formed in 4004BC, which made it only about 6000 years old. His calculations were believed, but this was far too short a time for evolution to have occurred. The Earth is now thought to be 4.6 billion years old.

Erasmus Darwin (1731–1802)

The grandfather of Charles Darwin, Erasmus Darwin put forward a firm theory of evolution. He believed that God was 'the first great cause', meaning that God had created the first simple organisms (Darwin called them 'filaments'). Thereafter, the organisms carried on under their own steam, evolving into the new species that we know today and from the fossil record. Erasmus Darwin recognized that related species could have evolved from a common ancestor, as shown on p. 206 for the major groups of animals. This would, of course, take millions of years – far longer than Bishop Ussher's calculations. By this time, however, a study of the Earth's surface was beginning to show that it had its origins hundreds of millions of years ago.

Lamarck (1744–1829)

Lamarck improved on Linnaeus's classification and put forward a comprehensive theory of evolution (he called it 'Transformism'). He suggested that animals had an innate tendency to increase in size and complexity. If this were the case, one might expect animals in the same species to follow identical pathways in becoming larger and more complex. But, in fact, there are wide-ranging variations between members of a species. Lamarck accounted for this variation by pointing out that some members of the species found themselves in a different environment from their relatives. The different environments made different demands on the population, which then evolved to meet these needs.

Lamark's theory suggested that, because of the environmental pressures, certain organs would be used more than others (e.g. for climbing or burrowing). Over many generations, this increasing use would cause these organs to become more highly developed and perhaps modified.

So a giraffe's neck, legs and tongue would become longer as a result of an environment where foliage could be reached only by stretching upwards; a mole's eyes were diminished in size and function through disuse in their underground habitat; a duck's feet became webbed because of their need to move efficiently on water.

For this kind of evolution to work, it would be essential for each tiny modification to be passed on to the offspring, with the modification becoming more pronounced with each generation.

This is known as the 'inheritance of acquired characteristics' – that is, characteristics acquired in the lifetime of an organism could be passed on to the offspring. A belief in the inheritance of acquired characteristics persisted until the 1900s.

Uniformitarianism vs catastrophism

These were two schools of thought about evolution in general. Geological studies showed that the Earth's crust was made up of distinct layers of rock, with different fossils in each layer.

Believers in catastrophism thought that this represented a series of world-wide disasters in which living organisms were wiped out, with new forms being created.

Uniformitarians, on the other hand, thought that the forces which shaped the Earth's crust in the past were no different from those in action today, e.g. mountain building, erosion, volcanic activity, and that they formed a continuous process.

The theory of evolution fitted the theory of uniformitarianism.

Charles Darwin (1809–82)

Charles Darwin tried studying first for medicine and then for the church but decided that he was suited for neither. His talent for natural history, however, was put to good use when he was appointed as naturalist to HMS *Beagle*. During his 5-year voyage he visited South America and noticed that species varied, little by little, as he travelled down the coast. In the Galapagos Islands he observed that each island had its own species of finch. These differed significantly from each other, and from finches on the mainland (Figure 40.4). He also made many more observations and kept records of animals, plants and fossils.

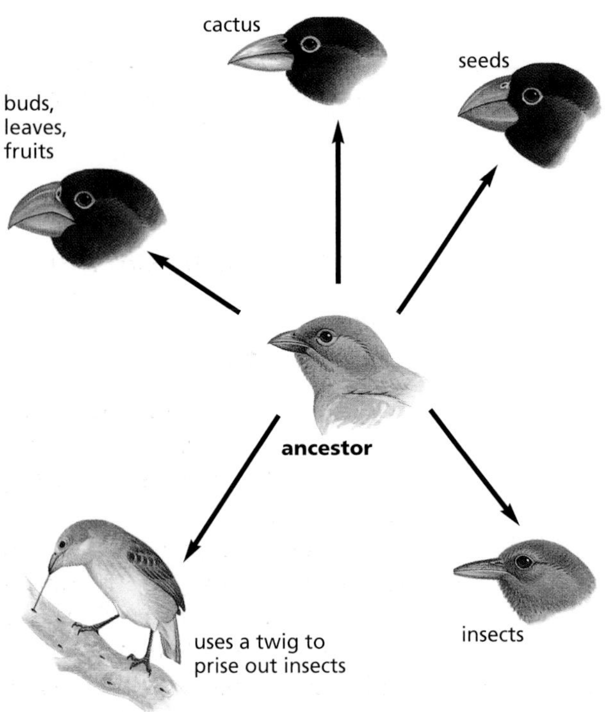

Figure 40.4 'Darwin's finches'. On each island, the finch species were quite different. Their beaks had become adapted to their feeding methods

Darwin bred pigeons as a hobby and noticed that the offspring in a single family differed slightly from each other. He reasoned that if you bred from one of the variants, e.g. one with a fan-tail, and selected its offspring showing the same characteristic, in a few generations you could get a distinctive subgroup of the species (Figure 40.5).

Figure 40.5 Varieties of pigeon produced by artificial selection (see also p. 211)

About this time he read 'An essay on the principles of population' by Thomas Malthus. Malthus pointed out that the human population increased faster than the food supply, so the numbers must be cut down by famine, disease or war. Darwin saw that this principle could apply to all organisms, but also that the varieties that survived were likely to be better adapted to their environment, could survive unfavourable periods and leave behind more offspring. The better-adapted varieties would be 'selected' by pressure of the environment.

Thus, the giraffe would have got its long neck as follows. Amongst the offspring of an antelope ancestor there occurred some individuals with slightly longer necks. When these grew to maturity they were able to browse off leaves that the other varieties could not reach. In conditions of food shortage, the short-necked varieties died and only the long-necked varieties survived to breed. Their first-generation offspring inherited the longer necks. When the first generation varieties grew up, mated and reproduced, some of their offspring had even longer necks than their parents or their siblings (brothers and sisters). In conditions of food shortage, these would have a better survival rate and leave more offspring. In each generation the longer-necked varieties would be selected and a new species of long-necked antelope would have been produced.

The principles of natural selection are set out on pp. 203–4.

Darwin first published his theory in a scientific journal in 1858, and his book '*The Origin of Species*' in the following year.

Principle differences between Lamarckism and Darwinism were:

Lamarck	Darwin
Believed in an 'in-built' drive towards greater complexity	Did not share this view
Thought the environment caused variations	Thought variations arose by chance and the environment 'weeded-out' the least efficient variants

However, they did agree on a number of points. Both realized that the variations had to be heritable. Both thought the process was continuous and took place in very small steps. Both rejected the 'fixity of species', i.e. the belief that all known species were created at the same time and have remained unchanged.

Neither knew how the variations were passed on to the offspring.

Although Darwin's book was widely read, his theory of natural selection was not accepted by the scientific world till about 1880 and is not accepted by some religious groups even today. Some of the reasons for rejection were that it contradicted religious belief in a single period of creation. It was seen as an attack on religion as a whole and denied that there was a 'fixity of species'. There were weaknesses in the theory, such as the problem of inheritance, and these were exploited by its opponents.

The theory was clinched only in the early 20th century (see 'Neo-Darwinism' below).

Alfred Russel Wallace (1823–1913)

Wallace's career was very similar to Darwin's. He tried surveying and architecture before turning to natural history. He was appointed as naturalist on voyages to the Amazon basin, Malaysia and the East Indies.

He observed the big difference between Asian and Australian animal species. He considered that the Australian species were more 'primitive' because, millions of years ago, Australia had split off from Asia before the more advanced species had evolved. Thus the Australian species were not out-competed by the Asian species.

Wallace, like Darwin, had read Malthus and formulated a theory of natural selection, which he sent to Darwin. At this time, Darwin had not published his theory but was spurred on to do so by the arrival of Wallace's manuscript. In the event, both scientists published jointly in 1858 and Wallace was given equal credit for the theory.

Neo-Darwinism

Neither Wallace nor Darwin knew how variations arose and were passed on to the offspring. Although Mendel's results (p. 348) were published during Darwin's lifetime, they appeared in an obscure journal. Even those scientists who read Mendel's paper dismissed his results as exceptions to the rule of blending inheritance.

In 1900, Mendel's papers were 're-discovered' and, at about the same time, mutations were observed in plants. It was then realized that mutation was the principal source of Darwin's 'variations' and Mendelian genetics could explain how they were inherited.

Today, when people talk about 'Darwinism' they usually mean 'Neo-Darwinism', which incorporates genetics and mutation into the theory of natural selection.

Francis Galton (1822–1911)

Galton was a cousin of Charles Darwin and was influenced by Darwin's theory of natural selection. Unlike Darwin, he lived long enough for his theories to benefit from Mendel's discoveries. Galton saw no reason why natural selection should not apply to humans, and made many studies of the inheritance of characteristics within families. These characteristics, including intelligence, he attributed almost entirely to inheritance, taking little account of environmental effects such as good diet, education, etc.

His ideas were supported by the study of identical twins who inherit the same genotype (p. 193) and who, even when separated at birth, still show a close correlation in nearly all their characteristics, including intelligence.

Galton believed that, by discouraging reproduction in 'inferior' individuals and encouraging marriage between 'gifted' people, it would be possible to improve the human race over the generations. This proposition was called 'eugenics'.

In fact, despite Galton's overall contribution to science, he speculated far beyond the available evidence, and eugenics fell into disrepute when, in addition to its scientific flaws, it was adopted by movements (e.g. the Nazi movement) seeking to establish so-called 'racial purity'.

The controversy about whether it is heredity or environment that contributes most to intelligence still goes on today.

Ideas about disease transmission and micro-organisms

Edward Jenner (1749–1823)

The history of immunization centres on the disease **smallpox**, which is caused by a virus. Only a few years ago it was a serious, world-wide disease causing hundreds of thousands of deaths.

It had long been noticed that people who had recovered from smallpox never caught the disease again. In the late 1600s this observation was exploited in countries such as Greece, Turkey, China and India. Fluid from the blisters, which characterized the disease, was introduced into healthy people through cuts in the skin. The patient suffered a mild form of smallpox but was, thereafter, immune to the disease. It was a risky practice, however, and some people developed smallpox and died as a result of the vaccination.

In the 1750s, a Suffolk surgeon, Robert Sutton, refined the technique with considerable success. Edward Jenner is usually given the credit for smallpox vaccination. While using Sutton's technique he noticed that milkmaids who had caught 'cowpox' from infected cows did not develop the mild symptoms of illness after vaccination.

In 1796, Jenner conducted a crucial, if somewhat risky, experiment. He took fluid from a blister on a milkmaid's hand and injected it into a young boy. Two months later, he inoculated the boy with smallpox and demonstrated that the boy was immune. After publication of the results, the practice spread widely throughout Europe, reducing deaths from smallpox by about two-thirds.

Jenner called his technique 'vaccination' to distinguish it from inoculation with smallpox. 'Vacca' is Latin for 'cow' and 'vaccinia' is the medical name for cowpox. We now know that viruses and bacteria often lose much of their virulence if they are allowed to pass through different animals or are cultured in a particular way. Such non-virulent microbes are said to be **attenuated**. Jenner and his contemporaries, of course, knew nothing about viruses or attenuation but their shrewd observations, logical deductions and bold experiments led to a massive reduction in suffering.

In 1967, the World Health Organization embarked on a programme to eradicate smallpox from the whole world. The strategy was to trace all cases of smallpox and isolate the patients so that they could not pass on the disease. Everyone at risk was then vaccinated. By 1987 the disease had been eradicated.

Louis Pasteur (1822–95)

Pasteur made outstanding contributions to chemistry, biology and medicine. In 1854, as professor of chemistry at the University of Lille, he was called in by the French wine industry to investigate the problem of wines going sour.

Under the microscope he observed the yeast cells that were present and proposed that these were responsible for the fermentation. Thus, he claimed, fermentation was the outcome of a living process in yeast and not caused solely by a chemical change in the grape juice. In time, Pasteur observed that the yeast cells were supplanted by microbes (which we now call 'bacteria') which appeared to change the alcohol into acetic and lactic acids.

Pasteur showed that souring was prevented by heating the wine to 120 °F (49 °C). He reasoned that this was because the microbes responsible for souring had been killed by the heat and, if the wine was promptly bottled, they could not return. This process is now called 'pasteurization' (p. 93).

Spontaneous generation

The micro-organisms in decaying products could be seen under the microscope, but where did they come from? Many scientists claimed that they were the *result* of decay rather than the *cause*; they had arisen 'spontaneously' in the decaying fluids.

In the 17th century, it was believed that organisms could be generated from decaying matter. The organisms were usually 'vermin' such as insects, worms and mice. To contest this notion, an experiment was conducted in 1668, comparing meat freely exposed to the air with meat protected from blowflies by a gauze lid on the container. Maggots appeared only in the meat to which blowflies had access.

This, and other experiments, laid to rest theories about spontaneous generation, as far as visible organisms were concerned, but the controversy about the origin of microbes continued into the 1870s.

It was already known that prolonged boiling, followed by enclosure, prevented liquids from putrifying. Exponents of spontaneous generation claimed that this was because the heat had affected some property of the air in the vessel. Pasteur designed experiments to put this to the test.

He made a variety of flasks, two of which are shown in Figure 40.6, and boiled meat broth in each of them. Fresh air was not excluded from the flask but could enter only through a tube which was designed to prevent 'dust' (and microbes) from reaching the liquid. The broths remained sterile until either the flask was opened or until it was tilted to allow some broth to reach the U-bend and then tipped back again.

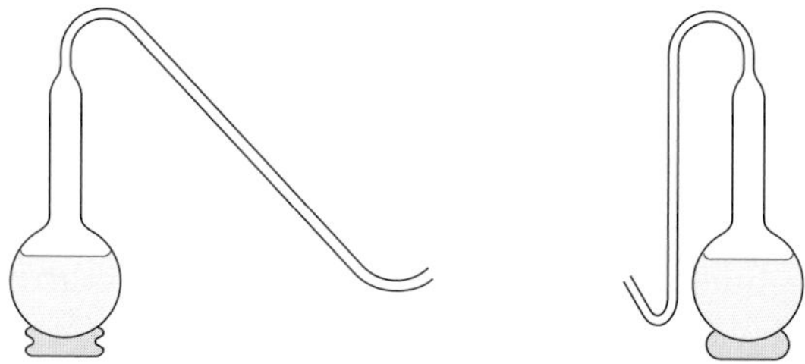

Figure 40.6 Two of Pasteur's flask shapes. The thin tubes admitted air but microbes were trapped in the U-bend

This series of experiments, and many others, supported the theory that micro-organisms *caused* decay and did not arise spontaneously in the liquids.

The germ theory of disease

In 1865, Pasteur was asked to investigate the cause of a disease of silkworms (silk-moth caterpillars) that was devastating the commercial production of silk. He observed that particular micro-organisms were present in the diseased caterpillars but not in the healthy ones. He demonstrated that, by removing all of the diseased caterpillars and moths, the disease could be controlled. This evidence supported the idea that the microbes passed from diseased caterpillars to healthy ones, thus causing the disease to spread.

He extended this observation to include many forms of transmissible disease, including anthrax. He also persuaded doctors to sterilize their instruments by boiling, and to steam-heat their bandages. In this way, the number of infections that followed surgery was much reduced.

Pasteur's discoveries led to the introduction of antiseptic surgery and also to the production of a rabies vaccine.

Ideas about antibiotics

Alexander Fleming (1881–1955)

Before 1934 there were few effective drugs. Some herbal preparations may have been useful; after all, many of our present-day drugs are derived from or based on plant products. Quinine, for example, was used for the treatment of malaria and was extracted from a specific kind of tree bark.

In 1935, a group of chemicals called **sulphanilamides** were found to be effective against some bacterial diseases such as blood poisoning, pneumonia and septic wounds.

Fleming had discovered penicillin in 1928, 7 years before the use of sulphanilamides, but he had been unable to purify it and test it on humans. Fleming was a bacteriologist working at St Mary's hospital in London. In 1928, he was studying different strains of *Staphylococcus* bacteria. He had made some cultures on agar plates and left them on the laboratory bench during a 4-week holiday. When he returned he noticed that one of the plates had been contaminated by a mould fungus and that round the margins of the mould there was a clear zone with no bacteria growing (Figure 40.7).

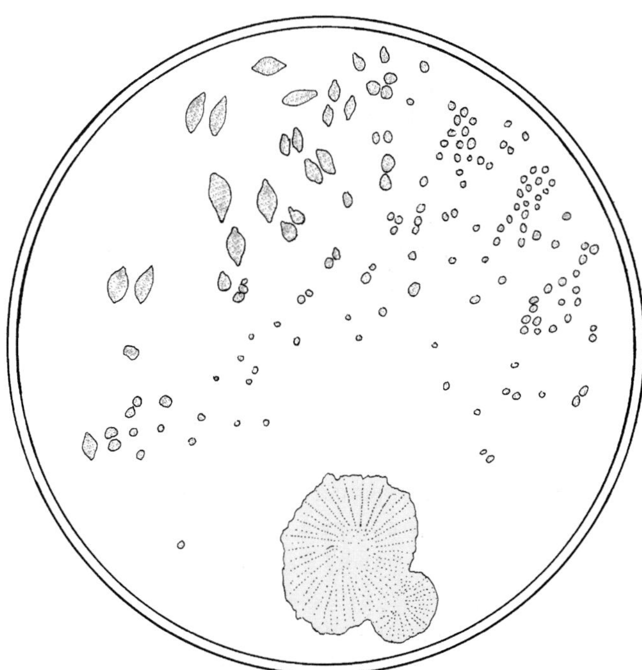

Figure 40.7 Appearance of the *Staphylococcus* colonies on Fleming's petri dish

Fleming reasoned that a substance had diffused out of the mould colony and killed the bacteria. The mould was identified as *Penicillium notatum* and the supposed anti-bacterial chemical was called penicillin. Fleming went on to culture the *Penicillium* on a liquid meat broth medium and showed that the broth contained penicillin, which suppressed the growth of a wide range of bacteria.

Two research assistants at St Mary's then tried to obtain a pure sample of penicillin, free from all the other substances in the broth. Although they succeeded, the procedure was cumbersome and the product was unstable. By this time, Fleming seemed to have lost interest and to assume that penicillin would be too difficult to extract and too unstable to be of medical value.

In 1939, **Howard Florey** (a pathologist) and **Boris Chain** (a biochemist), working at Oxford, succeeded in preparing reasonably pure penicillin and making it stable. Techniques of extraction had improved dramatically in 10 years and, in particular, freeze-drying enabled a stable water-soluble powder form of penicillin to be produced.

World War II was an urgent incentive for the production of penicillin in large quantities and this undoubtedly saved many lives that would otherwise have been lost as a result of infected wounds.

Once Boris Chain had worked out the molecular structure of penicillin, it became possible to modify it chemically and produce other forms of penicillin that attacked a different range of bacteria or had different properties. For example, ampicillin is a modified penicillin which can be taken by mouth rather than by injection.

Because penicillin was the product of a mould, chemists searched for other moulds, particularly those present in the soil, which might produce antibiotics. A large number of these were discovered, including streptomycin (for tuberculosis), chloramphenicol (for typhoid), aureomycin and terramycin (broad spectrum antibiotics, which attack a wide range of bacteria). The ideal drug is one which kills or suppresses the growth of harmful cells, such as bacteria or cancer cells, without damaging the body cells. Scientists have been trying for years to find a 'magic bullet' which 'homes in' exclusively on its target cells. For bacterial diseases, antibiotics come pretty close to the ideal, though the bacteria do seem able to develop resistant forms after a few years.

Ideas about DNA

The geneticists of the early 20th century interpreted Mendel's results in terms of heritable 'factors' carried in the gametes. These 'factors' (now called 'genes') were assumed to control the distinguishable characteristics of the organism in some way. The 'factors' were obviously carried in the gametes but what they were or how they achieved their effects was not known.

The microscope had been invented in the 1650s and cells had been studied. But because the cell contents were transparent it was difficult to make out internal details. This changed, however, when synthetic dyes were made in the late 19th century.

When cells were treated with the new dyes (stains), many internal structures showed up clearly. The staining techniques showed up structures in the nucleus. These are named 'chromosomes' and the behaviour of chromosomes during mitosis was observed and described. At about the same time, Mendel's work, which had remained in obscurity since 1860, was rediscovered.

The presence and behaviour of chromosomes at mitosis and meiosis could explain the way in which the hereditary 'factors' could be transmitted. However, one snag was that in humans, for example, there were only 46 chromosomes but thousands of heritable characteristics. So the thousands of 'factors' must be distributed along the length of each chromosome.

In this case, it might be expected that all the genes on any one chromosome would be inherited together. If this were so, Mendel's 'independent assortment' of characteristics could not happen.

Mendel had been lucky in unwittingly choosing characteristics which were controlled by genes on separate chromosomes. The eventual observation of chromosomes exchanging portions (crossing over p. 201) resolved this problem.

But what were genes? What did they consist of?

Nature of the gene

In 1869, a chemist working on cell chemistry, discovered a compound which contained nitrogen and phosphorus (as well as carbon). This was an unusual combination. The substance seemed to originate from nuclei and was at first called 'nuclein' and then 'nucleic acid'. Subsequent analysis revealed the bases adenine, thymine, cytosine and guanine in nucleic acid, together with a carbohydrate later identified as deoxyribose. In the early 1900s, the structure of nucleotides (base–sugar–phosphate, p. 187) was determined and also how they linked up to form deoxyribonucleic acid (DNA).

In the 1940s, a chemist, Chargaff, showed that, in a sample of DNA, the number of adenines was always the same as the number of thymines. Similarly, the amounts of cytosine and guanine were always equal. This information was to prove crucial to the work of Crick and Watson in determining the structure of DNA.

DNA

Francis Crick was a physicist, and **James Watson** (from the USA) a biologist. They worked together in the Cavendish Laboratory at Cambridge in the 1950s. They did not do chemical analyses or experiments, but used the data that was available from X-ray crystallography and the chemistry of nucleotides to try out different models for the structure of DNA.

The regular pattern of atoms in a crystal causes a beam of X-rays to be scattered in such a way that the structure of the molecules in the crystal can be determined (Figure 40.8a). The scattered X-rays are directed on to a photographic plate which, when developed, reveals images similar to the one in Figure 40.8b.

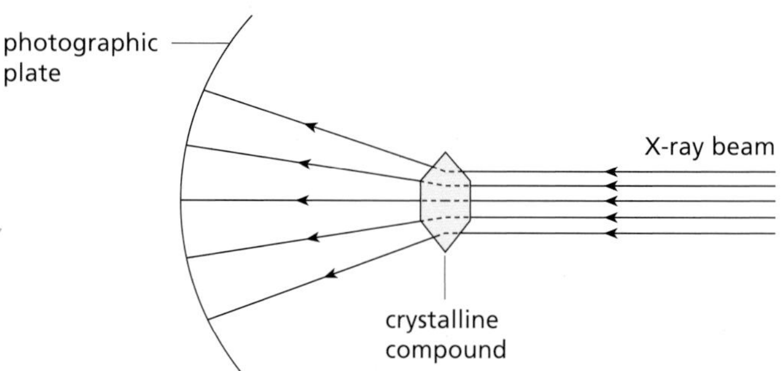

(a) Simplified representation of the scattering of X-rays by crystalline structures

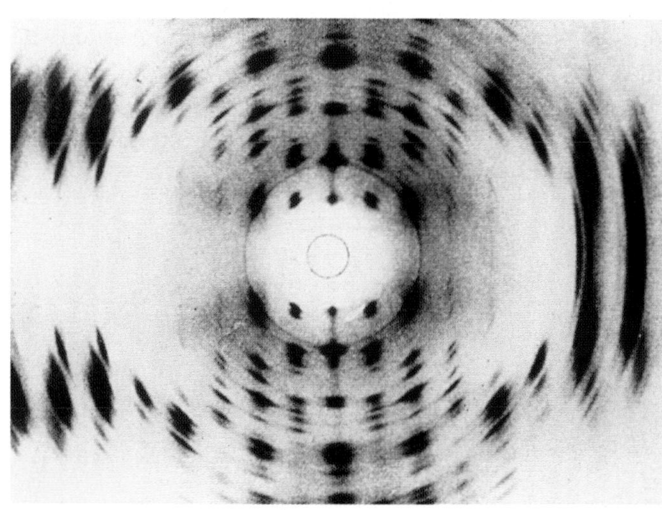

(b) One of the X-ray images produced by X-rays scattered by DNA. The number and positions of the dark areas allows the molecular structure to be calculated

Figure 40.8 X-ray crystallography

By precise measurements of the spots on the photograph and some very complex mathematics, the molecular structure of many compounds could be discovered.

It proved possible to obtain DNA in a crystalline form and subject it to X-ray analysis. Most of the necessary X-ray crystallography was carried out by **Maurice Wilkins** and **Rosalind Franklin** at Kings College, London.

Crick and Watson assembled models on a trial-and-error basis. The suitability of the model was judged by how well it conformed to the X-ray measurements and the chemical properties of the components.

The evidence all pointed to a helical structure (like a spiral staircase). At first they tried models with a core of three or four nucleotide chains twisted round each other and with the bases attached to the outside.

These models did not really fit the X-ray data or the chemical structures of the nucleotides. Watson tried a two-chain helical model with the bases pointing inwards. Initially he paired adenine with adenine, cytosine with cytosine, etc. But thymine and cytosine were smaller molecules than adenosine and guanine and this pairing would distort the double helix.

This is where Chargaff's work came to the rescue. If there were equal numbers of adenine and thymine, and equal numbers of cytosine and guanine, it was likely that this pairing of bases, large plus small, would fit inside the sugar–phosphate double helix without distortion.

The X-ray data confirmed that the diameter of the helix would allow this pairing and the chemistry of the bases would allow them to hold together. The outcome is the model of DNA shown on p. 188 and in Figure 40.9.

Crick, Watson and Wilkins were awarded the Nobel prize for medicine and physiology in 1962. Rosalind Franklin died in 1958, so her vital contribution was not formally rewarded.

Figure 40.9 Crick (right) and Watson with their model of the DNA molecule

Questions

1 Why was the connection between arteries and veins not known until after 1660?

2 Why were Harvey's observations on the circulation of blood originally dismissed?

3 **a** Describe briefly the current theory of the origin of fossils. (See p. 207.)
b State two of the theories held in the 17th century.

4 What are the advantages of the Linnaean system of naming organisms? (See p. 269).

5 **a** What conclusions did Mendel reach as a result of his experiments?
b Why did it take 30 years for Mendel's work to be appreciated?

6 **a** How do Lamarck's views on the origin of variation differ from Darwin's?
b What do their theories of evolution have in common?

7 What were the discoveries in the 1900s that supported the theory of natural selection?

8 Outline the experiments which disposed of 'spontaneous generation'
a of visible organisms,
b of micro-organisms?

9 In what way is Pasteur's work with silkworms relevant to the control of foot and mouth disease?

10 How did work by Chargaff in the 1940s influence Crick and Watson's work on DNA?

11 Why was the work of Maurice Wilkins and Rosalind Franklin crucial to the Crick and Watson model for the structure of DNA?

41 Observation and experiment

What sort of questions can be investigated scientifically?

Subjective and objective evidence
Subjective and objective observations. Anecdotal evidence.

Sampling
Populations; secondary sources of data; fieldwork.

Interpreting results
Probability.

Science progresses by a variety of methods. One important method is **hypothesis testing** (pp. 24 and 35). This depends on making observations, producing a hypothesis to explain them and testing the hypothesis by experiment. The experimental design usually involves **controlled experiments** (p. 23) and the results have to be interpreted (p. 362).

There are many other ways of conducting scientific investigations and there are also questions that cannot be answered by strictly scientific methods.

The following examples may help to distinguish what problems can and cannot be investigated by scientific methods.

What sort of questions can be investigated scientifically?

1. Are crop circles caused by aliens (Figure 41.1)?

Figure 41.1 Are crop circles caused by freak winds, aliens or misguided pranksters?

To investigate this, you could recruit all your friends to set up a round-the-clock watch (if you knew which fields to watch), but some of your friends might fall asleep or some might have vivid imaginations.

Using closed-circuit TV with infrared or thermal imaging would be more reliable and objective. But since there are no data on the physiology of aliens, these techniques could be quite unsuitable.

Given the enormous gap in our knowledge of aliens this is not really a suitable subject for scientific investigation; there is no way of obtaining evidence.

If you changed the question to 'Are crop circles made by humans?' you might have more chance of success.

2. Are fat people more cheerful than thin people?

To make the investigation scientific, you would need to have clear and measurable definitions of 'fat' and 'cheerful'.

'Fat', or better 'overweight', can be defined by measurements of weight and height, which are then fed into a formula to determine the **body mass index**.

$$\text{body mass index (BMI)} = \frac{\text{body mass (in kg)}}{\text{height (in m)}^2}$$

It is agreed that a body mass index of more than 25 is overweight. So, you could certainly collect objective evidence of this condition.

Defining 'cheerful' is more difficult (Figure 41.2). One way is to persuade the subjects to complete a questionnaire. For example:

Do you consider yourself to be cheerful?
always ☐ most of the time ☐
sometimes ☐ rarely ☐ never ☐
Tick the box.

Figure 41.2 On a scale of 1–10, how cheerful are they?

Self-assessment of this kind is unlikely to be very reliable, however.

Unless you can refine the question so that the evidence is objective, consistent and, preferably, measurable, this question cannot be investigated scientifically.

3. Are ASDA prices lower than Sainsbury's?

This is a better prospect. Prices of goods will provide measurable evidence. You will have to decide, however, which goods to compare. ASDA's baked beans might be 3p cheaper than Sainsbury's but their tinned soup could be 3p dearer, so cancelling each other out in your totals (Figure 41.3).

Figure 41.3 How might this affect your survey?

You would have to make a list of foodstuffs that most people would buy regularly and ensure that you were comparing the same quantities of the individual items. In other words, the items would have to be 'matched' for everything except price.

How many items would you choose? A recent government survey has been criticized for selecting only 56 items, so the more you have the better (up to a point). If the price difference of 150 items is little different from that of 100 items, why go beyond 100?

Although there are clearly problems with this investigation, it could be done scientifically. You could do it yourself for your own town or shopping centre, or you could go shopping on the internet for an even wider survey.

Alternatively, you could use 'second-hand' evidence. This is evidence (hopefully reliable) that has already been published, for example in a newspaper, in a government survey or on the internet. The advantage of second-hand evidence is that it is likely to be much more extensive than any you could gather for yourself.

4. Do athletes have a lower resting heart-rate than 'couch potatoes'?

This looks a more promising topic for scientific investigation. Resting heart rate is easily measured. More difficult is identification of the 'athletes'. Weight lifting, sprinting, discus throwing and cross-country running all make different demands on the body and could affect resting heart rate differently (Figure 41.4).

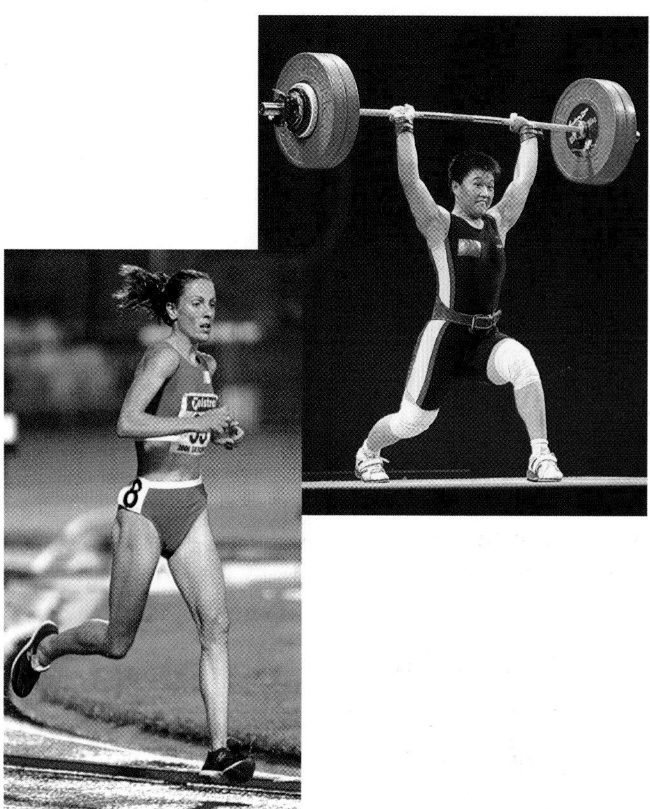

Figure 41.4 What kind of athlete will you select for your investigation?

Here then, is a case for refining the question to ask, for example, 'Do long-distance runners have a lower resting heart rate than couch potatoes?'

Now you will have to define your 'couch potato' as, for example, a person who takes less than 2 hours of exercise (also to be defined) per week. So long as you can come up with clearly defined characteristics, there is no reason why this should not be investigated scientifically (though you will meet criticism no matter how carefully you design your experiment).

Your next problem is to decide on your sample with regard to age and sex and numbers (see 'Sampling', p. 359).

5. Do 15-year-old boys weigh more on average than 15-year-old girls?

This should be easy to investigate. Scientifically you have to decide only on your sample size, what to allow for the different weights of clothing and the method of weighing. Whether it is worth doing is another matter.

Questions

The following are some questions suggested for a scientific investigation. In each case say: **a** whether or not the question could be investigated scientifically and if not, why not; **b** how you would refine the question, if necessary, to make it suitable for scientific investigation.

1 Do green algae grow better on the south side of tree-trunks than they do on the north side?
2 Do dogs make better companions than cats?
3 Can boys run faster than girls?
4 Do seeds need light in order to germinate?
5 Does taking vitamin C stop you getting colds?
6 Do robins in the north of England lay fewer eggs per clutch than robins in the south?

Subjective and objective evidence

You will be familiar with optical illusions such as Figure 41.5. The upper horizontal line looks shorter than the lower line. This is a **subjective** impression. If you measure the two lines with a ruler, they turn out to be the same length. This is an **objective** observation.

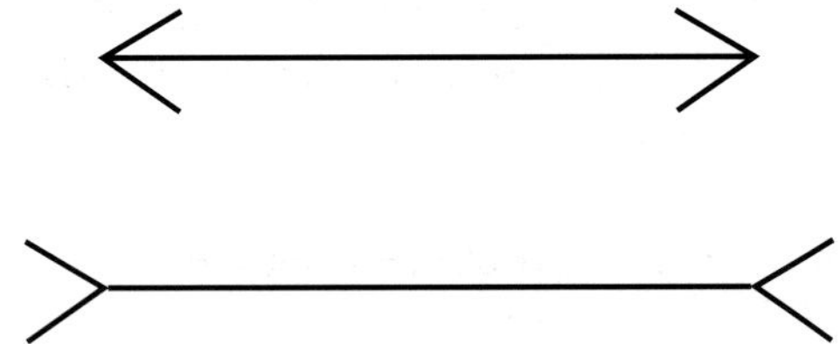

Figure 41.5 Which horizontal line is longer?

The subjective evidence depends on the impression of the 'subject' – that is, the person making the observation. It is thought that some African tribes who live in circular houses are less affected by perspective drawings and would see no difference between the two lines. Their 'subjective' observation would be different from yours.

Everybody would agree, however, that the 'objective' measurement of the two lines shows them to be identical in length.

Subjective and objective observations

Chris 'Have you noticed that there are far more buttercups in the lower half of this meadow than there are in the top half?'
Don 'I don't agree. It just looks like that because the top half of the field is in shadow from the trees.'

This argument about subjective observations could be resolved by objective measurements, for example by using metre quadrats (see Figure 41.10, p. 360) to sample the buttercup populations in each half of the field.

Subjective observations can be a useful starting point for scientific investigation even if they are unscientific on their own. It turns out that Chris was right; the buttercups were more abundant in the lower part of the field, probably because it was wetter there.

You could try designing a further investigation to see if there really was a correlation (pp. 121 and 129) between the wetness of the soil and the number of buttercups.

Anecdotal evidence

An anecdote is a very short story or an account of an event, sometimes funny, but not necessarily so.

'My grandad smoked 40 a day and lived to be 95' is an anecdote and may be seen, by some people, as evidence that smoking does you no harm. Good for grandad, but evidence from a single source is worthless scientifically. Even from a number of different sources, it is still **anecdotal** evidence. It might, however, lead to a scientific investigation – for example, what have all these people got in common that makes them resistant to the harmful effects of smoking?

Good scientific evidence on this subject involves the painstaking collection of data on smoking and health from a very large sample of smokers and non-smokers over a period of many years.

Anecdotal evidence often leads people to jump to conclusions about cause and effect (pp. 121 and 129). Stories about the incidence of leukaemia in the vicinity of a nuclear processing plant led people to assume that the disease was caused by radiation, but scientific investigations have not yet supported this 'cause and effect' assumption.

An extreme example of anecdotal evidence is the sighting of UFOs and abduction by aliens. The evidence about grandad's health was, at least, verifiable.

Nevertheless, people are greatly influenced by anecdotal evidence, possibly because it is usually more interesting than a reliable but rather dull scientific study. Would you prefer a gory story, or a dreary theory?

Some newspapers and other media often prefer anecdotal evidence because it has a greater impact on the reader or viewer (Figure 41.6). Once again, such evidence may spur authorities to set up a proper scientific enquiry.

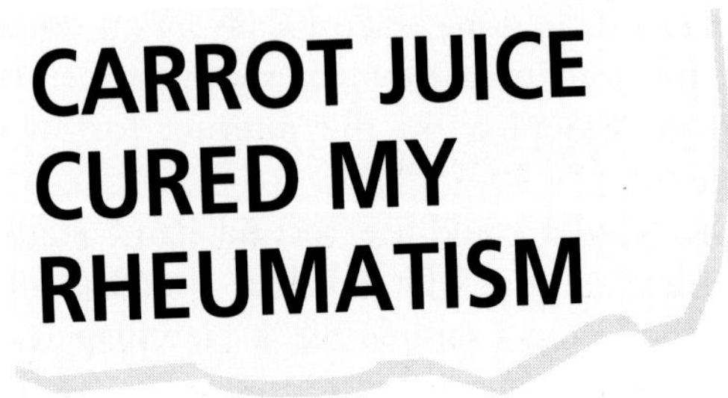

CARROT JUICE CURED MY RHEUMATISM

Figure 41.6 Can we be sure he wouldn't have got better anyway?

Questions

State which of the following you would consider to be anecdotal, subjective or objective evidence. (Don't worry, the boundaries are not always clear-cut.)

1 Smoking 20 cigarettes a day increases the chance of lung cancer by eight times.
2 My dog seems to understand everything I say to her.
3 'Flu? Mrs. Jones says there's a lot of it about.'
4 Adding nitrate fertilizer to a wheat crop increased the yield of grain by 20 per cent.

Sampling

Almost 18 million people watched 'Coronation Street' last Monday (Figure 41.7). How does ITV know this? They cannot ring up everybody in the country to ask. So, they need to ask a **sample** of the population.

The top ten programmes; a week's survey

	Millions		*Millions*
EastEnders	18.00	Casualty	11.98
Coronation Street	17.99	London's Burning	10.09
Heartbeat	14.65	Neighbours	9.89
Dinnerladies	13.05	Midsomer Murders	9.56
Emmerdale	12.13	Peak Practice	9.55

Figure 41.7 Daily viewing figures. But how do they know?

But how many people to ask? If you ask only one or two, it is possible that none of them watched on Monday. If you ask 100 people, you are very likely to find a substantial number of viewers, say 50 per cent, watched. If you then ask 1000 people, the figure might become 65 per cent. You then go on to ask 10 000 people but the result changes only to 65.5 per cent. Clearly, it is not worth asking a sample of more than 1000 in this case.

Which people do you ask? It is unlikely that many senior citizens will have watched 'Top of the Pops', so you will have to define the target individuals for your survey.

Sampling a population

The size of a small population of organisms confined to a limited area (e.g. rooks in a rookery (Figure 41.8) or oak trees in a copse) can probably be found by just counting the individuals. Mostly, however, you would need to take samples and estimate the population from the results.

Figure 41.8 Rookery. Count the nests; multiply by two for the adult population (assuming all the nests are occupied). Multiply by seven when the eggs hatch

One method for sampling small, mobile creatures is to capture as many as possible in a given time, mark them (e.g. with coloured nail varnish) and release them. Next day, you use the same method of capture and see what proportion of marked individuals appears in your sample. Using a formula called the Lincoln index (Figure 41.9), you can calculate the size of the whole population.

A sample of the population is captured.

5 individuals are captured, marked with cellulose paint and released.

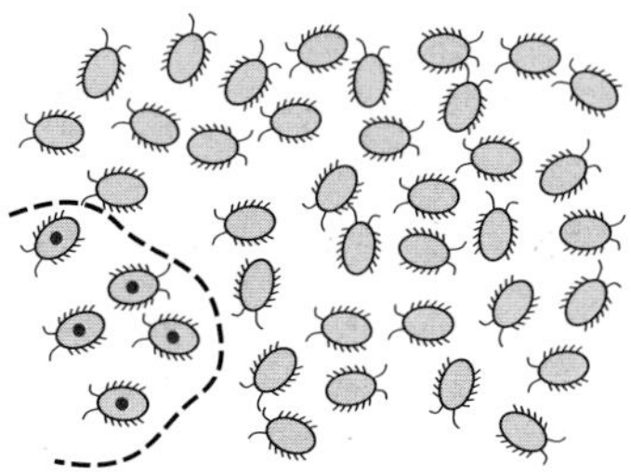

One day later, another sample is caught in the same locality. 8 individuals were captured, one of which was marked. So 1 in 8 of the recaptured population was marked, that is, ⅛ was marked. But 5 were initially marked, so five is one-eighth of the population. So population is $5 \times 8 = 40$.

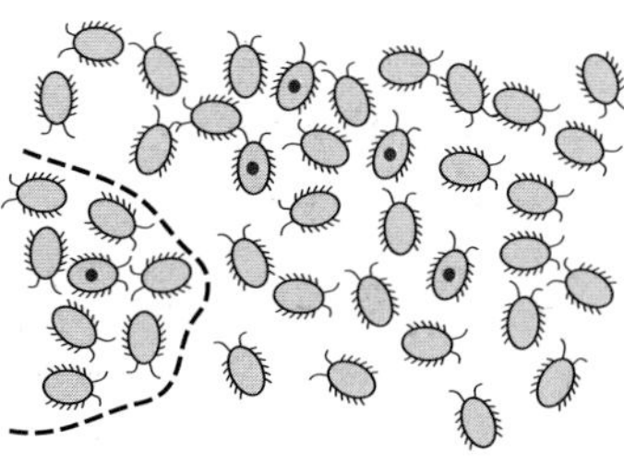

But note that if 2 marked indiviuals had been recaptured the population estimate would have been 20. So relatively large samples must be used; a recapture of one or two individuals is no good.

Figure 41.9 Lincoln index. The mark–release–recapture method of sampling. (The method assumes that marked individuals are not harmed, and that they distribute themslves randomly in the population)

For the number of birch trees in a wood, you could count the number in, say, a 50-metre square. If you then estimated the size of the entire wood, from a map perhaps, you could calculate the birch population for the whole wood (assuming the birch trees were evenly distributed).

Metre quadrats (Figure 41.10) are used to sample populations of small, herbaceous plants.

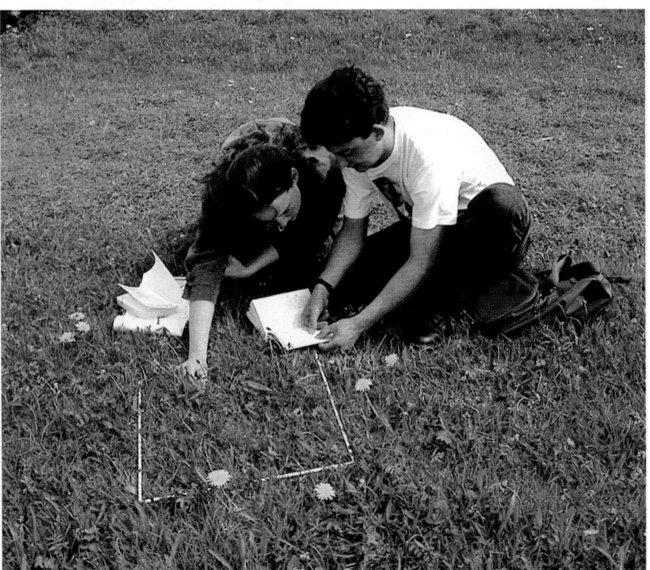

Figure 41.10 The metre quadrat. The frame is placed at random over an area of vegetation. The number of, for example, dandelion plants is counted

Secondary sources of data

Catch sizes in the fishing industry are a good indication of the population size. As over-fishing continues, not only are fewer fish caught with each trip, but their size is smaller. Eventually estimates of the fish stocks suggest that the population will crash. This has already happened with North Sea herring, Newfoundland cod and Southern bluefin tuna (Figure 41.11).

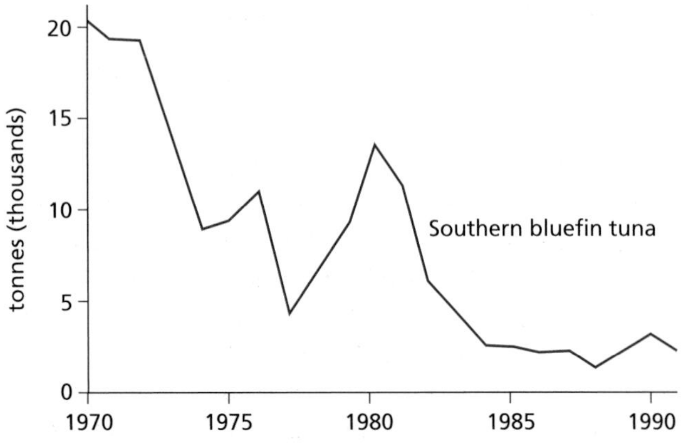

Figure 41.11 The fishing industry catches samples of a population, which is clearly declining

Sampling in fieldwork

Suppose you wanted to investigate the effect of a factory discharge on the organisms in a river. You would need to take samples of water above and below the factory's discharge point (Figure 41.12). You need to know what water qualities you are going to test for – pH, oxygen, heavy metals, nitrates, etc. You would need the appropriate equipment – pH meter, chemical test kits, etc. (Figure 41.13). You then have to decide how frequently you are going to take samples (the factory's operations may change from day to day).

Figure 41.12 Discharge of polluted effluent into a river

(a) Apparatus for measuring pH

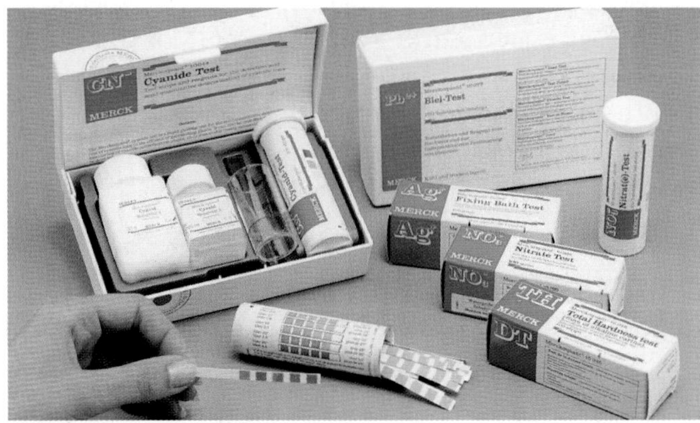

(b) Test kits for pollutants. The cyanide strips can detect 1 mg/l; the nitrate strips are sensitive to 10 mg/l

Figure 41.13 Sampling water quality

How far from the discharge point will you take your samples? Any pollutants may become progressively diluted as the river flows on. Perhaps you need two or three sampling sites at 500-metre intervals. Provided the pollutants are at a strength that your equipment can detect, the size of the sample will depend only on the design or capacity of your test apparatus.

Now, none of this may be necessary. If your samples of organisms above and below the discharge site showed little or no difference, there would not be much point in looking for pollutants. So, it would be best to start by sampling the organisms rather than the water.

If the river is shallow you could wade in with a sweep net. If it is deep you might have to be content with taking samples from the bank-side only. Since you are only *comparing* the populations at each sampling point, you need not try to estimate the actual population size. So, you decide on a method such as sweep-netting for 5 minutes, or ten sweeps at each site, and then you compare your catches (Figures 41.14 and 41.15). If your method produces too few organisms you may need to increase the time spent sweeping or the number of sweeps, provided you do this for all sites.

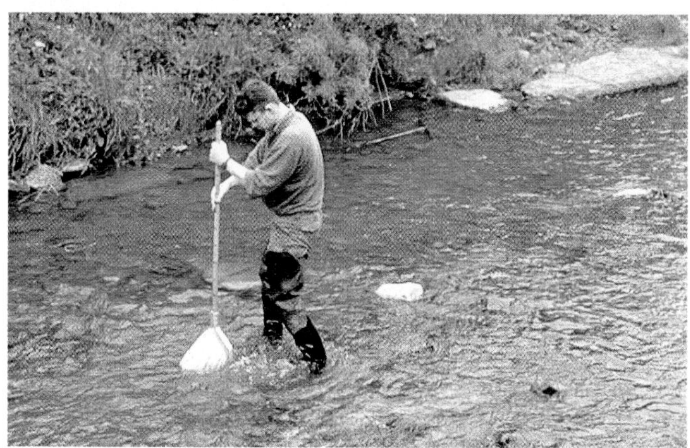

Figure 41.14 Sampling a river for invertebrates

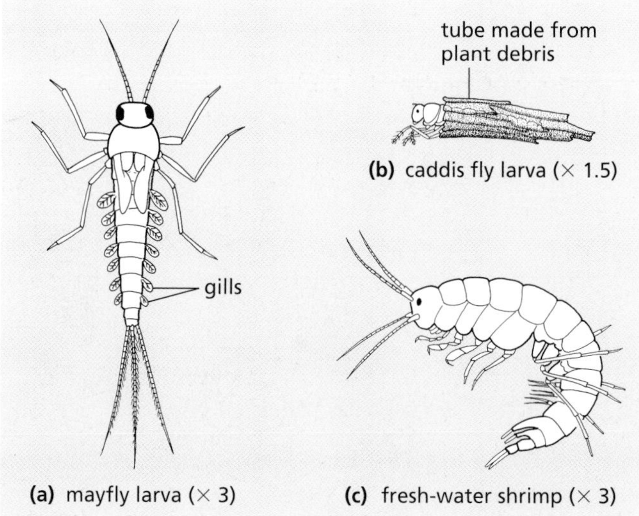

Figure 41.15 Good news for streams. If these animals are present, the stream is unlikely to be polluted

Interpreting results

A set of results might be interesting on their own but in order to contribute to a scientific investigation they need to be interpreted. There may be many possible interpretations, some more acceptable than others. For example, the result of Experiment 1 on p. 36 was that starch was formed, by photosynthesis, only in those parts of a leaf which had received light. The interpretation was that light was necessary for photosynthesis. An alternative interpretation might be that light was necessary for converting sugar, stored in the leaf, into starch. It is only because we already know quite a lot about the processes that the first interpretation is more acceptable.

Probability

Experimental results do not usually lead to certainty. The best they can do is to indicate that the probability of their interpretation being correct is, say, 95 per cent certain.

For example, suppose you wanted to investigate the effect of a fertilizer on plant growth. You might grow five plants in ordinary soil and compare their height, after a suitable interval, with the height of five plants (of the same kind), grown in soil with added fertilizer. A subjective observation might be that the plants with fertilizer *looked* taller than the others. To get objective evidence, however, you would need to measure all the plants.

The results might be as follows:

	Height of plant (shoot) in mm	
	A (no fertilizer)	*B (added fertilizer)*
	126	138
	123	129
	118	123
	104	131
	114	119
total	585	640
mean	585/5 = 117	640/5 = 128

difference between the means (128 − 117) = 11

The difference between the two means is 11 mm so it looks at first as if, on average, the plants with fertilizer have grown taller than those without. However, the differences *within* each group are greater than this. Subtracting the minimum from the maximum values gives: A 126 − 104 = 22, B 138 − 119 = 19.

If there is this much variation (about 20 mm) within each group, a mean difference of 11 cannot be regarded as significant. It could simply be due to chance variation.

If the two samples were much larger, the difference between the means might be much more significant.

A statistical test can be applied to decide whether the difference between the means is significantly greater than the difference within each group.

The results of such a test may then be expressed as a **probability**. For example, the odds against the difference between the means being due to chance alone might be so high that you can be confident that the fertilizer was effective.

Questions

1 Paul wants to estimate the population of dandelions on the football pitch by sampling, using a quadrat frame (see Figure 41.10) to see if it is worth spraying with weedkiller.

In some parts of the pitch, there are plenty of dandelions, so he places his quadrat frame over five areas, and in each counts the dandelion plants. He then multiplies his results by the ratio of the field area to the sample area to get the total population.

What fault can you find with his method of sampling?

2 Jane wants to check on Paul's results and also make a comparison between the two ends of the pitch. One way (not the best) of taking samples at random, is to throw your metre quadrat frame over your shoulder without looking (but making sure nobody is in the target area). Jane does this five times on the football pitch and estimates the dandelion population from her quadrat counts.

When she works on the other half of the pitch, she finds very few dandelions in the samples, so she throws the quadrat frame five more times.

Discuss whether this will make her comparison inaccurate.

3 You mark out a 10 m square of rough grassland and walk from corner to corner sweeping the grass with a butterfly net, making ten sweeps on each trip. You capture 35 grasshoppers, mark each one with a dab of white correcting fluid and release them.

Next day you repeat the operation and capture 27 grasshoppers, of which eight are marked.

Put these figures into the formula below to find an estimate of the population of grasshoppers in the 10 m square.

$$\text{estimated population} = \frac{\text{no. caught first time} \times \text{no. caught second time}}{\text{no. of marked individuals in second catch}}$$

a Criticize the design of this experiment.
b What might be the effect of the second day being much hotter than the first day?
c How might the marking process distort the results?
d If you had to anaesthetize the grasshoppers in order to mark them, how might this affect your result?

Glossary

(A) Scientific terms

Acid A sharp-tasting chemical, often a liquid. Some acids can dissolve metals and turn them into soluble salts. Nitric acid acts on copper and forms copper nitrate, which dissolves to give a blue-coloured solution. Plant and animal acids (amino acids, fatty acids) are weaker and do not dissolve metals. Amino acids and fatty acids are organic acids. Hydrochloric, sulphuric and nitric acids are called mineral acids or inorganic acids.

Agar A clear jelly extracted from one kind of seaweed. On its own it will not support the growth of bacteria or fungi, but will do so if food substances (e.g. potato juice or Bovril) are dissolved in it. Agar with different kinds of food dissolved in it is used to grow different kinds of micro-organism.

Alcohol Usually a liquid. There are many kinds of alcohol but the commonest is ethanol (or ethyl alcohol) which occurs in wines, spirits, beers, etc. It is produced by fermentation of sugar. Ethanol vaporises quickly and easily catches fire.

Alkali The opposite of an acid. An alkali can neutralise an acid and so remove its acid properties. Sodium hydroxide (NaOH) is an alkali. It neutralises hydrochloric acid (HCl) to form a salt, sodium chloride.

$$\mathrm{Na}\boxed{\mathrm{OH} + \mathrm{H}}\mathrm{Cl} \rightarrow \underset{\text{salt}}{\mathrm{NaCl}} + \underset{\text{water}}{\mathrm{H_2O}}$$

Atom The smallest possible particle of an element. Even a microscopic piece of iron would be made up of millions of iron atoms. When we write formulae, the letters represent atoms. So H_2O for water means two atoms of hydrogen joined to an atom of oxygen.

Carbon A black, solid non-metal which occurs as charcoal or soot, for example. Its atoms are able to combine together to make ring or chain molecules (see p.12). These molecules make up most of the chemicals of living organisms (see 'Organic'). One of the simplest compounds of carbon is carbon dioxide (CO_2).

Carbon dioxide A gas which forms 0.03 per cent (by volume) of the air. It is produced when carbon-containing substances burn ($C + O_2 \rightarrow CO_2$). It is also produced by the respiration of plants and animals. It is taken up by green plants to make food during photosynthesis.

Catalyst A substance which makes a chemical reaction go faster but does not get used up in the reaction. Platinum is a catalyst which speeds up the rate at which nitrogen and hydrogen combine to form ammonia, but does not get used up. Enzymes are catalysts for chemical reactions inside living cells.

Caustic A caustic substance can damage the skin and clothing and therefore should be handled with great care.

Compound Two or more elements joined together form a compound. Carbon dioxide, CO_2, is a compound of carbon and oxygen. Potassium nitrate, KNO_3, is a compound of potassium, nitrogen and oxygen.

Cubic centimetre (cm^3) This is a unit of volume. A tea-cup holds about 200 cm^3 liquid. One thousand cubic centimetres are called a cubic decimetre (dm^3) but this volume is also called a litre. Some measuring instruments are marked in millilitres (ml). A millilitre is a thousandth of a litre and therefore the same volume as a cubic centimetre. So $1\,cm^3 = 1\,ml$.

DCPIP The initials of an organic chemical called dichlorophenol–indophenol. It changes from blue to colourless in the presence of certain chemicals including vitamin C.

Density This is the weight (mass) of a given volume of a substance. Usually it is the weight in grams of one cubic centimetre of the substance, e.g. $1\,cm^3$ lead weighs 11 g, so its density is 11 g per cm^3. The density of water at 4°C is 1 g per cm^3.

Diffusion The random movement of molecules by which gases or dissolved substances move from a region of high concentration to a region of low concentration.

Dissolve A substance which mixes with a liquid and seems to 'disappear' in the liquid is said to dissolve. Sugar dissolves in water to make a solution.

Element An element is a substance which cannot be broken down into anything else. Sulphur is a non-metallic element. Iron is a metallic element. Oxygen and nitrogen are gaseous elements. Water (H_2O) is not an element because it can be broken down into hydrogen and oxygen.

Energy This can be heat, movement, light, electricity, etc. Anything which can be harnessed to do some kind of work is energy. Food consists of substances containing chemical energy. When food is turned into carbon dioxide and water by respiration, energy is released to do work, such as making muscles contract.

Filtrate The clear solution which passes through a filter; e.g. if a mixture of copper sulphate solution and sand is filtered, the blue copper sulphate solution which passes through the filter paper is called the filtrate.

Formula A way of showing the chemical composition of a substance. Letters are chosen to represent elements, and numbers show how many atoms of each element are present. The letter for carbon is C and for oxygen is O. A molecule of carbon dioxide is one atom of carbon joined to two atoms of oxygen and the formula is CO_2. There are more elements than letters in the alphabet, so some of the elements have two letters, e.g. Mg for magnesium. Other elements have letters standing for the latin name, e.g. sodium is Na (=natrium).

Gram (g) A unit of weight in the metric system.

A penny weighs 3.5 g.
A pack of butter weighs 225 g.
1000 g is a kilogram (kg).
One-thousandth of a gram is a milligram (mg).

Hydrogen Hydrogen is a gas which burns very readily. It is present in only tiny amounts in the air but forms part of many compounds such as water (H_2O), and organic compounds like carbohydrates (e.g. $C_6H_{12}O_6$ glucose) and fats.

Inorganic Substances like iron, salt, oxygen and carbon dioxide are inorganic. They do not have to come from a living organism. Salt is in the sea, iron is part of a mineral in the ground, oxygen is in the air. Inorganic substances can be made by industrial processes or extracted from minerals.

Insoluble An insoluble substance is one which will not dissolve. Sugar is soluble in water but insoluble in petrol.

Ion An atom or small group of atoms carrying an electric charge and having different properties from the uncharged atoms. A solution of sodium nitrate in water will consist of ions of positively charged sodium (Na^+) and negatively charged nitrate (NO_3^-).

Joule Just as a centimetre is a unit of length, a joule (J) is a unit of energy. For example 4.2 J of heat energy will raise the temperature of 1 g of water by 1 °C. A kilojoule (kJ) is 1000 J. The older unit of energy, the calorie, is often still used to express the energy value of food. 1 calorie = 4.2 J.

Lime water A weak solution of lime (calcium hydroxide) in water. When carbon dioxide bubbles through this solution, it reacts with the calcium hydroxide to form calcium carbonate (chalk) which is insoluble and forms a cloudy suspension. This makes lime water a good test of carbon dioxide.

$$Ca(OH)_2 + CO_2 \rightarrow CaCO_3 + H_2O$$

Manometer An instrument which measures pressure by the displacement of a liquid in a U-tube.

Mass This is the amount of matter in an object. The more mass an object has, the more it weighs, so mass can be measured by weighing something. However, if the force of gravity becomes less, as on the Moon, the same object will weigh less even though the amount of matter in it (its mass) has not changed. So mass and weight are related, but are not the same.

Molecule The smallest amount of a substance which you can have. For example, the water molecule is H_2O, that is, two atoms of hydrogen joined to one atom of oxygen. A drop of water consists of countless millions of molecules of H_2O moving about in all directions and with a lot of space between them.

Organic This usually refers to a substance produced by a living organism. Organic chemicals are things like carbohydrates, protein and fat. They have very large molecules and are often insoluble in water. Inorganic chemicals are usually quite simple substances like sodium chloride (salt) or carbon dioxide (CO_2).

Oxygen Oxygen is a gas which makes up about 20 per cent (by volume) of the air. It combines with other substances and oxidizes them, sometimes producing heat and light energy. In plants and animals it combines with food to release energy.

Permeable Allows liquids or gases to pass through. A cotton shirt is permeable to rain but a Gortex jacket is impermeable. Plant cell walls are permeable to water and dissolved substances.

pH This is a measure of how acid or alkaline a substance is. A pH of 7 is neutral. A pH in the range 8–11 is alkaline; a pH in the 6–2 range is acidic; pH 6 is slightly acidic; and pH 2 is very acidic.

Pigment A chemical which has a colour. Haemoglobin in blood is a red pigment; chlorophyll in leaves is a green pigment. A black pigment called melanin may give a dark colour to human skin, hair and eyes.

Reaction (chemical) A change which takes place when certain chemicals meet or are acted on by heat or light. The change results in the production of new substances. When paper burns, a reaction is taking place between the paper and the oxygen in the air.

Salt A salt is a compound formed from an acid and a metal. Salts have double-barrelled names like sodium chloride (NaCl) and potassium nitrate (KNO_3). The first name is usually a metal and the second name is the acid. Potassium (K) is a metal, and the nitrate (NO_3) comes from nitric acid (HNO_3).

Sodium hydrogencarbonate At one time this was called sodium bicarbonate. It is a salt which is used to make carbon dioxide in experiments. Its formula is $NaHCO_3$.

Sodium hydroxide (NaOH) An alkali with caustic properties, i.e. its solution will dissolve flesh, wood and fabrics.

Soluble A soluble substance is one which will dissolve in a liquid. Sugar is soluble in water.

Solution When something like sugar or salt dissolves in water it forms a solution. The molecules of the solid become evenly spread through the liquid.

Volume The amount of space something takes up, or the amount of space inside it. A milk carton has an internal volume of 1 pint. Your lungs have a volume of about 5 litres; they can hold up to 5 litres of air.

(B) Biological terms

(References in brackets are to pages)

Abdomen The part of the body below the diaphragm, which contains stomach, kidneys, liver etc.; in insects, it refers to the third region of the body.

Abiotic factors (255) Those features of an ecosystem which are not dependent on living organisms.

Accommodation (161) Changing the shape (and focal length) of the eye lens to focus on near or distant objects.

Active transport (28) The transport of a substance across a cell membrane with the expenditure of energy, often against a concentration gradient.

Adaptation (256) The development, during evolution of an organism, of structures or processes which make it more efficient in its environment.

Alleles (194) Alternative forms of a gene, occupying the same place on a chromosome and affecting the same characteristic but sometimes in different ways.

Anabolism (20) The building up of complex substances from simpler ones.

Antibiotic (329) An anti-bacterial drug derived from a fungus or bacterium.

Antigen (117) A substance which promotes the formation of antibodies.

Aseptic technique (289) Method of handling materials or apparatus so that unwanted micro-organisms are excluded.

Asexual reproduction (303) Reproduction without the involvement of gametes.

Assimilation (102) Absorption of substances which are then built into other compounds in the organism.

Autotroph (293) An organism which can build up its organic materials from inorganic substances.

Auxin (309) A chemical which affects the rate of growth in plants.

Basal metabolism (20) The minimum rate of chemical activity needed to keep an organism alive.

Biomass (225) The weight (mass) of all the organisms in a population, community or habitat.

Biosphere (254) That part of the Earth which contains living organisms.

Biotechnology (326) the use of living organisms or biological processes for industrial, agricultural or medical purposes.

Biotic factors (255) The effects on an ecosystem of the activities of living organisms.

Cardiac To do with the heart

Catabolism (20) The breakdown of complex substances to simpler substance in a cell, with release of energy.

Chromosomes (82) Thread-like structures in the nucleus, which carry the genes.

Clone (23) A population of organisms derived by asexual reproduction from a single individual.

Control (23) An experiment which is set up to ensure that only the condition being investigate has affected the result.

Co-ordination (163) The processes which make the different systems in an organism work effectively together.

Denature (11) Destroy the structure of a protein by means of heat or chemicals.

Detoxication (105) The process by which the liver makes poisonous chemicals harmless.

Dialysis (29) The separation of small molecules from large molecules by a selectively permeable membrane.

Ecosystem (29) A community of interdependent organisms and the environment in which they live.

Effector An organ which responds to a stimulus, nerve impulse or hormone.

Enzyme (13) A protein which acts as a catalyst for reactions in cells or organisms.

Eutrophic (238) An aquatic environment well supplied with nutrients for plant growth.

Fermentation (20) A form of anaerobic respiration in which carbohydrate is broken down to carbon dioxide and, in some cases, alcohol.

Gastric To do with the stomach.

Gene (186) sequence of nucleotides in a DNA molecule, which controls the development of a particular characteristic in an organism.

Genetic code (188) The sequence of bases in a molecule of DNA which specifies the order of amino acids in a protein.

Genetic engineering (213) Altering the genetic constitution of an organism by modifying its own genes or introducing genes from a different species.

Genetic fingerprinting (218) Identifying stretches of DNA specific to one, or very few individuals.

Genotype (193) The combination of genes present in an organism.

Genus (269) One of the categories in classification; a group of closely related species.

Gestation (146) The period of growth and development of a fetus in the uterus of a mammal.

GM organisms (215) Organisms whose genetic make-up has been changed by altering its own genes or introducing genes from a different species.

GM food (217) Food derived from GM (genetically modified) crops.

Hepatic (112) To do with the liver.

Heterotroph (293) An organism which feeds by taking in organic substances made by other organisms.

Heterozygous (193) Carrying a pair of contrasting alleles for any one heritable characteristic; will not breed true for this characteristic.

Homeostasis (105) Keeping the composition of the body fluids within narrow limits.

Homoiothermic (139) Animals whose body temperature is maintained at a constant level, usually above that of their surroundings; Sometimes called 'warm-blooded'.

Homologous chromosomes (184) A pair of corresponding chromosomes of the same shape and size; one from each parent.

Homozygous (193) Possessing a pair of identical alleles controlling the same characteristic; will breed true for this characteristic.

Hypothesis (24) A provisional explanation for an observation; it can be tested by experiment.

Immunity (117) Ability of an organism to resist infection, usually because it carries the appropriate antibodies in its blood.

Implantation (144) The process by which an embryo becomes attached to the lining of the uterus.

Inflorescence (69) A group of flowers on the same stalk.

Inhibit Slow down a process or prevent its happening.

Inoculation (118) Introduction to the body of a harmless micro-organism or substance to induce the formation of antibodies.

Laparoscopy (150) A method of examining the inside of the abdomen by inserting an optical instrument (endoscope) through the abdominal wall.

Laparotomy (151) Any operation which involves opening or entering the abdomen by surgery.

Limiting factor (41) A condition which limits the rate of a process, e.g. shortage of light for photosynthesis.

Metabolism (20) All the chemical changes going on in the cells of an organism which keep it alive.

Metamorphosis (307) The relatively sudden change by which the larval form of an insect or amphibian becomes an adult.

Monoculture (235) Growing a single species of crop plant, usually in the same ground for successive years.

Mutation (189) A spontaneous change in a gene or chromosome, which may affect the appearance or physiology of an organism.

Mutualism (296) A close and permanent relationship between two organisms of different species, from which both derive some benefit.

Osmoregulation (133) Regulation of the concentration of aqueous solutions within the cells of an organism.

Parasite (293) An organism living in or on another organism (the host). The parasite derives food from its host.

Pathogen (284) A parasite which causes disease.

Phenotype (193) The observable characteristics of an organism which are genetically controlled.

Plankton (22) The community of small plants and animals found in the surface waters of an aquatic environment.

Plasmid (213) Small circle of DNA in a bacterium. Often carries genes for antibiotic resistance. Used in genetic engineering to introduce foreign DNA

Poikilothermic (139) Having a body temperature which fluctuates with that of the environment. Sometimes referred to (inaccurately) as 'cold-blooded'.

Predator (224) An animal which kills and eats other animals.

Protoctista (268) Single-celled organisms which have a proper nucleus.

Protozoa (268) Those Protoctista which take in solid food and digest it.

Puberty (148) The period of growth during which humans become sexually mature.

Receptor A sense organ which detects a stimulus.

Recessive (192) A gene which, in the presence of its contrasting allele, is not expressed.

Recombinant (213) Refers to insertion , by genetic engineering, of foreign DNA into a genotype.

Recycling (227, 252) As a biological term, this means the return of matter to the soil, air or water, and its re-use by other organisms. In daily life it means the re-use of manufactured materials such as paper, glass and metals.

Renal (132) To do with the kidneys

Replication (182) Production of a duplicate set of chromosomes prior to cell division.

Restriction enzymes (213) Enzymes which 'cut' DNA at specific sites.

Section (2) A thin slice of tissue which can be examined under the microscope.

Sensitivity (317) The ability to detect and respond to a stimulus.

Sensor (322) A sensory cell or organ.

Specialization (6) Cells; able to do one particular job. Organisms; being closely adapted to a limited habitat or way of life.

Sphincter (101) A band of circular muscle which can contract to constrict or close a tubular organ

Spore (279) A cell or a small group of cells from plants fungi or bacteria, which can grow into a new organism.

Stem cells (218) Cells which have not become specialized themselves but can develop into a range of different specialized cells.

Stimulus (158) An event in the surroundings or internal anatomy of an organism, which provokes a response.

Stoma (52) A structure, in the epidermis of a plant, which consists of a pore enclosed by two guard cells. It permits gaseous exchange with the atmosphere.

Symbiosis (296) A very close association, between two unrelated organisms; it includes parasitism and mutualism.

Toxin (284) A poisonous protein produced by pathogenic bacteria.

Toxoid (118) A toxin which has been treated to make it harmless, but can still cause the body to make antibodies to the toxin.

Transgenic (214) An organism carrying DNA derived from an unrelated species.

Translocation (62) Transport of minerals and nutrients in a plant.

Trophic level (225) An organisms's position in a food chain, e.g. primary or secondary consumer.

Turgor (30) The pressure built up in a plant cell as a result of taking in water by osmosis.

Vascular To do with vessels; blood vessels or xylem and phloem.

Index

Page numbers in **bold type** show where a subject is introduced or most fully explained